Explorations in College Algebra

Kime • Clark • Michael

Fifth Edition

WILEY *Custom*
LEARNING SOLUTIONS

To order books or for customer service, please call 1(800)-CALL-WILEY (225-5945).

Printed in the United States of America.

ISBN 978-1-118-12400-0

Printed and bound by IPAK.

10 9 8 7 6 5 4 3 2

Brief Contents

Chapter 2	Rates of Change and Linear Functions	*60*
Chapter 3	When Lines Meet: Linear Systems	*160*
Chapter 4	The Laws of Exponents and Logarithms: Measuring the Universe	*212*
Chapter 5	Growth and Decay: An Introduction to Exponential Functions	*264*
Chapter 6	Logarithmic Links: Logarithmic and Exponential Functions	*346*
Chapter 8	Quadratics and the Mathematics of Motion	*468*
APPENDIX	Student Data Tables for Exploration	*617*
	Data Dictionary for FAM1000 Data	*621*
SOLUTIONS	For all Algebra Aerobics and Check Your Understanding problems; for odd-numbered problems in the Exercises and Chapter Reviews. All solutions are grouped by chapter.	*ANS-1*
INDEX		*I-1*

FIFTH EDITION

EXPLORATIONS IN COLLEGE ALGEBRA

LINDA ALMGREN KIME
JUDITH CLARK

University of Massachusetts, Boston, Retired

BEVERLY K. MICHAEL

University of Pittsburgh

in collaboration with

Norma M. Agras *Miami Dade College*

Meg Hickey *Massachusetts College of Art*

Sarah Hoffman *University of Arizona*

John A. Lutts *University of Massachusetts, Boston*

Peg Kem McPartland *Golden Gate University, Retired*

Software developed by

Hubert Hohn *Massachusetts College of Art*

Funded by a National Science Foundation Grant

WILEY

JOHN WILEY & SONS, INC.

Vice President and Publisher	Laurie Rosatone
Acquisitions Editor	Joanna Dingle
Project Editor	Ellen Keohane
Editorial Program Assistant	Beth Pearson
Production Services Manager	Dorothy Sinclair
Senior Production Editor	Janet Foxman
Marketing Manager	Jonathan Cottrell
Creative Director	Harry Nolan
Photo Manager	Hilary Newman
Senior Designer	Madelyn Lesure
Media Editor	Melissa Edwards
Media Assistant	Laura Abrams
Production Services	Ingrao Associates

Cover and chapter opener photo composite: (underwater scene in the tropics) © Gray Hardel/Corbis; (divers) Comstock/Getty Images, Inc.

Explorations photo feature: Stephen Frink/Stone/Getty Images, Inc.

This book was set in Times 10/12 by Aptara® Inc., and printed and bound by Courier/Westford. The cover was printed by Courier/Westford.

This book is printed on acid-free paper. ∞

Founded in 1807, John Wiley & Sons, Inc. has been a valued source of knowledge and understanding for more than 200 years, helping people around the world meet their needs and fulfill their aspirations. Our company is built on a foundation of principles that include responsibility to the communities we serve and where we live and work. In 2008, we launched a Corporate Citizenship Initiative, a global effort to address the environmental, social, economic, and ethical challenges we face in our business. Among the issues we are addressing are carbon impact, paper specifications and procurement, ethical conduct within our business and among our vendors, and community and charitable support. For more information, please visit our website: *www.wiley.com/go/citizenship*.

Library of Congress Cataloging in Publication Data:
Kime, Linda Almgren.
  Explorations in College Algebra/Linda Almgren Kime, Judy Clark, Beverly K. Michael. – 5th ed.
    p. cm.
  Includes index.
    ISBN 978-0-470-46644-5 (pbk.)
1. Algebra—Textbooks. I. Kime, Linda Almgren. II. Clark, Judy. III. Michael, Beverly K. IV. Title.
  QA152.3.K56 2011
  512.9—dc22   2010045241

Main Book ISBN   978-0-470-46644-5
Binder-Ready Version ISBN 978-0-470-91761-9

Printed in the United States of America

10 9 8 7 6 5 4 3 2

To our students, who inspired us.

My name is Lexi Fournier and I am a freshman here at Pitt. This semester I am enrolled in the Applied Algebra course using "Explorations in College Algebra." Before coming to Pitt, I had taken numerous math courses varying from algebra to calculus, all of which produced frustration, stress, and a detestation for math as a subject. When I was told that I was required to take a math course here, I was livid. I am a pre-law and creative writing major; why do I need math? My adviser calmed me by informing me of this new math class aimed at teaching non-math/science majors the basic skills they will need in everyday life.

At first I was skeptical, but I'm writing to you now to emphatically recommend this course. What I have learned thus far in this course have been realistic math skills presented in a "left brain" method that fosters confidence and motivation. For once in my career as a student, math is relatable. The concepts are clear and realistic (as opposed to the abstract, amorphous topics addressed in my earlier math classes). I look forward to this class. I enjoy doing my homework and projects because I feel that the lessons are applicable to my life and my future and because I feel empowered by my understanding.

This course is a vital addition to the math department. It has altered my view on the subject and stimulated an appreciation for what I like to call "everyday math."

It is my belief that many students will find the class as encouraging and helpful as I have. Thank you for your attention.

Sincerely,

Lexi Fournier
Student, University of Pittsburgh

PREFACE

This text was born from a desire to reshape the college algebra course, to make it relevant and accessible to all of our students. Our goal is to shift the focus from learning a set of discrete mechanical rules to exploring how algebra is used in the social and physical sciences, and in the world around you. By connecting mathematics to real-life situations, we hope students come to appreciate its power and beauty.

Guiding Principles

The following principles guided our work.

- Develop mathematical concepts using real-world data.
- Pose a wide variety of problems designed to promote mathematical reasoning in different contexts.
- Make connections among the multiple representations of functions.
- Emphasize communication skills, both written and oral.
- Facilitate the use of technology.
- Provide sufficient practice in skill building to enhance problem solving.

Evolution of "Explorations in College Algebra"

The fifth edition of *Explorations* is the result of an 18-year long process. Funding by the National Science Foundation enabled us to develop and publish the first edition, and to work collaboratively with a nationwide consortium of schools. Faculty from selected schools continued to work with us on the second, third, fourth, and now the fifth editions. During each stage of revision we solicited extensive feedback from our colleagues, reviewers and students.

Throughout the text, families of functions are used to model real-world phenomena. After an introductory chapter on data and functions, we first focus on linear and exponential functions, since these are the two most commonly used mathematical models. We then discuss logarithmic, power, quadratic, polynomial and rational functions. Finally we look at ways to extend and combine all these functions to create new functions and apply them in more complex situations.

The text adopts a problem-solving approach, where examples and exercises lie on a continuum from open-ended, non-routine questions to problems on algebraic skills. The materials are designed for flexibility of use and offer multiple options for a wide range of skill levels and departmental needs. The text is currently used in a variety of instructional settings including small classes, laboratory settings and large lectures, and in both two- and four-year institutions.

Special Features and Supplements

An instructor is free to choose among a number of special features. *The Instructor's Teaching and Solutions Manual* provides extensive teaching ideas and support for using these features. The manual also contains solutions to the even-numbered problems.[1]

Exploring Mathematical Ideas and Skill Building

NEW! Explore & Extend are short explorations that provide students and instructors with ideas for going deeper into topics or previewing new concepts. They can be found in almost every section.

Explorations are extensive problem-solving situations at the end of each chapter that can be used for small group or individual projects.

Algebra Aerobics are collections of skill-building practice problems found in each section. All of the answers are in the back of the text.

Check Your Understanding is a set of mostly true/false questions at the end of each chapter (answers are provided in the back of the text) that offer students a chance to assess their understanding of that chapter's mathematical ideas.

Chapter Review: Putting It All Together contains problems that apply all of the basic concepts in the chapter. The answers to the odd-numbered problems are in the back of the text.

60-Second Summaries are short writing assignments found in the exercises and the *Explore & Extend* problems that ask students to succinctly summarize their findings.

Readings are related to topics covered in the text and are available at *www.wiley.com/college/kimeclark* and at *www.**wileyplus**.com*.

Using Technology

Technology is not required to teach this course. However, we provide the following online resources on the course website *www.wiley.com/college/kimeclark* and at *www.**wileyplus**.com*.

Interactive Course Software provides illustrations of the properties of each function, simulations of concepts, and practice in skill building. Their modules can be used in the classroom or computer lab, or downloaded for student use.

Excel and TI Connect™ Graph Link Files contain all the major data sets used in the text and are available in Excel or TI Connect™ program formats.

Graphing Calculator Manual is coordinated with the chapters in the text and offers step-by-step instructions for using the TI-83/TI-84 family of calculators.

NEWLY UPDATED! WileyPLUS is an online course management and assessment system that provides resources for student learning, including short instructional videos. See Wiley's description at the end of the Preface.

---

[1]The manual is available for free to adopters online at *www.wiley.com/college/kimeclark* or at *www.**wileyplus**.com*. You can also contact your local Wiley sales representative to obtain a printed version of the manual.

The Fifth Edition

Overall Changes

Extensive faculty reviews guided our work on the fifth edition.

- The new *Explore & Extend* problems are short explorations imbedded within almost every section.
- The new *Chapter 9*: **Creating New Functions from Old** shows how to transform and combine functions to form new functions. It covers polynomials, rational functions, composition and inverse functions, and also provides more complex explorations.
- *Chapter 8* now covers quadratics and the mathematics of motion.
- **Data sets** were created or updated throughout the text.
- **Revisions** were made to many chapters for greater clarity.
- Many **new problems and exercises** were created, ranging from basic algebraic manipulations to real-world applications.
- **Extended Explorations were integrated into chapters.** The two Extended Explorations in the fourth edition have become Sections in Chapters 2 and 8.

Detailed Changes

CHAPTER 1: An Introduction to Data and Functions has more emphasis on the concepts of input and output, and a different presentation of domain and range. These topics are now more fully integrated throughout text.

CHAPTER 2: Rates of Change and Linear Functions now includes a discussion of the FAM1000 data set, regression lines and correlation coefficients (from 4th edition *Extended Exploration: Looking for Links between Education and Earnings*).

CHAPTER 3: When Lines Meet: Linear Systems now starts with graphs of non-linear systems to introduce intersection points in a real context and includes more real life examples of piecewise linear systems.

CHAPTER 4: The Laws of Exponents and Logarithms: Measuring the Universe merges two sections to generalize the properties of exponents sooner.

CHAPTER 5: Growth and Decay: An Introduction to Exponential Functions has earlier and more extensive coverage of doubling times and half-lives and now introduces *e* through continuous compounding.

CHAPTER 6: Logarithmic Links: Logarithmic and Exponential Functions The old section here on *e* and continuous compounding was moved to the end of Chapter 5.

CHAPTER 8: Quadratics and the Mathematics of Motion is dedicated to quadratic functions and now includes a discussion of freely falling bodies (from 4th edition *Extended Exploration: The Mathematics of Motion*).

CHAPTER 9: New Functions from Old discusses ways of combining and transforming all the functions we studied and contains sections on polynomial, rational and inverse functions, and composition of functions. It concludes with a more complex collection of Explore & Extend type problems.

Acknowledgments

We wish to express our appreciation to all those who helped and supported us during this extensive collaborative endeavor. We are grateful for the support of the National Science Foundation, whose funding made this project possible, and for the generous help of our program officers then, Elizabeth Teles and Marjorie Enneking. Our original Advisory Board, especially Deborah Hughes-Hallett and Philip Morrison, and our original editor, Ruth Baruth provided invaluable advice and encouragement.

Over the last 18 years, through seven versions (including a rough draft and preliminary and 1st through 5th editions), we worked with more faculty, students, teaching assistants, staff, and administrators than we can possibly list here. We are deeply grateful for supportive colleagues at our own universities. The generous support we received from Theresa Mortimer, Patricia Davidson, Mark Pawlak, Maura Mast, Dick Cluster, Anthony Beckwith, Bob Seeley, Randy Albelda, Art MacEwan, Rachel Skvirsky, and Brian Butler, among many others, helped to make this a successful project.

We are deeply indebted to Jennifer Blue, Celeste Hernandez, Paul Lorczak, Georgia Mederer, Ann Ostberg, and Sandra Zirkes for their dedicated search for mathematical errors in the text and solutions, and finding (we hope) all of them. A text designed around the application of real-world data would have been impossible without the time-consuming and exacting research done by Patrick Jarrett and Jie Chen. Edmond Tomastik, George Colpitts and Karl Schaffer were gracious enough to let us adapt some of their real-world examples in the text.

One of the joys of this project has been working with so many dedicated faculty who are searching for new ways to reach out to students. These faculty, their teaching assistants and students all offered incredible support, encouragement, and a wealth of helpful suggestions. In particular, our heartfelt thanks goes to members of our original consortium: Sandi Athanassiou and all the wonderful teaching assistants at the University of Missouri, Columbia; Natalie Leone, University of Pittsburgh; Peggy Tibbs and John Watson, Arkansas Technical University; Josie Hamer, Robert Hoburg, and Bruce King, all past and present faculty at Western Connecticut State University; Judy Stubblefield, Garden City Community College; Lida McDowell, Jan Davis, and Jeff Stuart, University of Southern Mississippi; Ann Steen, Santa Fe Community College; Leah Griffith, Rio Hondo College; Mark Mills, Central College; Tina Bond, Pensacola Junior College; and Curtis Card, Black Hills State University.

The following reviewers' thoughtful comments helped shape the fifth edition: Wendy Ahrendsen, South Dakota State University; Shemsi Alhaddad, University of South Carolina, Lancaster; Mathai Augustine, Cleveland State Community College; Said Bagherieh, Georgia Perimeter College, Dunwoody Campus; Teri Barnes, McLennan Community College; Nicoleta Virginia Bila, Fayetteville State University; Steven Brownstein, University of Arizona; Linda Buckwalter, Harrisburg Area Community College; Elizabeth Burns, Bowling Green State University; Rose Cavin, Chipola Junior College; Daniel P. Fahringer, Harrisburg Area Community College; Kenneth J. Frerichs, Columbus State University; Mark H. Goadrich, Centenary College of Louisiana; Linda Green, Santa Fe College; Lorraine Gregory, Lake Superior State University; Johanna Halsey, Dutchess Community College; Donald Harden, Georgia State University; Erick Hofacker, University of Wisconsin, River Falls; Heather Holley, Santa Fe College; Michael J. Johnson, Meredith College; Vicky Klima, Appalachian State University; Naomi Landau, Pima Community College, Downtown Campus; Xuhui Li, California State University, Long Beach; Gretchen H. Lynn, West Virginia Wesleyan College; Samuel Ofori, Cleveland State Community College; Mary Pearce, Wake Technical Community College; Timothy A. Redl, University of Houston-Downtown; Randy Scott, Santiago Canyon College; Niandong Shi, East Stroudsburg University; Brian A. Snyder, Lake Superior State University; Gilfred B. Swartz, Monmouth University; David E. Thomas, Centenary College; Sherri Wilson, Fort Lewis College; Christopher Yarrish, Harrisburg Area Community College; and Changyong Zhong, Georgia State University.

We are indebted to Laurie Rosatone at Wiley, whose gracious oversight helped to keep this project on track. Particular thanks goes to our new editors at Wiley, Joanna Dingle and especially to Ellen Keohane who, with the able editorial project assistant Beth Pearson, kept us on schedule. It has been a great pleasure, both professionally and personally, to work with Maddy Lesure on her creative cover designs and layouts of the text through multiple editions. Kudos to Sandra Dumas, Janet Foxman, Dorothy Sinclair, and Suzanne Ingrao in production for all their help in getting the text out. The accompanying media for *Explorations* would never have been produced without the experienced help from Melissa Edwards and Laura Abrams. Over the years many others at Wiley have been extraordinarily helpful in dealing with the myriad of endless details in producing a mathematics textbook. Our thanks goes to all of them.

Our families couldn't help but become caught up in this time-consuming endeavor. Judy's husband, Gerry, became our consortium lawyer, and her daughters, Rachel and Caroline were there when needed for support and to mail packages. Kristin and her husband John provided editorial help and more importantly produced two grandchildren: Nola, who asks "why" and Cordelia, who sings numbers. Beverly's husband, Dan, was patient and understanding about the amount of time this edition took. He tolerated Beverly working on Sundays and delaying their trip to Ireland. Beverly felt that without the support and encouragement from Dan, daughters Bridget and Megan, and new son-in-law Felipe Palamo, she couldn't have made it through this edition. Linda's husband, Milford, and her son Kristian provided invaluable scientific and, more importantly, emotional resources. Kristian and his wife Amy Mertl have just produced a marvelous grandchild, Evy, whom Linda believes already shows signs of mathematical acuity. All our family members ran errands, cooked meals, listened to our concerns, and gave us the time and space to work on the text. We offer our love and thanks to them.

Finally, we wish to thank all of our students. It is for them that this book was written.

Judy, Bev, and Linda

P.S. We've tried hard to write an error-free text, but we know that's impossible. You can alert us to any errors by sending an email to *math@wiley.com*. Be sure to reference *Explorations in College Algebra*. We would very much appreciate your input.

Since this text is a collaboration between authors and instructors, we encourage instructors to send new ideas and examples for the new *Explore & Extend* feature for possible future use. We'll put the best ideas on our website.

And last, but not least, we especially want to thank Dr. John Saber from Central Lake College for his kind and encouraging email: ". . . just wanted to thank you for writing this truly wonderful text." It made our day.

WileyPLUS

WileyPLUS is an innovative, research-based, online environment for effective teaching and learning.

What Do Students Receive with WileyPLUS?

A Research-Based Design. *WileyPLUS* provides an online environment that integrates relevant resources, including the entire digital textbook, in an easy-to-navigate framework that helps students study more effectively.

- *WileyPLUS* adds structure by organizing textbook content into smaller, more manageable "chunks."
- Related media, examples, and sample practice items reinforce the learning objectives.
- Innovative features such as calendars, visual progress tracking and self-evaluation tools improve time management and strengthen areas of weakness.

One-on-One Engagement. With *WileyPLUS* for *Explorations in College Algebra, Fifth Edition* students receive 24/7 access to resources that promote positive learning outcomes. Students engage with related examples (in various media) and sample practice items, including:

- Software
- Videos
- Readings
- Excel and TI-83/TI-84 Graph Link Files
- Guided Online (GO) Tutorial problems

Measurable Outcomes. Throughout each study session, students can assess their progress and gain immediate feedback. *WileyPLUS* provides precise reporting of strengths and weaknesses, as well as individualized quizzes, so that students are confident they are spending their time on the right things. With *WileyPLUS*, students always know the exact outcome of their efforts.

What Do Instructors Receive with WileyPLUS?

WileyPLUS provides reliable, customizable resources that reinforce course goals inside and outside of the classroom as well as visibility into individual student progress. Pre-created materials and activities help instructors optimize their time:

Customizable Course Plan: *WileyPLUS* comes with a pre-created Course Plan designed by a subject matter expert uniquely for this course. Simple drag-and-drop tools make it easy to assign the course plan as-is or modify it to reflect your course syllabus.

Pre-Created Activity Types Include:
- Questions
- Readings and Resources
- Presentation
- Print Tests
- Concept Mastery

Course Materials and Assessment Content:
- PowerPoint Slides
- Instructor's Solutions and Teaching Manual
- Readings
- Question Assignments: selected end-of-section and Chapter Review problems coded algorithmically with hints, links to text, whiteboard/show work feature and instructor controlled problem solving help.
- Computerized Test Bank
- Printable Test Bank

Gradebook: *WileyPLUS* provides instant access to reports on trends in class performance, student use of course materials, and progress toward learning objectives, helping inform decisions and drive classroom discussions.

WileyPLUS. Learn More. *www.wileyplus.com.*

Powered by proven technology and built on a foundation of cognitive research, *WileyPLUS* has enriched the education of millions of students, in over 20 countries around the world.

www.wileyplus.com

TABLE OF CONTENTS

CHAPTER 1

AN INTRODUCTION TO DATA AND FUNCTIONS

1.1 *Describing Single-Variable Data* 2
Visualizing Single-Variable Data 2
Numerical Descriptors: What Is "Average" Anyway? 4

1.2 *Describing Relationships between Two Variables* 10
Visualizing Two-Variable Data 10
Constructing a "60-Second Summary" 11
Using Equations to Describe Change 13

1.3 *An Introduction to Functions* 19
What Is a Function? 19
Representing Functions: Words, Tables, Graphs, and Equations 20
Input and Output: Independent and Dependent Variables 21
When Is a Relationship Not a Function? 21

1.4 *The Language of Functions* 26
Function Notation 26
Finding Output Values: Evaluating a Function 27
Finding Input Values: Solving Equations 27
Finding Input and Output Values from Tables and Graphs 28
Rewriting Equations Using Function Notation 28
Domain and Range 31

1.5 *Visualizing Functions* 36
Is There a Maximum or Minimum Value? 36
When is the Output of the Function Positive, Negative, or Zero? 37
Is the Function Increasing or Decreasing? 37
Is the Graph Concave Up or Concave Down? 38
Getting the Big Idea 39

CHAPTER SUMMARY 48
CHECK YOUR UNDERSTANDING 49
CHAPTER 1 REVIEW: PUTTING IT ALL TOGETHER 51
EXPLORATION 1.1 COLLECTING, REPRESENTING, AND ANALYZING DATA 56

CHAPTER 2

RATES OF CHANGE AND LINEAR FUNCTIONS

2.1 *Average Rates of Change* 60
Describing Change in the U.S. Population over Time 60
Defining the Average Rate of Change 61
Limitations of the Average Rate of Change 62

2.2 *Change in the Average Rate of Change* 67

2.3 *The Average Rate of Change Is a Slope* 72
Calculating Slopes 72

2.4 *Putting a Slant on Data* 78

Slanting the Slope: Choosing Different End Points 78
Slanting the Data with Words and Graphs 79

2.5 *Linear Functions: When Rates of Change Are Constant* 85

What If the U.S. Population Had Grown at a Constant Rate?
A Hypothetical Example 85
Real Examples of a Constant Rate of Change 86
The General Equation for a Linear Function 88

2.6 *Visualizing Linear Functions* 92

The Effect of b 93
The Effect of m 93

2.7 *Constructing Graphs and Equations of Linear Functions* 99

Finding the Graph 99
Finding the Equation 101

2.8 *Special Cases* 107

Direct Proportionality 107
Horizontal and Vertical Lines 110
Parallel and Perpendicular Lines 112

2.9 *Breaking the Line: Piecewise Linear Functions* 117

Piecewise Linear Functions 117
The absolute value and absolute value function 118
Step functions 120

2.10 *Constructing Linear Models of Data* 124

Fitting a Line to Data: The Kalama Study 124
Reinitializing the Independent Variable 127
Interpolation and Extrapolation: Making Predictions 128

**2.11 *Looking for Links between Education and Earnings:
A Case Study on Using Regression Lines*** 134

Using U.S. Census Data 134
Summarizing the Data: Regression Lines 135
Regression lines: How good a fit? 137
Interpreting Regression Lines: Correlation vs. Causation 138
Raising More Questions: Going Deeper 139

CHAPTER SUMMARY 146
CHECK YOUR UNDERSTANDING 147
CHAPTER 2 REVIEW: PUTTING IT ALL TOGETHER 149
EXPLORATION 2.1 HAVING IT YOUR WAY 154
EXPLORATION 2.2 A CASE STUDY ON EDUCATION AND
EARNINGS IN THE U.S. 156

CHAPTER 3

WHEN LINES MEET: LINEAR SYSTEMS

3.1 *Interpreting Intersection Points: Linear and Nonlinear Systems* 160

When Curves Collide: Nonlinear Systems 160
When Lines Meet: Linear Systems 163

3.2 *Visualizing and Solving Linear Systems*   171

Visualizing Linear Systems   **171**
Strategies for Solving Linear Systems   **171**
Systems with No Solution or Infinitely Many Solutions   **174**
Linear Systems in Economics: Supply and Demand   **175**

3.3 *Reading between the Lines: Linear Inequalities*   181

Above and Below the Line   **181**
Reading between the Lines   **182**
Manipulating Inequalities   **184**
Breakeven Points: Regions of Profit or Loss   **185**

3.4 *Systems with Piecewise Linear Functions: Tax Plans*   193

Graduated vs. Flat Income Tax   **193**
Comparing the Flat and Graduated Tax Plans   **195**

CHAPTER SUMMARY   **199**
CHECK YOUR UNDERSTANDING   **200**
CHAPTER 3 REVIEW: PUTTING IT ALL TOGETHER   **202**
EXPLORATION 3.1 FLAT VS. GRADUATED INCOME TAX: WHO BENEFITS?   **207**
EXPLORATION 3.2 A COMPARISON OF HYBRID AND CONVENTIONAL AUTOMOBILES   **209**

CHAPTER 4

THE LAWS OF EXPONENTS AND LOGARITHMS: MEASURING THE UNIVERSE

4.1 *The Numbers of Science: Measuring Time and Space*   212

Powers of 10 and the Metric System   **212**
Scientific Notation   **214**

4.2 *Positive Integer Exponents*   218

Exponent Rules   **219**
Common Errors   **221**
Estimating Answers   **222**

4.3 *Zero, Negative, and Fractional Exponents*   226

Zero and Negative Exponents   **226**
Evaluating $\left(\frac{a}{b}\right)^{-n}$ **227**
Fractional Exponents   **228**
Expressions of the Form $a^{\frac{1}{2}}$: Square Roots   **228**
nth Roots: Expressions of the Form $a^{\frac{1}{n}}$ **229**
Rules for Radicals   **230**
Expressions of the Form $a^{\frac{m}{n}}$ **232**

4.4 *Converting Units*   237

Converting Units within the Metric System   **237**
Converting between the Metric and English Systems   **238**
Using Multiple Conversion Factors   **238**

4.5 *Orders of Magnitude*   242

Comparing Numbers of Widely Differing Sizes   **242**
Orders of Magnitude   **242**
Graphing Numbers of Widely Differing Sizes: Log Scales   **243**

4.6 *Logarithms as Numbers* 247

Finding the Logarithms of Powers of 10 **247**
Finding the Logarithm of Any Positive Number **249**
Plotting Numbers on a Logarithmic Scale **250**

CHAPTER SUMMARY 255
CHECK YOUR UNDERSTANDING 256
CHAPTER 4 REVIEW: PUTTING IT ALL TOGETHER 256
EXPLORATION 4.1 THE SCALE AND THE TALE OF THE UNIVERSE 260

CHAPTER 5

GROWTH AND DECAY: AN INTRODUCTION TO EXPONENTIAL FUNCTIONS

5.1 *Exponential Growth* 264

The Growth of *E. coli* Bacteria **264**
The General Exponential Growth Function **265**
Doubling Time **266**
Looking at Real Growth Data for *E. coli* Bacteria **268**

5.2 *Exponential Decay* 271

The Decay of Iodine-131 **271**
The General Exponential Decay Function **272**
Half-Life **273**

5.3 *Comparing Linear and Exponential Functions* 278

Linear Functions **278**
Exponential Functions **278**
Identifying Exponential Functions in a Data Table **279**
A Linear vs. an Exponential Model through Two Points **280**
Comparing the Average Rates of Change **282**
In the Long Run, Exponential Growth Will Always Outpace Linear Growth **283**

5.4 *Visualizing Exponential Functions* 286

The Graphs of Exponential Functions **286**
The Effect of the Base *a* **286**
The Effect of the Initial Value *C* **287**
Horizontal Asymptotes **289**

5.5 *Exponential Functions: A Constant Percent Change* 292

Exponential Growth: Increasing by a Constant Percent **292**
Exponential Decay: Decreasing by a Constant Percent **293**
Revisiting Linear vs. Exponential Functions **295**

5.6 *More Examples of Exponential Growth and Decay* 301

Returning to Doubling Times and Half-Lives **302**
The Malthusian Dilemma **309**
Forming a Fractal Tree **311**

5.7 *Compound Interest and the Number e* 318

Compounding at Different Intervals **319**
Continuous Compounding Using *e* **321**
Continuous Compounding Formula **323**
Exponential Functions Base *e* **324**
Converting e^k into *a* **325**

5.8 Semi-Log Plots of Exponential Functions 331

CHAPTER SUMMARY 335
CHECK YOUR UNDERSTANDING 336
CHAPTER 5 REVIEW: PUTTING IT ALL TOGETHER 338
EXPLORATION 5.1: COMPUTER VIRUSES 342

CHAPTER 6

LOGARITHMIC LINKS: LOGARITHMIC AND EXPONENTIAL FUNCTIONS

6.1 Using Logarithms to Solve Exponential Equations 346
Estimating Solutions to Exponential Equations 346
Rules for Logarithms 347
Solving Exponential Equations Using Logarithms 352
Solving for Doubling Times and Half-Lives 353

6.2 Using Natural Logarithms to Solve Exponential Equations Base *e* 357
The Natural Logarithm 357
Returning to Doubling Times and Half-Lives 359
Converting Exponential Functions from Base *a* to Base *e* 361

6.3 Visualizing and Applying Logarithmic Functions 366
Logarithmic Growth 367
Inverse Functions: Logarithmic vs. Exponential 370
Applications of Logarithmic Functions 372
Measuring acidity: The pH scale 372
Measuring noise: The decibel scale 374

6.4 Using Semi-Log Plots to Construct Exponential Models for Data 379
Why Do Semi-Log Plots of Exponential Functions Produce Straight Lines? 379

CHAPTER SUMMARY 384
CHECK YOUR UNDERSTANDING 385
CHAPTER 6 REVIEW: PUTTING IT ALL TOGETHER 386
EXPLORATION 6.1: CHANGING BASES 390

CHAPTER 7

POWER FUNCTIONS

7.1 The Tension between Surface Area and Volume 394
Scaling Up a Cube 394
Size and Shape 396

7.2 Direct Proportionality: Power Functions with Positive Powers 399
Direct Proportionality 400
Properties of Direct Proportionality 401
Direct Proportionality with More Than One Variable 404

7.3 Visualizing Positive Integer Power Functions 408
The Graphs of $f(x) = x^2$ and $g(x) = x^3$ 408
Odd vs. Even Positive Integer Powers 410
The Effect of the Coefficient k 411

7.4 Comparing Power and Exponential Functions 416
Which Eventually Grows Faster, a Power Function or an Exponential Function? 416

7.5 *Inverse Proportionality: Power Functions with Negative Powers*   **421**

Inverse Proportionality   **421**
Properties of Inverse Proportionality   **423**
Inverse Square Laws   **426**

7.6 *Visualizing Negative Integer Power Functions*   **432**

The Graphs of $f(x) = x^{-1}$ and $g(x) = x^{-2}$ **433**
Odd vs. Even Negative Integer Powers   **434**
The Effect of the Coefficient k **436**

7.7 *Using Logarithmic Scales to Find the Best Functional Model*   **443**

Looking for Lines   **443**
Why Is a Log-Log Plot of a Power Function a Straight Line?   **444**
Translating Power Functions into Equivalent Logarithmic Functions   **445**
Analyzing Weight and Height Data   **447**
Allometry: The Effect of Scale   **450**

CHAPTER SUMMARY   **458**
CHECK YOUR UNDERSTANDING   **459**
CHAPTER 7 REVIEW: PUTTING IT ALL TOGTHER   **460**
EXPLORATION 7.1: SCALING OBJECTS   **464**

CHAPTER 8

QUADRATICS AND THE MATHEMATICS OF MOTION

8.1 *An Introduction to Quadratic Functions: The Standard Form*   **468**

The Simplest Quadratic   **468**
Designing Parabolic Devices   **469**
The Standard Form of a Quadratic   **470**
Properties of Quadratic Functions   **471**
Estimating the Vertex and Horizontal Intercepts   **473**

8.2 *Visualizing Quadratics: The Vertex Form*   **478**

Stretching and Compressing Vertically   **478**
Reflecting across the Horizontal Axis   **478**
Shifting Vertically and Horizontally   **480**
Using Transformations to Get the Vertex Form   **483**

8.3 *The Standard Form vs. the Vertex Form*   **487**

Finding the Vertex from the Standard Form   **487**
Converting between Standard and Vertex Forms   **489**

8.4 *Finding the Horizontal Intercepts: The Factored Form*   **496**

Using Factoring to Find the Horizontal Intercepts   **496**
Factoring Quadratics   **497**
Using the Quadratic Formula to Find the Horizontal Intercepts   **500**
   The discriminant   **501**
   Imaginary and complex numbers   **503**
The Factored Form   **504**
Standard, Factored and Vertex Forms   **506**

8.5 *The Average Rate of Change of a Quadratic Function*   510

8.6 *The Mathematics of Motion*   515
The Scientific Method   516
The Free Fall Experiment   516
Deriving an Equation Relating Distance and Time   516
Velocity: Change in Distance over Time   518
Acceleration: Change in Velocity over Time   520
Deriving an Equation for the Height of an Object in Free Fall   522
Working with an Initial Upward Velocity   525

CHAPTER SUMMARY   531
CHECK YOUR UNDERSTANDING   532
CHAPTER 8 REVIEW: PUTTING IT ALL TOGETHER   533
EXPLORATION 8.1: HOW FAST ARE YOU? USING A RULER TO MAKE A
     REACTION TIMER   536

CHAPTER 9

**NEW FUNCTIONS
FROM OLD**

9.1 *Transformations*   540
Transforming a Function   540

9.2 *The Algebra of Functions*   553

9.3 *Polynomials: The Sum of Power Functions*   561
Defining a Polynomial Function   562
Visualizing Polynomial Functions   564
Finding the Vertical Intercept   567
Finding the Horizontal Intercepts   567

9.4 *Rational Functions: The Quotient of Polynomials*   575
Building a Rational Function: Finding the Average Cost of an MRI Machine   575
Defining a Rational Function   576
Visualizing Rational Functions   577

9.5 *Composition and Inverse Functions*   585
Composing Two Functions   585
Composing More Than Two Functions   588
Inverse Functions: Returning the Original Value   589

9.6 *Exploring, Extending & Expanding*   599

CHAPTER SUMMARY   610
CHECK YOUR UNDERSTANDING   611
CHAPTER 9 REVIEW: PUTTING IT ALL TOGETHER   612

APPENDIX   ***Student Data Tables for Exploration 2.1***   617
                     Data Dictionary for FAM1000 Data   621

SOLUTIONS   *For all Algebra Aerobics and Check Your Understanding problems;
                     for odd-numbered problems in the Exercises and Chapter Reviews.
                     All solutions are grouped by chapter*   ANS-1

INDEX   I-1

CHAPTER 2
RATES OF CHANGE AND LINEAR FUNCTIONS

OVERVIEW

How does the U.S. population change over time? How do children's heights change as they age? Average rates of change provide a tool for measuring how change in one variable affects a second variable. When average rates of change are constant, the relationship is linear.

After reading this chapter, you should be able to

- calculate and interpret average rates of change
- understand how representations of data can be biased
- recognize that a constant rate of change denotes a linear relationship
- represent linear functions with equations, tables, graphs, or words
- derive by hand a linear model for a set of data
- use regression lines to summarize linear trends from scatter plots

2.1 *Average Rates of Change*

In Chapter 1 we looked at how change in one variable could affect change in a second variable. In this section we'll examine how to measure that change.

Describing Change in the U.S. Population over Time

We can think of the U.S. population as a function of time. Table 2.1 and Figure 2.1 are two representations of that function. They show the changes in the size of the U.S. population since 1790, the year the U.S. government conducted its first decennial census. Time, as usual, is the independent variable and population size is the dependent variable.

Population of the United States: 1790–2010

Year	Population in Millions
1790	3.9
1800	5.3
1810	7.2
1820	9.6
1830	12.9
1840	17.1
1850	23.2
1860	31.4
1870	39.8
1880	50.2
1890	63.0
1900	76.2
1910	92.2
1920	106.0
1930	123.2
1940	132.2
1950	151.3
1960	179.3
1970	203.3
1980	226.5
1990	248.7
2000	281.4
2010	309.2 (est.)

Table 2.1

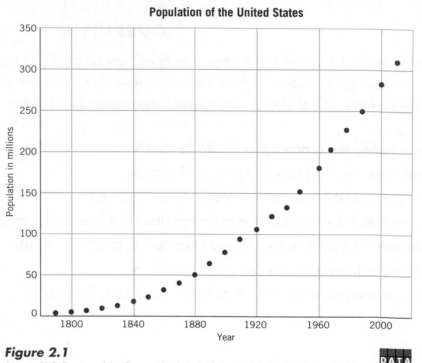

Figure 2.1
Source: U.S. Bureau of the Census, *Statistical Abstract of the United States: 2010.*

DATA
USPOP

Change in population

Figure 2.1 clearly shows that the size of the U.S. population has been growing over the last two centuries, and growing at what looks like an increasingly rapid rate. How can the change in population over time be described quantitatively? One way is to pick two points on the graph of the data and calculate how much the population has changed during the time period between them.

Suppose we look at the change in the population between 1900 and 1990. In 1900 the population was 76.2 million; by 1990 the population had grown to 248.7 million. How much did the population increase?

$$\text{change in population} = (248.7 - 76.2) \text{ million people}$$

$$= 172.5 \text{ million people}$$

This difference is portrayed graphically in Figure 2.2, at the top of the next page.

Change in time

Knowing that the population increased by 172.5 million tells us nothing about how rapid the change was; this change clearly represents much more dramatic growth if it

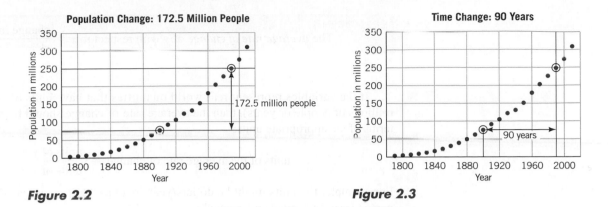

Figure 2.2   **Figure 2.3**

happened over 20 years than if it happened over 200 years. In this case, the length of time over which the change in population occurred is

$$\text{change in years} = (1990 - 1900) \text{ years}$$

$$= 90 \text{ years}$$

This interval is indicated in Figure 2.3 above.

Average rate of change

To find the *average rate of change* in population per year from 1900 to 1990, divide the change in the population by the change in years:

$$\text{average rate of change} \;=\; \frac{\text{change in population}}{\text{change in years}}$$

$$=\; \frac{172.5 \text{ million people}}{90 \text{ years}}$$

$$\approx\; 1.92 \text{ million people/year}$$

In the phrase "million people/year" the slash represents division and is read as "per." So our calculation shows that "on average," the population grew at a rate of 1.92 million people per year from 1900 to 1990. Figure 2.4 depicts the relationship between time and population increase.

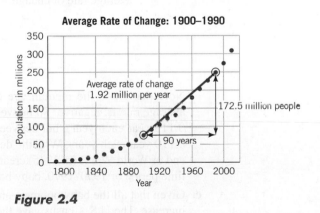

Figure 2.4

Defining the Average Rate of Change

The notion of average rate of change can be used to describe the change in any variable with respect to another. If you have a graph that represents a plot of data points of the form (x, y), then the average rate of change between any two points is the change in the y value divided by the change in the x value.

$$\text{The } \textit{average rate of change} \text{ of } y \text{ with respect to } x = \frac{\text{change in } y}{\text{change in } x}$$

If the variables represent real-world quantities that have units of measure (e.g., millions of people or years), then the average rate of change should be represented in terms of the appropriate units:

$$\text{units of the average rate of change} = \frac{\text{units of } y}{\text{units of } x}$$

For example, the units might be dollars/year (read as "dollars per year") or pounds/person (read as "pounds per person").

EXAMPLE 1

Average rate of change in median age

Between 1850 and 1950 the median age in the United States rose from 18.9 to 30.2, but by 1970 it had dropped to 28.0.

a. Calculate the average rate of change in the median age between 1850 and 1950.

b. Compare your answer in part (a) to the average rate of change between 1950 and 1970 in terms of numbers and societal context.

c. Would you expect the median age to increase or decrease after 1970?

SOLUTION

a. Between 1850 and 1950,

$$\text{average rate of change} = \frac{\text{change in median age}}{\text{change in years}}$$

$$= \frac{(30.2 - 18.9) \text{ years}}{(1950 - 1850) \text{ years}} = \frac{11.3 \text{ years (age)}}{100 \text{ years (calendar)}}$$

$$= 0.113 \text{ years/year}$$

The units are a little confusing. But the results mean that between 1850 and 1950 the median age increased an average of 0.113 years each calendar year.

b. Between 1950 and 1970,

$$\text{average rate of change} = \frac{\text{change in median age}}{\text{change in years}}$$

$$= \frac{(28.0 - 30.2) \text{ years}}{(1970 - 1950) \text{ years}} = \frac{-2.2 \text{ years}}{20 \text{ years}}$$

$$= -0.110 \text{ years/year}$$

Note that since the median age dropped in value between 1950 and 1970, the average rate is negative. So between 1850 and 1950 the median age increased (by 0.113 year each year), but between 1950 and 1970 the median age decreased (by 0.110 year each year). Why the decline? Americans started having babies after the end of World War II (in 1945) creating the "baby boom" lasting until 1964. So in our time period of 1950–1970, baby boomers lowered the overall median age.

c. Given that all the baby boomers are now aging, we would expect the median age to increase. The U.S. Census gave the U.S. median age as 28.0 in 1970 but as 36.9 in 2010, the highest ever.

Limitations of the Average Rate of Change

The average rate of change is an average. So it has the limitations of any average. For example, we calculated that the average rate of change of the U.S. population between

1900 and 1990 was 1.92 million people/year. But it is highly unlikely that each year the population grew by exactly 1.92 million people. Similarly, if the arithmetic average, or *mean*, height of students in your class is 67″, you wouldn't expect everyone (or possibly anyone) to be exactly 67″ tall.

The average rate of change depends upon the end points. If the data points do not all lie in a straight line, the average rate of change varies over different intervals.

E X A M P L E   2   **Average rate of change in population**
Calculate and graph the average rate of change of the U.S. population in these time periods:

a. 1840 to 1940

b. 1880 to 1980

S O L U T I O N   Table 2.2 and Figures 2.5 and 2.6 show the average rate of change calculations and graphs for parts (a) and (b). Note that the average rate of change for 1880–1980 is larger, and the corresponding line segment on the graph is steeper, than that for 1840–1940.

Time Interval	Change in Time	Change in Population	Average Rate of Change
1840–1940	100 yr	132.2 − 17.1 = 115.1 million	$\dfrac{115.1 \text{ million}}{100 \text{ yr}} \approx 1.15$ million/yr
1880–1980	100 yr	226.5 − 50.2 = 176.3 million	$\dfrac{176.3 \text{ million}}{100 \text{ yr}} \approx 1.76$ million/yr

Table 2.2

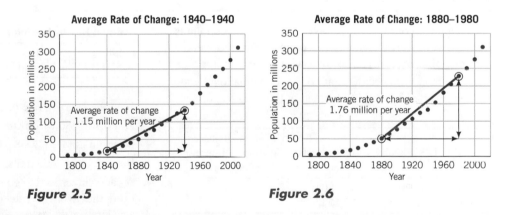

Figure 2.5   **Figure 2.6**

EXPLORE & EXTEND

The Federal Debt
The *federal debt* describes the *total* amount the government owes, accumulated over time (as opposed to the *yearly federal surplus or deficit,* see pages 11–12). The accompanying table and graph on the next page show the federal debt from 1945 to 2010. (*Note:* you can find the current debt at *www.brillig.com/debt_clock.*)

a. Has the gross federal debt decreased over any five-year interval between 1945 and 2010?

The federal debt has clearly been growing since 1950. But how quickly?

b. What was the average rate of change in the federal debt between 1945 and 1980? Between 1980 and 2010? How could you interpret these results?

Accumulated Gross Federal Debt

Year	Debt (billions of $)
1945	260
1950	257
1955	274
1960	291
1965	322
1970	381
1975	542
1980	909
1985	1,818
1990	3,207
1995	4,921
2000	5,674
2005	7,932
2010	12,130 (est.)

Source: www.census.gov.

Gross Federal Debt (billions of $)

c. Over what five-year interval does the largest rate of change occur? Calculate that rate of change. Show how you can use the debt at the starting point of this interval and the average rate of change to generate the debt at the end point.

d. Create a newspaper title to describe these data and the graph.

Algebra Aerobics 2.1

1. Suppose your weight five years ago was 135 pounds and your weight today is 143 pounds. Find the average rate of change in your weight with respect to time.

2. The table below shows data on U.S. international trade as reported by the U.S. Bureau of the Census.

Year	U.S. Exports (billions of $)	U.S. Imports (billions of $)	U.S. Trade Balance = Exports − Imports (billions of $)
2000	1,070.6	1,450.0	−379.4
2008	1,826.6	2,522.5	−695.9

a. What is the average rate of change between 2000 and 2008 for:
   i. Exports?
   ii. Imports?
   iii. The trade balance, the difference between what we sell abroad (exports) and buy from abroad (imports)?

b. What do these numbers tell us?

3. The following table indicates the number of deaths in motor vehicle accidents in the United States as listed by the U.S. Bureau of the Census.

Annual Deaths in Motor Vehicle Accidents (thousands)

1980	1990	2000	2008
52.1	44.6	41.9	43.3

Find the average rate of change:
a. From 1980 to 2000
b. From 2000 to 2008
Be sure to include units.

4. A car is advertised to go from 0 to 60 mph in 5 seconds. Find the average rate of change (i.e., the average acceleration) over that time.

5. According to the National Association of Insurance Commissioners, the mean cost for automobile insurance has gone from $689 in 2000 to $800 in 2009. What is the average rate of change from 2000 to 2009?

6. A football player runs for 1056 yards in 2006 and for 978 yards in 2010. Find the average rate of change in his performance from 2006 to 2010?

7. The African elephant is an endangered species, largely because poachers illegally kill elephants to sell the ivory from their tusks. In Kenya alone between 1972 and 1989, the elephant population fell from 140,000 to a mere 19,000. In 1989 the Convention on Trade in

Endangered Species introduced a ban on all ivory trade. But the estimated number of elephants in Kenya in 2009 is still only 30,000.

a. Calculate the average rate of change between 1972 and 1989 and then between 1989 and 2009. Describe what each rate means.

b. Using the annual average rate of change between 1989 and 2009, in how many years after 1989 would the number of elephants return to the 1972 size of 140,000?

Exercises for Section 2.1

1. If r is measured in inches, s in pounds, and t in minutes, identify the units for the following average rates of change:

a. $\dfrac{\text{change in } r}{\text{change in } s}$
b. $\dfrac{\text{change in } t}{\text{change in } r}$
c. $\dfrac{\text{change in } s}{\text{change in } r}$

2. Assume that R is measured in dollars, S in ounces, T in dollars per ounce, and V in ounces per dollar. Write a product of two of these terms whose resulting units will be:

a. Dollars
b. Ounces

3. Your car's gas tank is full and you take a trip. You travel 212 miles, then you fill your gas tank up again and it takes 10.8 gallons. If your change in distance is 212 miles and your change in gallons is 10.8, what is the average rate of change of gasoline used, measured in miles per gallon?

4. The gas gauge on your car is broken, but you know that the car averages 22 miles per gallon. You fill your 15.5-gallon gas tank and tell your friend, "I can travel 300 miles before I need to fill up the tank again." Why is this true?

5. The consumption of margarine (in pounds per person) decreased from 10.9 in 1990 to 4.6 in 2006. What was the annual average rate of change? (*Source: www.census.gov*)

6. The percentage of people who own homes in the United States has gone from 65.5% in 1980 to 67.6% in 2009. What is the average rate of change in percentage points per year?

7. The accompanying table shows females' SAT scores in 2000 and 2008.

Year	Average Female Verbal SAT	Average Female Math SAT
2000	504	498
2008	498	499

Source: www.collegeboard.com.

Find the average rate of change:

a. In the math scores from 2000 to 2008
b. In the verbal scores from 2000 to 2008

8. a. In 1992 the aerospace industry showed a net loss (negative profit) of $1.84 billion. In 2002 the industry had a net profit of $8.97 billion. Find the average annual rate of change in net profits from 1992 to 2002?

b. In 2007, aerospace industry net profits were $16.9 billion. Find the average rate of change in net profits:

i. From 1992 to 2007
ii. From 2002 to 2007

9. The U.S. Bureau of the Census provided the following data concerning computer use in all U.S. elementary and secondary schools.

Academic Year End	Total Number of Computers	Students per Computer
1985	630,000	84.1
2005	13,600,000	4.0
2006	14,165,000	3.9

a. Find the average rate of change from 1985 to 2005 in:
   i. The number of computers being used
   ii. The number of students per computer
b. Find the average rate of change from 2005 to 2006 in:
   i. The number of computers being used
   ii. The number of students per computer
c. Using the annual average rate of change between 2005 and 2006, in how many years after 2005 would there be a projection of two students per computer.
d. Look online to see if you can find more up-to-date data.

10. According to the U.S. Bureau of the Census, the percentage of persons 25 years old and over completing 4 or more years of college was 4.6 in 1940 and 29.4 in 2008.

a. Plot the data, labeling both axes and the coordinates of the points.
b. Calculate the average rate of change in percentage points per year.
c. Write a topic sentence summarizing what you think is the central idea to be drawn from these data.

11. According to a 2010 Kaiser Family Foundation report, during 2009 kids spent almost every waking hour outside of school using electronic devices such as iPhones, computers, or televisions. Those aged 8 to 18 spent an average seven and a half hours a day in 2009 versus six and a half hours in 2004. Calculate the annual average rate of change of electronic use for kids from 2004 to 2009.

12. The following graph was produced by the U.S. Department of Transportation to report actual and predicted highway crash fatalities (up until 2020). The left-hand axis shows the number of highway fatalities and the right-hand axis the percent of all crashes that were fatal.

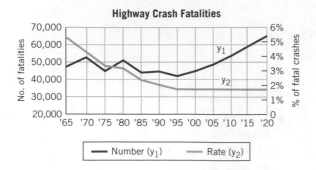

Highway Crash Fatalities

a. Estimate the number of fatalities (y_1) in 1965 and 1995 and calculate the annual average rate of change.

b. Estimate the number of fatalities (y_1) in 1995 and those predicted to 2020. Calculate the annual average rate of change.

c. The fatality *rate* (y_2) between 1995 and 2020 is predicted to be flat (just below 2%), while the *number* of fatalities is expected to increase. How could that be?

13. Use the information in the accompanying table to answer the following questions.

Percentage of Persons 25 Years Old and Over Who Have Completed 4 Years of High School or More

	1940	2008
All	24.5	86.6
White	26.1	87.1
Black	7.3	83.0
Asian/Pacific Islander	22.6	88.7

Source: U.S. Bureau of the Census, *Statistical Abstract of the United States: 2009.*

a. What was the average rate of change (in percentage points per year) of completion of 4 years of high school from 1940 to 2008 for whites? For blacks? For Asian/Pacific Islanders? For all?

b. If these rates continue, what percentages of whites, of blacks, of Asian/Pacific Islanders, and of all will have finished 4 years of high school in the year 2010? Check the Internet to see if your predictions are accurate.

c. If these rates continue, in what year will 100% of whites have completed 4 years of high school or more? In what year 100% of blacks? In what year 100% of Asian/Pacific Islanders? Do these projections make sense?

d. Write a 60-second summary describing the key elements in the high school completion data. Include rates of change and possible projections for the near future.

14. The data show U.S. consumption and exports of cigarettes.

Year	U.S. Consumption (billions)	Exports (billions)
1960	484	20
1980	631	82
2000	430	148
2007	364	102

Source: U.S. Department of Agriculture.

a. Calculate the average rates of change in U.S. cigarette consumption from 1960 to 1980, from 1980 to 2007, and from 1960 to 2007.

b. Compute the average rate of change for cigarette exports from 1960 to 2007. Does this give an accurate image of cigarette exports?

c. The total number of cigarettes consumed in the United States in 1960 was 484 billion, very close to the number consumed in 1995, 487 billion. Does that mean smoking was as popular in 1995 as it was in 1960? Explain your answer.

d. Write a paragraph summarizing what the data tell you about the consumption and exports of cigarettes since 1960, including average rates of change.

15. Examine the accompanying table on life expectancy.

Average Number of Years of Life Expectancy in the United States by Race and Sex Since 1900

Life Expectancy at Birth by Year	White Males	White Females	Black Males	Black Females
1900	46.6	48.7	32.5	33.5
1950	66.5	72.2	58.9	62.7
2000	74.8	80.0	68.2	74.9
2010	76.5	81.3	70.2	77.2

Source: U.S. National Center for Health Statistics, *Statistical Abstract of the United States*, 2010.

a. What group had the highest life expectancy in 1900? In 2010? What group had the lowest life expectancy in 1900? In 2010?

b. Which group had the largest average rate of change in life expectancy between 1900 and 2010?

c. Write a short summary of the patterns in U.S. life expectancy from 1900 to 2010 using average rates of change to support your points.

16. The table below shows U.S. marital status.

Marital Status of Population 15 Years Old and Older

Year	Number of Unmarried Males (in thousands)	Number of Unmarried Females (in thousands)
1950	17,735	19,525
1960	18,492	22,024
1970	23,450	29,618
1980	30,134	36,950
1990	36,121	43,040
2000	43,429	50,133
2008	51,383	57,756

Source: U.S. Bureau of the Census, *www.census.gov.*

a. Calculate the average rate of change in the number of unmarried males between 1950 and 2008. Interpret your results.

b. Calculate the average rate of change in the number of unmarried females between 1950 and 2008. Interpret your results.

c. Compare the two results.

d. What does this tell you, if anything, about the *percentages* of unmarried males and females?

2.2 *Change in the Average Rate of Change*

We can obtain an even better sense of patterns in the U.S. population if we look at how the average rate of change varies over time. One way to do this is to pick a fixed interval for time and then calculate the average rate of change for each successive time period. Since we have the U.S. population data in 10-year intervals, we can calculate the average rate of change for each successive decade. The third column in Table 2.3 shows the results of these calculations. Each entry represents the average population growth *per year* (the average annual rate of change) during the previous decade. A few of these calculations are worked out in the last column of the table.

Average Annual Rates of Change of U.S. Population: 1790–2010

Year	Population (millions)	Average Annual Rate of change for Prior Decade (millions/yr)	Sample Calculations
1790	3.9	Data not available	
1800	5.3	0.14	$0.14 = (5.3 - 3.9)/(1800 - 1790)$
1810	7.2	0.19	
1820	9.6	0.24	
1830	12.9	0.33	
1840	17.1	0.42	$0.42 = (17.1 - 12.9)/(1840 - 1830)$
1850	23.2	0.61	
1860	31.4	0.82	
1870	39.8	0.84	
1880	50.2	1.04	
1890	63.0	1.28	
1900	76.2	1.32	
1910	92.2	1.60	
1920	106.0	1.38	
1930	123.2	1.72	
1940	132.2	0.90	$0.90 = (132.2 - 123.2)/(1940 - 1930)$
1950	151.3	1.91	
1960	179.3	2.80	
1970	203.3	2.40	
1980	226.5	2.32	
1990	248.7	2.22	
2000	281.4	3.27	
2010	309.2 (est.)	2.78	

Table 2.3
Source: U.S. Bureau of the Census, *Statistical Abstract of the United States: 2010.*

What is happening to the average rate of change over time?

Start at the top of the third column and scan down the numbers. Notice that until 1910 the average rate of change increases every year. Not only is the population growing every decade until 1910, but it is growing at an increasing rate. It's like a car that is not only moving forward but also accelerating. A feature that was not so obvious in the original data is now evident: In the intervals 1910 to 1920, 1930 to 1940, and 1960 to

1990 we see an increasing population but a decreasing rate of growth. It's like a car decelerating—it is still moving forward but it is slowing down.

EXAMPLE 1

Changes in the rate of change in U.S. population
a. Graph and interpret the average rate of change of the population over time.
b. What does this tell you about the U.S. population?

SOLUTION

a. Figure 2.7 clearly shows how the average rate of change in population fluctuates over time.

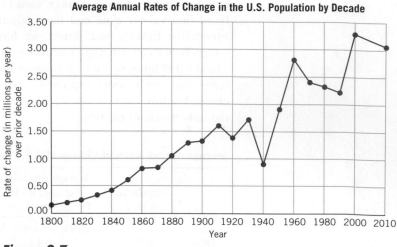

Average Annual Rates of Change in the U.S. Population by Decade

Figure 2.7

The first point, corresponding to the year 1800, shows an average rate of change of 0.14 million people/year for the decade 1790 to 1800. The rate 1.72, corresponding to the year 1930, means that from 1920 to 1930 the population was increasing at a rate of 1.72 million people/year.

b. The rate of growth was steadily increasing up until about 1910. Why did it change after that? A possible explanation for the slowdown in growth rate in the decade prior to 1920 might be World War I and the 1918 worldwide flu epidemic. By 1920 the flu killed nearly 20,000,000 people, including about 500,000 Americans.

The steepest decline in the average rate of change is between 1930 and 1940. One obvious suspect for the big slowdown in population growth in the 1930s is the Great Depression.

The average rate of change increases again between 1940 and 1960, an indication of the baby boomers born after World War II. It drops off from the 1960s through the 1980s, possibly due to the introduction of birth control. The rate increases once more in the 1990s. This latest surge in the growth rate is attributed partially to the "baby boom echo" (the result of baby boomers having children) and to a rise in birth rates and immigration.

EXPLORE & EXTEND

2.2

DATA

FEDDEBT

Changes in the Average Rate of Change in the Federal Debt
In Explore & Extend 2.1 (pages 63–64), the federal debt table and graph suggest a steady increase in the debt. But the average rate of change of the federal debt (in the accompanying table and graph) is a better gauge for showing how rapidly the debt is growing or slowing down.

Accumulated Gross Federal Debt

Year	Debt (billions of $)	Average Rate of Change (billions of $ per year)
1945	260	n.a.
1950	257	−1
1955	274	4
1960	291	3
1965	322	6
1970	381	12
1975	542	32
1980	909	73
1985	1,818	182
1990	3,207	278
1995	4,921	343
2000	5,674	151
2005	7,932	452
2010	12,130 (est.)	840

Source: www.census.gov.

Average Rate of Change of Federal Debt

a. What does it mean that the average rate of change between 1945 and 1950 is negative (i.e., −1)?

b. After 1950, the average rate of change is positive. What does that tell you about the national debt?

c. The average rate of change decreases between 1995 and 2000. Interpret this change.

d. Over what interval does the average rate of change dramatically increase? Identify what was happening in the U.S. economy then.

e. Create a 60-second summary about the average rate of change in the federal debt.

Algebra Aerobics 2.2

1. The accompanying table and graph show estimates for world population between 1800 and 2050.

   a. Fill in the third column of the table by calculating the annual average rate of change.

   b. Graph the annual average rate of change versus time.

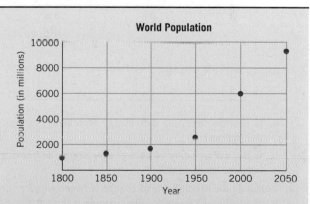

World Population

Year	Total Population (millions)	Annual Average Rate of Change (over prior 50 years)
1800	980	n.a.
1850	1260	
1900	1650	
1950	2520	
2000	6090	
2050	9317 (est.)	

Source: Population Division of the United Nations, *www.un.org/popin.*

c. During what 50-year period was the average annual rate of change the largest?

d. Describe in general terms what happened (and is predicted to happen) to the world population and its average rate of change between 1800 and 2050.

2. A graph illustrating a corporation's profits indicates a positive average rate of change between 2006 and 2007, another positive rate of change between 2007

and 2008, a zero rate of change between 2008 and 2009, and a negative rate of change between 2009 and 2010. Describe the graph and the company's financial situation over the years 2006–2010.

3. The data table shows educational data collected on 18- to 24-year-olds between 1960 and 2007 by the National Center for Educational Statistics. The table shows the number of students who graduated from high school or completed a GED (a high school equivalency exam) during the indicated year.

High School Completers

Year	Number (thousands)	Average Rate of Change (thousands per year)
1960	1679	n.a.
1970	2757	
1980	3089	
1990	2355	
2000	2756	
2007	2955	

a. Fill in the blank cells with the appropriate average rates of change for high school completers.

b. Describe the pattern in the number of high school completers between 1960 and 2007.

c. In this context what does it mean when the average rate of change is positive? Give a specific example from your data.

d. What does it mean when the average rate of change is negative? Give another specific example.

e. What does it mean when two adjacent average rates of change are positive, but the second one is smaller than the first?

Exercises for Section 2.2

Graphing program is optional for Exercises 4 and 11 and access to Internet recommended for Exercise 11.

1. Calculate the average rate of change between adjacent points for the following function. (The first few are done for you.)

x	$f(x)$	Average Rate of Change
0	0	n.a.
1	1	1
2	8	7
3	27	
4	64	
5	125	

a. Is the function $f(x)$ increasing, decreasing, or constant throughout?

b. Is the average rate of change increasing, decreasing, or constant throughout?

2. Calculate the average rate of change between adjacent points for the following function. The first one is done for you.

x	$f(x)$	Average Rate of Change
0	0	n.a.
1	1	1
2	16	
3	81	
4	256	
5	625	

a. Is the function $f(x)$ increasing, decreasing, or constant throughout?

b. Is the average rate of change increasing, decreasing, or constant throughout?

3. The accompanying table shows the number of registered motor vehicles in the United States.

Year	Registered Motor Vehicles (millions)	Annual Average Rate of Change (over prior decade)
1970	108	n.a.
1980	156	
1990	189	
2000	218	
2010	270 (est.)	

a. Fill in the third column in the table.

b. During which decade was the average rate of change the smallest?

c. During which decade was the average rate of change the largest?

d. Write a paragraph describing the change in registered motor vehicles between 1970 and 2010.

4. (Graphing program optional.) The accompanying table indicates the number of juvenile arrests (in thousands) in the United States for aggravated assault.

Year	Juvenile Arrests (thousands)	Annual Average Rate of Change over Prior 5 Years
1985	36.8	n.a.
1990	54.5	
1995	68.5	
2000	49.8	
2005	36.9	

a. Fill in the third column in the table by calculating the annual average rate of change.

b. Graph the annual average rate of change versus time.

c. During what 5-year period was the annual average rate of change the largest?

d. Describe the change in juvenile arrests during these years by referring both to the number and to the annual average rate of change.

5. Calculate the average rate of change between adjacent points for the following functions and place the values in a third column in each table. (The first entry is "n.a.")

x	f(x)	x	g(x)
0	5	0	270
10	25	10	240
20	45	20	210
30	65	30	180
40	85	40	150
50	105	50	120

a. Are the functions $f(x)$ and $g(x)$ increasing, decreasing, or constant throughout?

b. Is the average rate of change of each function increasing, decreasing, or constant throughout?

6. Calculate the average rate of change between adjacent points for each of the functions in Tables A–D and place the values in a third column in each table. (The first entry is "n.a.") Then for each function decide which statement best describes it.

Table A

x	f(x)
0	1
1	3
2	9
3	27
4	81
5	243

Table B

x	g(x)
0	200
15	155
30	110
45	65
60	20
75	−25

Table C

x	h(x)
0	50
10	55
20	60
30	65
40	70
50	75

Table D

x	k(x)
0	40
1	31
2	24
3	19
4	16
5	15

a. As x increases, the function increases at a constant rate.

b. As x increases, the function increases at an increasing rate.

c. As x increases, the function decreases at a constant rate.

d. As x increases, the function decreases at a decreasing rate.

7. Each of the following functions has a graph that is increasing. If you calculated the average rate of change between sequential equal-size intervals, which function can be said to have an average rate of change that is:

a. Constant?　　b. Increasing?　　c. Decreasing?

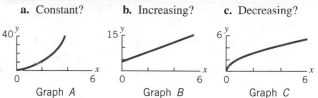

Graph *A*　　　Graph *B*　　　Graph *C*

8. Match each data table with its graph.

Table A

x	y
0	2
1	5
2	8
3	11
4	14
5	17
6	20

Table B

x	y
0	0
1	0.5
2	2
3	4.5
4	8
5	12.5
6	18

Table C

x	y
0	0
1	2
4	4
9	6
16	8
25	10
36	12

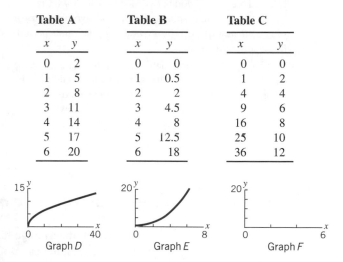

Graph *D*　　　Graph *E*　　　Graph *F*

9. Refer to the first two data tables (A and B) in Exercise 8. Insert a third column in each table and label the column "average rate of change." (Again the first entry is "n.a.")

a. Calculate the average rate of change over adjacent data points.

b. Identify whether the table represents an average rate of change that is constant, increasing, or decreasing.

c. Explain how you could tell this by looking at the corresponding graph in Exercise 8.

10. Following are data on the U.S. population over the time period 1830–1930 (extracted from Table 2.1).

U.S. Population

Year	Population (in millions)	Average Rate of Change (millions/yr)
1830	12.9	n.a.
1850	23.2	
1870		0.83
1890	63.0	1.16
1910	92.2	
1930		1.55

Source: U.S. Bureau of the Census, *www.census.gov*

a. Fill in the missing parts of the table.

b. Which 20-year interval experienced the largest average rate of change in population?

c. Which 20-year interval experienced the smallest average rate of change in population?

11. (Technology recommended.) The accompanying data give a picture of the two major methods of news communication in the United States. (See also Excel or graph link files NEWPRINT and ONAIRTV.)

a. Use the U.S. population numbers from Table 2.1 (at the beginning of this chapter) to calculate and compare the number of copies of newspapers *per person* in 1920 and in 2000.

b. Create a table that displays the annual average rate of change in TV stations for each decade between 1950 and 2000 and the 8-yr period between 2000 and 2008. Create a similar table that displays the annual average rate of change in newspapers published for the same period. Graph both of the results.

c. If new TV stations continue to come into existence at the same rate as from 2000 to 2008, how many would there be in the year 2010? If possible, check your predictions on the Internet. Is your estimate too high or too low?

d. What trends do you see in the dissemination of news as reflected in these data?

Number of U.S. Newspapers

Year	Newspapers (thousands of copies printed)	Number of Newspapers Published
1920	27,791	2042
1930	39,589	1942
1940	41,132	1878
1950	53,829	1772
1960	58,882	1763
1970	62,108	1748
1980	62,202	1745
1990	62,324	1611
2000	55,800	1480
2008	48,598	1408

Source: U.S. Bureau of the Census, *www.census.gov.*

Number of U.S. Commercial TV Stations

Year	Number of Commercial TV Stations
1950	98
1960	515
1970	677
1980	734
1990	1092
2000	1248
2008	1353

Source: U.S. Bureau of the Census, *www.census.gov.*

2.3 *The Average Rate of Change Is a Slope*

On a graph, the average rate of change is the *slope* of the line connecting two points. The slope is an indicator of the steepness of the line.

Calculating Slopes

The reading "Slopes" describes many of the practical applications of slopes, from cowboy boots to handicap ramps.

If (x_1, y_1) and (x_2, y_2) are two points, then the change in y equals $y_2 - y_1$. This difference is often denoted by Δy, read as "delta y," where Δ is the Greek letter capital D (think of D as representing difference): $\Delta y = y_2 - y_1$. Similarly, the change in x (delta x) can be represented by $\Delta x = x_2 - x_1$. See Figure 2.8.

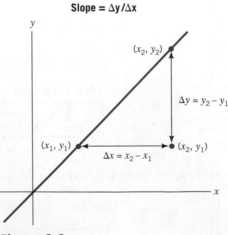

Slope = $\Delta y / \Delta x$

Figure 2.8

The average rate of change represents a *slope*. Given two points (x_1, y_1) and (x_2, y_2),

$$\text{average rate of change} = \frac{\text{change in } y}{\text{change in } x} = \frac{\Delta y}{\Delta x} = \frac{y_2 - y_1}{x_2 - x_1} = \text{slope}$$

When calculating a slope, it doesn't matter which point is first

Given two points, (x_1, y_1) and (x_2, y_2), it doesn't matter which one we use as the first point when we calculate the slope. In other words, we can calculate the slope between (x_1, y_1) and (x_2, y_2) as

$$\frac{y_2 - y_1}{x_2 - x_1} \quad \text{or as} \quad \frac{y_1 - y_2}{x_1 - x_2}$$

The two calculations result in the same value. Note: In calculating the slope, we need to be consistent in the order in which the coordinates appear in the numerator and the denominator. If y_1 is the first term in the numerator, then x_1 must be the first term in the denominator.

EXAMPLE 1

Slope through two points

Plot the two points $(-2, -6)$ and $(7, 12)$ and calculate the slope of the line passing through them.

SOLUTION

Treating $(-2, -6)$ as (x_1, y_1) and $(7, 12)$ as (x_2, y_2) (Figure 2.9), then

$$\text{slope} = \frac{y_2 - y_1}{x_2 - x_1} = \frac{12 - (-6)}{7 - (-2)} = \frac{18}{9} = 2$$

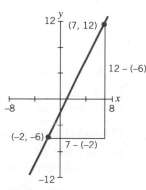

Figure 2.9

We could also have used -6 and -2 as the first terms in the numerator and denominator, respectively:

$$\text{slope} = \frac{y_1 - y_2}{x_1 - x_2} = \frac{-6 - 12}{-2 - 7} = \frac{-18}{-9} = 2$$

Either way we obtain the same answer.

EXAMPLE 2

Decrease in rural population

The percentage of the U.S. population living in rural areas decreased from 84.7% in 1850 to 21.0% in 2000. Plot the data, then calculate and interpret the average rate of change in the rural population over time.

SOLUTION

If we treat year as the input and percentage as the output, our given data can be represented by the points (1850, 84.7) and (2000, 21.0). (See Figure 2.10.)

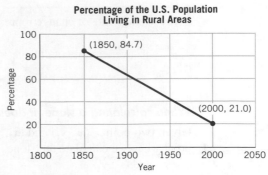

Figure 2.10

Source: U.S. Bureau of the Census, *www.census.gov.*

$$\text{The average rate of change} = \frac{\text{change in percentage points of rural population}}{\text{change in time}}$$

$$= \frac{(21.0 - 84.7) \text{ percentage points}}{(2000 - 1850) \text{ years}}$$

$$= \frac{-63.7 \text{ percentage points}}{150 \text{ years}}$$

$$\approx -0.42 \text{ percentage points per year}$$

The sign of the average rate of change is negative since the percentage of people living in rural areas was decreasing. (The negative slope of the graph in Figure 2.10 confirms this.) The value tells us that, on average, the percentage living in rural areas decreased by 0.42 percentage points (or about one-half of 1%) each year between 1850 and 2000. The change per year may seem small, but in a century and a half the rural population went from being the overwhelming majority (84.7%) to about one-fifth (21%) of the population.

If the slope of the line connecting any two data points on a graph is *positive,* then the output increases as the input increases.

If the slope is *negative,* the output decreases as the input increases.

If the slope is *zero,* the line is flat and the output is constant for all input.

EXAMPLE 3

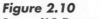

Plotting civil disturbances

a. Use Table 2.4 to plot the number of civil disturbances in U.S. cities (in three-month intervals) between 1967 and 1977, and then connect the points. Without doing any

Civil Disturbances in U.S. Cities

Year	Period	Number of Disturbances	Year	Period	Number of Disturbances
1968	Jan.–Mar.	6	1971	Jan.–Mar.	12
	Apr.–June	46		Apr.–June	21
	July–Sept.	25		July–Sept.	5
	Oct.–Dec.	3		Oct.–Dec.	1
1969	Jan.–Mar.	5	1972	Jan.–Mar.	3
	Apr.–June	27		Apr.–June	8
	July–Sept.	19		July–Sept.	5
	Oct.–Dec.	6		Oct.–Dec.	5
1970	Jan.–Mar.	26			
	Apr.–June	24			
	July–Sept.	20			
	Oct.–Dec.	6			

Table 2.4

Source: D.S. Moore and G.P. McCabe, *Introduction to the Practice of Statistics.*
Copyright © 1989 by W.H. Freeman and Company. Used with permission.

calculations, indicate on the graph when the average rate of change between adjacent points is positive (+), negative (−), and zero (0).

b. Describe the pattern of civil disturbances.

SOLUTION   **a.** The data are plotted in Figure 2.11. Each line segment is labeled +, −, or 0, indicating whether the average rate of change between adjacent points is positive, negative, or zero. The largest positive average rate of change, or steepest upward slope, seems to be between the January-to-March and April-to-June in 1968. The largest negative average rate of change, or steepest downward slope, appears later in the same year (1968) between the July-to-September and October-to-December.

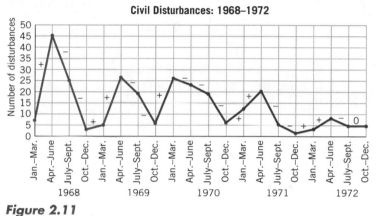

Figure 2.11

Source: D.S. Moore and G.P. McCabe, *Introduction to the Practice of Statistics.*
Copyright © 1989 by W.H. Freeman and Company. Used with permission.

b. Civil disturbances between 1968 and 1972 occurred in cycles: The largest numbers generally occurred in the spring months and the smallest in the winter months. The peaks decrease over time. What was happening in America that might correlate with the peaks? This was a tumultuous period in our history. Many previously silent factions of society were finding their voices. Recall that in April 1968 Martin Luther King was assassinated, and in March 1973 the last American troops were withdrawn from Vietnam, technically ending the Vietnam War. On April 30, 1975, frantic pro-U.S. Vietnamese tried a final escape via American helicopters on the U.S. embassy roof. The next day, ten Marines from the U.S. embassy were the last American soldiers to depart, concluding the U.S. presence in Vietnam.

Algebra Aerobics 2.3

1. a. Plot each pair of points and then calculate the slope of the line that passes through them.

   i. (4, 1) and (8, 11)

   ii. (−3, 6) and (2, 6)

   iii. (0, −3) and (−5, −1)

b. Recalculate the slopes in part (a), reversing the order of the points. Check that your answers are the same.

2. Specify the intervals on the graph of violent crimes for which the average rate of change between adjacent data points is approximately zero.

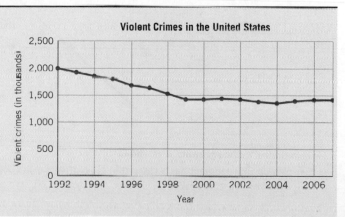

3. Specify the intervals on the graph of tornado deaths for which the average rate of change between adjacent data points appears positive, negative, or zero.

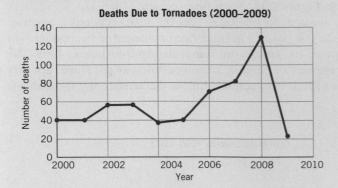

Deaths Due to Tornadoes (2000–2009)

4. What is the missing y-coordinate that would produce a slope of 4, if a line were drawn through the points $(3, -2)$ and $(5, y)$?

5. Find the slope of the line through the points $(2, 9)$ and $(2 + h, 9 + 2h)$ where $h \neq 0$.

6. Consider points $P_1 = (0, 0)$, $P_2 = (1, 1)$, $P_3 = (2, 4)$, and $P_4 = (3, 9)$.

   a. Verify that these four points lie on the graph of $y = x^2$.

   b. Find the slope of the line segments connecting P_1 and P_2, P_2 and P_3, and P_3 and P_4.

   c. What do these slopes suggest about the graph of the function within those intervals?

7. Show that the two forms are equivalent. (*Hint:* multiply by $\dfrac{-1}{-1}$.)

$$\frac{y_2 - y_1}{x_2 - x_1} \quad \text{and} \quad \frac{y_1 - y_2}{x_1 - x_2}$$

Exercises for Section 2.3

1. Find the slope of a straight line that goes through:

   a. $(-5, -6)$ and $(2, 3)$

   b. $(-5, 6)$ and $(2, -3)$

2. Find the slope of each line using the points where the graph intersects the x and y axes.

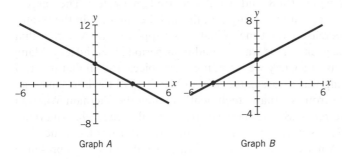

Graph *A*   Graph *B*

3. Find the slope of each line using the points where the line crosses the x- or y-axis.

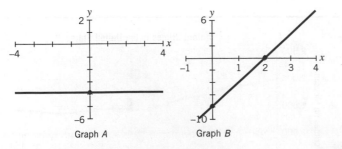

Graph *A*   Graph *B*

4. Examine the line segments A, B, and C.

   a. Which line segment has a slope that is positive? That is negative? That is zero?

b. Calculate the exact slope for each line segment A, B, and C.

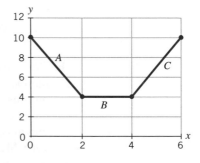

5. Given the following graph:

   a. Estimate the slope for each line segment A–F.

   b. Which line segment is the steepest?

   c. Which line segment has a slope of zero?

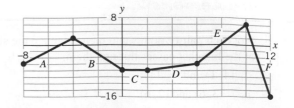

6. Plot each pair of points and calculate the slope of the line that passes through them.

   a. $(3, 5)$ and $(8, 15)$ **d.** $(-2, 6)$ and $(2, -6)$

   b. $(-1, 4)$ and $(7, 0)$ **e.** $(-4, -3)$ and $(2, -3)$

   c. $(5, 9)$ and $(-5, 9)$

7. The following problems represent calculations of the slopes of different lines. Solve for the variable in each equation.

   a. $\dfrac{150 - 75}{20 - 10} = m$ **c.** $\dfrac{182 - 150}{28 - x} = 4$

   b. $\dfrac{70 - y}{0 - 8} = 0.5$ **d.** $\dfrac{6 - 0}{x - 10} = 0.6$

8. Find the slope m of the line through the points $(0, b)$ and (x, y), then solve the equation for y.

9. Find the value of t if m is the slope of the line that passes through the given points.

   a. $(3, t)$ and $(-2, 1)$, $m = -4$

   b. $(5, 6)$ and $(t, 9)$, $m = \frac{2}{3}$

10. **a.** Find the value of x so that the slope of the line through $(x, 5)$ and $(4, 2)$ is $\frac{1}{3}$.

    b. Find the value of y so that the slope of the line through $(1, -3)$ and $(-4, y)$ is -2.

    c. Find the value of y so that the slope of the line through $(-2, 3)$ and $(5, y)$ is 0.

    d. Find the value of x so that the slope of the line through $(-2, 2)$ and $(x, 10)$ is 2.

    e. Find the value of y so that the slope of the line through $(-100, 10)$ and $(0, y)$ is $-\frac{1}{10}$.

    f. Find at least one set of values for x and y so that the slope of the line through $(5, 8)$ and (x, y) is 0.

11. Points that lie on the same straight line are said to be *collinear*. Determine if the following points are collinear.

    a. $(2, 3)$, $(4, 7)$, and $(8, 15)$

    b. $(-3, 1)$, $(2, 4)$, and $(7, 8)$

12. **a.** Find the slope of the line through each of the following pairs of points.

    i. $(-1, 4)$ and $(-2, 4)$

    ii. $(7, -3)$ and $(-7, -3)$

    iii. $(-2, -6)$ and $(5, -6)$

    b. Summarize your findings.

13. Graph a line through each pair of points and then calculate its slope.

    a. The origin and $(6, -2)$ **b.** The origin and $(-4, 7)$

14. Find some possible values of the y-coordinates for the points $(-3, y_1)$ and $(6, y_2)$ such that the slope $m = 0$.

15. Calculate the slope of the line passing through each of the following pairs of points.

    a. $(0, \sqrt{2})$ and $(\sqrt{2}, 0)$

    b. $(0, -\frac{3}{2})$ and $(-\frac{3}{2}, 0)$

    c. $(0, b)$ and $(b, 0)$

    d. What do these pairs of points and slopes all have in common?

16. Which pairs of points produce a line with a negative slope?

    a. $(-5, -5)$ and $(-3, -3)$ **d.** $(4, 3)$ and $(12, 0)$

    b. $(-2, 6)$ and $(-1, 4)$ **e.** $(0, 3)$ and $(4, -10)$

    c. $(3, 7)$ and $(-3, -7)$ **f.** $(4, 2)$ and $(6, 2)$

17. In the previous exercise, which pairs of points produce a line with a positive slope?

18. A study on numerous streams examined the effects of a warming climate. It found an increase in water temperature of about 0.7°C for every 1°C increase in air temperature.

    a. Find the rate of change in water temperature with respect to air temperature. What are the units?

    b. If the air temperature increased by 5°C, by how much would you expect the stream temperature to increase?

19. Handicapped Vietnam veterans successfully lobbied for improvements in the architectural standards for wheelchair access to public buildings.

    a. The old standard was a 1-foot rise for every 10 horizontal feet. What would the slope be for a ramp built under this standard?

    b. The new standard is a 1-foot rise for every 12 horizontal feet. What would the slope of a ramp be under this standard?

    c. If the front door is 3 feet above the ground, how long would the handicapped ramp be using the old standard? Using the new?

20. The following graph shows the relationship between education and health in the United States.

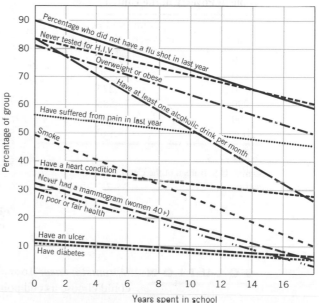

Years of Education vs. Health Issues

Percentage who did not have a flu shot in last year
Never tested for H.I.V.
Overweight or obese
Have at least one alcoholic drink per month
Have suffered from pain in last year
Smoke
Have a heart condition
Never had a mammogram (women 40+)
In poor or fair health
Have an ulcer
Have diabetes

(y-axis: Percentage of group; x-axis: Years spent in school)

Source: "A Surprising Secret to a Long Life, According to Studies: Stay in School," *New York Times,* January 3, 2007.

   a. Estimate the percentages for those with 8 years of education and for those with 16 years of education (college graduates) who:

   i. Are overweight or obese

   ii. Have at least one alcoholic drink per month

   iii. Smoke

   b. For each category in part (a), calculate the average rate of change with respect to years of education. What do they all have in common?

c. What does the graph suggest about the link between education and health? Could there be other factors at play?

21. Read the Anthology article "Slopes" on the course website and then describe two practical applications of slopes, one of which is from your own experience.

2.4 *Putting a Slant on Data*

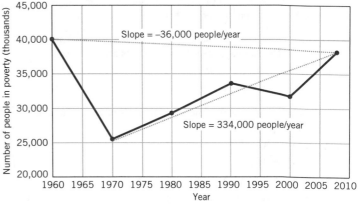

Exploration 2.1 gives you a chance to put your own "spin" on data.

Whenever anyone summarizes a set of data, choices are being made. One choice may not be more "correct" than another. But these choices can convey, either accidentally or on purpose, very different impressions.

Slanting the Slope: Choosing Different End Points

Within the same data set, one choice of end points may paint a rosy picture, while another choice may portray a more pessimistic outcome.

EXAMPLE 1   **Poverty in the United States**

The data in Table 2.5 and the scatter plot in Figure 2.12 show the number of people below the poverty level in the United States from 1960 to 2008. How could we use the information to make the case that the poverty level has decreased? Has increased?

People in Poverty in the United States

Year	Number of People in Poverty (in thousands)
1960	39,851
1970	25,420
1980	29,272
1990	33,585
2000	31,581
2008	38,110

Table 2.5
Source: U.S. Bureau of the Census, *www. census.gov.*

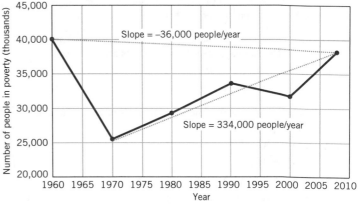

Figure 2.12 Number of people in poverty in the United States between 1960 and 2008.

SOLUTION   *Optimistic case:* To make an upbeat case that poverty numbers have decreased, we could choose as end points (1960, 39851) and (2008, 38110). Then

$$\text{average rate of change} = \frac{\text{change in no. of people in poverty (000s)}}{\text{change in years}}$$

$$= \frac{38,110 - 39,851}{2008 - 1960}$$

$$= \frac{-1741}{48}$$

$$\approx -36 \text{ thousand people/year}$$

So between 1960 and 2008 the number of impoverished individuals *decreased* on average by 36 thousand (or 36,000) each year. We can see this reflected in Figure 2.12 in the negative slope of the line connecting (1960, 39851) and (2008, 38110).

Pessimistic case: To make a depressing case that poverty numbers have risen, we could choose (1970, 25420) and (2008, 38110) as end points. Then

$$\text{average rate of change} = \frac{\text{change in no. of people in poverty (000s)}}{\text{change in years}}$$

$$= \frac{38{,}110 - 25{,}420}{2008 - 1970}$$

$$= \frac{12{,}690}{38}$$

$$\approx 334 \text{ thousand people/year}$$

So between 1970 and 2008 the number of impoverished individuals *increased* on average by 334 thousand (or 334,000) per year. This is reflected in Figure 2.12 in the positive slope of the line connecting (1970, 25420) and (2008, 38110). Both average rates of change are correct, but they give very different impressions of the changing number of people living in poverty in America.

Slanting the Data with Words and Graphs

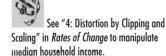 See "4: Distortion by Clipping and Scaling" in *Rates of Change* to manipulate median household income.

If we wrap data in suggestive vocabulary and shape graphs to support a particular viewpoint, we can influence the interpretation of information. In Washington, D.C., this would be referred to as "putting a spin on the data."

Take a close look at the following three examples. Each contains exactly the same underlying facts: the same average rate of change calculation and a graph with a plot of the same two data points (2000, 281.4) and (2010, 309.2), representing the U.S. population (in millions) in 2000 and in 2010.

The U.S. population increased by only 2.78 million/year between 2000 and 2010

Stretching the scale of the horizontal axis relative to the vertical axis makes the slope of the line look almost flat and hence minimizes the impression of change (Figure 2.13).

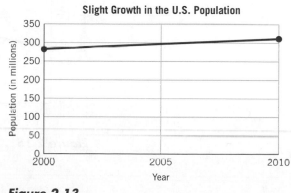

Figure 2.13

The U.S. population had an explosive growth of over 2.78 million/year between 2000 and 2010

Cropping the vertical axis (which now starts at 280 instead of 0) and stretching the scale of the vertical axis relative to the horizontal axis makes the slope of the line look steeper and strengthens the impression of dramatic change (Figure 2.14).

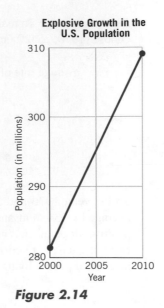

Figure 2.14

The U.S. population grew at a reasonable rate of 2.78 million/year between 2000 and 2010

Visually, the steepness of the line in Figure 2.15 seems to lie roughly halfway between the previous two graphs. *In fact, the slope of 2.78 million/year is precisely the same for all three graphs.*

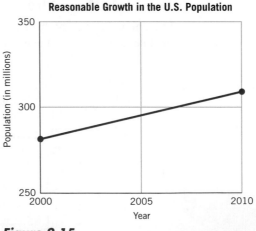

Figure 2.15

How could you decide upon a "fair" interpretation of the data? You might try to put the data in context by asking: How does the growth between 2000 and 2010 compare with other decades in the history of the United States? How does it compare with growth in other countries at the same time? Was this rate of growth easily accommodated, or did it strain national resources and overload the infrastructure?

A statistical claim is never completely free of bias. For every statistic that is quoted, others have been left out. This does not mean that you should discount all statistics. However, you will be best served by a thoughtful approach when interpreting the statistics you encounter daily. By getting in the habit of asking questions and then coming to your own conclusions, you will develop good sense about understanding data.

EXPLORE & EXTEND

2.4

Different Views of the Dow Jones

In Exercise 2, Section 1.2, we looked at one time interval of the Dow Jones Industrial Average, the best-known index of U.S. stocks. The following charts of the Dow Jones all end in August 2009 but begin in earlier and earlier years, each chart giving drastically different impressions. Describe the overall impression of each separate graph, using flamboyant and overblown language.

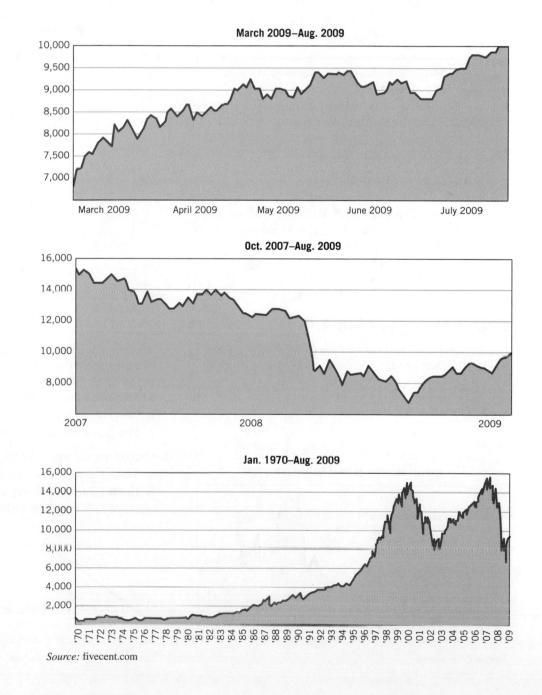

Source: fivecent.com

Algebra Aerobics 2.4

1. The table shows the *percentage* of the U.S. population in poverty between 1960 and 2008.

Year	% of U.S. Population in Poverty %
1960	22.2
1970	12.6
1980	13.0
1990	13.5
2000	11.3
2005	12.6
2006	12.3
2007	12.5
2008	13.2

Source: U.S. Bureau of the Census, *www.census.gov.*

Use the data for 2008 and for a different year from the table to make the case that poverty:

a. Has declined dramatically.

b. Has remained relatively stable.

c. Has increased substantially.

2. Assume you are the financial officer of a corporation whose stock earnings were $1.02 per share in 2008, and $1.12 per share in 2009, and $1.08 per share in 2010. How could you make a case for:

a. Dramatic growth?

b. Dramatic decline?

3. Sketch a graph and compose a few sentences to forcefully convey the views of the following persons.

a. You are an antiwar journalist reporting on American casualties during a war. In week 1 there were 17, in week 2 there were 29, and in week 3 there were 26.

b. You are the president's press secretary in charge of reporting war casualties (listed in part (a)) to the public.

4. The graph below appeared as part of an advertisement in the *Boston Globe*. Identify at least three strategies used to persuade you to buy gold.

Exercises for Section 2.4

Graphing program is optional for Exercise 10. Course software is required for Exercise 12.

1. Examine the following graph of 30-year fixed mortgage rates.

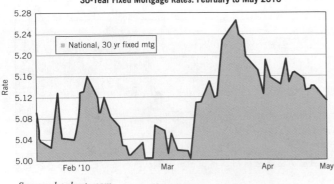

Source: bankrate.com.

Assuming the end date is at the beginning of May, estimate an earlier beginning date (different in each part (a)–(c)) to make the case that:

a. Mortgage rates have dramatically declined.

b. Mortgage rates have "exploded."

c. Mortgage rates have stayed the same.

2. Compare the accompanying graphs.

a. Which line appears to have the steeper slope?

b. Use the intercepts to calculate the slope. Which graph actually does have the steeper slope?

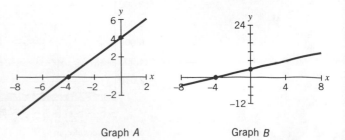

Graph A   Graph B

3. Compare the accompanying graphs.

   a. Which line appears to have the steeper slope?

   b. Use the intercepts to calculate the slope. Which graph has the more negative (and hence steeper) slope?

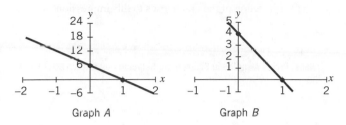

   Graph A   Graph B

4. The following graphs show the same function graphed on different scales.

   a. In which graph does Q appear to be growing at a faster rate?

   b. In which graph does Q appear to be growing at a near zero rate?

   c. Explain why the graphs give different impressions.

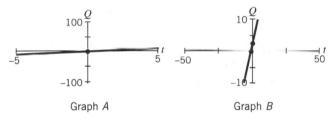

   Graph A   Graph B

5. The following graph shows the National Assessment of Educational Progress (NAEP) math scores for California students, grades 4 and 8, as well as national math scores.

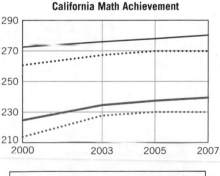

California Math Achievement

····· Grade 4 (CA)   ····· Grade 8 (CA)

——— National Grade 4   ——— National Grade 8

   Source: National Assessment of Educational Progress (NAEP).

   a. Summarize the message this graph conveys concerning achievement in math scores.

   b. What strategy is used to make the increase in math scores seem modest? How could you make the scores seem even more modest?

   c. What strategy could you use to make the increase more dramatic?

6. Generate two graphs and on each draw a line through the points (0, 3) and (4, 6), choosing x and y scales such that:

   a. The first line appears to have a slope of almost zero

   b. The second line appears to have a very large positive slope.

7. a. Generate a graph of a line through the points (0, −2) and (5, −2).

   b. On a new grid, choose different scales so that the line through the same points appears to have a large positive slope.

   c. What have you discovered?

8. What are three adjectives (like "explosive") that would imply rapid growth?

9. What are three adjectives (like "severe") that would imply rapid decline?

10. (Graphing program optional.) Examine the data given on women in the U.S. military forces.

Women in Uniform: Female Active-Duty Personnel

Year	Total	Army	Navy	Marine Corps	Air Force
1970	**41,479**	16,724	8,683	2,418	13,654
1980	**171,418**	69,338	34,980	6,706	60,394
1990	**227,018**	83,621	59,907	9,356	74,134
2000	**204,498**	72,021	53,920	9,742	68,815
2005	**189,465**	71,400	54,800	8,498	54,767
2008	**200,337**	73,902	50,008	12,290	64,137

Source: Department of Defense.

   a. Make the case with graphs and numbers that women are a growing presence in the U.S. military.

   b. Make the case with graphs and numbers that women are a declining presence in the U.S. military.

   c. Write a paragraph that gives a balanced picture of the changing presence of women in the military using appropriate statistics to make your points. What additional data would be helpful?

11. On the following page there is a graph on AIDS in America similar to one that appeared in Chapter 1. The left-hand axis shows both the number of new AIDS cases and deaths from AIDS. In this new version, the right-hand axis shows the number of people living with AIDS.

   a. Find something encouraging to say about these data by using numerical evidence, including estimated average rates of change.

   b. Find numerical support for something discouraging to say about the data.

   c. How might we explain the enormous increase in new AIDS cases reported from 1985 to a peak in 1992–1993, and the drop-off the following year?

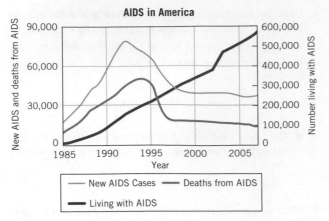

AIDS in America

Source: Centers for Disease Control.

12. (Course software required.) Open "L6: Changing Axis Scales" in *Linear Functions*. Generate a line in the upper left-hand box. The same line will appear graphed in the three other boxes but with the axes scaled differently. Describe how the axes are rescaled in order to create such different impressions.

13. The accompanying graphs show the same data on the median income of a household headed by a black person as a percentage of the median income of a household headed by a white person. Describe the impression each graph gives and how that was achieved. (Data from the U.S. Bureau of the Census.)

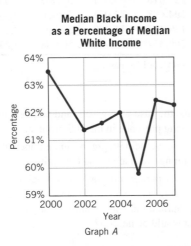

Median Black Income as a Percentage of Median White Income

Graph *A*

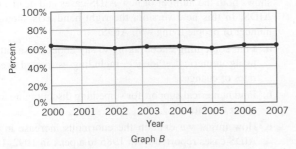

Median Black Income as a Percentage of Median White Income

Graph *B*

14. The accompanying graph shows the number of Nobel Prizes awarded in science for various countries between 1901 and 1974. It contains accurate information but gives the impression that the number of prize winners declined drastically in the 1970s, which was not the case. What flaw in the construction of the graph leads to this impression?

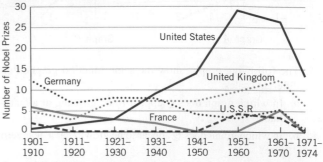

Nobel Prizes Awarded in Science, for Selected Countries, 1901–1974

Source: E. R. Tufte, *The Visual Display of Quantitative Information* (Cheshire, Conn.: Graphics Press, 1983).

15. The following graph shows the trend in National Assessment of Educational Progress (NAEP) math scores of 17-year-old students by gender from 1973 to 2008.

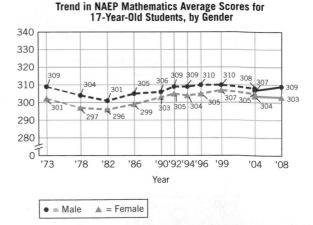

Trend in NAEP Mathematics Average Scores for 17-Year-Old Students, by Gender

● = Male   ▲ = Female

Source: nationsreportcard.gov.

a. Summarize the message this graph conveys concerning achievement in math scores by gender.

b. What strategy could be used to make the gap between genders seem greater?

2.5   *Linear Functions: When Rates of Change Are Constant*

In many of our examples so far, the average rate of change has varied depending on the choice of end points. Now we will examine the special case when the average rate of change remains constant.

What If the U.S. Population Had Grown at a Constant Rate? A Hypothetical Example

In Section 2.2 we calculated the average rate of change in the population between 1790 and 1800 as 0.14 million people/year. We saw that the average rate of change was different for different decades. What if the average rate of change had remained constant? What if in every decade after 1790 the U.S. population had continued to grow at the same rate of 0.14 million people/year? That would mean that starting with a population estimated at 3.9 million in 1790, the population would have grown by 0.14 million each year. The slopes of all the little line segments connecting adjacent population data points would be identical, namely 0.14 million people/year. The graph would be a straight line, indicating a constant average rate of change.

On the graph of actual U.S. population data, the slopes of the line segments connecting adjacent points are increasing, so the graph curves upward. Figure 2.16 compares the actual and hypothetical results.

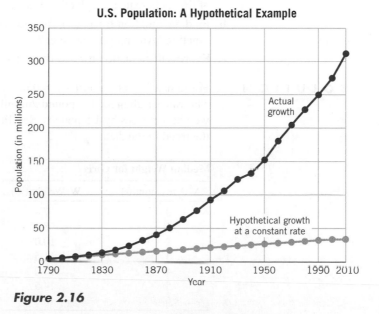

U.S. Population: A Hypothetical Example

Figure 2.16

Any function that has the same average rate of change on every interval has a graph that is a straight line. The function is called *linear*. This hypothetical example represents a linear function. When the average rate of change is constant, we can drop the word "average" and just say "rate of change."

A *linear function* has a constant rate of change. Its graph is a straight line.

EXPLORE & EXTEND

2.5

The Relationship between Velocity and Distance

Open up the course software "C3: Average Velocity and Distance" in *Rates of Change*.

a. Press and drag several velocity values. Describe the corresponding effect on the distance graph. What is the relationship between the two graphs?

b. Set all the velocity values to the same value (of your choice). What is the effect on the total distance graph?

c. If you set the velocities all at another fixed value, what would the distance graph look like now?

d. What could you conclude from parts (b) and (c)?

Real Examples of a Constant Rate of Change

Linear functions are frequently used to model real life situations.

EXAMPLE 1

Median weight for female infants

According to the standardized growth and development charts used by many American pediatricians, the median weight for girls during their first six months of life increases at an almost constant rate. Starting at 7.0 pounds at birth, female median weight increases by approximately 1.5 pounds per month. If we assume that the median weight for females, W, is increasing at a constant rate of 1.5 pounds per month, then W is a linear function of age in months, A.

a. Generate a table that gives the median weight for females, W, for the first six months of life and create a graph of W as a function of A.

b. Find an equation for W as a function of A. What is an appropriate domain for this function? An appropriate range?

c. Express the equation for part (b) using only units of measure.

SOLUTION

a. For female infants at birth ($A = 0$ months), the median weight is 7.0 lb ($W = 7.0$ lb). The rate of change, 1.5 pounds/month, means that as age increases by 1 month, weight increases by 1.5 pounds. See Table 2.6. The dotted line in Figure 2.17 shows the trend in the data.

Median Weight for Girls

A, Age (months)	W, Weight (lb)
0	7.0
1	8.5
2	10.0
3	11.5
4	13.0
5	14.5
6	16.0

Table 2.6
Source: Data derived from the Ross Growth and Development Program, Ross Laboratories, Columbus, OH.

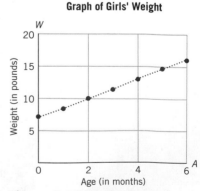

Figure 2.17 Median weight for girls.

b. To find a linear equation for W (median weight in pounds) as a function of A (age in months), we can study the table of values in Table 2.6.

$$W = \text{initial weight} + \text{weight gained}$$
$$= \text{initial weight} + \text{rate of growth} \cdot \text{number of months}$$
$$= 7.0 \text{ lb} + 1.5 \text{ lb/month} \cdot \text{number of months}$$

The equation would be

$$W = 7.0 + 1.5A$$

An appropriate domain would be $0 \leq A \leq 6$, and range $7.0 \leq W \leq 16.0$.

c. Since our equation represents quantities in the real world, each term in the equation has units attached to it. The median weight W is in pounds (lb), rate of change is in lb/month, and age, A, is in months. If we display the equation

$$W = 7.0 + 1.5A$$

showing only the units, we have

$$lb = lb + \left(\frac{lb}{month}\right)month$$

The rules for canceling units are the same as the rules for canceling numbers in fractions. So,

$$lb = lb + \left(\frac{lb}{\cancel{month}}\right)\cancel{month}$$

$$lb = lb + lb$$

This equation makes sense in terms of the original problem since pounds (lb) added to pounds (lb) should give us pounds (lb).

EXAMPLE   2   **Median weight for baby boys vs. baby girls**
a. If the median birth weight for baby boys is the same as for baby girls, but boys put on weight at a faster rate, which numbers in the previous weight model would change? Which would stay the same?
b. What would be different about the two graphs?

SOLUTION   a. The initial birth weight for males would be 7 lb (the same as for females), but the rate of growth (or slope) for males would be >1.5 lb/month (greater than that for females.)
b. Both graphs would have the same vertical intercept, but the male graph would be steeper.

EXAMPLE   3   **Depreciation of computer cost**
You spend $1200 on a computer and for tax purposes choose to depreciate it (or assume it decreases in value) to $0 at a constant rate over a 5-year period.

a. Calculate the rate of change of the assumed value of the equipment over 5 years. What are the units?
b. Create a table and graph showing the value of the equipment over 5 years.
c. Create a function for the value of the computer over time in years. Why is this a linear function?
d. What is an appropriate domain for this function? What is the range?

SOLUTION   a. After 5 years, your computer is worth $0. If $V(t)$ is the value of your computer in dollars as a function of t, the number of years you own the computer, then the rate of change of $V(t)$ from $t = 0$ to $t = 5$ is

$$\text{rate of change} = \frac{\text{change in value}}{\text{change in time}} = \frac{\Delta V}{\Delta t}$$

$$= \frac{\$1200}{5 \text{ years}} = -\$240/\text{year}$$

Thus, the worth of your computer drops at a constant rate of $240 per year. The rate of change in $V(t)$ is negative because the worth of the computer decreases over time. The units for the rate of change are dollars per year.

b. Table 2.7 and Figure 2.18 show the depreciated value of the computer.

Value of Computer Depreciated over 5 Years

Numbers of Years	Value of Computer ($)
0	1200
1	960
2	720
3	480
4	240
5	0

Table 2.7

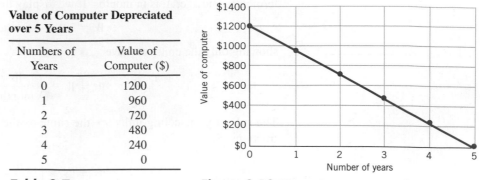

Figure 2.18 Value of computer over 5 years.

c. To find a linear function, $V(t)$, think about how we found the table of values.

$$\text{value of computer} = \text{initial value} + (\text{rate of decline}) \cdot (\text{number of years})$$

$$V(t) = \$1200 + (-\$240/\text{year}) \cdot t$$

$$V(t) = 1200 - 240t$$

This equation describes $V(t)$ as a function of t because for every value of t, there is one and only one value of $V(t)$. It is a linear function because the rate of change is constant.

d. The domain is $0 \le t \le 5$ and the range is $0 \le V(t) \le 1200$.

The General Equation for a Linear Function

The equations in Examples 1 and 3 can be rewritten in terms of the output (dependent variable) and the input (independent variable).

$$\text{weight} = \text{initial weight} + \text{rate of growth} \cdot \text{number of months}$$

$$\text{value of computer} = \text{initial value} + \text{rate of decline} \cdot \text{number of years}$$

$$\underbrace{\text{output}}_{y} = \underbrace{\text{initial value}}_{b} + \underbrace{\text{rate of change}}_{m} \cdot \underbrace{\text{input}}_{x}$$

So, the general linear equation can be written in the form

$$y = b + mx$$

where we use the traditional mathematical choices of y for the output (dependent variable) and x for the input (independent variable). We let m stand for the rate of change; so

$$m = \frac{\Delta \text{ output}}{\Delta \text{ input}} = \frac{\text{change in } y}{\text{change in } x} = \frac{\Delta y}{\Delta x}$$

$$= \text{slope of the graph of the line}$$

In mathematical models of the form $y = b + mx$, we often call b the initial or base value. Why? Many models exclude negative numbers (like the prior examples on

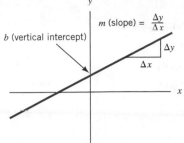

Figure 2.19 Graph of
$y = b + mx$.

infant weight or computer depreciation), so the smallest value in the domain is often 0. When $x = 0$, then

$$y = b + m \cdot 0$$
$$y = b$$

The point $(0, b)$ satisfies the equation and is the vertical intercept. Since the coordinate b tells us where the line crosses the y-axis, we often just refer to b as the *vertical* or *y-intercept* (see Figure 2.19).

A Linear Function

A function $y = f(x)$ is called *linear* if it can be represented by an equation of the form

$$y = b + mx$$

Its graph is a straight line where m is the *slope*, the rate of change of y with respect to x. So if (x_1, y_1) and (x_2, y_2) are any two distinct points on the line, then the

$$\text{slope} = \frac{y_2 - y_1}{x_2 - x_1} = \frac{\Delta y}{\Delta x} = m = \text{average rate of change}$$

b is the *vertical* or *y-intercept* and is the value of y when $x = 0$. If the function is a mathematical model, b typically represents an initial or starting value.

The equation $y = b + mx$ could, of course, be written in the equivalent form $y = mx + b$. This may be the traditional format that you first learned.

EXAMPLE 4 **Identifying the vertical intercept and slope**
For each of the following equations, identify the value of b and the value of m.

a. $y = -4 + 3.25x$ **c.** $g(x) = -4x + 3.25$
b. $f(x) = 3.25x - 4$ **d.** $y = 3.25 - 4x$

SOLUTION **a.** $b = -4$ and $m = 3.25$
b. $b = -4$ and $m = 3.25$
c. $b = 3.25$ and $m = -4$
d. $b = 3.25$ and $m = -4$

EXAMPLE 5 **Comparing legal costs**
In the following equations, $L(h)$ represents the legal fees (in dollars) charged by four different law firms and h represents the number of hours of legal advice.

$$L_1(h) = 500 + 200h \qquad L_3(h) = 800 + 350h$$
$$L_2(h) = 1000 + 150h \qquad L_4(h) = 500h$$

a. Which initial fee is the highest?
b. Which rate per hour is the highest?
c. If you need 5 hours of legal advice, which legal fee will be the highest?

SOLUTION **a.** L_2 has the highest initial fee of $1000.
b. L_4 has the highest rate of $500 per hour.

c. Evaluate each equation for $h = 5$ hours.

$$L_1(5) = 500 + 200(5) \qquad L_3(5) = 800 + 350(5)$$
$$= \$1500 \qquad\qquad = \$2550$$
$$L_2(5) = 1000 + 150(5) \qquad L_4(5) = 500(5)$$
$$= \$1750 \qquad\qquad = \$2500$$

For 5 hours of legal advice, L_3 has the highest legal fee.

Algebra Aerobics 2.5

1. From Figure 2.17, estimate the weight W of a baby girl who is 4.5 months old. Then use the equation $W = 7.0 + 1.5A$ to calculate the corresponding value for W. How close is your estimate?

2. From the same graph, estimate the age of a baby girl who weighs 11 pounds. Then use the equation to calculate the value for A.

3. **a.** If $C = 15P + 10$ describes the relationship between the number of persons (P) in a dining party and the total cost in dollars (C) of their meals, what is the unit of measure for 15? For 10?

   b. The equation $W = 7.0 + 1.5A$ (modeling weight, W, as a function of age, A) expressed in units of measure only is

   $$\text{lb} = \text{lb} + \left(\frac{\text{lb}}{\text{month}}\right)\text{months}$$

   Express $C = 15P + 10$ from part (a) in units of measure only.

4. (True story.) A teenager travels to Alaska with his parents and wins \$1200 in a rubber ducky race. (The race releases 5000 yellow rubber ducks marked with successive integers from one bridge over a river and collects them at the next bridge.) Upon returning home he opens up a "Rubber Ducky Savings Account," deposits his winnings, and continues to deposit \$50 each month. If D = amount in the account and M = months since the creation of the account, then $D = 1200 + 50M$ describes the amount in the account after M months.

   a. What are the units for 1200? For 50?

   b. Express the equation using only units of measure.

5. Assume $S = 0.8Y + 19$ describes the projected relationship between S, sales of a company (in millions of dollars), for Y years from today.

   a. What are the units of 0.8 and what does it represent?

   b. What are the units of 19 and what does it represent?

   c. What would be the projected company sales in three years?

   d. Express the equation using only units of measure.

6. Assume $C = 0.45N$ represents the total cost C (in dollars) of operating a car for N miles.

   a. What does 0.45 represent and what are its units?

   b. Find the total cost to operate a car that has been driven 25,000 miles.

   c. Express the equation using only units of measure.

7. The relationship between the balance B (in dollars) left on a mortgage loan (including interest) and N, the number of monthly payments, is given by $B = 302{,}400 - 840N$.

   a. What is the monthly mortgage payment?

   b. What does 302,400 represent?

   c. What is the balance on the mortgage after 10 years? 20 years? 30 years? (*Hint:* Remember there are 12 months in a year.)

8. Identify the slope, m, and the vertical intercept, b, of the line for each equation.

   a. $y = 5x + 3$ **e.** $f(x) = 7.0 - x$

   b. $y = 5 + 3x$ **f.** $h(x) = -11x + 10$

   c. $y = 5x$ **g.** $y = 1 - \frac{2}{3}x$

   d. $y = 3$ **h.** $2y + 6 = 10x$

9. If $f(x) = 50 - 25x$:

   a. Why does $f(x)$ describe a linear function?

   b. Evaluate $f(0)$ and $f(2)$.

   c. Use your answers in part (b) to verify that the slope is -25.

10. Identify the functions that are linear. For each linear function, identify the slope and the vertical intercept.

    a. $f(x) = 3x + 5$ **c.** $f(x) = 3x^2 + 2$

    b. $f(x) = x$ **d.** $f(x) = 4 - \frac{2}{3}x$

11. Write an equation for the line in the form $y = b + mx$ for the indicated values.

    a. $m = 3$ and $b = 4$

    b. $m = -1$ and passes through the origin

    c. $m = 0$ and $b = -3$

12. Write the equation of the graph of each of the four lines below in $y = b + mx$ form. Use the y-intercept and the slope indicated on each graph.

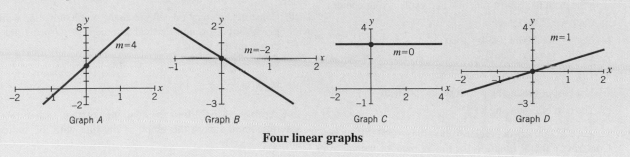

Graph A Graph B Graph C Graph D

Four linear graphs

Exercises for Section 2.5

You might wish to hone your algebraic mechanical skills with three programs in the course software; *Linear Functions:* "L1: Finding m & b", "L3: Finding a Line through 2 Points", "L4: Finding 2 Points on a Line" and "L5: Slope and Intercepts." They offer practice in predicting values for m and b, generating linear equations, and finding corresponding solutions.

1. Consider the equation $A = 300 + 50n$.

  a. Find the value of A for $n = 0, 1, 20$.

  b. Express your answers to part (a) as points with coordinates (n, A).

2. Consider the equation $B = 450 + 40n$.

  a. Find the value of B for $n = 0, 1, 20$.

  b. Express your answers to part (a) as points with coordinates (n, B).

3. Determine if any of the following points satisfy one or both of the equations in Exercises 1 and 2.

  a. $(300, 0)$ **b.** $(10, 850)$ **c.** $(15, 1050)$

4. Suppose during a 5-year period the profit $P(Y)$ (in billions of dollars) for a large corporation was given by $P(Y) = 7 + 2Y$, where Y represents the year.

  a. Fill in the chart.

Y	0	1	2	3	4
$P(Y)$					

  b. What are the units of $P(Y)$?

  c. What does the 2 in the equation represent, and what are its units?

  d. What was the initial profit?

5. Consider the equation $D = 3.40 + 0.11n$

  a. Find the values of D for $n = 0, 1, 2, 3, 4$.

  b. If D represents the average consumer debt, in thousands of dollars, over n years, what does 0.11 represent? What are its units?

  c. What does 3.40 represent? What are its units?

6. Suppose the equations $E(n) = 5000 + 1000n$ and $G(n) = 12,000 + 800n$ give the total cost of operating an electrical (E) versus a gas (G) heating/cooling system in a home for n years.

  a. Find the cost of heating a home using electricity for 10 years.

  b. Find the cost of heating a home using gas for 10 years.

  c. Find the initial (or installation) cost for each system.

  d. Determine how many years it will take before $40,000 has been spent in heating/cooling a home that uses:

  i. Electricity   **ii.** Gas

7. If the equation $E(n) = 5000 + 1000n$ gives the total cost of heating/cooling a home after n years, rewrite the equation using only units of measure.

8. Over a 5-month period at Acadia National Park in Maine, the mean night temperature increased on average 5 degrees Fahrenheit per month. If the initial temperature is 25 degrees Fahrenheit, create a function for the night temperature $N(t)$ for month t, where $0 \leq t < 4$.

9. A residential customer in the Midwest purchases gas from a utility company that charges according to the formula $C(g) = 11 + 10.50(g)$, where $C(g)$ is the cost, in dollars, for g thousand cubic feet of gas.

  a. Find $C(0)$, $C(5)$, and $C(10)$.

  b. What is the cost if the customer uses no gas?

  c. What is the rate per thousand cubic feet charged for using the gas?

  d. How much would it cost if the customer uses 96 thousand cubic feet of gas (the amount an average Midwest household consumes during the winter months)?

10. Create the formula for converting degrees centigrade, C, to degrees Fahrenheit, F, if for every increase of 5 degrees centigrade the Fahrenheit temperature increases by 9 degrees, with an initial point of $(C, F) = (0, 32)$.

11. Determine the vertical intercept and the rate of change for each of these formulas:

  a. $P = 4s$ **c.** $C = 2\pi r$

  b. $C = \pi d$ **d.** $C = \frac{5}{9}F - 17.78$

12. A hiker can walk 2 miles in 45 minutes.

  a. What is his average speed in miles per hour?

  b. What formula can be used to find the distance traveled, d, in miles in t hours?

13. The cost $C(x)$ of producing x items is determined by adding fixed costs to the product of the cost per item and the number of items produced, x. Below are three possible cost functions $C(x)$, measured in dollars. Match each description in parts (e) − (g) with the most likely cost function and explain your choices.

   a. $C(x) = 125,000 + 75x$

   b. $C(x) = 400,000 + 0.30x$

   c. $C(x) = 250,000 + 800x$

   e. The cost of producing a computer

   f. The cost of producing a college algebra text

   g. The cost of producing a CD

14. A new \$25,000 car depreciates in value by \$5000 per year. Construct a linear function for the value V of the car after t years.

15. The state of Pennsylvania has a 6% sales tax on taxable items. (*Note:* Clothes, food, and certain pharmaceuticals are not taxed in Pennsylvania.)

   a. Create a formula for the total cost (in dollars) of an item $C(p)$ with a price tag p. (Be sure to include the sales tax.)

   b. Find $C(9.50)$, $C(115.25)$, and $C(1899)$. What are the units?

16. **a.** If $S(x) = 20,000 + 1000x$ describes the annual salary in dollars for a person who has worked for x years for the Acme Corporation, what is the unit of measure for 20,000? For 1000?

   b. Rewrite $S(x)$ as an equation using only units of measure.

   c. Evaluate $S(x)$ for x values of 0, 5, and 10 years.

   d. How many years will it take for a person to earn an annual salary of \$43,000?

17. The following represent linear equations written using only units of measure. In each case supply the missing unit.

   a. inches = inches + (inches/hour) · (?)

   b. miles = miles + (?) · (gallons)

   c. calories = calories + (?) · (grams of fat)

18. Identify the slopes and the vertical intercepts of the lines with the given equations.

   a. $y = 3 + 5x$ **d.** $Q = 35t − 10$

   b. $f(t) = −t$ **e.** $f(E) = 10,000 + 3000E$

   c. $y = 4$

19. For each of the following, find the slope and the vertical intercept, then sketch the graph. (*Hint:* Find two points on the line.)

   a. $y = 0.4x − 20$ **b.** $P = 4000 − 200C$

20. Construct an equation and sketch the graph of its line with the given slope, m, and vertical intercept, b. (*Hint:* Find two points on the line.)

   a. $m = 2, b = −3$ **c.** $m = 0, b = 50$

   b. $m = −\frac{3}{4}, b = 1$

In Exercises 21 to 23 find an equation, generate a small table of solutions, and sketch the graph of the line with the indicated attributes.

21. A line that has a vertical intercept of −2 and a slope of 3.

22. A line that crosses the vertical axis at 3.0 and has a rate of change of −2.5.

23. A line that has a vertical intercept of 1.5 and a slope of 0.

24. Estimate b (the y-intercept) and m (the slope) for each of the accompanying graphs. Then, for each graph, write the corresponding linear function.

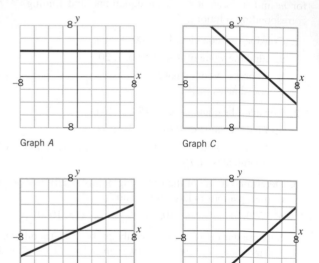

Graph A   Graph C

Graph B   Graph D

2.6 *Visualizing Linear Functions*

The values for b and m in the general form of the linear equation, $y = b + mx$, tell us about the graph of the function. The value for b tells us where to anchor the line on the y-axis. The value for m tells us whether the line climbs or falls and how steep it is.

EXPLORE & EXTEND

2.6

Making Predictions about the Effects of m and b

Each chapter in which a new function is introduced has a "Visualizing" section accompanied by course software called "sliders." For linear functions, open up the software "L1: m and b Sliders" to explore the effect of m and b on the graph of $y = mx + b$. (We recommend that you do this with a partner.)

a. Fix a value for *b*. Then construct and compare four graphs (of different colors) using the "*m*" slider. If you fix another value for *b* and construct four new graphs, do your previous conclusions still hold?

b. Fix a value for *m*. Then once again construct and compare four graphs (of different colors), now using the "*b*" slider. Pick a new *m* and repeat the process. What are your conclusions?

c. Write a 60-second summary on the effect of *m* and *b* on the graph of $y = mx + b$.

The Effect of *b*

In the equation $y = b + mx$, the value *b* is the vertical intercept, so it anchors the line at the point $(0, b)$ (see Figure 2.20).

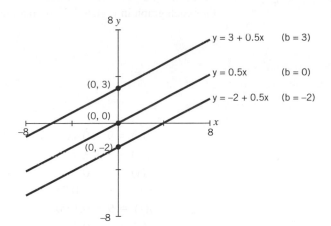

Figure 2.20 The effect of *b*, the vertical intercept.

EXAMPLE 1

Differing graphs
Explain how the graph of $y = 4 + 5x$ differs from the graphs of the following functions:

a. $y = 8 + 5x$
b. $y = 2 + 5x$
c. $y = 5x + 4$

SOLUTION

Although all of the graphs are straight lines with a slope of 5, they each have a different vertical intercept. The graph of $y = 4 + 5x$ has a vertical intercept at 4.

a. The graph of $y = 8 + 5x$ intersects the *y*-axis at 8, four units above the graph of $y = 4 + 5x$.

b. The graph of $y = 2 + 5x$ intersects the *y*-axis at 2, two units below the graph of $y = 4 + 5x$.

c. Since $y = 5x + 4$ and $y = 4 + 5x$ are equivalent equations, they have the same graph.

The Effect of *m*

The sign of *m*

The sign of *m* in the equation $y = b + mx$ determines whether the line climbs (slopes up) or falls (slopes down) as we move from left to right on the graph. If *m* is positive, the line climbs from left to right (as *x* increases, *y* increases). If *m* is negative, the line falls from left to right (as *x* increases, *y* decreases) (see Figure 2.21).

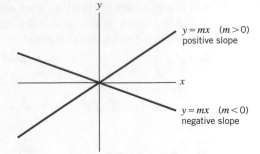

Figure 2.21 The effect of the sign of m.

EXAMPLE 2

Matching graphs and equations

Pair each graph in Figure 2.22 with a matching equation.

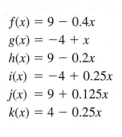

$f(x) = 9 - 0.4x$
$g(x) = -4 + x$
$h(x) = 9 - 0.2x$
$i(x) = -4 + 0.25x$
$j(x) = 9 + 0.125x$
$k(x) = 4 - 0.25x$

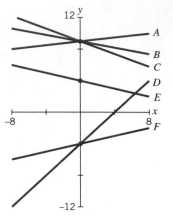

Figure 2.22 Graphs of multiple linear functions.

SOLUTION

A is the graph of $j(x)$.
B is the graph of $h(x)$.
C is the graph of $f(x)$.
D is the graph of $g(x)$.
E is the graph of $k(x)$.
F is the graph of $i(x)$.

The magnitude of m

The magnitude (absolute) value of m determines the steepness of the line. Recall that the absolute value of m is the value of m stripped of its sign; for example, $|-3| = 3$.

> The magnitude (absolute value) of any number x is written as $|x|$ and strips x of its sign. So $|x|$ is always ≥ 0.
>
> For example, $|-7| = |7| = 7$.

The greater the magnitude $|m|$, the steeper the slope of the line. This makes sense since m is the slope or rate of change of y with respect to x.

E X A M P L E   3

Comparing the steepness of slopes

a. Which function in Figure 2.23 (with positive slopes) has the steepest slope?

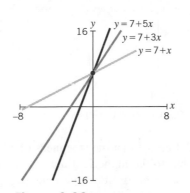

Figure 2.23 Graphs with positive values for m.

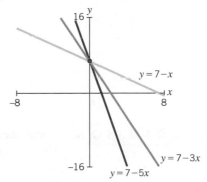

Figure 2.24 Graphs with negative values for m.

b. Which function in Figure 2.24 (with negative slopes) has the steepest slope?

c. Considering both Figures 2.23 and 2.24 which of all the lines is steepest?

S O L U T I O N

a. The steepness of the lines increases as the magnitude of m $(=|m|)$ increases. So in Figure 2.23 the slope ($m = 5$) of $y = 7 + 5x$ is steeper than both the slope ($m = 3$) of $y = 7 + 3x$ and the slope ($m = 1$) of $y = 7 + x$.

b. In Figure 2.24, the slope ($m = -5$) of $y = 7 - 5x$ is steeper than both the slope ($m = -3$) of $y = 7 - 3x$ and the slope ($m = -1$) of $y = 7 - x$, since $|-5| = 5 > |-3| = 3 > |-1| = 1$.

c. In Figure 2.24, the slopes (m) are all negative, but the steepness ($|m|$) depends upon the absolute value of the slope. So the lines $y = 7 - 5x$ and $y = 7 + 5x$ have the same steepness of 5 since $|-5| = |5| = 5$.

E X A M P L E   4

Finding the steepest slope from equations

Without graphing the following functions, how can you tell which graph will have the steepest slope?

a. $f(x) = 5 - 2x$

b. $g(x) = 5 + 4x$

c. $h(x) = 3 - 6x$

S O L U T I O N

The graph of the function h will be steeper than the graphs of the functions f and g since the magnitude of m is greater for h than for f or g. The greater the magnitude of m, the steeper the graph of the line. For $f(x)$, the magnitude of the slope is $|-2| = 2$. For $g(x)$, the magnitude of the slope is $|4| = 4$. For $h(x)$, the magnitude of the slope is $|-6| = 6$.

E X A M P L E   5   **Finding the steepest slope from graphs**
Which of the graphs in Figure 2.25 has the steeper slope?

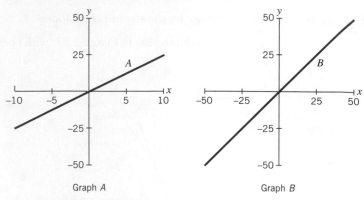

Figure 2.25 Comparing slopes.

S O L U T I O N   Remember from Section 2.4 that the steepness of a linear graph is not related to its visual impression, but to the numerical magnitude of the slope. The scales of the horizontal axes are different for the two graphs in Figure 2.25, so the impression of relative steepness is deceiving. Line A passes through $(0, 0)$ and $(10, 25)$, so its slope is $(25 - 0)/(10 - 0) = 2.5$. Line B passes through $(0, 0)$ and $(50, 50)$ so its slope is $(50 - 0)/(50 - 0) = 1$. So line A has a steeper slope than line B.

Domain and Range

The general linear function $y = b + mx$ has a defined domain (for the input) and range (for the output). Barring any constraints on the input or output then:

Domain: The domain is all real numbers; that is, all x in the interval $(-\infty, +\infty)$.

Range: The range is all real numbers; that is, all y in the interval $(-\infty, +\infty)$, except for the horizontal line $y = b$ (where $m = 0$), where the range is the single number b.

Specific examples however may place restrictions on the domain and range, such as in Section 2.5, Example 1 on the median weight of female infants and Example 3 on the depreciation of computer costs.

E X A M P L E   6   **Domain and Range**
Identify the domain and range for:

a. $y = 2.5 + 4x$ **b.** $y = 2.5 - 4x$ **c.** $y = 2.5$

S O L U T I O N   **a.** Domain: all x in the interval $(-\infty, +\infty)$. Range: all y in the interval $(-\infty, +\infty)$.
b. Domain: all x in the interval $(-\infty, +\infty)$. Range: all y in the interval $(-\infty, +\infty)$.
c. Domain: all x in the interval $(-\infty, +\infty)$. Range: $y = 2.5$

The Graph of a Linear Function

The graph of the linear function $y = b + mx$ is a straight line.

The y-intercept, b, tells us where the line crosses the y-axis.

The slope, m, tells us how fast the line is climbing or falling. The larger the magnitude (or absolute value) of m, the steeper the graph.

If the slope, m, is positive, the line climbs from left to right. If m is negative, the line falls from left to right.

Algebra Aerobics 2.6

1. Place these numbers in order from smallest to largest.

   $$|-12|, |-7|, |-3|, |-1|, 0, 4, 9$$

2. Without graphing the function, explain how the graph $y = 6x - 2$ differs from the graph of

   a. $y = 6x$ **c.** $y = -2 + 3x$

   b. $y = 2 + 6x$ **d.** $y = -2 - 2x$

3. Without graphing, order the graphs of the functions from least steep to steepest.

   a. $y = 100 - 2x$ **c.** $y = -3x - 5$

   b. $y = 1 - x$ **d.** $y = 3 - 5x$

4. On an x-y coordinate system, draw a line with a positive slope and label it $f(x)$.

   a. Draw a line $g(x)$ with the same slope but a y-intercept three units above $f(x)$.

   b. Draw a line $h(x)$ with the same slope but a y-intercept four units below $f(x)$.

   c. Draw a line $k(x)$ with the same steepness as $f(x)$ but with a negative slope.

5. Which function has the steepest slope? Create a table of values for each function and graph the function to show that this is true.

   a. $f(x) = 3x - 5$ **b.** $g(x) = 7 - 8x$

6. Create three functions with a y-intercept of 4 and three different negative slopes. Indicate which function has the steepest slope.

7. Match the four graphs below with the given functions.

   a. $f(x) = 2 + 3x$

   b. $g(x) = -2 - 3x$

   c. $h(x) = \frac{1}{2}x - 2$

   d. $k(x) = 2 - 3x$

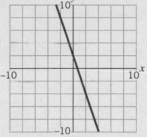

Graph A   Graph C

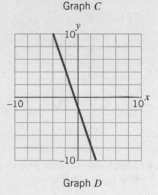

Graph B   Graph D

Exercises for Section 2.6

1. Assuming m is the slope, identify the graph(s) where:

   a. $m = 3$ **b.** $|m| = 3$ **c.** $m = 0$ **d.** $m = -3$

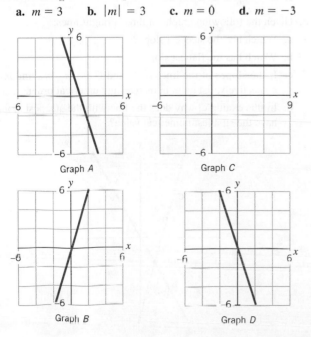

Graph A   Graph C

Graph B   Graph D

2. Which line(s), if any, have a slope m such that:

   a. $|m| = 2$?   **b.** $m = \frac{1}{2}$?

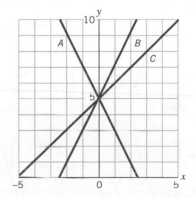

3. For each set of conditions, construct a linear equation and draw its graph

   a. A slope of zero and a y intercept of -3

   b. A positive slope and a vertical intercept of -3

   c. A slope of -3 and a vertical intercept that is positive

4. Construct a linear equation for each of the following conditions.

   a. A negative slope and a positive y-intercept

   b. A positive slope and a vertical intercept of -10.3

   c. A constant rate of change of \$1300/year

5. Match the graph with the correct equation.

   a. $y = x$ b. $y = 2x$ c. $y = \dfrac{x}{2}$ d. $y = 4x$

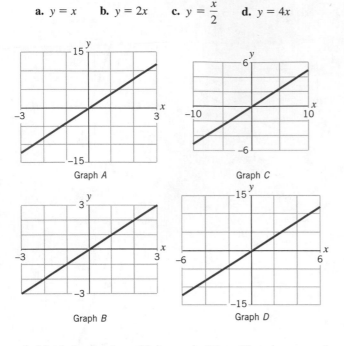

Graph A

Graph C

Graph B

Graph D

6. Match the function with its graph. (*Note:* There is one graph that has no match.)

   a. $f(x) = -4 + 3x$

   b. $g(x) = -3x + 4$

   c. $h(x) = 4 + 3x$

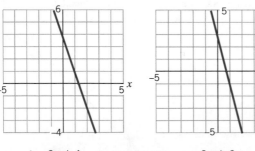

Graph A

Graph C

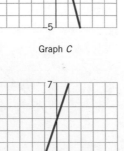

Graph B

Graph D

7. In each part construct three different linear equations that all have:

   a. The same slope

   b. The same vertical intercept

8. Which equation has the steepest slope?

   a. $y = 2 - 7x$

   b. $y = 2x + 7$

   c. $y = -2 + 7x$

9. Given the function $Q(t) = 13 - 5t$, construct a related function whose graph:

   a. Lies five units above the graph of $Q(t)$

   b. Lies three units below the graph of $Q(t)$

   c. Has the same vertical intercept as $Q(t)$

   d. Has the same slope as $Q(t)$

   e. Has the same steepness as $Q(t)$ but the slope is positive

10. Given the equation $C(n) = 30 + 15n$, construct a related equation whose graph:

    a. Is steeper

    b. Is flatter (less steep)

    c. Has the same steepness, but the slope is negative

11. On the same grid, graph (and label with the correct equation) three lines that go through the point $(0, 2)$ and have the following slopes:

    a. $m = \dfrac{1}{2}$

    b. $m = 2$

    c. $m = \dfrac{5}{6}$

12. On the same grid, graph (and label with the correct equation) three lines that go through the point $(0, 2)$ and have the following slopes:

    a. $m = -\dfrac{1}{2}$

    b. $m = -2$

    c. $m = -\dfrac{5}{6}$

13. Given the following graphs of three straight lines:

    a. Which has the steepest slope?

    b. Which has the flattest slope?

    c. If the slopes of the lines A, B, and C are m_1, m_2, and m_3, respectively, list them in increasing numerical order.

    d. In this example, why does the line with the steepest slope have the smallest numerical value?

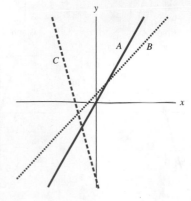

2.7 *Constructing Graphs and Equations of Linear Functions*

 The program "L4: Finding 2 Points on a Line" in *Linear Functions* creates a linear function and asks you to identify two points that lie on its graph.

There are many different strategies for finding the equations and graphs of linear functions.

Finding the Graph

E X A M P L E   1

Given the equation

A forestry study measured the diameter of the trunk of a red oak tree over 5 years. The scientists created a linear model $D(Y) = 1 + 0.13Y$, where $D(Y)$ = diameter in inches and Y = number of years from the beginning of the study.

a. What do the numbers 1 and 0.13 represent in this context?

b. Sketch a graph of the function model.

S O L U T I O N

a. The number 1 represents a starting diameter of 1 inch. The number 0.13 represents the annual growth rate of the oak's diameter (change in diameter/change in time), 0.13 inches per year.

b. The linear function $D(Y) = 1 + 0.13Y$ tells us that 1 is the vertical intercept, so the point (0, 1) lies on the graph. The graph represents solutions to the equation. So to find a second point, we can evaluate $D(Y)$ for any other value of Y. If we set $Y = 1$, then $D(Y) = 1 + (0.13 \cdot 1) = 1 + 0.13 = 1.13$. So (1, 1.13) is another point on the line. Since two points determine a line, we can sketch our line through (0, 1) and (1, 1.13) (see Figure 2.26).

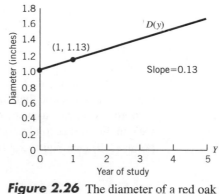

Figure 2.26 The diameter of a red oak over time.

E X A M P L E   2

Given the vertical intercept and the slope

The top speed a snowplow can travel on dry pavement is 40 miles per hour, which decreases by 0.8 miles per hour with each inch of snow on the highway.

a. Construct an equation describing the relationship between snowplow speed and snow depth.

b. Determine a reasonable domain and corresponding range, and then graph the function.

S O L U T I O N

a. If we think of the snow depth, D, as determining the snowplow speed, P, then we need an equation of the form $P = b + mD$. If there is no snow, then the snowplow can travel at its maximum speed of 40 mph; that is, when $D = 0$, then $P = 40$. So the point (0, 40) lies on the line, making the vertical intercept $b = 40$. The rate of change, m, (change in snowplow speed)/(change in snow depth) = -0.8 mph per inch of snow. So the desired equation is

$$P = 40 - 0.8D$$

b. Consider the snowplow as only going forward (i.e., not backing up). Then the snowplow speed does not go below 0 mph. So if we let $P = 0$ and solve for D, we have

$$0 = 40 - 0.8D$$
$$0.8D = 40$$
$$D = 50$$

So when the snow depth reaches 50 inches, the plow is no longer able to move. A reasonable domain then would be $0 \leq D \leq 50$. The corresponding range would be $0 \leq P \leq 40$. (See Figure 2.27.)

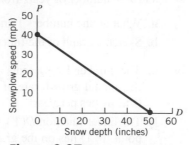

Figure 2.27 Snowplow speed versus snow depth.

EXAMPLE 3

Given a point off the *y*-axis and the slope
Given a point $(2, 3)$ and a slope of $m = -5/4$, describe at least two ways you could find a second point to plot a line with these characteristics without constructing the equation.

SOLUTION

Plot the point $(2, 3)$.

a. If we write the slope as $(-5)/4$, then a change of 4 in x corresponds to a change of -5 in y. So starting at $(2, 3)$, moving horizontally four units to the right (adding 4 to the x-coordinate), and then moving vertically five units down (subtracting 5 from the y-coordinate) gives us a second point on the line at $(2 + 4, 3 - 5) = (6, -2)$. Now we can plot our second point $(6, -2)$ and draw the line through it and our original point, $(2, 3)$ (see Figure 2.28).

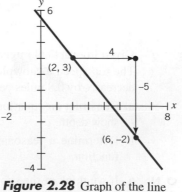

Figure 2.28 Graph of the line through $(2, 3)$ with slope $-5/4$.

b. If we are modeling real data, we are more likely to convert the slope to decimal form. In this case $m = (-5)/4 = -1.25$. We can generate a new point on the line by starting at $(2, 3)$ and moving 1 unit to the right and down (since m is negative) -1.25 units to get the point $(2 + 1, 3 - 1.25) = (3, 1.75)$.

EXAMPLE 4   **Given a general description**
A recent study reporting on the number of smokers showed:

a. A linear increase in Georgia
b. A linear decrease in Utah
c. A concave down decrease in Hawaii
d. A concave up increase in Oklahoma

Generate four rough sketches that could represent these situations.

SOLUTION   See Figure 2.29.

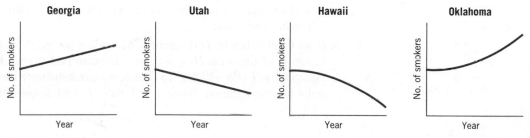

Figure 2.29 The change in the number of smokers over time in four states.

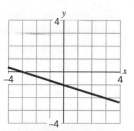
The program "L.3: Finding a Line Through 2 Points" in *Linear Functions* will give you practice in this skill.

Finding the Equation

To determine the equation of any particular linear function $y = b + mx$, we only need to find the specific values for the slope, m, and the vertical intercept, b.

EXAMPLE 5   **Given a graph**
Find the equation of the linear function graphed in Figure 2.30.

Figure 2.30 Graph of a linear function.

SOLUTION   We can use any two points on the graph to calculate m, the slope. If, for example, we take $(-3, 0)$ and $(3, -2)$, then

$$\text{slope} = \frac{\text{change in } y}{\text{change in } x} = \frac{-2 - 0}{3 - (-3)} = \frac{-2}{6} = \frac{-1}{3}$$

The y-intercept is -1, so $b = -1$.

Hence the equation is $y = -1 - \frac{1}{3}x$.

EXAMPLE 6

Given two points

Pediatric growth charts suggest a linear relationship between age (in years) and median height (in inches) for children between 2 and 12 years. Two-year-olds have a median height of 35 inches, and 12-year-olds have a median height of 60 inches.

a. Generate the average rate of change of height with respect to age. (Be sure to include units.) Interpret your result in context.

b. Generate an equation to describe height as a function of age. What is an appropriate domain? Range?

c. What would this model predict as the median height of 8-year-olds?

SOLUTION

a. Average rate of change $= \dfrac{\text{change in height}}{\text{change in age}} = \dfrac{60 - 35}{12 - 2} = \dfrac{25}{10} = 2.5$ inches/year

The chart suggests that, on average, children between the ages of 2 and 12 grow 2.5 inches each year.

b. If we think of height, H (in inches), depending on age, A (in years), then we want an equation of the form $H = b + m \cdot A$. From part (a) we know $m = 2.5$, so our equation is $H = b + 2.5A$. To find b, we can substitute the values for any known point into the equation. When $A = 2$, then $H = 35$. Substituting in, we get

$$H = b + 2.5A$$
$$35 = b + (2.5 \cdot 2)$$
$$35 = b + 5$$
$$30 = b$$

So the final form of our equation is

$$H = 30 + 2.5A$$

where the domain is $2 \le A \le 12$ and the range is $35 \le H \le 60$.

c. When $A = 8$ years, our model predicts that the median height is $H = 30 + (2.5 \cdot 8) = 30 + 20 = 50$ inches.

EXAMPLE 7

Given a table

a. Determine if the data in Table 2.8 represent a linear relationship between values of blood alcohol concentration and number of drinks consumed for a 160-pound person. (One drink is defined as 5 oz of wine, 1.25 oz of 80-proof liquor, or 12 oz of beer.)

D, Number of Drinks	A, Blood Alcohol Concentration
2	0.047
4	0.094
6	0.141
10	0.235

Table 2.8

Note that federal law requires states to have 0.08 as the legal BAC limit for driving drunk.

b. If the relationship is linear, determine the corresponding equation.

SOLUTION

a. We can generate a third column in the table that represents the average rate of change between consecutive points (see Table 2.9). Since the average rate of change of A with respect to D remains constant at 0.0235, these data represent a linear relationship.

b. The rate of change is the slope, so the corresponding linear equation will be of the form

$$A = b + 0.0235D \qquad (1)$$

D	A	Average Rate of Change
2	0.047	n.a.
4	0.094	$\dfrac{0.094 - 0.047}{4 - 2} = \dfrac{0.047}{2} = 0.0235$
6	0.141	$\dfrac{0.141 - 0.094}{6 - 4} = \dfrac{0.047}{2} = 0.0235$
10	0.235	$\dfrac{0.235 - 0.141}{10 - 6} = \dfrac{0.094}{4} = 0.0235$

Table 2.9

To find b, we can substitute any of the original paired values for D and A, for example, (4, 0.094), into Equation (1) to get

$$0.094 = b + (0.0235 \cdot 4)$$
$$0.094 = b + 0.094$$
$$0.094 - 0.094 = b$$
$$0 = b$$

So the final equation is

$$A = 0 + 0.0235D$$

or just

$$A = 0.0235D$$

So when D, the number of drinks, is 0, A, the blood alcohol concentration, is 0, which makes sense.

The slope m between an arbitrary point (x, y) and a specific point (x_1, y_1) is

$$m = \frac{y - y_1}{x - x_1}$$

If we solve this equation for y, we get $m(x - x_1) = y - y_1$ or

$$y = m(x - x_1) + y_1$$

This is called a linear function in *point-slope form*.

EXAMPLE 8   Using the point-slope formula to find the equation

a. If (2, 3) is a point on a line with slope 5, use the point-slope formula to find the equation of the line.

b. Confirm that the equation is equivalent to an equation in the form $y = mx + b$.

SOLUTION   **a.** Given the point (2, 3) on a line with slope 5, then

$$5 = \frac{y - 3}{x - 2}$$

$$5(x - 2) = y - 3$$

So in point slope form $\qquad y = 5(x - 2) + 3.$

b.
$$y = 5(x - 2) + 3$$
$$= 5x - 10 + 3$$

So in the form $y = mx + b$ $\qquad y = 5x - 7$

Algebra Aerobics 2.7

For Problems 1 and 2, find an equation, make an appropriate table, and sketch the graph of:

1. A line with slope 1.2 and vertical intercept −4.

2. A line with slope −400 and vertical intercept 300 (be sure to think about scales on both axes).

3. Write an equation for the line graphed below.

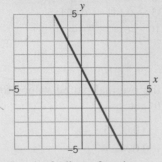

Graph of a linear function.

4. **a.** Find an equation to represent the current salary after x years of employment if the starting salary of an employee is $35,000 with annual increases of $3000.

   b. Create a small table of values and sketch a graph.

5. Complete this statement regarding the graph of $y = 6.2 + 3x$: Beginning with any point on the graph of the line, we could find another point by moving up ___ units for each unit that we move horizontally to the right.

6. Given the equation $y = 8 - 4x$, complete the following statements.

   a. Beginning with the vertical intercept, if we move one unit horizontally to the right, then we need to move down ___ units vertically to stay on the line and arrive at point (__, __).

   b. Beginning with the vertical intercept, if we move one unit horizontally to the left, then we need to move up ___ units vertically to stay on the line and arrive at point (__, __).

7. For both Graph A and Graph B, identify two points on each line, determine the slope, then write an equation for the line.

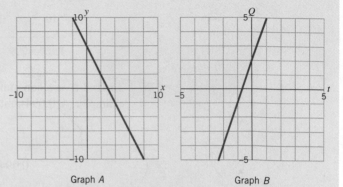

Graph A   Graph B

8. **a.** Write the equations of three lines each with vertical intercept of 6.

   b. Write the equations of three lines with slope −3.

9. Use the point-slope strategy to find linear equations through

   a. The point $(-2, 1)$ with a slope of 4

   b. The point $(3, 5)$ with a slope of $-2/3$

   c. The point $(-1, 3)$ with a slope of -10

   d. The point $(1.2, 4.5)$ with a slope of 2

10. Write an equation for the line:

   a. Through $(2, 5)$ and $(4, 11)$

   b. Through $(-3, 2)$ and $(6, 1)$

   c. Through $(4, -1)$ and $(-2, -7)$

Exercises for Section 2.7

Graphing program is optional in Exercises 8 and 10.

1. Write an equation for the line through $(-2, 3)$ that has slope:

   a. 5   **b.** $-\frac{3}{4}$ **c.** 0

2. Write an equation for the line through $(0, 50)$ that has slope:

   a. −20   **b.** 5.1   **c.** 0

3. Calculate the slope and write an equation for the linear function represented by each of the given tables.

a.

x	y
2	7.6
4	5.1

b.

A	W
5	12
7	16

4. Determine which of the following tables represents a linear function. If it is linear, write the equation for the linear function.

a.

x	y
0	3
1	8
2	13
3	18
4	23

b.

q	R
0	0.0
1	2.5
2	5.0
3	7.5
4	10.0

c.

x	$g(x)$
0	0
1	1
2	4
3	9
4	16

d.

t	r
10	5.00
20	2.50
30	1.67
40	1.25
50	1.00

e.

x	$h(x)$
20	20
40	−60
60	−140
80	−220
100	−300

f.

p	T
5	0.25
10	0.50
15	0.75
20	1.00
25	1.25

5. Plot each pair of points, then determine the equation of the line that goes through the points.

a. $(2, 3), (4, 0)$ **c.** $(2, 0), (0, 2)$

b. $(-2, 3), (2, 1)$ **d.** $(4, 2), (-5, 2)$

6. Find the equation for each of the lines A–C on the accompanying graph.

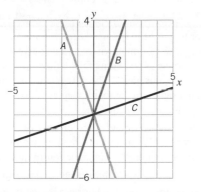

7. Put the following equations in $y = b + mx$ form, then identify the slope and the vertical intercept.

a. $2x - 3y = 6$ **d.** $2y - 3x = 0$

b. $3x + 2y = 6$ **e.** $6y - 9x = 0$

c. $\frac{1}{3}x + \frac{1}{2}y = 6$ **f.** $\frac{1}{2}x - \frac{2}{3}y = -\frac{1}{6}$

8. (Graphing program optional.) Solve each equation for y in terms of x, then identify the slope and the y-intercept. Graph each line by hand. Verify your answers with a graphing utility if available.

a. $-4y - x - 8 = 0$ **c.** $-4x - 3y = 9$

b. $\frac{1}{2}x - \frac{1}{4}y = 3$ **d.** $6x - 5y = 15$

9. Complete the table for each of the linear functions, and then sketch a graph of each function. Make sure to choose an appropriate scale and label the axes.

a.

x	$f(x) = 0.10x + 10$
-100	
0	
100	

b.

x	$h(x) = 50x + 100$
-0.5	
0	
0.5	

10. (Graphing program optional.) The equation $K = 4F - 160$ models the relationship between F, the temperature in degrees Fahrenheit, and K, the number of chirps per minute for the snow tree cricket.

a. Assuming F is the independent variable and K is the dependent variable, identify the slope and vertical intercept in the given equation.

b. Identify the units for K, 4, F, and -160.

c. What is a reasonable domain for this model?

d. Generate a small table of points that satisfy the equation. Be sure to choose realistic values for F from the domain of your model.

e. Calculate the slope directly from two data points. Is this value what you expected? Why?

f. Graph the equation, indicating the domain.

11. Find an equation to represent the cost of attending college classes if application and registration fees are $150 and classes cost $250 per credit.

12. a. Write an equation that describes the total cost to produce x items if the startup cost is $200,000 and the production cost per item is $15.

b. Why is the total average cost per item less if the item is produced in large quantities?

13. Your bank charges you a $2.50 monthly maintenance fee on your checking account and an additional $0.10 for each check you cash. Write an equation to describe your monthly checking account costs.

14. If a town starts with a population of 63,500 that declines by 700 people each year, construct an equation to model its population size over time. How long would it take for the population to drop to 53,000?

15. A teacher's union has negotiated a uniform salary increase for each year of service up to 20 years. If a teacher started at $26,000 and 4 years later had a salary of $32,000:

a. What was the annual increase?

b. What function would describe the teacher's salary over time?

c. What would be the domain for the function?

16. Your favorite aunt put money in a savings account for you. The account earns simple interest; that is, it increases by a fixed amount each year. After 2 years your account has $8250 in it and after 5 years it has $9375.

a. Construct an equation to model the amount of money in your account.

b. How much did your aunt put in initially?

c. How much will your account have after 10 years?

17. You read in the newspaper that the river is polluted with 285 parts per million (ppm) of a toxic substance, and local officials estimate they can reduce the pollution by 15 ppm each year.

a. Derive an equation that represents the amount of pollution, P, as a function of time, t.

b. The article states the river will not be safe for swimming until pollution is reduced to 40 ppm. If the cleanup proceeds as estimated, in how many years will it be safe to swim in the river?

18. The women's recommended weight formula from Harvard Pilgrim Healthcare says, "Give yourself 100 lb for the first 5 ft plus 5 lb for every inch over 5 ft tall."

a. Find a mathematical model for this relationship. Be sure you clearly identify your variables.

b. Specify a reasonable domain and range for the function and then graph the function.

c. Use your model to calculate the recommended weight for a woman 5 feet, 4 inches tall; and for one 5 feet, 8 inches tall.

19. In 1977 a math professor bought her condominium in Cambridge, Massachusetts, for $70,000. The value of the condo has risen steadily so that in 2007 real estate agents tell her the condo is now worth $850,000.

    a. Find a formula in point-slope form, to represent these facts about the value of the condo $V(t)$, as a function of time, t, in years.

    b. If she retires in 2010, what does your formula predict her condo will be worth then?

20. The y-axis, the x-axis, the line $x = 6$, and the line $y = 12$ determine the four sides of a 6-by-12 rectangle in the first quadrant (where $x > 0$ and $y > 0$) of the xy plane. Imagine that this rectangle is a pool table. There are pockets at the four corners and at the points $(0, 6)$ and $(6, 6)$ in the middle of each of the longer sides. When a ball bounces off one of the sides of the table, it obeys the "pool rule": The slope of the path after the bounce is the negative of the slope before the bounce. (*Hint:* It helps to sketch the pool table on a piece of graph paper first.)

    a. Your pool ball is at $(3, 8)$. You hit it toward the y-axis, along the line with slope 2.

        i. Where does it hit the y-axis?

        ii. If the ball is hit hard enough, where does it hit the side of the table next? And after that? And after that?

        iii. Show that the ball ultimately returns to $(3, 8)$. Would it do this if the slope had been different from 2? What is special about the slope 2 for this table?

    b. A ball at $(3, 8)$ is hit toward the y-axis and bounces off it at $\left(0, \frac{16}{3}\right)$. Does it end up in one of the pockets? If so, what are the coordinates of that pocket?

    c. Your pool ball is at $(2, 9)$. You want to shoot it into the pocket at $(6, 0)$. Unfortunately, there is another ball at $(4, 4.5)$ that may be in the way.

        i. Can you shoot directly into the pocket at $(6, 0)$?

        ii. You want to get around the other ball by bouncing yours off the y-axis. If you hit the y-axis at $(0, 7)$, do you end up in the pocket? Where do you hit the line $x = 6$?

        iii. If bouncing off the y-axis at $(0, 7)$ didn't work, perhaps there is some point $(0, b)$ on the y-axis from which the ball would bounce into the pocket at $(6, 0)$. Try to find that point.

21. Find the equation of the line shown on the accompanying graph. Use this equation to create two new graphs, taking care to label the scales on your new axes. For one of your graphs, choose scales that make the line appear steeper than in the original graph. For your second graph, choose scales that make the line appear less steep than in the original graph.

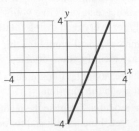

22. You are living in Paris but will be traveling in London. The exchange rate that a French bank gave in January 2010 was 1 euro (standard European currency) for 0.87 pound (currency in Great Britain). The bank charges 3 euros for any transaction.

    a. Write a general equation for changing euros into pounds (including transaction fees).

    b. Graph pounds as a function of euros.

    c. Would it make any sense to exchange 10 euros?

23. According to Moody's *www.econ.com.* in the fourth quarter of 2006, 6.5% of homeowners had negative equity (that is, the market value of the home falls below the outstanding debt to pay off the mortgage). The percentage increased at a roughly constant rate until the fourth quarter of 2008, gaining about 2.5 percentage points per quarter.

    a. Construct an equation describing the relationship of the percent of homeowners having negative equity vs. the number of quarters since the fourth quarter of 2006.

    b. Determine a reasonable domain and then graph the function.

    c. Given this model, what percentage of homeowners had negative equity in the fourth quarter of 2008?

24. **a.** Create a third column in Tables A and B, and insert values for the average rate of change. (The first entry will be "n.a.".)

t	d		t	d
0	400		0	1.2
1	370		1	2.1
2	340		2	3.2
3	310		3	4.1
4	280		4	5.2
5	250		5	6.1

        Table A   **Table B**

    b. In either table, is d a linear function of t? If so, construct a linear equation relating d and t for that table.

25. Adding minerals or organic compounds to water lowers its freezing point. Antifreeze for car radiators contains glycol (an organic compound) for this purpose. The accompanying table shows the effect of salinity (dissolved salts) on the freezing point of water. Salinity is measured in the number of grams of salts dissolved in 1000 grams of water. So our units for salinity are in parts per thousand, abbreviated ppt. Is the relationship between the freezing point and salinity linear? If so, construct an equation that models the relationship. If not, explain why.

Relationship between Salinity and Freezing Point

Salinity (ppt)	Freezing Point (°C)
0	0.00
5	−0.27
10	−0.54
15	−0.81
20	−1.08
25	−1.35

Source: Data adapted from P.R. Pinel, *Oceanography: An Introduction to the Planet Oceanus* (St. Paul, MN: West, 1992), p. 522.

26. The accompanying data show rounded mean values for blood alcohol concentration (BAC) for people of different weights, according to how many drinks (5 oz wine, 1.25 oz 80-proof liquor, or 12 oz beer) they have consumed.

Blood Alcohol Concentration for Selected Weights

Number of Drinks	100 lb	140 lb	180 lb
2	0.075	0.054	0.042
4	0.150	0.107	0.083
6	0.225	0.161	0.125
8	0.300	0.214	0.167
10	0.375	0.268	0.208

a. Examine the data on BAC for a 100-pound person. Are the data linear? If so, find a formula to express blood alcohol concentration, A, as a function of the number of drinks, D, for a 100-pound person.

b. Examine the data on BAC for a 140-pound person. Are the data linear? If they're not precisely linear, what might be a reasonable estimate for the average rate of change of blood alcohol concentration, A, with respect to number of drinks, D? Find a formula to estimate blood alcohol concentration, A, as a function of number of drinks, D, for a 140-pound person. Can you make any general conclusions about BAC as a function of number of drinks for all of the weight categories?

c. Examine the data on BAC for people who consume two drinks. Are the data linear? If so, find a formula to express blood alcohol concentration, A, as a function of weight, W, for people who consume two drinks. Can you make any general conclusions about BAC as a function of weight for any particular number of drinks?

2.8 Special Cases

Direct Proportionality

The simplest relationship between two variables is when one variable is equal to a constant multiple of the other. For instance, in the previous section (Example 7) we saw that $A = 0.0235D$; blood alcohol concentration A equals a constant, 0.0235, times D, the number of drinks. We say that A *is directly proportional to D*.

How to recognize direct proportionality

Linear functions of the form

$$y = mx \qquad (m \neq 0)$$

describe a relationship where y is directly proportional to x. If two variables are directly proportional to each other, the graph will be a straight line that passes through the point $(0, 0)$, the origin. Figure 2.31 shows the graphs of two relationships in which y is directly proportional to x, namely $y = 2x$ and $y = -x$.

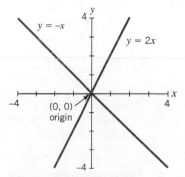

Figure 2.31 Graphs of two relationships in which y is directly proportional to x. Note that both graphs are lines that go through the origin.

> **Direct Proportionality**
>
> In a linear equation of the form
>
> $$y = mx \qquad (m \neq 0)$$
>
> we say that
>
> $$y \text{ is } \textit{directly proportional to } x.$$
>
> Its graph will go through the point (0, 0), the origin.

EXAMPLE 1

Braille vs print

Braille is a code, based on six-dot cells, that allows blind people to read. One page of regular print translates into 2.5 Braille pages. Construct a function describing this relationship. Does it represent direct proportionality? What happens if the number of print pages doubles? Triples?

SOLUTION

If P = number of regular pages and B = number of Braille pages, then $B = 2.5P$. So B is directly proportional to P. If the number of print pages doubles, the number of Braille pages doubles. If the number of print pages triples, the number of Braille pages will triple.

EXAMPLE 2

Monetary conversion

You are traveling to Canada and need to exchange American dollars for Canadian dollars. On that day the exchange rate is approximately 1 American dollar for 1.03 Canadian dollars.

a. Construct an equation for converting American to Canadian dollars. Does it represent direct proportionality?

b. Suppose the Exchange Bureau charges a $2 flat fee to change money. Alter your equation from part (a) to include the service fee. Does the new equation represent direct proportionality?

SOLUTION

a. If we let A = American dollars and C = Canadian dollars, then the equation

$$C = 1.03A$$

describes the conversion from American (the input) to Canadian (the output) dollars. The amount of Canadian money you receive is directly proportional to the amount of American money you exchange.

b. If there is a $2 service fee, you would have to subtract $2 from the American money you have before converting to Canadian. The new equation is

$$C = 1.03(A - 2)$$

or equivalently,

$$C = 1.03A - 2.06$$

where 2.06 is the service fee in Canadian dollars. Then C is no longer directly proportional to A.

EXAMPLE 3

University costs

A prominent midwestern university uses what it calls a linear model, charging $106 per credit hour for in-state students and $369 per credit hour for out-of-state students.

a. Is it correct to call this pricing scheme a linear model for in-state students? For out-of-state students? Why?

b. Generate equations and graphs for the cost of tuition for both in-state and out-of-state students. If we limit costs to one semester during which the usual maximum credit hours is 15, what would be a reasonable domain?

c. In each case is the tuition directly proportional to the number of credit hours?

SOLUTION   **a.** Yes, both relationships are linear since the rate of change is constant in each case: $106 per credit hour for in-state students and $369 per credit hour for out-of-state students.

b. Let N = number of credit hours, C_i = cost for an in-state student, and C_o = cost for an out-of-state student. In each case if the number of credit hours is zero ($N = 0$), then the cost would be zero ($C_i = 0 = C_o$). Hence both lines would pass through the origin $(0, 0)$, making the vertical intercept 0 for both equations. So the results would be of the form

$$C_i = 106N \quad \text{and} \quad C_o = 369N$$

which are graphed in Figure 2.32. A reasonable domain would be $0 \leq N \leq 15$.

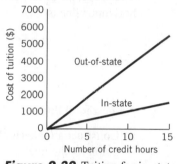

Figure 2.32 Tuition for in-state and out-of-state students at a midwestern university.

c. In both cases the tuition is directly proportional to the number of credit hours. The graphs verify this since both are straight lines going through the origin.

Algebra Aerobics 2.8a

1. Construct an equation and draw the graph of the line that passes through the origin and has the given slope.

a. $m = -1$

b. $m = 0.5$

2. For each of the tables below, determine whether x and y are directly proportional to each other. Represent each relationship with an equation.

a.

x	y
-2	6
-1	3
0	0
1	-3
2	-6

b.

x	y
0	5
1	8
2	11
3	14
4	17

3. In January 2010 the exchange rate was $1.00 U.S. to 0.70 euros, the common European currency.

a. Find a linear function that converts U.S. dollars to euros.

b. Find a linear function that converts U.S. dollars to euros with a service fee of $2.50.

c. Which function represents a directly proportional relationship and why?

4. Suppose you go on a road trip, driving at a constant speed of 60 miles per hour. Create an equation relating distance d in miles and time traveled t in hours. Does it represent direct proportionality? What happens to d if the value for t doubles? If t triples?

5. The total cost C for football tickets is directly proportional to the number of tickets purchased, N. If two tickets cost \$50, construct the formula relating C and N. What would the total cost of 10 tickets be?

6. Write a formula to describe each situation.

   a. y is directly proportional to x, and y is 4 when x is 12.

   b. d is directly proportional to t, and d is 300 when t is 50.

7. Write a formula to describe the following:

   a. The diameter, d, of a circle is directly proportional to the circumference, C.

   b. The amount of income tax paid, T, is directly proportional to income, I.

   c. The tip amount t, is directly proportional to the cost of the meal, c.

8. Assume that a is directly proportional to b. When $a = 10, b = 15$.

   a. Find a if b is 6.

   b. Find b if a is 4.

Horizontal and Vertical Lines

The slope, m, of any horizontal line is 0. So the general form for the equation of a horizontal line is

$$y = b + 0x$$

or just

$$y = b$$

E X A M P L E   4   **A horizontal line**

   a. Construct a table (with both positive and negative values for x) and corresponding graph for the line $y = 1$.

   b. Is y a function of x?

   c. Generate the slope of the line using two points from the table.

S O L U T I O N   **a.** Table 2.10 and Figure 2.33 show points that satisfy the equation $y = 1$.

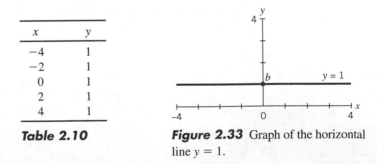

x	y
-4	1
-2	1
0	1
2	1
4	1

Table 2.10

Figure 2.33 Graph of the horizontal line $y = 1$.

   b. The graph of $y = 1$ passes the vertical line test, so y is a function of x. In this function every x is mapped to the same value of 1, so we can think of all the points on the line as of the form $(x, 1)$ where $y = 1$.

   c. If we calculate the slope between any two points in the table, for example $(-2, 1)$ and $(2, 1)$, we get

$$\text{slope} = \frac{1 - 1}{-2 - 2} = \frac{0}{-4} = 0$$

For a vertical line the slope, m, is undefined, so we can't use the standard $y = b + mx$ format. The graph of a vertical line (as in Figure 2.34) fails the vertical line test, so y is not a function of x. However, every point on a vertical line does have the same horizontal coordinate, which equals the coordinate of the horizontal intercept. Therefore, the general equation for a vertical line is of the form

$$x = c \qquad \text{where } c \text{ is a constant (the horizontal intercept)}$$

EXAMPLE 5   **A vertical line**

a. Construct a table (with both positive and negative values for y) and corresponding graph for the line $x = 1$.

b. Is y a function of x?

c. Generate the slope of the line using two points from the table.

SOLUTION   **a.** Table 2.11 and Figure 2.34 show points that satisfy the equation $x = 1$.

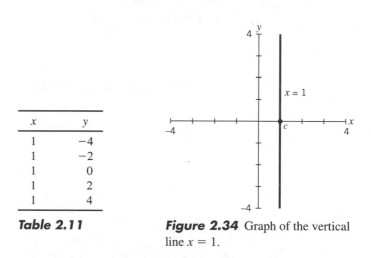

x	y
1	-4
1	-2
1	0
1	2
1	4

Table 2.11

Figure 2.34 Graph of the vertical line $x = 1$.

b. y is *not* a function of x since the graph fails the vertical line test.

c. If we calculate the slope between two points, say $(1, -4)$ and $(1, 2)$, we get

$$\text{slope} = \frac{-4 - 2}{1 - 1} = \frac{-6}{0} \quad \text{which is undefined.}$$

The general equation of a *horizontal line* is

$$y = b$$

where b is a constant (the vertical intercept) and the slope is 0.

The general equation of a *vertical line* is

$$x = c$$

where c is a constant (the horizontal intercept) and the slope is undefined.

E X A M P L E   6   **Finding an equation from a graph**
Find the equation for each line in Figure 2.35.

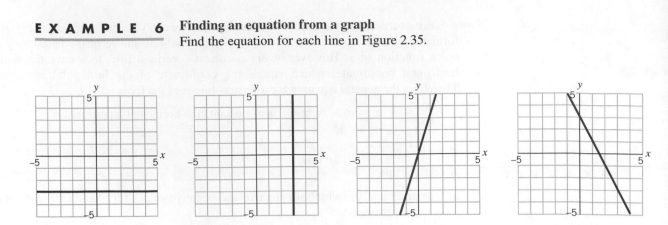

Figure 2.35 Four linear graphs.

S O L U T I O N   **a.** $y = -3$, a horizontal line

b. $x = 3$, a vertical line

c. $y = 3x$, a direct proportion, slope $= 3$

d. $y = -2x + 3$, a line with slope -2 and y-intercept 3

Parallel and Perpendicular Lines

Parallel lines have the same slope. So if the two equations $y = b_1 + m_1 x$ and $y = b_2 + m_2 x$ describe two parallel lines, then $m_1 = m_2$. For example, the two lines $y = 2.0 - 0.5x$ and $y = -1.0 - 0.5x$ each have a slope of -0.5 and thus are parallel (see Figure 2.36).

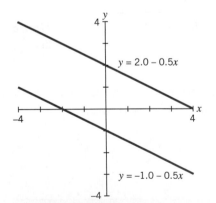

Figure 2.36 Two parallel lines have the same slope.

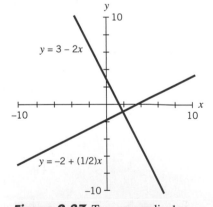

Figure 2.37 Two perpendicular lines have slopes that are negative reciprocals.

Two lines are perpendicular if their slopes are negative reciprocals. If $y = b_1 + m_1 x$ and $y = b_2 + m_2 x$ describe two perpendicular lines, then $m_1 = -1/m_2$. For example, in Figure 2.37 the two lines $y = 3 - 2x$ and $y = -2 + \frac{1}{2}x$ have slopes of -2 and $\frac{1}{2}$, respectively.

Since -2 is the negative reciprocal of $\frac{1}{2}$ (i.e., $-\dfrac{1}{\left(\frac{1}{2}\right)} = -1 \div \frac{1}{2} = -1 \cdot \frac{2}{1} = -2$), the two lines are perpendicular.

Why does this relationship hold for perpendicular lines?

Consider the line l_1 (in Figure 2.38) whose slope is

$$\frac{\text{vertical change}}{\text{horizontal change}} = \frac{v}{h}$$

Now imagine rotating the line 90 degrees clockwise to generate a second line, l_2, perpendicular to the original line, l_1. What would the slope of this new line be?

The positive vertical change, v, becomes a positive horizontal change. The positive horizontal change, h, becomes a negative vertical change. The slope of the original line is v/h, and the slope of the line rotated 90 degrees clockwise is $-h/v$. Note that $-h/v = -1/(v/h)$, which is the original slope inverted and multiplied by -1.

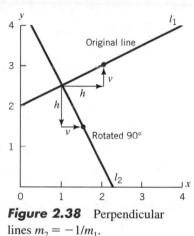

In general, the slope of a perpendicular line is the negative reciprocal of the slope of the original line. If the slope of a line is m_1, then the slope, m_2, of a line perpendicular to it is $-1/m_1$.

This is true for any pair of perpendicular lines for which slopes exist. It does not work for horizontal and vertical lines since vertical lines have undefined slopes.

Figure 2.38 Perpendicular lines $m_2 = -1/m_1$.

Parallel lines have the same slope.

Perpendicular lines have slopes that are negative reciprocals of each other.

EXAMPLE 7

Parallel, perpendicular or neither?
Determine from the equations which pairs of lines are parallel, perpendicular, or neither.

a. $F(x) = 2 + 7x$ and $G(x) = 7x + 3$ **c.** $y = 5 + 3x$ and $y = 5 - 3x$

b. $f(z) = 6 - z$ and $h(z) = 6 + z$ **d.** $y = 3x + 13$ and $3y + x = 2$

SOLUTION

a. The two lines are parallel since they share the same slope, 7.

b. The two lines are perpendicular since the negative reciprocal of -1 (the slope of first line) equals $-(1/(-1)) = -(-1) = 1$, the slope of the second line.

c. The lines are neither parallel nor perpendicular.

d. The lines are perpendicular. The slope of the first line is 3. If we solve the second equation for y, we get

$$3y + x = 2$$
$$3y = 2 - x$$
$$y = 2/3 - (1/3)x$$

So the slope of the second line is $-(1/3)$, the negative reciprocal of 3.

EXAMPLE 8

Equation for a horizontal line
Describe the equation for any line perpendicular to the horizontal line $y = b$. Is it a function?

SOLUTION

Any vertical line can be written in the form $x = c$ and is perpendicular to the horizontal line $y = b$. A vertical line is not a function because it clearly fails the vertical line test.

EXPLORE & EXTEND

2.8 Constructing Lines under Certain Constraints

Use the course software for "L1: *m* and *b* Sliders" or a graphing calculator or graphing paper to construct the following sets of lines. Be sure to write down your results. What generalizations can you make in each case? Are the slopes related in some way? What about the vertical intercepts?

1. **a.** Construct any line. Then construct another line that has a steeper slope, and then construct one that has a shallower slope.

   b. Construct three *parallel lines*.

   c. Construct three lines with the *same y-intercept*, the point where the line crosses the *y*-axis.

   d. Construct a pair of lines that are *horizontal*.

   e. Construct a pair of lines that go *through the origin*.

   f. Construct a pair of lines that are *perpendicular* to each other.

2. Write a 60-second summary of what you have learned about the equations of lines.

Algebra Aerobics 2.8b

1. In each case, find an equation for the horizontal line that passes through the given point.

   a. $(3, -5)$ **b.** $(5, -3)$ **c.** $(-3, 5)$

2. In each case, find an equation for the vertical line that passes through the given point.

   a. $(3, -5)$ **b.** $(5, -3)$ **c.** $(-3, 5)$

3. Construct the equation of the line that passes through the points.

   a. $(0, -7), (3, -7)$, and $(350, -7)$

   b. $(-4.3, 0), (-4.3, 8)$ and $(-4.3, -1000)$

4. Find the equation of the line that is parallel to $y = 4 - x$ and that passes through the origin.

5. Find the equation of the line that is parallel to $W = 360C + 2500$ and passes through the point where $C = 4$ and $W = 1000$.

6. Find the slope of a line perpendicular to each of the following.

   a. $y = 4 - 3x$ **c.** $y = 3.1x - 5.8$

   b. $y = x$ **d.** $y = -\frac{3}{5}x + 1$

7. **a.** Find an equation for the line that is perpendicular to $y = 2x - 4$ and passes through $(3, -5)$.

   b. Find the equations of two other lines that are perpendicular to $y = 2x - 4$ but do not pass through the point $(3, -5)$.

   c. How do the three lines from parts (a) and (b) that are perpendicular to $y = 2x - 4$ relate to each other?

   d. Check your answers by graphing the equations.

8. Find the slope of the line $Ax + By = C$ assuming that y is a function of x. (*Hint:* Solve the equation for y.)

9. Use the result of the previous exercise to determine the slope of each line described by the following linear equations (again assuming y is a function of x).

   a. $2x + 3y = 5$ **d.** $x = -5$

   b. $3x - 4y = 12$ **e.** $x - 3y = 5$

   c. $2x - y = 4$ **f.** $y = 4$

10. Solve the equation $2x + 3y = 5$ for y, identify the slope, then find an equation for the line that is parallel to the line $2x + 3y = 5$ and passes through the point $(0, 4)$.

11. Solve the equation $3x + 4y = -7$ for y, identify the slope, then find an equation for the line that is perpendicular to the line $3x + 4y = -7$ and passes through $(0, 3)$.

12. Solve the equation $4x - y = 6$ for y, identify the slope, then find an equation for the line that is perpendicular to the line $4x - y = 6$ and passes through $(2, -3)$.

13. Determine whether each equation could represent a vertical line, a horizontal line, or neither.

   a. $x + 1.5 = 0$

   b. $2x + 3y = 0$

   c. $y - 5 = 0$

14. Write an equation for the line perpendicular to $2x + 3y = 6$:

   a. That has a vertical intercept of 5

   b. That passes through the point $(-6, 1)$

15. Write an equation for the line parallel to $2x - y = 7$:

   a. That has a vertical intercept of 9

   b. That passes through $(4, 3)$

Exercises for Section 2.8

1. Using the general formula $y = mx$ that describes direct proportionality, find the value of m if:

   a. y is directly proportional to x and $y = 2$ when $x = 10$.

   b. y is directly proportional to x and $y = 0.1$ when $x = 0.2$.

   c. y is directly proportional to x and $y = 1$ when $x = \frac{1}{4}$.

2. For each part, construct an equation and then use it to solve the problem.

   a. Pressure P is directly proportional to temperature T, and P is 20 lb per square inch when T is 60 degrees Kelvin. What is the pressure when the temperature is 80 degrees Kelvin?

   b. Earnings E are directly proportional to the time T worked, and E is \$46 when T is 2 hours. How long has a person worked if she earned \$471.50?

   c. The number of centimeters of water depth W produced by melting snow is directly proportional to the number of centimeters of snow depth S. If W is 15.9 cm when S is 150 cm, then how many centimeters of water depth are produced by a 100-cm depth of melting snow?

3. In the accompanying table y is directly proportional to x.

Number of CDs purchased (x)	3	4	5
Cost of CDs (y)	42.69	56.92	

   a. Find the formula relating y and x, then determine the missing value in the table.

   b. Interpret the coefficient of x in this situation.

4. The electrical resistance R (in ohms) of a wire is directly proportional to its length l (in feet).

   a. If 250 feet of wire has a resistance of 1.2 ohms, find the resistance for 150 ft of wire.

   b. Interpret the coefficient of l in this context.

5. For each of the following linear functions, determine the independent and dependent variables and then construct an equation for each function.

   a. Sales tax is 6.5% of the purchase price.

   b. The height of a tree is directly proportional to the amount of sunlight it receives.

   c. The average salary for full-time employees of American domestic industries has been growing at an annual rate of \$1300/year since 1985, when the average salary was \$25,000.

6. On the scale of a map 1 inch represents a distance of 35 miles.

   a. What is the distance between two places that are 4.5 inches apart on the map?

   b. Construct an equation that converts inches on the map to miles in the real world.

7. Find a function that represents the relationship between distance, d, and time, t, of a moving object using the data in the table at the top of the next column. Is d directly proportional to t? Which is a more likely choice for the object, a person jogging or a moving car?

t (hours)	d (miles)
0	0
1	5
2	10
3	15
4	20

8. Determine which (if any) of the following variables (w, y, or z) is directly proportional to x:

x	w	y	z
0	1	0.0	0
1	2	2.5	$-\frac{1}{3}$
2	5	5.0	$-\frac{2}{3}$
3	10	7.5	-1
4	17	10.0	$-\frac{4}{3}$

9. Find the slope of the line through the pair of points, then determine the equation.

   a. $(2, 3)$ and $(5, 3)$ **c.** $(-3, 8)$ and $(-3, 4)$

   b. $(-4, -7)$ and $(12, -7)$ **d.** $(2, -3)$ and $(2, -1)$

10. Describe the graphs of the following equations.

    a. $y = -2$ **c.** $x = \frac{2}{5}$ **e.** $y = 324$

    b. $x = -2$ **d.** $y = \frac{x}{4}$ **f.** $y = \frac{2}{3}$

11. The accompanying figure shows the quantity of books (in millions) shipped by publishers in the United States between 2007 and 2010. Construct the equation of a horizontal line that would be a reasonable model for these data.

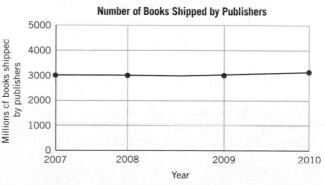

Number of Books Shipped by Publishers

Source: U.S. Bureau of the Census, *Statistical Abstract of the United States,* 2010.

12. An employee for an aeronautical corporation had a starting salary of \$25,000/year. After working there for 10 years and not receiving any raises, he decides to seek employment elsewhere. Graph the employee's salary as a function of time for the time he was employed with this corporation. What is the domain? What is the range?

13. For each of the given points write equations for three lines that all pass through the point such that one of the three lines is horizontal, one is vertical, and one has slope 2.

    a. $(1, -4)$ **b.** $(2, 0)$

14. Consider the function $f(x) = 4$.

    a. What is $f(0)$? $f(30)$? $f(-12.6)$?

    b. Describe the graph of this function.

    c. Describe the slope of this function's graph.

15. A football player who weighs 175 pounds is instructed at the end of spring training that he has to put on 30 pounds before reporting for fall training.

    a. If fall training begins 3 months later, at what (monthly) rate must he gain weight?

    b. Suppose that he eats a lot and takes several nutritional supplements to gain weight, but due to his metabolism he still weighs 175 pounds throughout the summer and at the beginning of fall training. Sketch a graph of his weight versus time for those 3 months.

16. **a.** Write an equation for the line parallel to $y = 2 + 4x$ that passes through the point $(3, 7)$.

    b. Find an equation for the line perpendicular to $y = 2 + 4x$ that passes through the point $(3, 7)$.

17. **a.** Write an equation for the line parallel to $y = 4 - x$ that passes through the point $(3, 7)$.

    b. Find an equation for the line perpendicular to $y = 4 - x$ that passes through the point $(3, 7)$.

18. Construct the equation of a line that goes through the origin and is parallel to the graph of the given equation.

    a. $y = 6$ **b.** $x = -3$ **c.** $y = -x + 3$

19. Construct the equation of a line that goes through the origin and is perpendicular to the given equation.

    a. $y = 6$ **b.** $x = -3$ **c.** $y = -x + 3$

20. Which lines are parallel to each other? Which lines are perpendicular to each other?

    a. $y = \frac{1}{3}x + 2$ **c.** $y = -2x + 10$ **e.** $2y + 4x = -12$

    b. $y = 3x - 4$ **d.** $y = -3x - 2$ **f.** $y - 3x = 7$

21. Because different scales may be used on the horizontal and vertical axes, it is often difficult to tell if two lines are perpendicular to each other. In parts (a) and (b), determine the equations of each pair of lines and show whether or not the paired lines are perpendicular to each other.

    a.   **b.**

    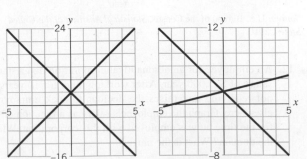

22. In each part construct the equations of two lines that:

    a. Are parallel to each other

    b. Intersect at the same point on the y-axis

    c. Both go through the origin

    d. Are perpendicular to each other

23. For each of the accompanying graphs you don't need to do any calculations or determine the actual equations. Using just the graphs, determine if the slopes for each pair of lines are the same. Are the slopes both positive or both negative, or is one negative and one positive? Do the lines have the same y-intercept?

    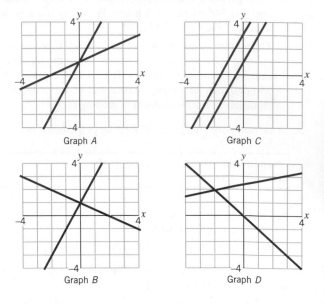

    Graph A   Graph C

    Graph B   Graph D

24. Find the equation of the line in the form $y = mx + b$ for each of the following sets of conditions. Show your work.

    a. Slope is \$1400/year and line passes through the point (10 yr, \$12,000).

    b. Line is parallel to $2y - 7x = y + 4$ and passes through the point $(-1, 2)$.

    c. Equation is $1.48x - 2.00y + 4.36 = 0$.

    d. Line is horizontal and passes through $(1.0, 7.2)$.

    e. Line is vertical and passes through $(275, 1029)$.

    f. Line is perpendicular to $y = -2x + 7$ and passes through $(5, 2)$.

25. In the equation $Ax + By = C$:

    a. Solve for y so as to rewrite the equation in the form $y = mx + b$.

    b. Identify the slope.

    c. What is the slope of any line parallel to $Ax + By = C$?

    d. What is the slope of any line perpendicular to $Ax + By = C$?

26. Use the results of Exercise 25, parts (c) and (d), to find the slope of any line that is parallel and then one that is perpendicular to the given lines.

    a. $5x + 8y = 37$ **b.** $7x + 16y = -14$ **c.** $30x + 47y = 0$

2.9 *Breaking the Line: Piecewise Linear Functions*

Piecewise Linear Functions

Some functions are not linear throughout but are made up of linear segments. They are called *piecewise linear functions*. For example, we could define a function $f(x)$ where:

$$f(x) = 3 + x \quad \text{for } x \leq 1$$
$$f(x) = 3 \qquad\quad \text{for } x > 1 \quad \text{or, more compactly,} \quad f(x) = \begin{cases} 3 + x & \text{for } x \leq 1 \\ 3 & \text{for } x > 1 \end{cases}$$

The graph of $f(x)$ in Figure 2.39 clearly shows the two distinct linear segments.

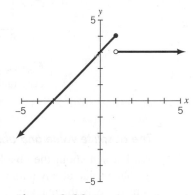

Figure 2.39 Graph of a piecewise linear function.

EXAMPLE 1

Gas consumption

Consider the amount of gas in your car during a road trip. You start out with 20 gallons and drive for 3 hours, leaving you with 14 gallons in the tank. You stop for lunch for an hour and then drive for 4 more hours, leaving you with 6 gallons.

a. Construct a piecewise linear function for the amount of gas in the tank as a function of time in hours.

b. Graph the results.

SOLUTION

a. Let t = time (in hours). For $0 \leq t \leq 3$, the average rate of change in gasoline over time is $(14-20)$ gallons/(3 hr) = $(-6$ gallons)/(3 hr) = -2 gallons/hr; that is, you are consuming 2 gallons per hour. The initial amount of gas is 20 gallons, so $G(t)$, the amount of gas in the tank at time t, is given by

$$G(t) = 20 - 2t \quad \text{for } 0 \leq t \leq 3$$

While you are at lunch for an hour, you consume no gasoline, so the amount of gas stays constant at 14 gallons. So

$$G(t) = 14 \quad \text{for } 3 < t \leq 4$$

At the end of lunch, $t = 4$ and $G(t) = 14$. You continue to drive for 4 more hours, ending up with 6 gallons. You are still consuming 2 gallons per hour since $(6-14)$ gallons/4 hr = -2 gallons/hr. So your equation will be of the form $G(t) = b - 2t$. Substituting in $t = 4$ and $G(t) = 14$, we have $14 = b - 2 \cdot 4$ and $b = 22$. So

$$G(t) = 22 - 2t \quad \text{for } 4 < t \leq 8$$

Writing $G(t)$ more compactly, we have

$$G(t) = \begin{cases} 20 - 2t & \text{if } 0 \leq t \leq 3 \\ 14 & \text{if } 3 < t \leq 4 \\ 22 - 2t & \text{if } 4 < t \leq 8 \end{cases}$$

b. The graph of $G(t)$ is shown in Figure 2.40.

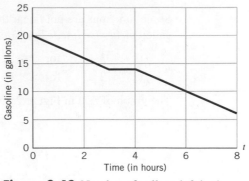

Figure 2.40 Number of gallons left in the car's tank.

The absolute value and absolute value function

We learned about the absolute value of the slope m in Section 2.6. In general, the absolute value of any number x is written as $|x|$ and strips x of its sign. That means we consider only the magnitude of x, so $|x|$ is never negative.[1]

- If x is positive (or 0), then $|x| = x$.
- If x is negative, then $|x| = -x$.

For example, $|-5| = -(-5) = 5$.

Absolute value inequalities in one variable are frequently used to describe an allowable range above or below a certain amount.

EXAMPLE 2

Margin of error of poll results

Poll figures are often given with a margin of error. For example, in January 2010 the CNN/Opinion Research Poll reported that 47% of Americans listed the economy as the most important issue facing the country today, with a margin of error of ±3 percentage points.

Construct an absolute value inequality that describes the range of percentages P that are possible within this poll. Restate this condition without using absolute values, and display it on a number line.

SOLUTION

$|P - 47| \leq 3$; that is, the poll takers are confident that the difference between the estimated percentage, 47%, and the actual percentage, P, is less than or equal to 3%.

Equivalently we could write $44 \leq P \leq 50$; that is, the actual percentage P is somewhere between 44% and 50% (see Figure 2.41).

Figure 2.41 Range of error around 47% is ±3 percentage points.

---

[1]Graphing calculators and spreadsheet programs usually have an absolute value function. It is often named *abs* where *abs* $(x) = |x|$. So *abs* $(-3) = |-3| = 3$.

We can construct the *absolute value function* as a piecewise linear function.

> **The absolute value function**
>
> If $f(x) = |x|$, then $f(x) = \begin{cases} x & \text{for } x \geq 0 \\ -x & \text{for } x < 0 \end{cases}$

E X A M P L E   3

The absolute value function
If $f(x) = |x|$, then:

a. What is $f(6)$? $f(0)$? $f(-6)$?
b. Graph the function $f(x)$ for x from -6 to 6.
c. What is the slope of the line segment when $x > 0$? When $x < 0$?

S O L U T I O N

a. $f(6) = 6$; $f(0) = 0$; $f(-6) = |-6| = 6$
b. See Figure 2.42.

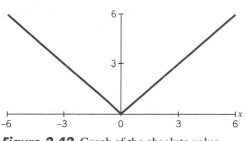

Figure 2.42 Graph of the absolute value function $f(x) = |x|$.

c. When $x > 0$, the slope is 1; when $x < 0$, the slope is -1.

E X A M P L E   4

Distance between a cell phone and cell tower
You are a passenger in a car, talking on a cell phone. The car is traveling at 60 mph along a straight highway and the nearest cell phone tower is 6 miles away.

a. How long will it take you to reach the cell phone tower (assuming it is right by the road)?
b. Construct a linear function $D(t)$ that describes your distance *to* the cell tower (in miles ≥ 0) or *from* the cell tower (in miles <0), where t is the number of hours traveled.
c. Graph your function using a reasonable domain for t.
d. What does $|D(t)|$ represent? Graph $|D(t)|$ on a separate grid and compare it with the graph of $D(t)$.

S O L U T I O N

a. Traveling at 60 miles per hour is equivalent to traveling at 1 mile per minute. So traveling 6 miles from the start will take you 6 minutes or 0.1 hr to reach the cell tower.
b. At the starting time $t = 0$ hours, the distance to the nearest cell phone tower is 6 miles, so $D(0) = 6$ miles. So, the vertical intercept is at $(0, 6)$. After $t = 0.1$ hr you are at the cell tower, so the distance between you and the cell tower is 0. Thus, $D(0.1) = 0$ miles and hence the horizontal intercept is $(0.1, 0)$. The slope of the line is $(0 - 6)/(0.1 - 0) = -60$. So the distance function is $D(t) = 6 - 60t$, where t is in hours and $D(t)$ is in miles (to or from the cell tower).
c. A reasonable domain might be $0 \leq t \leq 0.2$ hours. See Figure 2.43.

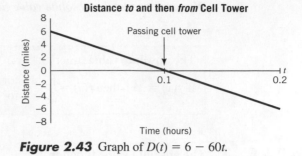

Figure 2.43 Graph of $D(t) = 6 - 60t$.

d. $|D(t)|$ describes the absolute value of the distance between you and the cell tower, indicating that the direction of travel no longer matters. Whether you are driving toward or away from the tower, the absolute value of the distance is always positive (or 0). For example, $|D(0.2)| = |6 - 60 \cdot 0.2| = |6 - 12| = |-6| = 6$ miles, which means that after 0.2 hours (or 12 minutes) you are 6 miles away from the tower. See Figure 2.44.

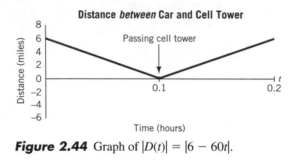

Figure 2.44 Graph of $|D(t)| = |6 - 60t|$.

Step functions

Some piecewise linear functions are called *step functions* because their graphs look like the steps of a staircase. Each "step" is part of a horizontal line.

E X A M P L E 5

Minimum wages

A federal minimum wage (as part of the Fair Labor Standards Act) was first set in 1938 in reaction to the Great Depression. Table 2.12 shows the recent minimum wage from 1997 up through 2009.

Federal Minimum Wage for 1997–2010

Year New Minimum Set	Minimum Hourly Wage
1997	$5.15
2007	$5.85
2008	$6.55
2009	$7.25

Table 2.12

a. Construct a step function $M(x)$, where $M(x)$ is the minimum wage at year x. What is the domain?

b. What is $M(1997)$? $M(2006)$? $M(2007)$?

c. Graph the step function.

d. Why do you think there was a lot of discussion in 2006 about raising the federal minimum wage? (*Note:* Individual states can set a higher minimum for their workers.)

SOLUTION

a. $M(x) = \begin{cases} 5.15 & \text{for } 1997 \le x < 2007 \\ 5.85 & \text{for } 2007 \le x < 2008 \\ 6.55 & \text{for } 2008 \le x < 2009 \\ 7.25 & \text{for } 2009 \le x < 2010 \end{cases}$

The domain is $1997 \le x < 2010$.

b. $M(1997) = \$5.15/\text{hr}; M(2006) = \$5.15/\text{hr}; M(2007) = \$5.85/\text{hr}$

c. See Figure 2.45.

Federal Minimum Wage

Figure 2.45 Step function for federal minimum wage between 1997 and 2010.

d. By the end of 2006 the federal minimum wage had stayed the same (\$5.15/hr) for 9 years. Inflation always erodes the purchasing power of the dollar, so \$5.15 in 2006 bought a lot less than \$5.15 in 1997. Many felt an increase in the minimum wage was way overdue.

Algebra Aerobics 2.9

1. Construct the graphs of the following piecewise linear functions. (Be sure to indicate whether each endpoint is included on or excluded from the graph.)

   a. $f(x) = \begin{cases} 1 & \text{for } 0 < x \le 1 \\ 0 & \text{for } 1 < x \le 2 \\ -1 & \text{for } 2 < x \le 3 \end{cases}$

   b. $g(x) = \begin{cases} x + 3 & \text{for } -4 \le x < 0 \\ 2 - x & \text{for } 0 \le x \le 4 \end{cases}$

2. Construct piecewise linear functions $Q(t)$ and $C(r)$ to describe the following two graphs.

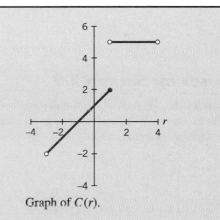

Graph of $C(r)$.

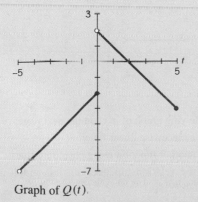

Graph of $Q(t)$.

3. Evaluate the following:

   a. $|-2|$

   b. $|6|$

   c. $|3 - 5|$

   d. $|3| - |5|$

   e. $-|3| \cdot |-5|$

4. Given the function $g(x) = |x - 3|$:

   a. What is $g(-3)$? $g(0)$? $g(3)$? $g(6)$?

   b. Sketch the graph of $g(x) = |x - 3|$ for $-6 \le x \le 6$.

c. Compare the graph of $g(x) = |x - 3|$ with the graph of $f(x) = |x|$.

d. Write $g(x)$ using piecewise linear notation.

5. Rewrite the following expressions without using an absolute value sign, and then describe in words the result.

a. $|t - 5| \leq 2$

b. $|Q - 75| < 6$

6. The optimal water temperature for trout is 55°F, but they can survive water temperatures that are 20° above or below that. Write an absolute value inequality that describes the temperature values, T, that lie within the trout survival temperature range. Then write an equivalent expression without the absolute value sign.

7. **a.** Sketch a piecewise linear graph of the *total* distance you would travel if you walked at a constant speed from home to a coffee shop, stopped for a cup, and then walked home at a faster pace.

b. For the scenario in part (a), sketch a piecewise linear graph that shows the distance *between* you and the coffee shop over time.

8. The federal funds rate is the short-term interest rate charged by the Federal Reserve for overnight loans to other federal banks. This is one of the major tools the Federal Reserve Board uses to stimulate the economy (with a rate decrease) or control inflation (with a rate increase). The table below shows the week of each rate change in 2008.

Federal Funds Rates During 2008

Week in 2008 When Rate Changed	Rate
Week 1	3.5%
Week 5	3.0%
Week 13	2.3%
Week 19	2.0%

Source: www.harpfinancial.com.

a. What was the federal funds rate in week 4? In week 52? What was the longest period in 2008 over which the federal funds rate remained the same?

b. In 2006 the federal fund rates were increasing (from 4.25% to 5.25%), but in 2009 the rates dropped to almost 0%. In 2006 was the Federal Reserve Board more concerned about stimulating growth or curbing inflation? What about in 2009?

c. Construct a piecewise linear function to describe the federal funds rate during 2008.

d. Sketch a graph of your function.

Exercises for Section 2.9

1. Construct the graphs of the following piecewise linear functions. Be sure to indicate whether an endpoint is included in or excluded from the graph.

a. $f(x) = \begin{cases} 2 & \text{for } -3 < x \leq 0 \\ 1 & \text{for } 0 < x \leq 3 \end{cases}$

b. $g(x) = \begin{cases} x + 3 & \text{for } -4 < x \leq 0 \\ 2 - x & \text{for } 0 < x \leq 4 \end{cases}$

2. Construct piecewise linear functions for the following graphs and determine the domain and range.

a. Graph of $f(x)$

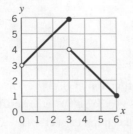

b. Graph of $g(x)$

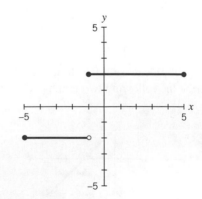

3. Given the following graph of $g(x)$ on the next page:

a. Construct a piecewise linear function of $g(x)$.

b. Construct another function using absolute values.

c. Describe the relationship between this graph and the graph of $f(x) = |x|$.

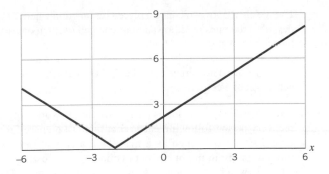

4. a. Normal human body temperature is often cited as 98.6°F. However, any temperature that is within 1°F more or less than that is still considered normal. Construct an absolute value inequality that describes normal body temperatures T that lie within that range. Then rewrite the expression without the absolute value sign.

b. The speed limit is set at 65 mph on a highway, but police do not normally ticket you if you go less than 5 miles above or below that limit. Construct an absolute value inequality that describes the speeds S at which you can safely travel without getting a ticket. Rewrite the expression without using the absolute value sign.

5. Assume two individuals, A and B, are traveling by car and initially are 400 miles apart on a highway. They travel toward each other, pass and then continue on.

a. If A is traveling at 60 miles per hour, and B is traveling at 40 mph, write two functions, $d_A(t)$ and $d_B(t)$, that describe the distance (in miles) that A and B each has traveled over time t (in hours).

b. Now construct a function for the distance $D_{AB}(t)$ *between* A and B at time t (in hours). Graph the function for $0 \le t \le 8$ hours.

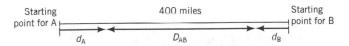

c. At what time will A and B cross paths? At that point, how many miles has each traveled?

d. What is the distance between them one hour before they meet? An hour after they meet? Interpret both values in context. (*Hint:* If they are traveling toward each other, the distance between them is considered positive. Once they have met and are traveling away from each other, the distance between them is considered negative.)

e. Now construct an absolute value function that describes the (positive) distance between A and B at any point, and graph your result for $0 \le t \le 8$ hours.

6. On Christmas Day 2009, there was a bomb scare aboard a Detroit-bound plane. Soon afterwards, a Gallup poll indicated that 39% of Americans were worried that they or a close friend or family member would become a victim of terrorism. The poll had a 4% margin of error. Create an absolute value inequality that describes the ranges of possible poll percentages and display it on a number line.

7. The greatest integer function $y = [x]$ is defined as the greatest integer $\le x$ (i.e., it rounds x down to the nearest integer at or below x).

a. What is [2] ? [2.5] ? [2.9999999] ?

b. Sketch a graph of the greatest integer function for $0 \le x < 5$. Be sure to indicate whether each endpoint is included or excluded.

[*Note:* A bank employee embezzled hundreds of thousands of dollars by inserting software to round down transactions (such as generating interest on an account) to the nearest cent, and siphoning the round-off differences into his account. He was eventually caught.]

8. The following table shows U.S. first-class stamp prices (per ounce) over time.

Year	Price for First-Class Stamp
2006	39 cents
2007	41 cents
2008	42 cents
2009	44 cents

a. Construct a step function describing stamp prices for 2006–2010.

b. Graph the function. Be sure to specify whether each of the endpoints is included or excluded.

c. In 2008 the post office created "forever stamps" that (if you bought them then, at 42 cents) would always be valid. How would this alter the graph if you used only "forever stamps"?

9. Sketch a graph for each of the following situations.

a. The amount in your savings account over a month, where you direct-deposit your paycheck each week and make one withdrawal during the month.

b. The amount of money in an ATM machine over one day, where the ATM is stocked with dollars at the beginning of the day, and then ATM withdrawals of various sizes are made.

10. A department store is offering a coupon for 20% off for any purchases *over* $100.

a. How much would you spend for a $100 purchase?

b. How much would you actually spend if the original purchase was $100.01? $101.00? $111.00?

c. Construct a piecewise function for the actual cost, C, of the purchase as a function of the original purchase price, P.

11. A name-brand cookie mix calls for baking cookies at a temperature of 350°F. However, for high altitudes (3500–6500 ft above sea level), baking at 375°F is suggested.

a. Construct a piecewise linear function for the recommended baking temperature, $T(s)$, as a function of feet above sea level, s.

b. What is a reasonable domain for this function? What is the range?

12. a. Create a piecewise function and sketch the graph of the following scenario where $D(t)$ is the total distance walked (in miles) as a function time t (in minutes): "You begin a walk from your home at a fast pace of 2 miles per hour for 30 minutes. You rest for 15 minutes, and then you continue walking at a leisurely pace of 1 mile per hour for 30 minutes."

b. How far have you walked at the end of 30, 45, and 60 minutes?

13. Create a piecewise linear function and sketch a graph for this scenario, where body temperature, $T(x)$, is a function of time x in hours since 6 a.m.

a. You are feeling fine at 6 a.m. and your temperature remains a normal 98.6°F. But by 8 a.m. you start to feel feverish. Your temperature begins to rise at a rate of 0.5° per hour for 4 hours. Your temperature remains at this level for 6 more hours and then gradually decreases at a rate of 0.25° per hour until it is back to normal.

b. What is your temperature at 8 a.m., noon, 6 p.m., and 10 p.m.?

14. To motivate customers, Puget Sound Energy explored ways to set lower electric rates for off-peak energy use and higher rates for hours of peak consumption.

Puget Sound Energy: Electric Rates vs. Actual Use

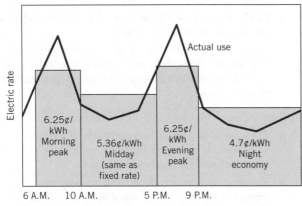

Source: http://energypriorities.com/.

Use the chart above to create a piecewise function where $E(h)$ describes the rates per kilowatt hour charged (written on each bar) and h = number of hours since 5 a.m. (*Note:* Use the times 6 a.m., 10 a.m., 5 p.m., and 9 p.m., to translate into values for h.) The jagged lines behind the bars indicate the actual use of electricity.

15. Interval training is a series of repetitions of work with a recovery period following each repetition. To prepare for a long-distance race such as a marathon, a typical interval training program might consist of running 1 mile at a fast rate, followed by a 3-minute recovery period of jogging. The runner repeats this 5 times.

A particular runner's training program is to run a mile at an average of 8 mph (or $1/8 = 0.125$ hr/mile), then jog at an average speed of 4 mph (or $1/4 = 0.25$ hr/mile). Create a piecewise function describing her total time, $T(d)$, for *one* repetition as a function of distance d traveled. Graph the function.

16. A 60-year-old U.S. Masters swimmer swam a 200-yard freestyle in 2 minutes. Each lap of the pool is 50 yards. The following table gives her splits (time to swim that 50-yard lap). Assume she swam at a constant rate during a given lap.

Laps (yards)	Total Distance (yards)	Time for Lap (seconds)	Cumulative Time (seconds)
First lap	50	27	27
Second lap	100	30	57
Third lap	150	31	88
Fourth lap	200	32	120

Create a piecewise linear function $S(y)$ for cumulative seconds as a function of distance, y, in yards.

2.10 *Constructing Linear Models of Data*

According to Edward Tufte in *Data Analysis of Politics and Policy*, "Fitting lines to relationships is the major tool of data analysis." Of course, when we work with actual data searching for an underlying linear relationship, the data points will rarely fall exactly in a straight line. However, we can model the trends in the data with a linear equation.

Linear relationships are of particular importance not because most relationships are linear, but because straight lines are easily drawn and analyzed. A human can fit a straight line by eye to a scatter plot almost as well as a computer. This paramount convenience of linear equations as well as their relative ease of manipulation and interpretation means that lines are often used as first approximations to patterns in data.

Fitting a Line to Data: The Kalama Study

Children's heights were measured monthly over several years as a part of a study of nutrition in developing countries. Table 2.13 and Figure 2.46 show data collected on the mean heights of 161 children in Kalama, Egypt.

Mean Heights of Kalama Children

Age (months)	Height (cm)
18	76.1
19	77.0
20	78.1
21	78.2
22	78.8
23	79.7
24	79.9
25	81.1
26	81.2
27	81.8
28	82.8
29	83.5

Table 2.13
Source: D. S. Moore and G. P. McCabe, *Introduction to the Practice of Statistics.* Copyright © 1989 by W. H. Freeman and Company. Used with permission.

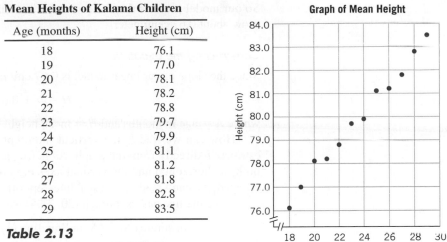

Graph of Mean Height

Figure 2.46

DATA
KALAMA

Sketching a line through the data

Although the data points do not lie exactly on a straight line, the overall pattern seems clearly linear. Rather than generating a line through two of the data points, try eyeballing a line that approximates all the points. A ruler or a piece of black thread laid down through the dots will give you a pretty accurate fit.

Figure 2.47 shows a line sketched that approximates the data points. (*Note:* This line does not necessarily pass through any of the original points.)

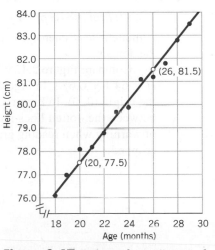

Figure 2.47 Estimated coordinates of two points on the line.

Finding the slope

Estimating the coordinates of two points *on the line*, say (20, 77.5) and (26, 81.5), we can calculate the slope, *m*, or rate of change, as

$$m = \frac{(81.5 - 77.5) \text{ cm}}{(26 - 20) \text{ months}}$$

$$= \frac{4.0 \text{ cm}}{6 \text{ months}}$$

$$\approx 0.67 \text{ cm/month}$$

So our model predicts that for each additional month an "average" Kalama child will grow about 0.67 centimeter.

Constructing the equation

Since the slope of our linear model is 0.67 cm/month, then our equation is of the form

$$H = b + 0.67A \qquad (1)$$

where A = age in months and H = mean height in centimeters.

How can we find b, the vertical intercept? We have to resist the temptation to estimate b directly from the graph. As is frequently the case in social science graphs, both the horizontal and the vertical axes are cropped. Because the horizontal axis is cropped, we can't read the vertical intercept off the graph. We'll have to calculate it.

Since the line passes through (20, 77.5) we can

substitute (20, 77.5) in Equation (1)	$77.5 = b + (0.67)(20)$
simplify	$77.5 = b + 13.4$
solve for b	$b = 64.1$

Having found b, we complete the linear model:

$$H = 64.1 + 0.67A$$

where A = age in months and H = height in centimeters. It offers a compact summary of the data.

What is the domain of this model? In other words, for what inputs does our model apply? The data were collected on children age 18 to 29 months. We don't know its predictive value outside these ages, so

the domain consists of all values of A for which $18 \leq A \leq 29$

The vertical intercept may not be in the domain

Although the H-intercept is necessary to write the equation for the line, it lies outside of the domain.

Compare Figure 2.47 with Figure 2.48. They show graphs of the same equation, $H = 64.1 + 0.67A$. In Figure 2.47 both axes are cropped, while Figure 2.48 includes the origin (0, 0). In Figure 2.48 the vertical intercept is visible, and the shaded area between the dotted lines indicates the region that applies to our model. So a word of warning when reading graphs: Always look carefully to see if the axes have been cropped.

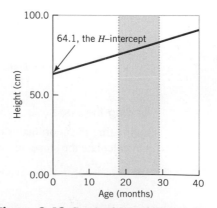

Figure 2.48 Graph of $H = 64.1 + 0.67A$ that includes the origin (0, 0). Shaded area shows the region that models the Kalama data.

Reinitializing the Independent Variable

When we model real data, it often makes sense to reinitialize the independent variable in order to have a reasonable vertical intercept. This is especially true for time series, as shown in the following example, where the independent variable is the year.

EXAMPLE 1

Time series
How can we find an equation that models the trend in smoking in the United States?

SOLUTION

The American Lung Association data are shown in Table 2.14 and in Figure 2.49. Show that although in some states smoking has increased, the overall trend is a steady decline in the percentage of adult smokers in the United States between 1965 and 2007.

Year	% of Adults Who Smoke
1965	42.4
1974	37.1
1979	33.5
1983	32.1
1985	30.1
1990	25.5
1995	24.7
2000	23.3
2005	20.9
2007	19.7

Table 2.14
Source: American Lung Association.

Percentage of U.S. Adults Who Smoke

Figure 2.49

The relationship appears reasonably linear. So the equation of a best-fit line could provide a fairly accurate description of the data. Since the horizontal axis is cropped, starting at the year 1965, the real vertical intercept would occur 1965 units to the left, at A.D. 0! If you drew a big enough graph, you'd find that the vertical intercept would occur at approximately (0, 1220). This nonsensical extension of the model outside its known values would say that in A.D. 0, 1220% of the adult population smoked. A better strategy would be to define the independent variable as the number of years *since* 1965. Table 2.15 shows the reinitialized values for the independent variable, and Figure 2.50 gives a sketched-in best-fit line.

Year	Years since 1965	% of Adults Who Smoke
1965	0	42.4
1974	9	37.1
1979	14	33.5
1983	18	32.1
1985	20	30.1
1990	25	25.5
1995	30	24.7
2000	35	23.3
2005	40	20.9
2007	42	19.7

Table 2.15

Percentage of Adult Smokers with Estimated Best-Fit Line

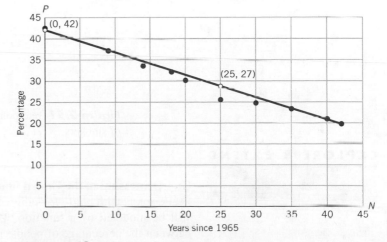

Figure 2.50

We can estimate the coordinates of two points, (0, 42) and (25, 27), on our best-fit line. (*Note:* In general the points on the best-fit line will not be from the actual data from the table.) Using them, we have

$$\text{vertical intercept} = 42 \quad \text{and} \quad \text{slope} = \frac{42 - 27}{0 - 25} = \frac{15}{-25} = -0.6$$

If we let N = the number of years since 1965 and P = percentage of adult smokers, then the equation for our best-fit line is

$$P = 42 - 0.6N$$

where the domain is $0 \le N \le 42$ (see Figure 2.50). This model says that starting in 1965, when about 42% of U.S. adults smoked, the percentage of the adult smokers has declined on average by 0.6 percentage points a year for 42 years.

What this model doesn't tell us is that (according to the U.S. Bureau of the Census) the total number of smokers during this time has remained fairly constant, at 50 million.

Interpolation and Extrapolation: Making Predictions

We can use this linear model on smokers to make predictions. We can *interpolate* or estimate new values between known ones. For example, in our smoking example we have no data for the year 1970. Using our equation we can estimate that in 1970 (when $N = 5$), $P = 42 - (0.6 \cdot 5) = 39\%$ of adults smoked. Like any other point on the best-fit line, this prediction is only an estimate and may, of course, be different from the actual percentage of smokers (see Figure 2.51).

We can also use our model to *extrapolate* or to predict beyond known values. For example, our model predicts that in 2010 (when $N = 45$), $P = 42 - (0.6 \cdot 45) = 15\%$ of adults will smoke (check the Internet to see if this is true). Extrapolation much beyond known values is risky. For 2035 (where $N = 70$) our model predicts that 0% will smoke, which seems unlikely. After 2035 our model would give the impossible answer that a negative percentage of adults will smoke.

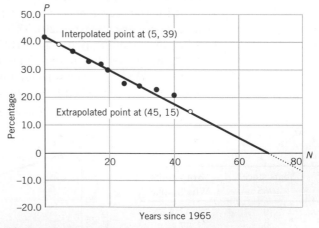

Figure 2.51 Interpolation and extrapolation of percentage of smokers.

EXPLORE & EXTEND

Gun Ownership in the United States

Scatter plots with *multiple outputs* for a single input that indicate an overall trend can still be modeled by a function. The following scatter plot shows the relationship between the percentage of people in a state who own a gun versus the percentage of people who live in rural areas.

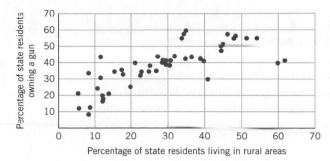

Source: The gun ownership statistics are from a 2001 survey by the
Behavioral Risk Factor Surveillance System (BRFSS). The percent
living in rural areas is from the 2000 report from the U.S. Census
Bureau. Each reports by state.

The relationship looks roughly linear. So:

a. Sketch in a best-fit line.

b. Pick two points on your line and calculate the slope.

c. Interpret the slope in terms of owning a gun and living in rural areas.

d. Create a title for your graph.

Algebra Aerobics 2.10

1. The scatter plot shows the total number of U.S. college graduates (age 25 or older) between 1960 and 2009.

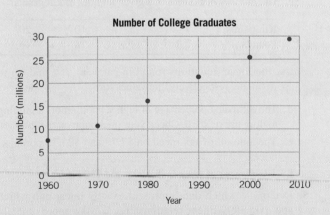

Number of College Graduates

Source: U.S. National Center for Education Statistics, *Digest of Education Statistics*, annual.

a. Estimate from the scatter plot the number of college graduates in 1960 and in 2009.

b. Since the data look fairly linear, sketch a line that would model the growth in U.S. college graduates.

c. Estimate the coordinates of two points *on your line* and use them to calculate the slope.

d. If x = number of years *since* 1960 and y = total number (in millions) of U.S. college graduates, what would the coordinates of your two points in part (c) be in terms of x and y?

e. Construct a linear equation using the x and y values defined in part (d).

f. What does your model tell you about the number of college graduates in the United States?

2. The scatter plot shows the percentage of adults who (according to the U.S. Census Bureau) had access to the Internet either at home or at work between 2000 and 2009.

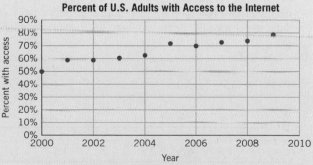

Percent of U.S. Adults with Access to the Internet

a. Since the data appear roughly linear, sketch a best-fit line. (This line need not pass through any of the data points.)

b. Reinitialize the years so that 2000 becomes year 0. Then identify the coordinates of any two points that lie on the line that you drew. Use these coordinates to find the slope of the line. What does this tell you about the percentage of adults with Internet access?

c. Give the approximate vertical intercept of the line that you drew (using the reinitialized value for the year).

d. Write an equation for your line.

e. What would you expect to happen to the percentage after 2009? Do you think your linear model will be a good predictor after 2009?

Exercises for Section 2.10

Graphing program is recommended for Exercises 3, 4, 13, and 15. Internet access optional for Exercises 7(b), 13(b), 14(f), and 18(c).

1. Match each equation with the appropriate table.

a. $y = 3x + 2$ **b.** $y = \frac{1}{2}x + 2$ **c.** $y = 1.5x + 2$

A.

x	y
0	2
2	3
4	4
6	5
8	6

B.

x	y
0	2
2	5
4	8
6	11
8	14

C.

x	y
0	2
2	8
4	14
6	20
8	26

2. Match each of the equations with the appropriate graph.

a. $y = 10 - 2x$ **b.** $y = 10 - 5x$ **c.** $y = 10 - 0.5x$

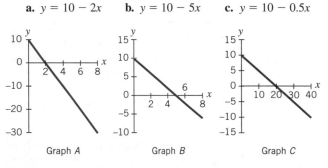

Graph A   Graph B   Graph C

3. (Graphing program recommended.) Identify which of the following data tables represent exact and which approximate linear relationships. For the one(s) that are exactly linear, construct the corresponding equation(s). For the one(s) that are approximately linear, generate the equation of a best-fit line; that is, plot the points, draw in a line approximating the data, pick two points on the line (not necessarily from your data) to generate the slope, and then construct the equation.

a.

x	−2	−1	0	1	2	3
y	−6.5	−5.0	−3.5	−2.0	−0.5	1.0

b.

t	−1	0	1	2	3	4
Q	8.5	6.5	3.0	1.2	−1.5	−2.0

c.

N	0	15	23	45	56	79
P	35	80	104	170	203	272

4. (Graphing program recommended.) Plot the data in each of the following data tables. Determine which data are exactly linear and which are approximately linear. For those that are approximately linear, sketch a line that looks like a best fit to the data. In each case generate the equation of a line that you think would best model the data.

a.

Year	2006	2007	2008	2009
% high school seniors admitting to drug use during previous year	36.5	35.9	36.6	36.5

Source: Centers for Disease Control and Prevention, National Center for Health-Related Statistics, 2009.

b. The amount of tax owed on a purchase price.

Price	$2.00	$5.00	$10.00	$12.00
Tax	$0.12	$0.30	$0.60	$0.72

c. The number of pounds in a given number of kilograms.

Kilograms	1	5	10	20
Pounds	2.2	11	22	44

Use the linear equations found in parts (a), (b), and (c) to approximate, respectively, the values for:

d. The estimated percentage of high school seniors admitting to drug use in 2010. Check your prediction on the Internet.

e. The amount of tax owed on $7.79 and $25.75 purchase prices.

f. The number of pounds in 15 kg and 150 kg.

5. Determine which data represent exactly linear and which approximately linear relationships. For the approximately linear data, sketch a line that looks like a best fit to the data. In each case generate the equation of a line that you think would best model the data.

a. The number of solar energy units consumed (in quadrillions of British thermal units, called Btus)

Year	1998	2000	2005	2006
Solar consumption (in quadrillions of Btus)	0.07	0.07	0.07	0.07

Source: U.S. Bureau of the Census, *Statistical Abstract*.

b.

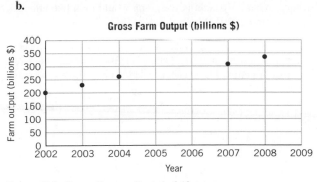

Gross Farm Output (billions $)

Source: U.S. Census Bureau, *Statistical Abstract.*

6. In 2002 the United States consumed 617 million gallons of wine and in 2007, about 745 million gallons.

a. Assuming the growth was linear, create a function that could model the trend in wine consumption.

b. Estimate the amount of wine consumed in 2000. The actual amount was about 568 million gallons worth of wine. How accurate was your approximation?

Source: U.S. Department of Agriculture, Economic Research Service, 2009.

7. The percentage of medical degrees awarded to women in the United States between 1970 and 2008 is shown in the accompanying graph.

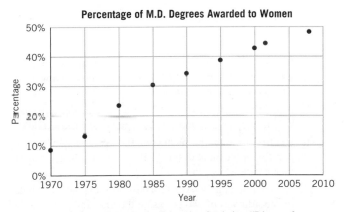

Percentage of M.D. Degrees Awarded to Women

Source: U.S. National Center for Education Statistics, "Digest of Education Statistics," annual; *Statistical Abstract of the United States.*

The data show that since 1970 the percentage of female doctors has been rising.

a. Sketch a line that best represents the data points. Use your line to estimate the rate of change of the percentage of M.D. degrees awarded to women.

b. If you extrapolate your line, estimate when 100% of doctors' degrees will be awarded to women.

c. It seems extremely unlikely that 100% of medical degrees will ever be granted to women. Comment on what is likely to happen to the rate of growth of women's degrees in medicine; sketch a likely graph for the continuation of the data into this century. If possible, check on the Internet for the current number of female medical degrees.

8. The accompanying graph shows the mortality rates (in deaths per 1000) for male and female infants in the United States from 1980 to 2009.

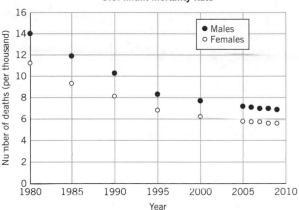

U.S. Infant Mortality Rate

Source: Centers for Disease Control and Prevention, *www.cdc.gov.*

a. Are male or female infants more likely to die? Estimate the infant mortality rates for both sexes in 1980 and in 2009. (Be sure to use the correct units.)

b. Both female and male infant mortality rates show a roughly linear decline between 1980 and 2000. Draw two best-fit lines (one female, one male) between 1980 and 2000, calculate their slopes, and interpret what each slope means.

c. Now draw two best-fit lines (one female, one male) for infant mortality between 2005 and 2009, calculate their slopes, and interpret what each means.

d. Describe the overall trends in U.S. infant mortality since 1980.

9. The accompanying scatter plot shows the relationship between literacy rate (the percentage of the population who can read and write) and infant mortality rate (infant deaths per 1000 live births) for 91 countries. The raw data are contained in the Excel or graph link file NATIONS and are described at the end of the Excel file. (You might wish to identify the outlier, the country with about a 20% literacy rate and a low infant mortality rate of about 40 per 1000 live births.) Construct a linear model. Show all your work and clearly identify the variables and units. Interpret your results.

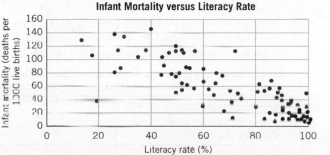

Infant Mortality versus Literacy Rate

10. The accompanying graph shows data for the men's Olympic 16-pound shot put from 1900 to 2008.

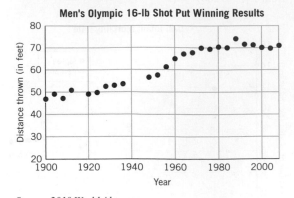

Men's Olympic 16-lb Shot Put Winning Results

Source: 2010 World Almanac.
Note: There were no Olympics in 1940 and 1944 due to World War II.

a. The shot put results are roughly linear between 1900 and 1988. Sketch a best-fit line for those years. Estimate the coordinates of two points on the line to calculate the slope. Interpret the slope in this context.

b. What is happening to the winning shot put results after 1988? Estimate the slope of the best-fit line for the years after 1988.

c. Letting x = years since 1900, construct a piecewise linear function $S(x)$ that describes the winning Olympic shot put results (in feet thrown) between 1900 and 2008.

11. Cell phones have become immensely popular, as the following chart shows.

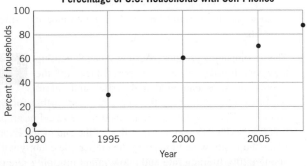

Percentage of U.S. Households with Cell Phones

Source: 2008 World Almanac and Book of Facts Forrester Reports.

a. A linear model seems reasonable between 1990 and 2007. Sketch a best-fit line for those dates, then pick two points on your line and calculate the slope. Interpret the slope in this context.

b. Letting x = years since 1990, construct a piecewise linear function $P(x)$ that would describe the percentage of U.S. households with cell phones between 1990 and 2007.

c. What do you think the graph would look like after 2015? After 2020? Why?

12. The accompanying graph shows the relationship between the age of a woman when she has her first child and her lifetime risk of getting breast cancer relative to a childless woman.

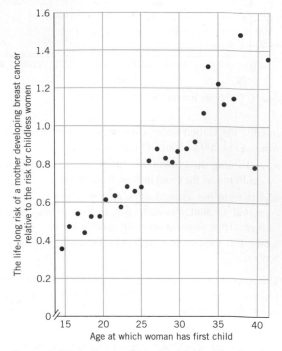

Source: J. Cairns, *Cancer: Science and Society* (San Francisco: W. H. Freeman, 1978), p. 49.

a. If a woman has her first child at age 18, approximately what is her risk of developing cancer relative to a woman who has never borne a child?

b. At roughly what age are the chances the same that a woman will develop breast cancer whether or not she has a child?

c. If a first-time mother is beyond the age you specified in part (b), is she more or less likely to develop breast cancer than a childless woman?

d. Sketch a line that looks like a best fit to the data, estimate the coordinates of two points on the line, and use them to calculate the slope.

e. Interpret the slope in this context.

f. Construct a linear model for these data, identifying your independent and dependent variables.

13. (Graphing program recommended.) The data in the table below show that health care is becoming more expensive and is taking a bigger share of the U.S. gross domestic product (GDP). The GDP is the market value of all goods and services that have been bought for final use.

Year	1960	1970	1980	1990	1995	2000	2005	2007
U.S. health care costs as a percentage of GDP	5.2	7.2	9.1	12.3	13.7	13.8	15.9	16.2
Cost per person, $	148	356	1100	2813	3783	4790	6649	7421

Source: U.S. Health Care Financing Administration and Centers for Medicare and Medicaid services. Numbers differ slightly from World Health Organizations (WHO) data.

a. Graph health care costs as a percentage of GDP versus year, with time on the horizontal axis. Measure time in years since 1960. Draw a straight line by eye that appears to be the closest fit to the data. Figure out the slope of your line and create a function $H(t)$ for health care's percentage of the GDP as a function of t, years since 1960.

b. What does your formula predict for health care as a percentage of GDP for the year 2010? Check your answer on the Internet if you have online access.

c. Why do you think the health care cost per person has gone up so much more dramatically than the health care percentage of the GDP?

14. Hybrid car sales peaked in early 2008, then showed a steady decline in the second half of 2008. The following graph shows the annual sales of hybrids between 2004 and 2009.

Number of U.S. Hybrid Sales

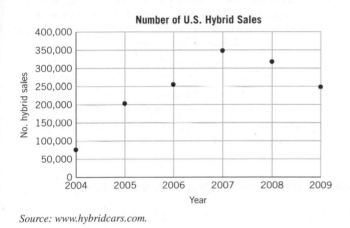

Source: www.hybridcars.com.

a. Relabel "years" as "years since 2004."

b. Sketch a line corresponding to increasing hybrid sales. Then calculate and interpret its slope in terms of car sales. Create a linear model, specifying its domain.

c. Sketch a line corresponding to decreasing hybrid sales. Then calculate and interpret its slope in terms of car sales. Create a linear model, specifying its domain.

d. Generate a piecewise linear function for hybrid sales.

e. In July 2009 there were 35,429 hybrids sold out of 999,890 total vehicle sales. What percentage of all vehicle sales were hybrids then?

f. If you have online access check on the Internet for 2011 hybrid sales to see if they have improved.

15. (Graphing program recommended.) The Gas Guzzler Tax is imposed on manufacturers on the sale of new-model cars (*not* minivans, sport utility vehicles, or pickup trucks) whose fuel economy fails to meet certain statutory regulations, to discourage the production of fuel-inefficient vehicles. The tax is collected by the IRS and paid by the manufacturer. The table (top right) shows the amount of tax that the manufacturer must pay for a vehicle's miles per gallon fuel efficiency.

Gas Guzzler Tax

MPG	Tax per Car
12.5	$6400
13.5	$5400
14.5	$4500
15.5	$3700
16.5	$3700
17.5	$2600
18.5	$2100
19.5	$1700
20.5	$1300
21.5	$1000
22.5	$0

Source: http://www.epa.gov/otaq/

a. Plot the data, verify that they are roughly linear, and add a line of best fit.

b. Choose two points on the line, find the slope, and then form a linear equation with x as the fuel efficiency in mpg and y as the tax in dollars.

c. What is the average rate of change of the amount of tax imposed on fuel-inefficient vehicles? Interpret the units.

16. A veterinarian's office displayed the following table comparing dog age (in dog years) to human age (in human years). The chart shows that the relationship is fairly linear.

Comparative Ages of Dogs and Humans

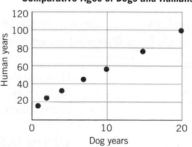

a. Draw a line that looks like a best fit to the data.

b. Estimate the coordinates and label two points on the line. Use them to find the slope. Interpret the slope in this context.

c. Using H for human age and D for dog age, identify which you are using as the independent and which you are using as the dependent variable.

d. Generate the equation of your line.

e. Use the linear model to determine the "human age" of a dog that is 17 dog years old.

f. Middle age in humans is 45–59 years. Use your model equation to find the corresponding middle age in dog years.

g. What is the domain for your model?

17. In general, heavier cars get lower or worse gas mileage than lighter cars. The graph on the following page shows the mpg (miles per gallon) for cars of given weights (in 1000 pounds).

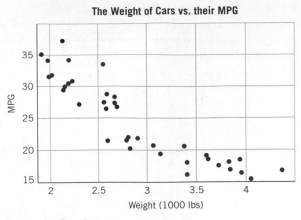

The Weight of Cars vs. their MPG

Source: StatCrunch.com.

18. The following scatter plot shows the relationship between the average January temperatures and latitude for 30 U.S. cities.

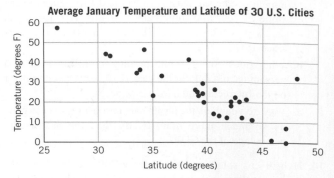

Average January Temperature and Latitude of 30 U.S. Cities

Source: http://mste.illinois.edu/malcz/DATA/WEATHER/ Temperatures.html.

a. A linear model seems reasonable for the mpg vs. weight of cars. Sketch a line corresponding to decreasing mpg. Then calculate and interpret its slope in terms of mpg and weight.

b. Letting w = weight (in 1000s of pounds), construct a linear function $M(w)$ that would describe the miles per gallon for cars of weights between 2 thousand and 4 thousand pounds. Create a linear model, specifying its domain.

c. Use your model to predict the mpg for a car weighing 3 thousand pounds.

d. Use your model to predict the weight of a car getting 30 mpg.

a. Sketch a line through the graph of the data that best represents the relationship of the average January temperature and latitude. Does the line seem to be a reasonable model for the data? What is the approximate slope of the line through two points? Show your work.

b. Construct a linear function $T(d)$ that describes the average January temperature as a function of d, the latitude.

c. Extra: Find the latitude and the average January temperature of your city and plot the point on the graph. How well did your model in part (b) predict your city's average January temperature?

2.11 *Looking for Links between Education and Earnings: A Case Study on Using Regression Lines*

How are earnings and education related? Does having more education give access to higher paying jobs? In this section we explore how a social scientist might start to answer these questions using regression lines and U.S. Census data.

Using U.S. Census Data

The data set we use in this section is a random sample from the March 2009 Current Population Survey of the U.S. Census that we call FAM1000. Our sample provides information on 1000 individuals and their households and is available in Excel on the course website. There is also a *data dictionary* with short definitions for each data category in the Appendix of this text.

Think of the data as a large array of rows and columns of facts. Each row represents all the information obtained from one particular respondent about his or her household. Each column contains the coded answers of all the respondents in the sample to one particular question. Table 2.16 shows one row of the 1000 rows in the FAM1000 data set.

Age	Sex	Region	Cencity	Marstat	Famsize	Edu	Edu (≥ 8 grade)	Occup	Hrswork	Wkswork	Yrft	Pearnings	Ptotinc	Faminc	Race	Hispanic
46	2	1	1	0	3	12	4	5	40	52	1	$55,000	$55,083	$74,087	1	0

Table 2.16

FAM1000

Referring to the data dictionary in the Appendix of this text, we learn that this respondent is a 46-year old male who lives in the Northeast in a city. He is married and lives in a family with 3 people. He has a high school education and works in an office as administrative support. He works 40 hours a week, 52 weeks a year, and so is considered a year-round full-time worker. He earns $55,000 a year. His personal total income $55,083, so perhaps the extra $83 is from interest on a savings account. His family income is $74,087. He is classified as white non-Hispanic.

Summarizing the Data: Regression Lines

In the social and life sciences it is usually difficult to tell whether one variable truly depends on another. For example, it is certainly plausible that a person's earnings depend in part on how much formal education he or she has had, since we may suspect that having more education gives access to higher-paying jobs, but many other factors also play a role. Some of these factors, such as the person's age or type of work, are measured in the FAM1000 data set; others, such as family background or good luck, may not have been measured or even be measurable. Despite this complexity, we attempt to determine as much as we can by first looking at the relationship between earnings and education alone.

Is there a relationship between education and earnings?

If we hypothesize that earnings depend on education, then the convention is to graph education on the horizontal axis. Each ordered pair of data values gives a point with coordinates in the form:

(education, personal earnings)

Using the FAM1000 data, we can create a scatter plot of earnings vs. education in Figure 2.52. The first coordinate gives years of education past grade eight (so zero

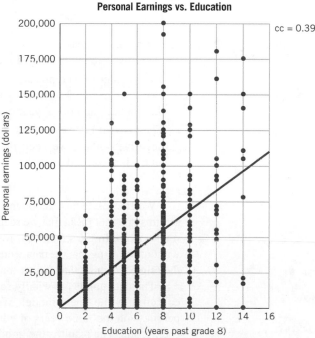

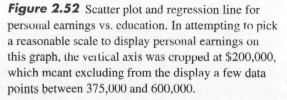

Figure 2.52 Scatter plot and regression line for personal earnings vs. education. In attempting to pick a reasonable scale to display personal earnings on this graph, the vertical axis was cropped at $200,000, which meant excluding from the display a few data points between 375,000 and 600,000.

represents an eighth-grade education or less) and the second coordinate gives the personal earnings.

How might we think of the relationship between these two variables? Clearly, personal earnings are not a function of education in the mathematical sense since people who have the same amount of education earn widely different amounts. The scatter plot obviously fails the vertical line test.

But suppose that, to form a simple description of these data, we were to insist on finding a simple functional description. And suppose we insist that this simple relationship be a linear function. In Section 2.10, we informally fit linear functions to data. A formal mathematical procedure called *regression analysis* lets us determine what linear function is the "best" approximation to the data; the resulting "best-fit" line is called a *regression line* and is similar in spirit to reporting only the mean of a set of single-variable data, rather than the entire data set. It can be a useful and powerful method of summarizing a set of data.

Figure 2.52 also shows the regression line determined from the data points. The equation of the line is:

$$\text{personal earnings} = -30 + 6840 \cdot \text{yrs. educ. (past grade 8)}$$

The reading "Linear Regression Summary" describes a standard technique for generating regression lines.

The calculations necessary to compute this regression line are tedious, although not difficult, and are easily carried out by computer software, such as the *FAM1000 Census Graphs*, Excel program, or graphing calculators.

E X A M P L E   1

Interpreting the regression line fit to all data

In Figure 2.52, the equation of the regression line for personal earnings vs. education is:

$$\text{personal earnings} = -30 + 6840 \cdot \text{yrs. educ. (past grade 8)}$$

a. What are the units for each term of the regression line?

b. Identify and interpret the slope in this context.

c. Identify and interpret the vertical intercept.

S O L U T I O N

a. Since personal earnings are dollars, the units for the term -30 must be dollars, and the units for 6840 must be dollars per years of education.

b. The number 6840 represents the slope of the regression line, or the average rate of change of personal earnings with respect to years of education. So this model predicts that for each additional year of education, personal earnings increase by \$6840.

c. The vertical intercept of the equation is -30. Since it doesn't make sense to earn negative dollars, we need to restrict our domain to perhaps 2 or more years of education (beyond grade 8).

We emphasize that, although we can construct an approximate linear model for a data set, this does not mean that we really believe that the data are truly represented by a linear relationship. In the same way, we may report the median to summarize a set of data, without believing that the data values are at the median. In both cases, there are features of the original data set that may or may not be important and that we do not report.

The data points are widely scattered about the line, for reasons that are clearly not captured by the linear model. We can eliminate the clutter by grouping together all people with the same years of education and plotting the median personal earnings of each group. The result is the graph in Figure 2.53, which includes a regression line for the new data.

For each year of education, only a single median earnings point has been graphed. For instance, the point corresponding to 12 years of education past grade 8 has a vertical value of approximately \$75,000; hence the median of the personal earnings of everyone in the FAM1000 data with 20 years of education is about \$75,000. The pattern is now clearer: An upward trend to the right is more obvious in this graph.

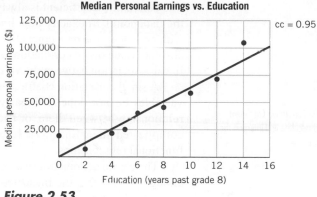

Figure 2.53

Every time we construct a simplified representation of an original data set, we should ask ourselves what information has been suppressed. In Figure 2.53 we have suppressed the spread of data in the vertical direction. For example, there are only 38 people with an eighth-grade education or less but 220 with 16 years of education. Yet each of these sets is represented by a single point.

EXAMPLE 2

Interpreting the regression line fit to median values

In Figure 2.53 the equation of the regression line for median personal earnings vs. education is:

$$\text{median personal earnings} = 1140 + 6250 \cdot \text{yrs. of educ (past grade 8)}$$

a. Identify and interpret the vertical intercept and slope for the regression line for median personal earnings vs. education.

b. What do the differences between the regression lines for all the FAM1000 data (in Example 1) and the median data (in Example 2) indicate?

SOLUTION

a. From the equation, the vertical intercept is 1140. So, this model predicts that median personal earnings for those with an eighth-grade education or less will be about $1140. The number 6250 represents the slope of the regression line, or the average rate of change of median personal earnings with respect to years of education. This model predicts that for each additional year of education, median personal earnings increase by $6250.

b. The regression line for all of the FAM1000 data (Figure 2.52) represents the fit to all of the data and predicts individual personal earnings while the regression line for the median data (Figure 2.53) represents the fit to the medians and predicts median personal earnings for the group, not the individual. Both lines are reasonable answers to the question "What straight line describes the relationship between education and earnings?" and the difference between them indicates the uncertainty in answering such a question. We may argue that the benefit in earnings for each year of education is $6250 or $6840, or something between these values.

Regression line: how good a fit?

The programs R1, R4 and R7 in *Linear Regression* can help you visualize the links among scatter plots, best-fit lines, and correlation coefficients.

Once we have determined a line that approximates our data, we must ask, "How good a fit is our regression line?" To help answer this, statisticians calculate a quantity called the *correlation coefficient*. This number can be computed by statistical software, and we have included it on our graphs and labeled it "cc." (See Figure 2.52 and Figure 2.53.)[2]

---

[2]We use the label "cc" for the correlation coefficient in the text and software to minimize confusion. In a statistics course the correlation coefficient is usually referred to as *Pearson's r* or just *r*.

The correlation coefficient is always between -1 (negative association with no scatter; the data points fit exactly on a line with a negative slope) and 1 (positive association with no scatter; the data points fit exactly on a line with a positive slope). The closer the absolute value of the correlation coefficient is to 1, the better the fit and the stronger the linear association between the variables.

A small correlation coefficient (with absolute value close to zero) indicates that the variables do not depend linearly on each other. This may be because there is no relationship between them, or because there is a relationship that is something more complicated than linear. In future chapters we discuss many possible nonlinear functional relationships.

There is no definitive answer to the question of when a correlation coefficient is "good enough" to say that the linear regression line is a good fit to the data. The cc for the regression line for all the data in Figure 2.52 is 0.39 and for the median data in Figure 2.53 is 0.95. Clearly the cc for the medians indicates a better fit. In general a fit to the graph of medians (or means) generally gives a higher correlation coefficient than a fit to the original data set because the scatter has been smoothed out. When in doubt, plot all the data along with the linear model and use your best judgment. The correlation coefficient is only a tool that may help you decide among different possible models or interpretations.

The reading "The Correlation Coefficient" explains how to calculate and interpret the correlation coefficient.

EXAMPLE 3

Correlation coefficient and scatter plots

a. Which graph in Figure 2.54 shows a positive linear correlation between x and y? A negative linear correlation? Zero correlation?

b. Which graph shows the closest linear correlation between x and y?

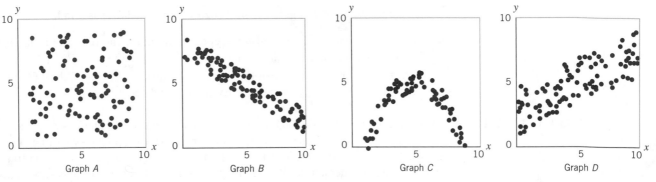

Figure 2.54 Four scatter plots.

SOLUTION

a. Graph D shows a positive linear correlation between x and y (when one variable increases, the other increases). Graph B shows a negative linear correlation (when one variable increases, the other decreases). Both graphs A and C show zero linear correlation between x and y. Even though graph C shows a pattern in the relationship between x and y, the pattern is not linear.

b. Graph B. The correlation coefficient of its regression line would be close to -1, almost a perfect (negative) correlation.

Interpreting Regression Lines: Correlation vs. Causation

One is tempted to conclude that increased education *causes* increased earnings. This may be true, but the model we have used does not offer conclusive proof. This model can show how strong or weak a relationship exists between variables but does not answer the question "Why are the variables related?" We need to be cautious in how we interpret our findings.

Regression lines show *correlation*, not *causation*. We say that two events are correlated when there is a statistical link. If we find a regression line with a correlation coefficient that is close to 1 in absolute value, a strong relationship is suggested. In our previous example, education is positively correlated with personal earnings. If education increases, personal earnings increase. Yet this does not prove that education causes an increase in personal earnings. The reverse might be true; that is, an increase in personal earnings might cause an increase in education. The correlation may be due to other factors altogether. It might occur purely by chance or be jointly caused by yet another variable. Perhaps both educational opportunities and earning levels are strongly affected by parental education or a history of family wealth. Thus a third variable, such as parental socioeconomic status, may better account for both more education and higher earnings. We call such a variable that may be affecting the results a *hidden variable*.

Figure 2.55 shows a clear correlation between the number of radios and the proportion of insane people in England between 1924 and 1937.

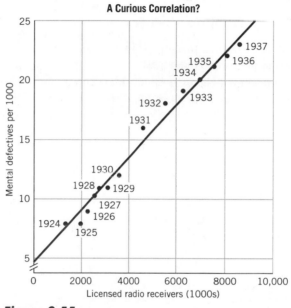

Figure 2.55

Source: E.A. Tufte, *Data Analysis for Politics and Policy*, p. 90.

Are you convinced that radios cause insanity? Or are both variables just increasing with the years? We tend to accept as reasonable the argument that an increase in education causes an increase in personal earnings, because the results seem intuitively possible and they match our preconceptions. But we balk when asked to believe that an increase in radios causes an increase in insanity. Yet the arguments are based on the same sort of statistical reasoning. The flaw in the reasoning is that statistics can show only that events occur together or are correlated, but *statistics can never prove that one event causes another.*

Raising More Questions: Going Deeper

Readings on the course website provide additional resources for examining the relationship between education and earnings.

When a strong link is found between variables, often the next step is to raise questions whose answers may provide more insight into the nature of the relationship such as:

Are there other variables that affect the relationship?

   Do earnings depend on age?

   Do earnings depend on gender?

Will the relationship still hold if we use other income measures?

What factors might be hidden or not taken into account?

What do the outliers tell us?

Questions about the methods used should also be raised, such as:

How good are the data?

How good is the analysis?

How can the evidence be strengthened?

 Exploration 2.2 provides an opportunity to generate your own conjectures and to use the course software to create regression lines to test those conjectures.

You may want to explore answers to the questions above or raise your own questions or conjectures. Your journey into exploratory data analysis is just beginning and you now have some tools to examine further the complex relationship between education and earnings.

Algebra Aerobics 2.11

1. **a.** Evaluate each of the following:

    $$|0.65| \quad |-0.68| \quad |-0.07| \quad |0.70|$$

    b. List the absolute values in part (a) in ascending order from the smallest to the largest.

2. Each of the four graphs shows a scatter plot with a regression line and its correlation coefficient (cc).

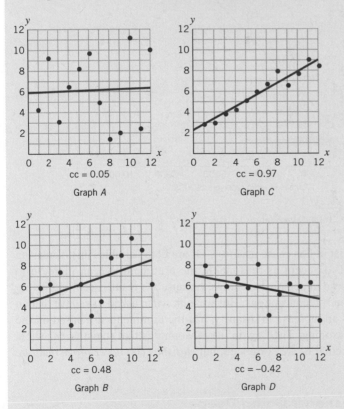

Graph A — cc = 0.05

Graph C — cc = 0.97

Graph B — cc = 0.48

Graph D — cc = −0.42

a. Which ones describe a positive correlation? A negative correlation?

b. Which one describes the weakest correlation? The strongest correlation?

3. We looked at the scatter plot of the *median* personal earnings in Figure 2.53. Below are the data and regression line for the *mean* personal earnings (with a cc of 0.96).

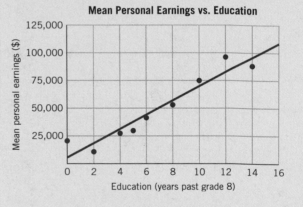

Mean Personal Earnings vs. Education

mean personal earnings =
$$5650 + 6430 \cdot \text{yrs. educ. past grade 8}$$

a. Identify and interpret the slope of the line.

b. What do the equations predict for both median (in Example 2) and mean personal earnings for 4 years of education past grade 8 (i.e., high school graduate)? For 8 years of education past grade 8 (i.e., college graduate)?

Exercises for Section 2.11

Technology is required for generating scatter plots and regression lines in Exercises 11, 12, and 18 and optional in Exercises 9, 13, 14, 16, and 17.

1. The figures (top left of next page) show regression lines and corresponding correlation coefficients for four different scatter plots.

a. Which of the lines describes the strongest linear relationship between the variables? Which of the lines describes the weakest linear relationship?

b. Which lines describe a positive relationship? A negative relationship?

Four regression lines with their correlation coefficients.

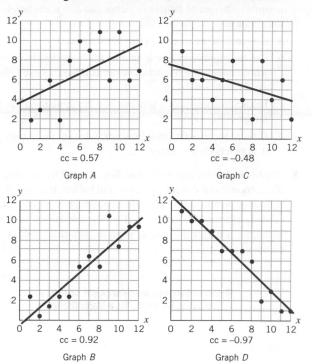

cc = 0.57

Graph *A*

cc = −0.48

Graph *C*

cc = 0.92

Graph *B*

cc = −0.97

Graph *D*

In Exercises 2 to 6, the data analyzed are from the FAM1000 data files and the equations can be generated using *FAM1000 Census Graphs*. Here, the income measure is personal total income, which includes personal earnings (from work) and other sources of unearned income, such as interest and dividends on investments.

2. The accompanying graph and regression line show median personal total income vs. years of education past grade 8.

Median Personal Total Income vs. Education

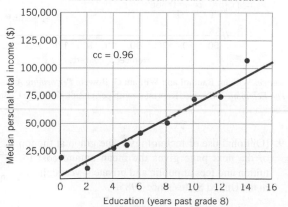

cc = 0.96

Median personal total income =
$$3230 + 6490 \cdot \text{yrs. educ. past grade 8}$$

a. What is the slope of the regression line?

b. Interpret the slope in this context.

c. By what amount does this regression line predict that median personal total income changes for 1 additional year of education? For 10 additional years of education?

d. What features of the data are not well described by the regression line?

3. The following equation represents the best-fit regression line for median personal earnings vs. years of education for the 304 people in the FAM1000 data set who live in the southern region of the United States.

$$S = 15820 + 2960E$$

where S = median personal earnings in the South and E = years of education past grade 8, and cc = 0.51.

a. Identify the slope of the regression line, the vertical intercept, and the correlation coefficient.

b. What does the slope mean in this context?

c. By what amount does this regression line predict that median personal earnings for those who live in the South change for 1 additional year of education? For 10 additional years of education?

4. The following equation represents a best-fit regression line for median personal total income of white males vs. years of education past grade eight:

median personal total income$_{\text{white males}}$ =
$$3240 + 7990 \cdot \text{yrs. educ. past grade 8}$$

The correlation coefficient is 0.93 and the sample size is 454 white males.

a. What is the rate of change of median personal total income with respect to years of education?

b. Generate three points that lie on this regression line. Use two of these points to calculate the slope of the regression line.

c. How does this slope relate to your answer to part (a)?

d. Sketch the graph.

5. From the FAM1000 data, the best-fit regression line for median personal total income of white females vs. years of education past grade 8 is:

median personal total income$_{\text{white females}}$ =
$$14,040 + 2310 \cdot \text{yrs. educ. past grade 8}$$

The correlation coefficient is 0.60 and the sample size is 356 white females.

a. Interpret the number 2310 in this equation.

b. Generate a small table with three points that lie on this regression line. Use two of these points to calculate the slope of the regression line.

c. How does this slope relate to your answer to part (a)?

d. Sketch the graph.

e. Describe some differences between median personal total income vs. education for white females and for white males (see Exercise 4). What are some of the limitations of the model in making this comparison?

6. We can look for correlations for personal earnings with other variables. Here is a scatter plot of mean personal earnings and age.

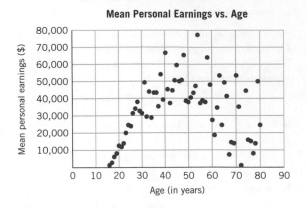

Mean Personal Earnings vs. Age

a. Is there a linear correlation between earnings and age? What would you expect the cc to be?

b. There does seem to be an underlying pattern in the scatter plot. Describe the pattern in terms of the personal earnings as a person ages.

7. The term "linear regression" was coined in 1903 by Karl Pearson as part of his efforts to understand the way physical characteristics are passed from generation to generation. He assembled and graphed measurements of the heights of fathers and their fully grown sons from more than a thousand families. The independent variable, F, was the height of the fathers. The dependent variable, S, was the mean height of the sons who all had fathers with the same height. The best-fit line for the data points had a slope of 0.516, which is much less than 1. If, on average, the sons grew to the same height as their fathers, the slope would equal 1. Tall fathers would have tall sons and short fathers would have equally short sons. Instead, the graph shows that whereas the sons of tall fathers are still tall, they are not (on average) as tall as their fathers. Similarly, the sons of short fathers are not as short as their fathers. Pearson termed this *regression;* the heights of sons *regress* back toward the height that is the mean for that population.

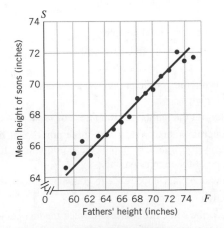

From Snedecor and Cochran, *Statistical Methods*, 8th ed. Reprinted with permission of John Wiley & Sons, Inc. Copyright © 1967.

The equation of this regression line is $S = 33.73 + 0.516F$, where F = height of fathers in inches and S = mean height of sons in inches.

a. Interpret the number 0.516 in this context.

b. Use the regression line to predict the mean height of sons whose fathers are 64 inches tall and of those whose fathers are 73 inches tall.

c. Predict the height of a son who has the same height as his father.

d. If there were over 1000 families, why are there only 17 data points on this graph?

8. The book *Performing Arts—The Economic Dilemma* studied the economics of concerts, operas, and ballets. It included the following scatter plot and corresponding regression line relating attendance per concert to the number of concerts given, for a major orchestra. What do you think were their conclusions?

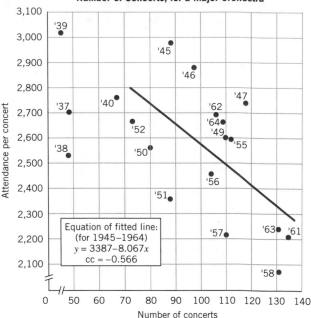

Relation Between Attendance per Concert and Number of Concerts, for a Major Orchestra

Source: William Baumol and William G. Bowen, *Performing Arts—The Economic Dilemma.* Reprinted with permission from The Century Foundation.

9. (Optional use of technology.) The graph at the top of the next page gives the mean annual cost for tuition and fees at public and private 4-year colleges in the United States since 1985.

EDUCOSTS

It is clear that the cost of higher education is going up, but public education is still less expensive than private. The graph suggests that costs of both public and private education versus time can be roughly represented as straight lines.

a. By hand, sketch two lines that best represent the data. Calculate the rate of change of education cost per year for public and for private education by estimating the coordinates of two points that lie on the line, and then estimating the slope.

Mean Cost of 4-Year Colleges, Public and Private

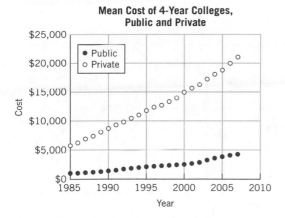

Source: U.S. Bureau of the Census, *Statistical Abstract of the United States: 2010*

b. Construct an equation for each of your lines in part (a). (Set 1985 as year 0.) If you are using technology, generate two regression lines from the Excel or graph link data file EDUCOSTS and compare these equations with the ones you constructed.

c. If the costs continue to rise at the same rates for both sorts of schools, what would be the respective costs for public and private education in the year 2010? Does this seem plausible to you? Check your answers on the Internet.

10. Stroke is the third-leading cause of death in the United States, behind heart disease and cancer. The accompanying graph shows the average neuron loss for a typical ischemic stroke.

Neuron Loss From a Typical Ischemic Stroke

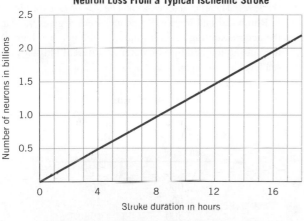

Source: American Heart Association.

a. Find the slope of the line by estimating the coordinates of two points on the line. Interpret the meaning of the slope in this context.

b. Construct an equation for the line, where n = number of neurons in billions and d = duration of stroke in hours.

c. The mean human forebrain has about 22 billion neurons and the mean stroke lasts about 10 hours. Find the percentage of neurons that are lost from a 10-hour stroke.

11. (Requires technology.) In Section 2.10 we examined the percentage of adult smokers. Here we break out smokers by sex—the percentage of males who smoke and the percentage of females who smoke.

SMOKERS

Percentage of Adult Smokers and by Sex (18 and over)

Year	% of Adult Population	% of All Males	% of All Females
1965	42.4	51.9	33.9
1974	37.1	43.1	32.1
1979	33.5	37.5	29.9
1983	32.1	35.1	29.5
1985	30.1	32.6	27.9
1990	25.5	28.4	22.8
1995	24.7	27.0	22.6
2000	23.3	25.7	21.0
2005	20.9	23.9	18.1
2007	19.7	22.0	17.5

Source: American Lung Association.

If you are male, use only the male data; if you are female, only the female data.

a. Construct a scatter plot of the percentage of smokers of your sex.

b. On your graph sketch a best-fit line. By estimating coordinates of points on your best-fit line, calculate the slope of the line. What does this tell you?

c. Reinitialize your data by setting 1965 as year 0. Now use technology to generate a linear regression line and correlation coefficient (cc) to your new data set. Record the equation and the cc. How good a fit is the regression line to the data? What does the vertical intercept tell you? The slope?

d. Write a 60-second summary about the smoking trends for your sex. If possible, compare your results to those of someone of the opposite sex.

12. (Requires technology.) The accompanying table shows the calories per minute burned by a 154-pound person moving at speeds from 2.5 to 12 miles/hour (mph). (*Note:* A fast walk is about 5 mph; faster than that is considered jogging or slow running.) Marathons, about 26 miles long, are now run in slightly over 2 hours, so the top distance runners are approaching a speed of 13 mph.

CALORIES

Speed (mph)	Calories per Minute	Speed (mph)	Calories per Minute
2.5	3.0	6.0	12.0
3.0	3.7	7.0	14.0
3.5	4.2	8.0	15.6
4.0	5.5	9.0	17.5
4.5	7.0	10.0	19.6
5.0	8.3	11.0	21.7
5.5	10.1	12.0	24.5

a. Plot the data.

b. Does it look as if the relationship between speed and calories per minute is linear? If so, generate a linear model. Identify the variables and a reasonable domain for the model, and interpret the slope and vertical intercept. How well does your line fit the data?

c. Describe in your own words what the model tells you about the relationship between speed and calories per minute.

13. (Optional use of technology.) The accompanying graph shows the winning running times in minutes for women in the Boston Marathon.

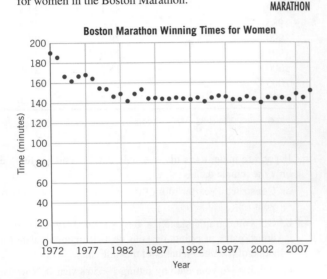

Boston Marathon Winning Times for Women

Source: www.bostonmarathon.org/BostonMarathon/PastChampions.asp.

a. The graph is roughly linear with negative slope between 1972 and 1982 but seems to flatten out after that. Sketch a piecewise linear function that best approximates the data.

b. Set 1972 as year 0 and create a piecewise linear function that describes winning times for women. If using technology, you can create 2 regression lines from 2 subsets of the data, using either the Excel or graph link file MARATHON.

c. Interpret the slopes of the 2 linear pieces.

d. What would you predict for the winning time in 2010? Since the 2010 Marathon has already occurred, check your prediction on the Internet.

e. Write a 60-second summary describing the Boston Marathon winning times for women.

14. (Optional use of technology.) The accompanying graph shows the world record times for the men's mile. As of January 2010, the 1999 record still stands. Note that several times the standing world record was broken more than once during a year.

a. Generate a line that approximates the data (by hand or, if using technology, with the data in the Excel or graph link file MENSMILE. (Set 1910 or 1913 as year 0). If you are using technology, specify the correlation coefficient. Interpret the slope of your line in this context.

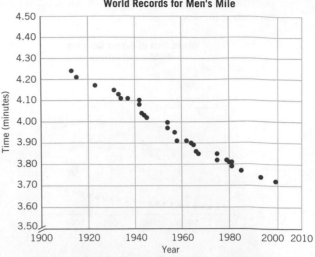

World Records for Men's Mile

Source: www.runnersworld.com.

b. If the world record times continue to change at the rate specified in your linear model, predict the record time for the men's mile in 2010. Check your answer on the Internet.

c. In what year would your linear model predict the world record to be 0 minutes? Since this is impossible, what do you think is a reasonable domain for your model? Describe what you think would happen in the years after those included in your domain.

d. Write a short paragraph summarizing the trends in the world record times for the men's mile.

15. The temperature at which water boils is affected by the difference in atmospheric pressure at different altitudes above sea level. The classic cookbook *The Joy of Cooking* by Irma S. Rombauer and Marion Rombauer Becker gives the data in the accompanying table (rounded to the nearest degree) on the boiling temperature of water at different altitudes above sea level.

Boiling Temperature of Water

Altitude (ft above sea level)	Temperature Boiling °F
0	212
2,000	208
5,000	203
7,500	198
10,000	194
15,000	185
30,000	158

a. Use the accompanying table for the following:

i. Plot boiling temperature in degrees Fahrenheit, °F, vs. the altitude. Find a formula to describe the boiling temperature of water, in °F, as a function of altitude.

ii. According to your formula, what is the temperature at which water will boil where you live? Can you verify this? What other factors could influence the temperature at which water will boil?

b. The highest point in the United States is Mount McKinley in Alaska, at 20,320 feet above sea level; the lowest point is Death Valley in California, at 285 feet below sea level. You can think of distances below sea level as negative altitudes from sea level. At what temperature in degrees Fahrenheit will water boil in each of these locations according to your formula?

c. Using your formula, find an altitude at which water can be made to boil at 32°F, the freezing point of water at sea level. At what altitude would your formula predict this would happen? (Note that airplane cabins are pressurized to near sea-level atmospheric pressure conditions in order to avoid unhealthy conditions resulting from high altitude.)

16. (Optional use of technology.) The accompanying graph shows the world distance records for the women's long jump. Several times a new long jump record was set more than once during a given year.

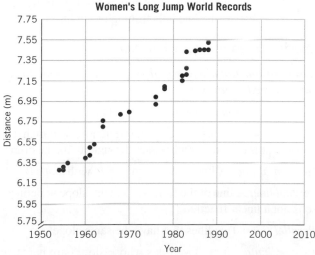

Women's Long Jump World Records

Note: As of January 2010, the 1988 long jump record still stands.
Source: Data extracted from the website at *http://www.uta.fi/-csmipe/sports/eng/mwr.html.*

a. Generate a line that approximates the data by hand or, if using technology, use the data from the Excel or graph link file LONGJUMP. (Set 1950 or 1954 as year 0). Interpret the slope of your line in this context. If you are using technology, specify the correlation coefficient.

b. If the world record distances continue to change at the rate described in your linear model, predict the world record distance for the women's long jump in the year 2010. Check on the Internet.

c. What would your model predict for the record in 1943? How does this compare with the actual 1943 record of 6.25 meters? What do you think would be a reasonable domain for your model? What do you think the data would look like for years outside your specified domain?

17. (Optional use of technology.) The accompanying graph shows the increasing number of motor vehicle registrations (cars and trucks) in the United States.

MOTOR

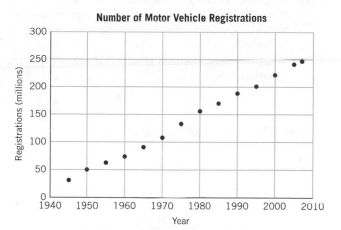

Number of Motor Vehicle Registrations

Source: U.S. Federal Highway Administration, *Highway Statistics,* annual.

a. Using 1945 as the base year, find a linear equation that would be a reasonable model for the data. If using technology, use the data in the Excel or graph link file MOTOR.

b. Interpret the slope of your line in this context.

c. What would your model predict for the number of motor vehicle registrations in 2007? How does this compare with the actual data value of 237.4 million?

d. Using your model, how many motor vehicles will be registered in the United States in 2010? Check your answer on the Internet.

18. (Requires technology.) Examine the following data on U.S. union membership from 1930 to 2009.

UNION

U.S. Union Membership, 1930–2009

Year	Labor Force* (thousands)	Union Members† (thousands)	Percentage of Labor Force
1930	29,424	3,401	11.6
1940	32,376	8,717	26.9
1950	45,222	14,267	31.5
1960	54,234	17,049	31.4
1970	70,920	19,381	27.3
1980	90,564	19,843	21.9
1990	103,905	16,740	16.1
2000	120,786	16,258	13.5
2005	125,889	15,685	12.5
2009	124,290	15,327	12.3

*Does not include agricultural employment; from 1985, does not include self-employed or unemployed persons.
†From 1930 to 1980, includes dues-paying members of traditional trade unions, regardless of employment status; from 1985, includes members of employee associations that engage in collective bargaining with employers.
Source: Bureau of Labor Statistics. U.S. Dept. of Labor.

a. Using the table, describe the overall trends in the size of the labor force, in the number of union members, and in the percentage of the labor force that belongs to a union.

b. Graph the percentage of the labor force in unions vs. time from 1950 to 2009. Measuring time in years since 1950, find a linear regression formula for these data using technology.

c. When does your formula predict that only 10% of the labor force will be unionized?

d. What data would you want to examine to understand why union membership is declining?

19. If a study shows that smoking and lung cancer have a high positive correlation, does this mean that smoking causes lung cancer? Explain your answer.

20. Parental income has been found to have a high positive correlation with their children's academic success. What are two different ways you could interpret this finding?

CHAPTER SUMMARY

The Average Rate of Change

The average rate of change of y with respect to x is

$$\frac{\text{change in } y}{\text{change in } x}$$

The units of the average rate of change $= \dfrac{\text{units of } y}{\text{units of } x}$

For example, the units might be dollars/year (read as "dollars per year") or pounds/person (read as "pounds per person").

The average rate of change between two points is the slope of the straight line connecting the points. Given two points (x_1, y_1) and (x_2, y_2),

$$\text{average rate of change} = \frac{\text{change in } y}{\text{change in } x}$$
$$= \frac{\Delta y}{\Delta x} = \frac{y_2 - y_1}{x_2 - x_1} = \text{slope}$$

If the slope, or average rate of change, of y with respect to x is *positive*, then the graph of the relationship rises when read from left to right. This means that as x increases in value, y increases in value.

If the slope is *negative*, the graph falls when read from left to right. As x increases, y decreases.

If the slope is *zero*, the graph is flat. As x increases, there is no change in y.

Linear Functions

A *linear function* has a constant average rate of change. It can be described by an equation of the form

$$\underbrace{\text{output}}_{y} = \underbrace{\text{initial value}}_{b} + \underbrace{\text{rate of change}}_{m} \cdot \underbrace{\text{input}}_{x}$$

or

$$y = b + mx$$

where b is the vertical intercept and m is the slope, or rate of change of y with respect to x.

The Graph of a Linear Function

The graph of the linear function $y = b + mx$ is a straight line.

The y-intercept, b, tells us where the line crosses the y-axis.

The slope, m, is the average rate of change, so it tells us how fast the line is climbing or falling. The larger the magnitude (or absolute value) of m, the steeper the graph.

If m is positive, then the line climbs from left to right. If m is negative, the line falls from left to right.

Special Cases of Linear Functions

Direct proportionality: y is *directly proportional to (or varies directly with)* x if

$$y = mx \quad \text{where the constant } m \neq 0$$

This equation represents a linear function in which the y-intercept is 0, so the graph passes through $(0, 0)$, the origin.

Horizontal line: A line of the form $y = b$, with slope 0.

Vertical line: A line of the form $x = c$, with slope undefined; so is not a linear function.

Parallel lines: Two lines that have the same slope.

Perpendicular lines: Two lines whose slopes are negative reciprocals.

Piecewise linear function: A function constructed from different linear segments. Some examples are the *absolute value function*, $f(x) = |x|$, and *step functions*, whose linear segments are all horizontal.

Fitting Lines to Data

We can draw by hand or use technology to find a line to fit data whose graph exhibits a linear pattern. A computer best-fit line is called a *regression line*, and its goodness of fit is called the *correlation coefficient*. The equation of a best-fit line offers an approximate but compact description of the data. Lines are of special importance in describing patterns in data because they are easily drawn and manipulated and give a quick first approximation of trends.

CHECK YOUR UNDERSTANDING

I. Is each of the statements in Problems 1–31 true or false? Give an explanation for your answer.

In Problems 1–4 assume that y is a function of x.

1. The graph of $y - 5x = 5$ is decreasing.

2. The graph of $2x + 3y = -12$ has a negative slope and negative vertical intercept.

3. The graph of $x - 3y + 9 = 0$ is steeper than the graph of $3x - y + 9 = 0$.

4. The accompanying figure is the graph of $5x - 3y = -2$.

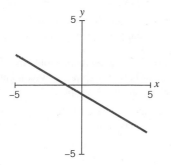

5. Health care costs between the years 1990 and 2010 would most likely show a positive average rate of change.

6. If we choose any two points on the graph in the accompanying figure, the average rate of change between them would be positive.

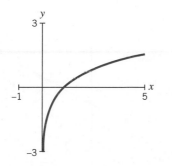

7. The average rate of change between two points (t_1, Q_1) and (t_2, Q_2) is the same as the slope of the line joining these two points.

8. The average rate of change of a variable M between the years 2000 and 2010 is the slope of the line joining two points of the form $(M_1, 2000)$ and $(M_2, 2010)$.

9. To calculate the average rate of change of a variable over an interval, you must have two distinct data points.

10. A set of data points of the form (x, y) that do not fall on a straight line will generate varying average rates of change depending on the choice of endpoints.

11. If the average rate of change of women's salaries from 2003 to 2010 is \$1200/year, then women's salaries increased by exactly \$1200 between 2003 and 2010.

12. If the average rate of change is positive, the acceleration (or rate of change of the average rate of change) may be positive, negative, or zero.

13. If the average rate of change is constant, then the acceleration (or rate of change of the average rate of change) is zero.

14. The average rate of change between (W_1, D_1) and (W_2, D_2) can be written as either $(W_1 - W_2)/(D_1 - D_2)$ or $(W_2 - W_1)/(D_2 - D_1)$.

15. If we choose any two distinct points on the line in the accompanying figure, the average rate of change between them would be the same negative number.

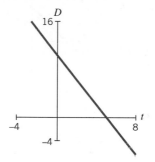

16. The average rate of change between (t_1, Q_1) and (t_2, Q_2) can be written as either $(Q_1 - Q_2)/(t_1 - t_2)$ or $(Q_2 - Q_1)/(t_2 - t_1)$.

17. On a linear graph, it does not matter which two distinct points on the line you use to calculate the slope.

18. The accompanying graph shows that the average rate of change of the percent of people who want smaller government and fewer services is positive from about October 2008 to October 2009.

Which Would You Rather Have...

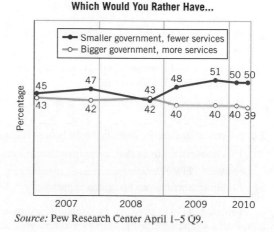

Source: Pew Research Center April 1–5 Q9.

19. If the distance a sprinter runs (measured in meters) is a function of the time (measured in minutes), then the units of the average rate of change are minutes per meter.

20. Every linear function crosses the horizontal axis exactly one time.

21. If a linear function in the form $y = mx + b$ has slope m, then increasing the x value by one unit changes the y value by m units.

22. If the units of the dependent variable y are pounds and the units of the independent variable x are square feet, then the units of the slope are pounds per square foot.

23. If the average rate of change between any two data points is increasing as you move from left to right, then the function describing the data is linear and is increasing.

24. The function in the accompanying figure has a slope that is increasing as you move from left to right.

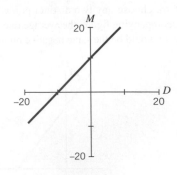

25. The slope of the function $f(x)$ is of greater magnitude than the slope of the function $g(x)$ in the accompanying figures.

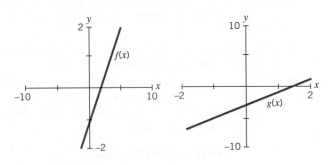

26. You can calculate the slope of a line that goes through any two points on the plane.

27. If $f(x) = y$ is decreasing throughout, then the y values decrease as the x values decrease.

28. Having a slope of zero is the same as having an undefined slope.

29. A vertical line can be described by a linear function.

30. Two nonvertical lines that are perpendicular must have slopes of opposite sign.

31. All linear functions can be written in the form $y = b + mx$.

II. In Problems 32–41, give an example of a function or functions with the specified properties. Express your answers using equations, and specify the independent and dependent variables.

32. Linear and decreasing with positive vertical intercept

33. Linear and horizontal with vertical intercept 0

34. Linear with positive horizontal intercept and negative vertical intercept

35. Linear and does not pass through the first quadrant (where $x > 0$ and $y > 0$)

36. Linear with average rate of change of 37 minutes/lap

37. Linear describing the value of a stock that is currently at $19.25 per share and is increasing exactly $0.25 per quarter

38. Two linear functions that are parallel, such that moving one of the functions horizontally to the right two units gives the graph of the other

39. Four distinct linear functions all passing through the point $(0, 4)$

40. Two linear functions where the slope of one is m and the slope of the second is $-1/m$, where m is a negative number

41. Five data points for which the average rates of change between consecutive points are positive and the sequential points are increasing at a decreasing rate

III. Is each of the statements in Problems 42–55 true or false? If a statement is true, explain how you know. If a statement is false, give a counterexample.

42. If the average rate of change between any two points of a data set is constant, then the data are linear.

43. If the slope of a linear function is negative, then the average rate of change decreases.

44. For any two distinct points, there is a linear function whose graph passes through them.

45. To write the equation of a specific linear function, one needs to know only the slope.

46. Function A in the accompanying figure is increasing at a faster rate than function B.

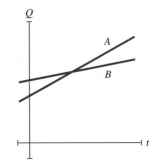

47. The graph of a linear function is a straight line.

48. Vertical lines are not linear functions because they cannot be written in the form $y = b + mx$, as they have undefined slope.

49. There exist linear functions that slant upward moving from left to right but have negative slope.

50. A constant average rate of change means that the slope of the graph of a function is zero.

51. All linear functions in x and y describe a relationship where y is directly proportional to x.

52. The range of the function $h(t) = |t - 2|$ is always positive for any value of t.

53. The function $h(t) = |t - 2|$ can be written as a piecewise linear function.

54. A linear model would be appropriate for Barry Bonds' graph of home runs vs. number of games.

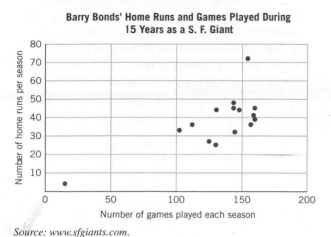

Barry Bonds' Home Runs and Games Played During 15 Years as a S. F. Giant

Source: www.sfgiants.com.

55. The graph below shows a step function.

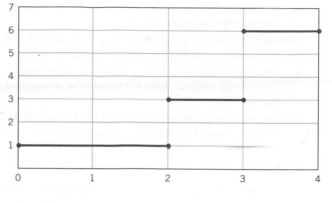

CHAPTER 2 REVIEW: PUTTING IT ALL TOGETHER

1. According to the National Association of Realtors, the median price for a single-family home fell from $217,900 in 2007 to $173,200 in 2009. Describe the change in the following ways.

a. The absolute change in dollars

b. The percent decrease

c. The annual average rate of change of the price with respect to year

(*Note:* Be sure to include units in all your answers.)

2. a. According to Apple Computer, sales of its iPod (the world's best-selling digital audio player) soared from 376,000 units in 2002 to an all-time high of 54,838,000 units in 2008. What was the average rate of change in iPods sold per year over that time period?

b. Apple's overall sales continue to rise (with iPhone sales), but iPod sales declined to 54,132,000 units in 2009. What was the average rate of change in iPod unit sales between 2008 and 2009?

3. Given the following graph of the function $h(t)$, identify any interval(s) over which:

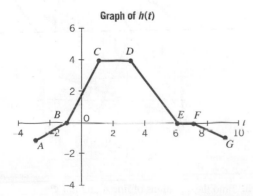

Graph of $h(t)$

a. The function is positive; is negative; is zero

b. The slope is positive; is negative; is zero

4. Which line has the steeper slope? Explain why.

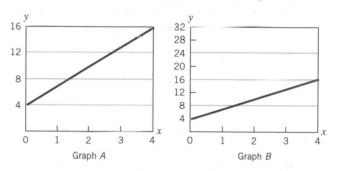

Graph A

Graph B

5. You are traveling abroad and realize that American, British, and French clothing sizes are different. The accompanying table shows the correspondence among female clothing sizes in these different countries.

Women's Clothing Sizes

U.S.	Britain	France
4	10	38
6	12	40
8	14	42
10	16	44
12	18	46
14	20	48

a. Are the British and French sizes both linear functions of the U.S. sizes? Why or why not?

b. Write a sentence describing the relationship between British sizes and U.S. sizes.

c. Construct an equation that describes the relationship between French and U.S. sizes.

6. A bathtub that initially holds 50 gallons of water starts draining at 10 gallons per minute.

   a. Construct a function $W(t)$ for the volume of water (in gallons) in the tub after t minutes.

   b. Graph the function and label the axes. Evaluate $W(0)$ and $W(5)$ and describe what they represent in this context.

   c. How would the original function and its graph change if the initial volume were 60 gallons? Call the new function $U(t)$.

   d. How would the original function and graph change if the drain rate were 12 gallons per minute? Call the new function $V(t)$.

   e. Add the graphs of $U(t)$ and $V(t)$ to the original graph of $W(t)$.

7. Cell phones have produced a seismic cultural shift. No other recent invention has incited so much praise—and criticism. In 2000 there were 109.4 million cell phone subscriptions in the United States; since then subscriptions steadily increased to reach 262.7 million in 2008. (*Note:* Some people had more than one subscription.)

   a. What was the average rate of change in millions of cell phone subscriptions per year between 2000 and 2008?

   b. Construct a linear function $C(t)$ for cell phone subscriptions (in millions) for t = years from 2000.

   c. If U.S. cell phone subscriptions continue to increase at the same rate, how many will there be in 2015? Does your result sound plausible?

8. The following graph shows average salaries for major league baseball players from 1990 to 2009.

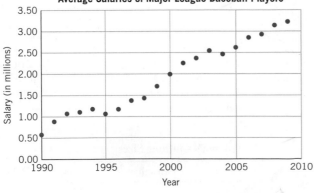

Average Salaries of Major League Baseball Players

Source: CBSsports.com.

   a. On the graph, sketch a straight line that is an approximate mathematical model for the data.

   b. Specify the coordinates of two points *on your line* and describe what they represent in terms of your model.

   c. Calculate the slope and then interpret it in terms of the salary in millions of dollars and the time interval.

   d. Let $S(x)$ be a linear approximation of the salaries where x = years since 2000. What is the slope of the graph of $S(x)$? The vertical intercept? What is the equation for $S(x)$?

   e. Use your model to predict the average salary for 2010. Then, if possible, check it on the Internet.

9. Match one or more of the following graphs with the following descriptions. Be sure to note the scale on each axis.

   a. Represents a constant rate of change of 3

   b. Has a vertical intercept of 10

   c. Is parallel to the line $y = 5 - 2x$

   d. Has a steeper slope than the line C

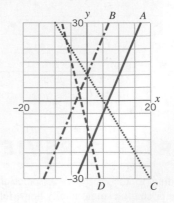

10. a. If you travel to China, the exchange rate at the Beijing airport is 1 yuan for 0.147 American dollar and at the hotel is 1 yuan for 0.145 American dollar. Which exchange rate should you use?

    b. Use the most favorable rate to construct a function Y to convert dollars, d, to yuan, $Y(d)$.

    c. What is $Y(50)$? $Y(200)$?

11. In May 2010 the Gallup poll reported that 46% of Americans said that they lived "with a lot of happiness/enjoyment without a lot of stress/worry." This poll had a 3% margin of error. Create an absolute value inequality that describes the ranges of possible poll percentages and display it on a number line.

12. Given the following graph:

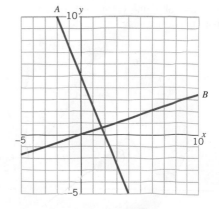

    a. Find the equations of line A and line B.

    b. Show that line A is (or is not) perpendicular to line B.

13. According to the Environmental Protection Agency's "National Coastal Conditions Report II," more than half of the U.S. population live in the narrow coastal fringes. Increasing population in these areas contributes to degradation (by runoff, sewage spills, construction, and overfishing) of the same resources that make the coasts desirable. In 2003 about 153 million people lived in coastal counties. This coastal population is currently increasing by an average of 3600 people per day.

   a. What is the average rate of change in millions of people per *year*?

   b. Let x = years since 2003 and construct a linear function $P(x)$ for the coast population (in millions).

   c. Graph the function $P(x)$ over a reasonable domain.

   d. What does your model give for the projected population in 2010? Check the actual population on the Internet.

14. Generate the equation for a line under each of the following conditions.

   a. The line goes through the points $(-1, -4)$ and $(4, 6)$.

   b. The line is parallel to the line in part (a) and goes through the point $(0, 5)$.

   c. The line is perpendicular to the line in part (a) and goes through the origin.

15. In 2005 a record 51.5% of paper consumed in the United States (51.3 million tons) was recycled. The American Forest and Paper Association states that its goal is to have 55% recovery by 2012.

   a. Plot two data points that represent the percentage of paper recycled at t number of years since 2005. (You may want to crop the vertical axis to start at 50%.) Connect the two points with a line and calculate its slope.

   b. Assuming linear growth, the line represents a model, $R(t)$, for the percentage of paper recycled at t years since 2005. Find the equation for $R(t)$.

   c. Find $R(0)$, $R(5)$, and $R(20)$. What does $R(20)$ represent?

16. Sketch a graph that could represent each of the following situations. Be sure to label your axes.

   a. The volume of water in a kettle being filled at a constant rate.

   b. The height of an airplane flying at 30,000 feet for 3 hours.

   c. The unlimited demand for a certain product. Assume a company produces 10,000 items and consumers will pay virtually any price for one of them. (Use the convention of economists, placing price on the vertical axis and quantity on the horizontal.)

   d. The price of subway tokens that periodically increase over time.

   e. The distance from your destination over time when you travel three subway stops to get to your destination.

17. Generate an equation for each line A, B, C, D, and E shown in the accompanying graph.

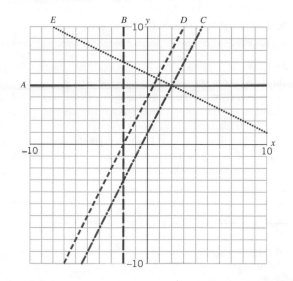

18. U.S. life expectancy has been steadily increasing (see Exercise 15 in Section 2.1 and data file for LIFEPXEC). The following graph shows the roughly linear trends for both males and females over more than three decades.

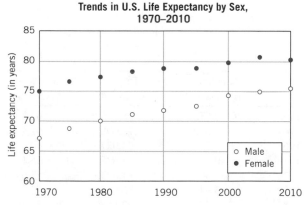

Trends in U.S. Life Expectancy by Sex, 1970–2010

Source: U.S. Bureau of the Census.

   a. According to these data, what is the life expectancy of a female born in 1970? Of a male born in 2005?

   b. Sketch a straight line to create a linear approximation of the data for your gender. Males and females will have different lines.

   c. Identify two points *on your line* and calculate the slope. What does your slope mean in this context? The male slope is steeper than the female slope. What does that imply?

   d. Generate a linear function for your line. (*Hint:* Let the independent variable be the number of years since 1970.)

   e. What would your model predict for your sex's life expectancy in 2010? How close does your model come to Census Bureau current predictions of life expectancies in 2010 of 75.6 years for males and 81.4 years for females?

19. For the function $g(x) = |x| + 2$:

   a. Generate a table of values for $g(x)$ using integer values of x between -3 and 3.

   b. Use the table to sketch a graph of $g(x)$.

   c. How does this graph compare with the graph of $f(x) = |x|$?

20. In Exercise 7 in Section 2.9 we met the greatest integer function $f(x) = [x]$, where $[x]$ is the greatest integer $\leq x$; (i.e., you round down to the nearest integer). So $[-1.5] = -2$ and $[3.99] = 3$. This function is sometimes written using the notation $\lfloor x \rfloor$ and called the *floor* function. There is a similar *ceiling* function, $g(x) = \lceil x \rceil$, where $\lceil x \rceil$ is the smallest integer $\geq x$; (i.e., here you round up to the nearest integer). So $\lceil -1.5 \rceil = -1$ and $\lceil 3.000001 \rceil = 4$.

   a. Complete the following table.

x	-2	-1.5	-1	-0.5	0	0.5	1	1.5	2
Floor $f(x) = \lfloor x \rfloor$									
Ceiling $g(x) = \lceil x \rceil$									

   b. Plot the ceiling function and the floor function on two separate graphs, being sure to specify whether the endpoints of each line segment are included or excluded.

   c. Why might a telephone company be interested in a ceiling function?

21. The table below gives historical data on voter turnout as a percentage of voting-age population (18 years and older) in U.S. presidential elections since 1960.

Year	Population (thousands)	Turnout (thousands)	% of Voting Age Pop.
1960	109,672	68,836	62.8
1964	114,090	70,098	61.4
1968	120,285	73,027	60.7
1972	140,777	77,625	55.1
1976	152,308	81,603	53.6
1980	163,945	86,497	52.8
1984	173,995	92,655	53.3
1988	181,956	91,587	50.3
1992	189,493	104,600	55.2
1996	196,789	96,390	49.0
2000	209,816	105,594	50.3
2004	219,726	122,349	55.6
2008	230,118	131,407	57.1

Using selected data from the table, graphs, and dramatically persuasive language, provide convincing arguments in a 60-second summary that during the years 1960 to 2008:

   a. Voter turnout has plummeted.

   b. Voter turnout has soared.

22. a. Construct the equation of a line y_1 whose graph is steeper than that of $8x + 2y = 6$ and does not pass through quadrant I (where both $x > 0$ and $y > 0$). Plot the lines for both equations on the same graph.

   b. Construct another equation of a line y_2 that is perpendicular to the original equation, $8x + 2y = 6$, and *does* go through quadrant I. Add the plot of this line to the graph in part (a).

23. Sweden has kept meticulous records of its population for many years. The following graph shows the mortality rate (as a percentage) for female and male children in Sweden over a 250-year period.

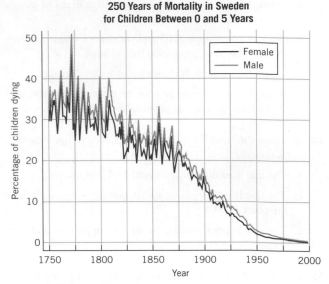

250 Years of Mortality in Sweden for Children Between 0 and 5 Years

Source: Statistics Sweden.

   a. Describe the overall trend, and the similarities and differences between the female and male graphs.

   b. Assuming the graphs are roughly linear between 1800 and 1950, draw a line approximating the female childhood mortality rate during this time. What is the slope of the line and what does it tell you in this context?

   c. Construct a linear function to model the deaths over this period.

   d. What is happening in the years after 1950?

Note: The large spike between 1750 and 1800 represents the more than 300,000 Swedish children (out of a total population of about 2 million) who died from smallpox, the most feared disease of that century.

24. According to the U.S. Environmental Protection Agency, carbon dioxide makes up 84.6% of the total U.S. greenhouse gas emissions. Carbon dioxide, or CO_2, arises from the combustion of coal, oil, and gas. Greenhouse gases trap heat on our planet, causing it to become warmer. Effects of this phenomenon are already being seen in increased melting of glaciers and permafrost. If this continues unchecked, water levels will rise, causing flooding in coastal communities, where the majority of the U.S. population lives. (See Exercise 13.)

   The graph at the top of the next page shows the average rate of change in the concentration of atmospheric CO_2 since 1990. (Note that before 1900 the average rate of change in CO_2 emissions was typically 0.2 ppm or below.)

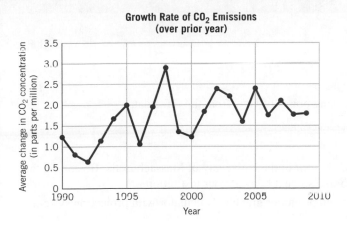

**Growth Rate of CO₂ Emissions
(over prior year)**

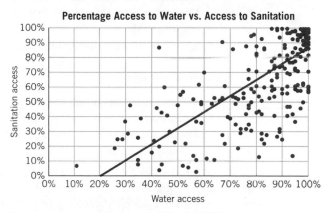

Farm Income from Safflower Oil

Source: Statistical Abstract of the United States, 2010.

a. Is the average rate of change always positive? If so, does that imply that the amount of atmospheric carbon dioxide is always increasing? Explain your answer.

b. What does it mean when the average rate of change was decreasing (though not negative)?

25. The following three graphs show U.S. farm income for three crops over the years 2000–2007.

a. Identify each scatter plot as showing a positive, negative, or no discernible correlation.

b. If you were a farmer and wanted to boost your farm income, which crop do you think you would plant based on the trend in these graphs? Check the Internet to see if you chose wisely.

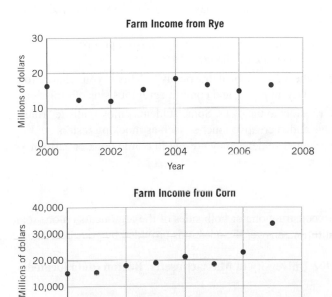

Farm Income from Rye

Farm Income from Corn

26. The World Health Organization (WHO) states: "Access to drinking water and improved sanitation is a fundamental need and a human right vital for the dignity and health of all people." Not surprisingly, there is a correlation between access to water and access to sanitation in the following scatter plot showing data for both urban and rural areas of 185 countries.

Percentage Access to Water vs. Access to Sanitation

Source: https://www.swivel.com/workbooks/20847.

The computer-generated regression line (from Excel using data file WATERSAN) is

% sanitation access = −0.224 + 1.096 · (% water access)

and has a cc = 0.75.

a. Identify the slope and interpret it in context.

b. Identify the vertical intercept of the equation. What would be a reasonable domain for the regression line and why?

c. How do you know the regression is a reasonable fit to the line?

d. Create a newspaper headline to summarize your results.

Having It Your Way

Objectives

- construct arguments supporting opposing points of view from the same data

Material/Equipment

- excerpts from the *Student Statistical Portrait* of the University of Massachusetts, Boston (in the Appendix) or from the equivalent for the student body at your institution
- computer with spreadsheet program and printer or graphing calculator with projection system (all optional)
- graph paper and/or overhead transparencies (and overhead projector)

Procedure

Working in Small Groups

Examine the data and graphs from the *Student Statistical Portrait* of the University of Massachusetts, Boston, or from your own institution. Explore how you would use the data to construct arguments that support at least two different points of view. Decide on the arguments you are going to make and divide up tasks among your team members.

Rules of the Game

- Your arguments needn't be lengthy, but you need to use graphs and numbers to support your position. You may use only legitimate numbers, but you are free to pick and choose those that support your case. If you construct your own graphs, you may, of course, use whatever scaling you wish on the axes.
- For any data that represent a time series, as part of your argument pick two appropriate endpoints and calculate the associated average rate of change.
- Use "loaded" vocabulary (e.g., "surged ahead," "declined drastically"). This is your chance to be outrageously biased, write absurdly flamboyant prose, and commit egregious sins of omission.
- Decide as a group how to present your results to the class. Some students enjoy realistic "role playing" in their presentations and have added creative touches such as mock protesters complete with picket signs. Visually oriented students may present in comic book or cartoon format–suitable for a student newspaper.

Suggested Topics

Your instructor might ask your group to construct one or both sides of the arguments on one topic. If you're using data from your own institution, answer the questions provided by your instructor.

Using the *Student Statistical Portrait* from the University of Massachusetts, Boston (data located in the Appendix of the text)

1. Use the table "Undergraduate Admissions Summary" to construct a persuasive case for each situation.
   a. You are the Provost, the chief academic officer of the university, arguing in front of the Board of Trustees that the university is becoming more appealing to students.
   b. You are a student activist arguing that the university is becoming less appealing to students.
2. Use the table "Trends in New Student Race/Ethnicity in the College of Liberal Arts" to make a convincing case for each of the following.

    a. You are the Affirmative Action Officer arguing that her office has done a terrific job.

    b. You are a reporter for the student newspaper criticizing the university for its neglect of minority students.

3. Use the table "SAT Scores of New Freshmen by College/Program" to "prove" each of the contradictory viewpoints.

    a. You are the Dean of the College of Liberal Arts arguing that you have brighter freshmen than those in the College of Management.

    b. You are the Dean of the College of Management arguing that your freshmen are superior.

Exploration-Linked Homework

With your partner or group prepare a short class presentation of your arguments, using, if possible, overhead transparencies or a projection panel. Then write individual 60-second summaries to hand in.

EXPLORATION 2.2

A Case Study on Education and Earnings in the U.S.

Objective

- to find relationships between education and earnings using the FAM1000 data

Materials/Equipment*

FAM1000

If using a computer

- course software called "F3: Regression with Multiple Subsets" in the "FAM1000 Census Graphs"
- Excel file with full FAM1000 data
- printer and overhead transparencies (optional)

If using a graphing calculator

- graphing calculator with graph link capabilities
- graph link files called FAM1000A to H
- printout of FAM1000 graph link files in Graphing Calculator Manual

Related Readings and Websites[3]

The Bureau of Labor Statistics created an interactive website where you can access all the data on earning and education from the most recent Current Population Survey at *http://www.bls.gov/data/*. This site allows you to search according to categories, access historical data, and create graphs.

Several relevant readings are on the course website.

Procedure

Working with a Partner

You may want to work with a partner so that you can discuss questions that may be worth pursuing and help each other interpret and analyze the findings. Also, it is easier to compare two or more regression lines using two computers or two graphing calculators.

Generating Conjectures

Your journey into exploratory data analysis is just beginning. You may want to explore answers to questions raised in the text (at the end of Section 2.11) or to generate your own questions and conjectures. For example, what other variables, such as gender, may affect the relationship between education and earnings? Will different income measures have different results? See the data dictionary in the Appendix for the variables included in the FAM1000 data.

Finding and Comparing Regression Lines

1. Finding regression lines:
   a. *Using a computer*: Open "F3: Regression with Multiple Subsets" in *FAM1000 Census Graphs*. This program allows you to find regression lines for education vs. earnings for different income variables and for different groups of people. Select at least two regression lines that would be interesting to compare (e.g., men vs. women, white vs. nonwhite, two or

---

[3]The course software, Excel, and graph link data files, Graphing Calculator Manual, and the readings can be downloaded from the web at *www.wiley.com/college/kimeclark* or on your class WileyPLUS site. The software provides easy-to-use interactive tools for analyzing the FAM1000 data.

more regions of the country). Print out your regression lines (on overhead transparencies if possible) so that you can present your findings to the class.

    b. *Using graphing calculators and graph link files:* FAM1000 graph link files A to H contain data for generating regression lines for several income variables as a function of years of education. The Graphing Calculator Manual contains descriptions and hardcopy of the files, as well as instructions for downloading and transferring them to TI-83 and TI-84 calculators. Decide on at least two regression lines that are interesting to compare.

2. Comparing regression lines:

With your partner, explore ways of comparing the regression lines. If possible, put the graphs you are comparing on the same grid; it is easier to make comparisons. Examine the regression lines for the categories you are comparing and ask if they make different predictions. How do the slopes compare? Is one group better off? Is that group better off no matter how many years of education they have? Where is the disparity most dramatic? What do the correlation coefficients tell you about the strength of these relationships? Are there outliers that may be affecting the regression line or the cc?

Discussion/Analysis

Were your original conjectures supported by your findings? What additional evidence could be used to support your analysis? Are your findings surprising in any way? If so, why? What factors might be hidden or not taken into account? What questions are raised by your analysis?

    You may wish to continue researching questions raised by your analysis by generating more regression lines, returning to the original FAM1000 data, or examining additional sources such as the related readings at *www.wiley.com/college/kimeclark* or on your class WileyPLUS site.

CHAPTER 3

WHEN LINES MEET: LINEAR SYSTEMS

OVERVIEW

When is solar heating cheaper than conventional heating? Will you pay more tax under a flat tax or a graduated tax plan? We can answer such questions using a system of linear equations or inequalities.

After reading this chapter, you should be able to

- interpret intersection points on graphs
- construct, graph, and interpret:
    - systems of linear equations
    - systems of linear inequalities
    - systems with piecewise linear functions
- use graphs and equations to find a solution for:
    - a system of two linear equations
    - a system of two linear inequalities
    - a system with piecewise linear functions

3.1 *Interpreting Intersection Points: Linear and Nonlinear Systems*

While the primary focus of this chapter is on linear systems, in this section we examine the intersection points of both linear and nonlinear systems. A system is a collection of two or more equations or graphs involving the same variables.

When Curves Collide: Nonlinear Systems

Figure 3.1 shows the percent of all college degrees conferred in the United States for males and for females from 1971 to 2007 with projections to 2017. The point of intersection indicates where the percent of college degrees conferred is the *same* for men and for women. We can estimate the coordinates of the point, P, where the graphs for males and females intersect.

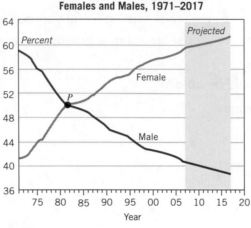

Percent of All College Degrees Conferred
Females and Males, 1971–2017

Coordinates of Point of Intersection

$$P = (\text{year, percent})$$
$$\approx (1981, 50)$$

In 1981, males and females each received about 50%, or half, of all the degrees conferred.

Figure 3.1
Source: Dept. of Education.

In Figure 3.2, the dotted lines from point P to both the x- and y-axes verify that our estimate is fairly accurate.

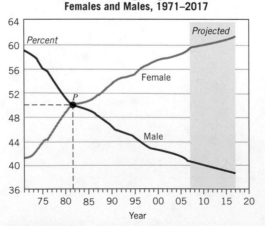

Percent of All College Degrees Conferred
Females and Males, 1971–2017

To the right of the intersection point, P, the percent of degrees conferred for females was greater than the percent for males.

To the left of the intersection point, P, the percent of degrees conferred for females was less than the percent for males.

Figure 3.2

To summarize, the percent of females receiving college degrees has been steadily increasing and in 1981 was roughly equal with the percent of males receiving degrees. By 2017, the percent of females is predicted to be about 62% vs. 38% for males.

EXAMPLE 1   **Global aging**

Figure 3.3 shows how the population of the world is aging. Predictions are shown with dotted lines that extend to 2050.

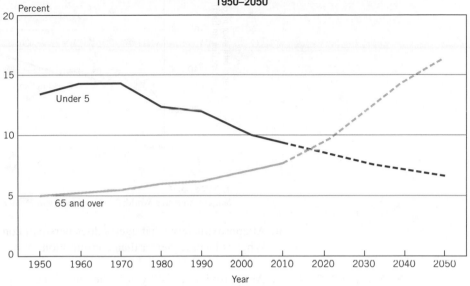

Figure 3.3

Source: An Aging World: 2008. *International Population Reports.* June 2009.

a. Estimate where the graph for young children (under 5 years) intersects with the graph for older people (65 years and older). Interpret the meaning of the intersection point.

b. Describe what happens to the percent of young children in relation to the percent of older people to the left and to the right of the intersection point.

c. Is the population under 5 years of age increasing?

SOLUTION   a. The coordinates of the point where the graphs intersect are approximately (2017, 8), where the units are years and percent, respectively. This means that around the year 2017 it is predicted that the percent of children (under 5 years) and the percent of older people (65 years and older) will be the same, each at approximately 8% of the total world population.

b. To the left of the intersection point, the percent of children in the world is greater than the percent of older people, while to right of the intersection point, the percent of children is less than the percent of older people. This means that although the world's population is aging, as of 2010, children still outnumber older people. Projections indicate that after 2017, older people will outnumber children for the first time since 1950 according to the graph and for the first time in history according to the National Institute on Aging.

c. The graph does not give the *size* of the total population, so we can't tell if the number of children under 5 years is increasing or decreasing.

EXAMPLE 2   **Challenges for aging societies**

A major challenge for aging societies is to provide for the needs of their older populations. Figure 3.4 on the next page shows the annual per capita income of a typical Thai worker and total annual per capita consumption in baht (unit of currency in Thailand) over the worker's life cycle (in years).

a. At what age is the annual per capita consumption approximately 40,000 baht?

b. Estimate the maximum per capita income and at what age it occurs.

c. Identify an interval in which the rate of change for consumption is approximately zero. Interpret your result.

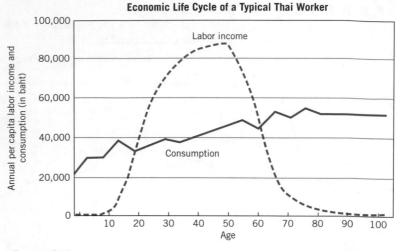

Figure 3.4
Source: An Aging World: 2008. *International Population Reports.* June 2009.

d. At approximately what age(s) does personal consumption equal personal income? When is income greater than consumption? When is income less than consumption?

SOLUTION

a. At approximately 40 years, the annual per capita consumption is approximately 40,000 baht.

b. The maximum per capita labor income of approximately 90,000 baht occurs at around 50 years of age.

c. From approximately 5 to 9 years and from 80 to 100 years, the rate of change for consumption is approximately zero, since the consumption graph is nearly horizontal on these intervals. This means the amount of consumption is almost constant.

d. At approximately 20 years and 61 years, per capita consumption equals per capita income. Per capita income is greater than per capita consumption from approximately 20 years to 61 years. Consumption is greater than income at less than 20 years and more than 61 years.

EXPLORE & EXTEND

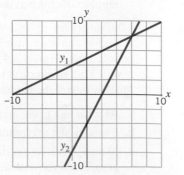

Using Equations to Find the Intersection Point
This *Explore & Extend* previews the discussion in the next section.

For linear systems we often know or can find the equations of the lines in the system. Find the equations for y_1 and y_2 in the Figure below.

Use the graph to estimate the coordinates of the intersection point for y_1 and y_2.
Use your equations to find the coordinates of the point of intersection.
Check to see if the coordinates of the point of intersection satisfy both equations.

When Lines Meet: Linear Systems

Using graphs, we found and interpreted the intersection points for nonlinear systems. We now focus on linear systems using both graphs and equations. We can manipulate linear equations to find the coordinates of the intersection points and, in some cases, find more accurate solutions than we can from graphs.

An economic comparison of solar vs. conventional heating systems

On a planet with limited fuel resources, heating decisions involve both monetary and ecological considerations. Typical costs for three different kinds of heating systems for a small one-bedroom housing unit are given in Table 3.1.

Typical Costs for Three Heating Systems

Type of System	Installation Cost ($)	Operating Cost ($/yr)
Electric	5,000	1,100
Gas	12,000	700
Solar	30,000	150

Table 3.1

Solar heating is clearly the most costly to install and the least expensive to run. Electric heating, conversely, is the cheapest to install and the most expensive to run.

Setting Up a System. By converting the information in Table 3.1 into equations, we can find out when the solar heating system begins to pay back its initially higher cost. If no allowance is made for inflation or changes in fuel price,[1] the general equation for the total cost, C, is

$$C = \text{installation cost} + (\text{annual operating cost})(\text{years of operation})$$

If we let n equal the number of years of operation and use the data from Table 3.1, we can construct the following linear equations:

$$C_{\text{electric}} = 5000 + 1100n$$
$$C_{\text{gas}} = 12{,}000 + 700n$$
$$C_{\text{solar}} = 30{,}000 + 150n$$

Together they form a *system of linear equations*, which is a set of two or more linear equations. Table 3.2 gives the cost data at 5-year intervals, and Figure 3.5 shows the costs over a 40-year period for the three heating systems.

Heating System Total Costs

Year	Electric ($)	Gas ($)	Solar ($)
0	5,000	12,000	30,000
5	10,500	15,500	30,750
10	16,000	19,000	31,500
15	21,500	22,500	32,250
20	27,000	26,000	33,000
25	32,500	29,500	33,750
30	38,000	33,000	34,500
35	43,500	36,500	35,250
40	49,000	40,000	36,000

Table 3.2

[1] A more sophisticated model might include many other factors, such as interest, repair costs, the cost of depleting fuel resources, risks of generating nuclear power, and what economists call opportunity costs.

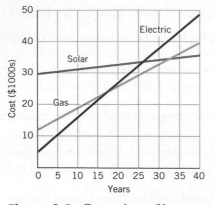

Figure 3.5 Comparison of home heating costs.

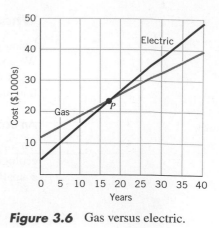

Figure 3.6 Gas versus electric.

Comparing Costs

Using Graphs. To compare the costs for gas and electric heat, we use the graphs of the equations for these two heating systems in Figure 3.6. The point of intersection, P, shows where the lines predict the *same* total cost for both gas and electricity, given a certain number of years of operation. From the graph, we can estimate the coordinates of the point P:

$$P = (\text{number of years of operation, cost})$$
$$\approx (17, \$24,000)$$

The total cost of operation is about $24,000 for both gas and electric after about 17 years of operation. We can compare the relative costs of each system to the left and right of the point of intersection. Figure 3.6 shows that gas is less expensive than electricity to the right of the intersection point and more expensive than electricity to the left of the intersection point.

Using Equations. Where the gas and electric lines intersect, the coordinates satisfy both equations, so both equations have the same value for the input and the same value for the output. At the point of intersection, the total cost of electric heat, C_{electric}, equals the total cost of gas heat, C_{gas}. We can set the cost equations equal to each other to find the exact values for the coordinates of the intersection point:

$$C_{\text{electric}} = 5000 + 1100n \qquad (1)$$
$$C_{\text{gas}} = 12,000 + 700n \qquad (2)$$

Set (1) equal to (2)	$C_{\text{electric}} = C_{\text{gas}}$
substitute	$5000 + 1100n = 12,000 + 700n$
subtract 5000 from each side	$1100n = 7000 + 700n$
subtract 700n from each side	$400n = 7000$
divide each side by 400	$n = 17.5 \text{ years}$

When $n = 17.5$ years, the total cost for electric or gas heating is the same. To find the total cost, substitute 17.5 for n into Equation (1) or (2):

Substitute 17.5 for n in Equation (1) $C_{\text{electric}} = 5000 + 1100(17.5)$

$$= 5000 + 19,250$$

$$C_{\text{electric}} = \$24,250$$

Since we claim that the pair of values (17.5, $24,250) satisfies both equations, we need to check, when $n = 17.5$ years, that C_{gas} is also $24,250:

Substitute 17.5 for n in Equation (2)

$$C_{gas} = 12,000 + 700(17.5)$$
$$= 12,000 + 12,250$$
$$C_{gas} = \$24,250$$

The coordinates (17.5, $24,250) satisfy both equations.

After 17.5 years, a total of $24,250 could have been spent on heat for either an electric or a gas heating system, close to our estimated values.

Solutions to a System. When $n = 17.5$ years, then $C_{electric} = C_{gas} = \$24,250$. The point (17.5, $24,250) is a solution to both equations and is called a *solution to the system* of these two equations.

A Solution to a System of Equations

A pair of real numbers is a *solution* to a system of equations in two variables if and only if the pair of numbers is a solution to each equation in the system.

We started with a system of three linear equations in two variables (cost and years), as shown in Figure 3.5. This system has three intersection points but no intersection point for all three equations; so, there is no solution for the system of all three equations. Figure 3.6 represents a system of two linear equations, and the intersection point is a solution to this system of two equations.

EXAMPLE 3

When solar becomes the cheapest option
Using the graphs in Figures 3.7 and 3.8, estimate when the cost will be the same for each of the following systems:

a. Electric vs. solar heating

b. Gas vs. solar heating

You are asked to find more accurate solutions in the exercises.

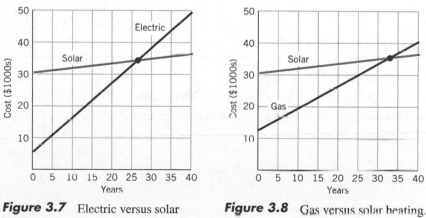

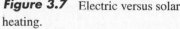

Figure 3.7 Electric versus solar heating.

Figure 3.8 Gas versus solar heating.

SOLUTION

a. The graphs of electric and solar costs intersect at approximately (26, $34,000). That means the costs of electric and solar heating are both approximately $34,000 at about 26 years of operation. Solar becomes less expensive than electric after 26 years.

b. Gas and solar heating costs are equivalent at approximately (33, $35,000), or in 33 years at a cost of $35,000. Solar becomes less expensive than gas after 33 years.

EXPLORE & EXTEND

3.1b

Predicting Sales

Over the past decade, warehouse club stores such as Sam's Club and Costco have been gaining ground on their grocery store counterparts. Consider what the following regression equations suggest about the future of these grocery options.

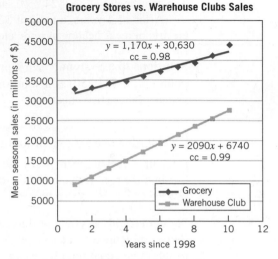

Grocery Stores vs. Warehouse Clubs Sales

$y = 1,170x + 30,630$
$cc = 0.98$

$y = 2090x + 6740$
$cc = 0.99$

Mean seasonal sales (in millions of $)

Years since 1998

— Grocery
— Warehouse Club

Source: http://www.economagic.com/cenret.htm.

a. What do the slopes of the regression equations predict about the sales growth for grocery stores? For warehouse clubs?

b. If this pattern continues, what is likely to happen?

c. Use the system of linear equations:

$$\text{grocery store sales} = 1,170x + 30,630$$
$$\text{warehouse club sales} = 2090x + 6,740$$

to estimate when sales in grocery stores are likely to equal sales in warehouse clubs.

d. Are there any risks to extrapolating our data so far into the future?

Algebra Aerobics 3.1

1. For each of the following graphs:
   a. Estimate the coordinates of the point of intersection.
   b. Estimate the interval(s) for x when $y_1 > y_2$ and when $y_1 < y_2$.

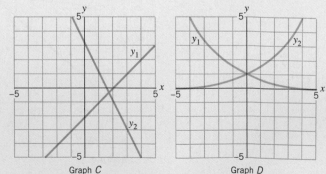

Graph C

Graph D

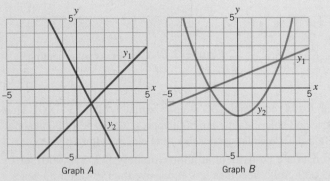

Graph A

Graph B

2. The following graphs show the oil production and the oil imports in the United States between 1920 and 2005. Estimate the coordinates of the point of intersection of the two graphs. What information does this point give us? During what years did the United States produce more oil than it imported? During what years did the United States import more oil than it produced?

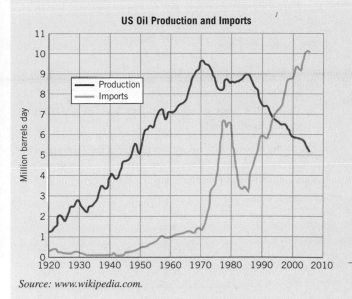

Source: *www.wikipedia.com.*

3. Our model tells us that electric systems are cheaper than gas for the first 17.5 years of operation. Using Figure 3.5, estimate the interval(s) over which gas is the cheapest of the three heating systems.

4. For the system of equations

$$4x + 3y = 9$$
$$5x + 2y = 13$$

   a. Determine whether $(3, -1)$ is a solution.
   b. Show why $(1, 4)$ is not a solution for this system.

5. Estimate the solution(s) for each of the systems of equations in the graphs below.

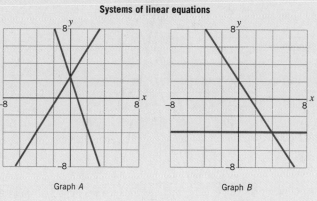

Exercises for Section 3.1

1. a. Estimate the coordinates of the point of intersection on the accompanying graph.
   b. Describe what happens to the population of Pittsburgh in relation to the population of Las Vegas to the right and to the left of this intersection point.

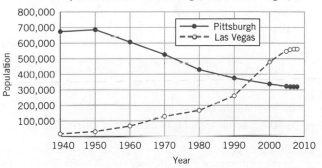

2. a. Determine whether $(-2, 3)$ is a solution for the following system of equations:

$$3x + y = -3$$
$$x + 2y = 4$$

   b. Explain why $(3, -2)$ is not a solution for the system in part (a).

3. a. Estimate the solution to the system of linear equations graphed in the accompanying figure.

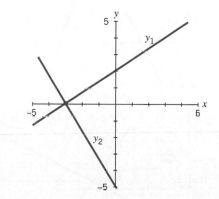

   b. Construct equations for y_1 and y_2. Use your equations to find a solution for this system. How does your answer compare to your estimate for part (a)?

4. Determine which of the given ordered pairs is a solution of the system of equations:

   a. $(-1, 3)$ or $(0, 1)$ for $2x + y = 1$ and $-3x + 2y = 9$
   b. $(6, 0)$ or $(3, 2)$ for $3x - 4y = 1$ and $2x + 3y = 12$

5. a. Find the equation of a line with slope 3 that goes through the point (0, 17).

b. Find the equation of a line that goes through the points (0, 1) and (5, 0).

c. Graph both lines and find the point that lies on both lines.

6. Examine the graphs of three systems below.

a. What do the three systems have in common?

b. Which of the graphs is the correct set of graphs for the following system of equations:

$$y = 2x + 2 \quad \text{and} \quad y = -3x + 12$$

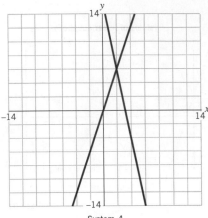

System A

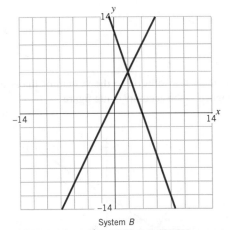

System B

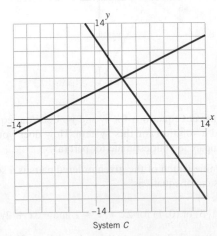

System C

7. Construct a sketch of each system by hand and then estimate the solution(s) to the system (if any).

a. $x + 2y = 1$
   $x + 4y = 3$

b. $x + y = 9$
   $2x - 3y = -2$

8. For the linear system $\begin{aligned} x - y &= 5 \\ 2x + y &= 1 \end{aligned}$

a. Graph the system. Estimate the solution for the system and then find the exact solution.

b. Check that your solution satisfies both of the original equations.

9. Create the system of equations that produced the accompanying graph. Estimate the solution for the system from the graph and then confirm using your equations.

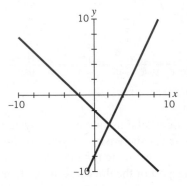

10. The U.S. Bureau of Labor Statistics reports on the percent of all men in the civilian work force, and the corresponding percent of all women. Lines of best-fit on the accompanying graph are used to make predictions about the future of the labor force.

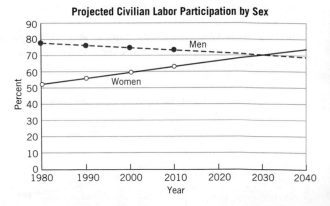

Projected Civilian Labor Participation by Sex

a. Estimate the point of intersection.

b. Describe the meaning of the point of intersection.

11. The figure on the next page shows graphs of population growth in selected areas of the world over the 500 years from 1500 to 2000. Use these graphs to answer the following questions.

a. Estimate the decade when the populations of Africa and of Latin America both surpassed the population of Western Europe.

b. Describe what is happening to the populations of China and India between 1600 and 1700. What happens to the population of China in relation to that of India after 1725?

11. (continued)

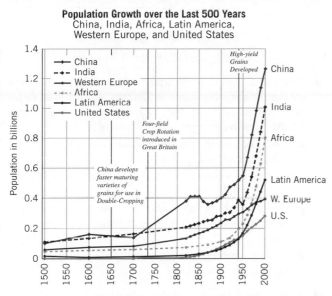

Population Growth over the Last 500 Years
China, India, Africa, Latin America,
Western Europe, and United States

Source: Data from Angus Maddison, University of Groningen. Graph from VisualizingEconomics.com.

12. The accompanying figure shows the heights of two balloons at t hours.[2]

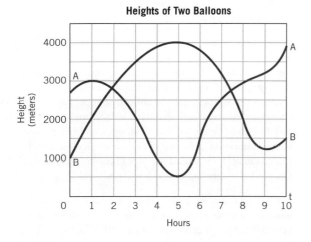

Heights of Two Balloons

a. What is the vertical distance between the balloons at 8 hours?

b. When is balloon A 2000 meters above balloon B?

c. What is the maximum vertical distance between the balloons?

d. At approximately what hour(s) will the heights of the two balloons be the same?

e. Describe the behavior of the balloons to the left and right of each of the points of intersection.

13. The graph at top of the next column shows changes in how health care spending was financed from 1970 to 2009.

²Problem adapted from Swenson, Carl. *The Precalculus Source Book.* Compiled by the Washington State Precalculus Revitalization Project.

a. Estimate when the percentage paid for health care by individuals was the same as the percentage paid by private insurance.

b. Compare the percentage of health care spending that is paid by individuals to the percentage paid by private insurance to the left and right of the intersection point.

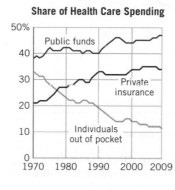

Share of Health Care Spending

By the numbers

$2.38 trillion
Estimated amount spent in the U.S. on health care in 2008

$239 billion
Effect of House health-care bill on U.S. deficit by 2019*

$235.4 billion
Estimated amount spent in the U.S. on prescription drugs in 2008

45.7 million
Number of uninsured people in the U.S. in 2007

15.3%
Portion of the U.S. population that was uninsured in 2007

*Congressional Budget Office Estimate

Sources: Centers for Medicare and Medicaid Services, Census Bureau

14. The accompanying graphs show net debt as a percentage of gross domestic product (GDP) for Belgium, Italy, Japan, and the United States from 1991 to 2009, with projections for the United States to 2015.

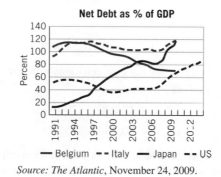

Net Debt as % of GDP

— Belgium -- Italy — Japan -- US

Source: The Atlantic, November 24, 2009.

a. Estimate where the graph for Belgium intersects with the graph for Japan. Interpret the meaning of the intersection point.

b. Describe changes in debt over time for Belgium as compared to changes for Japan.

c. Estimate where the graph for the United States intersects with the graph for Japan. Interpret the meaning of the intersection point.

d. Describe changes in debt over time for the United States as compared to changes in debt for Japan.

15. Two companies offer starting employees incentives to stay with the company after they are trained for their new jobs. Company A offers an initial hourly wage of $7.25, then increases the hourly wage by $0.15 per month. Company B offers an initial hourly wage of $7.70, then increases the hourly wage by $0.10 per month.

15. (continued)

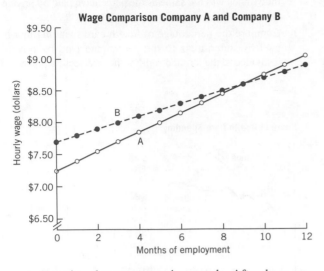

Wage Comparison Company A and Company B

a. Examine the accompanying graph. After how many months does it appear that the hourly wage will be the same for both companies?

b. Estimate that hourly wage.

c. Form two linear functions for the hourly wages in dollars of $W_A(m)$ for company A and $W_B(m)$ for company B after m months of employment.

d. Does your estimated solution from part (a) satisfy both equations? If not, find the exact solution.

e. What is the exact hourly wage when the two companies offer the same wage?

f. Describe the circumstances under which you would rather work for company A and for company B.

16. In the text the following cost equations were given for gas and solar heating:

$$C_{gas} = 12,000 + 700n$$
$$C_{solar} = 30,000 + 150n$$

where n represents the number of years since installation and the cost represents the total accumulated costs up to and including year n.

a. Sketch the graph of this system of equations.

b. What do the coefficients 700 and 150 represent on the graph, and what do they represent in terms of heating costs?

c. What do the constant terms 12,000 and 30,000 represent on the graph? What does the difference between 12,000 and 30,000 say about the costs of gas vs. solar heating?

d. Label the point on the graph where the total accumulated gas and solar heating costs are equal. Make a visual estimate of the coordinates, and interpret what the coordinates mean in terms of heating costs.

e. Use the equations to find a better estimate for the intersection point. To simplify the computations, you may want to round values to two decimal places. Show your work.

f. When is the total cost of solar heating more expensive than gas? When is the total cost of gas heating more expensive than solar?

17. Answer the questions in Exercise 16 (with suitable changes in wording and values) for the following cost equations for electric and solar heating:

$$C_{electric} = 5000 + 1100n$$
$$C_{solar} = 30,000 + 150n$$

18. Consider the following job offers. At Acme Corporation, you are offered a starting salary of $20,000 per year, with raises of $2500 annually. At Boca Corporation, you are offered $25,000 to start and raises of $2000 annually.

a. Find an equation to represent your salary, $S_A(n)$, after n years of employment with Acme.

b. Find an equation to represent your salary, $S_B(n)$, after n years of employment with Boca.

c. Create a table of values showing your salary at each of these corporations for integer values of n up to 12 years.

d. In what year of employment would the two corporations pay you the same salary?

19. a. Solve the following system algebraically:

$$S = 20,000 + 2500n$$
$$S = 25,000 + 2000n$$

b. Graph the system in part (a) and use the graph to estimate the solution to the system. Check your estimate with your answer in part (a).

20. While solar energy powered home systems are quite expensive to install, adding some passive energy features to conventional construction can make a substantial reduction in heating and cooling costs. Passive solar requires no machinery and uses the building itself to collect, store and distribute solar energy. An excellent passive solar investment is the installation of better quality windows with high resistance to heat transmission, provided with exterior shading on the south side to cut off high angled sun rays in cooling season and let in low angled winter sun during heating season.

For a typical conventional housing unit improved with better windows and south shading we can see the effect on heating, ventilating and air conditioning costs (HVAC):

Cost of conventional HVAC system	$11,500
Annual cost for HVAC in conventional house	$1,240
Extra installation cost for better windows and shading	$2,700
Annual cost for HVAC in improved house	$860

a. Create equations showing total accumulated HVAC cost $C_{conventional}$ over n years for the conventional house, and total cost $C_{improved}$ over n years for the improved house.

Generate data from your equations and graph them on the same axes.

b. After how many years are the total accumulated HVAC costs the same? When do the passive solar improvements result in smaller total cost?

21. In a previous exercise in Section 2.7, you read of a math professor who purchased her condominium in Cambridge, MA, for $70,000 in 1977. Its assessed value has climbed at a steady rate so that it was worth $850,000 as of 2007. Alas, one of her colleagues has not been so fortunate. He bought a house in that same year for $160,000. Not long after his family moved in, rumors began to circulate that the housing complex had been built on the site of a former toxic dump. Although never substantiated, the rumors adversely affected the value of his home, which has steadily decreased in value over the years and in 2007 was worth a meager $40,000. In what year would the two homes have been assessed at the same value?

3.2   *Visualizing and Solving Linear Systems*

In this section we examine more strategies for solving a system of linear equations graphically and algebraically.

Visualizing Linear Systems

When a system consists of two linear functions using the same variables, there are three possible ways the graphs of the equations can relate to each other, as shown in Figure 3.9. There may be one point of intersection giving one solution, no point of intersection if the lines are parallel giving no solution, or an infinite number of points if the two equations represent the same line giving an infinite number of solutions.

One Solution: Two lines intersect at a single point

No Solution: Parallel lines never intersect

Infinitely Many Solutions: Two equations representing the same line

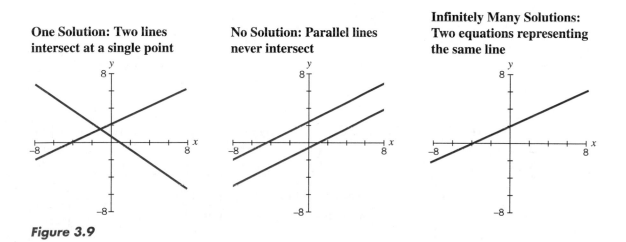

Figure 3.9

The Number of Solutions for a Linear System with Two Equations

On the graph of a system of two linear equations, a *solution* is a point where the two lines intersect. There can be

One solution, if the lines intersect once

No solution, if the lines are parallel and distinct

Infinitely many solutions, if the two lines are the same

Strategies for Solving Linear Systems

The program "L1: m + b sliders" can be used to create systems of linear equations.

A system of two linear equations can be solved in several ways. The form of the equations can help determine the most effective strategy. Two of the most common methods are *substitution* and *elimination*.

Substitution method

Using the substitution method with a system of two equations, we *substitute* the expression for one output from one equation into other equation. In Section 3.1, we used substitution to solve the following system of equations:

$$C_{\text{electric}} = 5{,}000 + 1{,}110n$$
$$C_{\text{gas}} = 12{,}000 + 700n$$

In this case both equations were in function form, and we set them equal to each other to find the coordinates of the point of intersection. We can also say we substituted the formula for C_{electric} for C_{gas} or vice versa.

What if the equations are not in function form?

EXAMPLE 1 **Solving a system when only one equation is in function form**
a. Find the point (if any) where the graphs of the following two linear equations intersect:

$$6x + 7y = 25 \tag{1}$$
$$y = 15 + 2x \tag{2}$$

b. Graph the two equations on the same grid, labeling their intersection point.

SOLUTION **a.** In Equation (1) substitute the expression for y from Equation (2):

$$6x + 7(\mathbf{15 + 2x}) = 25$$

multiply $\qquad\qquad 6x + 105 + 14x = 25$

simplify $\qquad\qquad\qquad\qquad 20x = -80$

$$x = -4$$

We can use one of the original equations to find the value for y when $x = -4$. Using Equation (2),

$$y = 15 + 2x \tag{2}$$

substitute -4 for $x \qquad y = 15 + 2(-4)$

multiply $\qquad\qquad\quad y = 15 - 8$

$$y = 7$$

Try double-checking your answer in Equation (1).

b. Figure 3.10 shows a graph of the two equations and the intersection point at $(-4, 7)$.

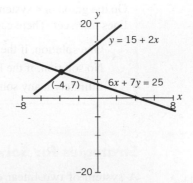

Figure 3.10 Graphs of $6x + 7y = 25$ and $y = 15 + 2x$.

Algebra Aerobics 3.2a

1. Solve for the indicated variable.
   a. $2x + y = 7$ for y
   b. $3x + 5y = 6$ for y
   c. $x - 2y = -1$ for x
2. Determine the number of solutions without solving the system. Justify your answer.
   a. $y = 3x - 5$ b. $y = 2x - 4$
   $y = 3x + 8$ $y = 3x - 4$
3. Solve the following systems of equations using the substitution method.
   a. $y = x + 4$
   $y = -2x + 7$
   b. $y = -1700 + 2100x$
   $y = 4700 + 1300x$
   c. $F = C$
   $F = 32 + \frac{9}{5}C$
   [Part (c) was a question on the TV program *Who Wants to Be a Millionaire?*]

4. Solve the following systems of equations using the substitution method.
   a. $y = x + 3$ c. $x = 2y - 5$
   $5y - 2x = 21$ $4y - 3x = 9$
   b. $z = 3w + 1$ d. $r - 2s = 5$
   $9w + 4z = 11$ $3r - 10s = 13$
5. Consider the system $2x + 3y = 9$ and $2x + y = 3$. Solve each equation for y and explain how you could find the point of intersection by inspection.

Elimination method

Another method, called *elimination*, can be useful when neither equation is in function form. The strategy is to modify the equations (through multiplication or rearrangement) so that adding (or subtracting) the modified equations eliminates one variable.

EXAMPLE 2

Solving a system using the elimination method

Assume you have $10,000 to invest in an "up market" when the economy is booming. You want to split your investment between conservative bonds and riskier stocks. The bonds will stay fixed in value but return a guaranteed 7% per year in dividends. The stocks pay no dividends, but your return is from the increase in stock value, predicted to be 14% per year. Overall you want a 12% or $1200 return on your $10,000 at the end of one year.

a. How much should you invest in bonds and how much in stocks?

b. What if the economy has a drastic downturn, as it did between 2007 and 2010? Assuming you split your investment as recommended in part (a), what will your return be after one year if your stock value decreased by 10%?

SOLUTION

a. If $B = $ $ invested in bonds and $S = $ $ invested in stocks, then

$$B + S = \$10,000 \tag{1}$$

The expected return on your investments after one year is

$$(7\% \text{ of } B) + (14\% \text{ of } S) = \$1200$$

or

$$0.07B + 0.14S = \$1200 \tag{2}$$

We can solve the system by eliminating one variable, in this case B, from both equations. Given the two equations

$$B + S = \$10,000 \tag{1}$$
$$0.07B + 0.14S = \$1200 \tag{2}$$

If we multiply both sides of Equation (1) by 0.07 we get an equivalent Equation (1)*, which has the same coefficient for B as Equation (2). We do this so that we can subtract the equations and eliminate B.

Given $0.07B + 0.07S = \$700$ (1)*

subtract Equation (2) $-(0.07B + 0.14S = \$1200)$ (2)

to eliminate B $-0.07S = -\$500$

Dividing both sides by -0.07, we have $S \approx \$7143$.

So to achieve your goal of a $1200 return on $10,000 you would need to invest $7143 in stocks and $10,000 − $7143 = $2857 in bonds.

b. If the economy turns sour at the end of the year and the value of your stock drops 10%, then

your return = (gain of 7% of $2857 from bonds) and (loss of 10% of $7143 from stocks)

$$\approx \ \$200 - \$714$$
$$= -\$514$$

So you would lose over $500 on your $10,000 investment that year.

Systems with No Solution or Infinitely Many Solutions

At the beginning of Section 3.2, we described three possible cases for the solutions to a linear system of two equations: one solution when lines intersect, no solution when lines are parallel, and infinitely many solutions when the equations are of the same line. How can we tell whether or not the system of equations falls into one of these categories?

EXAMPLE 3   A system with no solution: Parallel lines

a. Solve the following system of two linear equations:

$$y = 20,000 + 1500x \tag{1}$$
$$2y - 3000x = 50,000 \tag{2}$$

b. Graph your results.

SOLUTION   **a.** Since the first equation is in function form, we can use the substitution method. Substitute the expression for y from Equation (1) into Equation (2):

substitute $2(20,000 + 1500x) - 3000x = 50,000$

simplify $40,000 + 3000x - 3000x = 50,000$

false statement $40,000 = 50,000 \ (???)$

What could this possibly mean? Where did we go wrong? If we return to the original set of equations and solve Equation (2) for y in terms of x, we can see why there is no solution for this system of equations.

given $2y - 3000x = 50,000$ $\qquad$ (2)

add 3000x to both sides $2y = 50,000 + 3000x$

divide by 2 $y = 25,000 + 1500x$ $\qquad$ (2)*

Equations (1) and (2)* (rewritten form of the original Equation (2)) have the same slope of 1500 but different y-intercepts.

$$y = 20,000 + 1500x \tag{1}$$
$$y = 25,000 + 1500x \tag{2)*}$$

Written in this form, we can see that we have two parallel lines, so the lines never intersect (see Figure 3.11). Our initial premise, that the two lines intersected and so the two y-values were equal at some point, was incorrect.

b. Figure 3.11 shows the graphs of the two lines.

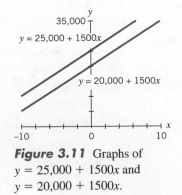

Figure 3.11 Graphs of $y = 25,000 + 1500x$ and $y = 20,000 + 1500x$.

EXAMPLE 4   **A system with infinitely many solutions: Equivalent equations**
Solve the following system:

$$45x = -y + 33 \tag{1}$$
$$2y + 90x = 66 \tag{2}$$

SOLUTION   As always, there are multiple ways of solving the system. One strategy is to put both equations in function form:

Solve Equation (1) for y	$45x = -y + 33$	(1)
add y to both sides	$y + 45x = 33$	
add $-45x$ to both sides	$y = -45x + 33$	

Solve Equation (2) for y	$2y + 90x = 66$	(2)
add $-90x$ to both sides	$2y = -90x + 66$	
divide by 2	$y = -45x + 33$	

The two original equations really represent the same line, $y = -45x + 33$, so any of the infinitely many points on the line is a solution to the system (Figure 3.12).

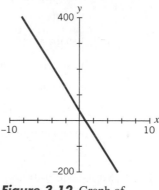

Figure 3.12 Graph of
$y = -45x + 33$.

Linear Systems in Economics: Supply and Demand

Economists study the relationship between the price p of an item and the quantity q, the number of items produced. Economists traditionally place quantity q on the horizontal axis and price p on the vertical axis.[3] From the consumer's point of view, an increase in price decreases the quantity demanded. So the consumer's *demand curve* would slope downward (see Figure 3.13).

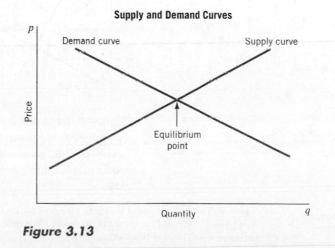

Figure 3.13

[3]This can be confusing since we usually think of quantity as a function of price.

From the manufacturer's (or supplier's) point of view, an increase in price motivates them to increase the quantity they supply. So the manufacturer's *supply curve* slopes upward. The intersection point between the demand and supply curves is called the *equilibrium point*. At this point supply equals demand, so both suppliers and consumers are happy with the quantity produced and the price charged.

E X A M P L E   5

Milk supply and demand curves

Loren Tauer[4] studied the U.S. supply and demand curves for milk. If q = billions of pounds of milk and p = dollars per cwt (where 1 cwt or "hundred weight" = 100 lb), he estimated that the demand function for milk is $p = 55.9867 - 0.2882q$ and the supply function is $p = 0.0865q$.

a. Find the equilibrium point.

b. What will happen if the price of milk is higher than the equilibrium price?

S O L U T I O N

a. The equilibrium point occurs where

$$\text{supply} = \text{demand}$$

substituting for p $\qquad 0.0865q = 55.9867 - 0.2882q$

solving for q $\qquad 0.3747q = 55.9867$

we have $\qquad\qquad q \approx 149.42$ billions of pounds of milk

If we use the supply function to find p, we have

the supply function $\qquad p = 0.0865q$

so when $q = 149.42$ $\qquad p = 0.0865 \cdot 149.42 \approx \$12.92$ per cwt

If we use the demand function to find p we would also get

the demand function $\qquad p = 55.9867 - 0.2882q$

so when $q = 149.42$ $\qquad p = 55.9867 - (0.2882 \cdot 149.42) \approx \$12.92$ per cwt

So the equilibrium point is (149.42, $12.92); that is, when the price is $12.92 per cwt, manufacturers are willing to produce and consumers are willing to buy 149.42 billion pounds of milk.

b. If the price of milk rises above $12.92 per cwt to, say, p_1, then, as shown in Figure 3.14, consumers would buy less than 149.42 billion pounds (amount q_1) while manufacturers would be willing to produce more than 149.42 billion pounds (amount q_2).

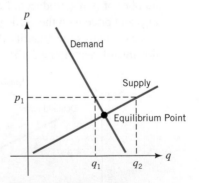

Figure 3.14 At price p_1, consumers would buy q_1 billion pounds of milk, but producers would manufacture q_2 billion pounds.

[4]Loren W. Tauer, "The value of segmenting the milk market into bST-produced and non-bST produced milk," *Agribusiness 10*(1): 3–12 as quoted in Edmond C. Tomastik, *Calculus: Applications and Technology,* 3rd ed. (Belmont, CA: Thomson Brooks/Cole, 2004).

So there would be a surplus of $q_2 - q_1$ billion pounds of milk, which would drive down the price of milk toward the equilibrium point.

EXPLORE & EXTEND

3.2

Constructing Systems under Constraints

Find the missing coefficient of x such that there will be an infinite number of solutions to the system of linear equations:

$$y = 2x + 4$$
$$??x = -2y + 8$$

Is this the only coefficient of x that would work to give an infinite number of solutions to this system? Explain your answer.

Now find a different coefficient for x such that there will be only one solution to this system of linear equations. Are there other coefficients for x that would work to give only one solution to this system? Explain your answer.

Algebra Aerobics 3.2b

1. Solve each system of equations using the method you think is most efficient.

   a. $2y - 5x = -1$
   $3y + 5x = 11$

   c. $t = 3r - 4$
   $4t + 6 = 7r$

   b. $3x + 2y = 16$
   $2x - 3y = -11$

   d. $z = 2000 + 0.4(x - 10,000)$
   $z = 800 + 0.2x$

2. Solve each system of equations. If technology is available, check your answers by graphing each system.

   a. $y = 2x + 4$
   $y = -x + 4$

   c. $y = 1500 + 350x$
   $2y = 700x + 3500$

   b. $5y + 30x = 20$
   $y = -6x + 4$

3. Construct a system of two linear equations in two unknowns that has no solution.

4. Determine the number of solutions without solving the system. Explain your reasoning.

   a. $2x + 5y = 7$
   $3x - 8y = -1$

   c. $2x + 3y = 1$
   $4x + 6y = 2$

   b. $3x + y = 6$
   $6x + 2y = 5$

   d. $3x + y = 8$
   $3x + 2y = 8$

5. Solve each of the following systems of equations.

   a. $y = x + 4$
   $\frac{x}{2} + \frac{y}{3} = 3$

   b. $0.5x + 0.7y = 10$
   $30x + 50y = 1000$

6. A paint dealer has determined that the demand function for interior white paint is $4p + 3q = 240$, where $p =$ dollars/gallon of paint and $q =$ number of gallons.

   a. Find the demand for white paint when the price is $39.00 per gallon.

   b. If consumer demand is for 20 gallons of paint, what price would these consumers be willing to pay?

   c. Sketch the demand function, placing q on the horizontal axis and p on the vertical axis.

   d. The supply function for interior white paint is $p = 0.85q$. Sketch the supply curve on the same graph as the demand curve.

   e. Find the equilibrium point and interpret its meaning.

   f. At a price of $39.00 per gallon of paint, is there a surplus or shortage of supply?

7. Solve each equation for y and then determine which system has no solutions or infinitely many solutions. Explain your answers.

   a. $3y - 2x = -3$ and $-2x = -3y + 6$

   b. $5y + x = 15$ and $10y - 30 = -2x$

Exercises for Section 3.2

1. a. Match each system of linear equations with the graph of the system.

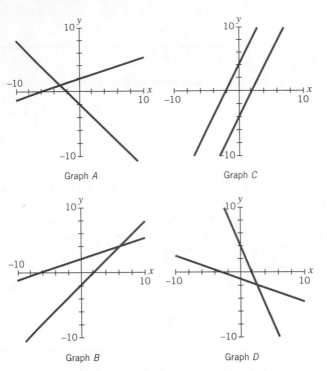

Graph A

Graph C

Graph B

Graph D

  i. $y = -x - 2$ $y = \frac{1}{3}x + 2$
  ii. $y - 2x = 4$ $y - 2x = -4$
  iii. $y - x = -2$ $3y - x = 6$
  iv. $3y + x = -3$ $y = -2x + 4$

  b. Find the point of intersection in each system if there is one.
  c. Verify that the point of intersection satisfies both equations.

2. Predict the number of solutions to each of the following systems. Give reasons for your answer. You don't need to find any actual solutions.

  a. $y = 20{,}000 + 700x$ $y = 15{,}000 + 800x$
  b. $y = 20{,}000 + 700x$ $y = 15{,}000 + 700x$
  c. $y = 20{,}000 + 700x$ $y = 20{,}000 + 800x$

3. For each system:
  a. Indicate whether the substitution or elimination method might be easier for finding a solution to the system of equations.

  i. $y = \frac{1}{3}x + 6$ **iv.** $y = 2x - 3$
  $\quad y = \frac{1}{3}x - 4$ $\quad 4y - 8x = -12$

  ii. $2x - y = 5$ **v.** $-3x + y = 4$
  $\quad 5x + 2y = 8$ $\quad -3x + y = -2$

  iii. $3x + 2y = 2$ **vi.** $3y = 9$
  $\quad x = 7y - 30$ $\quad x + 2y = 11$

  b. Using your chosen method, find the solution(s), if any, of each system.

4. For each graph, construct the equations for each of the two lines in the system, and then solve the system using your equations.

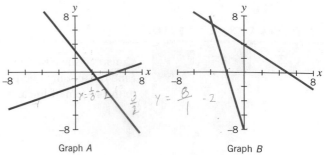

Graph A

Graph B

5. a. Solve the following system algebraically:

$$x + 3y = 6$$
$$5x + 3y = -6$$

  b. Graph the system of equations in part (a) and estimate the solution to the system. Check your estimate with your answer in part (a).

6. Calculate the solution(s), if any, to each of the following systems of equations. Use any method you like.

  a. $y = -1 - 2x$ **c.** $y = 2200x - 700$
  $\quad y = 13 - 2x$ $\quad y = 1300x + 4700$

  b. $t = -3 + 4w$ **d.** $3x = 5y$
  $\quad -12w + 3t + 9 = 0$ $\quad 4y - 3x = -3$

In some of the following examples you may wish to round off your answers:

  e. $y = 2200x - 1800$ **h.** $2x + 3y = 13$
  $\quad y = 1300x - 4700$ $\quad 3x + 5y = 21$

  f. $y = 4.2 - 1.62x$ **i.** $xy = 1$
  $\quad 1.48x - 2y + 4.36 = 0$ $\quad x^2y + 3x = 2$

  g. $4r + 5s = 10$ (A nonlinear system!
  $\quad 2r - 4s = -3$ *Hint:* Solve $xy = 1$ for y
  $\qquad$ and use substitution.)

7. Assume you have $2000 to invest for 1 year. You can make a safe investment that yields 4% interest a year or a risky investment that yields 8% a year. If you want to combine safe and risky investments to make $100 a year, how much of the $2000 should you invest at the 4% interest? How much at the 8% interest? (*Hint:* Set up a system of two equations in two variables, where one equation represents the total amount of money you have to invest and the other equation represents the total amount of money you want to make on your investments.)

8. Two investments in high-technology companies total $1000. If one investment earns 10% annual interest and the other earns 20%, find the amount of each investment if the total interest earned is $140 for the year.

9. Solve the following systems:

  a. $\frac{x}{3} + \frac{y}{2} = 1$ **b.** $\frac{x}{4} + y = 9$
  $\quad x - y = \frac{4}{3}$ $\quad y = \frac{x}{2}$

10. For each of the following systems of equations, describe the graph of the system and determine if there is no solution, an infinite number of solutions, or exactly one solution.

    a. $2x + 5y = -10$ **b.** $3x + 4y = 5$ **c.** $2x - y = 5$
         $y = -0.4x - 2$ $3x - 2y = 5$ $6x - 3y = 4$

11. If $y = b + mx$, solve for values of m and b by constructing two linear equations in m and b for the given sets of ordered pairs.

    a. When $x = 2$, $y = -2$ and when $x = -3$, $y = 13$.
    b. When $x = 10$, $y = 38$ and when $x = 1.5$, $y = -4.5$.

12. The following are formulas predicting future raises for four different groups of union employees. N represents the number of years from the start date of all the contracts. Each equation represents the salary that will be earned after N years.

Group A:	Salary $= 49{,}000 + 1500N$
Group B:	Salary $= 49{,}000 + 1800N$
Group C:	Salary $= 43{,}000 + 1500N$
Group D:	Salary $= 37{,}000 + 2100N$

    a. Will group A ever earn more per year than group B? Explain.
    b. Will group C ever catch up to group A? Explain.
    c. How much total salary would an individual in each group have earned 3 years after the contract?
    d. Will group D ever catch up to group C? If so, after how many years and at what salary?
    e. Which group will be making the highest yearly salary in 5 years? How much will that salary be?

13. The supply and demand equations for a particular bicycle model relate price per bicycle, p (in dollars) and q, the number of units (in thousands). The two equations are

    $$p = 250 + 40q \qquad \text{Supply}$$
    $$p = 510 - 25q \qquad \text{Demand}$$

    a. Sketch both equations on the same graph. On your graph identify the supply equation and the demand equation.
    b. Find the equilibrium point and interpret its meaning.

14. For a certain model of DVD player, the following supply and demand equations relate price per player, p (in dollars) and number of players, q (in thousands).

    $$p = 50 + 2q \qquad \text{Supply}$$
    $$p = 155 - 5q \qquad \text{Demand}$$

    a. Find the point of equilibrium.
    b. Interpret this result.

15. Explain what is meant by "two equivalent equations." Give an example of two equivalent equations.

16. Without graphing each system of equations (at the top of the next column), how can you tell if the graphs of the equations intersect? If they don't intersect, explain why. If they do intersect, what is the intersection point?

a. $y = 5$ and $x = -3$
b. $y = -2$ and $y = 2$
c. $x = 7$ and $y = -2x + 3$
d. $y = \dfrac{2}{3}x + 1$ and $y = -\dfrac{3}{2}x + 1$

17. A restaurant is located on ground that slopes up 1 foot for every 20 horizontal feet. The restaurant is required to build a wheelchair ramp starting from an entry platform that is 3 feet above ground. Current regulations require a wheelchair ramp to rise up 1 foot for every 12 horizontal feet, (See accompanying figure where H = height in feet, d = distance from entry in feet, and the origin is where the H-axis meets the ground.) Where will the new ramp intersect the ground?

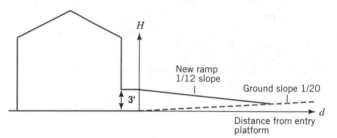

18. A house attic as shown has a roofline with a slope of 5″ up for every 12″ of horizontal run; this is a slope of 5/12. Since the roofline starts at 4′ above the floor, it is not possible to stand in much of the attic space. The owner wants to add a dormer with a 2/12 slope, starting at a 7′ height, to increase the usable space.

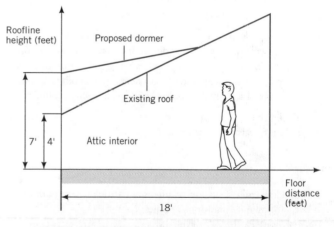

a. With height and floor distance coordinates as shown on the house sketch, find a formula for the original roofline, using R for roof height and F for floor distance from the wall. Also find a formula for the dormer roofline, using D for dormer roof height and F for floor distance from the wall.
b. Make a graph showing the R and D roof height lines for floor distances F from $0'$ to $18'$.
c. At what height and floor distance will the dormer roofline intersect the existing roofline?
d. How far do you need to measure along the horizontal floor distance to give 6′6″ of head room in the original roofline? What percent of the horizontal floor distance of 18 ft allows less than 6′6″ of head room?

19. Solve the following system of three equations in three variables, using the steps outlined below:

$$2x + 3y - z = 11 \qquad (1)$$
$$5x - 2y + 3z = 35 \qquad (2)$$
$$x - 5y + 4z = 18 \qquad (3)$$

   a. Use Equations (1) and (2) to eliminate one variable, creating a new Equation (4) in two variables.

   b. Use Equations (1) and (3) to eliminate the same variable as in part (a). You should end up with a new Equation (5) that has the same variables as Equation (4).

   c. Equations (4) and (5) represent a system of two equations in two variables. Solve the system.

   d. Find the corresponding value for the variable eliminated in part (a).

   e. Check your work by making sure your solution works in all three original equations.

20. Using the strategy described in Exercise 19, solve the following system:

$$2a - 3b + c = 4.5 \qquad (1)$$
$$a - 2b + 2c = 0 \qquad (2)$$
$$3a - b + 2c = 0.5 \qquad (3)$$

21. a. Construct a system of linear equations in two variables that has no solution.

   b. Construct a system of linear equations in two variables that has exactly one solution.

   c. Solve the system of equations you constructed in part (b) by using two different algebraic strategies and by graphing the system of equations. Do your answers all agree?

22. Nenuphar wants to invest a total of $30,000 into two savings accounts, one paying 6% per year in interest and the other paying 9% per year in interest (a more risky investment). If after 1 year she wants the total interest from both accounts to be $2100, how much should she invest in each account?

23. When will the following system of equations have no solution? Justify your answer.

$$y = m_1 x + b_1$$
$$y = m_2 x + b_2$$

24. a. Construct a system of linear equations where both of the following conditions are met:

   The coordinates of the point of intersection are (2, 5).

   One of the lines has a slope of −4 and the other line has a slope of 3.5.

   b. Graph the system of equations you found in part (a). Verify that the coordinates of the point of intersection are the same as the coordinates specified in part (a).

25. Life-and-death travel problems are dealt with by air traffic computers and controllers who are trying to prevent collisions of planes traveling at various speeds in three-dimensional space. To get a taste of what is involved, consider this situation: Airplanes A and B are traveling at the same altitude on the paths shown on the position plot. (See figure at top of next column.)

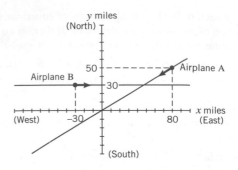

   a. Construct two equations that describe the positions of airplanes A and B in x- and y-coordinates. Use y_A and y_B to denote the north/south coordinates of airplanes A and B, respectively.

   b. What are the coordinates of the intersection of the airplanes' paths?

   c. Airplane A travels at 2 miles/minute and airplane B travels at 6 miles/minute. Clearly, their paths will intersect if they each continue on the same course, but will they arrive at the intersection point at the same time? How far does plane A have to travel to the intersection point? How many minutes will it take to get there? How far does plane B have to travel to the intersection point? How many minutes will it take to get there? (*Hint:* Recall the rule of Pythagoras for finding the hypotenuse of a right triangle: $a^2 + b^2 = c^2$, where c is the hypotenuse and a and b are the other sides.) Is this a safe situation?

26. a. Examine the accompanying figure, where the demand curve has been moved to the right. Does the new demand curve represent an increase or decrease in demand. Why? (*Hint:* Pick an arbitrary price, and see if consumers would want to buy more or less at that price.)

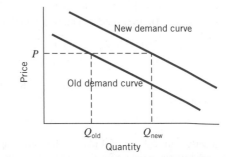

   b. Sketch in a possible supply curve identifying the old and new equilibrium points. What does the shift from the old equilibrium point to the new mean for both consumers and suppliers?

27. a. Examine the accompanying figure at the top of the next page. Does the new supply curve represent an increase or decrease in supply. Why? (Again, try picking an arbitrary price and see if at that price, the supplier would want to increase or decrease production.)

   b. Sketch in a possible demand curve, and label the old and new equilibrium points. What does the shift from the old to the new equilibrium point mean for both consumers and suppliers?

27. (continued)

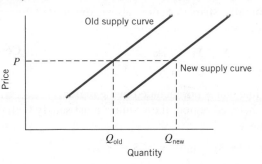

28. In studying populations (human or otherwise), the two primary factors affecting population size are the birth rate and the death rate. There is abundant evidence that, other things being equal, as the population density increases, the birth rate tends to decrease and the death rate tends to increase.[5]

a. Generate a rough sketch showing birth rate as a function of population density. Note that the units for population density on the horizontal axis are the number of individuals for a given area. The units on the vertical axis represent a rate, such as the number of individuals born in a year per 1000 people. Now add to your graph a rough sketch of the relationship between death rate and population density. In both cases assume the relationship is linear.

b. At the intersection point of the two lines the growth of the population is zero. Why? (*Note:* We are ignoring all other factors, such as immigration.)

The intersection point is called the *equilibrium point*. At this point the population is said to have stabilized, and the size of the population that corresponds to this point is called the *equilibrium number*.

c. What happens to the equilibrium point if the overall death rate decreases, that is, at each value for population density the death rate is lower? Sketch a graph showing the birth rate and both the original and the changed death rates. Label the graph carefully. Describe the shift in the equilibrium point.

d. What happens to the equilibrium point if the overall death rate increases? Analyze as in part (c).

29. Use the information in Exercise 28 to answer the following questions:

a. What if the overall birth rate increases (that is, if at each population density level the birth rate is higher)? Sketch a graph showing the death rate and both the original and the changed birth rates. Be sure to label the graph carefully. Describe the shift in the equilibrium point.

b. What happens if the overall birth rate decreases? Analyze as in part (a).

[5]See E. O. Wilson and W. H. Bossert, *A Primer of Population Biology.* Sunderland, MA: Sinauer Associates, 1971, p. 104.

3.3 *Reading between the Lines: Linear Inequalities*

Above and Below the Line

Sometimes we are concerned with values that lie above or below a line, or that lie between two lines. To describe these regions we need some mathematical conventions.

Terminology for describing regions

The two linear functions $y_1 = 1 + 3x$ and $y_2 = 5 - x$ are graphed in Figure 3.15. How would you describe the various striped regions?

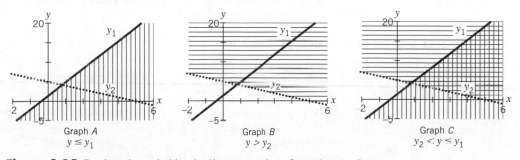

Figure 3.15 Regions bounded by the lines $y_1 = 1 + 3x$ and $y_2 = 5 - x$.

A solid line indicates that the points on the line are included in the area. A dotted line indicates that the points on the line are *not* included in the area. In Graph A the vertical-striped region below the solid line y_1 can be described as all points (x, y) that satisfy the inequality

$$y \leq y_1 \qquad \text{Condition (1)}$$

or equivalently $\qquad y \leq 1 + 3x$

The equation $y_1 = 1 + 3x$ is a *boundary line* that is included in the region.

In Graph B, the horizontally striped region above the dotted line y_2 can be described as all points (x, y) that satisfy the inequality

$$y > y_2 \qquad\qquad \text{Condition (2)}$$
$$\text{or} \qquad y > 5 - x$$

So $y_2 = 5 - x$ is a boundary line that is not included in the region.

In Graph C, in the cross-hatched region, the y values must satisfy both conditions (1) and (2); that is, we must have

$$y \leq 1 + 3x \quad \text{and} \quad y > 5 - x, \text{so}$$
$$5 - x < y \leq 1 + 3x$$
$$\text{or equivalently} \qquad y_2 < y \leq y_1$$

This is called a *compound inequality*. We could describe this inequality by saying that y is greater than $5 - x$ and less than or equal to $1 + 3x$. So the region can be described as all points (x, y) that satisfy the compound inequality

$$5 - x < y \leq 1 + 3x$$

Reading between the Lines

EXAMPLE 1

Healthy weight zones

The U.S. Department of Agriculture recommends healthy weight zones for adults based on their height. For men between 60 and 84 inches tall, the recommended lowest weight, W_{lo} (in lb), is

$$W_{lo} = 105 + 4.0H$$

and the recommended highest weight, W_{hi} (in lb), is

$$W_{hi} = 125.4 + 4.6H$$

where H is the number of inches above 60 inches (5 feet).

a. Graph and label the two boundary equations and indicate the underweight, healthy, and overweight zones.

b. Give a mathematical description of the healthy weight zone for men.

c. Two men each weigh 180 lb. One is 5′ 10″ tall and the other 6′ 1″ tall. Is either within the healthy weight zone?

d. A 5′ 11″ man weighs 135 lb. If he gains 2 lb a week, how long will it take him to reach the healthy range?

SOLUTION **a.** See Figure 3.16.

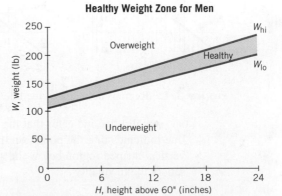

Healthy Weight Zone for Men

Figure 3.16 Graph of men's healthy weight zone between recommended low (W_{lo}) and high (W_{hi}) weights.

b. Assuming that men's heights generally run between 5 feet (60″) and 7 feet (84″), then H (the height in inches above 60 inches) is bounded by

$$0'' \leq H \leq 24''$$

The recommendations say that a man's weight W (in lb) should be more than or equal to (W_{lo}) and should be less than or equal to W_{hi}. So we have the computed inequality

$$W_{lo} \leq W \leq W_{hi}$$

or $\qquad\qquad 105 + 4.0H \leq W \leq 125.4 + 4.6H$

c. For a man who is 5′ 10″ (or 70″) tall, $H = 10''$ and his maximum recommended weight, W_{hi}, is $125.4 + (4.6 \cdot 10) = 171.4$ lb. So if he weighs 180 lb, he is overweight. For a man who is 6′ 1″ (or 73″) tall, $H = 13''$. His minimum recommended weight, W_{lo}, is $105 + (4.0 \cdot 13) = 157$ lb. His maximum recommended weight, W_{hi}, is $125.4 + (4.6 \cdot 13) = 185.2$ lb. So his weight of 180 lb would place him in the healthy zone.

d. If a man is 5′ 11″, his recommended minimum weight, W_{lo} is $105 + (4.0 \cdot 11) = 149$ lb. If he currently weighs 135 lb, he would need to gain at least $149 - 135 = 14$ lb. If he gained 2 lb a week, it would take him 7 weeks to reach the minimum recommended weight of 149 lb.

EXAMPLE 2

Army sleeping bags

The U.S. Army recommends that sleeping bags, which will be used in temperatures between −40° and +40° Fahrenheit, have a thickness of 2.5 inches minus 0.025 times the number of degrees Fahrenheit.

a. Construct a linear equation that models the recommended sleeping bag thickness as a function of degrees Fahrenheit. Identify the variables and domain of your function.

b. Graph the model (displaying it over its full domain) and shade in the area where the sleeping bag is not warm enough for the given temperature.

c. What symbolic expressions would describe the shaded area?

d. If a manufacturer submitted a sleeping bag to the Army with a thickness of 2.75 inches, would it be suitable for

i. −20° Fahrenheit? **ii.** 0° Fahrenheit? **iii.** +20° Fahrenheit?

SOLUTION

a. If we let F = number of degrees Fahrenheit and T = thickness of the sleeping bag in inches, then

$$T = 2.5 - 0.025F \qquad \text{where the domain is } -40° \leq F \leq 40°$$

b. See Figure 3.17.

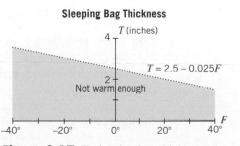

Sleeping Bag Thickness

Figure 3.17 T, sleeping bag thickness as a function of F, degrees Fahrenheit.

When the thickness of the sleeping bag, T, is less than the recommended thickness, then the bag will not be warm enough.

c. The shaded region can be described as

$$0 < \text{thickness of sleeping bag} < \text{recommended thickness}$$

$$0 < \qquad\qquad T \qquad\qquad < 2.5 - 0.025F$$

where $-40° \leq F \leq 40°$.

We could rephrase this to say that the region is bounded by four lines:

$$T = 0, \quad T = 2.5 - 0.025F, \quad F = -40°, \text{ and } F = 40°.$$

Note that the dotted line indicates that the line itself is not included.

d. If the sleeping bag thickness T is 2.75, we can find the corresponding recommended temperature by solving our equation for F.

Substitute 2.75 for T	$2.75 = 2.5 - 0.025F$
simplify to get	$0.25 = -0.025F$
or	$F = -10°$

So the sleeping bag would not be thick enough for $-20°F$, since the point $(-20°, 2.75'')$ lies in the shaded area below the Army's recommended values. It would be more than thick enough for $0°F$ or $20°F$ since the points $(0°, 2.75'')$ and $(20°, 2.75'')$ both lie above the shaded area.

Manipulating Inequalities

Recall that any term may be added to or subtracted from both sides of an inequality without changing the direction of the inequality. The same holds for multiplying or dividing by a positive number. However, multiplying or dividing by a negative number reverses the inequality. For example,

Given the previous inequality	$5 - x < y$
if we wanted to solve for x we could subtract 5 from both sides	$-x < y - 5$
then multiply both sides by -1	$x > -y + 5$

Note that subtracting 5 (or equivalently adding -5) to both sides preserved the inequality, but multiplying by -1 on both sides reversed the inequality. So "$<$" in the first two inequalities became "$>$" in the last inequality.

EXAMPLE 3

An overlapping region

Given the inequalities $2x - 3y \leq 12$ and $x + 2y < 4$:

a. Solve each for y.

b. On the same graph, plot the boundary line for each inequality (indicating whether it is solid or dotted) and then shade the region described by each inequality.

c. Write a compound inequality describing the overlapping region.

SOLUTION

a. To solve for y in the first inequality $\qquad 2x - 3y \leq 12$

add $-2x$ to both sides $\qquad -3y \leq -2x + 12$

divide both sides by -3, reversing the inequality symbol $\qquad \dfrac{-3y}{-3} \geq \dfrac{-2x}{-3} + \dfrac{12}{-3}$

then simplify $\qquad y \geq \dfrac{2}{3}x - 4$

Solving for y in the second inequality $\qquad x + 2y < 4$

add $-x$ to both sides $\qquad 2y < -x + 4$

divide both sides by 2 $\qquad \dfrac{2y}{2} < \dfrac{-x}{2} + \dfrac{4}{2}$

then simplify $\qquad y < -\dfrac{1}{2}x + 2$

b. See Figure 3.18.

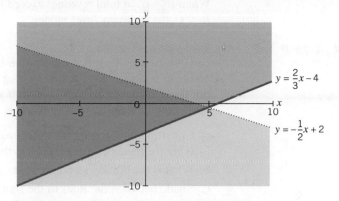

Figure 3.18 The darkest shaded region lies between
$y < -\frac{1}{2}x + 2$ and $y \geq \frac{2}{3}x - 4$.

c. The overlapping region consists of all ordered pairs (x, y) such that $\frac{2}{3}x - 4 \leq y < -\frac{1}{2}x + 2$. We could describe the compound inequality by saying "y is greater than or equal to $\frac{2}{3}x - 4$ and less than $-\frac{1}{2}x + 2$."

EXAMPLE 4

Four boundary lines
Describe the shaded region in Figure 3.19.

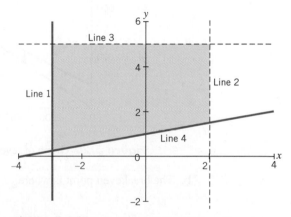

Figure 3.19 A shaded region with four boundary lines.

SOLUTION

We need to find the equation for each of the four boundary lines. The two vertical lines are the easiest: Line 1 is the line $x = -3$ (a solid line, so included in the region); Line 2 is the line $x = 2$ (dotted, so excluded from the region). Line 3 is the horizontal line $y = 5$ (dotted, so it is excluded). Line 4 has a vertical intercept of 1. It passes through $(0, 1)$ and $(-4, 0)$ so its slope is $(0 - 1)/(-4 - 0) = 1/4$ or 0.25. So the equation for Line 4 is $y = 1 + 0.25x$. The region can be described as all pairs (x, y) that satisfy both compound inequalities:

$$-3 \leq x < 2 \quad \text{and} \quad 1 + 0.25x \leq y < 5$$

Breakeven Points: Regions of Profit or Loss

A simple model for the total cost C to a company producing n units of a product is

$$C = \text{fixed costs} + (\text{cost per unit}) \cdot n$$

A corresponding model for the total revenue R is

$$R = (\text{price per unit}) \cdot n$$

The breakeven point occurs when $C = R$, or the total cost is equal to the total revenue. When $R > C$, or total revenues exceed total cost, the company makes a profit and when $R < C$, the company loses money.

EXAMPLE 5

Cost versus revenue

In 1996 two professors from Purdue University reported on their study of fertilizer plants in Indiana.[6] They estimated that for a large-sized fertilizer manufacturing plant the fixed costs were about $450,000 and the additional cost to produce each ton of fertilizer was about $210. The fertilizer was sold at $270 per ton.

a. Construct and graph two equations, one representing the total cost $C(n)$, and the other the revenue $R(n)$, where n is the number of tons of fertilizer produced.

b. What is the breakeven point, where costs equal revenue?

c. Shade between the lines to the right of the breakeven point. What does the region represent?

SOLUTION

a. $C(n) = 450,000 + 210n$ and $R(n) = 270n$. See Figure 3.20.

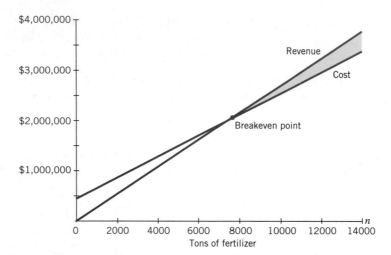

Figure 3.20 Graph of revenue vs. cost with breakeven point.

b. The breakeven point is where $\qquad\qquad$ cost = revenue

or $\qquad\qquad\qquad\qquad\qquad\qquad C(n) = R(n)$

substituting for $C(n)$ and $R(n)$ $\qquad 450,000 + 210n = 270n$

solving for n $\qquad\qquad\qquad\qquad\qquad 450,000 = 60n$

we get $\qquad\qquad\qquad\qquad\qquad\qquad n = 7500$ tons of fertilizer

Substituting 7500 tons into the revenue equation, we have

$$R(7500) = 270 \cdot 7500 = \$2,025,000$$

We could have substituted 7500 tons into the cost equation to get the same value.

$$C(7500) = 450,000 + 210 \cdot 7500 = \$2,025,000$$

So the breakeven point is (7500, $2,025,000). At this point the cost of producing 7500 tons of fertilizer and the revenue from selling 7500 tons both equal $2,025,000.

c. In the shaded area between $R(n)$ and $C(n)$ to the right of the breakeven point, $R(n) > C(n)$, so revenue exceeds costs, and the manufacturers are making money. Economists call this the *region of profit*.

[6]Duane S. Rogers and Jay T. Aldridge, "Economic impact of storage and handling regulation on retail fertilizer and pesticide plants," *Agribusiness* 12(4): 327–337, July/August 1996. Copyright Wiley Periodicals, Inc. This material is reproduced with permission of John Wiley & Sons, Inc.

3.3

The "Fortunate 400"

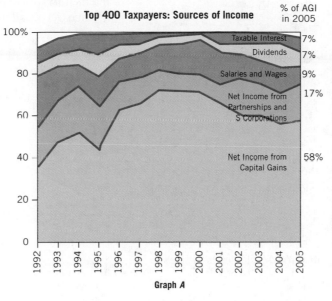

Top 400 Taxpayers: Sources of Income

% of AGI in 2005

Taxable Interest — 7%

Dividends — 7%

Salaries and Wages — 9%

Net Income from Partnerships and S Corporations — 17%

Net Income from Capital Gains — 58%

Graph *A*

Graph A describes the sources of income for the very wealthy from 1992 to 2005. The "Top 400" are the 400 tax returns with the highest adjusted gross income (AGI) reported to the IRS. They accounted for 1.15% of total income reported in the United States in 2005, more than twice as large as their 0.49% share in 1995.

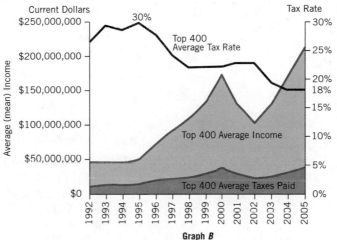

Top 400 Taxpayers: Income and Taxes

Current Dollars

Tax Rate

Top 400 Average Tax Rate

Top 400 Average Income

Top 400 Average Taxes Paid

Graph *B*

Graph B shows the average (mean) income (left vertical axis) reported to the IRS and the mean tax rate (right vertical axis) paid by the "Top 400" from 1992 to 2005.

Source: Data from *Internal Revenue Statistics*; Graphs from *www.VisualizingEconomics.com*.

Use these graphs to estimate the answers to the following questions about the "Top 400" taxpayers.

a. What was the difference between the percentage of income from capital gains and the percentage of income from salaries and wages in 2005? What was this difference in 1992?

b. How did the amount of income after taxes (in 2005 dollars) change from 1992 to 2005?

c. What are two striking trends for income in relation to taxes from 1992 to 2005 for the "Top 400" taxpayers?

d. Check the Internal Revenue Department statistics website at *www.irs.gov/taxstats/* for the income and taxes for the "Top 400" taxpayers in 2010. Are there any differences between 2005 and 2010?

Algebra Aerobics 3.3

1. Solve graphically each set of conditions.

a. $y \geq 2x - 1$
$y \geq 4 - x$

b. $y \geq 2x - 1$
$y \leq 3 - x$

c. $y \geq 400 + 10x$
$y \geq 200 + 20x$
$x \geq 0$
$y \geq 0$

d. $y \geq 0$
$y < -0.5x + 2$
$y < 0.5x + 2$

e. $y \leq 0$
$x \geq 0$
$y \geq -100 + x$

f. $0 \leq x \leq 200$
$y > 2x - 400$
$y < 100 - 1.5x$

2. Determine which of the following points (if any) satisfy the system of inequalities $y > 2x - 3$ and $y \leq 3x + 8$

a. $(2, 3)$ **c.** $(0, 8)$ **e.** $(20, -8)$
b. $(-4, 7)$ **d.** $(-4, -6)$ **f.** $(1, -1)$

3. Determine each inequality for the shaded area in each of the following graphs.

Graphs of Four Linear Inequalities

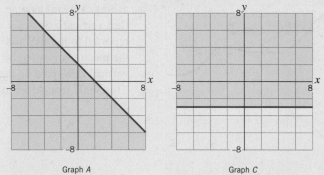

Graph A

Graph C

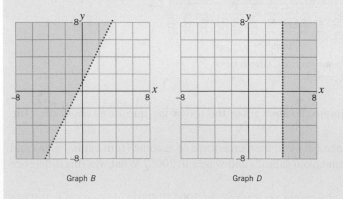

Graph B

Graph D

4. A small company sells dulcimer[7] music books on the Internet. Examine the graph below of the cost (C) and revenue (R) equations for selling n books.

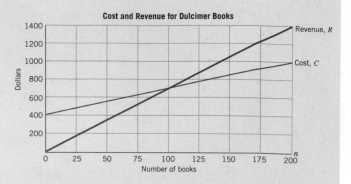

a. Estimate the breakeven point and interpret its meaning.

b. Shade in the region corresponding to losses for the company.

c. What are the fixed costs for selling dulcimer music books?

d. Another company buys dulcimer music books for $3.00 each and sells them for $7.00 each. The fixed cost for this company is $400. Form a system of inequalities that represents when this company would make a profit. Use C_1 for cost and R_1 for revenue for n books.

5. Suppose that the two professors from Example 5 (p. 186) estimated 6 years later that for a large-sized fertilizer manufacturing plant the fixed costs were about $500,000 and the additional cost to produce each ton of fertilizer increased to $235. However, market conditions and competition caused the company to continue to sell the fertilizer at $270 per ton.

a. Form the cost function $C(n)$ and revenue function $R(n)$ for n tons of fertilizer.

b. Find the breakeven point and interpret its meaning.

c. Graph the two functions and shade in the region corresponding to profits for the company.

---

[7]A mountain dulcimer is an Appalachian string instrument, usually with four strings, commonly played on the lap by strumming or plucking.

Exercises for Section 3.3

A graphing program is required for Exercises 19 and 20 and recommended for Exercises 15, 16, and 17. Access to the Internet is required in Exercise 14(c).

1. On different grids, graph and shade in the areas described by the following linear inequalities.

   a. $y < 2x + 2$ **c.** $y < 4$

   b. $y \geq -3x - 3$ **d.** $y \geq \frac{2}{3}x - 4$

2. On different grids, graph and shade in the areas described by the following linear inequalities.

   a. $x - y < 0$ **c.** $3x + 2y > 6$

   b. $x \leq -2$ **d.** $5x - 2y \leq 10$

3. Use inequalities to describe each shaded region.

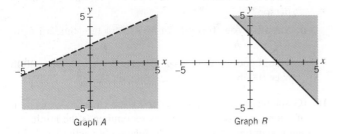

   Graph A   Graph B

4. Use inequalities to describe each shaded region.

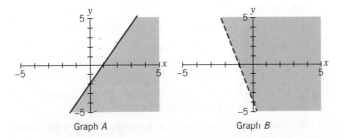

   Graph A   Graph B

5. On different grids, graph each inequality (shading in the appropriate area) and then determine whether or not the origin, the point $(0, 0)$, satisfies the inequality.

   a. $-2x + 6y < 4$ **c.** $y > 3x - 7$

   b. $x \geq 3$ **d.** $y - 3 > x + 2$

6. Determine whether or not the point $(-1, 3)$ satisfies the inequality.

   a. $x - 3y > 6$ **c.** $y \leq 3$

   b. $x < 3$ **d.** $y \leq -\frac{1}{2}x + 3$

7. Explain how you can tell if the region described by the inequality $3x - 5y < 15$ is above or below the boundary line of $3x - 5y = 15$.

8. Shade the region bounded by the inequalities

   $$x + 3y \leq 15$$
   $$2x + y \leq 15$$
   $$x \geq 0$$
   $$y \geq 0$$

9. Match each description in parts (a) to (e) with the appropriate compound inequality in parts (f) to (j).

   a. y is greater than 4 and less than $x - 3$.

   b. y is greater than or equal to $x - 3$ and less than -6.

   c. y is less than $2x + 5$ and greater than -6.

   d. y is greater than or equal to $2x + 5$ and less than -6.

   e. y is less than or equal to $x - 3$ and greater than $2x + 5$.

   f. $2x + 5 \leq y < -6$

   g. $4 < y < x - 3$

   h. $2x + 5 < y \leq x - 3$

   i. $x - 3 \leq y < -6$

   j. $-6 < y < 2x + 5$

10. For the inequalities $y > 4x - 3$ and $y \leq -3x + 4$:

    a. Graph the two boundary lines and indicate with different stripes the two regions that satisfy the individual inequalities.

    b. Write the compound inequality for y. Indicate the double-hatched region on the graph that satisfies both inequalities.

    c. What is the point of intersection for the boundary lines?

    d. If $x = 3$, are there any corresponding y values in the region defined in part (b)?

    e. Is the point $(1, 4)$ part of the double-hatched region?

    f. Is the point $(-1, 4)$ part of the double-hatched region?

11. Examine the shaded region in the graph.

    a. Create equations for the boundary lines l_1 and l_2 using y as a function of x.

    b. Determine the compound inequality that created the shaded region.

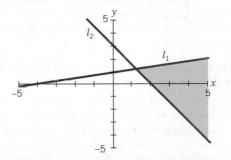

12. Examine the shaded region in the graph. Determine the compound inequality of y in terms of x that created the shaded region.

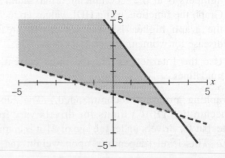

13. The Food and Drug Administration labels suntan products with a sun protection factor (SPF) typically between 2 and 45. Multiplying the SPF by the number of unprotected minutes you can stay in the sun without burning, you are supposed to get the increased number of safe sun minutes. For example, if you can stay unprotected in the sun for 30 minutes without burning and you apply a product with a SPF of 10, then supposedly you can sun safely for $30 \cdot 10 = 300$ minutes or 5 hours.

Assume that you can stay unprotected in the sun for 20 minutes without burning.

   a. Write an equation that gives the maximum safe sun time T as a function of S, the sun protection factor (SPF).

   b. Graph your equation. What is a suggested domain for S?

   c. Write an inequality that suggests times that would be unsafe to stay out in the sun.

   d. Shade in and label regions on the graph that indicate safe and unsafe regions. (Use two different shadings and remember to include boundaries.)

   e. How would the graph change if you could stay unprotected in the sun for 40 minutes?

Note that you should be cautious about spending too much time in the sun. Factors such as water, wind, and sun intensity can diminish the effect of SPF products.

14. (Access to Internet required for part (c).) Doctors measure two kinds of cholesterol in the body: low-density lipoproteins, LDL, called "bad cholesterol" and high-density lipoproteins, HDL, called "good cholesterol" because it helps to remove the bad cholesterol from the body. Rather than just measuring the total cholesterol, TC, many doctors use the ratio of TC/HDL to help control heart disease. General guidelines have suggested that men should have TC/HDL of 4.5 or below, and women should have TC/HDL of 4 or below.

On the graphs you construct, place HDL on the horizontal and TC on the vertical axis.

   a. Construct an equation for men that describes TC as a function of HDL assuming that the ratio of the two numbers is at the recommended maximum for men. Graph the function, using HDL values up to 75. Label on the graph higher-risk and lower-risk areas for heart disease for men.

   b. Construct a similar equation for women that describes TC as a function of HDL assuming that the ratio of the two numbers is at the recommended maximum for women. Graph the function, using HDL values up to 75. Label on the graph higher-risk and lower-risk areas for heart disease for women.

   c. Use the Internet to find the most current cholesterol guidelines.

15. (Graphing program recommended.) The blood alcohol concentration (BAC) limits for drivers vary from state to state, but for drivers under the age of 21 it is commonly set at 0.02. This level (depending upon weight and medication levels) may be exceeded after drinking only one 12-oz can of beer. The formula

$$N = 6.4 + 0.0625(W - 100)$$

gives the number of ounces of beer, N, that will produce a BAC legal limit of 0.02 for an average person of weight W. The formula works best for drivers weighing between 100 and 200 lb.

   a. Write an inequality that describes the condition of too much blood alcohol for drivers under 21 to legally drive.

   b. Graph the corresponding equation and label the areas that represent legally safe to drive, and not legally safe to drive conditions.

   c. How many ounces of beer is it legally safe for a 100-lb person to consume? A 150-lb person? A 200-lb person?

   d. Simplify your formula in part (b) to the standard form $y = mx + b$.

   e. Would you say that "6 oz of beer + 1 oz for every 20 lb over 100 lb" is a legally safe rule to follow?

16. (Graphing program recommended.) The Ontario Association of Sport and Exercise Sciences recommends the minimum and maximum pulse rates P during aerobic activities, based on age A. The maximum recommended rate, P_{max}, is $0.87(220 - A)$. The minimum recommended pulse rate, P_{min}, is $0.72(220 - A)$.

   a. Convert these formulas to the $y = mx + b$ form.

   b. Graph the formulas for ages 20 to 80 years. Label the regions of the graph that represent too high a pulse rate, the recommended pulse rate, and too low a pulse rate.

   c. What is the maximum pulse rate recommended for a 20-year-old? The minimum for an 80-year-old?

   d. Construct an inequality that describes too low a pulse rate for effective aerobic activity.

   e. Construct an inequality that describes the recommended pulse range.

17. (Graphing program recommended.) We saw in Example 1 in this section the U.S. Department of Agriculture recommendations for healthy weight zones for men based on height. There are comparable recommendations for women between 5′ (or 60″) and 6′3″ (or 75″) tall. For women the recommended lowest weight W_{lo} (in lb) is

$$W_{lo} = 100 + 3.5H$$

and the recommended highest weight W_{hi} (in lb) is

$$W_{hi} = 118.2 + 4.2H$$

where H is the number of inches above 60″ (or 5 feet).

   a. Graph and label the two equations and indicate the underweight, healthy weight, and overweight zones.

   b. Give a mathematical description of the healthy weight zone for women.

17. (continued)

    c. Two women have the same weight of 130 lb. One is 5'2" and the other is 5'5". Does either one lie within the healthy weight zone? Why?

    d. A 5'4" woman weighs 165 lb. If her doctor puts her on a weight-loss diet of 1.5 lb per week, how many weeks would it take for her to reach the healthy range?

18. Cotton and wool fabrics, unless they have been preshrunk, will shrink when washed and dried the first time at high temperatures. If washed in cold water and drip-dried they will shrink a lot less, and if dry-cleaned, they will not shrink at all. A particular cotton fabric has been found to shrink 8% with a hot wash/dry, and 3% with a cold wash/drip dry.

    a. Find a formula to express the hot wash length, H, as a function of the original length, L. Then find a formula for the cold wash length, C, as a function of the original length, L. What formula expresses dry-clean length, D, as a function of the original length?

    b. Make a graph with original length, L, on the horizontal axis, up to 60 inches. Plot three lines showing cold wash length, C, hot wash length, H, and dry-clean length, D. Label the lines.

    c. If you buy trousers with an original inseam length of 32", how long will the inseam be after a hot wash? A cold wash?

    d. If you need 3 yards of fabric (a yard is 3') to make a well-fitted garment, how much would you have to buy if you plan to hot-wash the garment?

19. (Graphing program required.) Two professors from Purdue University reported that for a typical small-sized fertilizer plant in Indiana the fixed costs were $235,487 and it cost $206.68 to produce each ton of fertilizer.

    a. If the company planned to sell the fertilizer at $266.67 per ton, find the cost, C, and revenue, R, equations for x tons of fertilizer.

    b. Graph the cost and revenue equations on the same graph and calculate and interpret the breakeven point.

    c. Indicate the region where the company would make a profit and create the inequality to describe the profit region.

20. (Graphing program required.) A company manufactures a particular model of DVD player that it sells to retailers for $85. It costs $55 to manufacture each DVD player, and the fixed manufacturing costs are $326,000.

    a. Create the revenue function $R(x)$ for selling x number of DVD players.

    b. Create the cost function $C(x)$ for manufacturing x DVD players.

    c. Plot the cost and revenue functions on the same graph. Estimate and interpret the breakeven point.

    d. Shade in the region where the company would make a profit.

    e. Shade in the region where the company would experience a loss.

    f. What is the inequality that represents the profit region?

21. Describe the shaded region in each graph with the appropriate inequalities.

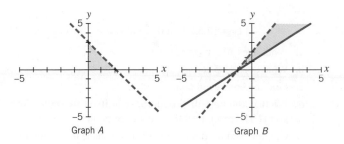

Graph *A*   Graph *B*

22. Describe the shaded region.

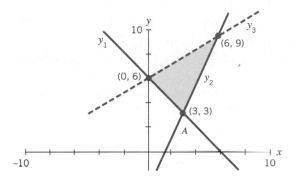

23. A financial advisor has up to $30,000 to invest, with the stipulation that at least $5000 is used to purchase Treasury bonds and at most $15,000 in corporate bonds.

    a. Construct a set of inequalities that describes the relationship between buying corporate vs. Treasury bonds, where the total amount invested must be less than or equal to $30,000. (Let C be the amount of money invested in corporate bonds, and T the amount invested in Treasury bonds.).

    b. Construct a feasible region of investment; that is, shade in the area on a graph that satisfies the spending constraints on both corporate and Treasury bonds. Label the horizontal axis "Amount invested in Treasury bonds" and the vertical axis "Amount invested in corporate bonds."

    c. Find all of the intersection points (corner points) of the bounded investment feasibility region and interpret their meanings.

24. A Texas oil supplier ships at most 10,000 barrels of oil per week. Two distributors need oil. Southern Oil needs at least 2000 barrels of oil per week and Regional Oil needs at least 5000 barrels of oil per week.

    a. Let S be the number of barrels of oil shipped to Southern Oil and let R be the number of barrels shipped to Regional Oil per week. Create a system of inequalities that describes all of the conditions.

    b. Graph the feasible region of the system.

    c. Choose a point inside the region and describe its meaning.

25. A small T-shirt company created the following cost and revenue equations for a line of T-shirts, where cost C is in dollars for producing x units and revenue R is in dollars from selling x units:

$$C = 12.5x + 360 \quad \text{and} \quad R = 15.5x$$

a. What does 12.5 represent?

b. What does 15.5 represent?

c. Find the breakeven point.

d. What is the cost of producing x units at the breakeven point? The revenue at the breakeven point?

e. Graph C and R on the same grid and shade the region that represents profit.

26. A large wholesale nursery sells shrubs to retail stores. The cost $C(x)$ and revenue $R(x)$ equations (in dollars) for x shrubs are

$$C(x) = 15x + 12{,}000 \quad \text{and} \quad R(x) = 18x$$

a. Find the breakeven point.

b. Explain the meaning of the coordinates for the breakeven point.

c. Graph $C(x)$ and $R(x)$ on the same grid and shade the region that represents loss.

27. The accompanying graph compares the total energy production, $P(t)$, and consumption, $C(t)$, in the United States for 1980–2008, with projections to 2035.

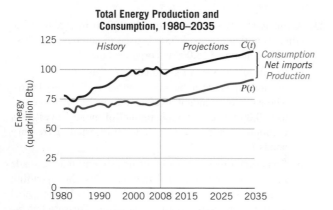

Total Energy Production and Consumption, 1980–2035

Source: U.S. Energy Information Administration.

a. On the graph, identify the region that is $\leq C(t)$ and the region that is $\geq P(t)$.

b. Write a compound inequality describing the overlapping region called net imports.

c. Write a 60-second summary about the consumption and production of energy in the United States and projections by the U.S. Energy Information Administration.

28. The accompanying graph shows supply, $S(q)$, and demand curves, $D(q)$, for a quantity at a market price, P.

a. Let P_1 be the price at equilibrium. How would you describe the equilibrium point in terms of supply and demand?

b. Match the following regions shown on the graph at the top of the next column to the corresponding system of inequalities describing it.

Region on graph: a. Consumer surplus b. Producer surplus

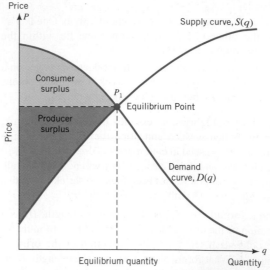

Source: http://en.wikipedia.org/wiki/Economic_surplus.

System of inequalities:

  i. $P \leq S(q)$ and $P \geq P_1$ **iii.** $P \geq S(q)$ and $P \leq P_1$

  ii. $P \leq D(q)$ and $P \geq P_1$

c. Economists define the *consumer surplus* as the amount that consumers benefit by being able to purchase a product for a price that is less than the most that they would be willing to pay. The *producer surplus* is the amount that producers benefit by selling at a price that is higher than the least that they would be willing to sell for. Do these definitions correspond to the system of inequalities that you found in part (b)?

29. The following graph represents the percent of land in California affected by different levels of drought conditions from June 2008 to April 2010.

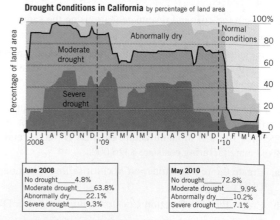

Less of California in a Drought
Since January, drought conditions have affected a much smaller portion of the state's land area, according to data compiled by the National Drought Mitigation Center.

Drought Conditions in California by percentage of land area

June 2008	
No drought	4.8%
Moderate drought	63.8%
Abnormally dry	22.1%
Severe drought	9.3%

May 2010	
No drought	72.8%
Moderate drought	9.9%
Abnormally dry	10.2%
Severe drought	7.1%

Source: Data from the National Drought Mitigation Center. San Francisco Chronicle by Todd Trumbull. Copyright May 16, 2010 by San Francisco Chronicle. Reproduced with permission of the San Francisco Chronicle in the format textbook via Copyright Clearance Center.

a. If $D(t)$ represents the percentage of land area affected by drought and $S(t)$ represents percentage of land area affected by severe drought, what would $S(t) < P < D(t)$ represent?

b. In what months were there severe or moderate drought conditions for 80% or more of California's land area?

3.4 *Systems with Piecewise Linear Functions: Tax Plans*

In this section, different types of tax plans are modeled. Comparisons among the plans are made from a mathematical point of view, while the readings on the course website contrast different political views.

Graduated vs. Flat Income Tax

Income taxes may be based on either a flat or a graduated tax rate. With a flat tax rate, no matter what the income level, everyone is taxed at the same percentage. Flat taxes are often said to be unfair to those with lower incomes, because paying a fixed percentage of income is considered more burdensome to someone with a low income than to someone with a high income.

A graduated tax rate means that people with higher incomes are taxed at higher rates. Such a tax is called *progressive* and is generally less popular with those who have high incomes. Whenever the issue appears on the ballot, the pros and cons of the graduated vs. flat tax rate are hotly debated in the news media and paid political broadcasting. Of the 43 states with a broad-based income tax, 36 had a graduated income tax in 2010.

The *New York Times* article "How a Flat Tax Would Work for You and for Them" and other articles on the course website discuss the trade-offs in using a flat tax.

In Exploration 3.1 you can examine the effects of the 2010 U.S. tax rate on individuals in different income brackets.

The taxpayer

For the taxpayer there are two primary questions in comparing the effects of flat and graduated tax schemes. For what income level will the taxes be the same under both plans? And, given a certain income level, how will taxes differ under the two plans?

Taxes are influenced by many factors, such as filing status, exemptions, source of income, and deductions. For our comparisons of flat and graduated income tax plans, we examine one filing status and assume that exemptions and deductions have already been subtracted from income.

E X A M P L E 1

Visualizing different tax plans
Match each graph in Figure 3.21 with the appropriate description.

a. Income taxes are a flat rate of 5% of your income.

b. Income taxes are graduated, with a rate of 5% for first $100,000 of income and a rate of 8% for any additional income $> \$100,000$.

c. Sales taxes are 5% for $0 \leq$ purchases $\leq \$100,000$ and a flat fee of $5000 for purchases $> \$100,000$.

Graphs of Different Tax Plans

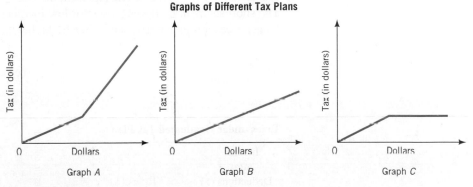

Graph *A*   Graph *B*   Graph *C*

Figure 3.21

S O L U T I O N

Graph *A* matches description (b).
Graph *B* matches description (a).
Graph *C* matches description (c).

A flat tax model

Flat taxes are a fixed percentage of income. If the flat tax rate is 15% (or 0.15 in decimal form), then flat taxes can be represented as

$$f(i) = 0.15i$$

where i = income. This flat tax plan is represented in Figure 3.22.

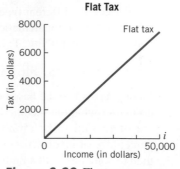

Figure 3.22 Flat tax at a rate of 15%.

A graduated tax model

A Piecewise Linear Function.   Under a graduated income tax, the tax rate changes for different income levels. We can use a piecewise linear function to model a graduated tax. Let's consider a graduated tax where the first $10,000 of income is taxed at 10% and any income over $10,000 is taxed at 20%. For example, if your income after deductions is $30,000, then your taxes under this plan are

$$\text{graduated tax} = (10\% \text{ of } \$10,000) + (20\% \text{ of income over } \$10,000)$$
$$= 0.10(\$10,000) + 0.20(\$30,000 - \$10,000)$$
$$= 0.10(\$10,000) + 0.20(\$20,000)$$
$$= \$1000 + \$4000$$
$$= \$5000$$

If an income, i, is over $10,000, then under this plan,

$$\text{graduated tax} = 0.10(\$10,000) + 0.20(i - \$10,000)$$
$$= \$1000 + 0.20(i - \$10,000)$$

The Graph of a Piecewise Linear Function.   This graduated tax plan is represented in Table 3.3 and Figure 3.23. The graph of the graduated tax is the result of piecing together two different line segments that represent the two different formulas used to find taxes. The short segment represents taxes for low incomes between $0 and $10,000, and the longer, steeper segment represents taxes for higher incomes that are greater than $10,000.

Taxes under Graduated Tax Plan

Income after Deductions ($)	Taxes ($)	
0		0
5,000		500
10,000		1,000
20,000	1,000 + 2,000	= 3,000
30,000	1,000 + 4,000	= 5,000
40,000	1,000 + 6,000	= 7,000
50,000	1,000 + 8,000	= 9,000

Table 3.3

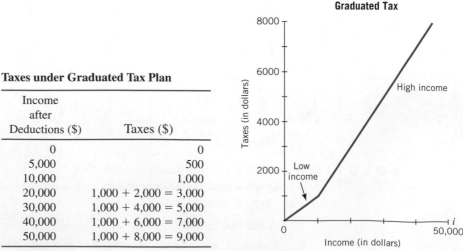

Figure 3.23

The Equations for a Piecewise Linear Function.   To find an algebraic expression for the graduated tax, we need to use different formulas for different levels of income. Functions that use different formulas for different intervals of the domain are called *piecewise defined*. Since each income determines a unique tax, we can define the graduated tax as a piecewise function, g, of income i as

$$g(i) = \begin{cases} 0.10i & \text{for } 0 \le i \le \$10{,}000 \\ 1000 + 0.20(i - 10{,}000) & \text{for } i > \$10{,}000 \end{cases}$$

The value of i (the input or independent variable) determines which formula to use to evaluate the function. This function is called a *piecewise linear function,* since each piece consists of a different linear formula. The formula for incomes equal to or below $10,000 is different from the formula for incomes above $10,000.

E X A M P L E   2   **Evaluating piecewise functions**
Use the piecewise function defined above for $g(i)$ to evaluate:

a. $g(\$8{,}000)$ b. $g(\$40{,}000)$

S O L U T I O N   **a.** To find $g(\$8000)$, the value of $g(i)$ when $i = \$8000$, use the first formula in the definition since income, i, in this case is less than $10,000:

For $i \le \$10{,}000$ $\qquad\qquad\qquad\qquad$ $g(i) = 0.10i$
substituting $8000 for i $\qquad\qquad$ $g(\$8000) = 0.10(\$8000)$
$\qquad\qquad\qquad\qquad\qquad\qquad\qquad = \$800$

b. To find $g(\$40{,}000)$, we use the second formula in the definition, since in this case income is greater than $10,000:

For $i > \$10{,}000$ $\qquad\qquad\qquad\qquad$ $g(i) = \$1000 + 0.20(i - \$10{,}000)$
substituting $40,000 for i $\qquad$ $g(\$40{,}000) = \$1000 + 0.20(\$40{,}000 - \$10{,}000)$
$\qquad\qquad\qquad\qquad\qquad\qquad\qquad = \$1000 + 0.20(\$30{,}000)$
$\qquad\qquad\qquad\qquad\qquad\qquad\qquad = \$1000 + \$6000$
$\qquad\qquad\qquad\qquad\qquad\qquad\qquad = \$7000$

Comparing the Flat and Graduated Tax Plans

Using graphs

In Figure 3.24 we compare the two different tax plans by plotting the flat and graduated tax equations on the same graph.

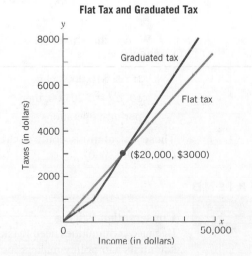

In Exploration 3.2, you compare the cost of hybrid and conventional automobiles using piecewise functions.

Figure 3.24

The intersection points indicate the incomes at which the amount of tax is the same under both plans. From the graph, we can estimate the coordinates of the two points as (0, 0) and ($20,000, $3000). That is, under both plans, with $0 income you pay $0 taxes, and with approximately $20,000 in income you would pay approximately $3000 in taxes. Individual voters want to know what impact these different plans will have on their taxes. From the graph in Figure 3.24, we can see that to the left of the intersection point at ($20,000, $3000), the flat tax is *greater* than the graduated tax for the same income. To the right of this intersection point, the flat tax is *less* than the graduated tax for the same income. So for incomes *less than* $20,000, taxes are *greater* under the flat tax plan, and for incomes *greater than* $20,000, taxes will be *less* under the flat tax plan.

Using equations

To verify the accuracy of our estimates for the coordinates of the point(s) of intersection, we can set $f(i) = g(i)$. We know that $f(i) = 0.15i$. Which of the two expressions do we use for $g(i)$? The answer depends upon what value of income, i, we consider. For $i \leq \$10,000$, we have $g(i) = 0.10i$.

If	$f(i) = g(i)$
and $i \leq \$10,000$, then	$0.15i = 0.10i$
This can only happen when	$i = 0$

If $i = 0$, both $f(i)$ and $g(i)$ equal 0; therefore, one intersection point is indeed (0, 0).

For $i > \$10,000$, we use $g(i) = \$1000 + 0.20(i - \$10,000)$ and again set $f(i) = g(i)$.

If	$f(i) = g(i)$
and $i > \$10,000$, then	$0.15i = \$1000 + 0.20(i - \$10,000)$
apply the distributive property	$0.15i = \$1000 + 0.20i - (0.20)(\$10,000)$
multiply	$0.15i = \$1000 + 0.20i - \$2000$
combine terms	$0.15i = -\$1000 + 0.20i$
add $-0.20i$ to each side	$-0.05i = -\$1000$
divide by -0.05	$i = -\$1000/(-0.05)$
	$i = \$20,000$

So each plan results in the same tax for an income of $20,000. How much tax is required? We can substitute $20,000 for i into either the flat tax formula or the graduated tax formula for incomes over $10,000 and solve for the tax. Given the flat tax function,

$$f(i) = 0.15i$$
$$\text{if } i = \$20,000, \text{ then} \quad f(i) = (0.15)(\$20,000)$$
$$= \$3000$$

We can check to make sure that when $i = \$20,000$, the graduated tax, $g(i)$, will also be $3000:

If $i > \$10,000$, then	$g(i) = \$1000 + 0.20(i - \$10,000)$
so, if $i = \$20,000$, then	$g(i) = \$1000 + 0.20(\$20,000 - \$10,000)$
perform operations	$= \$3000$

These calculations confirm that the other intersection point is, as we estimated, ($20,000, $3000).

EXPLORE & EXTEND

3.4 **Conservation Pricing**

Because communities need to balance growth with protection of their natural resources, scientists and policy makers look closely at the problem of balancing water use and available water resources. One approach is to use conservation pricing to decrease the demand.

In the early 1990s Santa Barbara, California, experienced a severe drought and used conservation pricing as an incentive to decrease demand for water. The water rate structures are given below. Let x = water usage per month in HCF (hundred cubic feet) where 1 HCF = 100 cubic feet.

Time Period	Rate/HCF per Month	Service Charge per Month
Before the drought (6/86–7/88)	$0.89	$4.10
During drought (3/90–10/90)	$1.09/HCF for $0 \leq x \leq 4$	$1.47
	$3.27/HCF for $4 < x \leq 8$	
	$9.81/HCF for $8 < x \leq 14$	
	$29.43/HCF for $x > 14$	
After drought (after 8/95)	$2.10/HCF for $0 \leq x \leq 4$	$5.50
	$3.50/HCF for $4 < x \leq 20$	
	$3.70/HCF for $x > 20$	

Source: Public Works Department, City of Santa Barbara.

a. Create three piecewise functions, $b(x)$, $d(x)$, and $a(x)$, the cost/month before, during, and after the drought, for x = number of HCF/month, where $0 \leq x \leq 50$ and cost = service charge + (rate · number HCF).

b. Graph the three functions.

c. For what water usage is the cost the same before the drought and during the drought?

d. For what water usage is the cost the same during the drought and after the drought?

Algebra Aerobics 3.4

1. Graph the following piecewise functions.

a. $f(x) = \begin{cases} -x - 1 & \text{for } x \leq 0 \\ \frac{1}{2}x - 1 & \text{for } x > 0 \end{cases}$

b. $g(x) = \begin{cases} 4 & \text{for } 0 \leq x \leq 5 \\ 2x - 6 & \text{for } x > 5 \end{cases}$

c. $k(x) = \begin{cases} 0 & \text{for } 0 \leq x \leq 10 \\ 3x - 30 & \text{for } x > 10 \end{cases}$

2. Use equations to describe the piecewise linear function on each graph.

Two Piecewise Linear Functions

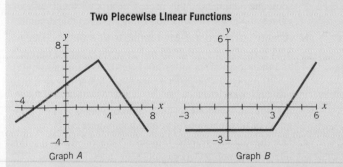

Graph A

Graph B

3. Evaluate each of the following piecewise defined functions at $x = -5, 0, 2,$ and 10.

a. $P(x) = \begin{cases} 3 & \text{for } x \leq 1 \\ 1 - 2x & \text{for } x > 1 \end{cases}$

b. $W(x) = \begin{cases} x - 4 & \text{for } x < 2 \\ x + 4 & \text{for } x \geq 2 \end{cases}$

4. a. Construct a graduated tax function, where $A(i)$ = the amount of tax, i = income, and the tax is 5% on the first $50,000 of income and 15% on income in excess of $50,000.

b. Construct a flat tax function, $B(i)$, where the tax rate is 10% of income.

c. Find the solution to the system of equations $A(i)$ and $B(i)$.

Exercises for Section 3.4

Graphing program recommended for Exercise 12.

1. Match each function with its graph.

a. $f(x) = \begin{cases} x & \text{if } x \leq 2 \\ -x + 4 & \text{if } x > 2 \end{cases}$

b. $f(x) = \begin{cases} -x & \text{if } x \leq 0 \\ x & \text{if } 0 < x \leq 2 \\ -x + 4 & \text{if } x > 2 \end{cases}$

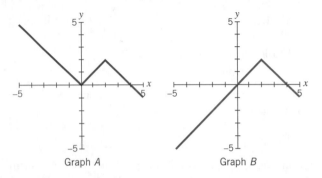

Graph A   Graph B

2. Construct a piecewise linear function for each of the accompanying graphs.

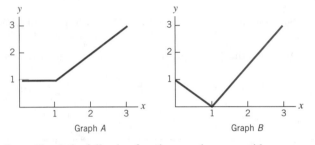

Graph A   Graph B

3. a. Graph the following functions on the same grid.

$$h(x) = \begin{cases} 20 - 2x & \text{if } 0 \leq x < 5 \\ 10 & \text{if } 5 \leq x \leq 10 \\ 10 + 2(x - 10) & \text{if } x > 10 \end{cases}$$

$j(x) = 5 + x$

b. Estimate from your graphs any intersection points and confirm with your equations.

4. a. Construct a graduated tax function where the tax is 10% on the first $30,000 of income, then 20% on any income in excess of $30,000.

b. Construct a flat tax function where the tax is 15% of income.

c. Calculate the tax for both the flat tax function from part (b) and the graduated tax function from part (a) for each of the following incomes: $10,000, $20,000, $30,000, $40,000, and $50,000.

d. Graph the graduated and flat tax functions on the same grid and estimate the coordinates of the points of intersection. Interpret the points of intersection.

5. You are thinking about replacing your long-distance telephone service. A cell phone company charges a monthly fee of $40 for the first 450 minutes and then charges $0.45 for every minute after 450. Every call is considered a long-distance call.

Your local phone company charges you a fee of $60 per month and then $0.05 per minute for every long-distance call.

a. Assume you will be making only long-distance calls. Create two functions, $C(x)$ for the cell phone plan and $L(x)$ for the local telephone plan, where x is the number of long-distance minutes.

After how many minutes would the two plans cost the same amount?

b. Describe when it is more advantageous to use your cell phone for long-distance calls.

c. Describe when it is more advantageous to use your local phone company to make long-distance calls.

6. (Graphing program recommended). In 2004, Missouri had a graduated tax plan (which it still had in 2010) but it was considering adopting a flat-rate tax of 4% on income after deductions. The tax rate for single people under the graduated plan is shown in the accompanying table. For what income levels would a single person pay less tax under the flat tax plan than under the graduated tax plan?

Missouri State Tax for a Single Person

Income after Deductions ($)	Tax Rate (%)
≤1000	1.50
1001–2000	2.00
2001–3000	2.50
3001–4000	3.00
4001–5000	3.50
5001–6000	4.00
6001–7000	4.50
7001–8000	5.00
8001–9000	5.50
>9000	6.00

7. Heart health is a prime concern, because heart disease is the leading cause of death in the United States. Aerobic activities such as walking, jogging, and running are recommended for cardiovascular fitness, because they increase the heart's strength and stamina.

a. A typical training recommendation for a beginner is to walk at a moderate pace of about 3.5 miles/hour (or approximately 0.0583 miles/minute) for 20 minutes. Construct a function that describes the distance traveled $D_{beginner}$, in miles, as a function of time, T, in minutes, for someone maintaining this pace. Construct a small table of values and graph the function using a reasonable domain.

b. A more advanced training routine is to walk at a pace of 3.75 miles/hour (or 0.0625 miles/minute) for 10 minutes and then jog at 5.25 miles/hour (or 0.0875 miles/minute) for 10 minutes. Construct a piecewise linear function that gives the total distance, $D_{advanced}$, as a function of time, T, in minutes. Generate a small table of values and plot the graph of this function on your graph in part (a).

c. Do these two graphs intersect? If so, what do the intersection point(s) represent?

8. A graduated income tax is proposed in Borduria to replace an existing flat rate of 8% on all income. The new proposal states that persons will pay no tax on their first $20,000 of income, 5% on income over $20,000 and less than or equal to $100,000, and 10% on their income over $100,000. (*Note:* Borduria is a fictional totalitarian state in the Balkans that figures in the adventures of Tintin.)

a. Construct a table of values that shows how much tax persons will pay under both the existing 8% flat tax and the proposed new tax for each of the following incomes: $0, $20,000, $50,000, $100,000, $150,000, $200,000.

b. Construct a graph of tax dollars vs. income for the 8% flat tax.

c. On the same graph plot tax dollars vs. income for the proposed new graduated tax.

d. Construct a function that describes tax dollars under the existing 8% tax as a function of income.

e. Construct a piecewise function that describes tax dollars under the proposed new graduated tax rates as a function of income.

f. Use your graph to estimate the income level for which the taxes are the same under both plans. What plan would benefit people with incomes below your estimate? What plan would benefit people with incomes above your estimate?

g. Use your equations to find the coordinates that represent the point at which the taxes are the same for both plans. Label this point on your graph.

h. If the median income in the state is $27,000 and the mean income is $35,000, do you think the new graduated tax would be voted in by the people? Explain your answer.

9. It is often said that 1 year of a dog's life is equivalent to 7 years of a human's life. A more accurate veterinarian's estimate is that for the first 2 years of a dog's life, each dog year is equivalent to 10.5 years of a human's life, and after 2 years each dog year is equivalent to 4 years of a human's life.

a. Write formulas to describe human-equivalent years as a function of dog years, D, for the two methods of relating dog years to human years. Use H for the popular method and H_v for the veterinary method.

b. At what dog age do the 2 methods give the same human years, and what human age is that?

10. You check around for the best deal on your prescription medicine. At your local pharmacy it costs $4.39 a bottle; by mail-order catalog it costs $3.85 a bottle, but there is a flat shipping charge of $4.00 for any size order; by a source found on the Internet it costs $3.99 a bottle and shipping costs $1.00 for each bottle, but for orders of ten or more bottles it costs $3.79 a bottle, and handling is $2.50 per order plus shipping costs of $1.00 for each bottle.

a. Find a formula to express each of these costs if N is the number of bottles purchased and C_p, C_c, and C_I are the respective costs for ordering from the pharmacy, catalog, and Internet.

b. Graph the costs for purchases up to twelve bottles at a time. Which is the cheapest source if you buy fewer than ten bottles at a time? If you buy more than ten bottles at a time? Explain.

CHAPTER SUMMARY

Systems of Linear Equations

A pair of real numbers is a *solution* to a system of linear equations in two variables if and only if the pair of numbers is a solution of each equation. A system of two linear equations in two variables may have one solution, no solutions, or infinitely many solutions.

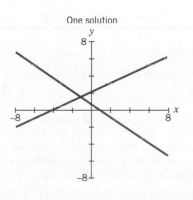

One solution

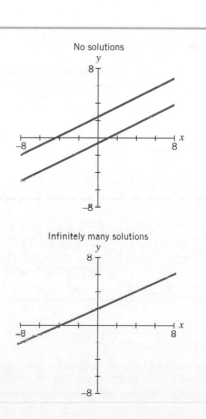

No solutions

Infinitely many solutions

Systems of Linear Inequalities

The solutions to a system of linear inequalities in two variables are pairs of real numbers that satisfy all the inequalities. On a graph, the solutions typically form a region of the plane and may be represented by a compound inequality such as

$$y_1 < y \le y_2$$
$$5 - x < y \le 1 + 3x$$

The shaded area in the figure represents the solution area for this system.

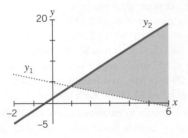

Systems with Piecewise Linear Functions

Systems with piecewise linear functions are often used to model graduated tax plans.

The accompanying figure shows the graphs of a system with a linear piecewise function g of income i, where

$$g(i) = \begin{cases} 0.10i & \text{for } i \le \$10,000 \\ 1000 + 0.20(i - 10,000) & \text{for } i > \$10,000 \end{cases}$$

and a linear function f of income i where $f(i) = 0.15i$.

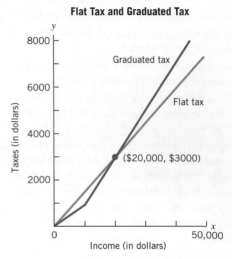

CHECK YOUR UNDERSTANDING

I. Are the statements in Problems 1–30 true or false? Give an explanation for your answer.

Questions 1 and 2 refer to the accompanying graph that shows predictions for the 2008 Michigan Republican primary.

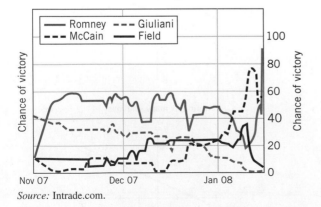

Source: Intrade.com.

1. Around the first week in January 2008, the chances of Romney and McCain winning the 2008 Michigan Republican primary were the same, at approximately 40%.

2. Before January 2008, the chances of McCain winning the Michigan Republican primary were higher than the chances of Romney winning.

3. Assuming y is a function of x, the number pair $(-5, 3)$ is a solution to the following system of equations:

$$2x - 3y = 21 \quad \text{and} \quad x = 2y + 3$$

4. Assuming y is a function of x, the number pair $\left(\frac{16}{5}, \frac{2}{5}\right)$ is a solution to the following system of equations:

$$y = 10 - 3x \quad \text{and} \quad 2x - y = 6$$

5. The system $x - \dfrac{5}{y} = 2$ and $\dfrac{x}{3} + 2y = 1$ is not a linear system of equations.

6. The system of linear equations $2y - \dfrac{x}{3} = -1$ and $\dfrac{6y}{5} - \dfrac{x}{5} = \dfrac{-3}{5}$ has no solution(s).

7. A system of linear equations in two variables either has a pair of numbers that is a unique solution or has no solution.

8. A pair of numbers is a solution to a system of linear equations in two variables if the pair is a solution to at least one of the equations.

Questions 9 and 10 refer to the accompanying graphs of systems of two equations in two variables.

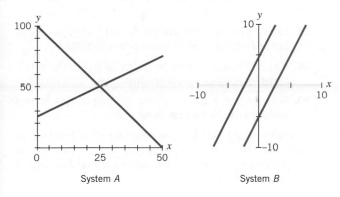

System *A*   System *B*

9. System *A* has a unique solution at approximately (50, 25).

10. System *B* appears to have no solution.

Questions 11 and 12 refer to the supply and demand curves in the accompanying graph.

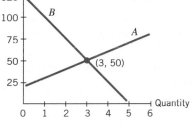

11. Graph *B* is the demand curve because as price increases, the quantity demanded decreases.

12. If three items are produced and sold at a price of $50, the quantity supplied will be equal to the quantity demanded.

13. The shaded region in the accompanying graph appears to satisfy the linear inequality $12 - x < 3y$.

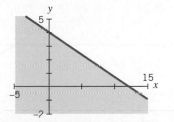

14. The darkest shaded region in the graph below appears to satisfy the system of linear inequalities

$$y \le 4 - \frac{x}{3} \quad \text{and} \quad y \le 2$$

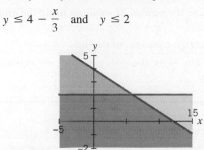

15. The linear inequalities $y \le 5x - 3$ and $y < 5x - 3$ have exactly the same solutions.

Questions 16 and 17 refer to the accompanying graph of supply and demand, where q = quantity and p = price:

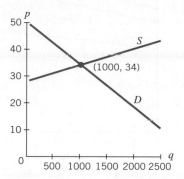

16. The equilibrium point (1000, 34) means that at a price of $1000, suppliers will supply 34 items and consumers will demand 34 items.

17. If the price is less than the equilibrium price, the quantity supplied is less than the quantity demanded and therefore competition will cause the price to increase toward the equilibrium price.

Questions 18–22 refer to the accompanying graph, where *R* represents total revenue and *C* represents total costs.

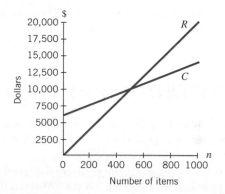

18. The estimated breakeven point is approximately (500, $10,000).

19. The profit is zero at the breakeven point.

20. If 200 units are produced and sold, the total revenue is larger than the total cost.

21. If the revenue per unit increases, and hence, *R*, the revenue line, becomes steeper, the breakeven point moves to the right.

22. If 700 units are produced and sold, the company is making about a $5000 profit.

Questions 23–25 refer to the accompanying graph of linear approximations for the body mass index (BMI) of girls age 7–15. B_U gives the minimal recommended BMI and B_N the maximum recommended BMI. B_0 is the dividing line between being somewhat overweight and being obese.

The National Center For Health Statistics calculates the BMI as

$$\text{BMI} = \left[\frac{\text{weight (lb)}}{\text{height (in)}}\right] \cdot \left[\frac{703}{\text{height (in)}}\right]$$

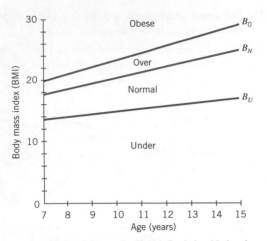

Sources: National Center for Health Statistics, National Center for Chronic Disease Prevention and Health Promotion (2000), *www.cdc.gov/growthcharts.*

23. A 15-year-old girl who weighs 110 pounds and is 55 inches tall has a BMI that is within the normal zone.

24. A 10-year-old girl with BMI = 12 is underweight for her height and age.

25. A mathematical description of the normal BMI zone for girls age 7–15 is $B_U \leq$ BMI $\leq B_N$.

Questions 26 and 27 refer to the following system of linear inequalities: $2x + 3 \leq y$ and $y < 5 + x$

26. The boundary line $y = 5 + x$ is not included in the solution region.

27. The pair of numbers (0, 0) is a solution to the system.

28. Linear systems of equations in two variables always have exactly one pair of numbers as a solution.

29. Any solution(s) to a linear system of equations in two variables is either a point or a line.

30. Solutions to a linear system of inequalities in two variables can be a region in the plane.

II. In Problems 31–35, give an example of a function or functions with the specified properties. Express your answer using formulas, and specify the independent and dependent variables.

31. A system of two linear equations in two variables that has no solution.

32. A system of two linear equations in two variables that has an infinite number of solutions, including the pairs (2, 3) and (−1, 9).

33. A system of two linear inequalities in two variables that has no solution.

34. A system of two linear equations in variables c and r with the ordered pair (−1, 0) as a solution, where c is a function of r.

35. A cost and revenue function that has a breakeven point at (100, $5000).

CHAPTER 3 REVIEW: PUTTING IT ALL TOGETHER

1. The following graph shows worldwide production and consumption of grain in millions of metric tons (MMT). It is based on data from 2005 to 2010.

a. Estimate the maximum production of grain in this time period. In what year did it occur? What was the minimum production and in what year did it occur?

b. Explain the meaning of the intersection point(s) in this context.

c. In what year was the difference between consumption and production the largest? Was there a surplus or deficit?

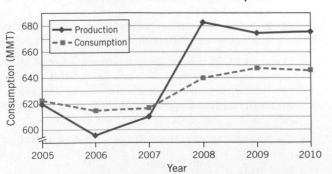

World Grain Production and Consumption

Source: Foreign Agricultural Service Circular Series, FG 05-06, May 2006, *www.fas.usda.gov/grain/circular/2006/05-06/graintoc.htm.*

2. The accompanying graphs show two systems of linear equations.

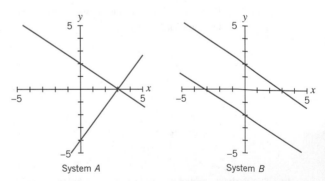

System *A*   System *B*

For each system:

a. Determine the number of solutions using the graph of the system.

b. Construct the equations of the lines.

c. If possible, solve the system using your equations from part (b).

3. a. Construct a system of linear equations where both of the following conditions are met:

  i. The coordinates of the point of intersection are (4, 10).

  ii. The two lines are perpendicular to each other and one of the lines has a slope of −4.

b. Graph the system of equations you found in part (a). Estimate the coordinates of the point of intersection on your graph. Does your estimate confirm your answer for part (a)?

4. New York City taxi fares are listed in the following table:

Initial fare	$2.50
Each 1/5 mile (4 blocks)	$0.40
Each 1 minute idle	$0.40
Peak surcharge	$1.00 (after 4 p.m. until 8 p.m. Mon–Fri)
Night surcharge	$0.50 (after 8 p.m. until 6 a.m.)
NY state tax surcharge	$0.50 (ride)
Tolls	extra
Additional riders	FREE

Source: http://www.ny.com/transportation/taxis/.

a. Create and graph on the same grid an equation for the following functions:

C_d, the cost of a day time taxi ride for m miles

C_p, the cost of a peak time taxi ride for m miles

C_n, the cost of a night time taxi ride for m miles

(Assume that you are never idling for more than one minute and no tolls are needed.)

b. Describe how the graphs compare.

c. The bus or subway fare in NYC is $2.25 (each way). Four friends want to go out to a place that is 2 miles away. Which mode of transportation, taxi or bus/subway, makes more economic sense for them? Explain when it would not make economical sense for the option you have chosen for this group.

5. Graph and shade the region bounded by the following inequalities:

$$y < 2$$
$$y \geq -2x + 3$$
$$x \leq 5$$

6. Determine the inequalities that describe the shaded region in the following graph.

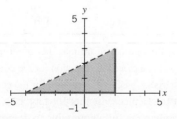

7. A musician produces and sells CDs on her website. She estimates fixed costs of $10,000, with an additional cost of $7 to produce each CD. She currently sells the CDs for $12 each.

a. Create the cost equation, C, and revenue equation R, in terms of x number of CDs produced and sold.

b. Find the breakeven point.

c. How much should the musician charge for each CD if she wants to break even when producing and selling 1600 CDs, assuming other costs remain the same?

d. By how much would she need to reduce fixed costs if she wants to break even when producing and selling 1600 CDs, assuming other costs and prices remain the same as originally stated?

8. Data from the U.S. Department of Energy show that gasoline prices vary immensely for different countries. These prices include taxes and are in U.S. dollars/gallon of unleaded regular gasoline, standardized on the U.S. gallon.

Year	Germany	Japan	United States
1990	2.65	3.16	1.16
1992	3.26	3.59	1.13
1994	3.52	4.39	1.11
1996	3.94	3.65	1.23
1998	3.34	2.83	1.06
2000	3.45	3.65	1.51
2002	3.67	3.15	1.36
2004	5.24	3.93	1.88
2006	6.03	4.47	2.58
2008	8.60	4.49	3.26
2010	7.19	5.19	2.84

a. Compute the average annual rate of change of gas prices for each country using data for 1990 and 2010.

b. Compute the average (mean) gas price from 1990 to 2010 for each country.

c. Examine the accompanying graph of the data in the table. What information is more obvious in the graph? Using the information found in parts (a) and (b) and information from the graph below, write a 60-second summary comparing the gas prices in the United States, Germany, and Japan over the time period given.

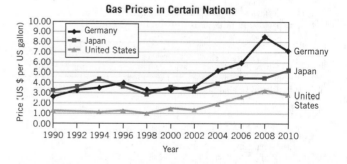

9. The graph below shows the production and consumption of oil from 1986 to 2008 by our neighbor to the north, Canada.

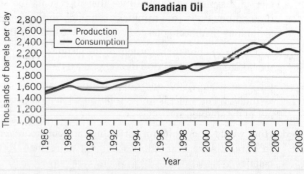

Source: EIA International Petroleum Monthly.

9. (continued)

   a. Estimate the amount of oil consumption and production in Canada in 1990 and 2008. What was the net difference amount between production and consumption? What does this mean?

   b. Estimate in which year(s) the consumption and production were the same? Explain what this means.

   c. Over what years did Canada have to import oil? Over what years was Canada able to export oil?

10. You keep track of how much gas your car uses and estimate that it gets 32 miles/gallon.

    a. If you start out with a full tank of 14 gallons, write a formula for how many gallons of gas G are left in the tank after you have gone M miles.

    b. You consider borrowing your friend's larger car, which gets 18 miles/gallon and holds 20 gallons. Write a formula for how many gallons of gas G_F are left in the tank after you have gone M miles. Generate a data table and graph the gas remaining (vertical axis) versus miles (horizontal axis) for both cars.

    c. Which car has the longer mileage range?

    d. Is there a distance at which they both have the same amount of gas left? If so, what is it?

11. a. Solve algebraically each of the following systems of linear equations.

    i. $4x + 21y = -5$ ii. $7 = 2.5a - b$
    $3x + 7y = -10$ $b = \dfrac{a}{6}$

    b. Create a system of linear equations for which there is no solution.

The accompanying table is for problems 12 and 13.

Activity	Calories Burned per Minute
Running (moderate pace)	10
Swimming laps	8

Let r and s represent the number of minutes spent running and swimming, respectively.

12. Use the table to answer the following questions. Your goal is to run and swim so that you burn at least 800 calories but no more than 1200 calories.

    a. Create a system of inequalities that represents the set of combinations of minutes running and minutes swimming that meet your goal. (*Note: $r \geq 0$ and $s \geq 0$.*)

    b. Graph the solution to the system of inequalities using s on the horizontal axis and r on the vertical axis.

    c. Give an example of one combination of running and swimming times that is in the solution set and one that is not in the solution set.

    d. How would your solution set change if your goal were to burn at least 600 calories but not more than 1000 calories?

13. Use the table to answer the following questions. Your goal is to run and swim so that you spend no more than 60 minutes exercising but burn at least 560 calories.

    a. Create a system of inequalities that represents the set of combinations of minutes running and minutes swimming that meet your goal. (*Note: $r \geq 0$ and $s \geq 0$.*)

    b. Graph the solution to the system of inequalities with s on the horizontal axis and r on the vertical axis.

    c. Give an example of one combination of running and swimming times that is in the solution set and one that is not in the solution set.

    d. How would your solution set change if you spend no more than 70 minutes exercising but still burn at least 560 calories?

14. In a 400-meter relay swim, each team has four swimmers. In sequence, the swimmers each swim 100 meters. The total time (cumulative) and rates for team A are provided in the accompanying table.

Swimmer	Total Time, Cumulative (seconds)	Rate (seconds/meter), Rounded	Total Distance, Cumulative (meters)
1	99	0.99	100
2	201	1.02	200
3	279	0.78	300
4	341	0.62	400

    a. Which swimmer is the fastest? The slowest?

    b. The total swim time function can be written as

$$t(m) = \begin{cases} 0.99m & \text{for } 0 \leq m \leq 100 \\ 99 + 1.02(m - 100) & \text{for } 100 < m \leq 200 \\ 201 + 0.78(m - 200) & \text{for } 200 < m \leq 300 \\ 279 + 0.62(m - 300) & \text{for } 300 < m \leq 400 \end{cases}$$

Find and interpret the following: $t(50)$, $t(125)$, $t(250)$, $t(400)$.

15. The accompanying table lists the monthly charge, the number of minutes allowed, and the charge per additional minute for three wireless phone plans.

Cell Phone Plan	Monthly Charge	Number of Minutes	Overtime Charge/Minute
A	$39.99	450	$0.45
B	$59.99	900	$0.40
C	$79.99	1350	$0.35

Let m be the number of minutes per month, $0 \leq m \leq 2500$.

    a. Construct a piecewise function for the cost $A(m)$ for plan A.

    b. Construct a piecewise function for the cost $B(m)$ for plan B.

    c. Construct a piecewise function for the cost $C(m)$ for plan C.

    d. Complete the following table to determine the best plan for each of the estimated number of minutes per month.

Number of Minutes/Month	Cost		
	Plan A	Plan B	Plan C
500			
800			
1000			

16. Suppose a flat tax amounts to 10% of income. Suppose a graduated tax is a fixed $1000 for any income ≤ $20,000 plus 20% of any income over $20,000.

a. Construct functions for the flat tax and the graduated tax.

b. Construct a small table of values:

Income	Flat Tax	Graduated Tax
$0		
$10,000		
$20,000		
$30,000		
$40,000		

c. Graph both tax plans and estimate any point(s) at which the two plans would be equal.

d. Use the function definitions to calculate any point(s) at which the two plans would be equal.

e. For what levels of income would the flat tax be more than the graduated tax? For what levels of income would the graduated tax be more than the flat tax.

f. Are there any conditions in which an individual might have negative income under either of the above plans, that is, the amount of taxes would exceed an individual's income? This is not as strange as it sounds. For example, many states impose a minimum corporate tax on a company, even if it is a small, one-person operation with no income in that year.

17. The time series at the bottom of the page shows the price per barrel of crude oil from 1861 to 2009 in both dollars actually spent (*nominal*) and dollars adjusted for inflation (*real* 2009 dollars). Write a 60-second summary about crude oil prices.

18. Older toilets use about 7 gallons per flush. Since using this much pure water to transport human waste is especially undesirable in areas of the country where water is scarce, new toilets now must meet a water conservation standard and are designed to use 1.6 gallons or less per flush.

a. An old toilet that leaks about 2 cups of water an hour from the tank to the bowl and uses 7 gallons/flush is replaced with a new toilet, which does not leak and uses 1.6 gallons/flush. There are 16 cups in a gallon. For each toilet write an equation for daily water use, W (gallons), as a function of number of flushes per day, F.

b. Graph the equations in part (a) on the same plot with F on the horizontal axis, where $0 \leq F \leq 25$. From your graph, estimate the amount of water used by each system for 20 flushes a day, and check your estimates with your equations. What is the net difference? What does this mean?

c. If a family flushes the toilet an average of 10 times a day, how much water do they save every day by replacing their leaky old toilet with a new water saver toilet? How much water would be saved over a year?

d. In England during World War II the citizens were asked to flush their toilets only once every five times the toilet was used, in order to save water. A pencil was hung near the toilet, and each user made a vertical mark on the wall, the fifth user made a horizontal line through the last four marks and then flushed the toilet. If this method were used today for the family with the old leaky toilet, would the amount saved be greater than the amount saved with the new toilet if the new toilet is flushed after every use?

19. Regular aerobic exercise at a target heart rate is recommended for maintaining health. Someone starting an exercise program might begin at an intensity level of 60% of the target rate and work up to 70%; athletes need to work at 85% or higher. The American College of Sports Medicine method to compute target heart rate, H (in beats per minute), is based on maximum heart rate, H_{max}, age in years, A, and exercise intensity level, I, where

$$H_{max} = 220 - A$$

and $$H = I \cdot H_{max}$$

thus $$H = I \cdot (220 - A)$$

a. Write a formula for beginners' target heart rate H_b if the intensity level, I, is 60%.

b. Write a formula for intermediate target heart rate H_i if the intensity level, I, is 70%.

c. Write a formula for athletes' target heart rate H_a if the intensity level, I, is 85%.

d. Construct a graph showing all three heart rate levels, H_b, H_i, H_a, and the maximum rate, H_{max}, for ages 15 to 75 years. Put age on the horizontal axis. Mark the zone in which athletes should work.

e. Compute the target heart rate for a 20-year-old to work at all three levels. What is H_{max} for a 20-year-old?

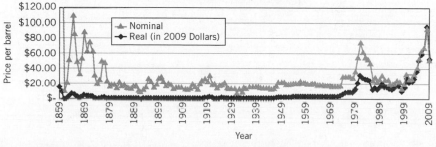

Crude Oil Prices, 1859–2009

Source: EIA International Petroleum Monthly; CPI conversion factors from oregonstate.edu/cla/polisci/faculty-research/sahr/cv2009.xls.

f. If a 65-year-old is working at a heart rate of 134 beats/minute, what is her intensity level? Has she exceeded the maximum heart rate for her age? If her 45-year-old son is working at the same heart rate, what is his intensity level? If her 25-year-old granddaughter is also working at the same heart rate, what is her intensity level?

20. In the graph to the right, the Energy Information Administration of the Department of Energy projects the energy consumption by sector from 1980 to 2008 with projections into the future.

a. In 2008 what sector was the largest consumer of energy? Estimate the amount of energy consumed.

b. In what year did the industrial and transportation sectors consume the same amount of energy? Estimate that amount. Explain what was true before that year and what is projected after that year.

c. Let $T(t)$ and $I(t)$ represent transportation and industrial consumption, respectively, for t years. Specify the years t, when

$$T(t) < I(t)$$

and when

$$T(t) > I(t)$$

d. Projecting even further into the future, which sectors appear to intersect? If their rates of increase continue as projected, which sector will be consuming more energy after that intersection point?

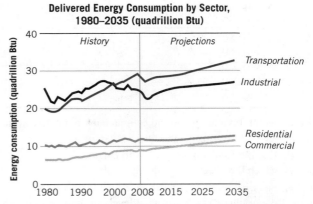

Source: http://www.eia.doe.gov/oiaf/aeo/overview.html.

Flat vs. Graduated Income Tax: Who Benefits?

Objectives

- compare the effects of different tax plans on individuals in different income brackets
- interpret intersection points

Material/Equipment

- spreadsheet or graphing calculator (optional). If using a graphing calculator, see examples on graphing piecewise functions in the Graphing Calculator Manual.
- graph paper

Related Readings

"Your Real Tax Rate," *www.msnmoney.com*, Feb. 21, 2007, and other readings can be found on the course website.

Procedure

A variety of tax plans were debated in all recent elections for president. (See related readings.) One plan recommended a flat tax of 19% on income after exemptions and deductions. In this exploration we examine who benefits from this flat tax as opposed to the current graduated tax plan. The questions we explore are

- For what income will taxes be equal under both plans?
- Who will benefit under the graduated income tax plan compared with the flat tax plan?
- Who will pay more taxes under the graduated plan compared with the flat tax plan?

In a Small Group or with a Partner

The table on the next page gives the 2010 federal graduated tax rates on income after deductions for single people.

1. Construct a function for flat taxes of 19%, where income, i, is income after deductions.
2. Construct a piecewise linear function for the graduated federal tax for single people in 2010, where income, i, is income after deductions.
3. Graph your two functions on the same grid. Estimate from your graph any intersection points for the two functions.
4. Use your equations to calculate more accurate values for the points of intersection.
5. (Extra credit.) Use your results to make changes in the tax plans. Decide on a different income for which taxes will be equal under both plans. You can use what you know about the distribution of income in the United States from the FAM1000 data to make your decision. Alter one or both of the original functions such that both tax plans will generate the same tax given the income you have chosen.

2010 Tax Rate Schedule for Single Persons

Schedule X—Use if your filing status is **Single**

If the amount on form 1040, line 40, is: Over—	*But not over—*	Enter on 1040, line 41:	*of the amount over—*
$0	$8,375	10%	$0
$8,375	$34,000	$837.50 + 15%	$8,375
$34,000	$82,400	$4,681.25 + 25%	$34,000
$82,400	$171,850	$16,781 + 28%	$82,400
$171,850	$373,650	$41,827 + 33%	$171,850
$373,650	No limit	$108,421 + 35%	$373,650

Source: www.irs.gov.

Analysis

- Interpret your findings. Assume deductions are treated the same under both the flat tax plan and the graduated tax plan. What do the intersection points tell you about the differences between the tax plans?
- What information would be useful in deciding on the merits of each of the plans?
- What if the amount of deductions that most people can take under the flat tax is less than the graduated tax plan? How will your analysis be affected?

Exploration-Linked Homework

Reporting Your Results

Take a stance for or against a flat federal income tax. Using supportive quantitative evidence, write a 60-second summary for a voters' pamphlet advocating your position. Present your arguments to the class.

EXPLORATION 3.2

A Comparison of Hybrid and Conventional Automobiles

Objectives

- compare the costs of hybrid and conventional automobiles
- interpret intersection points
- construct and solve linear systems of two equations

Material/Equipment

- graph paper
- graphing calculator or computer with graphing program

Related Readings/Links

"Compare Hybrids—Side by Side" at *http://www.fueleconomy.gov/feg/hybrid_sbs.shtml*. Other related readings and links are on the course website.

Procedure

Hybrid cars, such as the Toyota Prius, are becoming more popular as gas prices increase. However, the price of hybrid cars is often more than the price of conventional cars. In this exploration, you will create models for three cars (one hybrid and two conventional), where the total cost is represented as a function of purchase price and cost of gas/miles driven. Some questions to explore are:

- For what mileage at given gasoline prices will the cost of the cars be equal?
- How does change in the gasoline price affect the intersection(s) of the cost lines?
- How does location (for example, city vs. highway driving) affect the intersection(s) of the cost lines?

The accompanying table compares the 2010 Hybrid Prius II and conventional cars, the Corolla and the Camry.

	Purchase Price	Average MPG (city)	Average MPG (highway)
2010 Prius (Hybrid)	$22,800	51	48
2010 Corolla	$16,200	26	35
2010 Camry	$20,145	22	33

The graph on the next page gives the average gas price per gallon in the United States from February 15, 2009, to February 15, 2010.

Source: http://gasbuddy.com/gb_retail_price_chart.aspx.

Working with a Partner

If you work with a partner or in a small group, you can divvy up the work and discuss issues and your findings. Included below are possible ways to compare the cost of the cars.

1. Construct cost equations for each of the three cars, using the purchase price and the number of city miles driven, x. Estimate the average minimum price of gasoline between February 15, 2009, and February 15, 2010, from the above graph. Use this minimum price to solve the following equation:

$$C_{prius}(x) = \text{purchase price} + (\text{cost/mile})x$$

[*Hint*: cost/mile = (1/miles/gal)(cost/gallon)]

2. On the same grid, graph the three cost equations for the Prius, Corolla, and Camry at the minimum gasoline price. Find and interpret any points of intersection.

3. Now estimate the maximum price for gasoline from the graph. Use this value to construct three new cost equations and find any new intersection points. Compare these points to the intersection points derived in part (2).

4. Compare the three cars in terms of the number of highway miles driven.

Class Presentation

Present your findings to your class in the form of a 60-second summary. Include any surprising findings as well as graphs and equations to support your analysis. If a projector is available, put your visuals on overhead transparencies.

Further Exploring

What questions were raised by your analysis? What else would you want to know if you were thinking about buying a Hybrid car?

CHAPTER 4
THE LAWS OF EXPONENTS AND LOGARITHMS: MEASURING THE UNIVERSE

OVERVIEW

Most of the examples we've studied so far have come from the social sciences. In order to delve into the physical and life sciences, we need to compactly describe and compare the extremes in deep time and deep space. In this chapter, we introduce the tools that scientists use to represent very large and very small quantities.

After reading this chapter, you should be able to

- write expressions in scientific notation
- convert between English and metric units
- simplify expressions using the rules of exponents
- compare numbers of widely differing sizes
- calculate logarithms base 10 and plot numbers on a logarithmic scale

4.1 *The Numbers of Science: Measuring Time and Space*

On a daily basis we encounter quantities measured in tenths, tens, hundreds, or perhaps thousands. Finance or politics may bring us news of "1.3 billion people living in China" or "a federal debt of over $12 trillion." In the physical sciences the range of numbers encountered is much larger. *Scientific notation* was developed to provide a way to write numbers compactly and to compare the sizes found in our universe, from the largest object we know—the observable universe—to the tiniest—the minuscule quarks oscillating inside the nucleus of an atom. We use examples from deep space and deep time to demonstrate powers of 10 and the use of scientific notation.

Powers of 10 and the Metric System

The international scientific community and most of the rest of the world use the *metric system,* a system of measurements based on the meter (which is about 39.37 inches, a little over 3 feet). In daily life Americans have resisted converting to the metric system and still use the *English system* of inches, feet, and yards. Table 4.1 shows the conversions for three standard metric units of length: the meter, the kilometer, and the centimeter. For a more complete conversion table, see the inside back cover.

Conversions from Metric to English for Some Standard Units

Metric Unit	Abbreviation	In Meters	Equivalent in English Units	Informal Conversion
meter	m	1 m	3.28 ft	The width of a twin bed, a little more than a yard
kilometer	km	1000 m	0.62 mile	A casual 12-minute walk, a little over half a mile
centimeter	cm	0.01 m	0.39 in	The length of a black ant, a little under half an inch

Table 4.1

Deep space

The Observable Universe.   Current measurements with the most advanced scientific instruments generate a best guess for the radius of the observable universe at about 100,000,000,000,000,000,000,000,000 meters, or "one hundred trillion trillion meters." Obviously, we need a more convenient way to read, write, and express this number. To avoid writing a large number of zeros, exponents can be used as a shorthand:

For an appreciation of the size of things in the universe, we highly recommend the video by Charles and Ray Eames and related book by Philip and Phylis Morrison titled *Powers of Ten: About the Relative Size of Things.*

10^{26} can be written as a 1 with twenty-six zeros after it.

10^{26} means: $10 \cdot 10 \cdot 10 \cdot \cdots \cdot 10$, the product of twenty-six 10s.

10^{26} is read as "10 to the twenty-sixth" or "10 to the twenty-sixth power."

So the estimated size of the radius of the observable universe is 10^{26} meters. The sizes of other relatively large objects are listed in Table 4.2.[1]

The Relative Sizes of Large Objects in the Universe

Object	Radius (in meters)
Milky Way	$1,000,000,000,000,000,000,000 = 10^{21}$
Our solar system	$1,000,000,000,000 = 10^{12}$
Our sun	$1,000,000,000 = 10^{9}$
Earth	$10,000,000 = 10^{7}$

Table 4.2

[1]The rough estimates for the sizes of objects in the universe in this section are taken from Timothy Ferris, *Coming of Age in the Milky Way* (New York: Doubleday, 1988). Copyright © 1988 By Timothy Ferris. Reprinted by permission of Harper Collins Publishers.

Us.   Human beings are roughly in the middle of the scale of measurable objects in the universe. Human heights, including children's, vary from about one-third of a meter to 2 meters. In the wide scale of objects in the universe, a rough estimate for human height is 1 meter.

To continue the system of writing all sizes using powers of 10, we need a way to express 1 as a power of 10. Since $10^3 = 1000$, $10^2 = 100$, and $10^1 = 10$, a logical way to continue would be to say that $10^0 = 1$. Since reducing a power of 10 by 1 is equivalent to dividing by 10, the following calculations give justification for defining 10^0 as equal to 1.

$$10^2 = \frac{10^3}{10} = \frac{(10)(10)(\cancel{10})}{\cancel{10}} = 100$$

$$10^1 = \frac{10^2}{10} = \frac{(10)(\cancel{10})}{\cancel{10}} = 10$$

$$10^0 = \frac{10^1}{10} = \frac{\cancel{10}}{\cancel{10}} = 1$$

By using negative exponents, we can continue to use powers of 10 to represent numbers less than 1. For consistency, reducing the power by 1 should remain equivalent to dividing by 10. So, continuing the pattern established above, we define $10^{-1} = 1/10$, $10^{-2} = 1/10^2$, and so on. For any positive integer, n, we define

$$10^{-n} = \frac{1}{10^n}$$

DNA Molecules.   A DNA strand provides genetic information for a human being. It is made up of a chain of building blocks called nucleotides. The chain is tightly coiled into a double helix, but stretched out it would measure about 0.01 meter in length. How does this DNA length translate to a power of 10? The number 0.01, or one-hundredth, equals $1/10^2$. We can write $1/10^2$ as 10^{-2}. So a DNA strand, uncoiled and measured lengthwise, is approximately 10^{-2} meter, or one centimeter.

Table 4.3 shows the sizes of some objects relative to the size of human beings.

The Relative Sizes of Small Objects in the Universe

Object	Radius (in meters)	
Human beings	$1 = \frac{10}{10} =$	10^0
DNA molecules	$0.01 = \frac{1}{100} = \frac{1}{10^2} =$	10^{-2}
Living cells	$0.000\ 01 = \frac{1}{100,000} = \frac{1}{10^5} =$	10^{-5}
Atoms	$0.000\ 000\ 000\ 1 = \frac{1}{10,000,000,000} = \frac{1}{10^{10}} =$	10^{-10}

Table 4.3

The following box gives the definition for various powers of 10.

Powers of 10

When n is a positive integer:

$10^n = \underbrace{10 \cdot 10 \cdot 10 \cdot \, \cdots \, \cdot 10}_{n \text{ factors}}$ or a 1 followed by n zeros.

$10^0 = 1$

$10^{-n} = \dfrac{1}{10^n}$ or a decimal point followed by $n-1$ zeros and a 1.

Multiplying by 10^n is equivalent to moving the decimal point to the right n places.

Multiplying by 10^{-n} is equivalent to dividing by 10^n, or moving the decimal point to the left n places.

The metric language

By international agreement, standard prefixes specify the power of 10 that is attached to a specific unit of measure. They indicate the number of times the basic unit has been multiplied or divided by 10. Usually these prefixes are attached to metric units of measure, but they are occasionally used with the English system. Table 4.4 gives prefixes and their abbreviations for certain powers of 10. A more complete table is on the inside back cover.

Prefixes for Powers of 10

pico-	p	10^{-12}	(unit)		10^0
nano-	n	10^{-9}	kilo-	k	10^3
micro-	μ	10^{-6}	mega-	M	10^6
milli-	m	10^{-3}	giga-	G	10^9
centi-	c	10^{-2}	tera-	T	10^{12}

Table 4.4

EXAMPLE 1

Understanding units of measure
Indicate the number of meters in each unit of measure: cm, mm, Gm.

SOLUTION

$$1 \text{ cm} = 1 \text{ } centi\text{meter} = 10^{-2} \text{ m} = \frac{1}{10^2} \text{ m} = \frac{1}{100} \text{ m} = 0.01 \text{ meter}$$

$$1 \text{ mm} = 1 \text{ } milli\text{meter} = 10^{-3} \text{ m} = \frac{1}{10^3} \text{ m} = \frac{1}{1000} \text{ m} = 0.001 \text{ meter}$$

$$1 \text{ Gm} = 1 \text{ } giga\text{meter} = 10^9 \text{ m} = 1{,}000{,}000{,}000 \text{ meters}$$

EXAMPLE 2

Interpreting prefixes
Translate the following underlined expressions.

A standard CD holds about 700 megabytes of information.

Translation: $700 \cdot 10^6$ bytes or 700,000,000 bytes

A calculator takes about one millisecond to add or multiply two 10-digit numbers.

Translation: $1 \cdot 10^{-3}$ second or 0.001 second

In Tokyo on January 11, 1999, the NEC company announced that it had developed a picosecond pulse emission, optical communications laser.

Translation: $1 \cdot 10^{-12}$ second or 0.000 000 000 001 second

It takes a New York City cab driver one nanosecond to beep his horn when the light changes from red to green.

Translation: $1 \cdot 10^{-9}$ second or 0.000 000 001 second

Scientific Notation

In the previous examples we estimated the sizes of objects to the nearest power of 10 without worrying about more precise measurements. For example, we used a gross estimate of 10^7 meters for the measure of the radius of Earth. A more accurate measure is 6,368,000 meters. This number can be written more compactly using *scientific notation* as

$$6.368 \cdot 10^6 \text{ meters}$$

The number 6.368 is called the *coefficient*. The absolute value of the coefficient must always lie between 1 and 10. The power of 10 tells us how many places to shift the decimal point of the coefficient in order to get back to standard decimal form. Here, we would multiply 6.368 times 10^6, which means we would move the decimal place six places to the right, to get 6,368,000 meters.

Any nonzero number, positive or negative, can be written in scientific notation, that is, written as the product of a coefficient N multiplied by 10 to some power, where $1 \le |N| < 10$. Thus 2 million, 2,000,000, and $2 \cdot 10^6$ are all equivalent representations of the same number. The one you choose depends on the context.

Scientific Notation

A number is in *scientific notation* if it is in the form

$$N \cdot 10^n \qquad \text{where}$$

N is called the *coefficient* and $1 \le |N| < 10$

n is an integer

In the following examples, you learn how to write numbers in scientific notation. Later we use scientific notation to simplify operations with very large and very small numbers.

EXAMPLE 3   **Using scientific notation for very large numbers**
The distance to Andromeda, our nearest neighboring galaxy, is 15,000,000,000,000,000,000,000 meters. Express this number in scientific notation.

SOLUTION   The coefficient needs to be a number between 1 and 10. We start by identifying the first nonzero digit and then placing a decimal point right after it to create the coefficient of 1.5. The original number written in scientific notation will be of the form

$$1.5 \cdot 10^?$$

What power of 10 will convert this expression back to the original number? The original number is larger than 1.5, so the exponent will be positive. If we move the decimal place 22 places to the right, we will get back 1.5,000,000,000,000,000,000,000.

This is equivalent to multiplying 1.5 by 10^{22}. So, in scientific notation, 15,000,000,000,000,000,000,000 is written as

$$1.5 \cdot 10^{22}$$

EXAMPLE 4   **Using scientific notation for very small numbers**
The radius of a hydrogen atom is 0.000 000 000 052 9 meter. Express this number in scientific notation.

SOLUTION   The coefficient is 5.29. The original number written in scientific notation will be of the form

$$5.29 \cdot 10^?$$

What power of 10 will convert this expression back to the original number? The original number is smaller than 5.29, so the exponent will be negative. If we move the decimal place 11 places to the left, we will get back 0.<u>000 000 000 05</u>.2 9. This is equivalent to dividing 5.29 by 10^{11} or multiplying it by 10^{-11}:

$$0.000\ 000\ 000\ 052\ 9 = \frac{5.29}{10^{11}}$$

$$= 5.29 \left(\frac{1}{10^{11}} \right)$$

$$= 5.29 \cdot 10^{-11}$$

This number is now in scientific notation.[2]

E X A M P L E   5

Using scientific notation for negative numbers
Express $-0.000\ 000\ 000\ 052\ 9$ in scientific notation.

S O L U T I O N

In this case the coefficient, -5.29, is negative. Notice that the absolute value of the coefficient, $|-5.29|$, is equal to 5.29, which is between 1 and 10. In scientific notation, $-0.000\ 000\ 000\ 052\ 9$ is written as

$$-5.29 \cdot 10^{-11}$$

Converting from Standard Decimal Form to Scientific Notation

Place a decimal point to the right of the first nonzero digit, creating the coefficient N, where

$$1 \le |N| < 10.$$

Determine n, the power of 10 needed to convert the coefficient back to the original number.

Write in the form $N \cdot 10^n$, where the exponent n is an integer.

Examples: $346,800,000 = 3.468 \cdot 10^8$ and $0.000\ 008\ 4 = 8.4 \cdot 10^{-6}$

The poem "Imagine" offers a creative look at the Big Bang.

Deep time

The Big Bang. In 1929 the American astronomer Edwin Hubble published an astounding paper claiming that the universe is expanding. Most astronomers and cosmologists now agree with his once-controversial theory and believe that approximately 13.7 billion years ago the universe began as an explosive expansion from an infinitesimally small point. This event is referred to as the "Big Bang," and the universe has been expanding ever since it occurred.[3] Scientific notation can be used to record the progress of the universe since the Big Bang Theory, as shown in Table 4.5.

The Tale of the Universe in Scientific Notation

Object	Age (in years)
Universe	13.7 billion = 13,700,000,000 = $1.37 \cdot 10^{10}$
Earth	4.6 billion = 4,600,000,000 = $4.6 \cdot 10^9$
Human life	100 thousand = 100,000 = $1.0 \cdot 10^5$

Table 4.5

---

[2]Most calculators and computers automatically translate a number into scientific notation when it is too large or small to fit into the display. The notation is often slightly modified by using the letter E (short for "exponent") to replace the expression "times 10 to some power." So $3.0 \cdot 10^{26}$ may appear as 3.0 E+26. The number after the E tells how many places, and the sign (+ or −) indicates in which direction to move the decimal point of the coefficient.

[3]Depending on its total mass and energy, the universe will either expand forever or collapse back upon itself. However, cosmologists are unable to estimate the total mass or total energy of the universe, since they are in the embarrassing position of not being able to find about 90% of either. Scientists call this missing mass *dark matter* and missing energy *dark energy,* which describes not only their invisibility but also the scientists' own mystification.

Carl Sagan's video *Cosmos* condenses the life of the universe into one calendar year. The reading "The Universe in One Year", inspired by Carl Sagan, visualizes this concept.

Table 4.5 tells us that humans, *Homo sapiens sapiens,* first walked on Earth about 100,000 or $1.0 \cdot 10^5$ years ago. In the life of the universe, this is almost nothing. If all of time, from the Big Bang to today, were scaled down into a single year, with the Big Bang on January 1, our early human ancestors would not appear until less than 4 minutes before midnight on December 31, New Year's Eve.

EXPLORE & EXTEND

 **4.1**

What Comes Next?
Using what you know about powers of 10, generalize the exponent rules to other bases, *a.*

Algebra Aerobics 4.1

1. Express as a power of 10:
  a. 10,000,000,000
  b. 0.000 000 000 000 01
  c. 100,000
  d. $\dfrac{1}{100,000}$

2. Express in standard notation (without exponents):
  a. 10^{-8} **c.** 10^{-4}
  b. 10^{13} **d.** 10^7

3. Express as a power of 10 and then in standard notation:
  a. A nanosecond in terms of seconds
  b. A kilometer in terms of meters
  c. A gigabyte in terms of bytes

4. Rewrite each measurement in meters, first using a power of 10 and then using standard notation:
  a. 7 cm
  b. 9 mm
  c. 5 km

5. Avogadro's number is $6.02 \cdot 10^{23}$. A mole of any substance is defined to be Avogadro's number of particles of that substance. Express this number in standard notation.

6. The distance between Earth and our moon is 384,000,000 meters. Express this in scientific notation.

7. An angstrom (denoted by Å), a unit commonly used to measure the size of atoms, is 0.000 000 01 cm. Express its size using scientific notation.

8. The width of a DNA double helix is approximately 2 nanometers, or $2 \cdot 10^{-9}$ meter. Express the width in standard notation.

9. Express in standard notation:
  a. $-7.05 \cdot 10^8$ **c.** $5.32 \cdot 10^6$
  b. $-4.03 \cdot 10^{-5}$ **d.** $1.021 \cdot 10^{-7}$

10. Express in scientific notation:
  a. $-43,000,000$ **c.** 5,830
  b. $-0.000 008 3$ **d.** 0.000 000 024 1

Exercises for Section 4.1

The inside of the back cover of the text contains metric prefixes and conversion facts.

1. Write each expression as a power of 10.
  a. $10 \cdot 10 \cdot 10 \cdot 10 \cdot 10 \cdot 10$
  b. $\dfrac{1}{10 \cdot 10 \cdot 10 \cdot 10 \cdot 10}$
  c. one billion
  d. one-thousandth
  e. 10,000,000,000,000
  f. 0.000 000 01

2. Express in standard decimal notation (without exponents):
  a. 10^{-7} **c.** -10^8 **e.** 10^{-3}
  b. 10^7 **d.** -10^{-5} **f.** 10^5

3. Express each in meters, using powers of 10.
  a. 10 cm   **c.** 3 terameters
  b. 4 km   **d.** 6 nanometers

4. Express each unit using a metric prefix.
  a. 10^{-3} seconds   **b.** 10^3 grams   **c.** 10^2 meters

5. Computer storage is often measured in gigabytes and terabytes. Write these units as powers of 10.

6. Express each of the following using powers of 10.

  a. 10,000,000,000,000   **d.** $\dfrac{1}{10 \cdot 10 \cdot 10 \cdot 10}$

  b. 0.000 000 000 001   **e.** one million

  c. $10 \cdot 10 \cdot 10 \cdot 10$ **f.** one-millionth

7. Write each of the following in scientific notation:

  a. 0.000 29   **d.** 0.000 000 000 01   **g.** -0.0049

  b. 654.456   **e.** 0.000 002 45

  c. 720,000   **f.** $-1{,}980{,}000$

8. Why are the following expressions *not* in scientific notation? Rewrite each in scientific notation.

  a. $25 \cdot 10^4$ **c.** $0.012 \cdot 10^{-2}$

  b. $0.56 \cdot 10^{-3}$ **d.** $-425.03 \cdot 10^2$

9. Write each of the following in standard decimal form:

  a. $7.23 \cdot 10^5$ **d.** $1.5 \cdot 10^6$

  b. $5.26 \cdot 10^{-4}$ **e.** $1.88 \cdot 10^{-4}$

  c. $1.0 \cdot 10^{-3}$ **f.** $6.78 \cdot 10^7$

10. Express each in scientific notation. (Refer to the chart in Exploration 4.1, p. 261.)

  a. The age of the observable universe

  b. The size of the first living organism on Earth

  c. The size of Earth

  d. The age of Pangaea

  e. The size of the first cells with a nucleus

11. Determine if the expressions are true or false. If false, change the right-hand side to make the expression true.

  a. $0.00\,756 = 7.56 \cdot 10^{-2}$

  b. $3.432 \cdot 10^5 = 343{,}200$

  c. 49 megawatts $= 4.9 \cdot 10^6$ watts

  d. $1{,}596{,}000{,}000 = 1.5 \cdot 10^9$

  e. 5 megapixels $= 5 \cdot 10^6$ pixels

  f. 6 picoseconds $= 6 \cdot 10^{12}$ seconds

12. Express each quantity in scientific notation.

  a. The mass of an electron is about 0.000 000 000 000 000 000 000 000 001 67 gram.

  b. One cubic inch is approximately 0.000 016 cubic meter.

  c. The radius of a virus is about 0.000 000 05 meter.

13. Evaluate:

  a. $|9|$ **c.** $|-1000|$

  b. $|-9|$ **d.** $-|-1000|$

14. The accompanying amusing graph shows a roughly linear relationship between the "scientifically" calculated age of Earth and the year the calculation was published. For instance, in about 1935 Ellsworth calculated that Earth was about 2 billion years old. The age is plotted on the horizontal axis and the year the calculation was published on the vertical axis. The triangle on the horizontal coordinate represents the presently accepted age of Earth.

  a. Who calculated that Earth was less than 1 billion years old? Give the coordinates of the points that give this information.

  b. In about what year did scientists start putting the age of Earth at over a billion years? Give the coordinates that represent this point.

  c. On your graph sketch an approximation of a best-fit line for these points. Use two points on the line to calculate the slope of the line.

  d. Interpret the slope of that line in terms of the year of calculation and the estimated age of Earth.

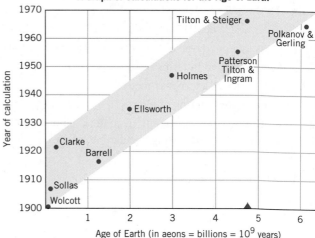

A Graph of Calculations for the Age of Earth

Reprinted by permission of American Scientist, magazine of Sigma Xi, The Scientific Research Society.

4.2 *Positive Integer Exponents*

No matter what the base, whether it is 10 or any other number, repeated multiplication leads to *exponentiation*. For example,

$$3 \cdot 3 \cdot 3 \cdot 3 = 3^4$$

Here 4 is the *exponent* of 3, and 3 is called the *base*. In general, if a is a real number and n is a positive integer, then we define a^n as the product of n factors of a.

> **Definition of a^n**
>
> In the expression a^n, the number a is called the *base* and n is called the *exponent* or *power*.
>
> If n is a positive integer, then
>
> $$a^n = \underbrace{a \cdot a \cdot a \cdots a}_{n \text{ factors}} \qquad \text{(the product of } n \text{ factors of } a\text{)}$$

Exponent Rules

In this section we see how the rules for manipulating expressions with exponents make sense if we remember what the exponent tells us to do to the base. First we focus on cases where the exponents are positive integers and then apply the rules to cases where the exponents can be any rational number, such as a negative integer or a fraction.

> **Rules for Exponents**
>
> **1.** $a^n \cdot a^m = a^{(n+m)}$
>
> **2.** $\dfrac{a^n}{a^m} = a^{(n-m)}$ where $a \neq 0$
>
> **3.** $(a^m)^n = a^{(m \cdot n)}$
>
> **4.** $(ab)^n = a^n b^n$
>
> **5.** $\left(\dfrac{a}{b}\right)^n = \dfrac{a^n}{b^n}$ where $b \neq 0$

We show below how Rules 1, 3, and 5 make sense and leave Rules 2 and 4 for you to justify in the exercises.

Rule 1. To justify this rule, think about the total number of times a is a factor when a^n is multiplied by a^m:

$$a^n \cdot a^m = \underbrace{a \cdot a \cdot a \cdots a}_{n \text{ factors}} \cdot \underbrace{a \cdot a \cdots a}_{m \text{ factors}} = \underbrace{a \cdot a \cdot a \cdots a}_{n+m \text{ factors}} = a^{(n+m)}$$

Rule 3. First think about how many times a^m is a factor when we raise it to the nth power:

$$(a^m)^n = \underbrace{\overbrace{a^m \cdot a^m \cdots a^m}^{n \text{ terms}}}_{n \text{ factors of } a^m}$$

Use Rule 1: $\qquad\qquad = a^{(m+m+\cdots+m)}$

Represent adding m
n times as $m \cdot n$ $\qquad = a^{(m \cdot n)}$

Rule 5. Remember that the exponent n in the expression $(a/b)^n$ applies to the whole expression within the parentheses:

$$\left(\frac{a}{b}\right)^n = \underbrace{\left(\frac{a}{b}\right) \cdot \left(\frac{a}{b}\right) \cdots \left(\frac{a}{b}\right)}_{n \text{ factors of } a/b}$$

$$= \frac{\overbrace{a \cdot a \cdots a}^{n \text{ factors of } a}}{\underbrace{b \cdot b \cdots b}_{n \text{ factors of } b}}$$

$$= \frac{a^n}{b^n}$$

EXAMPLE 1 Simplify and write as an expression with exponents:

$$7^3 \cdot 7^2 = 7^{3+2} = 7^5 \qquad\qquad (x^5)^3 = x^{5 \cdot 3} = x^{15}$$

$$w^3 \cdot w^5 = w^{3+5} = w^8 \qquad\qquad (11^2)^4 = 11^{2 \cdot 4} = 11^8$$

$$\frac{10^8}{10^3} = 10^{8-3} = 10^5 \qquad\qquad \frac{z^8}{z^3} = z^{8-3} = z^5$$

EXAMPLE 2 Simplify:

$$(3a)^4 = 3^4 a^4 = 81a^4$$

$$(-5x)^3 = (-5)^3 x^3 = -125x^3$$

$$\left(\frac{2}{3}\right)^3 = \frac{2^3}{3^3} = \frac{8}{27}$$

EXAMPLE 3 Simplify:

$$\left(\frac{-2a}{3b}\right)^3 = \frac{(-2a)^3}{(3b)^3} = \frac{(-2)^3 a^3}{3^3 b^3} = \frac{-8a^3}{27b^3}$$

$$\frac{-5(x^3)^2}{(2y^2)^3} = \frac{-5x^6}{8y^6}$$

EXAMPLE 4 **Using scientific notation to simplify calculations**
Deneb is 1600 light years from Earth. How far is Earth from Deneb when measured in miles?

SOLUTION The distance that light travels in 1 year, called a *light year,* is approximately 5.88 trillion miles.

Since 1 light year = 5,880,000,000,000 miles

then the distance from Earth to Deneb is

$$1600 \text{ light years} = (1600) \cdot (5{,}880{,}000{,}000{,}000 \text{ miles})$$

$$= (1.6 \cdot 10^3) \cdot (5.88 \cdot 10^{12} \text{ miles})$$

$$= (1.6 \cdot 5.88) \cdot (10^3 \cdot 10^{12}) \text{ miles}$$

$$\approx 9.4 \cdot 10^{3+12} \text{ miles}$$

$$\approx 9.4 \cdot 10^{15} \text{ miles}$$

Using ratios to compare sizes of objects

In comparing two objects of about the same size, it is common to subtract one size from the other and say, for instance, that one person is 6 inches taller than another. This method of comparison is not effective for objects of vastly different sizes. To say that the difference between the estimated radius of our solar system (1 terameter, or 1,000,000,000,000 meters) and the average size of a human (about 10^0 or 1 meter) is 1,000,000,000,000 − 1 = 999,999,999,999 meters is not particularly useful. In fact, since our measurement of the solar system certainly isn't accurate to within 1 meter, this difference is meaningless. As shown in the following example, a more useful method for comparing objects of wildly different sizes is to calculate the ratio of the two sizes.

E X A M P L E   5   **Comparing objects using ratios**
How many times larger is the sun than Earth?

S O L U T I O N   1   The radius of the sun is approximately 10^9 meters and the radius of Earth is about 10^7 meters. One way to answer the question "How many *times* larger is the sun than Earth?" is to form the ratio of the two radii:

$$\frac{\text{radius of the sun}}{\text{radius of Earth}} = \frac{10^9 \text{ m}}{10^7 \text{ m}}$$

$$= \frac{10^9 \text{ m}}{10^7 \text{ m}} = 10^{9-7} = 10^2$$

The units cancel, so 10^2 is unitless. The radius of the sun is approximately 10^2, or 100, times larger than the radius of Earth.

S O L U T I O N   2   Another way to answer the question is to compare the volumes of the two objects. The sun and Earth are both roughly spherical. The formula for the volume V of a sphere with radius r is $V = \frac{4}{3}\pi r^3$.

The radius of the sun is approximately 10^9 meters and the radius of Earth is about 10^7 meters. The ratio of the two volumes is

$$\frac{\text{volume of the sun}}{\text{volume of Earth}} = \frac{(4/3)\pi(10^9)^3 \text{ m}^3}{(4/3)\pi(10^7)^3 \text{ m}^3}$$

$$= \frac{(10^9)^3}{(10^7)^3} \quad (\textit{Note: } \tfrac{4}{3}\pi \text{ and m}^3 \text{ cancel.})$$

$$= \frac{10^{27}}{10^{21}}$$

$$= 10^6$$

So while the radius of the sun is 100 times larger than the radius of Earth, the *volume* of the sun is approximately $10^6 = 1,000,000$, or 1 million, times larger than the volume of Earth!

Common Errors

The first question to ask in evaluating expressions with exponents is: To what does the exponent apply? Consider the following expressions:

1. $-a^n = -(a^n)$ **but** $-a^n \neq (-a)^n$ **(unless n is odd)**

For example, in the expression -2^4, the exponent 4 applies only to 2, not to -2. The order of operations says to compute the power first, before applying the negation sign. So $-2^4 = -(2^4) = -16$. If we want to raise -2 to the fourth power, we write $(-2)^4 = (-2)(-2)(-2)(-2) = 16$.

In the expression $(-3b)^2$, everything inside the parentheses is squared. So $(-3b)^2 = (-3b)(-3b) = 9b^2$. But in the expression $-3b^2$, the exponent 2 applies only to the base b.

In the case where n is an *odd integer*, then $(-a)^n$ will equal $-(a)^n$. For example, $(-2)^3 = (-2)(-2)(-2) = -8 = -2^3$.

2. $ab^n = a(b^n)$ **and** $-ab^n = -a(b^n)$ **but** $ab^n \neq (ab)^n$

Remember, the exponent applies only to the variable to which it is attached. In the expressions ab^n and $-ab^n$, only b is raised to the nth power.

For example,

$$2 \cdot 5^3 = 2 \cdot 125 = 250 \qquad \text{but} \quad (2 \cdot 5)^3 = (10)^3 = 1000$$

$$-2 \cdot 5^3 = -2 \cdot 125 = -250 \qquad \text{but} \quad (-2 \cdot 5)^3 = (-10)^3 = -1000$$

You can use parentheses ( ) to indicate when more than one variable is raised to the nth power.

3. $(ab)^n = a^n b^n$ **but** $(a + b)^n \neq a^n + b^n$ **(if $n \neq 1$)**

For example,

$$(2 \cdot 5)^3 = 2^3 \cdot 5^3 \qquad \text{but} \qquad (2 + 5)^3 \neq 2^3 + 5^3$$
$$(10)^3 = 8 \cdot 125 \qquad\qquad\qquad (7)^3 \neq 8 + 125$$
$$1000 = 1000 \qquad\qquad\qquad 343 \neq 133$$

4. $a^n \cdot a^m = a^{n+m}$ **but** $a^n + a^m \neq a^{n+m}$

For example,

$$10^2 \cdot 10^3 = 10^5 \qquad \text{but} \qquad 10^2 + 10^3 \neq 10^5$$
$$100 \cdot 1000 = 100{,}000 \qquad\qquad 100 + 1000 \neq 100{,}000$$

Algebra Aerobics 4.2a

1. Simplify where possible, leaving the answer in a form with exponents:
   a. $10^5 \cdot 10^7$ d. $5^5 \cdot 6^7$ g. $3^4 + 7 \cdot 3^4$
   b. $8^6 \cdot 8^{14}$ e. $7^3 + 7^3$ h. $2^3 + 2^4$
   c. $z^5 \cdot z^4$ f. $5 \cdot 5^6$ i. $2^5 + 5^2$

2. Simplify (if possible), leaving the answer in exponent form:
   a. $\dfrac{10^{15}}{10^7}$ c. $\dfrac{3^5}{3^4}$ e. $\dfrac{5}{5^6}$ g. $\dfrac{2^3 \cdot 3^4}{2 \cdot 3^2}$
   b. $\dfrac{8^6}{8^4}$ d. $\dfrac{5}{6^7}$ f. $\dfrac{3^4}{3}$ h. $\dfrac{6}{2^4}$

3. Write each number as a power of 10, then perform the indicated operation. Write your final answer as a power of 10.
   a. $100{,}000 \cdot 1{,}000{,}000$
   b. $1{,}000 \cdot 0.000\,001$
   c. $0.000\,000\,000\,01 \cdot 0.000\,01$
   d. $\dfrac{1{,}000{,}000{,}000}{10{,}000}$
   e. $\dfrac{1{,}000{,}000}{0.001}$
   f. $\dfrac{0.000\,01}{0.0001}$
   g. $\dfrac{0.000\,001}{10{,}000}$

4. Simplify:
   a. $(10^4)^5$ d. $(2x)^4$ g. $(-3x^2)^3$
   b. $(7^2)^3$ e. $(2a^4)^3$ h. $((x^3)^2)^4$
   c. $(x^4)^5$ f. $(-2a)^3$ i. $(-5y^2)^3$

5. Simplify:
   a. $\left(\dfrac{-2x}{4y}\right)^3$ c. -5^2 e. $(-3yz^2)^4$
   b. $(-5)^2$ d. $-3(yz^2)^4$ f. $(-3yz^2)^3$

6. A DVD optical disk has a storage capacity of about 4.7 gigabytes ($4.7 \cdot 10^9$ bytes). If a hard drive has a capacity of 3 terabytes ($3.0 \cdot 10^{12}$ bytes), how many DVDs would it take to equal the storage capacity of the hard drive?

7. Write as a single number with no exponents:
   a. $(3 + 5)^3$ c. $3 \cdot 5^2$
   b. $3^3 + 5^3$ d. $-3 \cdot 5^2$

Estimating Answers

By rounding off numbers and using scientific notation and the rules for exponents, we can often make quick estimates of answers to complicated calculations. In this age of calculators and computers we need to be able to roughly estimate the size of an answer, to make sure our calculations with technology make sense.

EXAMPLE 6   **Using scientific notation to do calculations**
Estimate the value of

$$\frac{(382{,}152) \cdot (490{,}572{,}261)}{(32{,}091) \cdot (1942)}$$

Express your answer in both scientific and standard notation.

SOLUTION Round each number:

$$\frac{(382,152) \cdot (490,572,261)}{(32,091) \cdot (1942)} \approx \frac{(400,000) \cdot (500,000,000)}{(30,000) \cdot (2000)}$$

rewrite in scientific notation $\approx \dfrac{(4 \cdot 10^5) \cdot (5 \cdot 10^8)}{(3 \cdot 10^4) \cdot (2 \cdot 10^3)}$

group the coefficients and the powers of 10 $\approx \left(\dfrac{4 \cdot 5}{3 \cdot 2}\right) \cdot \left(\dfrac{10^5 \cdot 10^8}{10^4 \cdot 10^3}\right)$

simplify each expression $\approx \dfrac{20}{6} \cdot \dfrac{10^{13}}{10^7}$

we get in scientific notation $\approx 3.33 \cdot 10^6$

or in standard notation $\approx 3,330,000$

Using a calculator on the original problem, we get a more precise answer of 3,008,200.

EXAMPLE 7

Farmable land

As of 2010 the world population was approximately 6.848 billion people. There are roughly 57.9 million square miles of land on Earth, of which about 22% are favorable for agriculture. Estimate how many people per square mile of farmable land there are as of 2010.

SOLUTION

$$\frac{\text{size of world population}}{\text{amount of farmable land}} = \frac{6.848 \text{ billion people}}{22\% \text{ of } 57.9 \text{ million square miles}}$$

rewrite as powers of 10 $= \dfrac{6.848 \cdot 10^9 \text{ people}}{(0.22) \cdot (57.9) \cdot 10^6 \text{ mile}^2}$

round each number $\approx \dfrac{6.8 \cdot 10^9 \text{ people}}{(0.2) \cdot 60 \cdot 10^6 \text{ mile}^2}$

simplify $\approx \dfrac{6.8 \cdot 10^9 \text{ people}}{12 \cdot 10^6 \text{ mile}^2}$

we get in scientific notation $\approx 0.57 \cdot 10^3 \text{ people/mile}^2$

or in standard notation $\approx 570 \text{ people/mile}^2$

So there are roughly 570 people/mile^2 of farmable land in the world. Using a calculator and the original numbers, we get a more accurate answer of 538 people/mile^2 of farmable land, which is close to our estimate.

EXPLORE & EXTEND

 4.2

Finding Specific and General Cases

Why does $(a + b)^2 \neq a^2 + b^2$?

a. Find a specific case for values of a and b where these expressions are not equal.

Are there any cases when these expressions would be equal?

b. Generalize by rewriting $(a + b)^2 = (a + b)(a + b) = $ _____.

Algebra Aerobics 4.2b

1. Estimate the value of:

   a. $(0.000\ 297\ 6) \cdot (43{,}990{,}000)$

   b. $\dfrac{453{,}897 \cdot 2{,}390{,}702}{0.004\ 38}$

   c. $\dfrac{0.000\ 000\ 319}{162{,}000}$

   d. $28{,}000{,}000 \cdot 7{,}629$

   e. $0.000\ 021 \cdot 391{,}000{,}000$

2. Evaluate the following without the aid of a calculator:

   a. $(3.0 \cdot 10^3)(4.0 \cdot 10^2)$ c. $\dfrac{2.0 \cdot 10^5}{5.0 \cdot 10^3}$

   b. $\dfrac{(5.0 \cdot 10^2)^2}{2.5 \cdot 10^3}$ d. $(4.0 \cdot 10^2)^3 \cdot (2.0 \cdot 10^3)^2$

3. The radius of Jupiter, the largest of the planets in our solar system, is approximately $7.14 \cdot 10^4$ km. (If r is the radius of a sphere, the sphere's surface area equals $4\pi r^2$ and its volume equals $\frac{4}{3}\pi r^3$.) Assuming Jupiter is roughly spherical,

   a. Estimate the surface area of Jupiter.

   b. Estimate the volume of Jupiter.

   Express your answers in scientific notation.

4. Only about three-sevenths of the land favorable for agriculture is actually being farmed. Using the facts in Example 7, estimate the number of people per square mile of farmable land that is being used. Should your estimate be larger or smaller than the ratio of people to farmable land? Explain. (Round your answer to the nearest integer.)

Exercises for Section 4.2

1. Simplify, when possible, writing your answer as an expression with exponents:

   a. $10^4 \cdot 10^3$ d. $x^5 \cdot x^{10}$ g. $\dfrac{z^7}{z^2}$ j. $4^5 \cdot (4^2)^3$

   b. $10^4 + 10^3$ e. $(x^5)^{10}$ h. 256^0

   c. $10^3 + 10^3$ f. $4^7 + 5^2$ i. $\dfrac{3^5 \cdot 3^2}{3^8}$

2. Simplify:

   a. $(-1)^4$ c. $(a^4)^3$ e. $(2a^4)^3$ g. $(10a^2b^3)^3$

   b. $-(1)^4$ d. $-(2a^2)^3$ f. $(-2a^4)^3$ h. $(2ab)^2 - 3a^2b^2$

3. Simplify:

   a. $(-2a)^4$ e. $(2x^4)^5$

   b. $-2(a)^4$ f. $(-4x^3)^2 + x^3(2x^3)$

   c. $(-x^5)^3$ g. $(50a^{10})^2$

   d. $(-2ab^2)^3$ h. $(3ab)^3 + ab$

4. Simplify and write each variable as an expression with positive exponents:

   a. $-\left(\dfrac{3}{5}\right)^2$ b. $\left(\dfrac{-5a^3}{a^2}\right)^4$ c. $\left(\dfrac{10a^3}{5b}\right)^2$ d. $\left(\dfrac{-2x^3}{3y^2}\right)^3$

5. Simplify and write each variable as an expression with positive exponents:

   a. $-\left(\dfrac{5}{8}\right)^2$ b. $\left(\dfrac{3x^3}{5y^2}\right)^3$ c. $\left(\dfrac{-10x^5}{2b^2}\right)^4$ d. $\left(\dfrac{-x^5}{x^2}\right)^3$

6. Evaluate and express your answer in standard decimal form:

   a. $-2^4 + 2^2$ e. $10^3 + 2^3$

   b. $-2^3 + (-4)^2$ f. $2 \cdot 10^3 + 10^3 + 10^2$

   c. $2 \cdot 3^2 - 3(-2)^2$ g. $2 \cdot 10^3 + (-10)^3$

   d. $-10^4 + 10^5$ h. $(1000)^0$

7. Convert each number into scientific notation then perform the indicated operation. Leave your answer in scientific notation.

   a. $2{,}000{,}000 \cdot 4000$

   b. 1.4 million $\div\ 7000$

   c. 50 billion $\cdot\ 60$ trillion

   d. 2500 billion $\div\ 500$ thousand

8. Convert each number into scientific notation and then perform the operation without a calculator.

   a. $60{,}000{,}000{,}000 + 40{,}000{,}000{,}000$

   b. $\dfrac{(20{,}000)^6}{(400)^3}$

   c. $(2{,}000{,}000) \cdot (40{,}000)$

9. Simplify each expression using the properties of exponents.

   a. $(x^5y)(x^6)(x^2y^3)$ c. $\left(\dfrac{-2x^5y^5}{x^2y^2}\right)^3$ e. $(3x^2y^5)^4$

   b. $\dfrac{5x^6y^3}{x^2y^2}$ d. $(x^2)^5 \cdot (2y^2)^4$ f. $\left(\dfrac{3x^3y}{5xy}\right)^2$

10. Each of the following simplifications contains an error made by students on a test. Find the error and correct the simplification so that the expression becomes true.

   a. $[(x^2)^3]^5 = [x^5]^5 = x^{25}$

   b. $\dfrac{7x^2y^6}{(xy)^2} = \dfrac{7x^2y^6}{x^2y^2} = 7x^4y^8$

   c. $\left(\dfrac{4x^3y^5}{6xy^4}\right)^3 = \left(\dfrac{2x^2y}{3}\right)^3 = \frac{2}{3}x^6y^3$

   d. $(1.1 \cdot 10^6) \cdot (1.1 \cdot 10^4) = 1.1 \cdot 10^6$

   e. $\dfrac{4 \cdot 10^6}{8 \cdot 10^3} = 0.5 \cdot 10^3 = 5.0 \cdot 10^4$

   f. $6 \cdot 10^3 + 7 \cdot 10^5 = 13 \cdot 10^8$

11. Express your answer as a power of 10 and in standard decimal form. (Refer to table on p. 214.)

    a. How many times larger is a gigabyte of memory than a megabyte?

    b. How many times farther is a kilometer than a dekameter?

    c. How many times heavier is a kilogram than a milligram?

    d. How many times longer is a microsecond than a nanosecond?

12. In 2010 the People's Republic of China was estimated to have about 1,338,000,000 people, and Monaco about 33,000. Monaco has an area of 0.75 miles^2, and China has an area of 3,705,000 miles^2.

    a. Express the populations and geographic areas in scientific notation.

    b. By calculating a ratio, determine how much larger China's population is than Monaco's.

    c. What is the population density (the number of people per square mile) for each country?

    d. Write a paragraph comparing and contrasting the population size and density for these two nations.

13. a. In 2010 Japan had a population of approximately 127.4 million people and a total land area of about 146.9 thousand square miles. What was the population density (the number of people per square mile)?

    b. In 2010 the United States had a population of approximately 309.3 million people and a total land area of about 3720 thousand square miles. What was the population density of the United States?

    c. Compare the population densities of Japan and the United States.

14. The distance that light travels in 1 year (a light year) is $5.88 \cdot 10^{12}$ miles. If a star is $2.4 \cdot 10^8$ light years from Earth, what is this distance in miles?

15. On April 20th, 2010, an explosion and fire on an oil rig operated by British Petroleum (BP) caused the worst oil spill in history. On May 20th, the Associated Press reported, "oil has been pouring into the Gulf from a blown-out undersea well at a rate of at least 210,000 gallons per day." Unfortunately, this turned out to be a best-case estimate. On June 20, 2010, Associated Press reported that a confidential memo by BP gave a worst-case estimate of 4.2 million gallons per day pouring into the ocean.

    a. What is the difference in gallons per day between the worst- and best-case scenarios? Use scientific notation to express your answer.

    b. Using orders of magnitude, how many times larger is the worst-case scenario than the best-case?

    c. After 105 days, a temporary cap on the pipe began to stop the spill. Using the best-case scenario, estimate how much oil was spilled over 105 days. Using the worst-case, how much was spilled?

16. Change each number into scientific notation, then perform the indicated calculation without a calculator.

    a. A $600,000 lottery jackpot is divided among 300 people. What are the winnings per person?

    b. A total of 2500 megawatts are used over 500 hours. What is the rate in watts per hour?

    c. If there were 6 million births in 30 years, what is the birth rate per year?

17. a. For any nonzero real number a, what can we say about the sign of the expression $(-a)^n$ when n is an even integer? What can we say about the sign of $(-a)^n$ when n is an odd integer?

    b. What is the sign of the resulting number if a is a positive number? If a is a negative number? Explain your answer.

18. Round off the numbers and then estimate the value of each of the following expressions without using a calculator. Show your work, writing your answers in scientific notation. If available, use a calculator to verify your answers.

    a. $(2,968,001,000) \cdot (189,000)$

    b. $(0.000\ 079) \cdot (31,140,284,788)$

    c. $\dfrac{4,083,693 \cdot 49,312}{213 \cdot 1945}$

19. Simplify each expression using two different methods, and then compare your answers.

    Method I: Simplify inside the parentheses first, and then apply the exponent rule outside the parentheses.

    Method II: Apply the exponent rule outside the parentheses, and then simplify.

    a. $\left(\dfrac{m^2 n^3}{mn}\right)^2$ b. $\left(\dfrac{2a^2 b^3}{ab^2}\right)^4$

20. Verify that $(a^2)^3 = (a^3)^2$ using the rules of exponents.

21. Verify that $\left(\dfrac{2a^3}{5b^2}\right)^4 = \dfrac{16a^{12}}{625b^8}$ using the rules of exponents.

22. As of May 22, 2010, the U.S. national debt was estimated at $12,995,989,428,360 and the estimated population of the United States was 308,422,640. Convert each number into scientific notation and then estimate the debt per person in the U.S. Check your estimate using a calculator. You can find an estimate of the current debt and population at *www.brillig.com/debt_clock/* to see how much the debt/person has changed.

23. Justify Rule 4 for exponents when n is an integer ≥ 0:

    If a and b are any nonzero real numbers, then

    $$(ab)^n = a^n b^n$$

24. Justify Rule 2 for exponents when $n \geq m$ and m and n are integers > 0:

    If a is any nonzero real number, then

    $$\frac{a^n}{a^m} = a^{(n-m)}$$

25. In 2008 the United Kingdom generated approximately 53 terawatt-hours of nuclear energy[4] for a population of about 60 million on 94,000 square miles. In the same year the United States generated approximately 805 terawatt-hours of nuclear energy for a population of about 309 million on about 3,720,000 miles^2. A terawatt is 10^{12} watts.

[4]*www.world-nuclear.org.*

25. (continued)

a. How many terawatt-hours is the United Kingdom generating per person? How many terawatt-hours is it generating per square mile? Express each in scientific notation.

b. How many terawatt-hours is the United States generating per person? How many terawatt-hours are we generating per square mile? Express each in scientific notation.

c. How much nuclear energy is being generated in the United Kingdom per square mile relative to the United States?

d. Write a brief statement comparing the relative magnitude of generation of nuclear power per person in the United Kingdom and the United States.

4.3 Zero, Negative, and Fractional Exponents

In this section we show how to apply the rules for exponents to expressions with zero, negative, and fractional exponents. First we define each type of exponent and then show how to apply the rules of exponents.

Zero and Negative Exponents

The definitions for raising any base to the zero power or a negative power follow a logic that is similar to the one used to define 10^0 and 10^{-n}.

Zero and Negative Exponents

If n is a positive integer and $a \neq 0$, then

$$a^0 = 1 \qquad \text{For example, } 100^0 = 1$$

$$a^{-n} = \frac{1}{a^n} \qquad \text{For example, } a^{-100} = \frac{1}{a^{100}}$$

In the following examples, we show how the rules for exponents are applied when the exponents are negative integers or zero.

EXAMPLE 1

Simplifying expressions with exponents

Simplify the following expressions and express your answers with positive exponents.

a. $x^2 \cdot x^{-5}$ **b.** $\dfrac{(-5)^2}{(-5)^6}$ **c.** $\dfrac{x^{-2}}{x^4}$ **d.** $(13^{-8})^3$ **e.** $\dfrac{v^{-2}(w^5)^2}{(v^{-1})^4 w^{-3}}$

SOLUTION

a. Using Rule 1 for exponents,

$$x^2 \cdot x^{-5} = x^{2+(-5)} = x^{-3} = \frac{1}{x^3}$$

b. Using Rule 2 for exponents,

$$\frac{(-5)^2}{(-5)^6} = (-5)^{2-6} = (-5)^{-4} = \frac{1}{(-5)^4} = \frac{1}{(-1)^4(5)^4} = \frac{1}{5^4}$$

c. Using Rule 2 for exponents,

$$\frac{x^{-2}}{x^4} = x^{-2-4} = x^{-6} = \frac{1}{x^6}$$

d. Using Rule 3 for exponents,

$$(13^{-8})^3 = 13^{(-8)3} = 13^{-24} = \frac{1}{13^{24}}$$

e. Apply Rule 3 twice: $\dfrac{v^{-2}(w^5)^2}{(v^{-1})^4 w^{-3}} = \dfrac{v^{-2} w^{10}}{v^{-4} w^{-3}}$

Apply Rule 2 twice: $= v^{-2-(-4)} w^{10-(-3)}$

                                      $= v^2 w^{13}$

Evaluating $\left(\frac{a}{b}\right)^{-n}$

The rule for applying negative powers is the same whether a is an integer or a fraction:

$$a^{-n} = 1/a^n \quad \text{where } a \neq 0$$

For example,

$$\left(\frac{1}{2}\right)^{-1} = \frac{1}{(1/2)^1} = 1 \div \left(\frac{1}{2}\right) = 1 \cdot \left(\frac{2}{1}\right) = 2$$

In general, if a and b are nonzero, then

$$\left(\frac{a}{b}\right)^{-n} = \frac{1}{(a/b)^n} = 1 \div \left(\frac{a}{b}\right)^n = 1 \cdot \left(\frac{b}{a}\right)^n = \left(\frac{b}{a}\right)^n = \frac{b^n}{a^n}$$

EXAMPLE 2   Simplify:

a. $\left(\frac{1}{2}\right)^{-11} \cdot \left(\frac{1}{2}\right)^{-2}$ **b.** $\left(\frac{a}{b}\right)^{3} \cdot \left(\frac{a}{b}\right)^{-5}$

SOLUTION   **a.** Using Rule 1 for exponents and the definition of a^{-n},

$$\left(\frac{1}{2}\right)^{-11} \cdot \left(\frac{1}{2}\right)^{-2} = \left(\frac{1}{2}\right)^{-11+(-2)}$$

$$= \left(\frac{1}{2}\right)^{-13} = \left(\frac{2}{1}\right)^{13} = 2^{13} = 8192$$

b. Using Rules 1 and 5 for exponents and the definition of a^{-n},

$$\left(\frac{a}{b}\right)^{3} \cdot \left(\frac{a}{b}\right)^{-5} = \left(\frac{a}{b}\right)^{3+(-5)}$$

$$= \left(\frac{a}{b}\right)^{-2} = \left(\frac{b}{a}\right)^{2} = \frac{b^2}{a^2}$$

Algebra Aerobics 4.3a

1. Simplify and express, if possible, with a single positive exponent.

a. $10^5 \cdot 10^{-7}$

b. $\dfrac{11^6}{11^{-4}}$

c. $\dfrac{3^{-5}}{3^{-4}}$

d. $\dfrac{5^5}{6^7}$

e. $\dfrac{7^3}{7^3}$

f. $a^{-2} \cdot a^{-3}$

g. $3^4 \cdot 3^3$

h. $(2^2 \cdot 3)(2^6)(2^4 \cdot 3)$

2. A typical TV signal, traveling at the speed of light, takes $3.3 \cdot 10^{-6}$ second to travel 1 kilometer. Estimate how long it would take the signal to travel across the United States (a distance of approximately 4300 kilometers).

3. Distribute and simplify:

a. $x^{-2}(x^5 + x^{-6})$

b. $-a^2(b^2 - 3ab + 5a^2)$

4. Simplify:

a. $(10^4)^{-5}$

b. $(7^{-2})^{-3}$

c. $(2a^3)^{-2}$

d. $\left(\dfrac{8}{x}\right)^{-2}$

e. $(2x^{-2})^{-1}$

f. $2(x^{-2})^{-1}$

g. $\left(\dfrac{3}{2y^2}\right)^{-4}$

h. $\dfrac{3}{(2y^2)^{-4}}$

5. Simplify:

a. $\dfrac{t^{-3}t^0}{(t^{-4})^3}$

b. $\dfrac{v^{-3}w^7}{(v^{-2})^3w^{-10}}$

c. $\dfrac{7^{-8}x^{-1}y^2}{7^{-5}xy^3}$

d. $\dfrac{a(5b^{-1}c^3)^2}{5ab^2c^{-6}}$

Fractional Exponents

Before we can apply the rules of exponents to expressions in the form of $a^{m/n}$ where the exponent is a fraction, we need to consider what such an expression means. The expression m/n can also be written as $m \cdot (1/n)$ or $(1/n) \cdot m$. For the rules of exponents to be consistent, then

$$a^{m/n} = (a^m)^{1/n} = (a^{1/n})^m$$

For example, if we apply Rule 3 for exponents to the expression $(a^{1/3})^2$, then the following is true:

$$(a^{1/3})^2 = a^{(1/3)2} = a^{2/3}$$

Expressions of the Form $a^{1/2}$: Square Roots

Applying Rule 3 for exponents to $a^{1/2}$, we have:

$$(a^{1/2})^2 = a^{2/2} = a^1 = a$$

This means $a^{1/2}$ is a number that when squared is equal to a.

The expression $a^{1/2}$ is often written as $\sqrt{a}$ and is called the *square root* of a. The symbol $\sqrt{}$ is called a *radical*. If $a = 4$, both $(2)^2$ and $(-2)^2$ are equal to 4, but $\sqrt{4}$ is defined as the *positive* square root. When we want to consider both 2 and -2, we write $\pm\sqrt{4}$, so

$$\sqrt{4} = 2 \quad \text{and} \quad \pm\sqrt{4} = \pm2$$

When you solve $x^2 = 4$, the solution is 2 and -2.

In the real numbers, $\sqrt{a}$ is not defined when a is negative. For example, $\sqrt{-4}$ is undefined, since there is no real number b such that $b^2 = -4$.

The Square Root

For $a \geq 0$,

$$a^{1/2} = a^{0.5} = \sqrt{a}$$

where $\sqrt{a}$ is the nonnegative number b such that $b^2 = a$.

For example, $25^{1/2} = \sqrt{25} = 5$, since $5^2 = 25$.

Estimating Square Roots. A number is called *a perfect square* if its square root is an integer. For example, 25 and 36 are both perfect squares since $25 = 5^2$ and $36 = 6^2$, so $\sqrt{25} = 5$ and $\sqrt{36} = 6$. If we don't know the square root of some number x and don't have a calculator handy, we can estimate the square root by bracketing it between two perfect squares, a and b, for which we do know the square roots. If $0 \leq a < x < b$, then $\sqrt{a} < \sqrt{x} < \sqrt{b}$. For example, to estimate $\sqrt{10}$,

we know $\qquad 9 < 10 < 16 \qquad$ where 9 and 16 are perfect squares

so $\qquad \sqrt{9} < \sqrt{10} < \sqrt{16}$

and $\qquad 3 < \sqrt{10} < 4$

Therefore $\sqrt{10}$ lies somewhere between 3 and 4, probably closer to 3 because 10 is closer to 9 than to 16. According to a calculator, $\sqrt{10} \approx 3.16$.

Using a calculator

Many calculators and spreadsheet programs have a square root function, often labeled $\sqrt{}$ or perhaps "SQRT." You can also calculate square roots by raising a number to the $\frac{1}{2}$ or 0.5 power using the ^ key, as in 4 ^ 0.5. Try using a calculator to find $\sqrt{4}$ and $\sqrt{9}$.

In any but the simplest cases where the square root is immediately obvious, you will probably use the calculator. For example, use your calculator to find

$$8^{1/2} = \sqrt{8} \approx 2.8284$$

Double-check the answer by verifying that $(2.8284)^2 \approx 8$.

EXAMPLE 3 Estimate $\sqrt{27}$.

SOLUTION We know $25 < 27 < 36$

therefore $\sqrt{25} < \sqrt{27} < \sqrt{36}$

and $5 < \sqrt{27} < 6$

So $\sqrt{27}$ lies somewhere between 5 and 6. We would expect $\sqrt{27}$ to be closer to 5 than 6, since 27 is closer to 25 than 36. Check your answer with a calculator.

EXAMPLE 4 **Calculating square roots**
The function $S = \sqrt{30d}$ describes the relationship between S, the speed of a car in miles per hour, and d, the distance in feet a car skids after applying the brakes on a dry tar road. Use a calculator to estimate the speed of a car that:

a. Leaves 40-foot-long skid marks on a dry tar road.
b. Leaves 150-foot-long skid marks.

SOLUTION **a.** If $d = 40$ feet, then $S = \sqrt{30 \cdot 40} = \sqrt{1200} \approx 35$, so the car was traveling at about 35 miles per hour.
b. If $d = 150$ feet, then $S = \sqrt{30 \cdot 150} = \sqrt{4500} \approx 67$, so the car was traveling at almost 70 miles per hour.

*n*th Roots: Expressions of the Form $a^{1/n}$

The term $a^{1/n}$ denotes the *n*th root of a, often written as $\sqrt[n]{a}$. For $a \geq 0$, the *n*th root of a is the nonnegative number whose *n*th power is a.

$8^{1/3} = \sqrt[3]{8} = 2$ since $2^3 = 8$ (we call 2 the third or cube root of 8)
$16^{1/4} = \sqrt[4]{16} = 2$ since $2^4 = 16$ (we call 2 the fourth root of 16)

For $a < 0$, if n is odd, $\sqrt[n]{a}$ is the negative number whose *n*th power is a. Note that if n is even, then $\sqrt[n]{a}$ is not a real number when $a < 0$.

$(-8)^{1/3} = \sqrt[3]{-8} = -2$ since $(-2)^3 = -8$
$(-27)^{1/3} = \sqrt[3]{-27} = -3$ since $(-3)^3 = -27$
$(-16)^{1/4} = \sqrt[4]{-16}$ is not a real number

The *n*th Root

If a is a real number and n is a positive integer,

$$a^{1/n} = \sqrt[n]{a},\qquad \text{the } n\text{th root of } a$$

For $a \geq 0$,

$\sqrt[n]{a}$ is the nonnegative number b such that $b^n = a$.

For $a < 0$,

If n is odd, $\sqrt[n]{a}$ is the negative number b such that $b^n = a$.
If n is even, $\sqrt[n]{a}$ is not a real number.

If the nth root exists, you can find its value on a calculator. For example, to determine a fifth root, raise the number to the $\frac{1}{5}$ or the 0.2 power. So

$$3125^{1/5} = \sqrt[5]{3125} = 5$$

Double-check your answer by verifying that $5^5 = 3125$.

E X A M P L E   5   **Simplifying expressions with fractional exponents**
Simplify:

a. $625^{1/4}$ b. $(-625)^{1/4}$ c. $125^{1/3}$ d. $(-125)^{1/3}$

S O L U T I O N   a. $625^{1/4} = 5$ since $5^4 = 625$

b. $(-625)^{1/4}$ does not have a real-number solution

c. $125^{1/3} = 5$ since $5^3 = 125$

d. $(-125)^{1/3} = -5$ since $(-5)^3 = -125$

E X A M P L E   6   **The volume of spheres and cubes**

a. The volume of a sphere is given by the equation $V = \frac{4}{3}\pi r^3$. Rewrite the formula, solving for the radius as a function of the volume.

b. If the volume of a sphere is 370 cubic inches, what is its radius? What common object might have that radius?

c. What are the dimensions of a cube that contains this volume?

S O L U T I O N   a. Given:

$$V = \tfrac{4}{3}\pi r^3$$

multiply both sides by 3

$$3V = 4\pi r^3$$

divide by 4π

$$\frac{3V}{4\pi} = r^3$$

take the cube root and switch sides

$$r = \sqrt[3]{\frac{3V}{4\pi}}$$

b. Substituting 370 for V and 3.14 for π in our derived formula in part (a), we have

$$r \approx \sqrt[3]{\frac{3 \cdot 370}{4 \cdot 3.14}} \approx 4.45 \text{ inches}$$

A regulation-size soccer ball is basically a sphere with a diameter of about 8.9 inches, so its radius is about 4.45 inches.

c. A cube has the same length on all three sides. If s is the side length, then the volume of the desired cube is $s^3 = 370$ cubic inches. So $s = \sqrt[3]{370} \approx 7.18$ inches. A cube of length 7.18 inches on each side would give a volume equivalent to a sphere with a radius of 4.45 inches.

Rules for Radicals

The following rules can help you compute with radicals. They represent extensions of the rules for integer exponents. In the following table we assume that n is a positive integer and that $\sqrt[n]{a}$ and $\sqrt[n]{b}$ exist.

> **Rules for Radicals**
>
> Example
>
> 1. $\sqrt[n]{a} \cdot \sqrt[n]{b} = \sqrt[n]{ab}$
>
> $\sqrt{3} \cdot \sqrt{2} = \sqrt{6}$
>
> 2. $\dfrac{\sqrt[n]{a}}{\sqrt[n]{b}} = \sqrt[n]{\dfrac{a}{b}}$ $(b \neq 0)$
>
> $\dfrac{\sqrt[4]{125}}{\sqrt[4]{25}} = \sqrt[4]{\dfrac{125}{25}} = \sqrt[4]{5}$
>
> 3. $(\sqrt[n]{a})^n = \sqrt[n]{a^n} = a$ $(a > 0)$
>
> $(\sqrt{7})^2 = \sqrt{7^2} = 7$

EXAMPLE 7   **Measuring Earth**

Assuming that the surface area of Earth is approximately 200 million square miles, estimate the radius of Earth.

SOLUTION   **Step 1.** Find the formula for the radius of a sphere, using the surface area formula.

If we assume that Earth is roughly spherical, we can solve for the radius r in the formula for the surface area of a sphere, $S = 4\pi r^2$. We get

$$r = \sqrt{\frac{S}{4\pi}} = \sqrt{\frac{1}{4}} \cdot \sqrt{\frac{S}{\pi}} = \frac{1}{2}\sqrt{\frac{S}{\pi}}$$

Step 2. Estimate the radius of Earth.

Given that Earth's surface area is approximately 200,000,000 square miles, we can use our derived formula to estimate Earth's radius. Substituting for S, we get

$$r = \frac{1}{2}\sqrt{\frac{200,000,000 \text{ miles}^2}{\pi}}$$

$$\approx \frac{1}{2}\sqrt{63,661,977 \text{ miles}^2}$$

$$\approx \frac{1}{2} \cdot 7979 \text{ miles} \approx 3989 \text{ miles}$$

So Earth has a radius of about 4000 miles.

EXAMPLE 8   **Simplifying radicals**

Simplify the following radical expressions. Assume all variables are nonnegative real numbers.

a. $\sqrt[3]{625x^4}$ **b.** $3\sqrt{48} - 5\sqrt{27}$

SOLUTION   **a.** Factor 625 $\sqrt[3]{625x^4} = \sqrt[3]{5^4 \cdot x^4}$

rewrite using perfect cube factors $= \sqrt[3]{5^3 \cdot 5 \cdot x^3 \cdot x}$

use Rule 1 for radicals $= \sqrt[3]{5^3 x^3} \cdot \sqrt[3]{5x}$

extract the perfect cubes (Rule 3), leaving the remaining factors under the radical $= 5x \cdot \sqrt[3]{5x}$

b. Find the largest perfect square factors $3\sqrt{48} - 5\sqrt{27} = 3\sqrt{16 \cdot 3} - 5\sqrt{9 \cdot 3}$

extract the perfect squares (Rule 3) $= 3 \cdot 4\sqrt{3} - 5 \cdot 3\sqrt{3}$

multiply $= 12\sqrt{3} - 15\sqrt{3}$

use distributive law $= (12 - 15)\sqrt{3}$

$= -3\sqrt{3}$

Algebra Aerobics 4.3b

1. Evaluate each of the following without a calculator.
   a. $81^{1/2}$ c. $36^{1/2}$ e. $(-36)^{1/2}$
   b. $144^{1/2}$ d. $-49^{1/2}$

2. Assume that all variables represent nonnegative quantities. Then simplify and rewrite the following without radical signs. (Use fractional exponents if necessary.)
   a. $\sqrt{9x}$ c. $\sqrt{36x^2}$ e. $\sqrt{\dfrac{49}{x^2}}$
   b. $\sqrt{\dfrac{x^2}{25}}$ d. $\sqrt{\dfrac{9y^2}{25x^4}}$ f. $\sqrt{\dfrac{4a}{169}}$

3. Use the formula $S = \sqrt{30d}$ (in Example 4, p. 229) to estimate the following:
   a. The speed of a car that leaves 60-foot-long skid marks on a dry tar road.
   b. The speed of a car that leaves 200-foot-long skid marks on a dry tar road.

4. Without a calculator, find two consecutive integers between which the given square root lies.
   a. $\sqrt{29}$ c. $\sqrt{117}$ e. $\sqrt{39}$
   b. $\sqrt{92}$ d. $\sqrt{79}$

5. Evaluate each of the following without a calculator:
   a. $27^{1/3}$ c. $8^{-1/3}$ e. $27^{-1/3}$ g. $\left(\dfrac{8}{27}\right)^{-1/3}$
   b. $16^{1/4}$ d. $32^{1/5}$ f. $25^{-1/2}$ h. $\left(\dfrac{1}{16}\right)^{1/2}$

6. Evaluate:
   a. $\sqrt[3]{-27}$ c. $(-1000)^{1/3}$ e. $(-8)^{1/3}$
   b. $(-10,000)^{1/4}$ d. $-16^{1/4}$ f. $\sqrt{2500}$

7. Estimate the radius of a spherical balloon with a volume of 2 cubic feet. Volume of a sphere $= \frac{4}{3}\pi r^3$.

8. Simplify if possible.
   a. $\sqrt{9+16}$ c. $\sqrt[3]{-125}$
   b. $-\sqrt{49}$ d. $\sqrt{45} - 3\sqrt{125}$

9. Change each radical expression into exponent form, then simplify. Assume all variables are nonnegative.
   a. $\sqrt{36}$ b. $\sqrt[3]{27x^6}$ c. $\sqrt[4]{81a^4b^{12}}$

10. Solve for the indicated variable. Assume all variables represent nonnegative quantities.
   a. $V = \pi r^2 h$ for r d. $c^2 = a^2 + b^2$ for a
   b. $V = \frac{1}{3}\pi r^2 h$ for r e. $S = 6x^2$ for x
   c. $V = s^3$ for s

Expressions of the Form $a^{m/n}$

In this section, we apply the rules of exponents and the rules of radicals to expressions of the form $a^{m/n}$. Recall that we used Rule 3 for exponents to show that we can write $a^{m/n}$ either as $(a^m)^{1/n}$ or $(a^{1/n})^m$. For example,

$$2^{3/2} = (2^3)^{1/2}$$
$$= (8)^{1/2}$$

using a calculator ≈ 2.8284

Equivalently, $2^{3/2} = (2^{1/2})^3$
$$\approx (1.414)^3$$
$$\approx 2.8284$$

We could, of course, use a calculator to compute $2^{3/2}$ (or $2^{1.5}$) directly by raising 2 to the $\frac{3}{2}$ or 1.5 power.

If $a \geq 0$ and m and n are integers ($n \neq 0$), then using radical notation,

$$a^{m/n} = \left(\sqrt[n]{a}\right)^m$$

or equivalently $= \sqrt[n]{a^m}$

EXAMPLE 9   **Calculating with radicals**
Find the product of $(\sqrt{5}) \cdot (\sqrt[3]{5})$, leaving the answer in exponent form.

SOLUTION $\left(\sqrt{5}\right) \cdot \left(\sqrt[3]{5}\right) = 5^{1/2} \cdot 5^{1/3} = 5^{(1/2)+(1/3)} = 5^{5/6}$

EXAMPLE 10

The diameter of nails

According to McMahon and Bonner in *On Size and Life*,[5] common nails range from 1 to 6 inches in length. The weight varies even more, from 11 to 647 nails per pound. Longer nails are relatively thinner than shorter ones. A good approximation of the relationship between length and diameter is given by the equation

$$d = 0.07L^{2/3}$$

where d = diameter and L = length, both in inches. Estimate the diameters of nails that are 1, 3, and 6 inches long.

SOLUTION

When $L = 1$ inch, the diameter $d = 0.07 \cdot (1)^{2/3} = 0.07 \cdot 1 = 0.07$ inch.

When $L = 3$ inches, then $d = 0.07 \cdot (3)^{2/3} \approx 0.07 \cdot 2.08 \approx 0.15$ inch.

When $L = 6$ inches, then $d = 0.07 \cdot (6)^{2/3} \approx 0.07 \cdot 3.30 \approx 0.23$ inch.

Summary of Zero, Negative, and Fractional Exponents

If m and n are positive integers and $a \neq 0$, then

$$a^0 = 1$$

$$a^{-n} = \frac{1}{a^n}$$

$$a^{1/n} = \sqrt[n]{a}$$

$$a^{m/n} = \sqrt[n]{a^m} = \left(\sqrt[n]{a}\right)^m \qquad a > 0$$

EXPLORE & EXTEND

4.3

Patterns in the positions and motions of the planets

Four hundred years ago, Johannes Kepler discovered a law that relates the periods of planets to their average (mean) distances from the sun. (A period of a planet is the time it takes the planet to complete one orbit around the sun.) Kepler's strong belief that harmonious laws governed the solar system led him to discover an interesting pattern, called Kepler's Third Law:

cube of the average distance from sun = square of the orbital period of the planet

using only units $\qquad (\text{A.U.})^3 = \text{years}^2$

At the time of his work, Kepler did not know the distance from the sun to each planet in terms of measures of distance such as the kilometer. But he was able to determine the distance from each planet to the sun in terms of the distance from Earth to the sun. This distance from Earth to the sun is now called the astronomical unit, or A.U. for short.

a. The planets Uranus and Neptune were discovered after Kepler made his discovery. Check to see whether the above relationship Kepler found holds true for these two planets, using the table at the top of the next page.

---

[5]A. McMahon and J. Tyler Bonner, *On Size and Life* (New York: Scientific American Library, Scientific American Books, 1983).

Planet	Average (mean) Distance from Sun (A.U.)	Cube of the Distance (A.U.)^3	Orbital Period (years)	Square of the Orbital Period (years^2)
Uranus	19.1911		84.0086	
Neptune	30.0601		164.7839	

Source: Data from S. Parker and J. Pasachoff, *Encyclopedia of Astronomy,* 2nd ed. (New York: McGraw-Hill, 1993), Table 1, Elements of Planetary Orbits. Copyright © 1993 by McGraw-Hill, Inc. Reprinted with permission.

b. Construct an equation representing the relationship between distance from the sun and orbital periods for the planets. Solve the equation for distance from the sun. Then solve the equation for the orbital period.

c. Use your equations to find the following information about the planets:

  i. Saturn's orbital period is 29.4557 years. What is the average distance from the sun to Saturn?

  ii. Mars is 1.5233 astronomical units (A.U.) from the sun. Find the orbital period for Mars.

Algebra Aerobics 4.3c

Assume all variables represent positive quantities. Problem 3 requires a calculator that can evaluate powers.

1. Find the product expressed in exponent form:

  a. $\sqrt{2} \cdot \sqrt[3]{2}$ **c.** $\sqrt{3} \cdot \sqrt[3]{9}$ **e.** $\sqrt[4]{x^3} \cdot \sqrt{x}$

  b. $\sqrt{5} \cdot \sqrt[4]{5}$ **d.** $\sqrt[4]{x} \cdot \sqrt[3]{x}$ **f.** $\sqrt[3]{xy^2} \cdot \sqrt{xy}$

2. Find the quotient by representing the expression in exponent form. Leave the answer in positive exponent form.

  a. $\dfrac{\sqrt{2}}{\sqrt[3]{2}}$ **b.** $\dfrac{2}{\sqrt[4]{2}}$ **c.** $\dfrac{\sqrt[4]{5}}{\sqrt[3]{5}}$ **d.** $\dfrac{\sqrt{x}}{\sqrt[4]{x^3}}$ **e.** $\dfrac{\sqrt[3]{xy^2}}{\sqrt{xy}}$

3. McMahon and Bonner give the relationship between chest circumference and body weight of adult primates as

$$c = 17.1w^{3/8}$$

where w = weight in kilograms and c = chest circumference in centimeters. Estimate the chest circumference of a:

  a. 0.25-kg tamarin   **b.** 25-kg baboon

4. Simplify each expression by removing all possible factors from the radical.

  a. $\sqrt{20x^2}$ **c.** $\sqrt[3]{16x^3y^4}$

  b. $\sqrt{75a^3}$ **d.** $\dfrac{\sqrt[4]{32x^4y^6}}{\sqrt[4]{81x^8y^5}}$

5. Change each radical expression into a form with fractional exponents, then simplify.

  a. $\sqrt{4a^2b^6}$ **c.** $\sqrt[3]{8.0 \cdot 10^{-9}}$

  b. $\sqrt[4]{16x^4y^6}$ **d.** $\sqrt{8a^{-4}}$

Exercises for Section 4.3

Exercises 30, 36, 37 and 38 require a calculator that can evaluate powers.

1. Simplify and express your answer using positive exponents. Check your answers by applying the rules for exponents and doing the calculations.

  a. $10^3 \cdot 10^{-2}$ **c.** $(10^{-3})^2$

  b. $\dfrac{10^{-2}}{10^3}$ **d.** $\dfrac{10^3}{10^{-2}}$

2. Simplify and express your answer with positive exponents:

  a. $(x^{-3}) \cdot (x^4)$

  b. $(x^{-3}) \cdot (x^{-2})$

  c. $(x^2)^{-3}$

  d. $(n^{-2})^{-3}$

  e. $(2n^{-2})^{-3}$

  f. $n^{-4}(n^5 - n^2) + n^{-3}(n - n^4)$

3. Simplify where possible. Express your answer with positive exponents.

a. $\dfrac{x^4y^7}{x^3y^{-5}}$

c. $\dfrac{(x+y)^4}{(x+y)^{-7}}$

b. $\dfrac{x^{-2}y}{xy^3}$

d. $\dfrac{a^{-2}bc^{-5}}{(ab^2)^{-3}c}$

4. Simplify where possible. Express your answer with positive exponents.

a. $(3 \cdot 3^8)^{-2}$

c. $2x^{-3} + 3x(x^{-4})$

b. $x^3 \cdot x^{-4} \cdot x^{12}$

d. $10^{-5} + 5^{-2} + 10^{10}$

5. Evaluate without using technology and write the result using scientific notation:

a. $(2.3 \cdot 10^4)(2.0 \cdot 10^6)$

d. $\dfrac{3.25 \cdot 10^8}{6.5 \cdot 10^{15}}$

b. $(3.7 \cdot 10^{-5})(1.0 \cdot 10^8)$

e. $(4.0 \cdot 10^{52})^3$

c. $\dfrac{8.2 \cdot 10^{23}}{4.1 \cdot 10^{12}}$

f. $(5.0 \cdot 10^{-11})^2$

6. Write each of the following in scientific notation:

a. $725 \cdot 10^{23}$

c. $\dfrac{1}{725 \cdot 10^{23}}$

e. $-725 \cdot 10^{-23}$

b. $725 \cdot 10^{-23}$

d. $-725 \cdot 10^{23}$

7. Change each number to scientific notation, then simplify using rules of exponents. Show your work, recording your final answer in scientific notation.

a. 10% of 0.000 01

d. $\dfrac{8000}{0.000\ 8}$

b. $\dfrac{0.000\ 05}{50,000}$

e. $\dfrac{0.006\ 4}{8000}$

c. $\dfrac{3}{0.006}$

f. $5,000,000 \cdot 40,000$

8. Use scientific notation and the rules of exponents to perform the indicated operation without a calculator. Show your work, recording your answer in decimal form.

a. $\dfrac{20}{200,000}$

c. $200 \cdot 0.000\ 007\ 5$

e. $0.06 \cdot 600$

b. $\dfrac{0.006}{60,000}$

d. $\dfrac{10,000,000}{25,000}$

f. 10% of 0.000 05

9. Evaluate without a calculator:

a. $100^{1/2}$

c. $100^{-1/2}$

e. $-1000^{1/3}$

b. $-100^{1/2}$

d. $-100^{-1/2}$

f. $(-1000)^{1/3}$

10. Evaluate without a calculator:

a. $\sqrt{10,000}$

c. $625^{1/2}$

e. $\left(\dfrac{1}{9}\right)^{1/2}$

b. $\sqrt{-25}$

d. $100^{1/2}$

f. $\left(\dfrac{625}{100}\right)^{1/2}$

11. Assume that all variables represent positive quantities and simplify.

a. $\sqrt{\dfrac{a^2b^4}{c^6}}$

b. $\sqrt{36x^4y}$

c. $\sqrt{\dfrac{49x}{y^6}}$

d. $\sqrt{\dfrac{x^4y^2}{100z^6}}$

12. Simplify by removing all possible factors for each radical. Assume all variable quantities are positive.

a. $\sqrt{125a}$

b. $\sqrt{\dfrac{x^2}{4x^4y^6}}$

c. $\sqrt{8x^3y^2}$

d. $\sqrt{64x^4y^5}$

13. Evaluate each expression without using a calculator.

a. $\sqrt{36 \cdot 10^6}$

c. $\sqrt[4]{625 \cdot 10^{20}}$

b. $\sqrt[3]{8 \cdot 10^9}$

d. $\sqrt{1.0 \cdot 10^{-4}}$

14. Calculate the following:

a. $4^{1/2}$ **c.** $27^{1/3}$ **e.** $8^{2/3}$ **g.** $16^{1/4}$

b. $-4^{1/2}$ **d.** $-27^{1/3}$ **f.** $-8^{2/3}$ **h.** $16^{3/4}$

15. Calculate:

a. $\left(\dfrac{1}{100}\right)^{1/2}$ **b.** $25^{-1/2}$ **c.** $\left(\dfrac{9}{16}\right)^{-1/2}$ **d.** $\left(\dfrac{1}{1000}\right)^{1/3}$

16. Evaluate:

a. $27^{2/3}$ **b.** $16^{-3/4}$ **c.** $25^{-3/2}$ **d.** $81^{-3/4}$

17. Simplify the following expressions by using properties of exponents. Write your final answers with only positive exponents.

a. $\dfrac{(-2x^3y^{-1})^{-3}}{(x^2y^{-2})^0}$

c. $\left(\dfrac{3x^2y^{-5}}{5x^3y^4}\right)^{-1}$

b. $\dfrac{(-2x^3y^{-1})^{-2}}{(x^2y^{-2})^{-1}}$

d. $\left[(3x^{-1}z^4)^{-2}\right]^{-3}$

18. Each of the following simplifications is false. In each case identify the error and correct it.

a. $x^{-2}x^5 = x^{10}$

b. $\dfrac{2^{-1}x^2y^{-2}}{x^{-1}y^5} = \dfrac{x^2x^{-1}}{2y^{-2}y^5} = \dfrac{x}{2y^3}$

c. $(3x^{-1})^2 = \left(\dfrac{1}{3x}\right)^2 = \dfrac{1}{9x^2}$

d. $(x+y)^{-1} = \dfrac{1}{x} + \dfrac{1}{y}$

19. Evaluate when $x = 2$:

a. $(-x)^2$ **c.** $x^{1/2}$ **e.** $x^{3/2}$

b. $-x^2$ **d.** $(-x)^{1/2}$ **f.** x^0

20. Determine if the following statements are true or false.

a. $\sqrt[4]{(3x^2)^4} = 3x^2$

b. $\sqrt[3]{(x+1)^4} = (x+1)(\sqrt[3]{x+1})$

c. $\sqrt[3]{\dfrac{9}{25}} = \dfrac{\sqrt[3]{45}}{5}$

d. $\sqrt{15} - \sqrt{3} = \sqrt{12}$

21. Use $<$, $>$, or $=$ to make each statement true.

a. $\sqrt{3} + \sqrt{7}$? $\sqrt{3+7}$

b. $\sqrt{3^2 + 2^2}$? 5

c. $\sqrt{5^2 - 4^2}$? 2

22. Fill in the missing forms in the table.

Radical Form	Rational Exponent Form
a. $\sqrt[3]{64} = 4$	
b.	$-(144)^{1/2} = -12$
c. $\left(\sqrt[4]{81x}\right)^3 = 27 \cdot \sqrt[4]{x^3}$	
d.	$(-243)^{1/5} = -3$
e.	$16^{5/4} = 32$

23. A TV signal traveling at the speed of light takes about $8 \cdot 10^{-5}$ second to travel 15 miles. How long would it take the signal to travel a distance of 3000 miles?

24. Estimate the length of a side, s, of a cube with volume, V, of 6 cm^3 (where $V = s^3$).

25. Estimate the radius of a spherical balloon that has a volume of 4 ft^3.

26. *Constellation.* Reduce each of the following expressions to the form $u^a \cdot m^b$; then plot the exponents as points with coordinates (a, b) on graph paper. Do you recognize the constellation?

a. $\dfrac{(u^2)^2 \cdot m}{u^2 \cdot m^{-4}}$

b. $\dfrac{u^{-9/5} \cdot m^3}{(umu^2)^1 \cdot m^{-1}}$

c. $\dfrac{u^2 \cdot u^{-4}}{u^3 \cdot (m^{-2})^3}$

d. $\dfrac{(um^2)^3 \cdot u^2}{(um)^4}$

e. $\dfrac{u^{-3/2} \cdot u^{-7/2} \cdot m^1 \cdot (m^3)^3}{(um)^2}$

f. $\dfrac{1}{u^{12} \cdot m^{-9}}$

g. $\dfrac{(mu)^0 \cdot (u^{10})^{-1} \cdot m^{1/4}}{(m^{-3} \cdot u^{-1/3})^3}$

27. An equilateral triangle has sides of length 8 cm.

    a. Find the height of the triangle. (*Hint:* Use the Pythagorean theorem on the inside back cover.)

    b. Find the area A of the triangle if $A = \dfrac{1}{2}bh$.

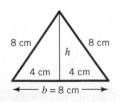

28. An Egyptian pyramid consists of a square base and four triangular sides. A model of a pyramid is constructed using four equilateral triangles each with a side length of 30 inches. Find the surface area of the pyramid model, including the base. (*Note:* Surface area is the sum of the areas of the four triangular sides and the rectangular base. The previous exercise gives the formula for finding the area of a triangle.)

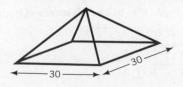

29. Simplify each expression by removing all possible factors from the radical, then combining any like terms.

    a. $2\sqrt{50} + 12\sqrt{8}$ **c.** $10\sqrt{32} - 6\sqrt{18}$

    b. $3\sqrt{27} - 2\sqrt{75}$ **d.** $2\sqrt[3]{16} + 4\sqrt[3]{54}$

30. (Requires a calculator that can evaluate powers.) Perform the following calculations using technology, then write the answer in standard scientific notation rounded to three places.

    a. $\left(\dfrac{9}{5}\right)^{50}$ **d.** $\dfrac{7}{6^{15}}$

    b. 2^{35} **e.** $(5)^{-10}(2)^{10}$

    c. $(0.000025)^{1/2}$ **f.** $(8{,}000{,}000)^{2/3}$

31. Describe at least three different methods for entering $5.23 \cdot 10^{-3}$ into a calculator or spreadsheet.

32. Using rules of exponents, show that $\dfrac{9^5}{27^{-7}} = 3^{31}$.

33. Using rules of exponents, show that $\dfrac{1}{x^{-n}} = x^n$.

34. Write an expression that shows the calculation(s) necessary to answer the question. Then use scientific notation and exponent rules to determine the answer.

    a. Find the number of nickels in $500.00.

    b. The circumference of Earth is about 40.2 million meters. Find the radius of Earth, in kilometers, using the formula $C = 2\pi r$.

35. The time it takes for one complete swing of a pendulum is called the *period* of its motion. The period T (in seconds) of a swinging pendulum is found using the formula $T = 2\pi\sqrt{\dfrac{L}{32}}$, where L is the length of the pendulum in feet and 32 is the acceleration of gravity in feet per second^2.

    a. Find the period of a pendulum whose length is 2 ft 8 in.

    b. How long would a pendulum have to be to have a period of 2 seconds?

36. (Requires the use of a calculator that can evaluate powers.) A wheelchair ramp is constructed at the end of a porch, which is 4 ft off the ground. The base of the ramp is 48 ft from the porch. How long is the ramp? (*Hint:* Use the Pythagorean theorem on the inside back cover.)

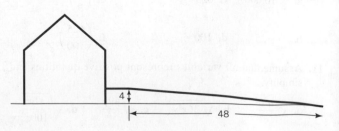

37. (Requires the use of a calculator that can evaluate powers.) The breaking strength S (in pounds) of a three-strand manila rope is a function of its diameter, D (in inches). The relationship can be described by the equation $S = 1700D^{1.9}$. Calculate the breaking strength when D equals:

a. 1.5 in.   **b.** 2.0 in.

38. (Requires the use of a calculator that can evaluate powers.) If a rope is wound around a wooden pole, the number of pounds of frictional force, F, between the pole and the rope is a function of the number of turns, N, according to the equation $F = 14 \cdot 10^{0.70N}$. What is the frictional force when the number of turns is:

a. 0.5   **b.** 1   **c.** 3

4.4   *Converting Units*

Exploration 4.1 will help you understand the relative ages and sizes of objects in our universe and give you practice in scientific notation and unit conversion.

Problems in science constantly require converting back and forth between different units of measure. To do so, we need to be comfortable with the laws of exponents and the basic metric and English units (see Table 4.1 or a more complete table on the inside back cover). The following unit conversion examples describe a strategy based on *conversion factors*.

Converting Units within the Metric System

EXAMPLE 1

Conversion factors

Light travels at a speed of approximately $3.00 \cdot 10^5$ kilometers per second (km/sec). Describe the speed of light in meters per second (m/sec).

SOLUTION

The prefix *kilo* means thousand. One kilometer (km) is equal to 1000 or 10^3 meters (m):

$$1 \text{ km} = 10^3 \text{ m} \tag{1}$$

Dividing both sides of Equation (1) by 1 km, we can rewrite it as

$$1 = \frac{10^3 \text{ m}}{1 \text{ km}}$$

If instead we divide both sides of Equation (1) by 10^3 m, we get

$$\frac{1 \text{ km}}{10^3 \text{ m}} = 1$$

The ratios $\frac{10^3 \text{ m}}{1 \text{ km}}$ and $\frac{1 \text{ km}}{10^3 \text{ m}}$ are called *conversion factors*, because we can use them to convert between kilometers and meters.

What is the right conversion factor? If units in $\frac{\text{km}}{\text{sec}}$ are multiplied by units in meters per kilometer, we have

$$\frac{\cancel{\text{km}}}{\text{sec}} \cdot \frac{\text{m}}{\cancel{\text{km}}}$$

and the result is in meters per second. So multiplying the speed of light in km/sec by a conversion factor in m/km will give us the correct units of m/sec.

A conversion factor always equals 1. So we will not change the value of the original quantity by multiplying it by a conversion factor. In this case, we use the conversion factor of $\frac{10^3 \text{ m}}{1 \text{ km}}$.

$$3.00 \cdot 10^5 \text{ km/sec} = 3.00 \cdot 10^5 \frac{\cancel{\text{km}}}{\text{sec}} \cdot \frac{10^3 \text{ m}}{1 \cancel{\text{km}}}$$

$$= 3.00 \cdot 10^5 \cdot 10^3 \text{ m/sec}$$

$$= 3.00 \cdot 10^8 \text{ m/sec}$$

Hence light travels at approximately $3.00 \cdot 10^8$ m/sec.

EXAMPLE 2

Finding the right conversion factor
Check your answer in Example 1 by converting $3.00 \cdot 10^8$ m/sec back to km/sec.

SOLUTION

Here we use the same strategy, but now we need to use the other conversion factor. Multiplying $3.00 \cdot 10^8$ m/sec by $(1 \text{ km})/(10^3 \text{ m})$ gives us

$$3.00 \cdot 10^8 \frac{\text{m}}{\text{sec}} \cdot \frac{1 \text{ km}}{10^3 \text{ m}} = 3.00 \cdot \frac{10^8 \text{ km}}{10^3 \text{ sec}}$$

$$= 3.00 \cdot 10^5 \text{ km/sec}$$

which was the original value given for the speed of light.

Converting between the Metric and English Systems

EXAMPLE 3

Distance to Toronto
You're touring Canada, and you see a sign that says it is 130 kilometers to Toronto. How many miles is it to Toronto?

SOLUTION

The crucial question is, "What conversion factor should be used?" From Table 4.1 we know that

$$1 \text{ km} \approx 0.62 \text{ mile}$$

This equation can be rewritten in two ways:

$$1 \approx \frac{0.62 \text{ mile}}{1 \text{ km}} \quad \text{or} \quad 1 \approx \frac{1 \text{ km}}{0.62 \text{ mile}}$$

It produces two possible conversion factors:

$$\frac{0.62 \text{ mile}}{1 \text{ km}} \quad \text{and} \quad \frac{1 \text{ km}}{0.62 \text{ mile}}$$

Which one will convert kilometers to miles? We need one with kilometers in the denominator and miles in the numerator, namely $\frac{0.62 \text{ mile}}{1 \text{ km}}$, so that the km will cancel when we multiply by 130 km:

$$130 \text{ km} \cdot \frac{0.62 \text{ mile}}{1 \text{ km}} = 80.6 \text{ miles}$$

So it is a little over 80 miles to Toronto.

Using Multiple Conversion Factors

EXAMPLE 4

How fast is the speed of light?
Light travels at $3.00 \cdot 10^5$ km/sec. How many kilometers does light travel in one *year*?

SOLUTION

Here our strategy is to use more than one conversion factor to convert from seconds to years. Use your calculator to perform the following calculations:

$$3.00 \cdot 10^5 \frac{\text{km}}{\text{sec}} \cdot \frac{60 \text{ sec}}{1 \text{ min}} \cdot \frac{60 \text{ min}}{1 \text{ hr}} \cdot \frac{24 \text{ hr}}{1 \text{ day}} \cdot \frac{365 \text{ days}}{1 \text{ year}} = 94{,}608{,}000 \cdot 10^5 \text{ km/year}$$

$$\approx 9.46 \cdot 10^7 \cdot 10^5 \text{ km/year}$$

$$= 9.46 \cdot 10^{12} \text{ km/year}$$

So a light year, the distance light travels in one year, is approximately equal to $9.46 \cdot 10^{12}$ kilometers.

EXPLORE & EXTEND

4.4

Liters vs. Kilograms

On July 23, 1983, Air Canada Flight 143 had to make an emergency landing at Gimli, Manitoba Industrial Park Airport after completely running out of jet fuel. The investigation into the incident revealed that the amount of fuel needed for the flight was miscalculated due to a conversion error between the metric and English systems. (Source: *http://www.answers.com/topic/gimli glider*)

Air Canada Flight 143 needed 22,300 kilograms of fuel to safely make the trip from Montreal to Edmonton. A check of the fuel tank revealed that the plane had only 7682 liters of fuel before take-off from Montreal. In the following exercise, you will calculate the additional amount of fuel needed to safely complete the flight and identify the crew's critical error.

a. A liter of jet fuel weighs 0.803 kilogram. Using this information, calculate how many liters of fuel should have been added to the tank before takeoff.

b. Given that 2.2 pounds is equal to 1 kilogram, determine the number of pounds in a liter of jet fuel.

c. The crew mistakenly assumed that a liter of fuel weighs 1.77 kilograms. Explain how their error affected how much jet fuel was added to the tank.

d. How many kilograms of jet fuel short was Air Canada Flight 143 when it took off?

Algebra Aerobics 4.4

The inside of the back cover of the text contains metric prefixes and conversion facts. Round your answers to two decimal places.

1. Coca-Cola Classic comes in a 2-liter container. How many fluid ounces is that? (*Note:* There are 32 ounces in a quart.)

2. A child's height is 120 cm. How tall is she in inches?

3. Convert to the desired unit:

   a. 12 inches = _____ cm

   b. 100 yards = _____ meters

   c. 20 kilograms = _____ pounds

   d. \$40,000 per year = \$_____ per hour (assume a 40-hour work week for 52 weeks)

   e. 24 hr/day = _____ sec/day

   f. 1 gallon = _____ ml

   g. 1 mph = _____ ft/sec

4. The distance between the sun and the moon is $3.84 \cdot 10^8$ meters. Express this in kilometers.

5. The mean distance from our sun to Jupiter is $7.8 \cdot 10^8$ kilometers. Express this distance in meters.

6. A light year is about $5.88 \cdot 10^{12}$ miles. Verify that $9.46 \cdot 10^{12}$ kilometers $\approx 5.88 \cdot 10^{12}$ miles.

7. 1 angstrom (Å) = 10^{-8} cm. Express 1 angstrom in meters.

8. If a road sign says the distance to Quebec is 218 km, what is the distance in miles?

9. The distance from Earth to the sun is about 93,000,000 miles. There are 5280 feet in a mile, and a dollar bill is approximately 6 inches long. Estimate how many dollar bills would have to be placed end to end to reach from Earth to the sun.

10. Fill in the missing parts of the following conversion.

$$\frac{2560 \text{ mi}}{4.2 \text{ hrs}} = \frac{2560 \text{ mi}}{4.2 \text{ hrs}} \cdot \frac{1.6 \text{ km}}{?} = \frac{?}{?}$$

$$= \frac{?}{?} \cdot \frac{?}{60 \text{ min}} = \frac{? \text{ km}}{? \text{ min}}$$

11. The Harvard Bridge, which connects Cambridge to Boston along Massachusetts Avenue, is literally marked off in units called *Smoots*. A Smoot is equal to about 5.6 feet, the height of an M.I.T. fraternity pledge named Oliver Smoot. The bridge is approximately 364 Smoots long. How long is the bridge in feet? Show all units when doing your conversion.

Exercises for Section 4.4

The inside of the back cover of the text contains metric prefixes and conversion facts.

1. Change the following English units to the metric units indicated.
   a. 50 miles to kilometers
   d. 12 inches to centimeters
   b. 3 feet to meters
   e. 60 feet to meters
   c. 5 pounds to kilograms
   f. 4 quarts to liters

2. Change the following metric units to the English units indicated.
   a. 25 kilometers to miles
   d. 50 grams to ounces
   b. 700 meters to yards
   e. 10 kilograms to pounds
   c. 250 centimeters to inches
   f. 10,000 milliliters to quarts

3. For the following questions, make an estimate and then check your estimate using the conversion table on the inside back cover:
   a. One foot is how many centimeters?
   b. One foot is what part of a meter?

4. A football field is 100 yards long. How many meters is this? What part of a kilometer is this?

5. How many droplets of water are in a river that is 100 km long, 250 m wide, and 25 m deep? Assume a droplet is 1 milliliter. (*Note:* one liter = one cubic decimeter and 10 decimeters = 1 meter.)

6. a. A roll of aluminum foil claims to be 50 sq ft or 4.65 m². Show the conversion factors that would verify that these two measurements are equivalent.
   b. One cm³ of aluminum weighs 2.7 grams. If a sheet of aluminum foil is 0.003 8 cm thick, find the weight of the roll of aluminum foil in grams.

7. If a falling object accelerates at the rate of 9.8 meters per second every second, how many feet per second does it accelerate each second?

8. Convert the following to feet and express your answers in scientific notation.
   a. The radius of the solar system is approximately 10^{12} meters.
   b. The radius of a proton is approximately 10^{-15} meter.

9. The speed of light is approximately $1.86 \cdot 10^5$ miles/sec.
   a. Write this number in decimal form and express your answer in words.
   b. Convert the speed of light into meters per year. Show your work.

10. The average distance from Earth to the sun is about 150,000,000 km, and the average distance from the planet Venus to the sun is about 108,000,000 km.
    a. Express these distances in scientific notation.
    b. Divide the distance from Venus to the sun by the distance from Earth to the sun and express your answer in scientific notation.

c. The distance from Earth to the sun is called 1 astronomical unit (1 A.U.) How many astronomical units is Venus from the sun?
    d. Pluto is 5,900,000,000 km from the sun. How many astronomical units is it from the sun?

11. The distance from Earth to the sun is approximately 150 million kilometers. If the speed of light is $3.00 \cdot 10^5$ km/sec, how long does it take light from the sun to reach Earth? If a solar flare occurs right now, how long would it take for us to see it?

12. Earth travels in an approximately circular orbit around the sun. The average radius of Earth's orbit around the sun is $9.3 \cdot 10^7$ miles. Earth takes one year, or 365 days, to complete one orbit.
    a. Use the formula for the circumference of the circle to determine the distance Earth travels in one year.
    b. How many hours are in one year?
    c. Speed is distance divided by time. Find the orbital speed of Earth in miles per hour.

13. A barrel of U.S. oil is 42 gallons. A barrel of British oil is 163.655 liters. Which barrel is larger and by how much?

14. A barrel of wheat is 3.2812 bushels (U.S. dry) or 4.0833 cubic feet.
    a. How many cubic feet are in a bushel of wheat?
    b. How many cubic inches are in a barrel?
    c. How many cubic centimeters are in a bushel?

15. In the United States, land is measured in acres and one acre is 43,560 sq ft.
    a. If you buy a one-acre lot that is in the shape of a square, what would be the length of each side in feet?
    b. A newspaper advertisement states that all lots in a new housing development will be a minimum of one and a half acres. Assuming the lot is rectangular and has 150 ft of frontage, how deep will the minimal-size lot be? If the new home owner wants to fence in the lot, how many yards of fencing would be needed?
    c. The metric unit for measuring land is the square hectometer. (A hectometer is a length of 100 meters.) Find the size of a one-acre lot if it were measured in square hectometers.
    d. A hectare is 100 acres. How many one-acre lots can fit in a square mile? How many hectares is that?

16. Estimate the number of heartbeats in a lifetime. Explain your method.

17. A nanosecond is 10^{-9} second. Modern computers can perform on the order of one operation every nanosecond. Approximately how many feet does an electrical signal moving at the speed of light travel in a computer in 1 nanosecond?

18. Since light takes time to travel, everything we see is from the past. When you look in the mirror, you see yourself not as you are, but as you were nanoseconds ago.
    a. Suppose you look up tonight at the bright star Deneb. Deneb is 1600 light years away. When you look at Deneb, how old is the image you are seeing?

18. (continued)

b. Even more disconcerting is the fact that what we see as simultaneous events do not necessarily occur simultaneously. Consider the two stars Betelgeuse and Rigel in the constellation Orion. Betelgeuse is 300 light years away and Rigel 500. How many years apart were the images generated that we see simultaneously?

19. Use the conversion factor of 1 light year $= 9.46 \cdot 10^{12}$ kilometers or $5.88 \cdot 10^{12}$ miles to determine the following.

a. Alpha Centauri, the nearest star to our sun, is 4.3 light years away. What is the distance in kilometers? How many miles away is it?

b. The radius of the Milky Way is 10^8 light years. How many meters is that?

c. Deneb is a star 1600 light years from Earth. How far is that in feet?

20. A homeowner would like to spread shredded bark (mulch) over her flowerbeds. She has three flowerbeds measuring 25 ft by 3 ft, 15 ft by 4 ft, and 30 ft by 1.5 ft. The recommended depth for the mulch is 4 inches, and the shredded bark costs $27.00 per one cubic yard. How much will it cost to cover all of the flowerbeds with shredded bark? (*Note:* You cannot buy a portion of a cubic yard of mulch.)

21. A circular swimming pool is 18 ft in diameter and 4 ft deep.

a. Determine the volume of the pool in gallons if one gallon is 231 cubic inches.

b. The pool's filter pump can circulate 2500 gal per hour. How many hours do you need to run the filter in order to filter the number of gallons contained in the pool?

c. One pound of chlorine shock treatment can treat 10,000 gal. How much of the shock treatment should you use?

22. An angstrom, Å, is a metric unit of length equal to one ten billionth of a meter. It is useful in specifying wavelengths of electromagnetic radiation (e.g., visible light, ultraviolet light, X-rays, and gamma rays).

a. The visible-light spectrum extends from approximately 3900 angstroms (violet light) to 7700 angstroms (red light) Write this range in centimeters using scientific notation.

b. Some gamma rays have wavelengths of 0.0001 angstrom. Write this number in centimeters using scientific notation.

c. The nanometer (nm) is 10 times larger than the angstrom, so 1 nm is equal to how many meters?

23. The National Institutes of Health guidelines suggest that adults over 20 should have a body mass index, or BMI, under 25. This index is created according to the formula

$$\text{BMI} = \frac{\text{weight in kilograms}}{(\text{height in meters})^2}$$

a. Given that 1 kilogram = 2.2 pounds, and 1 meter = 39.37 inches, calculate the body mass index of President Obama, who is 6′ 1″ tall and in 2010 weighed 180 pounds. According to the guidelines, how would you describe his weight?

b. Most Americans don't use the metric system. So in order to make the BMI easier to use, convert the formula to an equivalent one using weight in pounds and height in inches. Check your new formula by using President Obama's weight and height, and confirm that you get the same BMI.

c. An article about weight gain in a prominent U.S. newspaper tells readers how to estimate their body mass index. In order to do so, the author of the article directs readers to convert their weight in pounds into kilograms by multiplying pounds by 0.45. Then they are told to find their height in inches and convert it to meters by multiplying inches by 0.0254. The author tells them to take that number and multiply it by itself, then divide the results into their weight in kilograms. Can you do a better job of describing the process?

d. A college professor reads the article on weight gain mentioned in part c of this exercise and writes a letter to the newspaper in response. The professor argues that the math used in the article makes calculating BMI unnecessarily complicated. Instead, a simple formula is all that's needed. He says that readers just need to multiply their weight in pounds by 703 and then divide by the square of their height (in inches). If the resulting number is higher than 25, then the reader is overweight. Is the author of this letter correct?

24. Computer technology refers to the storage capacity for information with its own special units. Each minuscule electrical switch is called a "bit" and can be off or on. As the information capacity of computers has increased, the industry has developed some much larger units based on the bit:

1 byte = 8 bits

1 kilobit = 2^{10} bits, or 1024 bits (a kilobit is sometimes abbreviated Kbit)

1 kilobyte = 2^{10} bytes, or 1024 bytes (a kilobyte is sometimes abbreviated Kbyte)

1 megabit = 2^{20} bits, or 1,048,576 bits

1 megabyte = 2^{20} bytes, or 1,048,576 bytes

1 gigabyte = 2^{30} bytes, or 1,073,741,824 bytes

a. How many kilobytes are there in a megabyte? Express your answer as a power of 2 and in scientific notation.

b. How many bits are there in a gigabyte? Express your answer as a power of 2 and in scientific notation.

25. If 1 angstrom, Å, = 10^{-10} meter, determine the following values.

a. The radius of a hydrogen atom is 0.5 angstrom. How many meters is the radius?

b. The radius of a cell is 10^5 angstroms. How many meters is the cell's radius?

c. A radius of a proton is 0.00001 angstrom. Express the proton's radius in meters.

26. Anthrax spores, which were inhaled by postal workers, causing severe illness and death, are no larger than 5 microns in diameter. Using ratios, compare the diameter of a spore to the diameter of the tip of a pencil, which is approximately 1 millimeter. (*Note:* A micron is the same as a micrometer, μm.)

4.5 *Orders of Magnitude*

Comparing Numbers of Widely Differing Sizes

In Section 4.2, we learned that a useful method of comparing two objects of widely different sizes is to calculate the ratio rather than the difference of the sizes. The ratio can be estimated by computing *orders of magnitude,* the number of times we would have to multiply or divide by 10 to convert one size to the other. Each factor of 10 represents one order of magnitude.

For example, the radius of the observable universe is approximately 10^{26} meters and the radius of our solar system is approximately 10^{12} meters. To compare the radius of the observable universe to the radius of our solar system, calculate the ratio

$$\frac{\text{radius of the universe}}{\text{radius of our solar system}} \approx \frac{10^{26} \text{ meters}}{10^{12} \text{ meters}}$$

$$\approx 10^{26-12}$$

$$\approx 10^{14}$$

Orders of Magnitude

The radius of the universe is roughly 10^{14} times larger than the radius of the solar system; that is, we would have to multiply the radius of our solar system by 10 fourteen times in order to obtain the radius of the universe. Since each factor of 10 is counted as a single order of magnitude, the radius of the universe is *fourteen orders of magnitude larger* than the radius of our solar system. Equivalently, we could say that the radius of our solar system is *fourteen orders of magnitude smaller* than the radius of the universe.

When something is one order of magnitude larger than a *reference object,* it is 10 times larger. You *multiply* the *reference size* by 10 to get the other size. If the object is two orders of magnitude larger, it is 100 or 10^2 times larger, so you would multiply the reference size by 100. If it is one order of magnitude smaller, it is 10 times smaller, so you would *divide* the reference size by 10. Two orders of magnitude smaller means the reference size is divided by 100 or 10^2.

E X A M P L E   1

Comparing the radius of the sun to the radius of the hydrogen atom
The radius of the sun (10^9 meters) is how much larger than the radius of a hydrogen atom (10^{-11} meter)?

S O L U T I O N

$$\frac{\text{radius of sun}}{\text{radius of the hydrogen atom}} \approx \frac{10^9 \text{ meters}}{10^{-11} \text{ meter}}$$

$$\approx 10^{9-(-11)}$$

$$\approx 10^{20}$$

So the radius of the sun is 10^{20} times, or twenty orders of magnitude, larger than the radius of the hydrogen atom.

E X A M P L E   2

Comparing the size of a DNA strand to the size of a living cell
Compare the length of an unwound DNA strand (10^{-2} meter) with the size of a living cell (radius of 10^{-5} meter).

S O L U T I O N

$$\frac{\text{length of DNA strand}}{\text{radius of the living cell}} \approx \frac{10^{-2} \text{ meter}}{10^{-5} \text{ meter}}$$

$$\approx 10^{-2-(-5)}$$

$$\approx 10^{-2+5}$$

$$\approx 10^3$$

Surprisingly enough, the average width of a living cell is approximately three orders of magnitude *smaller* than one of the single strands of DNA it contains, if the DNA is uncoiled and measured lengthwise.

The Richter Scale

The reading "Earthquake Magnitude Determination" describes how earthquake tremors are measured.

The *Richter scale,* designed by the American Charles Richter in 1935, allows us to compare the magnitudes of earthquakes throughout the world. The Richter scale measures the maximum ground movement (tremors) as recorded on an instrument called a seismograph. Earthquakes vary widely in severity, so Richter designed the scale to measure order-of-magnitude differences. The scale ranges from less than 1 to over 8. Each increase of one unit on the Richter scale represents an increase of ten times, or one order of magnitude, in the maximum tremor size of the earthquake. So an increase from 2.5 to 3.5 indicates a 10-fold increase in maximum tremor size. An increase of two units from 2.5 to 4.5 indicates an increase in maximum tremor size by a factor of 10^2 or 100.

Description of the Richter Scale

Richter Scale Magnitude	Description
2.5	Generally not felt, but recorded on seismographs
3.5	Felt by many people locally
4.5	Felt by all locally; slight local damage may occur
6	Considerable damage in ordinary buildings; a destructive earthquake
7	"Major" earthquake; most masonry and frame structures destroyed; ground badly cracked
8 and above	"Great" earthquake; a few per decade worldwide; total or almost total destruction; bridges collapse, major openings in ground, tremors visible

Table 4.6

Table 4.6 contains some typical values on the Richter scale along with a description of how humans near the center (called the *epicenter*) of an earthquake perceive its effects. There is no theoretical upper limit on the Richter scale. The U.S. Geological Survey reported in 2010 that the largest measured earthquake in the United States was in Prince William Sound, Alaska, in 1964 (magnitude 9.2), and the largest in the world was in Chile in 1960 (magnitude 9.5).[6]

Graphing Numbers of Widely Differing Sizes: Log Scales

Exploration 4.1 asks you to construct a graph using logarithmic scales on both axes.

If the sizes of various objects in our solar system are plotted on a standard linear axis, we get the uninformative picture shown in Figure 4.1. The largest value stands alone, and all the others are so small when measured in terameters that they all appear to be zero. When objects of widely different orders of magnitude are compared on a linear scale, the effect is similar to pointing out an ant in a baseball stadium.

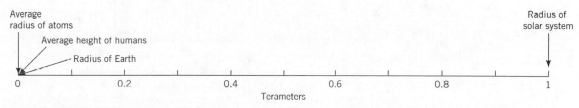

Figure 4.1 Sizes of various objects in the universe on a linear scale.
(*Note:* One terameter $= 10^{12}$ meters.)

[6]See the National Earthquake Information Center website at *http://earthquake.usgs.gov/eqcenter/historic_eqs.php.*

A more effective way of plotting sizes with different orders of magnitude is to use an axis that has powers of 10 evenly spaced along it. This is called a *logarithmic* or *log scale*. The plot of the previous data graphed on a logarithmic scale is much more informative (see Figure 4.2).

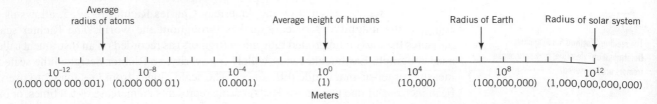

Figure 4.2 Sizes of various objects in the universe on an order-of-magnitude (logarithmic) scale.

Reading log scales

Graphing sizes on a log scale can be very useful, but we need to read the scales carefully. When we use a linear scale, each move of one unit to the right is equivalent to *adding* one unit to the number, and each move of k units to the right is equivalent to *adding k* units to the number (Figure 4.3).

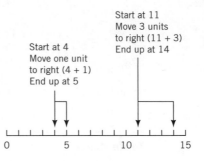

Figure 4.3 Linear scale.

When we use a log scale (see Figure 4.4), we need to remember that one unit of length now represents a change of one order of magnitude. Moving one unit to the right is equivalent to *multiplying by 10*. So moving from 10^4 to 10^5 is equivalent to multiplying 10^4 by 10. Moving three units to the right is equivalent to *multiplying* the starting number by 10^3, or 1000. In effect, a linear scale is an "additive" scale and a logarithmic scale is a "multiplicative" scale.

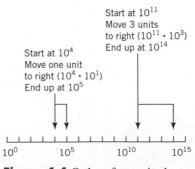

Figure 4.4 Order-of-magnitude (logarithmic) scale.

EXPLORE & EXTEND

4.5

Comparing Energy Released by Earthquakes

The Richter scale derives from the maximum ground movement of an earthquake (see p. 243). In February 2010, Chile experienced an earthquake with a magnitude of 8.8 on the Richter scale, and the month before, Haiti experienced one with magnitude of 7.0.

The Associated Press reported that the difference in magnitude of energy released by the Chilean earthquake was about 500 times greater than that of the one in Haiti, although the subsequent damage and death toll in Haiti were orders of magnitudes greater due to the differences in building structures.

A formula for the energy released by an earthquake is $(10^n)^{3/2}$ where n is the magnitude on the Richter scale (which measures maximum ground movement). So, an increase of 1.0 unit (or one order of magnitude) on the Richter scale is equivalent to a factor of $(10^{1.0})^{(3/2)}$, or 31.6, times more energy released. An increase of n units is equivalent to a factor of $(10^n)^{3/2}$ energy released.

Calculate the difference in energy released by the Chilean and Haiti earthquakes. How do your results compare to what was reported by the Associated Press?

Algebra Aerobics 4.5

1. In 1987 Los Angeles had an earthquake that measured 5.9 on the Richter scale. In 1988 Armenia had an earthquake that measured 6.9 on the Richter scale. Compare the sizes of the two earthquakes using orders of magnitude.

2. On July 15, 2003, Little Rock, Arkansas, had an earthquake that measured 6.5 on the Richter scale. Compare the size of this earthquake to the largest ever recorded as of 2010, 9.5 in Chile in 1960.

3. If my salary is $100,000 per year and you make an order of magnitude more, what is your salary? If Henry makes two orders of magnitude less money than I do, what is his salary?

4. For each of the following pairs, determine the order-of-magnitude difference:

    a. The radius of the sun (10^9 meters) and the radius of the Milky Way (10^{21} meters)

    b. The radius of a hydrogen atom (10^{-11} meter) and the radius of a proton (10^{-15} meter)

5. Joe wants to move from Wyoming to California, but he has been advised that houses in California cost an order of magnitude more than houses in Wyoming.

    a. If Joe's house in Wyoming is worth $400,000, how much would a similar house cost in California?

    b. If a house in California sells for $1,250,000, how much would it cost in Wyoming?

6. How many orders of magnitude greater is a kilometer than a meter? Than a millimeter?

7. By rounding the number to the nearest power of 10, find the approximate location of each of the following on the logarithmic scale in Figure 4.2 on page 244.

    a. The radius of the sun, at approximately 1 billion meters

    b. The radius of a virus, at 0.000 000 7 meter

    c. An object whose radius is two orders of magnitude smaller than that of Earth

Exercises for Section 4.5

1. What is the order-of-magnitude difference between the following units? (Refer to table on inside back cover.)

    a. A millimeter and a gigameter

    b. A second and a day

    c. A square centimeter and an acre (1 acre $= 43,560$ ft^2)

    d. A microfarad and a picofarad

2. Fill in the blanks to make each of the following statements true.

    a. Attaching the prefix "micro" to a unit _____ the size by _____ orders of magnitude.

    b. Attaching the prefix "kilo" to a unit _____ the size by _____ orders of magnitude.

    c. Scientists and engineers have designated prefix multipliers from septillionths (10^{-24}) to septillions (10^{24}), a span of _____ orders of magnitude.

3. Compare the following numbers using orders of magnitude.

    a. 5.261 and 52.61

    b. 5261 and 5.261

    c. $5.261 \cdot 10^6$ and 526.1

4. An ant is roughly 10^{-3} meter in length and the average human roughly one meter. How many times longer is a human than an ant?

5. Refer to the chart on p. 261 in Exploration 4.1.

    a. How many orders of magnitude larger is the Milky Way than the first living organism on Earth?

    b. How many orders of magnitude older is the Pleiades (a cluster of stars) than the first *Homo sapiens sapiens*?

6. Water boils (changes from a liquid to a gas) at 373 kelvins. The temperature of the core of the sun is 20 million kelvins. By how many orders of magnitude is the sun's core hotter than the boiling temperature of water?

7. An electron weighs about 10^{-27} gram, and a raindrop weighs about 10^{-3} gram. How many times heavier is a raindrop than an electron? How many times lighter is an electron than a raindrop? What is the order-of-magnitude difference?

8. Determine the order-of-magnitude difference in the sizes of the radii for:

   a. The solar system (10^{12} meters) compared with Earth (10^7 meters)

   b. Protons (10^{-15} meter) compared with the Milky Way (10^{21} meters)

   c. Atoms (10^{-10} meter) compared with neutrons (10^{-15} meter)

9. To compare the sizes of different objects, we need to use the same unit of measure.

   a. Convert each of these to meters:

   i. The radius of the moon is approximately 1,922,400 yards.

   ii. The radius of Earth is approximately 6400 km.

   iii. The radius of the sun is approximately 432,000 miles.

   b. Determine the order-of-magnitude difference between:

   i. The surface areas of the moon and Earth

   ii. The volumes of the sun and the moon

10. The pH scale measures the hydrogen ion concentration in a liquid, which determines whether the substance is acidic or alkaline. A strong acid solution has a hydrogen ion concentration of 10^{-1} M. One M equals $6.02 \cdot 10^{23}$ particles per liter, or 1 mole per liter.[7] A strong alkali solution has a hydrogen ion concentration of 10^{-14} M. Pure water, with a concentration of 10^{-7} M, is neutral. The pH value is the power without the minus sign, so pure water has a pH of 7, acidic substances have a pH less than 7, and alkaline substances have a pH greater than 7.

   a. Tap water has a pH of 5.8. Before the industrial age, rain water commonly had a pH of about 5. With the spread of modern industry, rain in the northeastern United States and parts of Europe now has a pH of about 4, and in extreme cases the pH is about 2. Lemon juice has a pH of 2.1. If acid rain with a pH of 3 is discovered in an area, how much more acidic is it than preindustrial rain?

   b. Blood has a pH of 7.4; wine has a pH of about 3.4. By how many orders of magnitude is wine more acidic than blood?

11. Which is an additive scale? Explain why. Which is a multiplicative or logarithmic scale? Explain why.

   a.

   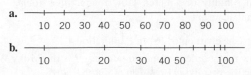

   b.

   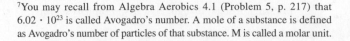

---

[7]You may recall from Algebra Aerobics 4.1 (Problem 5, p. 217) that $6.02 \cdot 10^{23}$ is called Avogadro's number. A mole of a substance is defined as Avogadro's number of particles of that substance. M is called a molar unit.

12. Graph the following on a power-of-10 (logarithmic) scale. (See sample log scale at the end of the exercises.)

   a. 1 meter   **c.** 1000 kilometers

   b. 10 meters   **d.** 10 gigameters

13. (Refer to the chart on p. 261 in Exploration 4.1.) Describe how to plot on a logarithmic scale an object whose radius is:

   a. Five orders of magnitude larger than the radius of the first atoms

   b. Twenty orders of magnitude smaller than the radius of the sun

14. **a.** Read "The Universe in One Year", inspired by Carl Sagan, the late great astronomer.

   b. Carl Sagan tried to give meaning to the cosmic chronology by imagining the almost 15 billion–year lifetime of the universe compressed into the span of one calendar year. To get a more personal perspective, consider your date of birth as the time at which the Big Bang took place. Map the following five cosmic events onto your own life span:

   i. The Big Bang   **iv.** First *Homo sapiens*

   ii. Creation of Earth   **v.** American Revolution

   iii. First life on Earth

   Once you have done the necessary mathematical calculations and placed your results on either a chart or a timeline, form a topic sentence and write a playful paragraph about what you were supposedly doing when these cosmic events took place. Hand in your calculations along with your writing.

15. Graph the following on a power-of-10 (logarithmic) scale. (See sample log scale at the end of the exercises.)

   a. 1 watt   **c.** 100 billion kilowatts

   b. 10 kilowatts   **d.** 1000 terawatts

16. Radio waves, sent from a broadcast station and picked up by the antenna of your radio, are a form of electromagnetic (EM) radiation, as are microwaves, X-rays, and visible, infrared, and ultraviolet light. They all travel at the speed of light. Electromagnetic radiation can be thought of as oscillations like the vibrating strings of a violin or guitar or like ocean swells that have crests and troughs. The distance between the crest or peak of one wave and the next is called the wavelength. The number of times a wave crests per minute, or per second for fast-oscillating waves, is called its frequency. Wavelength and frequency are inversely proportional: the longer the wavelength, the lower the frequency, and vice versa—the faster the oscillation, the shorter the wavelength. For radio waves and other EM, the number of oscillations per second of a wave is measured in hertz, after the German scientist who first demonstrated that electrical waves could transmit information across space. One cycle or oscillation per second equals 1 hertz (Hz).

   You may see the notation kHz beside the AM band and MHz beside the FM band on old radios. AM radio waves oscillate at frequencies measured in the kilohertz range, and FM radio waves oscillate at frequencies measured in the megahertz range.

16. (continued)

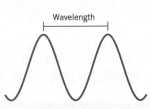

Wavelength

a. The Boston FM rock station WBCN transmits at 104.1 MHz. Write its frequency in hertz using scientific notation.

b. The Boston AM radio news station WBZ broadcasts at 1030 kHz. Write its frequency in hertz using scientific notation.

The wavelength λ (Greek lambda) in meters and frequency μ (Greek mu) in oscillations per second are related by the formula $\lambda = \dfrac{c}{\mu}$ where c is the speed of light in meters per second.

c. Estimate the wavelength of the WBCN FM radio transmission.

d. Estimate the wavelength of the WBZ AM radio transmission.

e. Compare your answers in parts (c) and (d), using orders of magnitude, with the length of a football field (approximately 100 meters).

$10^0 \quad 10^1 \quad 10^2 \quad 10^3 \quad 10^4 \quad 10^5 \quad 10^6 \quad 10^7 \quad 10^8 \quad 10^9 \quad 10^{10} \quad 10^{11} \quad 10^{12} \quad 10^{13} \quad 10^{14} \quad 10^{15}$

Sample Log Scale

4.6 *Logarithms as Numbers*

In Section 4.5 we used a logarithmic scale to graph numbers of widely disparate sizes. We labeled the axis with integer powers of 10, so it was easy to plot numbers such as $100,000 = 10^5$. But how would we plot a number such as $4,600,000,000 = 4.6 \cdot 10^9$, the approximate age of Earth in years? To create logarithmic (or orders of magnitude) scales, we need to learn about logarithms as numbers. We return to this scale repeatedly, starting in the next chapter. Later in Chapter 6, we learn about logarithms as functions.

Finding the Logarithms of Powers of 10

For handling very large or very small numbers, it is often easier to write the number using powers of 10. For example,

$$100,000 = 10^5$$

We say that

100,000 equals the base 10 raised to the fifth power

But we could rephrase this as

5 is the exponent of the base 10 that is needed to produce 100,000

The more technical way to say this is

5 is the *logarithm* base 10 of 100,000

In symbols we write

$$5 = \log_{10} 100,000$$

So the expressions

$$100,000 = 10^5 \quad \text{and} \quad 5 = \log_{10} 100,000$$

are two ways of saying the same thing. The key point to remember is that a logarithm is an exponent.

Definition of Logarithm

The *logarithm base 10 of* x is the exponent of 10 needed to produce x:

$$\log_{10} x = c \quad \text{means} \quad 10^c = x$$

So to find the logarithm of a number, write it as 10 to some power. The power is the logarithm of the original number.

E X A M P L E 1   **Logarithms are exponents**

Find log (1,000,000,000) without using a calculator.

SOLUTION

Since $\qquad 1{,}000{,}000{,}000 = 10^9$

then $\qquad \log_{10} 1{,}000{,}000{,}000 = 9$

and we say that the logarithm base 10 of 1,000,000,000 is 9. The logarithm of a number tells us the exponent of the number when written as a power of 10. Here the logarithm is 9, so that means that when we write 1,000,000,000 as a power of 10, the exponent is 9.

EXAMPLE 2

What is log 1?
Find log 1 without using a calculator.

SOLUTION

Since $\qquad 1 = 10^0$

then $\qquad \log_{10} 1 = 0$

and we say that the logarithm base 10 of 1 is 0. Since logarithms represent exponents, this says that when we write 1 as a power of 10, the exponent is 0.

EXAMPLE 3

Logarithms of very small numbers
How do we calculate the logarithm base 10 of decimals such as 0.000 01?

SOLUTION

Since $\qquad 0.000\,01 = 10^{-5}$

then $\qquad \log_{10} 0.000\,01 = -5$

and we say that the logarithm base 10 of 0.000 01 is -5.

When is $\log_{10} x$ not defined?

In the previous example, we found that the log (short for "logarithm") of a number can be negative. This makes sense if we think of logarithms as exponents, since exponents can be any real number. But we cannot take the log of a negative number or zero; that is, $\log_{10} x$ is not defined when $x \leq 0$. Why? If $\log_{10} x = c$, where $x \leq 0$, then $10^c = x$ (a number ≤ 0). But 10 to any power will never produce a number that is negative or zero, so $\log_{10} x$ is not defined if $x \leq 0$.

$$\log_{10} x \text{ is not defined when } x \leq 0.$$

Table 4.7 gives a sample set of values for x and their associated logarithms base 10. To find the logarithm base 10 of x, we write x as a power of 10, and the logarithm is just the exponent.

Most scientific calculators and spreadsheet programs have a LOG function that calculates logarithms base 10. Try using technology to double-check some of the numbers in Table 4.7.

Logarithms of Powers of 10

x	Exponential Notation	$\log_{10} x$
0.0001	10^{-4}	-4
0.001	10^{-3}	-3
0.01	10^{-2}	-2
0.1	10^{-1}	-1
1	10^0	0
10	10^1	1
100	10^2	2
1000	10^3	3
10,000	10^4	4

Table 4.7

Logarithms base 10 are used frequently in our base 10 number system and are called *common logarithms*. We write $\log_{10} x$ as log x.

> **Common Logarithms**
>
> Logarithms base 10 are called *common logarithms*.
>
> $$\log_{10} x \text{ is written as } \log x.$$

Algebra Aerobics 4.6a

1. Without using a calculator, find the logarithm base 10 of:
   a. 10,000,000   c. 10,000   e. 1000   g. 1
   b. 0.000 000 1   d. 0.0001   f. 0.001

2. Rewrite the following expressions in an equivalent form using powers of 10:
   a. $\log 100,000 = 5$ c. $\log 10 = 1$
   b. $\log 0.000\ 000\ 01 = -8$ d. $\log 0.01 = -2$

3. Evaluate without using a calculator. Find a number if its log is:
   a. 3   c. 6   e. -2
   b. -1 d. 0

4. Find c and then rewrite as a logarithm:
   a. $10^c = 1000$ d. $10^c = 0.000\ 01$
   b. $10^c = 0.001$ e. $10^c = 1,000,000$
   c. $10^c = 100,000$ f. $10^c = 0.000\ 001$

5. Find the value of x that makes the statement true.
   a. $10^{x-3} = 10^2$ c. $\log (x - 2) = 1$
   b. $10^{2x-1} = 10^4$ d. $\log 5x = -1$

Finding the Logarithm of Any Positive Number

Scientific calculators have a LOG function that will calculate the log of any positive number. However, it's easy to make errors typing in numbers, so it's important not to rely solely on technology-generated answers. To verify that the calculated number is the right order of magnitude, you should estimate the answer without using technology.

EXAMPLE 4   **Estimating, then using technology to calculate logs**

a. Estimate the size of log (2000) and log (0.07).

b. Use a calculator to find the logarithm of 2000 and 0.07.

SOLUTION   a. i. If we place 2000 between the two closest integer powers of 10, we have

$$1000 < 2000 < 10,000$$

Rewriting 1000 and 10,000 as powers of 10 gives

$$10^3 < 2000 < 10^4$$

Taking the log of each term preserves the inequality, so we would expect

$$3 < \log 2000 < 4$$

ii. If we place 0.07 between the two closest integer powers of 10, we have

$$0.01 < 0.07 < 0.10$$

Rewriting 0.01 and 0.10 as powers of 10 gives

$$10^{-2} < 0.07 < 10^{-1}$$

Taking the log of each term preserves the inequality, so we would expect

$$-2 < \log 0.07 < -1$$

b. Using a calculator, we have

i. $\log 2000 \approx 3.301$ ii. $\log (0.07) \approx -1.155$.

So our estimates were correct.

EXAMPLE 5   **Calculating with logs of very large or small numbers**
Find the logarithm of

a. 3.7 trillion

b. A Planck length of 0.000 000 000 000 000 000 000 000 000 000 000 016 meter

SOLUTION Our strategy in each case is to

- Write the number in scientific notation.
- Then convert the number into a single power of 10.

The resulting exponent is the desired log.

a. In scientific notation 3.7 trillion is $3.7 \cdot 10^{12}$. To convert the entire expression into a single power of 10, we need to first convert the coefficient 3.7 to a power of 10. Using a calculator, we have

$$\log 3.7 \approx 0.568$$

so $$3.7 \approx 10^{0.568}$$

If we substitute for 3.7, $$3.7 \cdot 10^{12} \approx 10^{0.568} \cdot 10^{12}$$

and use rules for exponents, $$= 10^{0.568+12}$$

we have $$= 10^{12.568}$$

So the exponent 12.568 is the desired logarithm.

b. In scientific notation a Planck length is $1.6 \cdot 10^{-35}$ meter. We need to convert 1.6 to a power of 10. Using a calculator, we have $\log 1.6 \approx 0.204$, so $1.6 = 10^{0.204}$. If we

substitute for 1.6 $$1.6 \cdot 10^{-35} \approx 10^{0.204} \cdot 10^{-35}$$

use rules for exponents $$= 10^{0.204-35}$$

and subtract, we get $$= 10^{-34.796}$$

So the exponent -34.796 is the desired log.

   So far we have dealt with finding the logarithm of a given number. Logarithms can, of course, occur in expressions involving variables.

EXAMPLE 6 **Finding the number given the log**
Rewrite the following expressions using exponents, and then solve for x without using a calculator.

a. $\log x = 3$ b. $\log x = 0$ c. $\log x = -2$

SOLUTION a. If $\log x = 3$, then $10^3 = x$, so $x = 1000$.

b. If $\log x = 0$, then $10^0 = x$, so $x = 1$.

c. If $\log x = -2$, then $10^{-2} = x$, so $x = 1/10^2 = 1/100 = 0.01$.

EXAMPLE 7 **Rewriting exponents using logarithms**
Rewrite the following expressions using logarithms and then solve for x using a calculator. Round off to three decimal places.

a. $10^x = 11$ b. $10^x = 0.5$ c. $10^x = 0$

SOLUTION a. If $10^x = 11$, then $\log 11 = x$. Using a calculator gives $x \approx 1.041$.

b. If $10^x = 0.5$, then $\log 0.5 = x$. Using a calculator gives $x \approx -0.301$.

c. There is no power of 10 that equals 0. Hence there is no solution for x.

Plotting Numbers on a Logarithmic Scale

We are finally prepared to answer the question posed at the very beginning of this section: How can we plot on a logarithmic (or order-of-magnitude) scale a number such as 4.6 billion years, the estimated age of Earth?

A Strategy for Plotting Numbers on a Logarithmic Scale

- Write the number in scientific notation.
- Then convert the number into a single power of 10.
- Use the exponent to help plot the number.

In scientific notation the age of Earth equals 4.6 billion $= 4,600,000,000 = 4.6 \cdot 10^9$ years. To plot this number on a logarithmic scale, we need to convert 4.6 into a power of 10. Using a calculator, we have $\log 4.6 \approx 0.663$, so $4.6 \approx 10^{0.663}$. If we

substitute for 4.6 $4.6 \cdot 10^9 \approx 10^{0.663} \cdot 10^9$

and use rules of exponents $= 10^{0.663+9}$

we have $= 10^{9.663}$

The power of 10 seems reasonable since $10^9 < 4.6 \cdot 10^9 < 10^{10}$.

Having converted our original number $4.6 \cdot 10^9$ into $10^{9.663}$, we can plot it on an order-of-magnitude graph between 10^9 and 10^{10} (see Figure 4.5).

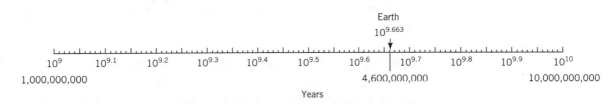

Figure 4.5 Age of Earth plotted on an order-of-magnitude (or logarithmic) scale.

E X A M P L E   8   **Plotting numbers on a logarithmic scale**
Plot the numbers 100, 200, 300, 400, 500, 600, 700, 800, 900, and 1000 on a logarithmic scale.

S O L U T I O N   Using our log plotting strategy, we first convert each number to a single power of 10. We have

$$100 = 10^2 \qquad\qquad 600 \approx 10^{2.778}$$
$$200 \approx 10^{2.301} \qquad 700 \approx 10^{2.845}$$
$$300 \approx 10^{2.477} \qquad 800 \approx 10^{2.903}$$
$$400 \approx 10^{2.602} \qquad 900 \approx 10^{2.954}$$
$$500 \approx 10^{2.699} \qquad 1000 = 10^3$$

We can now use the exponents of each power of 10 to plot the numbers directly onto a logarithmic scale (see Figure 4.6).

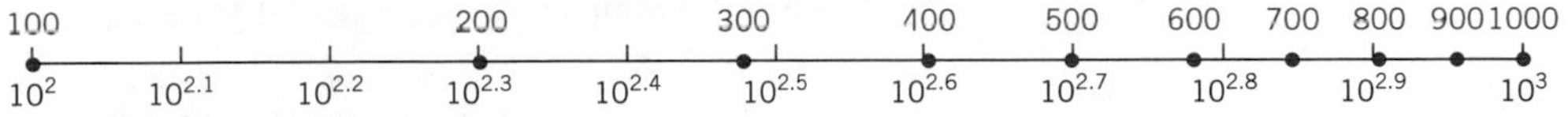

Figure 4.6 A logarithmic plot of the numbers 100, 200, 300, . . . , 1000.

Note that on the logarithmic scale in Figure 4.6 the point halfway between 10^2 and 10^3 is at $10^{2.5} = 316$.

When are numbers evenly spaced on a logarithmic scale?

On a linear (additive) scale the numbers 100, 200, . . . , 1000 would be evenly spaced, since you *add* a constant amount to move from one number to the next. On a logarithmic (multiplicative) scale, numbers that are evenly spaced are generated by *multiplying* by a constant amount to get from one number to the next. For example, the integer powers of 10 are all evenly spaced on a log plot since you multiply each number by 10 to get the next number in the sequence. Similarly, the numbers 100, 200, 400, and 800 are evenly spaced in Figure 4.6 since you multiply by the constant 2 to get from one number in the sequence to the next. The sequence 100, 200, 300, . . . , 1000 is not evenly spaced, since there is not a constant factor that you could multiply one number by to get to the next.

Labeling using only the exponent

Instead of labeling the axis using powers of 10, we can label it using just the exponents of 10 as in Figure 4.7. Remember that exponents are logarithms, which is why we call the scale logarithmic.

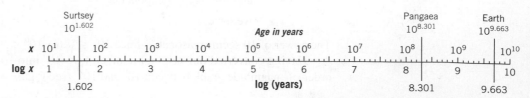

Figure 4.7 The age of Surtsey, Pangaea, and Earth plotted using an order-of-magnitude or logarithmic scale.

EXAMPLE 9

Relating powers of ten and logarithms on a logarithmic scale

Identify on the logarithmic plot in Figure 4.7 the following numbers:

a. 200 million, the number of years since all of Earth's continents collided to form one giant land mass called Pangaea

b. 40, the number of years since the volcanic island of Surtsey, Earth's newest land mass, emerged near Iceland

SOLUTION

a. In scientific notation 200 million $= 200{,}000{,}000 = 2.0 \cdot 10^8$. Using a calculator, we have $\log 2.0 \approx 0.301$, so $2.0 \approx 10^{0.301}$. Hence $2.0 \cdot 10^8 \approx 10^{0.301} \cdot 10^8 = 10^{8.301}$ is the age of Pangaea in a form easily plotted on a logarithmic scale.

b. In scientific notation $40 = 4.0 \cdot 10^1$. Using a calculator, we have $\log 4.0 \approx 0.602$, so $4.0 \approx 10^{0.602}$. Therefore $4.0 \cdot 10^1 \approx 10^{0.602} \cdot 10^1 = 10^{1.602}$ is the age of Surtsey in a form easily plotted on a log scale.

EXPLORE & EXTEND

4.6

The Dow Jones Industrial Average

The graph below shows the trends in the Dow Jones Industrial Average (an indicator of stock market prices) from 1901 to 2010 using a powers-of-ten or log scale on the vertical axis and a linear scale for years on the horizontal axis.

Use orders of magnitude to describe three striking changes over time in the Dow Jones Industrial Average.

Source: Observations *http://observationsandnotes.blogspot.com/.*

Algebra Aerobics 4.6b

Most of these problems require a calculator that can evaluate logs.

1. Use a calculator to estimate each of the following:
   a. log 3   b. log 6   c. log 6.37

2. Use the answers from Problem 1 to estimate values for:
   a. log 3,000,000   b. log 0.006
   Then use a calculator to check your answers.

3. Write each of the following as a power of 10:
   a. 0.000 000 7 m (the radius of a virus)
   b. 780,000,000 km (the mean distance from our sun to Jupiter)
   c. 0.0042
   d. 5,400,000,000

4. Rewrite the following equations using exponents instead of logarithms. Estimate the solution for x.

Check your estimate with a calculator. Round the value of x to the nearest integer.
   a. log x = 4.125
   b. log x = 5.125
   c. log x = 2.125

5. Rewrite the following equations using logs instead of exponents. Estimate a solution for x and check your estimate with a calculator. Round the value of x to three decimal places.
   a. $10^x = 250$ c. $10^x = 0.075$
   b. $10^x = 250,000$ d. $10^x = 0.000\ 075$

6. Write each number as a power of 10 and then plot them all on the logarithmic scale below.
   a. 57   c. 25,000
   b. 182   d. 7,200,000,000

Exercises for Section 4.6

Many of the problems in this section require the use of a calculator that can evaluate logs.

1. Rewrite in an equivalent form using logarithms:
   a. $10^4 = 10,000$ c. $10^0 = 1$
   b. $10^{-2} = 0.01$ d. $10^{-5} = 0.00001$

2. Use your calculator to evaluate to two decimal places:
   a. $10^{0.4}$ c. $10^{0.6}$ e. $10^{0.8}$
   b. $10^{0.5}$ d. $10^{0.7}$ f. $10^{0.9}$

3. Express the number 375 in the form 10^x.

4. Estimate the value of each of the following:
   a. log 4000   b. log 5,000,000   c. log 0.0008

5. Rewrite the following statements using logs:
   a. $10^2 = 100$ b. $10^7 = 10,000,000$ c. $10^{-3} = 0.001$
   Rewrite the following statements using exponents:
   d. log 10 = 1   e. log 10,000 = 4   f. log 0.0001 = −4

6. Evaluate the following without a calculator.
   a. Find the following values:
      I. log 100
      ii. log 1000
      iii. log 10,000,000
      What is happening to the values of log x as x gets larger?
   b. Find the following values:
      i. log 0.1   ii. log 0.001   iii. log 0.000 01
      What is happening to the values of log x as x gets closer to 0?

c. What is log 0?
d. What is log(−10)? What do you know about log x when x is any negative number?

7. Rewrite the following equations using exponents instead of logs. Estimate a solution for x and then check your estimate with a calculator. Round the value of x to the nearest integer.
   a. log x = 1.255
   b. log x = 3.51
   c. log x = 4.23
   d. log x = 7.65

8. Rewrite the following equations using exponents instead of logs. Estimate a solution for x and then check your estimate with a calculator. Round the value of x to the nearest integer.
   a. log x = 1.079
   b. log x = 0.699
   c. log x = 2.1
   d. log x = 3.1

9. Rewrite the following equations using logs instead of exponents. Estimate a solution for x and then check your estimate with a calculator. Round the value of x to three decimal places.
   a. $10^x = 12,500$
   b. $10^x = 3,526,000$
   c. $10^x = 597$
   d. $10^x = 756,821$

10. Rewrite the following equations using logs instead of exponents. Estimate a solution for x and then check your estimate with a calculator. Round the value of x to three decimal places.

   a. $10^x = 153$
   c. $10^x = 0.125$
   b. $10^x = 153{,}000$
   d. $10^x = 0.001\,25$

11. Solve for x. (*Hint:* Rewrite each expression so that you can use a calculator to solve for x.)

   a. $\log x = 0.82$
   c. $\log x = 0.33$
   b. $10^x = 0.012$
   d. $10^x = 0.25$

12. Without using a calculator, show how you can solve for x.

   a. $10^{x-2} = 100$
   c. $10^{2x-3} = 1000$
   b. $\log(x - 4) = 1$
   d. $\log(6 - x) = -2$

13. Without using a calculator show how you can solve for x.

   a. $10^{x-5} = 1000$
   c. $10^{3x-1} = 0.0001$
   b. $\log(2x + 10) = 2$
   d. $\log(500 - 25x) = 3$

14. Find the value of x that makes the equation true.

   a. $\log x = -2$ b. $\log x = -3$ c. $\log x = -4$

15. Without using a calculator, for each number in the form $\log x$, find some integers a and b such that $a < \log x < b$. Justify your answer. Then verify your answers with a calculator.

   a. $\log 11$ b. $\log 12{,}000$ c. $\log 0.125$

16. Use a calculator to determine the following logs. Double-check each answer by writing down the equivalent expression using exponents, and then verify this equivalence using a calculator.

   a. $\log 15$ b. $\log 15{,}000$ c. $\log 1.5$

17. On a logarithmic scale, what would correspond to moving over to the right:

   a. 0.001 unit   b. $\frac{1}{2}$ unit   c. 2 units   d. 10 units

18. The difference in the noise levels of two sounds is measured in decibels, where $\text{decibels} = 10 \log \left(\dfrac{I_2}{I_1} \right)$ and I_1 and I_2 are the intensities of the two sounds. Compare noise levels when $I_1 = 10^{-15}$ watt/cm^2 and $I_2 = 10^{-8}$ watt/cm^2.

19. The concentration of hydrogen ions in a water solution typically ranges from 10 M to 10^{-15} M. (One M equals $6.02 \cdot 10^{23}$ particles, such as atoms, ions, molecules, etc., per liter or 1 mole per liter.) Because of this wide range, chemists use a logarithmic scale, called the pH scale, to measure the concentration (see Exercise 10 of Section 4.5). The formal definition of pH is $\text{pH} = -\log[\text{H}^+]$, where $[\text{H}^+]$ denotes the concentration of hydrogen ions. Chemists use the symbol H$^+$ for hydrogen ions, and the brackets [ ] mean "the concentration of."

   a. Pure water at 25°C has a hydrogen ion concentration of 10^{-7} M. What is the pH?

   b. In orange juice, $[\text{H}^+] \approx 1.4 \cdot 10^{-3}$ M. What is the pH?

   c. Household ammonia has a pH of about 11.5. What is its $[\text{H}^+]$?

   d. Does a higher pH indicate a lower or a higher concentration of hydrogen ions?

   e. A solution with a pH > 7 is called basic, one with a pH $= 7$ is called neutral, and one with a pH < 7 is called acidic. Identify pure water, orange juice, and household ammonia as either acidic, neutral, or basic. Then plot their positions on the accompanying scale, which shows both the pH and the hydrogen ion concentration.

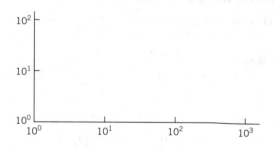

20. a. Place the number 50 on the *additive* scale below.

   b. Place the number 50 on the *multiplicative* scale below.

21. The coordinate system below uses multiplicative or log scales on both axes. Position the point whose coordinates are (708, 25).

22. Change each number to a power of 10, then plot the numbers on a power-of-10 scale. (See sample log scale below.)

   a. 125   b. 372   c. 694   d. 840

23. Compare the times listed below by plotting them on the same order-of-magnitude scale. (*Hint:* Start by converting all the times to seconds.)

   a. The time of one heartbeat (1 second)
   b. Time to walk from one class to another (10 minutes)
   c. Time to drive across the country (7 days)
   d. One year (365 days)
   e. Time for light to travel to the center of the Milky Way (38,000 years)
   f. Time for light to travel to Andromeda, the nearest large galaxy (2.2 million years)

10^0 10^1 10^2 10^3 10^4 10^5 10^6 10^7 10^8 10^9 10^{10} 10^{11} 10^{12} 10^{13} 10^{14} 10^{15}

Sample Log Scale

CHAPTER SUMMARY

Powers of 10

If n is a positive integer, we define

$$10^n = \underbrace{10 \cdot 10 \cdot 10 \cdot \,\cdots\, \cdot 10}_{n \text{ factors}}$$

$$10^0 = 1$$

$$10^{-n} = \frac{1}{10^n}$$

Scientific Notation

A number is in scientific notation if it is in the form

$$N \cdot 10^n$$

where N is called the *coefficient*, $1 \leq |N| < 10$, and n is an integer.

Example: In scientific notation 67,000,000 is written as $6.7 \cdot 10^7$ and $-0.000\,000\,000\,008\,1$ is written as $-8.1 \cdot 10^{-12}$.

Powers of a

In the expression a^n, a is called the *base* and n is called the *exponent* or *power*.

If a is nonzero real number and n is a positive integer, then

$$a^n = \underbrace{a \cdot a \cdot a \cdots a}_{n \text{ factors}}$$

$$a^0 = 1$$

$$a^{-n} = \frac{1}{a^n}$$

If m and n are positive integers and the base, a, is restricted to values for which the power is defined, then

$$a^{1/2} = \sqrt{a}$$
$$a^{1/n} = \sqrt[n]{a}$$
$$a^{m/n} = (a^m)^{1/n} = (a^{1/n})^m$$
$$\qquad = \sqrt[n]{a^m} = (\sqrt[n]{a})^m$$

Rules of Exponents

If a and b are nonzero, then

1. $a^m \cdot a^n = a^{(m+n)}$ **4.** $(ab)^n = a^n b^n$

2. $\dfrac{a^n}{a^m} = a^{(n-m)}$ **5.** $\left(\dfrac{a}{b}\right)^n = \dfrac{a^n}{b^n}$

3. $(a^m)^n = a^{(m \cdot n)}$

Orders of Magnitude

We use *orders of magnitude* when we compare objects of widely different sizes. Each *factor* of 10 is counted as a single order of magnitude.

Example: The radius of the universe is 10^{14} times or fourteen orders of magnitude larger than the radius of the solar system. And vice versa: The radius of the solar system is fourteen orders of magnitude smaller than the radius of the universe.

Logarithms

The *logarithm base 10 of x* is the exponent of 10 needed to produce x. So

$$\log_{10} x = c \qquad \text{means} \qquad 10^c = x$$

We say that c is the logarithm base 10 of x.

Example: $\log_{10} 6{,}370{,}000 \approx 6.804$ means that $10^{6.804} = 10^6 \cdot 10^{0.804} \approx 10^6 \cdot 6.37 \approx 6{,}370{,}000$.

Logarithms base 10 are called *common logarithms*. We usually write $\log_{10} x$ as $\log x$. When $x \leq 0$, $\log x$ is not defined.

Plotting Numbers on a Logarithmic Scale

Logarithmic or powers-of-10 scales are used to graph objects of widely differing sizes. We can plot a number on a log scale by converting the number to a power of 10.

Age of Earth Plotted on a Log Scale

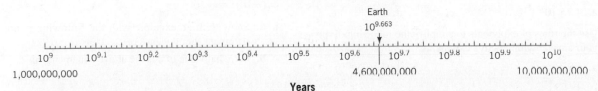

CHECK YOUR UNDERSTANDING

I. Are the statements in Problems 1–24 true or false? Give an explanation for your answer.

1. A distance of 10 miles is longer than a distance of 10 kilometers.

2. There are 39 centimeters in 1 inch.

3. 10^{15} is 10 followed by fifteen zeros.

4. $10^0 = 0$.

5. $\dfrac{1}{10^{-m}} = 10^m$.

6. $-0.000\,005\,62 = -5.62 \cdot 10^{-6}$.

7. $15 \cdot 10^4$ is correct scientific notation for the number 150,000.

8. The age of the universe $(1.37 \cdot 10^{10}$ years) is about three times the age of Earth $(4.6 \cdot 10^9$ years).

9. On February 22, 2010, the population of the world (about 6,804,397,000) was approximately three orders of magnitude larger than the population of the United States (308,738,000).

10. $-8^2 = (-8)^2$.

11. $\left(\dfrac{5}{3}\right)^{-3} = \dfrac{-15}{-9}$.

12. $10^2 + 10^1 + 10^5 = 10^{2+1+5} = 10^8$.

13. To convert a distance D in kilometers to miles, you could multiply D by $\dfrac{1\ \text{km}}{0.62\ \text{mile}}$.

14. The units of $300\,\dfrac{\text{km}}{\text{hr}} \cdot \dfrac{1\ \text{hr}}{60\ \text{min}} \cdot \dfrac{1\ \text{min}}{60\ \text{sec}} \cdot \dfrac{10^3\ \text{m}}{1\ \text{km}}$ are meters per second.

15. $-9 < -\sqrt{75} < -8$.

16. $8^{1/2} = 8^{0.5}$.

17. $\log_{10} 0.0001 = 10^{-4}$.

18. $\log_{10} 1821$ is not defined since 1821 is not a power of 10.

19. $(81)^{1/2} = \pm 9$ because $(9)^2 = 81$ and $(-9)^2 = 81$.

20. $\log 0 = 1$.

21. $-4 < \log 0.00015 < -3$.

22. $\log 0.143 \approx -0.845$ means that $10^{-0.845} \approx 0.143$.

23. If $P > 0$, $\log P = Q$ means that $10^P = Q$.

24. The following figure illustrates the number 7,500,000 plotted correctly on a logarithmic scale.

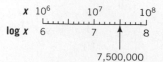

II. In Problems 25–30, give examples with the specified properties.

25. Populations of two cities A and B, where the population of city A is two orders of magnitude larger than that of city B.

26. A number x such that $\log x$ lies between 8 and 9.

27. A number x such that $\log x$ is a negative number.

28. A positive number b such that $\sqrt{b} > b$.

29. A non-zero number b such that $b^m = b^n$ for any numbers m and n.

30. A number b such that $|b| = -b$.

III. Are the statements in Problems 31–40 true or false? If a statement is true, explain how you know. If a statement is false, give a counterexample.

31. If one quantity is four orders of magnitude larger than a second quantity, it is four times as large as the second quantity.

32. $|c| = c$ for any real number c.

33. $\log x$ is defined only for numbers $x > 0$.

34. Raising a number to the $\frac{1}{3}$ power is the same as taking the cube root of that number.

35. $b^m \cdot b^m = b^{m^2}$.

36. $(b^p)^q = b^{p+q}$.

37. $(b + c)^m = b^m + c^m$.

38. $(-b)^q = b^q$.

39. $b^m \cdot c^n = (b \cdot c)^{m+n}$.

40. If n is odd, $\sqrt[n]{b}$ can be positive, negative, or zero depending on the value of b.

CHAPTER 4 REVIEW: PUTTING IT ALL TOGETHER

Exercise 23 requires a calculator that can evaluate powers.

1. Evaluate each of the following without a calculator.

   a. $4.2 \cdot 10^3$ **b.** $(-5)2^3$ **c.** -4^2 **d.** $100^{-1/2}$ **e.** $\dfrac{3^5}{3^2}$

2. Use the rules of exponents to simplify the following. Express your answer with positive exponents.

   a. $x^4 x^3$ **c.** $(-2xy^2)^3$ **e.** $(x^{-3}y)(x^2y^{-1/2})$

   b. $\dfrac{10x^2y^4}{5xy^3}$ **d.** $(x^{-1/2})^2$

3. For what integer values of x will the following statement be true?

   $(-10)^x = -10^x$

4. **a.** Show with an example why the following is not a true statement for all values of x: $x^3 + x^5 = x^8$.

   b. For what value of x is the above statement true?

5. Elephant seals can weigh as much as 5000 lb (for males) and 2000 lb (for females). On land, these seals can travel short distances quite quickly, as much as 20 feet in 3 seconds. How many miles per hour is this?

6. An NFL regulation playing field for football is 120 yd (110 m) long including the end zones, and 53 yd 1 ft (48.8 m) wide. An acre is 4840 square yards, and 1 yard = 3 feet.

   a. Which is larger, a football field (including the end zones) or an acre? By how much?

   b. If you bought a house on a square lot that measured half an acre, what would the dimensions of the lot be in feet?

7. According to a special report in 2010 by *Forbes,* the highest paid chief executive officer (CEO) in the U.S. was Lawrence Culp Jr., CEO for Danaher. His total compensation was $141.36 million, which included salary, bonuses, perks, stock grants, and stock gains. By what order of magnitude is his salary greater than that of a minimum-wage worker in the same state as Danaher's corporate headquarters, making $8.25/hr working 40 hours/week for 50 weeks/year? How many years would the minimum-wage worker have to work to earn what Lawrence Culp Jr. made in 1 year?

8. The Yangtze River (China) is 6380 km long. The Colorado River is 1400 miles long.

   a. Which river is longer?

   b. Compare the lengths of these rivers using orders of magnitude.

9. Use the accompanying table to answer the following questions.

Country	Area
Russia	17,075,200 km^2
Chile	290,125 mi^2
Canada	3,830,840 mi^2
South Africa	1,184,825 km^2
Norway	323,895 km^2
Monaco	0.5 mi^2

   a. Which country has the largest area? The smallest?

   b. Using scientific notation, arrange the countries from largest area to smallest area.

   c. What is the order-of-magnitude difference between the country with the largest area and the country with the smallest area?

10. The Energy Information Administration of the U.S. Department of Energy estimates that in 2010 the world energy use will be 470.8 quadrillion Btu (British thermal units), where 1 Btu = 0.000 293 1 kWh (kilowatt-hours.)

    a. Express 470.8 quadrillion Btu and 0.000 293 1 kWh in scientific notation.

    b. How many kilowatt-hours are there in 470.8 quadrillion Btu? Give your answer in scientific notation.

11. Is the following statement true or false? "An increase in one order of magnitude is the same as an increase of 100%." If true, explain why. If false, revise the statement to make it true.

12. Three projections for the world population (high, medium, and low) made by the United Nations are shown in the graph. Use this graph to estimate answers to the following questions:

a. What is the difference in millions of people between the high and low projections for world population in 2060? Now compare these projections using ratios and orders of magnitude.

b. What is the difference in millions of people between the high and medium projections for world population in 2100? Now compare these projections using ratios and orders of magnitude.

c. Compare the difference in the low projections for the world population between 2060 and 2100 using three measures: amount in millions, orders of magnitude, and rate of change.

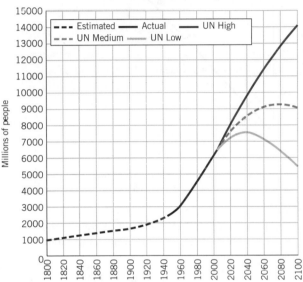

Source: World population from 1800 to 2100, based on United Nations projections and historical estimates, *http://en.wikipedia.org/wiki/World_population.*

13. Hubble's Law states that galaxies are receding from one another at velocities directly proportional to the distances separating them. The graph on the next page illustrates that Hubble's Law holds true across the known universe. The plot includes ten major clusters of galaxies. The boxed area at the lower left represents the galaxies observed by Hubble when he discovered the law. The easiest way to understand this graph is to think of Earth as being at the center of the universe (at 0 distance) and not moving (at 0 velocity). In other words, imagine Earth at the origin of the graph (a favorite fantasy of humans). Think of the horizontal axis as measuring the distance of the galaxy from Earth, and the vertical as measuring the velocity at which a galaxy cluster is moving away from Earth (the recession velocity). Then answer the following questions.

   a. Identify the coordinates of two data points that lie on the regression line drawn on the graph.

   b. Use the coordinates of the points in part (a) to calculate the slope of the line. That slope is called the *Hubble constant.*

   c. What does the slope mean in terms of distance from Earth and recession velocity?

   d. Construct an equation for our line in the form $y = mx + b$. Show your work.

13. (continued)

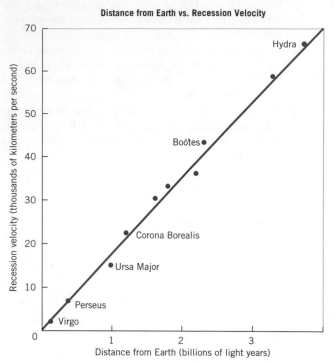

Distance from Earth vs. Recession Velocity

Source: T. Ferris, *Coming of Age in the Milky Way* (New York: William Morrow, 1988). Copyright © 1988 by Timothy Ferris. Reprinted by permission of Harper Collins Publishers

14. Temperature can affect the speed of sound. The speed of sound, S (in feet/second), at an air temperature of T (in degrees Celsius) is

$$S = \frac{1087(273 + T)^{0.5}}{16.52}$$

a. Express T in terms of S.

b. The speed of sound is often given as 1120 feet/second. At what temperature in degrees Celsius would that be? At what temperature in degrees Fahrenheit would that be? (Recall that degrees Fahrenheit = 1.8 · degrees Celsius + 32.)

15. The radius of Earth is about $6.3 \cdot 10^6$ m and its mass is approximately $5.97 \cdot 10^{24}$ kg. Find its density in kg/m³ (density = mass/volume).

16. Objects that are less dense than water will float; those that are more dense than water will sink. The density of water is 1.0 g/cm³. A brick has a mass of 2268 g and a volume of 1230 cm³. Show that the brick will sink in water (recall density = mass/volume).

17. An adult patient weighs 130 lb. The prescription for a drug is 5 mg per kg of the patient's weight per day. This drug comes in 100-mg tablets. What daily dosage should be prescribed?

18. On March 2, 2007, the *Boston Globe* reported the following:

An exabyte is 1 quintillion bytes. In 2006 alone, the human race generated 161 exabytes of digital information. So? Well, that's about 3 million times the information in all the books ever written or the equivalent of 12 stacks of books, each extending more than 93 million miles from Earth to the sun.

a. Use scientific notation to represent the amount of digital information generated in 2006. (One quintillion is 1 followed by eighteen zeros.)

b. Compare the amount of digital information generated in 2006 with the amount of information in all the books ever written, using orders of magnitude.

c. Estimate how many miles of books are needed to hold the equivalent information in 161 exabytes. Express your answer in scientific notation.

d. In May 2010, the International Data Corporation predicted the amount of digital information generated in 2010 would grow to 1.2 zettabytes (10^{21} bytes). Using orders of magnitude, compare this prediction to the amount of digital information generated in 2006.

19. Find the logarithm of each of the following numbers:
a. 1   **b.** 1 billion   **c.** 0.000 001

20. Estimate the following by placing the log between the two closest integer powers of 10.
a. log 3000   **b.** log 150,000   **c.** log 0.05

21. One way of defining the energy unit *joule* (J) is the amount of the energy required to lift a small apple weighing 102 grams one meter above Earth's surface. The table below lists the estimated energy in joules for different situations.

Item	Value	Value in Scientific Notation
Mass-energy of electron	0.000 000 000 000 051 J	
The kinetic energy of a flying mosquito	0.000 000 160 2 J	
An average person swinging a baseball bat	80 J	
Energy received from the sun at Earth's orbit on one square meter in one second	1,360 J	
Energy released by one gram of TNT	4,184 J	
Energy released by metabolism of one gram of fat	38,000 J	
Approximate annual power usage of a standard clothes dryer	320,000,000 J	

Source: http://en.wikipedia.org.

21. (continued)

Use the table to answer the following questions

a. Write each value in scientific notation.

b. A year's use of a clothes dryer requires how many times the energy of swinging a baseball bat once?

c. Metabolizing one gram of fat releases how many times the kinetic energy of a flying mosquito?

22. Rewrite each number as a power of 10, then create a logarithmic scale and estimate the location of the number on that scale. (*Hint:* log 2 = 0.301.)

a. 10   b. 100   c. 200   d. 20,000

23. (Requires a calculator that can evaluate powers.)

"For many drugs, it is important to adjust the dose for individual patients. A patient's body weight or body surface area (BSA) is sometimes used for estimating drug dosages for children and for estimating the starting dosage for chemotherapy for cancer patients. BSA is difficult to measure directly, but it may be calculated by the following formula based on height and weight."[8] $BSA = 71.84W^{0.425}H^{0.725}$, where BSA is measured in square centimeters, W is weight in kilograms, and H is height in centimeters.

A patient weighs 180 lb and is 6 feet tall. His dosage of a particular drug is 15 mg/m²/day (that is, 15 mg per square meter of body surface area per day). What is his daily dosage in mg?

[8]*Drug and Therapeutics Bulletin,* 2010, 48, 33–36.

EXPLORATION 4.1

The Scale and the Tale of the Universe

Objective

- gain an understanding of the relative sizes and relative ages of objects in the universe using scientific notation and unit conversions

Materials/Equipment

- tape, pins, paper, and string to generate a large wall graph (optional)
- enclosed worksheet and conversion table on inside back cover

Related Readings/Videos

"Powers of Ten" and "The Universe in One Year".
Videos: *Powers of Ten* and *The Cosmic Calendar* in the PBS series *Cosmos*

Related Software

"E1: Tale and Scale of the Universe" in *Exponential & Log Functions*

Procedure

Work in small groups. Each group should work on a separate subset of objects on the accompanying worksheet.

1. Convert the ages and sizes of objects so they can be compared. You can refer to the conversion table that shows equivalences between English and metric units on the inside back cover. In addition, 1 light year $\approx 9.46 \cdot 10^{12}$ km.
2. Generate on the blackboard or on the wall (with string) a blank graph whose axes are marked off in orders of magnitude (integer powers of 10), with the units on the vertical axis representing age of object, ranging from 10^0 to 10^{11} years, and the units on the horizontal axis representing size of object, ranging from 10^{-12} to 10^{27} meters.
3. Each small group should plot the approximate coordinates of their selected objects (size in meters, age in years) on the graph. You might want to draw and label a small picture of your object to plot on your graph.

Discussion/Analysis

- Scan the plotted objects from left to right, looking only at relative sizes. Now scan the plotted objects from top to bottom, considering only relative ages. Does your graph make sense in terms of what you know about the relative sizes and ages of these objects?
- Describe the scale and the tale of the universe.

In Scientific Notation

Object	Age (in years)	Size (of radius)	Age (in years)	Size (in meters)
Observable universe	13.7 billion	10^{26} meters		
Surtsey (Earth's newest land mass)	40 years	0.5 mile		
Pleiades (a galactic cluster)	100 million	32.6 light years		
First living organisms on Earth	4.6 billion	0.000 05 meter		
Pangaea (Earth's prehistoric supercontinent)	200 million	4500 miles		
First *Homo sapiens sapiens*	100 thousand	100 centimeters		
First *Tyrannosaurus rex*	200 million	20 feet		
Eukaryotes (first cells with nuclei)	2 billion	0.000 05 meter		
Earth	4.6 billion	6400 kilometers		
Milky Way galaxy	14 billion	50,000 light years		
First atoms	13.7 billion	0.000 000 0001 meter		
Our sun	5 billion	1 gigameter		
Our solar system	5 billion	1 terameter		

CHAPTER 5
GROWTH AND DECAY: AN INTRODUCTION TO EXPONENTIAL FUNCTIONS

OVERVIEW

Exponential and linear functions are used to describe quantities that change over time. Exponential functions represent quantities that are multiplied by a constant factor during each time period. Linear functions represent quantities to which a fixed amount is added (or subtracted) during each time period. Exponential functions can model such diverse phenomena as bacteria growth, radioactive decay, compound interest rates, inflation, musical pitch, and family trees.

After reading this chapter, you should be able to

- recognize the properties of exponential functions and their graphs
- understand the differences between exponential and linear growth
- model growth and decay phenomena with exponential functions
- represent exponential functions using percentages, factors, or rates
- use semi-log plots to determine if data can be modeled by an exponential function
- use base *e* to understand and model continuous or instantaneous compounding

5.1 *Exponential Growth*

 See course software "E2: Exponential Growth & Decay" for a dramatic visualization of growth and decay.

The Growth of *E. coli* Bacteria

Measuring and predicting growth is of concern to population biologists, ecologists, demographers, economists, and politicians alike. The growth of bacteria provides a simple model that scientists can generalize to describe the growth of other phenomena such as cells, countries, or capital.

Bacteria are very tiny, single-celled organisms that are by far the most numerous organisms on Earth. One of the most frequently studied bacteria is *E. coli,* a rod-shaped bacterium approximately 10^{-6} meter (or 1 micrometer) long that inhabits the intestinal tracts of humans and other mammals.[1] The cells of *E. coli* reproduce by a process called fission: The cell splits in half, forming two "daughter cells."

Under ideal conditions *E. coli* divide every 20 minutes. If we start with an initial population of 100 *E. coli* bacteria that doubles every 20-minute time period, we generate the data in Table 5.1. The initial 100 bacteria double to become 200 bacteria at the end of the first time period, double again to become 400 at the end of the second time period, and so on. At the end of the twenty-fourth time period (at $24 \cdot 20$ minutes $= 480$ minutes, or 8 hours), the initial 100 bacteria in our model have grown to over 1.6 billion bacteria!

Because the numbers become astronomically large so quickly, we run into the problems we saw in Chapter 4 when graphing numbers of widely different sizes. Figure 5.1 shows a graph of the data in Table 5.1 for only the first ten time periods. We can see from the graph that the relationship between number of bacteria and time is not linear. The number of bacteria seems to be increasing more and more rapidly over time.

Growth of *E. coli* Bacteria

Time Periods (of 20 minutes each)	Number of *E. coli* Bacteria
0	100
1	200
2	400
3	800
4	1,600
5	3,200
6	6,400
7	12,800
8	25,600
9	51,200
10	102,400
11	204,800
12	409,600
13	819,200
14	1,638,400
15	3,276,800
16	6,553,600
17	13,107,200
18	26,214,400
19	52,428,800
20	104,857,600
21	209,715,200
22	419,430,400
23	838,860,800
24	1,677,721,600

Table 5.1

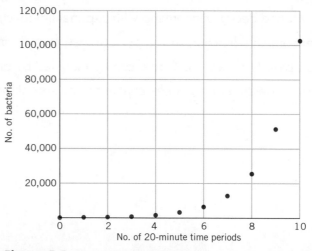

Figure 5.1 Growth of *E. coli* bacteria.

A mathematical model for E. coli growth

Table 5.1 shows us that the initial number of 100 bacteria repeatedly doubles. If we record in a third column (Table 5.2) the number of times we multiply 2 times the original value of 100, we begin to see a pattern emerge.

---

[1]Most types of *E. coli* are beneficial to humans, aiding in digestion. A few types are lethal. You may have read about deaths resulting from people eating certain deadly strains of *E. coli* bacteria in undercooked hamburgers or tainted spinach. The explosive nature of exponential growth shows how a few dangerous bacteria can multiply rapidly to become a deadly quantity in a very short time.

Pattern in *E. coli* Growth

Number of Time Periods	Number of *E. coli* Bacteria	Generalized Expression
0	100	$100 = 100 \cdot 2^0$
1	200	$100 \cdot 2 = 100 \cdot 2^1$
2	400	$100 \cdot 2 \cdot 2 = 100 \cdot 2^2$
3	800	$100 \cdot 2 \cdot 2 \cdot 2 = 100 \cdot 2^3$
4	1,600	$100 \cdot 2 \cdot 2 \cdot 2 \cdot 2 = 100 \cdot 2^4$
5	3,200	$100 \cdot 2 \cdot 2 \cdot 2 \cdot 2 \cdot 2 = 100 \cdot 2^5$
6	6,400	$100 \cdot 2 \cdot 2 \cdot 2 \cdot 2 \cdot 2 \cdot 2 = 100 \cdot 2^6$
7	12,800	$100 \cdot 2 \cdot 2 \cdot 2 \cdot 2 \cdot 2 \cdot 2 \cdot 2 = 100 \cdot 2^7$
8	25,600	$100 \cdot 2 \cdot 2 \cdot 2 \cdot 2 \cdot 2 \cdot 2 \cdot 2 \cdot 2 = 100 \cdot 2^8$
9	51,200	$100 \cdot 2 \cdot 2 \cdot 2 \cdot 2 \cdot 2 \cdot 2 \cdot 2 \cdot 2 \cdot 2 = 100 \cdot 2^9$
10	102,400	$100 \cdot 2 \cdot 2 \cdot 2 \cdot 2 \cdot 2 \cdot 2 \cdot 2 \cdot 2 \cdot 2 \cdot 2 = 100 \cdot 2^{10}$

Table 5.2

Remembering that 2^0 equals 1 by definition, we can describe the relationship by

$$\text{number of } \textit{E. coli} \text{ bacteria} = 100 \cdot 2^{\text{number of time periods}}$$

If we let N = number of bacteria and T = number of time periods, we can write the equation more compactly as

$$N = 100 \cdot 2^T$$

Since each value of t determines one and only one value for N, the equation represents N as a function of T. The number 100 is the *initial bacteria population*. This function is called *exponential* since the input or independent variable, t, occurs in the exponent of the base 2. The base 2 is the *growth factor*, or the multiple by which the population grows during each time period. The bacteria double every time period, so the population increases at a rate of 100% during each time period. If *E. coli* grew unchecked at this pace, the offspring from one cell would cover Earth with a layer a foot deep in less than 36 hours!

The General Exponential Growth Function

The *E. coli* growth equation

$$N = 100 \cdot 2^T$$

is in the form

$$\text{output} = (\text{initial quantity}) \cdot (\text{growth factor})^{\text{input}}$$

Such an equation, with the input in the exponent and a growth factor >1, describes an *exponential growth function*.

Exponential Growth Function

An exponential growth function can be represented by an equation of the form

$$y = Ca^x \quad (a > 1, \quad C > 0) \qquad \text{where}$$

a, the base, is the *growth factor*, the amount by which y is multiplied when x increases by 1.

C is the *initial value* or y-intercept.

E X A M P L E 1

Constructing an exponential function

Construct an exponential function to model 200 cells that triple during every time period T.

S O L U T I O N

Let N represent the number of cells; then

$$N = 200 \cdot 3^T$$

E X A M P L E 2

Interpreting an exponential equation

Interpret the numbers in the following equation:

$$Q = (4 \cdot 10^6) \cdot 2.5^T$$

S O L U T I O N

$4 \cdot 10^6$, or 4,000,000, represents an initial population that grows by a factor of 2.5 each time period, T. This means the initial population is multiplied by 2.5 each time period, T.

E X A M P L E 3

To the edge of the universe

Take a piece of ordinary paper, about 0.1 mm in thickness. Now start folding it in half, then in half again, and so on. Assume you could continue indefinitely.

a. How would you model the growth in thickness of the folded paper?

b. How many folds would it take to reach:

   i. The height of a human?

   ii. The height of the Matterhorn, a famous mountain in Switzerland, which is 4478 meters?

   iii. The sun?

   iv. The edge of the known universe?

S O L U T I O N

a. Each time we fold the paper, we double the thickness. If F = number of folds, the initial thickness = 0.1 mm, and P = thickness of the folded paper (in mm), then

$$P = (0.1) \cdot 2^F$$

b. If you start plugging in values for F (this is where technology is useful), it takes 14 folds to reach the height of an average person (well, a short person). At 26 folds, the paper is higher than the Matterhorn. At 51 folds, the paper would reach the sun, and at 54 folds the edge of our solar system. It would take only 84 folds to reach the limits of the Milky Way. A mere 100 folds takes you to the edge of the known universe!

Doubling Time

The amount of time it takes for a quantity to double is called its *doubling time*. In the equation for *E. coli* bacteria, $N = 100 \cdot 2^T$, where T = number of 20-minute intervals. The quantity N doubles every 20 minutes, then 20 minutes represents the *doubling time*. Every exponential growth function has a fixed doubling time.

One of the most common applications of exponential functions is compound interest on savings and on loans. In the following example, we determine whether functions representing simple and compound interest rates have a fixed doubling time.

E X A M P L E 4

Doubling times?

Simple interest is calculated on the same amount, the principal, each year. The amount of money, S, in a savings account that pays simple interest at 4% annually and starts with an initial value of \$10,000 can be modeled with the linear function

$$S = 10,000 + (0.04 \cdot 10,000)n \quad \text{where } n \text{ represents the number of years}$$

Compound interest is calculated on both the principal and accumulated interest each time period, rather than just the principal as in simple interest calculations.

The amount of money, C, in a savings account that pays compound interest at 4% annually and starts with an initial value of $10,000 can be modeled with the exponential function

$$C = 10,000 \cdot 1.04^n \quad \text{where } n \text{ represents the number of years}$$

a. Graph both functions on the same grid. What happens over time to the difference in the amount of money earned from these two types of savings accounts?

b. Use your graphs to estimate when your investment will double from $10,000 to $20,000 in each account.

c. Use the equations for each function to check your estimates from the graphs.

d. How many years will it take for each account to double from $20,000 to $40,000? Which functions have a fixed doubling time?

SOLUTION   a.

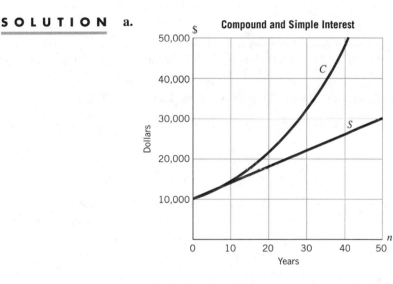

The graphs show that the difference between C and S in the amount of interest earned increases over time. Savings based on compound interest (Graph C) increase each time period, but simple interest yields the same interest each year (Graph S).

b. From the graphs, you can see that the savings account with simple interest will double from $10,000 to $20,000 in about 25 years and the account with compound interest will double in about 18 years.

c. Substitute $n = 25$ in the equation for the amount of money in the account with simple interest:

$$S = 10,000 + 0.04(25) \cdot 10,000$$
$$= 10,000 + 1 \cdot 10,000$$
$$= \$20,000$$

So our estimate of 25 years for the account with simple interest to double from $10,000 to $20,000 is accurate.

Substitute $n = 18$ in the equation for the amount of money in the account with compound interest:

$$C = 10,000(1.04)^{18}$$

use a calculator to apply exponent $\approx 10,000 \cdot 2.026$

multiply $\approx \$20,260$

So our estimate of 18 years for the doubling time of the account with compound interest is fairly accurate.

d. Using our equation for S, the linear function representing simple interest, we find that it takes another 50 years for the account to double from $20,000 to $40,000, so the time it takes for the account to double is not fixed. For the linear function there is not a constant doubling time.

The doubling time for an exponential function is fixed, so it takes the same amount of time, or about 18 years, for C, the account with compound interest, to double from $20,000 to $40,000, as shown in the graph for C. You can confirm this by evaluating C for $20,000 and $40,000.

Doubling Time

The *doubling time* of an exponentially growing quantity is the time required for the quantity to double in size.

We return to doubling times and half-lives in Sec. 5.6 and compound interest in Sec. 5.7, where we focus on finding equivalent equations to model these phenomena.

Looking at Real Growth Data for *E. coli* Bacteria

Figure 5.2 shows a plot of real *E. coli* growth over 24 time periods (8 hours). The growth appears to be exponential for the first 12 time periods (4 hours). We can use technology to generate a best-fit exponential function for that section of the curve. The function is

$$N = (1.37 \cdot 10^7) \cdot 1.5^T$$

where N = number of cells per milliliter and T = the number of 20-minute periods. So the initial quantity is $1.37 \cdot 10^7$ (or 13.7 million) cells. More important, the growth factor is 1.5. Every 20 minutes the number of cells is multiplied by 1.5 (a 50% increase), as opposed to the doubling (a 100% increase) in our first, idealized example.

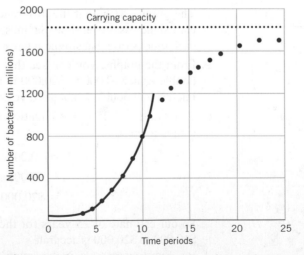

Figure 5.2 Sigmoid curve of real *E. coli* growth over 24 time periods and the best-fit exponential for 12 time periods.

The bacterial growth rate in the laboratory is half that of the idealized data. But even this rate can't be sustained for long. Conditions are rarely ideal; bacteria die, nutrients are used up, space to grow is limited. In the real world, growth that starts out as exponential must eventually slow down (see Figure 5.2). The curve flattens out as the number of bacteria reaches its maximum size, called the *carrying capacity*. The overall shape is called *sigmoid*. The arithmetic of exponentials leads to the inevitable conclusion that in the long term—for bacteria, mosquitoes, or humans—the rate of growth for populations must approach zero.

EXPLORE & EXTEND

5.1

The Water Lily Pond

French children are told a story in which they imagine having a pond with water lily leaves floating on the surface. The lily population doubles in size every day and if left unchecked will smother the pond in 30 days, killing all the other living things in the water. Day after day the leaf cover seems relatively small, so it is decided to let it grow until it half-covers the pond before cutting it back. On what day will this occur? How would you explain these results to a child?

Algebra Aerobics 5.1

A calculator that can evaluate powers is recommended for Problems 3 and 4.

1. Identify the initial value and the growth factor in each of the following exponential growth functions:

   a. $y = 350 \cdot 5^x$

   b. $Q = 25{,}000 \cdot 1.5^t$

   c. $P = (7 \cdot 10^3) \cdot 4^t$

   d. $N = 5000 \cdot 1.025^t$

2. Write an equation for an exponential growth function where:

   a. The initial population is 3000 and the growth factor is 3.

   b. The initial population is $4 \cdot 10^7$ and the growth factor is 1.3.

   c. The initial population is 75 and the population quadruples during each time period.

   d. The initial amount is \$30,000 and the growth factor is 1.12.

3. The population growth of a small country is described by the function $P = 28 \cdot 1.065^t$, where t is in years and P is in millions.

   a. Determine P when $t = 0$. What does that quantity represent?

   b. Determine P when $t = 10$, 20, and 30.

   c. Estimate the value of t that would result in a doubling of the population to 56,000,000. If available, use technology to check your estimate.

4. Fill in a table of values for the amount $A = 80 \cdot 1.06^t$ for values of the time period t of 0, 1, 10, 15, and 20, and then complete the following statements.

   a. The initial amount is ___.

   b. During the first time period, the amount grew from 80 to ___.

   c. Based on this table, the amount doubles between the values $t = $ ___ and $t = $ ___.

   d. Estimate the number of time periods it will take the amount to double.

5. A substance, Q, doubles in quantity every 5 years. Construct a function for the amount of Q as a function of T, where $T = $ number of 5-year periods and the initial amount is 30 grams. Find the amount of Q after 5, 10, and 15 years.

6. Estimate the doubling time and construct a function for each of the following examples.

   a.

   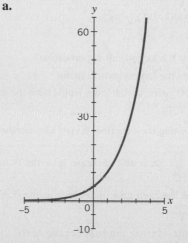

   b. In the following table, $T = $ number of decades and $G = $ number of gallons.

T	G
0	300
1	424
2	600
3	848
4	1200

Exercises for Section 5.1

A calculator that can evaluate powers is recommended for Exercises 6 to 11.

1. The following exponential functions represent population growth. Identify the initial population and the growth factor.
   a. $Q = 275 \cdot 3^T$ d. $A = 25(1.18)^t$
   b. $P = 15,000 \cdot 1.04^t$ e. $P(t) = 8000(2.718)^t$
   c. $y = (6 \cdot 10^8) \cdot 5^x$ f. $f(x) = 4 \cdot 10^5(2.5)^x$

2. Write the equation of each exponential growth function in the form $y = Ca^x$ where:
   a. The initial population is 350 and the growth factor is $\frac{4}{3}$.
   b. The initial population is $5 \cdot 10^9$ and the growth factor is 1.25.
   c. The initial population is 150 and the population triples during each time period.
   d. The initial population is 2, and it quadruples every time period.

3. Fill in the missing parts of the table.

Initial Value C	Growth Factor a	Exponential Function $y = Ca^x$
1600	1.05	
		$y = 6.2 \cdot 10^5(2.07)^x$
	3.25	$y = 1400(\underline{\hphantom{xx}})^x$

4. The populations of four towns for time t, in years, are given by:
   $$P_1(t) = 12,000(1.05)^t$$
   $$P_2(t) = 6000(1.07)^t$$
   $$P_3(t) = 100,000(1.01)^t$$
   $$P_4(t) = 1000(1.9)^t$$
   a. Which town has the largest initial population?
   b. Which town has the largest growth factor?
   c. At the end of 10 years, which town would have the largest population?

5. Find C and a such that the function $f(x) = Ca^x$ satisfies the given conditions.
   a. $f(0) = 6$ and for each unit increase in x, the output is multiplied by 1.2.
   b. $f(0) = 10$ and for each unit increase in x, the output is multiplied by 2.5.

6. In the United States during the first decade of the 2000s, live births to unmarried mothers, $B(t)$, grew according to the exponential model $B(t) = 1.307 \cdot 10^6(1.033)^t$, where t is the number of years after 2000.
   a. What does the model give as the number of live births to unwed mothers in 2000?
   b. What was the growth factor?
   c. What does the model predict for the number of live births to unwed mothers in 2005? In 2010?

7. A cancer patient's white blood cell count grew exponentially after she had completed chemotherapy treatments. The equation $C(d) = 63(1.17)^d$ describes $C(d)$, her white blood cell count per milliliter, d days after the treatment was completed.
   a. What is the white blood cell count growth factor?
   b. What was the initial white blood cell count?
   c. Create a table of values that shows the white blood cell counts from day 0 to day 10 after the chemotherapy.
   d. From the table of values, approximate when the number of white blood cells doubled.

8. $5,000 is invested in an institution that promises to double the investment every 8 years.
   a. Assuming no withdrawals, find the balance after 8, 16, and 24 years.
   b. Write a function, $f(T)$, that gives the balance using T to represent 8-year periods.
   c. Use this function to determine the balance after 40 years.

9. A tuberculosis culture increases by a factor of 1.185 each hour.
   a. If the initial concentration is $5 \cdot 10^3$ cells/ml, construct an exponential function to describe its growth over time.
   b. What will the concentration be after 8 hours?

10. An ancient king of Persia was said to have been so grateful to one of his subjects that he allowed the subject to select his own reward. The clever subject asked for a grain of rice on the first square of a chessboard, two grains on the second square, four on the next, and so on.
    a. Construct a function that describes the number of grains of rice, G, as a function of the square, n, on the chessboard. (*Note:* There are 64 squares.)
    b. Construct a table recording the numbers of grains of rice on the first ten squares.
    c. Sketch your function.
    d. How many grains of rice would the king have to provide for the 64th (and last) square?

11. Suppose you accumulate $2000 in debt on your credit card at an interest rate of 3.25% compounded monthly. If you are not able to pay off this debt, the amount of money, A, owed to the credit card company after m months is $A = 2,000 \cdot 1.0325^m$. Graph A and use the graph to explain why it is unwise to accumulate credit card debt without paying off some of the debt each month. In approximately how many months would the interest accumulated be equal to the original amount of your debt?

12. Using the following table, construct an equation to represent Q as a function of s.

s	Q
-1	125
0	250
1	500
2	1000
3	2000
4	4000

13. Does the following graph approximate an exponential function? Why or why not?

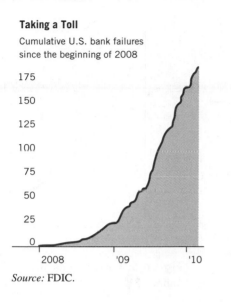

Taking a Toll

Cumulative U.S. bank failures since the beginning of 2008

Source: FDIC.

14. A new species of fish is introduced into a pond. The size of the fish population can be modeled by the accompanying graph of the function $P(t)$, where t is time in months.

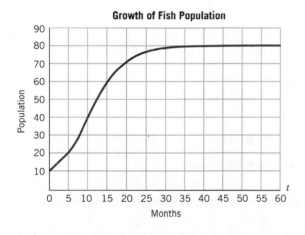

Growth of Fish Population

a. What is the initial population of the fish?

b. How long did it take for the initial fish population to double?

c. What was the maximum sustainable fish population (carrying capacity) of the pond?

d. Estimate the number of months it took for the fish population to reach its sustainable size.

15. The Northern Wildlife Prairie Research Center in Jamestown, North Dakota, measured the weights of three duckling species (wild mallard, gadwall, and blue-winged teal). Each one-day-old duckling weighed about 32 grams. Weights were tracked and are graphed on the following chart. The weight curves appear sigmoidal, each approaching a maximum sustainable (or mature) weight.

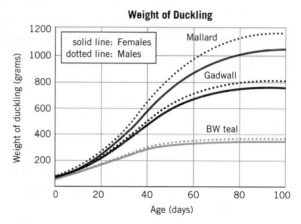

Weight of Duckling

solid line: Females
dotted line: Males

Source: http://www.npwrc.usgs.gov.

a. Estimate the mature weight for the female ducklings in each species.

b. Estimate the mature weight for the male ducklings in each species.

c. Approximately how long did it take for each species of ducklings to attain the mature weight?

5.2 *Exponential Decay*

An exponential growth function is of the form $y = Ca^x$, where $a > 1$. Each time x increases by 1, the population is multiplied by a, a number greater than 1, so the population increases. But what if a were positive and less than 1, that is, $0 < a < 1$? Then each time x increases by 1 and the population is multiplied by a, the population size would be reduced. We would have *exponential decay*.

The Decay of Iodine-131

Iodine-131 is one of the radioactive isotopes of iodine that is used in nuclear medicine for diagnosis and treatment. It can be used to test how well the thyroid gland is functioning. An exponential function can be used to describe the decay of iodine-131. After being generated by a fission reaction, iodine-131 decays into a nontoxic, stable isotope. Every 8 days the amount of iodine-131 remaining is cut in half. For example, we show in Table 5.3 and Figure 5.3 how 160 milligrams (mg) of iodine-131 decays over four time periods (or $4 \cdot 8 = 32$ days).

Time Periods (8 days each)	Amount of Iodine-131 (mg)		General Expression
0	160	=	$160 = 160 \left(\frac{1}{2}\right)^0$
1	80	=	$160 \left(\frac{1}{2}\right) = 160 \left(\frac{1}{2}\right)^1$
2	40	=	$160 \left(\frac{1}{2}\right)\left(\frac{1}{2}\right) = 160 \left(\frac{1}{2}\right)^2$
3	20	=	$160 \left(\frac{1}{2}\right)\left(\frac{1}{2}\right)\left(\frac{1}{2}\right) = 160 \left(\frac{1}{2}\right)^3$
4	10	=	$160 \left(\frac{1}{2}\right)\left(\frac{1}{2}\right)\left(\frac{1}{2}\right)\left(\frac{1}{2}\right) = 160 \left(\frac{1}{2}\right)^4$

Table 5.3

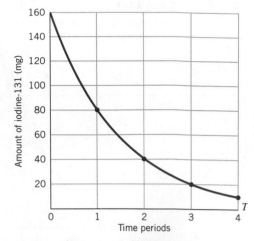

Figure 5.3 Exponential decay of iodine-131.

We can describe the decay of iodine-131 with the equation

$$Q = 160 \cdot \left(\frac{1}{2}\right)^T$$

where Q = quantity of iodine-131 (in mg) and T = number of time periods (of 8 days each). The number 160 is the initial amount of iodine-131 and $\frac{1}{2}$ is the *decay factor*.

The General Exponential Decay Function

The equation for Q is in the form

$$\text{output} = (\text{initial quantity}) \cdot (\text{decay factor})^{\text{input}}$$

This is the standard format for an exponential decay function.

Exponential Decay Function

An *exponential decay function* $y = f(x)$ can be represented by an equation of the form

$$y = Ca^x \qquad (0 < a < 1 \text{ and } C > 0) \qquad \text{where}$$

a is the *decay factor,* the amount by which y is multiplied when x increases by 1. C is the *initial value* or y-intercept.

EXAMPLE 1   **Estimating how much remains**
   a. How would you model the decay of 500 mg of iodine-131?
   b. How many milligrams would be left after four time periods (or 32 days)?

SOLUTION   **a.** The initial value is 500 mg, and we know from the previous discussion that the decay factor is $\frac{1}{2}$ for each time period T (of 8 days). So the decay function is $Q = 500 \cdot \left(\frac{1}{2}\right)^T$.
   b. When $T = 4$, then $Q = 500 \cdot \left(\frac{1}{2}\right)^4 = 500 \cdot \left(\frac{1}{16}\right) = 31.25$. So after four time periods (32 days) there would be 31.25 mg of iodine-131 remaining.

E X A M P L E   2   **Identifying and interpreting the decay factor**
Identify the initial value and the decay factor for the function $N = (3 \cdot 10^4) \cdot (0.25)^T$.
Interpret the decay factor.

S O L U T I O N   The initial value is $3 \cdot 10^4$ (or 30,000) and the decay factor is 0.25 or $\left(\text{or } \frac{1}{4}\right)$. So each
time T increases by 1, the value for N is one-fourth of its previous value.

E X A M P L E   3   **When is it growth and when is it decay?**
Which of the following exponential functions represent growth and which represent
decay? Identify the growth or decay factor.

a. $P(t) = 2000(1.05)^t$ **c.** $N(t) = 16\left(\frac{2}{3}\right)^t$

b. $Q(t) = 25(0.75)^t$ **d.** $f(x) = 5(4)^{-x}$

S O L U T I O N   **a.** $P(t)$ represents exponential growth because $1.05 > 1$; the term 1.05 is the growth
factor.

b. $Q(t)$ represents exponential decay because $0 < 0.75 < 1$; the term 0.75 is the decay
factor.

c. $N(t)$ represents exponential decay because $0 < \frac{2}{3} < 1$; the term $\frac{2}{3}$ is the decay factor.

d. We can rewrite the function as $f(x) = 5\left(\frac{1}{4}\right)^x$. Thus $f(x)$ represents exponential decay
because $0 < \frac{1}{4} < 1$; the term $\frac{1}{4}$ is the decay factor.

Half-Life

Just as every exponential growth function has a fixed doubling time, every exponential
decay function has a fixed half-life. The amount of time it takes for a quantity to reduce
to one-half of its initial value is called its *half-life*. We saw that the half-life of iodine-131
is 8 days. In this section we use data tables and graphs to examine other exponential
decay functions and their half-lives.

E X A M P L E   4   **Using a data table to solve an exponential decay problem**
Each year the Dance Association holds a dance marathon where 256 dancers perform
solos. After each 25-minute round of dancing, the judges eliminate half of the participants.
How many rounds would it take before there is a winner? How long would this take?

S O L U T I O N   After each 25-minute round, the number of competitors is cut in half, so the half-life is
25 minutes. Setting T = number of 25-minute intervals and letting $N(T)$ = number of
dancers on the floor, then

$$N(T) = 256 \cdot \left(\frac{1}{2}\right)^T$$

If only one dancer is left, we need to solve the following equation for T:

$$1 = 256 \cdot \left(\frac{1}{2}\right)^T$$

In Chapter 6 we learn how to solve this type of equation, but for now we can use a table
to find an answer.

Number of rounds (in 25-minute intervals)	0	1	2	3	4	5	6	7	8
Number of people remaining on dance floor	256	256(1/2) = 128	128(1/2) = 64	64(1/2) = 32	32(1/2) = 16	16(1/2) = 8	8(1/2) = 4	4(1/2) = 2	2(1/2) = 1

Table 5.4

From Table 5.4, we can see that after 8 rounds of dancing there will be only one person, the winner, left on the dance floor. So it would take $8 \cdot 25 = 200$ minutes, or 3 hours and 20 minutes, for a winner to be declared.

Half-Life

The *half-life* of an exponentially decaying quantity is the time required for the quantity to be reduced by a half.

E X A M P L E   5

Using a graph to estimate the half-life of an exponential function
Radionuclides are radioactive substances often used in medical procedures since they have a relatively short half-life, so the amount of radioactivity in the body diminishes rapidly. Bone scans routinely require patients to receive an injection containing Tc-99m radionuclides that will briefly concentrate in the patient's bones and show up in a gamma camera image or scanner.

a. Figure 5.4 is a graph showing the decay of Tc-99m over time, t, in hours. From the graph, estimate the half-life of the radioactivity level of Tc-99m. Check your estimate using the Internet.

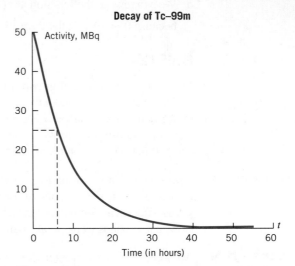

Figure 5.4

Note: MBq, or megabecquerel, is 10^6 Bq, where the unit Bq is defined as the activity of a quantity of radioactive material in which one nucleus decays per second.

b. Create an exponential function for the amount of radioactivity of Tc-99m remaining, where T = number of half-life intervals.

c. What fraction of the radioactivity of Tc-99m remains after 1 day?

S O L U T I O N

a. On the graph, the initial radioactivity level of Tc-99m is 50 MBq. When $t \approx$ 6 hours, then radioactivity level $\approx$ 25 MBq, or half of the initial level. According to the U.S. Environmental Protection Agency, the radioactivity of Tc-99m has a half-life of 6.01 hours, so our estimate is pretty good.[2]

b. For an initial level of radioactivity of 50 MBq and a half-life of 6.01 hours, an exponential function for the amount of radioactivity of Tc-99m, A, is:

$$A = 50 \cdot \left(\frac{1}{2}\right)^T \qquad \text{where } T = \text{number of 6.01-hour intervals}$$

---

[2]Since processes in the body can speed up the decay of Tc-99m, the amount of a radioactive substance remaining in the body is called the biological half-life and is slightly less than the half-life outside of the body.

c. In 1 day, or a little less than 4 T (where T = number of 6.01-hour intervals), there will be about $\left(\frac{1}{2}\right)^4$, or 1/16, of the original substance remaining in the body.

EXAMPLE 6

Representing caffeine levels with equivalent equations
When you drink an 8-ounce cup of coffee, virtually all the caffeine is absorbed in your gut and passes through the liver and into your bloodstream, acting as a stimulant. A peak blood level of caffeine, 120 milligrams, is reached in about 30 minutes—and then the blood levels begin to fall exponentially. Five hours after the peak, your blood contains 60 milligrams of caffeine.

a. Construct a function to model the caffeine decrease in your bloodstream over time.

b. Find an equivalent equation to model the decay of caffeine, using a = growth factor.

c. Generate a sketch of the caffeine levels in your bloodstream over time, where time is measured in hours.

SOLUTION

a. Five hours after the peak level of caffeine (120 mg), your blood contains only half the caffeine (60 mg) of the peak level. So the half-life is 5 hours.[3] If we let C = amount of caffeine in your blood (in milligrams), C_0 = peak amount of caffeine or 120 milligrams, and T = a half-life of 5-hour intervals, an exponential decay function to model the amount of caffeine in your blood is

$$C = C_0 \cdot (0.5)^T \qquad \text{where } T = \text{number of 5-hour intervals}$$
$$= 120 \cdot (0.5)^T$$

b. We know that approximately every five hours the amount of caffeine decreases by a half. This means that the growth factor, a, can be represented as:

$$a^5 \approx \frac{1}{2}$$

taking the fifth root on each side
$$(a^5)^{(1/5)} = \left(\frac{1}{2}\right)^{(1/5)}$$

using a calculator
$$a \approx 0.8705$$

Using the growth factor, an equivalent equation for the amount of caffeine, C, after t hours is:

$$C = 120 \cdot 0.8705^t$$

c. Figure 5.5 shows a graph for C, the decay of caffeine from coffee, after t hours.

Blood Caffeine Levels after Drinking 8 oz of Coffee

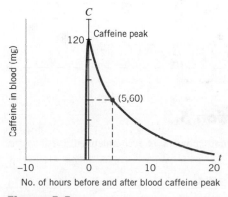

Figure 5.5

[3]Note to coffee drinkers: Consumption of 250 mg of caffeine (2 to 3 cups) is considered a moderate amount; over 800 mg of caffeine is considered excessive. Various factors can influence the rate of caffeine decay. For a pregnant woman or a woman on oral contraceptives, the decrease in caffeine slows down considerably; a little caffeine takes a long time to pass through the body. For smokers, the decrease in caffeine is speeded up, so they can actually drink more coffee without feeling its side effects.

EXPLORE & EXTEND

5.2

Half-Life of Atomic Weapons

Plutonium, the fuel for atomic weapons, has a half-life of about 24,400 years. An atomic weapon is usually designed with a 1% mass margin: that is, it will remain functional until the original fuel has decayed by more than 1%, leaving less than 99% of the original amount. Estimate how many years a plutonium bomb would remain functional. Why do you think there is concern over stored caches of atomic weapons?

Algebra Aerobics 5.2

1. Which of the following exponential functions represent growth and which represent decay?

   a. $y = 100 \cdot 3^x$ **d.** $g(r) = (2 \cdot 10^6) \cdot (1.15)^t$

   b. $f(t) = 75 \cdot \left(\frac{2}{3}\right)^t$ **e.** $y = (7 \cdot 10^9) \cdot (0.20)^z$

   c. $w = 250 \cdot (0.95)^t$ **f.** $h(x) = 150 \cdot \left(\frac{5}{2}\right)^x$

2. Write an equation for an exponential decay function where:

   a. The initial population is 2300 and the decay factor is $\frac{1}{3}$.

   b. The initial population is $3 \cdot 10^9$ and the decay factor is 0.35.

   c. The initial population is 375 and the population drops to one-tenth its previous size during each time period.

3. For each of the following, identify the half-life and construct an exponential function. Specify the units for each variable.

   a.

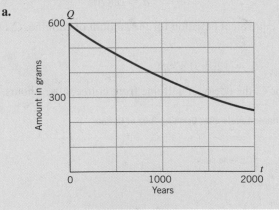

b.

Number of 500-year intervals	0	1	2	3	4	5	6
Amount of substance remaining (grams)	60.0	47.6	37.8	30.0	23.8	18.9	15.0

4. A substance, R, reduces to half its size every 88 years. Construct a function that models the amount of the substance remaining, R, where $T =$ number of 88-year periods, and the initial amount is 6 mg. Find the amount remaining of this substance after 88, 176, and 440 years.

5. Does the exponential function $y = 12 \cdot (5)^{-x}$ represent growth or decay? (*Hint:* Rewrite the function in the standard form $y = Ca^x$.)

6. Rewrite each of the following using the general form and indicate whether the function represents growth or decay.

   a. $y = 23(2.4)^{-x}$

   b. $f(x) = 8000(0.5)^{-x}$

   c. $P = 52{,}000(1.075)^{-t}$

Exercises for Section 5.2

A calculator that can evaluate powers is recommended for Exercises 8, 10, and 16.

1. Identify and interpret the decay factor for each of the following functions:

   a. $P = 450(0.43)^t$ **b.** $f(t) = 3500(0.95)^t$ **c.** $y = 21(3)^{-x}$

2. Which of the following exponential functions represent growth and which decay?

   a. $N = 50 \cdot 2.5^T$ **b.** $y = 264(5/2)^x$

   c. $R = 745(1.001)^t$ **e.** $f(T) = (1.5 \cdot 10^{11}) \cdot (0.35)^T$

   d. $g(z) = (3 \cdot 10^5) \cdot (0.8)^z$ **f.** $h(x) = 2000\left(\frac{2}{3}\right)^x$

3. Write an equation for an exponential decay function $f(t)$, for a given time, t, where:

   a. The initial population is 10,000 and the decay factor is $\frac{2}{5}$.

   b. The initial population is $2.7 \cdot 10^{13}$ and the decay factor is 0.27.

   c. The initial population is 219 and the population drops to one-tenth its previous size during each time period.

4. The accompanying tables show approximate values for the four exponential functions: $f(x) = 5(2^x)$, $g(x) = 5(0.7^x)$, $h(x) = 6(1.7^x)$, and $j(x) = 6(0.6^x)$. Which table is associated with each function?

Function A

x	y
-2	16.67
-1	10.00
0	6.00
1	3.60
2	2.16

Function C

x	y
-2	1.25
-1	2.50
0	5.00
1	10.00
2	20.00

Function B

x	y
-2	10.2
-1	7.1
0	5.0
1	3.5
2	2.5

Function D

x	y
-2	2.1
-1	3.5
0	6.0
1	10.2
2	17.3

5. Determine which of the following functions are exponential. For each exponential function, identify the growth or decay factor and the vertical intercept.

   a. $y = 5(x^2)$

   b. $y = 100 \cdot 2^{-x}$

   c. $P = 1000(0.999)^t$

6. Determine which of the following functions are exponential. Identify each exponential function as representing growth or decay and find the vertical intercept.

   a. $A = 100(1.02^t)$ d. $y = 100x + 3$

   b. $f(x) = 4(3^x)$ e. $M = 2^p$

   c. $g(x) = 0.3(10^x)$ f. $y = x^2$

7. Fill in the missing parts of the table.

Initial Value, C	Decay Factor, a	Exponential Function $y = Ca^x$
500	0.95	
		$y = (1.72 \cdot 10^6)(0.75)^x$
	0.25	$y = 1600(\_\_\_)^x$

8. Complete the following table, where x represents the number of years. The first row is finished for you.

Function	Decay Factor, a, where $y = Ca^x$	Half-life in years
$y = 100\left[\left(\dfrac{1}{2}\right)^{1/5}\right]^x$	$y = 100(0.8706)^x$	5 years
$y = 100 \cdot (\quad)^x$		2 years
$y = 2.5 \cdot 10^3(\quad)^x$		20 years

9. Match each function with its graph.

   a. $y = 100(0.8)^x$ b. $y = 100(0.5)^x$

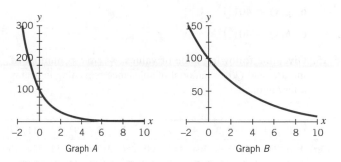

10. The per capita (per person) consumption of milk was 28 gallons in 1980 and has been steadily decreasing by an annual decay factor of 0.99.

   a. Form an exponential function for per capita milk consumption $M(t)$ for year t after 1980.

   b. According to your function, what was the per capita consumption of milk in 2010? If available, use the Internet to check your predictions.

11. A 2.5-gram sample of an isotope of strontium-90 was formed in a 1960 explosion of an atomic bomb at Johnson Island in the Pacific Test Site. The half-life of strontium-90 is 28 years. In what year will only 0.625 gram of this strontium-90 remain?

12. From the graph below, estimate the half-life for the population, P, then create an exponential function to model the population where t is the time in years.

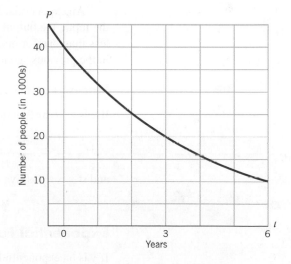

13. a. A linear function $f(t) = b + mt$ has a slope of -4 and a vertical intercept of 20. Find its equation.

   b. An exponential function $g(t) = Ca^t$ has a decay factor of $1/4$ and an initial value of 20. Find its equation.

   c. Plot both functions on the same grid.

14. Which of the following functions (if any) are equivalent? Explain your answer.

    a. $f(x) = 40(0.625)^x$

    b. $g(x) = 40\left(\frac{5}{8}\right)^x$

    c. $h(x) = 40\left(\frac{8}{5}\right)^{-x}$

15. Given the following table of values, where t = number of minutes and $Q(t)$ = amount of substance remaining in grams after t minutes:

t	0	1	2	3	4	5	6	7	8	9	10	11	12
$Q(t)$	800	713	635	566	504	449	400	356	317	283	252	224	200

    a. What is the initial value of $Q(t)$?

    b. What is the half-life?

    c. Construct an exponential decay function $Q(t)$, where t is measured in minutes.

    d. What is $Q(5)$? What does it represent?

16. The half-life of francium is 21 minutes. Starting with $4 \cdot 10^{18}$ atoms of francium, how many atoms would disintegrate in 1 hour and 45 minutes? What fraction of the original sample remains?

17. Examine the following graph on the decay of fossil fuel emissions. Does it represent exponential decay? Why or why not?

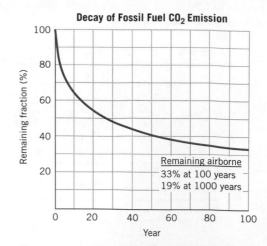

Source: www.tececo.com.au/images/graphics/climate_change/.

18. It takes 1.31 billion years for radioactive potassium-40 to drop to half its original size.

    a. Construct a function to describe the decay of potassium-40.

    b. Approximately what amount of the original potassium-40 would be left after 4 billion years? Justify your answer.

5.3 Comparing Linear and Exponential Functions

A linear function represents *additive* change. For each unit increase in the input, a constant amount is *added* to the output. In Chapter 2 we learned this constant amount is the average rate of change (which is positive when the output is increasing as the input increases, and negative when decreasing).

An exponential function represents *multiplicative* change. For each unit increase in the input, the output is *multiplied* by a constant factor (greater than one for growth and less than one for decay). This constant factor is the growth (or decay) factor discussed in the previous sections.

Linear Functions

If y is a linear function of x, then

$$y = b + mx$$
$$y = \text{initial value} + (\text{rate of change}) \cdot x$$

and if x is a positive integer $\quad y = \quad b \quad + \underbrace{(m + m + m + \cdots + m)}_{x \text{ times}}$

Exponential Functions

If y is an exponential function of x, then

$$y = C \cdot a^x$$
$$y = \text{initial value} \cdot (\text{base})^x$$

and if x is a positive integer $\quad y = C \cdot \underbrace{(a \cdot a \cdot a \cdot a \cdots \cdot a)}_{x \text{ times}}$

> **Linear vs. Exponential Functions**
>
> A linear function represents a quantity to which a constant amount is added for each unit increase in the input.
>
> An exponential function represents a quantity that is multiplied by a constant factor for each unit increase in the input.

Identifying Exponential Functions in a Data Table

We can examine data tables for repeated additions or multiplications to identify linear or exponential functions. If the difference in consecutive input values is constant, the data represent a linear function if the *difference* between consecutive output values is constant, whereas the data represent an exponential function if the *ratio* of consecutive output values is constant.

EXAMPLE 1 Do the data represent a linear or an exponential function? Use the data in Table 5.5 to show the following:

a. $f(x)$ is a linear function and $g(x)$ is not a linear function.

b. $g(x)$ is an exponential function.

x	$f(x)$	$g(x)$
0	25	100
2	50	196
4	75	384.16
6	100	752.95
8	125	1475.79
10	150	2892.55

Table 5.5

SOLUTION **a.** The function $f(x)$ is a linear function, since the difference between consecutive output values is constant when $\Delta x = 2$. Each time x increases by 2, $f(x)$ increases by 25.

The function $g(x)$ is not linear since each time x increases by 2, the difference between consecutive outputs is *not* constant:

$$196 - 100 = 96$$
$$384.16 - 196 = 188.16$$

b. The function $g(x)$ is exponential since the ratio of consecutive output values is constant at 1.96 when $\Delta x = 2$.

$$\frac{196}{100} = 1.96; \quad \frac{384.16}{196} = 1.96; \quad \frac{752.95}{384.16} = 1.96; \quad \frac{1475.79}{752.95} = 1.96; \quad \frac{2892.55}{1475.79} = 1.96$$

> **For a Table Describing y as a Function of x, Where Δx Is Constant**
>
> If the *difference* between consecutive y values is constant, the function is linear.
>
> If the *ratio* of consecutive y values is constant, the function is exponential.

A Linear vs. an Exponential Model through Two Points

A town's population increased from 20,000 to 24,000 over a 5-year period. The town council is concerned about the rapid population growth and wants to predict future population size. The functions most commonly used to predict growth patterns over time are linear and exponential.

If we let t = number of years and P = population size, then when $t = 0$, $P = 20,000$ and when $t = 5$, $P = 24,000$. This gives us two points, (0, 20000) and (5, 24000), to construct our models.

Constructing a linear model: Adding a constant amount each year

Assuming linear growth, our function is $P = b + mt$, with initial population, $b = 20,000$. The average rate of change or slope, m, through the two points is

$$m = \frac{\text{change in population}}{\text{change in time}} = \frac{24,000 - 20,000}{5 - 0} = \frac{4000}{5} = 800 \text{ people/yr}$$

So the equation is

$$P = 20,000 + 800t$$

This linear model says that the original population of 20,000 is increasing by the constant amount of 800 people each year.

Constructing an exponential model: Multiplying by a constant factor each year

Assuming exponential growth and an initial population of 20,000, our function is

$$P = 20,000 \cdot a^t$$

where P is the population of the town and t the number of years. To complete our model, we need a value for a, the annual growth factor.

Finding the Growth Factor from Two Data Points.   We have enough information to find the 5-year growth factor. The inputs 0 and 5 years give us two outputs, 20,000 and 24,000 people, respectively. Exponential growth is multiplicative, so for each additional year, the initial population is multiplied by the growth factor, a. After 5 years, the initial population of 20,000 is multiplied by a^5 to get 24,000.

Given $\qquad\qquad\qquad 20,000 \cdot a^5 = 24,000$

divide by 20,000 $\qquad\qquad a^5 = \frac{24,000}{20,000} = 1.2$

So the 5-year growth factor, a^5, is 1.2. Using a calculator, we can find the annual growth factor, a.

Take the fifth root of both sides $\qquad (a^5)^{1/5} = (1.2)^{1/5}$

$$a^1 \approx 1.0371$$

Now we can represent our exponential function as

$$P = 20,000 \cdot (1.0371)^t$$

where t = number of years and P = population size. Our exponential model tells us the initial population of 20,000 is multiplied by 1.0371 each year. We used a similar method to find the growth factor for the decay of caffeine in Example 6, p. 275.

> **Translating the Growth (or Decay) Factor from *n* Time Units to One Time Unit**
>
> If *a* is the growth (or decay) factor for *n* time units, then
>
> $$a^{1/n}$$
>
> is the growth (or decay) factor for one time unit.
>
> *Examples:* If 2.3 is the *10-year* growth factor, then $2.3^{1/10} \approx 1.0869$ is the *annual* growth factor.
>
> If 0.7 is the *5-month* decay factor, then $0.7^{1/5} \approx 0.9311$ is the *one-month* decay factor.

EXAMPLE 2

Making predictions with our models

For our problem on population growth, the population size is the same in both models for year 0 and year 5.

a. What would the two models predict for years 10, 15, 20, and 25? Construct a table and graph to show these predictions.

b. What are some limitations in using these models to make predictions?

SOLUTION

a.

Town Population

Year	Linear Model $P = 20{,}000 + 800t$	Exponential Model $P = 20{,}000(1.0371)^t$
0	20,000	20,000
5	24,000	24,000
10	28,000	28,790
15	32,000	34,540
20	36,000	41,440
25	40,000	49,720

Table 5.6

As shown in Table 5.6 and Figure 5.6, for year 10, the linear and exponential predictions are fairly close: 28,000 versus approximately 28,790. Beyond year 10, the exponential predictions exceed the linear by an increasingly greater amount. By year 25, the linear model predicts a population of 40,000 whereas the exponential model predicts almost 50,000.

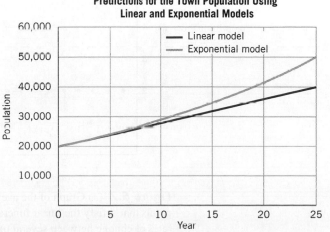

Predictions for the Town Population Using Linear and Exponential Models

Figure 5.6

b. Both models should be considered as generating only crude future estimates, particularly since we constructed the models using only two data points. Clearly, the further out we try to predict, the more unreliable the estimates become.

Comparing the Average Rates of Change

Another way to compare linear and exponential functions is to examine average rates of change. Recall that if N is a function of t, then

$$\text{average rate of change} = \frac{\text{change in } N}{\text{change in } t} = \frac{\Delta N}{\Delta t}$$

Examine Table 5.7, which contains average rates of change for both the linear and exponential functions, where $\Delta t = 1$. For all linear functions, we know that the average rate of change is constant. For the linear function $N = 100 + 2t$, we can tell from the equation, Table 5.7, and Figure 5.7(a) that the (average) rate of change is constant at two units. Average rates of change for the exponential function $N = 100 \cdot 2^t$ are calculated in Table 5.7 and then graphed in Figure 5.7(b). These suggest that the average rates of change of an exponential growth function grow *exponentially*.

Comparing Average-Rate-of-Change Calculations

	Linear Function			Exponential Function	
	$N = 100 + 2t$	Average Rate of Change (when $\Delta t = 1$)		$N = 100 \cdot 2^t$	Average Rate of Change (when $\Delta t = 1$)
t	N			N	
0	100	n.a.		100	n.a.
1	102	2		200	100
2	104	2		400	200
3	106	2		800	400
4	108	2		1,600	800
5	110	2		3,200	1,600
6	112	2		6,400	3,200
7	114	2		12,800	6,400
8	116	2		25,600	12,800
9	118	2		51,200	25,600
10	120	2		102,400	51,200

Table 5.7

Graphs of Average Rates of Change

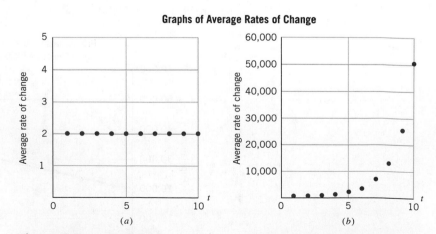

(a) *(b)*

Figure 5.7 (*a*) Graph of the average rates of change between several pairs of points that satisfy the linear function $N = 100 + 2t$. (*b*) Graph of the average rates of change between several pairs of points that satisfy the exponential function $N = 100 \cdot 2^t$.

In the Long Run, Exponential Growth Will Always Outpace Linear Growth

When we examined the average rate of change for an exponential growth function, we saw that it mirrors an exponential function and increases at an exponential rate. Since the average rate of change of a linear function is constant, any exponential growth function will eventually overtake any linear growth function. Sooner or later the graph of the exponential function will permanently lie above the linear graph and will continue to grow faster and faster (Figure 5.8).

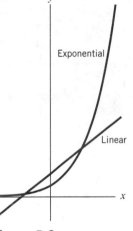

Linear vs. Exponential Growth

Figure 5.8

EXPLORE & EXTEND

5.3

Working with Unequal Input Intervals
Do either of the following tables represent exponential functions? Justify your answers.

x	f(x)		x	g(x)
0	4.00		0	3.0
2	2.13		2	12.0
5	0.83		5	25.5
10	0.17		10	48.0
11	0.13		11	52.5

Algebra Aerobics 5.3

1. Fill in the following table and sketch the graph of each function.

t	$N = 10 + 3t$	$N = 10 \cdot 3^t$
0		
1		
2		
3		
4		

2. **a.** Create a table of values (from $t = 0$ to $t = 5$) for the functions $f(t)$, with a vertical intercept at 200 and a constant rate of change of 20, and $g(t)$, with a vertical intercept at 200 and a growth factor of 1.20.

   b. Sketch the graphs of g and f on the same grid.

   c. Compare the graphs.

3. Given the values in the table on the next page, determine which functions (if any) are linear, exponential, or neither. Justify your answer based on common differences or common ratios.

x	0	3	6	9
$f(x)$	2	6.5	11.0	15.5
$g(x)$	13	12.4	11.8	11.2
$F(x)$	10	13.3	17.7	23.6
$G(x)$	12	11.0	9.0	6.0

4. Write an equation for the linear function and the exponential function that pass through the given points.

   a. $(0, 500)$ and $(1, 620)$ **b.** $(0, 3)$ and $(1, 3.2)$

5. In each of the following situations, assume growth is exponential, find the growth factor, and then construct an exponential function that models the situation.

   a. The initial population, P, is 1500 and one time unit, t, later, the population grew to 3750.

   b. The initial amount, A, is \$80,000 and one time unit, t, later, the amount grew by \$2300.

c. A quantity, Q, grew from 30 mg to 32.7 mg after one time unit, t.

6. In the table below, assume that Q is an exponential function of t.

t (years)	Q
0	10
2	20
4	
6	
8	

   a. By what factor is Q multiplied when t increases by 2 years? This is the 2-year growth factor.

   b. Fill in the rest of the Q values in the table.

   c. What is the annual growth factor?

   d. Construct an equation to model the relationship between Q and t.

Exercises for Section 5.3

A graphing program is recommended for Exercises 3 and 11.

1. Determine which of the following functions are linear, which are exponential, and which are neither. In each case identify the vertical intercept.

   a. $C(t) = 3t + 5$

   b. $f(x) = 3(5)^x$

   c. $y = 5x^2 + 3$

   d. $Q = 6\left(\frac{3}{2}\right)^t$

   e. $P = 7(1.25)^t$

   f. $T = 1.25n$

2. Create a linear or exponential function based on the given conditions.

   a. A function with an average rate of change of 3 and a vertical intercept of 4.

   b. A function with growth factor of 3 and vertical intercept of 4.

   c. A function with slope of 4/3 and initial value of 5.

   d. A function with initial value of 5 and growth factor of 4/3.

3. (Graphing program recommended.) A small village has an initial size of 50 people at time $t = 0$, with t in years.

   a. If the population increases by 5 people per year, find the formula for the population size $P(t)$.

   b. If the population increases by a factor of 1.05 per year, find a new formula $Q(t)$ for the population size.

   c. Plot both functions on the same graph over a 30-year period.

   d. Estimate the coordinates of the point(s) where the graphs intersect. Interpret the meaning of the intersection point(s).

4. A herd of deer has an initial population of 10 at time $t = 0$, with t in years.

   a. If the size of the herd increases by 8 per year, find the formula for the population of deer, $P(t)$, over time.

   b. If the size of the herd increases by a factor of 1.8 each year, find the formula for the deer population, $Q(t)$, over time.

   c. For each model create a table of values for the deer population for a 10-year period.

   d. Using the table, estimate when the two models predict the same population size.

5. Construct both a linear and an exponential function that go through the points $(0, 6)$ and $(1, 9)$.

6. Construct both a linear and an exponential function that go through the points $(0, 200)$ and $(10, 500)$.

7. Find the equation of the linear function and the exponential function that are sketched through two points on each of the following graphs.

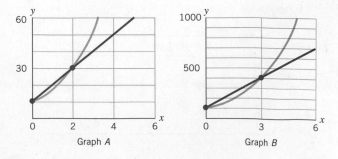

Graph A   Graph B

8. Match the description to any appropriate graph(s) shown below.

   a. When *x* increases by 10, *y* increases by 5.

   b. When *x* increases by 10, *y* is multiplied by a factor of 5.

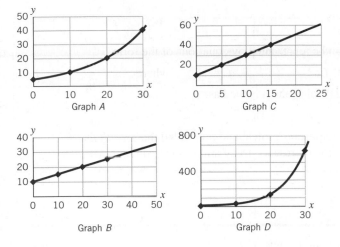

Graph A

Graph C

Graph B

Graph D

9. Each of the following tables contains values representing either linear or exponential functions. Find the equation for each function.

   a.

x	−2	−1	0	1	2
$f(x)$	1.12	2.8	7	17.5	43.75

   b.

x	−2	−1	0	1	2
$g(x)$	0.1	0.3	0.5	0.7	0.9

10. Each table has values representing either linear or exponential functions. Find the equation for each function. (Values for $j(x)$ are rounded to the nearest tenth.)

   a.

x	−2	−1	0	1	2
$h(x)$	160	180	200	220	240

   b.

x	0	10	20	30	40
$j(x)$	200	230	264.5	304.2	349.8

11. (Graphing program recommended.) Create a table of values for the following functions, then graph the functions.

   a. $f(x) = 6 + 1.5x$

   b. $g(x) = 6(1.5)^x$

   c. $h(x) = 1.5(6)^x$

12. Suppose you are given a table of values of the form (x, y) where Δx, the distance between two consecutive x values, is constant. Why is calculating $y_2 - y_1$, the distance between two consecutive y values, equivalent to calculating the average rate of change between consecutive points?

13. Mute swans were imported from Europe in the nineteenth century to grace ponds. Now there is concern that their population is growing too rapidly, edging out native species. During the 2000s, many states created mute swan controls.

The goal was to reduce the mute swan count from 14,313 in 2002 to 3000 swans by 2013.

   a. Using a linear model, what would be the average rate of change per year needed to bring the swan population to that level?

   b. Using an exponential model, what would be the decay factor needed?

14. The price of a home in Carnegie, PA was $60,000 in 1980 and rose to $120,000 in 2010.

   a. Create two models, $f(t)$ assuming linear growth and $g(t)$ assuming exponential growth, where t = number of years after 1980.

   b. Fill in the following table representing linear growth and exponential growth for t years after 1980.

t	Linear Growth $f(t)$ = Price (in thousands of dollars)	Exponential Growth $g(t)$ = Price (in thousands of dollars)
0	60,000	60
10	80,000	
20	100,000	
30	120	120
40	140,000	
50	160,000	

   c. Which model do you think is more realistic?

15. *The Mass Media*, a student publication at the University of Massachusetts–Boston, reported on a proposed parking fee increase. The university administration recommended gradually increasing the daily parking fee on this campus from $7.00 in the year 2004, by an increase of 5% every year after that. Call this plan A. Several other plans were also proposed; one of them, plan B, recommended that every year after 2004 the rate be increased by 50 cents.

   a. Let $t = 0$ for year 2004 and fill in the chart for parking fees under plans A and B.

Years after 2004	Parking Fee under Plan A	Parking Fee under Plan B
0	$7.00	$7.00
1		
2		
3		
4		

   b. Write an equation for parking fees F_A as a function of t (years since 2004) for plan A and an equation F_B for plan B.

   c. What will the daily parking fee be by the year 2025 under each plan? (Show your calculations.)

   d. Imagine that you are the student representative to the UMass Board of Trustees. Which plan would you recommend for adoption? Explain your reasons and support your position using quantitative arguments.

5.4 *Visualizing Exponential Functions*

We can summarize what we've learned so far about exponential functions:

Exponential Functions

Exponential functions can be represented by equations of the form

$$y = Ca^x \qquad (a > 0, a \neq 1) \qquad \text{where}$$

C is the initial value and a is the base

Assuming $C > 0$, then if

$a > 1$, the function represents growth and a is called the *growth factor*.
$0 < a < 1$, the function represents decay and a is called the *decay factor*.

EXPLORE & EXTEND

5.4

Making Predictions About the Effects of *a* and *C*
This *Explore & Extend* previews this section. Using the course software "E3: $y = Ca^x$ Sliders," you can make your own observations on the effects of changing the values of a and C on the graph of the function.

What effect does changing the value of a have on the graph? Start with the simplest case where $C = 1$. Make predictions when $a > 1$ and when $0 < a < 1$. Test your predictions with the software. Now consider cases where $C \neq 1$. Do your observations about a still hold?

What effect does changing the value of C have on the graph? Test your predictions with the software. Set $a = 2$. Describe your graphs when $C > 0$ and when $C < 0$. Do your observations about C hold for different values of a?

Summarize your observations with a 60-second summary.

The Graphs of Exponential Functions

In this section we examine what happens to the graph of an exponential functions when the value of a or C changes and what happens as x approaches $\pm\infty$. The graphs of exponential growth and decay functions in the form $y = Ca^x$, where the initial value $C > 0$, lie above the horizontal or x-axis.

Domain and range

Domain: all real numbers, that is all values of the input in the interval $(-\infty, +\infty)$.

Range: restricted to all positive real numbers, that is values of the output >0 when C or the initial value is >0.

The Effect of the Base *a*

Changes in value of the base a affect the steepness of the graph of an exponential function.

Exponential growth: *a* > 1

Given an exponential function $y = Ca^x$, when $a > 1$ (and $C > 0$), the function represents growth. The graph of an exponential growth function is concave up and increasing. Figure 5.9 shows how the value of a affects the steepness of the graphs of the following functions:

$$y = 100 \cdot 1.2^x \qquad y = 100 \cdot 1.3^x \qquad y = 100 \cdot 1.4^x$$

Each function has the same initial value of $C = 100$ but different values of a, that is, 1.2, 1.3, and 1.4, respectively. When $a > 1$, the larger the value of a, the more rapid the growth and the more rapidly the graph rises. So as the values of x increase, the values of y not only increase, but increase at an increasing rate.

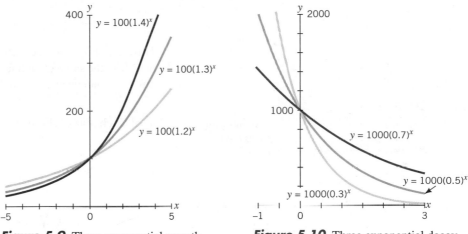

Figure 5.9 Three exponential growth functions.

Figure 5.10 Three exponential decay functions.

Exponential decay: $0 < a < 1$

In an exponential function $y = Ca^x$, when a is positive and less than 1 (and $C > 0$), we have decay. The graph of an exponential decay function is concave up but decreasing.

The smaller the value of a, the more rapid the decay. For example, if $a = 0.7$, after each unit increase in x, the value of y would be multiplied by 0.7. So y would drop to 70% of its previous size—a loss of 30%. But if $a = 0.5$, then when x increases by 1, y would be multiplied by 0.5, equivalent to dropping to 50% of its previous size—a loss of 50%.

Figure 5.10 shows how the value of a affects the steepness of the graph. Each function has the same initial population of 1000 but different values of a, that is, 0.3, 0.5, and 0.7, respectively. When $0 < a < 1$, the smaller the value of a, the more rapid the decay and the more rapidly the graph falls.

The Effect of the Initial Value C

In the exponential function $y = Ca^x$, the initial value C is the vertical intercept. When $x = 0$, then

$$y = Ca^0$$
$$= C \cdot 1$$
$$= C$$

Figure 5.11 compares the graphs of three exponential functions with the same base a of 1.1, but with different C values of 50, 100, and 250, respectively.

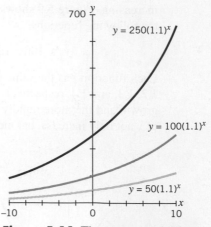

Figure 5.11 Three exponential functions with the same growth factor but different y-intercepts, or initial values.

Changing the value of C changes where the graph of the function crosses the vertical axis.

EXAMPLE 1

Matching graphs and equations
Match each of the graphs (A to D) in Figure 5.12 to one of the following equations and explain your answers.

a. $f(x) = 1.5(2)^x$ **c.** $j(x) = 5(0.6)^x$

b. $g(x) = 1.5(3)^x$ **d.** $k(x) = 5(0.8)^x$

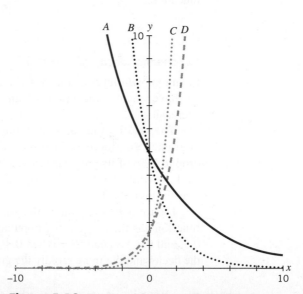

Figure 5.12 Graphs of four exponential functions.

SOLUTION

A is the graph of $k(x) = 5(0.8)^x$. B is the graph of $j(x) = 5(0.6)^x$.

Reasoning: Graphs A and B both have a vertical intercept of 5 and represent exponential decay. The steeper graph (B) must have the smaller decay factor (in this case 0.6).

C is the graph of $g(x) = 1.5(3)^x$. D is the graph of $f(x) = 1.5(2)^x$.

Reasoning: Graphs C and D both have a vertical intercept of 1.5 and both represent exponential growth. The steeper graph (C) must have the larger growth factor (in this case 3).

Horizontal Asymptotes

In the exponential decay function $y = 1000 \cdot (\frac{1}{2})^x$, graphed in Figure 5.13, the initial population of 1000 is cut in half each time x increases by 1. So when $x = 0, 1, 2, 3, 4, 5, 6, \ldots$, the corresponding y values are 1000, 500, 250, 125, 62.5, 31.25, 15.625, ... As x gets larger and larger, the y values come closer and closer to, but never reach, zero. As x approaches positive infinity ($x \to +\infty$), y approaches zero ($y \to 0$).

The graph of the function $y = 1000(\frac{1}{2})^x$ is said to be *asymptotic* to the x-axis, or we say the x-axis is a *horizontal asymptote* to the graph. In general, a horizontal asymptote is a horizontal line that the graph of a function approaches for extreme values of the input or independent variable.

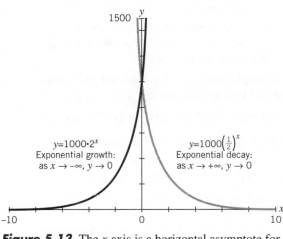

Figure 5.13 The x-axis is a horizontal asymptote for both exponential growth and decay functions of the form $y = Ca^x$.

Exponential decay functions of the form $y = Ca^x$ are asymptotic to the x-axis. Similarly, for exponential growth functions, as x approaches negative infinity ($x \to -\infty$), y approaches zero ($y \to 0$). Examine, for instance, the exponential growth function $y = 1000 \cdot 2^x$, also graphed in Figure 5.13. When $x = 0, -1, -2, -3, \ldots$, the corresponding y values are 1000, $1000 \cdot 2^{-1} = 500$, $1000 \cdot 2^{-2} = 250$, $1000 \cdot 2^{-3} = 125, \ldots$. So as $x \to -\infty$, the y values come closer and closer to but never reach zero. So exponential growth functions are also asymptotic to the x axis.

Graphs of Exponential Functions

For functions in the form $y = Ca^x$,

The value of C tells us where the graph crosses the y-axis.

The value of a affects the steepness of the graph.

Exponential growth ($a > 1$, $C > 0$). The larger the value of a, the more rapid the growth and the more rapidly the graph rises.

Exponential decay ($0 < a < 1$, $C > 0$). The smaller the value of a, the more rapid the decay and the more rapidly the graph falls.

The graphs are *asymptotic* to the x-axis.

Exponential growth: As $x \to -\infty$, $y \to 0$
Exponential decay: As $x \to +\infty$, $y \to 0$

Algebra Aerobics 5.4

1. **a.** Draw a rough sketch of each of the following functions all on the same graph:

$$y = 1000(1.5)^x$$
$$y = 1000(1.1)^x$$
$$y = 1000(1.8)^x$$

   b. Do the three curves intersect? If so, where?

   c. In the first quadrant (where $x > 0$ and $y > 0$), which curve should be on the top? Which in the middle? Which on the bottom?

   d. In the second quadrant (where $x < 0$ and $y > 0$), which curve should be on the top? Which in the middle? Which on the bottom?

2. **a.** Draw a rough sketch of each of the following functions all on the same graph:

$$Q = 250(0.6)^t \quad Q = 250(0.3)^t \quad Q = 250(0.2)^t$$

   b. Do the three curves intersect? If so, where?

   c. In the first quadrant, which curve should be on the top? Which in the middle? Which on the bottom?

   d. In the second quadrant, which curve should be on the top? Which in the middle? Which on the bottom?

3. **a.** Draw a rough sketch of each of the following functions on the same graph:

$$P = 50 \cdot 3^t \quad \text{and} \quad P = 150 \cdot 3^t$$

   b. Do the curves intersect anywhere?

   c. Describe when and where one curve lies above the other.

4. Identify any horizontal asymptotes for the following functions:

   a. $Q = 100 \cdot 2^t$ **c.** $y = 100 - 15x$

   b. $g(t) = (6 \cdot 10^7) \cdot (0.95)^t$

5. Which of these functions has the most rapid growth? Which the most rapid decay?

   a. $f(t) = 100 \cdot 1.06^t$
   b. $g(t) = 25 \cdot 0.89^t$
   c. $h(t) = 4000 \cdot 1.23^t$
   d. $r(t) = 45.9 \cdot 0.956^t$
   e. $P(t) = 32{,}000 \cdot 1.092^t$

6. Examine the graphs of the exponential growth functions $f(x)$ and $g(x)$ in the figure below.

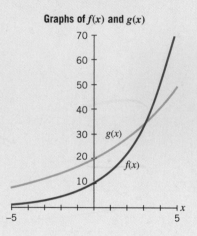

Graphs of $f(x)$ and $g(x)$

   a. Which graph has the larger initial value?

   b. Which graph has the larger growth factor?

   c. As $x \to -\infty$, which graph approaches zero more rapidly?

   d. Approximate the point of intersection. After the point of intersection, as $x \to +\infty$, which function has larger values?

Exercises for Section 5.4

A graphing program is required for Exercises 9 and 10 and recommended for Exercises 11, 15, and 16.

1. Each of the following three exponential functions is in the standard form $y = Ca^x$.

$$y = 2^x \quad y = 5^x \quad y = 10^x$$

   a. In each case identify C and a.

   b. Specify whether each function represents growth or decay. In particular, for each unit increase in x, what happens to y?

   c. Do all three curves intersect? If so, where?

   d. In the first quadrant, which curve should be on top? Which in the middle? Which on the bottom?

   e. Describe any horizontal asymptotes.

   f. For each function, generate a small table of values.

   g. Graph the three functions on the same grid and verify that your predictions in part (d) are correct.

2. Repeat Exercise 1 for the functions

$$y = (0.5)2^x, \ y = 2 \cdot 2^x, \text{ and } y = 5 \cdot 2^x.$$

3. Repeat Exercise 1 for the functions

$$y = 3^x, \ y = \left(\tfrac{1}{3}\right)^x, \text{ and } y = 3 \cdot \left(\tfrac{1}{3}\right)^x.$$

4. Match each equation with its graph (at the top of the next page).

$$f(x) = 30 \cdot 2^x \qquad h(x) = 100 \cdot 2^x$$
$$g(x) = 30 \cdot (0.5)^x \qquad j(x) = 50 \cdot (0.5)^x$$

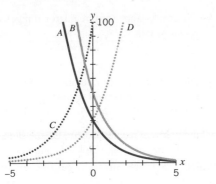

5. Below are graphs of four exponential functions. Match each function with its graph.

$$P = 5 \cdot (0.7)^x \qquad R = 10 \cdot (1.8)^x$$
$$Q = 5 \cdot (0.4)^x \qquad S = 5 \cdot (3)^x$$

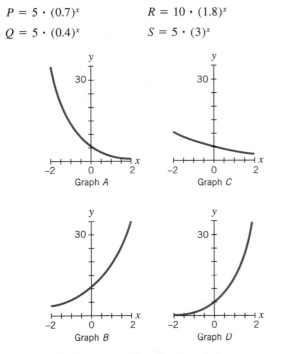

Graph A   Graph C

Graph B   Graph D

6. Examine the accompanying graphs of three exponential growth functions.

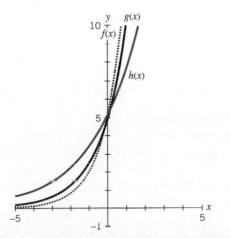

a. Order the functions from smallest growth factor to largest.

b. What point do all of the functions have in common? Will they share any other points?

c. As $x \to -\infty$, will the function with the largest growth factor approach zero more slowly or more quickly than the other functions?

d. For $x > 0$, the graph of which function remains on top?

7. Examine the accompanying graphs of three exponential functions.

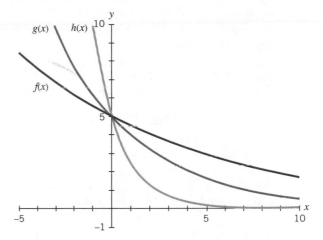

a. Order the functions from smallest decay factor to largest.

b. What point do all of the functions have in common? Will they share any other points?

c. As $x \to +\infty$, will the function with the largest decay factor approach zero more slowly or more quickly than the other functions?

d. When $x < 0$, the graph of which function remains on top?

8. Generate quick sketches of each of the following functions, without the aid of technology.

$$f(x) = 4(3.5)^x \qquad g(x) = 4(0.6)^x$$
$$h(x) = 4 + 3x \qquad k(x) = 4 - 6x$$

a. As $x \to +\infty$, which function(s) approach $+\infty$?

b. As $x \to +\infty$, which function(s) approach 0?

c. As $x \to -\infty$, which function(s) approach $-\infty$?

d. As $x \to -\infty$, which function(s) approach 0?

9. (Graphing program required.) Graph the functions $f(x) = 30 + 5x$ and $g(x) = 3(1.6)^x$ on the same grid. Supply the symbol $<$ or $>$ in the blank that would make the statement true.

a. $f(0)$ _____ $g(0)$ **e.** $f(-6)$ _____ $g(-6)$

b. $f(6)$ _____ $g(6)$ **f.** As $x \to +\infty$, $f(x)$ _____ $g(x)$

c. $f(7)$ _____ $g(7)$ **g.** As $x \to -\infty$, $f(x)$ _____ $g(x)$

d. $f(-5)$ _____ $g(-5)$

10. (Graphing program required.) Graph the functions $f(x) = 6(0.7)^x$ and $g(x) = 6(1.3)^x$ on the same grid. Supply the symbol $<$, $>$, or $=$ in the blank that would make the statement true.

a. $f(0)$ _____ $g(0)$

b. $f(5)$ _____ $g(5)$

c. $f(-5)$ _____ $g(-5)$

d. As $x \to +\infty$, $f(x)$ _____ $g(x)$

e. As $x \to -\infty$, $f(x)$ _____ $g(x)$

11. (Graphing program recommended.)

    a. As $x \to +\infty$, which function will dominate,
    $f(x) = 100 + 500x$ or $g(x) = 2(1.005)^x$?

    b. Determine over which x interval(s) $g(x) > f(x)$.

12. Two cities each have a population of 1.2 million people. City A is growing by a factor of 1.15 every 10 years, while city B is decaying by a factor of 0.85 every 10 years.

    a. Write an exponential function for each city's population $P_A(t)$ and $P_B(t)$ after t years.

    b. For each city's population function generate a table of values for $t = 0$, 10, 20, 30, 40 and 50, then sketch a graph of each town's population on the same grid.

13. Which function has the steepest graph?

    $$F(x) = 100(1.2)^x$$
    $$G(x) = 100(0.8)^x$$
    $$H(x) = 100(1.2)^{-x}$$

14. Examine the graphs of the following functions.

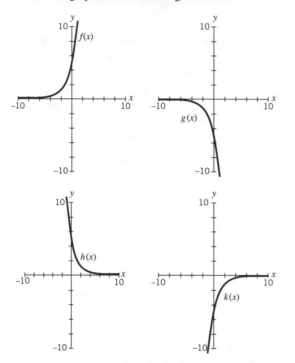

    a. Which graphs appear to be mirror images of each other across the y-axis?

    b. Which graphs appear to be mirror images of each other across the x-axis?

    c. Which graphs appear to be mirror images of each about the origin (i.e., you could translate one into the other by reflecting first about the y-axis, then about the x-axis)?

15. (Graphing program recommended.) On the same graph, sketch $f(x) = 3(1.5)^x$, $g(x) = -3(1.5)^x$, and $h(x) = 3(1.5)^{-x}$.

    a. Which graphs are mirror images of each other across the y-axis?

    b. Which graphs are mirror images of each other across the x-axis?

    c. Which graphs are mirror images of each other about the origin (i.e., you could translate one into the other by reflecting first about the y-axis, then about the x-axis)?

    d. What can you conclude about the graphs of the two functions $f(x) = Ca^x$ and $g(x) = -Ca^x$?

    e. What can you conclude about the graphs of the two functions $f(x) = Ca^x$ and $g(x) = Ca^{-x}$?

16. (Graphing program recommended.) Make a table of values and plot each pair of functions on the same coordinate system.

    a. $y = 2^x$ and $y = 2x$ for $-3 \le x \le 3$

    b. $y = (0.5)^x$ and $y = 0.5x$ for $-3 \le x \le 3$

    c. Which of the four functions that you drew in parts (a) and (b) represent growth?

    d. How many times did the graphs that you drew for part (a) intersect? Find the coordinates of any points of intersection.

    e. How many times did the graphs that you drew for part (b) intersect? Find the coordinates of any points of intersection.

5.5 Exponential Functions: A Constant Percent Change

We defined exponential functions in terms of multiplication by a constant (growth or decay) factor. In this section we show how these factors can be translated into a constant percent increase or decrease. While growth and decay factors are needed to construct an equation, journalists are more likely to use percentages.

Exponential Growth: Increasing by a Constant Percent

In Section 5.1 we modeled the growth of *E. coli* bacteria with the exponential function $N = (1.37 \cdot 10^7) \cdot 1.5^t$. The growth factor, 1.5, tells us that for each unit increase in time, the initial number of *E. coli* is multiplied by 1.5. How can we translate this into a statement involving percentages?

If Q is the number of bacteria at any point in time, then one time period later,

$$\text{number of } E. \ coli = Q \cdot (1.5)$$

Rewrite 1.5 as a sum	$= Q(1 + 0.5)$
Use the distributive property	$= 1 \cdot Q + (0.5) \cdot Q$
Rewrite 0.5 as a percentage	$= Q + 50\% \text{ of } Q$

So for each unit increase in time, the number of bacteria increases by 50%. The term 50% is called the *growth rate in percentage form* and 0.5, the equivalent in decimal form, is called the *growth rate in decimal form.*

A quantity that increases by a constant percent over each time interval represents exponential growth. We can represent this growth in terms of growth factors or growth rates. Exponential growth requires a growth factor > 1. If the exponential growth factor $= 1$, the object would never grow, but would stay constant through time. So,

$$\text{growth factor} = 1 + \text{growth rate}$$
$$\text{growth factor} = 1 + r$$

where r is the growth rate in decimal form.

EXAMPLE 1

Converting the growth rate to a growth factor

The U.S. Bureau of the Census has made a number of projections for the first half of this century. For each projection, convert the growth rate into a growth factor.

a. Disposable income is projected to grow at an annual rate of 2.9%.

b. Nonagriculture employment is projected to grow at 1% per year.

c. Employment in manufacturing is projected to grow at 0.2% per year.

SOLUTION

a. A growth rate of 2.9% is 0.029 in decimal form. The growth factor is $1 + 0.029 = 1.029$.

b. A growth rate of 1% is 0.01 in decimal form. The growth factor is $1 + 0.01 = 1.01$.

c. A growth rate of 0.2% is 0.002 in decimal form. The growth factor is $1 + 0.002 = 1.002$.

EXAMPLE 2

Urban population projection

According to UN statistics, in the year 2010 there were an estimated 3.4 billion people living in urban areas. The number of people living in urban areas is projected to grow at a rate of 1.9% per year. If this projection is accurate, how many people will be living in urban areas in 2030?

SOLUTION

If the urban population is increasing by a constant percent each year, the growth is exponential. To construct an exponential function modeling this growth, we need to find the growth factor. We know that the growth rate $= 1.9\%$ or 0.019. So,

$$\text{growth factor} = 1 + 0.019 = 1.019$$

Given an initial population of 3.4 billion in 2010, our model is

$$U(n) = 3.4 \cdot (1.019)^n$$

where $n =$ number of years since 2010 and $U(n) =$ urban population (in billions).

Using our model, the urban population in 2030, $U(20) \approx 5.0$ billion people.

Exponential Decay: Decreasing by a Constant Percent

In Section 5.2, we used the exponential function $C = C_0(0.871)^t$ (where $C_0 =$ the peak amount of caffeine and $t =$ number of hours since peak caffeine level) to model the amount of caffeine in the body after drinking a cup of coffee. The decay factor, 0.871, tells us that for each unit increase in time, the amount of caffeine is multiplied by 0.871. As with exponential growth, we can translate this statement into one that involves percentages.

If Q is the amount of caffeine at any point in time, then one time period later,

$$\text{amount of caffeine} = Q \cdot (0.871)$$

rewrite 0.871 as a difference	$= Q \cdot (1 - 0.129)$
use the distributive property	$= 1 \cdot Q - (0.129) \cdot Q$
rewrite 0.129 as a percentage	$= Q - 12.9\% \text{ of } Q$

So for each unit increase in time, the amount of caffeine decreases by 12.9%. The term 12.9% is called the *decay rate in percentage form* and the equivalent decimal form, 0.129, is called the *decay rate in decimal form.*

A quantity that decreases by a constant percent over each time interval represents exponential decay. We can represent this decay in terms of decay factors or decay rates. Exponential decay requires a decay factor between 0 and 1.

$$\text{decay factor} = 1 - \text{decay rate}$$
$$\text{decay factor} = 1 - r$$

where r is the decay rate in decimal form.

Factors, Rates, and Percentages

For an exponential function in the form $y = Ca^x$, where $C > 0$,

The *growth factor $a > 1$* can be represented as

$$a = 1 + r$$

where r is the *growth rate,* the decimal representation of the percent rate of change.

The *decay factor a*, where $0 < a < 1$, can be represented as

$$a = 1 - r$$

where r is the *decay rate,* the decimal representation of the percent rate of change.

Examples: If a quantity is growing by 3% per month, the growth rate r in decimal form is 0.03, so the growth factor $a = 1 + 0.03 = 1.03$.

If a quantity is decaying by 3% per month, the decay rate r in decimal form is 0.03, so the decay factor $a = 1 - 0.03 = 0.97$.

E X A M P L E 3

Constructing an exponential decay function

Sea ice that survives the summer and remains year round, called perennial sea ice, is melting at the alarming rate of 9% per decade, according to a 2006 report by the National Oceanic and Atmospheric Administration. Assuming the percent decrease per decade in sea ice remains constant, construct a function that represents this decline.

S O L U T I O N

If the percent decrease remains constant, then the melting of sea ice represents an exponential function with a decay rate of 0.09 in decimal form. So,

$$\text{decay factor} = 1 - 0.09 = 0.91$$

If d = number of decades since 2006 and A = amount of sea ice (in millions of acres) in 2006, then we can model the amount of sea ice, $S(d)$, by the exponential function:

$$S(d) = A(0.91)^d$$

EXAMPLE 4   Growth or decay?

In the following functions, identify the growth or decay factors and the corresponding growth or decay rates in both percentage and decimal form.

a. $f(x) = 100(0.01)^x$ **c.** $p = 5.34(0.015)^n$

b. $g(x) = 230(3.02)^x$ **d.** $y = 8.75(2.35)^x$

SOLUTION   **a.** The decay factor is 0.01 and if r is the decay rate in decimal form, then

$$\text{decay factor} = 1 - \text{decay rate}$$
$$0.01 = 1 - r$$
$$r = 1 - 0.01$$
$$r = 0.99$$

The decay rate is 0.99, or 99% as a percentage.

b. The growth factor is 3.02 and if r is the growth rate in decimal form, then

$$\text{growth factor} = 1 + \text{growth rate}$$
$$3.02 = 1 + r$$
$$3.02 - 1 = r$$
$$r = 2.02$$

The growth rate in decimal form is 2.02, or 202% as a percentage.

c. The decay factor is 0.015 and the decay rate is 0.985 in decimal form, or 98.5% as a percentage.

d. The growth factor is 2.35 and the growth rate is 1.35 in decimal form, or 135% as a percentage.

Revisiting Linear vs. Exponential Functions

In Section 5.3 we compared linear vs. exponential functions as additive versus multiplicative. We now have another way to compare them.

Linear vs. Exponential Functions

Linear functions represent quantities that increase or decrease by a constant amount. Exponential functions represent quantities that increase or decrease by a constant percent.

EXAMPLE 5   Identifying linear vs. exponential growth

According to industry sources, U.S. wireless data service revenues are growing by 34% annually. India is the biggest growth market, increasing by about 8 million new cell phone subscribers every month. Which of these two statements represents linear and which represents exponential growth? Justify your answers.

SOLUTION   Quantities that increase by a constant amount represent linear growth. So, an increase of 8 million cell phones every month describes linear growth. Quantities that increase by a constant percent represent exponential growth. So, an increase of 34% annually describes exponential growth.

EXAMPLE 6   Using constant amount and constant percent to construct functions

Construct an equation for each description of the population of four different cities.

a. In 1950, city A had a population of 123,000 people. Each year since 1950, the population decreased by 0.8%.

b. In 1950, city B had a population of 4,500,000 people. Since 1950, the population has grown by approximately 2000 people per year.

c. In 1950, city C had a population of 625,000. Since 1950, the population has declined by about 5000 people each decade.

d. In 1950, city D had a population of 2.1 million people. Each decade since 1950, the population has increased by 15%.

SOLUTION For each of the following equations, let P = population (in thousands) and t = years since 1950.

a. $P_A = 123(0.992)^t$

b. $P_B = 4500 + 2t$

where the unit for rate of change is people (in thousands) per year.

c. $P_C = 625 - 0.5t$

where the unit for rate of change is people (in thousands) per year. The rate of change is -0.5 people (in thousands) per *year*, since the population declined by 5000 people per *decade*, or 500 people per *year*.

d. growth rate per decade $= 0.15$
growth factor per decade $= 1 + 0.15 = 1.15$
growth factor per year $= (1.15)^{1/10} \approx 1.014$
$P_D = 2100(1.014)^t$

EXAMPLE 7 **When the average rate of change is not constant**
You are offered a job with a salary of $30,000 a year and annual raises of 5% for the first 10 years of employment. Show why the average rate of change of your salary in dollars per year is not constant.

SOLUTION Let n = number of years since you signed the contract and $S(n)$ = your salary in dollars after n years. When $n = 0$, your salary is represented by $S(0) = \$30,000$. At the end of your first year of employment, your salary increases by 5%, so

$$S(1) = 30,000 + 5\% \text{ of } 30,000$$
$$= 30,000 + 0.05 \,(30,000)$$
$$= 30,000 + 1500$$
$$= \$31,500$$

At the end of your second year, your salary increases again by 5%, so

$$S(2) = 31,500 + 5\% \text{ of } 31,500$$
$$= 31,500 + 0.05(31,500)$$
$$= 31,500 + 1575$$
$$= \$33,075$$

The average rate of change between years 0 and 1 is

$$\frac{S(1) - S(0)}{1} = \$31,500 - \$30,000 = \$1500$$

The average rate of change between years 1 and 2 is

$$\frac{S(2) - S(1)}{1} = \$33,075 - \$31,500 = \$1575$$

So, the average rates of change differ. This is to be expected, since in the first year the 5% raise applies only to the initial salary of $30,000, whereas in the second year the 5% raise applies to both the initial $30,000 salary and to the $1500 raise from the first year.

EXPLORE & EXTEND

5.5   Madmen or Economists

"Anyone who believes exponential growth can go on forever in a finite world is either a madman or an economist"—Kenneth Boulding

World population growth has been a source of hot debate, as past data and the graph on the left indicate the world population is growing at an exponential rate. Can the world's population continue to grow at this rate? Limitations in natural resources suggest not. In fact, it is suggested that world population has already started not to follow an exponential trend.

Consider the world population growth rates:

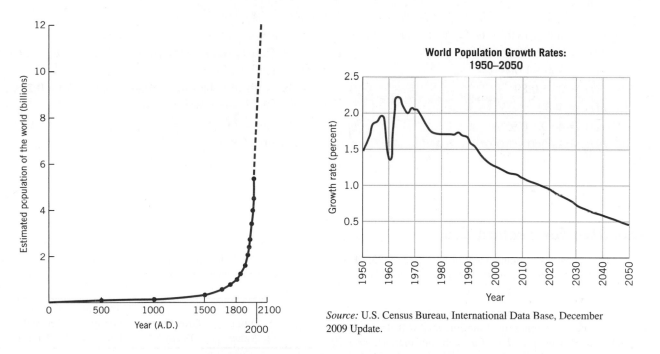

Source: U.S. Census Bureau, International Data Base, December 2009 Update.

a. The dip in the growth rate during the late 1950s and early 1960s was due primarily to China's Great Leap Forward, which triggered a widespread famine and resulted in the deaths of possibly 20 million or more Chinese people. Why did this have such a noticeable effect on world growth rates?

b. If the world population were growing exponentially, what would we expect to see in the growth rates?

c. What do these data suggest about world population growth as we approach 2050 and beyond? Based on your analysis, what would the graph of the estimated world population look like?

Algebra Aerobics 5.5

A calculator that can evaluate powers is recommended for Problem 4.

1. Complete the following statements.

   a. A growth factor of 1.22 corresponds to a growth rate of _____.

   b. A growth rate of 6.7% corresponds to a growth factor of _____.

   c. A decay factor of 0.972 corresponds to a decay rate of _____.

   d. A decay rate of 12.3% corresponds to a decay factor of _____.

2. Fill in the table below.

Exponential Function	Initial Value	Growth or Decay?	Growth or Decay Factor	Growth or Decay Rate (as %)
$A = 4(1.03)^t$				
$A = 10(0.98)^t$				
	1000		1.005	
	30		0.96	
	$50,000	Growth		7.05%
	200 grams	Decay		49%

3. Determine the growth or decay factor and the growth or decay rate in percentage form.

a. $f(t) = 100 \cdot 1.06^t$

b. $g(t) = 25 \cdot 0.89^t$

c. $h(t) = 4000 \cdot 1.23^t$

d. $r(t) = 45.9 \cdot 0.956^t$

e. $P(t) = 32,000 \cdot 1.092^t$

4. Find the percent increase (or decrease) if:

a. Hourly wages grew from $1.65 per hour to $6.00 per hour in 30 years.

b. A player's batting average dropped from 0.299 to 0.167 in one season.

c. The number of new flu cases grew from 17 last month to 23 this month.

Exercises for Section 5.5

A graphing program is recommended for Exercises 15, 23, 24, and 26. Access to the Internet is required for Exercises 8 (d), 16 (d), and 26 (e).

1. Assume that you start with 1000 units of some quantity Q. Construct an exponential function that will describe the value of Q over time T if, for each unit increase in T, Q increases by:

a. 300%   **b.** 30%   **c.** 3%   **d.** 0.3%

2. Given each of the following exponential growth functions, identify the growth rate in percentage form:

a. $Q = 10,000 \cdot 1.5^T$ **d.** $Q = 10,000 \cdot 2^T$

b. $Q = 10,000 \cdot 1.05^T$ **e.** $Q = 10,000 \cdot 2.5^T$

c. $Q = 10,000 \cdot 1.005^T$ **f.** $Q = 10,000 \cdot 3^T$

3. Given the following exponential decay functions, identify the decay rate in percentage form.

a. $Q = 400(0.95)^t$ **d.** $y = 200(0.655)^x$

b. $A = 600(0.82)^t$ **e.** $A = 10(0.996)^T$

c. $P = 70,000(0.45)^t$ **f.** $N = 82(0.725)^T$

4. What is the growth or decay factor for each given time period?

a. Weight increases by 0.2% every 5 days.

b. Mass decreases by 6.3% every year.

c. Population increases 23% per decade.

d. Profit increases 300% per year.

e. Blood alcohol level decreases 35% per hour.

5. Fill in the following chart and then construct exponential functions for each part (a) to (g).

	Initial Value	Growth or Decay?	Growth or Decay Factor	Growth or Decay Rate (% form)
a.	600		2.06	
b.	1200			200%
c.	6000	Decay		75%
d.	1.5 million	Decay		25%
e.	1.5 million	Growth		25%
f.	7		4.35	
g.	60		0.35	

6. Match the statements (a) through (d) with the correct exponential function in (e) through (h). Assume time t is measured in the unit indicated.

a. Radon-222 decays by 50% every day and t represents the number of days.

b. Money in a savings account increases by 2.5% per year.

c. The population increases by 25% per decade.

d. The pollution in a stream decreases by 25% every year.

e. $A = 1000(1.025)^t$

f. $A = 1000(0.75)^t$

g. $A = 1000(\frac{1}{2})^t$

h. $A = 1000(1.25)^t$

7. Each of two towns had a population of 12,000 in 1990. By 2000 the population of town A had increased by 12% while the population of town B had decreased by 12%. Assume these growth and decay rates continued.

   a. Write two exponential population models $A(T)$ and $B(T)$ for towns A and B, respectively, where T is the number of decades since 1990.

   b. Write two new exponential models $a(t)$ and $b(t)$ for towns A and B, where t is the number of *years* since 1990.

   c. Now find $A(2)$, $B(2)$, $a(20)$, and $b(20)$ and explain what you have found.

8. [Access to the Internet is required for part (d).] On May 18, 2010, United Envirotech reported that it had a net profit of $14.9 million, which represented an increase of 305.4% for the 12-month period ending on March 31, 2010.

   a. Estimate the net profit for the previous year ending on March 31, 2009.

   b. Assume continued growth at this rate and construct an exponential model for this data. Indicate what your variables represent and the units for each variable.

   c. Do you think an exponential model is appropriate for this set of data? Explain your reasoning.

   d. Check the current profits of United Envirotech on the Internet. Do their current profits provide support for your answers? Why or why not?

Each of the tables in Exercises 9–12 represents an exponential function. Construct that function and then identify the corresponding growth or decay rate in percentage form. Values are rounded to two decimal places.

9.

x	0	1	2	3
y	500.00	425.00	361.25	307.06

10.

x	0	1	2	3
y	500.00	575.00	661.25	760.44

11.

x	0	1	2	3
y	225.00	228.38	231.80	235.28

12.

x	0	1	2	3
y	225.00	221.63	218.30	215.03

13. Generate equations that represent the pollution levels, $P(t)$, as a function of time, t (in years), such that $P(0) = 150$ and:

   a. $P(t)$ triples each year.

   b. $P(t)$ decreases by twelve units each year.

   c. $P(t)$ decreases by 7% each year.

   d. The annual average rate of change of $P(t)$ with respect to t is a constant value of 1.

14. Given an initial value of 50 units for parts (a)–(d) below, in each case construct a function that represents Q as a function of time t. Assume that when t increases by 1:

   a. $Q(t)$ doubles

   b. $Q(t)$ increases by 5%

   c. $Q(t)$ increases by ten units

   d. $Q(t)$ is multiplied by 2.5

15. (Graphing program recommended.) Between 1970 and 2000, the United States grew from about 200 million to 280 million, an increase of approximately 40% in this 30-year period. If the population continues to expand by 40% every 30 years, what will the U.S. population be in the year 2030? In 2060? Estimate when the population of the United States would reach 1 billion.

16. [Access to the Internet is required for part (d).]

   The global economic crisis hit the tourism industry around the world. In 2008 almost 13 million foreign tourists visited Egypt, but in 2009 there were only 10.9 million. Assume this decline continues.

   a. Construct two models, one linear and one exponential, for the number of foreign tourists to Egypt.

   b. What would your linear model predict for number of tourists in 2010?

   c. What would your exponential model predict for number of tourists in 2010?

   d. Use the Internet to check which model is more accurate.

17. A pollutant was dumped into a lake, and each year its amount in the lake is reduced by 25%.

   a. Construct a general formula to describe the amount of pollutant after n years if the original amount is A_0.

   b. How long will it take before the pollution is reduced to below 1% of its original level? Justify your answer.

18. A swimming pool is initially shocked with chlorine to bring the chlorine concentration to 3 ppm (parts per million). Chlorine dissipates in reaction to bacteria and sun at a rate of about 15% per day. Above a chlorine concentration of 2 ppm, swimmers experience burning eyes, and below a concentration of 1 ppm, bacteria and algae start to proliferate in the pool environment.

   a. Construct an exponential decay function that describes the chlorine concentration (in parts per million) over time.

   b. Construct a table of values that corresponds to monitoring chlorine concentration for at least a 2-week period.

   c. How many days will it take for the chlorine to reach a level tolerable for swimmers? How many days before bacteria and algae will start to grow and you will need to add more chlorine? Justify your answers.

19. a. If the inflation rate is 0.7% a month, what is it per year?

   b. If the inflation rate is 5% a year, what is it per month?

20. Rewrite each expression so that no fraction appears in the exponent and each expression is in the form a^x.

   a. $3^{x/4}$ b. $2^{x/3}$ c. $\left(\frac{1}{2}\right)^{x/4}$ d. $\left(\frac{1}{4}\right)^{x/2}$

   Hint: Remember that $a^{x/n} = (a^{1/n})^x$.

21. a. Show that the following two functions are roughly equivalent.

$$f(x) = 15{,}000(1.2)^{x/10} \quad \text{and} \quad g(x) = 15{,}000(1.0184)^x$$

b. Create a table of values for each function for $0 < x \leq 25$.

c. Would it be correct to say that a 20% increase over 10 years represents a 1.84% annual increase? Explain your answer.

22. The exponential function $Q(T) = 600(1.35)^T$ represents the growth of a species of fish in a lake, where T is measured in 5-year intervals.

a. Determine $Q(1)$, $Q(2)$, and $Q(3)$.

b. Find another function $q(t)$, where t is measured in years.

c. Determine $q(5)$, $q(10)$, and $q(15)$.

d. Compare your answers in parts (a) and (c) and describe your results.

23. (Graphing program recommended.) Blood alcohol content (BAC) is the amount of alcohol present in your blood as you drink. It is calculated by determining how many grams of alcohol are present in 100 milliliters (1 deciliter) of blood. So if a person has 0.08 grams of alcohol per 100 milliliters of blood, the BAC is 0.08 g/dl. After a person has stopped drinking, the BAC declines over time as his or her liver metabolizes the alcohol. Metabolism proceeds at a steady rate and is impossible to speed up. For instance, an average (150-lb) male metabolizes about 8 to 12 grams of alcohol an hour (the amount in one bottle of beer[4]). The behavioral effects of alcohol are closely related to the blood alcohol content. For example, if an average 150-lb male drank two bottles of beer within an hour, he would have a BAC level of 0.05 and could suffer from euphoria, inhibition, loss of motor coordination, and overfriendliness. The same male after drinking four bottles of beer in an hour would be legally drunk with a BAC of 0.10. He would likely suffer from impaired motor function and decision making, drowsiness, and slurred speech. After drinking twelve beers in one hour, he will have attained the dosage for stupor (0.30) and possibly death (0.40).

The following table gives the BAC of an initially legally drunk person over time (assuming he doesn't drink any additional alcohol).

Time (hours)	0	1	2	3	4
BAC (g/dl)	0.100	0.067	0.045	0.030	0.020

a. Graph the data from the table (be sure to carefully label the axes).

b. Justify the use of an exponential function to model the data. Then construct the function where $B(t)$ is the BAC for time t in hours.

c. By what percentage does the BAC decrease every hour?

d. What would be a reasonable domain for your function? What would be a reasonable range?

e. Assuming the person drinks no more alcohol, when does the BAC reach 0.005 g/dl?

[4]One drink is equal to $1\frac{1}{4}$ oz of 80-proof liquor, 12 oz of beer, or 4 oz of table wine.

24. (Graphing program recommended.) If you have a heart attack and your heart stops beating, the amount of time it takes paramedics to restart your heart with a defibrillator is critical. According to a medical report on the evening news, each minute that passes decreases your chance of survival by 10%. From this wording it is not clear whether the decrease is linear or exponential. Assume that the survival rate is 100% if the defibrillator is used immediately.

a. Construct and graph a linear function that describes your chances of survival. After how many minutes would your chance of survival be 50% or less?

b. Construct and graph an exponential function that describes your chances of survival. Now after how many minutes would your chance of survival be 50% or less?

25. Algeria is a developing country on the northern coast of Africa. Its infant mortality rates (number of deaths per 1000 live births) declined from 94 in 1980 to 33 in 2006.

a. Assume that the infant mortality rate is declining linearly over time. Construct an equation modeling the relationship between infant mortality rate and time, where time is measured in years since 1980. Make sure you have clearly identified your variables.

b. Assuming that the infant mortality rate is declining exponentially over time, construct an equation modeling the relationship, where time is measured in years since 1980.

c. What would each of your models predict for the infant mortality rate in 2010? If you have access, check the Internet to see which prediction is more accurate.

26. [Part (e) requires use of the Internet and graphing program to find a best-fit function.] A "rule of thumb" used by car dealers is that the trade-in value of a car decreases by 30% each year.

a. Is this decline linear or exponential?

b. Construct a function that would express the value of the car as a function of years owned.

c. Suppose you purchase a car for $15,000. What would its value be after 2 years?

d. Explain how many years it would take for the car in part (c) to be worth less than $1000. Explain how you arrived at your answer.

e. *Internet search:* Go to the Internet site for the Kelley Blue Book (*www.kbb.com*).

   i. Enter the information about your current car or a car you would like to own. Specify the actual age and mileage of the car. What is the Blue Book value?

   ii. Keeping everything else the same, assume the car is 1 year older and increase the mileage by 10,000. What is the new Blue Book value?

   iii. Find a best-fit exponential function to model the value of your car as a function of years owned. What is the annual decay rate?

   iv. According to this function, what will the value of your car be 5 years from now?

5.6 *More Examples of Exponential Growth and Decay*

Here we take a look at some examples showing the wide range of applications of exponential functions.

EXAMPLE 1

The rabbit epidemic

In 1859, Thomas Austin released 24 rabbits onto his farm in western Australia in order to have a small population to hunt. He didn't realize the population would flourish, having no natural predators, and grow by a factor of 2.5 every 6 months.

a. Construct an equation for the number of rabbits, R, as a function of the number of 6-month time periods, T, that have passed since Thomas Austin first released the rabbits.

b. Create three tables, one for the number of rabbits as a function of 6-month time periods; one as a function of the number of months, m; and one for number of years, x, since the first release of the rabbits. Show how you find the new growth factor when changing time units.

c. Construct equivalent equations for the number of rabbits, R, as a function of the number of months, m, and as a function of the number of years, x.

d. In 1926, even after various measures were taken to reduce the size of the population,[5] there were an estimated 10 billion rabbits. Using a calculator, estimate how many years your model predicts it would take to reach 10 billion rabbits.

SOLUTION

a. We know that the initial value is 24 and that the growth factor = 2.5 when time T = number of 6-month periods, so an equation that models the number of rabbits, R, is

$$R = 24(2.5)^T \qquad \text{where } T = \text{number of 6-month periods}$$

b. One year is equal to two 6-month periods, so the growth factor for one year is

$$2.5^2 = 6.25$$

One month is equal to 1/6 of a 6-month period, so the growth factor for one month is

$$2.5^{1/6} = 1.165$$

Using these growth factors, we can create Tables 5.8, 5.9, and 5.10.

Number of 6-Month Periods	Number of Rabbits	Number of Years	Number of Rabbits	Number of Months	Number of Rabbits
0	24	0	24	0	24
1	60	1	150	1	28
2	150	2	938	2	33
3	375	3	5,859	3	38
4	938	4	36,621	4	44
5	2,344	5	228,882	5	52
6	5,859	6	1,430,511	6	60

Table 5.8   *Table 5.9*   *Table 5.10*

[5]Between 1865 and 1866, Thomas and his guests killed more than 34,000 rabbits on his property. From January to August 1887, more than 10 million rabbits were killed in one colony alone. In the early 1900s, Australians created 3200 km of fences to block the spread of the rabbits, but the rabbits simply burrowed under. In the 1950s, they released viruses that reduced the rabbit population by 99%, but the rabbits that remained developed resistance to the viruses and began to multiply. By 1990, there were 600 million rabbits, and another virus was developed to target the rabbits. This virus reduced the population of rabbits to 100 million.
Source: http://australian-history.suite101.com/article.cfm/wild-rabbits.

We can check the monthly growth factor by seeing if the number of rabbits we get for 6 months (in Table 5.10 on the previous page) is the same as what we get for one 6-month period (in Table 5.8). Similarly the number of rabbits after 3 years (in Table 5.9) should be the same as the number of rabbits we get for six 6-month periods (in Table 5.8), which it is.

c. With initial value = 24 and monthly growth factor = 1.165, an equation that models the number of rabbits, R, is

$$R = 24(1.165)^m \qquad \text{where } m = \text{number of months}$$

With initial value = 24 and yearly growth factor = 6.25, an equation that models the number of rabbits, R, is

$$R = 24(6.25)^x \qquad \text{where } x = \text{number of years}$$

d. Using a calculator and our model with a yearly growth rate, we get

$$R = 24(6.25)^{10} \approx 2{,}182{,}787{,}284$$

$$R = 24(6.25)^{11} \approx 13{,}642{,}420{,}527$$

So the number of rabbits reaches 10 billion after almost 11 years.

EXAMPLE 2

China's one-child policy

China's one-child policy, introduced in 1979 and implemented in the 1980s, was a response to concerns that China's population was growing too fast. The goal was to reduce China's population to 700 million by 2050. In 2010, the 30th anniversary of the first significant implementation of the policy, China's population was approximately 1.3 billion. Even though China's population grew over these 30 years, in 2009, its rate of growth was 0.494% and the population was expected to inevitably begin to decline in the near future. Assume that China's population will start to decline at a rate of 0.5% per year in 2010. Will this rate be sufficient to meet the original goal?

SOLUTION

If the decay rate = 0.5% per year, then the decay factor is $1 - 0.005 = 0.995$ per year. Using this decay factor and an initial population of 1.3 billion, or 1300 million, people, we can use an exponential function to model the population, P (in millions), after n years:

$$P = 1{,}300 \cdot 0.995^n \qquad \text{where } n = \text{number of years}$$

For the year 2050, $n = 2050 - 2010 = 40$ years:

$$P = 1300 \cdot 0.995^{40}$$

using a calculator and rounding,

$$P \approx 1300 \cdot 0.818$$

multiplying, rounding, and adding unit

$$P \approx 1064 \text{ million people}$$

So a decay rate of 0.5% is not enough to reduce the population to 700 million people in 2050.

Returning to Doubling Times and Half-Lives

Previously we saw that every exponential growth function has a fixed doubling time, and every exponential decay function has a fixed half-life. Here we consider general strategies for constructing equivalent exponential equations and learn how to estimate the doubling time and half-life of quantities using the "rule of 70."

EXAMPLE 3   **Constructing exponential functions given the doubling time or half-life**
In each situation construct an exponential function, where t = number of years.

a. The doubling time is 3 years; the initial quantity is 1000.

b. The half-life is 5 years; the initial quantity is 50.

SOLUTION   **a.** A doubling time of 3 years describes an exponentially growing quantity that doubles every 3 years. We can construct an exponential function,

$$P(t) = 1000a^t \qquad \text{where } a \text{ is the yearly} \qquad (1)$$
$$\text{growth factor and}$$
$$t = \text{number of years.}$$

Since the doubling time is 3 years, then $2000 = 1000a^3$

dividing and reversing sides $a^3 = 2$

taking the cube root of each side $a = 2^{1/3}$

substituting in (1) $P(t) = 1000 \cdot (2^{1/3})^t$

using Rule 3 for exponents $P(t) = 1000 \cdot 2^{t/3}$

b. A half-life of 5 years means an exponentially decaying quantity that decreases by half every 5 years. We can construct an exponential decay function,

$$Q(t) = 50a^t \qquad \text{where } a \text{ is the yearly} \qquad (2)$$
$$\text{decay factor and}$$
$$t = \text{number of years.}$$

Since the half-life is 5 years, then $25 = 50a^5$

dividing and reversing sides $a^5 = \frac{1}{2}$

taking the fifth root of each side $a = \left(\frac{1}{2}\right)^{1/5}$

substituting in (2) and
using Rule 3 for exponents $Q(t) = 50 \cdot \left(\frac{1}{2}\right)^{t/5}$

Finding Equivalent Equations Using the Doubling Time (or Half-Life)

If an exponential growth function $G(t) = Ca^t$ has a doubling time of n time units, and t = number of time units, then

$$a^n = 2$$
$$a = 2^{1/n}$$
so, $$G(t) = C \cdot 2^{t/n}$$

If an exponential decay function $D(t) = Ca^t$ has a half-life of n, then

$$a^n = \frac{1}{2}$$
$$a = \left(\frac{1}{2}\right)^{1/n}$$
so, $$D(t) = C \cdot \left(\frac{1}{2}\right)^{t/n}$$

For example, if $G(t) = 500a^t$ has a doubling time of 6 hours, then $G(t) = 500 \cdot 2^{t/6}$, where t = number of hours.

If $D(t) = 25a^t$ has a half-life of 3 minutes, then $D(t) = 25 \cdot \left(\frac{1}{2}\right)^{t/3}$, where t = number of minutes.

EXAMPLE 4

Half-life: Radioactive decay

One of the toxic radioactive by-products of nuclear fission is strontium-90. A nuclear accident, like the one in Chernobyl, can release clouds of gas containing strontium-90. The clouds deposit the strontium-90 onto vegetation eaten by cows, and humans ingest strontium-90 from the cows' milk. The strontium-90 then replaces calcium in bones, causing cancer and birth defects. Strontium-90 is particularly insidious because it has a *half-life* of approximately 28 years. That means that every 28 years about half (or 50%) of the existing strontium-90 has decayed into nontoxic, stable zirconium-90, but the other half still remains.

a. Construct a model for the decay of 100 mg of strontium-90 as a function of 28-year time periods.

b. Construct a new model that describes the decay as a function of the number of years. Generate a corresponding table and graph.

SOLUTION

a. Strontium-90 decays by 50% every 28 years. If we define our basic time period T as 28 years, then the decay rate is 0.5 and the decay factor is $1 - 0.5 = 0.5$ (or 1/2). Assuming an initial value of 100 mg, the function

$$S = 100 \cdot (0.5)^T$$

gives S, the remaining milligrams of strontium-90 after T time periods (of 28 years).

b. The number 0.5 is the decay factor over a 28-year period. If we let $a = $ *annual* decay factor, then

$$a^{28} = 0.5$$

To find the value for a we can take the 28th root of both sides.

$$(a^{28})^{1/28} = (0.5)^{1/28}$$

or
$$a \approx 0.976$$

So if $t = $ number of *years,* the *yearly* decay factor is 0.976, making our function

$$S \approx 100 \cdot (0.976)^t$$

We now have an equation that describes the amount of strontium-90 as a function of years, t. The decay factor is 0.976, so the decay rate is $1 - 0.976 = 0.024$, or 2.4%. So each year there is 2.4% less strontium-90 than the year before.

Decay of Strontium-90

Time T (28-year time periods)	Time t (years)	Strontium-90 (mg)
0	0	100
1	28	50
2	56	25
3	84	12.5
4	112	6.25
5	140	3.125

Table 5.11

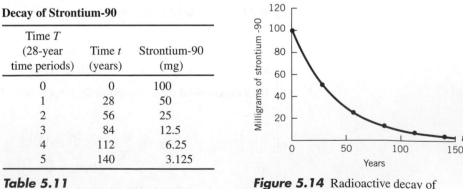

Figure 5.14 Radioactive decay of strontium-90.

Table 5.11 and Figure 5.14 show the decay of 100 mg of strontium-90 over time. After 28 years, half (50 mg) of the original 100 mg still remains. It takes an additional 28 years for the remaining amount to halve again, still leaving 25 mg out of the original 100 after 56 years.

EXAMPLE 5   **Identifying the doubling time or half-life**
Assuming t is in years, identify the doubling time or half-life for each of the following exponential functions.

a. $Q(t) = 80(2)^{t/12}$ **b.** $P(t) = 5 \cdot 10^6 \left(\frac{1}{2}\right)^{4t}$ **c.** $R(t) = 130(0.5)^{3t}$

SOLUTION   **a.** When $t = 12$ years, the quantity $Q(12) = 80 \cdot 2^{12/12} = 80 \cdot 2^1 = 160$. So the quantity has doubled from the initial value of 80 and the doubling time is 12 years.

b. When $t = 1/4$, the quantity $P(1/4) = (5 \cdot 10^6)(1/2)^{4(1/4)} = (5 \cdot 10^6)(1/2)^1$. So the quantity is half the initial value of $5 \cdot 10^6$ and the half-life is 1/4 years or 3 months.

c. When $t = 1/3$, the quantity $R\left(\frac{1}{3}\right) = 130(0.5)^{3(1/3)} = 130(0.5)^1 = 65$. So the quantity is half the initial value of 130 and the half-life is 1/3 year or 4 months.

The "rule of 70": A rule of thumb for calculating doubling or halving times

A simple way to understand the significance of constant percent growth rates is to compute the doubling time. A rule of thumb is that a quantity growing at $R\%$ per year has a doubling time of approximately $70/R$ years. If the quantity is growing at $R\%$ per *month*, then $70/R$ gives its doubling time in *months*. The same reasoning holds for any unit of time.

The rule of 70 is easy to apply. For example, if a quantity is growing at a rate of 2% a year, then the doubling time is about $70/2$ or approximately 35 years. The rule of 70 also applies when R represents the percentage at which some quantity is decaying. In these cases, $70/R$ equals the half-life.

For now we'll take the rule of 70 on faith. It provides good approximations, especially for smaller values of R (those under 10%). When we return to logarithms in Chapter 6, we find out why this rule works.

The Rule of 70

The *rule of 70* states that if a quantity is growing (or decaying) at $R\%$ per time period, then the doubling time (or half-life) is approximately

$$\frac{70}{R}$$

provided R is not much bigger than 10.

> *Examples:* If a quantity is growing at 7% per year, the doubling time is approximately $70/7 = 10$ years.
>
> If a quantity is decaying at 5% per minute, the half-life is approximately $70/5 = 14$ minutes.

EXAMPLE 6   **Using the rule of 70**
Suppose that at age 23 you invest $1000 in a retirement account that grows at 5% per year.

a. Roughly how long will it take your investment to double?

b. If you retire at age 65, approximately how much will you have in your account?

SOLUTION

a. Using the rule of 70, we get 70/5 = 14, so your investment will double approximately every 14 years.

b. If you retire at 65, then 42 years or three doubling periods will have elapsed (3 · 14 = 42). So your $1000 investment (disregarding inflation) will have increased by a factor of 2^3 and be worth about $8000.

EXAMPLE 7

How long will it take for the population of Saudi Arabia to double?
The CIA Factbook lists the population of Saudi Arabia as 28.7 million in 2009 and growing at an estimated rate of 1.85% per year. If this rate continues, estimate when the population would double.

SOLUTION

Using the rule of 70, the approximate doubling time for a 1.85% annual growth rate would be 70/1.85 ≈ 38 years. If this rate of growth continues, the population of Saudi Arabia will be over 57 million people in 2047.

EXAMPLE 8

What is plutonium's annual decay rate?
Plutonium, the fuel for atomic weapons, has an extraordinarily long half-life, about 24,400 years. Once the radioactive element plutonium is created from uranium, 24,400 years later half the original amount still remains. Use the rule of 70 to estimate plutonium's annual decay rate.

SOLUTION

The rule of 70 says	70/R = 24,400
multiply by R	70 = 24,400R
divide and simplify	R = 70/24,400
	≈ 0.003

Remember that R is already in percentage, not decimal, form. So the annual decay rate is a tiny 0.003%, or three-thousandths of 1%.

Algebra Aerobics 5.6a

A calculator that can evaluate powers is recommended for Problem 2.

1. For each of the following exponential functions (with t in days) determine whether the function represents exponential growth or decay, then estimate the doubling time or half-life. For each function find the initial amount and the amount after one day, one month, and one year. (*Note:* To make calculations easier, use an approximation of 360 days in one year.)

   a. $f(t) = 300 \cdot 2^{t/30}$ c. $P = 32,000 \cdot 0.5^t$

   b. $g(t) = 32 \cdot 0.5^{t/2}$ d. $h(t) = 40,000 \cdot 2^{t/360}$

2. Estimate the doubling time using the rule of 70.

   a. $g(x) = 100(1.02)^x$, where x is in years.

   b. $M = 10,000(1.005)^t$, where t is in months.

   c. The annual growth rate is 8.1%.

   d. The annual growth factor is 1.065.

3. Use the rule of 70 to approximate the growth rate when the doubling time is:

   a. 10 years

   b. 5 minutes

   c. 25 seconds

4. Use the rule of 70 to estimate the time it will take an initial quantity to drop to half its value when:

   a. $h(x) = 10(0.95)^x$, where x is in months.

   b. $K = 1000(0.75)^t$, where t is in seconds.

   c. The annual decay rate is 35%.

E X A M P L E   9   **White blood cell counts**

On September 27 a patient was admitted to Brigham and Women's Hospital in Boston for a bone marrow transplant. The transplant was needed to cure myelodysplastic syndrome, in which the patient's own marrow fails to produce enough white blood cells to fight infection.

The patient's bone marrow was intentionally destroyed using chemotherapy and radiation, and on October 3 the donated marrow was injected. Each day, the hospital carefully monitored the patient's white blood cell count to detect when the new marrow became active. The patient's counts are plotted in Figure 5.15. Normal counts for a healthy individual are between 4000 and 10,000 cells per milliliter.

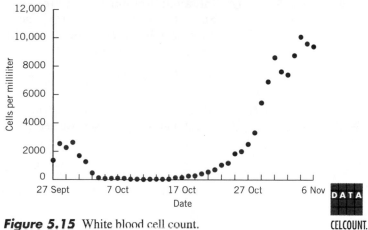

Figure 5.15 White blood cell count.

DATA

CELCOUNT.

What was the daily percent increase and estimated doubling time during the period of exponential growth?

S O L U T I O N   Clearly, the whole data set does not represent exponential growth, but we can reasonably model the data between, say, October 15 and 31 with an exponential function. Figure 5.16 shows a plot of this subset of the original data, along with a computer-generated best-fit exponential function, where t = number of days after October 15.

$$W(t) = 105(1.32)^t$$

The fit looks quite good.

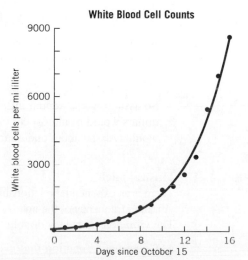

White Blood Cell Counts

Figure 5.16 White blood cell counts with a best-fit exponential function between October 15 and October 31.

The initial quantity of 105 is the white blood cell count (per milliliter) for the model, not the actual white blood cell count of 70 measured by the hospital. This discrepancy is not unusual; remember that a best-fit function may not necessarily pass through any of the specific data points.

The growth factor is 1.32, so the growth rate is 0.32. So between October 15 and 31, the number of white blood cells was increasing at a rate of 32% a day. According to the rule of 70, the number of white blood cells was doubling roughly every $70/32 = 2.2$ days!

EXAMPLE 10

Inflation: How much will your money be worth in 20 years?

Suppose you have a $100,000 nest egg hidden under your bed, awaiting your retirement. In 20 years, how much will your nest egg be worth in today's dollars if the annual inflation rate is:

a. 3%? **b.** 7%?

SOLUTION

a. Given a 3% inflation rate, what costs $1 today will cost $1.03 next year. Using a ratio to compare costs, we get

$$\frac{\text{cost now}}{\text{cost in 1 year}} = \frac{\$1.00}{\$1.03} \approx 0.971$$

Rewriting $\qquad \text{cost now} \approx 0.971 \cdot (\text{cost in 1 year})$

Our equation shows that next year's dollars need to be multiplied by 0.971 to reduce them to the equivalent value in today's dollars. An annual inflation rate of 3% means that in one year, $1 will be worth only $0.971 \cdot (\$1) = \$0.971$, or about 97 cents in today's dollars. Each additional year reduces the value of your nest egg by a factor of 0.971; in other words, 0.971 is the decay factor. So value of your nest egg in real dollars is given by

$$V_{3\%}(n) = \$100,000 \cdot 0.971^n$$

where $V_{3\%}(n)$ is the real value (measured in today's dollars) of your nest egg after n years with 3% annual inflation. When $n = 20$, we have

$$V_{3\%}(20) = \$100,000 \cdot 0.971^{20} \approx \$100,000 \cdot 0.555 = \$55,500$$

b. Similarly, since $\$1.00/\$1.07 \approx 0.935$, then

$$V_{7\%}(n) = \$100,000 \cdot 0.935^n$$

where $V_{7\%}(n)$ is the real value (measured in today's dollars) of your nest egg after n years with 7% annual inflation. When $n = 20$, we have

$$V_{7\%}(20) = \$100,00 \cdot 0.935^{20} \approx \$100,000 \cdot 0.261 = \$26,100$$

So assuming 3% annual inflation, after 20 years your nest egg of $100,000 in real dollars would be almost cut in half. With a 7% inflation your nest egg in real dollars would only be about one-quarter of the original amount.

EXAMPLE 11

Musical pitch

There is an exponential relationship between musical octaves and vibration frequency. The vibration frequency of the note A above middle C is 440 cycles per second (or 440 hertz). The vibration frequency doubles at each octave.

a. Construct a function that gives the vibration frequency as a function of octaves. Construct a corresponding table and graph.

b. Rewrite the function as a function of individual note frequencies. (*Note:* There are twelve notes to an octave.)

SOLUTION   **a.** Let F be the vibration frequency in hertz (Hz) and N the number of octaves above or below the chosen note A. Since the frequency doubles over each octave, the growth factor is 2. If we set the initial frequency at 440 Hz, then we have

$$F = 440 \cdot 2^N$$

Table 5.12 and the graph in Figure 5.17 show a few of the values for the vibration frequency.

Musical Octaves

Number of Octaves above or below A (at 440 Hz), N	Vibration Frequency (Hz), F
−3	55
−2	110
−1	220
0	440
1	880
2	1760
3	3520

Table 5.12

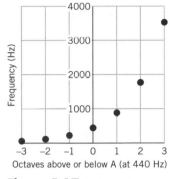

Figure 5.17 Octaves versus frequency.

"E6: Musical Keyboard Frequencies" in *Exponential & Log Functions* offers a multimedia demonstration of this function.

b. There are twelve notes in each octave. If we let n = number of notes above or below A, then we have $n = 12N$. Solving for N, we have $N = n/12$. By substituting this expression for N, we can define F as a function of n.

$$F = 440 \cdot 2^{n/12}$$
$$= 440 \cdot 2^{(1/12)\cdot n}$$
$$= 440 \cdot (2^{1/12})^n$$
$$\approx 440 \cdot (1.059)^n$$

So each note on the "even-tempered" scale has a frequency 1.059 times the frequency of the preceding note.

The Malthusian Dilemma

The most famous attempt to predict growth mathematically was made by a British economist and clergyman, Thomas Robert Malthus, in an essay published in 1798. He argued that the growth of the human population would overtake the growth of food supplies, because the population size was *multiplied* by a fixed amount each year, whereas food production only increased by *adding* a fixed amount each year. In other words, he assumed populations grew exponentially and food supplies grew linearly as functions of time. He concluded that humans are condemned always to breed to the point of misery and starvation, unless the population is reduced by other means, such as war or disease.

EXAMPLE 12   **Constructing the Malthusian equations**
Malthus believed that the population of Great Britain, then about 7,000,000, was growing by 2.8% per year. He counted food supply in units that he defined to be enough food for one person for a year. At that time the food supply was adequate, so he assumed that Britons were producing 7,000,000 food units. He predicted that food production would increase by about 280,000 units a year.

a. Construct functions modeling Britain's population size and the amount of food units over time.

b. Plot the functions from part (a) on the same grid. Estimate when the population would exceed the food supply.

SOLUTION

a. Since the population is assumed to increase by a constant *percent* each year, an exponential model is appropriate. Given an initial size of 7,000,000 and an annual growth rate of 2.8%, the exponential function

$$P(t) = 7{,}000{,}000 \cdot (1.028)^t$$

describes the population growth over time, where t is the number of years and $P(t)$ is the corresponding population size.

Since the food units are assumed to grow by a constant *amount* each year, a linear model is appropriate. Given an initial value of 7,000,000 food units and an annual constant rate of change of 280,000, the linear function

$$F(t) = 7{,}000{,}000 + 280{,}000t$$

describes the food unit increase over time, where t = number of years and $F(t)$ = number of food units in year t.

b. Figure 5.18 reveals that if the formulas were good models, then after about 25 years the population would start to exceed the food supply, and some people would starve.

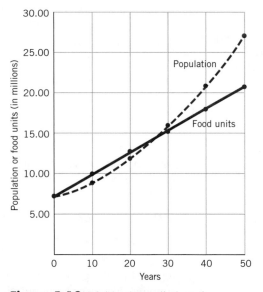

Figure 5.18 Malthus's predictions for population and food.

The two centuries since Malthus published his famous essay have not been kind to his theory. The population of Great Britain in 2002 was about 58 million, whereas Malthus's model predicted over 100 million people before the year 1900. Improved food production techniques and the opening of new lands to agriculture have kept food production in general growing faster than the population. The distribution of food is a problem and famines still occur with unfortunate regularity in parts of the world, but the mass starvation Malthus predicted has not come to pass.

Forming a Fractal Tree

Tree structures offer another useful way of visualizing exponential growth. A computer program can generate a tree by drawing two branches at the end of a trunk, then two smaller branches at the ends of each of those branches, and two smaller branches at the ends of the previous branches, and so on until a branch reaches twig size. This kind of structure produced from self-similar repeating scaled graphic operations is called a *fractal* (Figure 5.19). There are many examples of fractal structures in nature, such as ferns, coastlines, and human lungs.

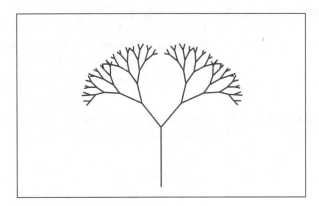

Figure 5.19 A fractal tree.

The fractal tree shown is drawn in successive levels (Figure 5.20):

- At level 0 the program draws the trunk, one line.
- At level 1 it draws two branches on the previous one, for a total of two new lines.
- At level 2 it draws two branches on each of the previous two, for a total of four new lines.
- At level 3 it draws two branches on each of the previous four, for a total of eight new lines.

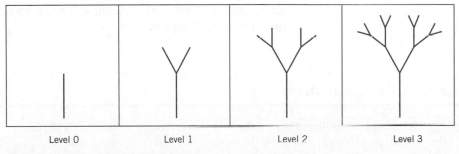

| Level 0 | Level 1 | Level 2 | Level 3 |

Figure 5.20 Forming a fractal tree.

Table 5.13 on the next page shows the relationship between the level, L, and the number of new lines, N, at each level. The formula

$$N = 2^L$$

describes the relationship between level, L, and the number of new lines, N. For example, at the fifth level, there would be $2^5 = 32$ new lines.

Level, L	New Lines, N	N as Power of 2
0	1	$1 = 2^0$
1	$2 \cdot 1 = 2$	$2 = 2^1$
2	$2 \cdot 2 = 4$	$4 = 2 \cdot 2 = 2^2$
3	$2 \cdot 4 = 8$	$8 = 2 \cdot 2 \cdot 2 = 2^3$
4	$2 \cdot 8 = 16$	$16 = 2 \cdot 2 \cdot 2 \cdot 2 = 2^4$

Table 5.13

EXAMPLE 13

Your family tree

You can think of Figure 5.20 on the previous page and Table 5.13 as depicting your family tree. Each level represents a generation. The trunk is you (at level 0). The first two branches are your parents (at level 1). The next four branches are your grandparents (at level 2), and so on. How many ancestors do you have ten generations back?

SOLUTION

The answer is $2^{10} = 1024$; that is, you have 1024 great-great-great-great-great-great-great-great-grandparents.

EXAMPLE 14

A phone tree

An information system that is similar to this process is an emergency phone tree, in which one person calls two others, each of whom calls two others, until everyone in the organization has been called. How many levels of phone calls would be needed to reach an organization with 8000 people?

SOLUTION

If we think of Figure 5.20 and Table 5.13 as representing this phone tree, then each new line (or branch) represents a person. We need to count not just the number of new people N at each level L, but also all the previous people called.

At level 0, there is the one person who originates the phone calls. At level 1, there is the original person plus the two he or she called, for a total of $1 + 2 = 3$ people. At level 2, there are $1 + 2 + 4 = 7$ people, etc. At level 11, there are $1 + 2 + 4 + 8 + 16 + 32 + 64 + 128 + 256 + 512 + 1024 + 2048 = 4095$ people who have been called. At level 12, there would be $2^{12} = 4096$ new people called, for a total of $4095 + 4096 = 8191$ people called. So it would take twelve levels of the phone tree to reach 8000 people.

Algebra Aerobics 5.6b

1. For each of the equations in parts (a) to (d), determine the initial investment, the growth factor, and the growth rate, and estimate the time it will take to double the investment when $t =$ number of years.

   a. $A = \$10,000 \cdot 1.065^t$ **c.** $A = \$300,000 \cdot 1.11^t$
   b. $A = \$25 \cdot 1.08^t$ **d.** $A = \$200 \cdot 1.092^t$

2. Fill in the table below.

Function	Initial Investment	Growth Factor	Growth Rate	Amount 1 Year Later	Doubling Time
	$50,000		7.2%		
	$100,000			$106,700	
			5.8%	$52,500	
	$3,000	1.13			

3. Identify the value of x that would make each of the following equations a true statement.

 a. $2^x = 32$ **e.** $2^x = 1$

 b. $2^x = 256$ **f.** $2^x = \frac{1}{2}$

 c. $2^x = 1024$ **g.** $2^x = \frac{1}{8}$

 d. $2^x = 2$ **h.** $2^x = \sqrt{2}$

4. Assume the tree-drawing process was changed to draw three branches at each level.

 a. Draw a trunk and at least two levels of the tree.

 b. What would the general formula be for N, the number of new lines, as a function of L, the level?

5. In an emergency phone tree in which one person calls three others, each of whom calls three others, and so on, until everyone in the organization has been called, how many levels of phone calls are required for this phone tree to reach an organization of 8000 people?

Exercises for Section 5.6

Technology for finding a best-fit function is required for Exercises 7, 9, 10, 17, 18, and 24. Internet access is required for Exercises 7 (c), 9 (c), 10 (c), 22 (b), 26 (e), and 27 (c). Graphing program is recommended for Exercise 12.

1. Which of the following functions have a fixed doubling time? A fixed half-life?

 a. $y = 6(2)^x$ **d.** $A = 10(2)^{t/5}$

 b. $y = 5 + 2x$ **e.** $P = 500 - \frac{1}{2}T$

 c. $Q = 300\left(\frac{1}{2}\right)^r$ **f.** $N = 50\left(\frac{1}{2}\right)^{t/20}$

2. Identify the doubling time or half-life of each of the following exponential functions. Assume t is measured in years. [*Hint:* What value of t would give you a growth (or decay) factor of 2 (or 1/2)?]

 a. $Q = 70(2)^t$ **d.** $Q = 100\left(\frac{1}{2}\right)^{t/250}$

 b. $Q = 1000(2)^{t/50}$ **e.** $N = 550(2)^{t/10}$

 c. $Q = 300\left(\frac{1}{2}\right)^t$ **f.** $N = 50\left(\frac{1}{2}\right)^{t/20}$

3. Fill in the following chart. (The first column is done for you.)

	a.	**b.**	**c.**	**d.**
Initial Value	50	1000	4	5000
Doubling time	30 days	7 years	25 minutes	18 months
Exponential function $(f(t) = Ca^{t/n})$	$f(t) = 50(2)^{t/30}$			
Growth factor per unit of time	$2^{1/30} = 1.0234$ per day			
Growth rate per unit of time (in percentage form)	2.34% per day			

4. Make a table of values for the dotted points on each graph. Using your table, determine if each graph has a fixed doubling time or a fixed half-life. If so, create a function formula for that graph.

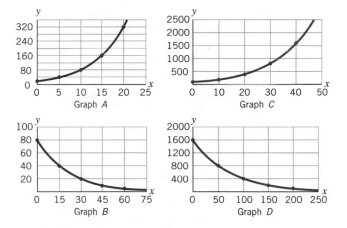

Graph A

Graph B

Graph C

Graph D

5. Insert the symbol $>$, $<$, or $\approx$ (approximately equal) to make the statement true. Assume $x > 0$.

 a. $3(2)^{x/5}$ _____ $3(1.225)^x$

 b. $50\left(\frac{1}{2}\right)^{x/20}$ _____ $50(0.9659)^x$

 c. $200(2)^{x/8}$ _____ $200(1.0905)^x$

 d. $750\left(\frac{1}{2}\right)^{x/165}$ _____ $750(0.911)^x$

6. Lead-206 is not radioactive, so it does not spontaneously decay into lighter elements. Radioactive elements heavier than lead undergo a series of decays, each time changing from a heavier element into a lighter or more stable one. Eventually, the element decays into lead-206 and the process stops. So, over billions of years, the amount of lead in the universe has increased because of the decay of numerous radioactive elements produced by supernova explosions.

 Radioactive uranium-238 decays sequentially into thirteen other lighter elements until it stabilizes at lead-206. The half-lives of the fifteen different elements in this decay chain vary from

6. (continued)

0.000 164 seconds (from polonium-214 to lead-210) all the way up to 4.47 billion years (from uranium-238 to thorium-234).

a. Find the decay rate per billion years for uranium-238 to decay into thorium-234.

b. Find the decay rate per second for polonium-214 to decay into lead-210.

7. (Requires data file and technology to find a best-fit function.) We have seen the accompanying table and graph of the U.S. population at the beginning of Chapter 2.

a. Find a best-fit exponential function. (You may want to set 1790 as year 0.) Be sure to clearly identify the variables and their units for time and population. What is the annual growth factor? The annual growth rate? The estimated initial population?

b. Graph your function and the actual U.S. population data on the same grid. Describe how the estimated population size differs from the actual population size. In what ways is this exponential function a good model for the data? In what ways is it flawed?

c. What would your model predict the population to be in the year 2010? In 2025? Check your results for the year 2010 on the Internet.

8. (Requires results from Exercise 7.) According to a letter published in the Ann Landers column in the *Boston Globe* on Friday, December 10, 1999, "When Elvis Presley died in 1977, there were 48 professional Elvis impersonators. In 1996, there were 7,328. If this rate of growth continues, by the year 2012, one person in every four will be an Elvis impersonator."

a. What was the growth factor in the number of Elvis impersonators for the 19 years between 1977 and 1996?

b. What would be the *annual* growth factor in the number of Elvis impersonators between 1977 and 1996?

c. Construct an exponential function that describes the growth in the number of Elvis impersonators since 1977.

d. Use your function to estimate the number of Elvis impersonators in 2012.

e. Use your model for the U.S. population from Exercise 7 to determine if in 2012 one person out of every four will be an Elvis impersonator. Explain your reasoning.

9. (Requires data file and technology to find a best-fit function and use of the Internet.) The International Telecommunications Union tracks market information about telecom services. The following table and graph use its data about the number of worldwide cell phone subscribers.

Population of the United States, 1790–2000

Year	Millions	Year	Millions
1790	3.9	1910	92.2
1800	5.3	1920	106.0
1810	7.2	1930	123.2
1820	9.6	1940	132.2
1830	12.9	1950	151.3
1840	17.1	1960	179.3
1850	23.2	1970	203.3
1860	31.4	1980	226.5
1870	39.8	1990	248.7
1880	50.2	2000	281.4
1890	63.0	2010	309.2 (est)
1900	76.2		

Source: U.S. Bureau of the Census, *www.census.gov.*

Worldwide Cell Phone Subscribers

Year	Number of Subscribers (in millions)
1997	215
1998	318
1999	490
2000	738
2001	961
2002	1157
2003	1417
2004	1763
2005	2219
2006	2759
2007	3305
2008	4100 (est)

Source: International Telecommunications Union, *www.itu.int.*

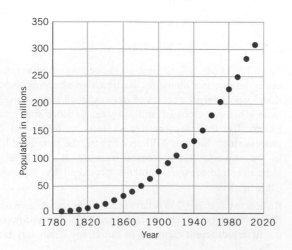

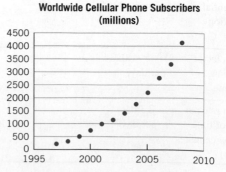

9. (continued)

   a. Use the data file to create a best-fit exponential function. (You may want to reinitialize the years.)

   b. What is the growth factor? The growth rate?

   c. Use your function to predict the number of worldwide cell phone subscribers in 2010. Check on the Internet to see if the numbers are accurate.

10. (Requires data file and technology to find a best-fit function and use of the Internet.) Medicare is a federal program that provides health care for nearly all people age 65 years and older. **MEDICARE**

Medicare Expenses

Year	Years since 1970	Medicare Expenses (billions of dollars)
1970	0	7.7
1975	5	16.3
1980	10	37.1
1985	15	71.5
1990	20	109.5
1995	25	184.4
2000	30	224.3
2005	35	339.8
2008	38	469.2

Source: www.census.gov.

   a. Use the data file for the following table to graph a best-fit function, with time, *t*, equal to years since 1970.

   b. Estimate the annual percentage increase in Medicare expenses.

   c. What does your model predict for Medicare expenses in the year 2010. Check your results on the Internet.

11. Tritium, the heaviest form of hydrogen, is a critical element in a hydrogen bomb. It decays exponentially with a half-life of about 12.3 years. Any nation wishing to maintain a viable hydrogen bomb has to replenish its tritium supply roughly every 3 years, so world tritium supplies are closely watched. Construct an exponential function that shows the remaining amount of tritium as a function of time as 100 grams of tritium decays (about the amount needed for an average size bomb). Be sure to identify the units for your variables.

12. (Graphing program recommended.) Cosmic ray bombardment of the atmosphere produces neutrons, which in turn react with nitrogen to produce radioactive carbon-14. Radioactive carbon-14 enters all living tissue through carbon dioxide (via plants). As long as a plant or animal is alive, carbon-14 is maintained in the organism at a constant level. Once the organism dies, however, carbon-14 decays exponentially into carbon-12. By comparing the amount of carbon-14 to the amount of carbon-12, one can determine approximately how long ago the organism died. (Willard Libby won a Nobel Prize for developing this technique for use in dating archaeological specimens.) The half-life of

carbon-14 is about 5730 years. In answering the following questions, assume that the initial quantity of carbon-14 is 500 milligrams.

   a. Construct an exponential function that describes the relationship between *A*, the amount of carbon-14 in milligrams, and *t*, the number of 5730-year time periods.

   b. Generate a table of values and plot the function. Choose a reasonable set of values for the domain. Remember that the objects we are dating may be up to 50,000 years old.

   c. From your graph or table, estimate how many milligrams are left after 15,000 years and after 45,000 years.

   d. Now construct an exponential function that describes the relationship between *A* and *T*, where *T* is measured in years. What is the annual decay factor? The annual decay rate?

   e. Use your function in part (d) to calculate the number of milligrams that would be left after 15,000 years and after 45,000 years.

13. The body eliminates drugs by metabolism and excretion. To predict how frequently a patient should receive a drug dosage, the physician must determine how long the drug will remain in the body. This is usually done by measuring the half-life of the drug, the time required for the total amount of drug in the body to diminish by one-half.

   a. Most drugs are considered eliminated from the body after five half-lives, because the amount remaining is probably too low to cause any beneficial or harmful effects. After five half-lives, what percentage of the original dose is left in the body?

   b. The accompanying graph shows a drug's concentration in the body over time, starting with 100 milligrams.

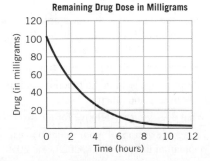

Remaining Drug Dose in Milligrams

Use the given graph to answer the following questions.

   i. Estimate the half-life of the drug.

   ii. Construct an equation that approximates the curve. Specify the units of your variables.

   iii. How long would it take for five half-lives to occur? Approximately how many milligrams of the original dose would be left then?

   iv. Write a 60 second summary describing your results to a prospective buyer of the drug.

14. Estimate the doubling time using the rule of 70 when:
    a. $P = 2.1(1.0475)^t$, where t is in years
    b. $Q = 2.1(1.00475)^T$, where T is in years

15. Use the rule of 70 to approximate the growth rate when the doubling time is:
    a. 5730 years
    b. 11,460 years
    c. 5 seconds
    d. 10 seconds

16. Estimate the time it will take an initial quantity to drop to half its value when:
    a. $P = 3.02(0.998)^t$, with t in years
    b. $Q = 12(0.75)^T$, with T in decades

17. (Requires technology to find a best-fit function.) Estimates for world population vary, but the data in the accompanying table are reasonable estimates.

World Population

Year	Total Population (millions)
1800	980
1850	1260
1900	1650
1950	2520
1970	3700
1980	4440
1990	5270
2000	6080
2005	6480
2010	6850 (est)

Source: United Nations Population Division, *www.undp.org/popin.*

a. Enter the data table into a graphing program (you may wish to enter 1800 as 0, 1850 as 50, etc.) or use the data file WORLDPOP in Excel or in graph link form.

b. Generate a best-fit exponential function.

c. Interpret each term in the function, and specify the domain and range of the function.

d. What does your model give for the growth rate?

e. Using the graph of your function, estimate the world population in 1790, 1920, 2025, and 2050.

f. Estimate the length of time your model predicts it takes for the population to double from 4 billion to 8 billion people.

18. (Requires data file and technology to find a best-fit function.) In 1911, reindeer were introduced to St. Paul Island, one of the Pribilof Islands, off the coast of Alaska in the Bering Sea. There was plenty of food and no hunting or reindeer predators. The size of the reindeer herd grew rapidly for a number of years, as given in the accompanying table.

Population of Reindeer Herd

Year	Population Size	Year	Population Size
1911	17	1925	246
1912	20	1926	254
1913	42	1927	254
1914	76	1928	314
1915	93	1929	339
1916	110	1930	415
1917	136	1931	466
1918	153	1932	525
1919	170	1933	670
1920	203	1934	831
1921	280	1935	1186
1922	229	1936	1415
1923	161	1937	1737
1924	212	1938	2034

Source: V.B. Scheffer, "The rise and fall of a reindeer herd," *Scientific Monthly,* 73:356–362, 1951.

a. Use the reindeer data file (in Excel or graph link form) to plot the data.

b. Find a best-fit exponential function.

c. How does the predicted population from part (b) differ from the observed ones?

d. Does your answer in part (c) give you any insights into why the model does not fit the observed data perfectly?

e. Estimate the doubling time of this population.

19. In medicine and biological research, radioactive substances are often used for treatment and tests. In the laboratories of a large East Coast university and medical center, any waste containing radioactive material with a half-life under 65 days must be stored for 10 half-lives before it can be disposed of with the non-radioactive trash.

a. By how much does this policy reduce the radioactivity of the waste?

b. Fill out the accompanying chart and develop a general formula for the amount of radioactive pollution at any period, given an initial amount, A_0.

Number of Half-Life Periods	Pollution Amount
0	A_0, original amount
1	$A_1 = 0.5A_0$
2	A_2
3	
4	
Period n	

20. It is now recognized that prolonged exposure to very loud noise can damage hearing. The accompanying table gives the permissible daily exposure hours to very loud noises as recommended by OSHA, the Occupational Safety and Health Administration.

20. (continued)

Sound Level, D (decibels)	Maximum Duration, H (hours)
120	0
115	0.25
110	0.5
105	1
100	2
95	4
90	8

a. Examine the data for patterns. How is D progressing? How is H progressing? Do the data represent a growth or a decay phenomenon? Explain your answer.

b. Find a formula for H as a function of D. Fit the data as closely as possible. Graph your formula and the data on the same grid.

21. It takes 3 months for a malignant lung tumor to double in size. At the time a lung tumor was detected in a patient, its mass was 10 grams.

a. If untreated, determine the size in grams of the tumor at each of the listed times in the table. Find a formula to express the tumor mass M (in grams) at any time t (in months).

t, Time (months)	M, Mass (g)
0	10
3	
6	
9	
12	

b. Lung cancer is fatal when a tumor reaches a mass of 2000 grams. If a patient diagnosed with lung cancer went untreated, estimate how long he or she would survive after the diagnosis.

c. By what percentage of its original size has the 10-gram tumor grown when it reaches 2000 grams?

22. According to the *Arkansas Democrat Gazette* (February 27, 1994):

Jonathan Holdeen thought up a way to end taxes forever. It was disarmingly simple. He would merely set aside some money in trust for the government and leave it there for 500 or 1000 years. Just a penny, Holdeen calculated, could grow to trillions of dollars in that time. But the stash he had in mind would grow much bigger—to quadrillions or quintillions—so big that the government, one day, could pay for all its operations simply from the income. Then taxes could be abolished. And everyone would be better off.

a. Holdeen died in 1967, leaving a trust of $2.8 million that is being managed by his daughter, Janet Adams. In 1994, the trust was worth $21.6 million. The trust was debated in Philadelphia Orphans' Court. Some lawyers who were trying to break the trust said that it is dangerous to let it go on, because "it would sponge up all the money in the world." Is this possible?

b. After 500 years, how much would the trust be worth? Would this be enough to pay off the current national debt (can be found on Internet at *www.brillig.com/debt_clock*)? What about after 1000 years? Describe the model you used to make your predictions.

23. Describe how a 6% inflation rate will erode the value of a dollar over time. Approximately how long would it take for a dollar to be worth only 50 cents? This is the half life of the dollar's buying power under 6% inflation.

24. (Graphing program required.) The average female adult *Ixodes scapularis* (a deer tick that can carry Lyme disease) lives only a year but can lay up to 10,000 eggs right before she dies. Assume that the ticks all live to adulthood, and half are females that reproduce at the same rate and half the eggs are male.

a. Describe a formula that will tell you how many female ticks there will be in n years, if you start with one impregnated tick.

b. How many male ticks will there be in n years? How many total ticks in n years?

c. If the surface of Earth is approximately $5.089 \cdot 10^{14}$ square meters and an adult tick takes up 0.5 square centimeters of land, approximately how long would it take before the total number of ticks would cover the surface of Earth?

25. MCI, a phone company that provides long-distance service, introduced a marketing strategy called "Friends and Family." Each person who signed up received a discounted calling rate to ten specified individuals. The catch was that the ten people also had to join the "Friends and Family" program.

a. Assume that one individual agrees to join the "Friends and Family" program and that this individual recruits ten new members, who in turn each recruit ten new members, and so on. Write a function to describe the number of new people who have signed up for "Friends and Family" at the nth round of recruiting.

b. Now write a function that would describe the total number of people (including the originator) signed up after n rounds of recruiting.

c. How many "Friends and Family" members, stemming from this one person, will there be after five rounds of recruiting? After ten rounds?

d. Write a 60-second summary of the pros and cons of this recruiting strategy. Why will this strategy eventually collapse?

26. In a chain letter one person writes a letter to a number of other people, N, who are each requested to send the letter to N other people, and so on. In a simple case with $N = 2$, let's assume person A1 starts the process.

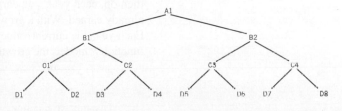

26. (continued)

A1 sends to B1 and B2; B1 sends to C1 and C2; B2 sends to C3 and C4; and so on. A typical letter has listed in order the chain of senders who sent the letters. So D7 receives a letter that has A1, B2, and C4 listed.

If these letters request money, they are illegal. A typical request looks like this:

- When you receive this letter, send $10 to the person on the top of the list.
- Copy this letter, but add your name to the bottom of the list and leave off the name at the top of the list.
- Send a copy to two friends within 3 days.

For this problem, assume that all of the above conditions hold.

a. Construct a mathematical model for the number of new people receiving letters at each level L, assuming $N = 2$ as shown in the above tree.

b. If the chain is not broken, how much money should an individual receive?

c. Suppose A1 sent out letters with two additional phony names on the list (say A1a and A1b) with P.O. box addresses she owns. So both B1 and B2 would receive a letter with the list A1, A1a, A1b. If the chain isn't broken, how much money would A1 receive?

d. If the chain continued as described in part (a), how many new people would receive letters at level 25?

e. *Internet search:* Chain letters are an example of a "pyramid growth" scheme. A similar business strategy is multilevel marketing. This marketing method uses the customers to sell the product by giving them a financial incentive to promote the product to potential customers or potential salespeople for the product. (See Exercise 31.) Sometimes the distinction between multilevel marketing and chain letters gets blurred. Search the U.S. Postal Service website (*www.usps.gov*) for "pyramid schemes" to find information about what is legal and what is not. Report what you find.

27. [Technology to find a best-fit function is recommended. Internet access required for part (c).] The following data show the total government debt for the United States from 1945 to 2010.

Year	Debt ($ billions)
1945	260
1950	257
1955	274
1960	291
1965	322
1970	381
1975	542
1980	909
1985	1818
1990	3207
1995	4921
2000	5674
2005	7932
2010	12,130 (est)

DATA

FEDDEBT

a. By hand or with technology, plot the data in the accompanying table and sketch a curve that approximates the data.

b. Construct an exponential function that models the data. What would your model predict for the current debt?

c. Use the "debt clock" at *www.brillig.com/debt_clock* to find the current debt. How accurate was your prediction in part (b)?

28. If you look back at your family tree forty generations ago (roughly 800 years), you had 2^{40} ancestors. This number is larger than all the people that ever lived on the surface of the earth. How can that be?

5.7 Compound Interest and the Number *e*

We have seen that exponential functions occur frequently in finance. One of the most common applications is compound interest. In Section 5.1, we learned that compound interest means that you earn interest not only on your investment but also on the interest you have already earned. Compounding is an example of exponential growth.

If you have $500 in a savings account that returns 1% interest *compounded annually,* at the end of the first year, you earn 1% interest on the initial $500, called the *principal*. From then on, each year, you earn 1% interest on your principal and on the interest you have already earned. With a growth rate of 1%, or 0.01, the growth factor $= 1 + 0.01 = 1.01$. Each year, the current value of your savings account is multiplied by 1.01. The following function P models the growth:

$$P = 500 \cdot (1.01)^t$$
$$= 500 \cdot (1 + 0.01)^t$$

$$\text{value of account} = (\text{original investment}) \cdot (1 + \text{interest rate})^{\text{no. of years}}$$

Compounding Interest Annually

If

$$P_0 = \text{original investment or principal}$$
$$r = \text{annual interest rate (in decimal form)}$$
$$t = \text{number of years}$$

the resulting value, P of the investment after t years is given by the formula

$$P = P_0 \cdot (1 + r)^t$$

EXAMPLE   1   **Comparing investments**
Suppose you have $10,000 that you could put in either a checking account that earns no interest, a mutual fund that earns 5% a year, or a risky stock investment that you hope will return 15% a year. Model the potential growth of each investment. Compare the investments after 10, 20, 30, and 40 years.

Assuming you make no withdrawals, the money in the checking account will remain constant at $10,000. For both the mutual fund and the stock investment, you are expecting a constant annual percent rate, so you can think of them as compounding annually. The mutual fund's predicted annual growth rate is 5%, so its growth factor would be 1.05. The stock's annual growth rate (you hope) will be 15%, so its growth factor would be 1.15. The initial value is $10,000, so after t years

Table 5.14 shows the value of each investment over time.

Number of Years	Checking Account (0% per year)	Mutual Fund (5% per year)	Stock (15% per year)
0	$10,000	$10,000	$10,000
10	$10,000	$16,289	$40,456
20	$10,000	$26,533	$163,665
30	$10,000	$43,219	$662,118
40	$10,000	$70,400	$2,678,635

Table 5.14

Forty years from now, your checking account will still have $10,000, but your mutual fund will be worth $70,400. If you were lucky in your stock investment and it gave a 15% interest rate compounded annually, your $10,000 would become well over $2.5 million in 40 years.

Next we investigate investments that receive interest more frequently than once a year.

Compounding at Different Intervals

Suppose we invest a principal of $100 in a bank account that pays interest of 6% per year. To compute the amount of money we have at the end of 1 year, we must also know how often the interest is credited (added) to our account, that is, how often it is *compounded*.

Compounding annually

If we invest $100 in a bank account that pays 6% interest per year, then at the end of 1 year we would have $100(1.06) = $106. When the interest is applied to your account once a year, we say that the interest is *compounded* annually.

Compounding twice a year

Now suppose that the interest is compounded twice a year. This means that instead of applying the annual rate of 6% once, it is divided by 2 and applied twice, at the end of each 6-month period. At the end of the first 6 months we earn 6%/2, or 3%, interest, so our balance is $100(1.03) = $103. At the end of the second 6 months we earn 3% interest on our new balance of $103. So, after 1 year we have

$$\begin{aligned} \$100 \text{ at } 6\% \text{ interest compounded twice a year} &= \$100 \cdot (1.03) \cdot (1.03) \\ &= \$100 \cdot (1.03)^2 \\ &= \$106.09 \end{aligned}$$

We earn 9 cents more when interest is credited twice per year than when it is credited once per year. The difference is a result of the interest earned during the second half-year on the $3 in interest credited at the end of the first half-year. In other words, we're starting to earn "interest on interest." To earn the same amount with only annual compounding, we would need an interest rate of 6.09%.

When 6% interest is compounded twice a year, then

> 6%   is the *nominal interest rate* (in name only)
>
>   or the annual percentage rate (APR)

and

> 6.09%   is the *effective interest rate*
>
>   or the annual percentage yield (APY)

which tells you how much interest you earn (or pay) after one year. Banks and credit card companies are required by law to list both the nominal (the APR) and effective (the APY) interest rates.

EXAMPLE 2

Compounding four times a year
Suppose that interest is compounded quarterly, or four times per year. What is the effective interest rate? What is an investment of $100 worth at the end of a year?

SOLUTION

In each quarter, we receive one-quarter of 6% or 6%/4 = 1.5% interest. Each quarter, our investment is multiplied by $1 + 0.015 = 1.015$ and, after the first quarter, we earn interest on the interest we have already received. At the end of 1 year we have received interest four times, so our initial $100 investment has become

$$\$100 \cdot (1.015)^4 \approx \$106.14$$

The nominal rate is still 6%, but the effective interest rate (or annual percentage yield) is now about 6.14%.

Compounding n times a year

We may imagine dividing the year into smaller and smaller time intervals and computing the interest earned at the end of 1 year. The nominal rate remains at 6%, but the effective interest rate is slightly more each time (Table. 5.15 on the next page).

Investing $100 for One Year at an Annual Interest Rate of 6%

Number of Times Interest Computed During the Year	Value of $100 at End of One Year ($)			Effective Annual Interest Rate (%)
1	$100(1 + 0.06) =$	$100(1.06) =$	106.00	6.00
2	$100(1 + 0.06/2)^2 =$	$100(1.03)^2 = 100(1.0609) = 106.09$		6.09
4	$100(1 + 0.06/4)^4 =$	$100(1.015)^4 \approx 100(1.0614) = 106.14$		6.14
6	$100(1 + 0.06/6)^6 =$	$100(1.010)^6 \approx 100(1.0615) = 106.15$		6.15
12	$100(1 + 0.06/12)^{12} =$	$100(1.005)^{12} \approx 100(1.0617) = 106.17$		6.17
24	$100(1 + 0.06/24)^{24} =$	$100(1.025)^{24} \approx 100(1.0618) = 106.18$		6.18
.				
.				
.				
n	$100(1 + 0.06/n)^n$			

Table 5.15

In general, if we calculate the interest on $100 n times a year for one year when the annual interest rate is 6%, we get

$$\$100 \left(1 + \frac{0.06}{n}\right)^n$$

At the end of t years

What if we invest $100 for t years at an annual interest rate of 6% compounded n times a year? The annual growth factor is $(1 + 0.06/n)^n$, that is, every year the $100 is multiplied by $(1 + 0.06/n)^n$. After t years the $100 is multiplied by $(1 + 0.06/n)^n$ a total of t times or, equivalently, multiplied by $[(1 + 0.06/n)^n]^t = (1 + 0.06/n)^{nt}$. So $100 will be worth

$$\$100(1 + 0.06/n)^{nt}$$

We can generalize our results for any annual rate, r.

> **Compounding n Times a Year for t Years**
>
> The value of P_0 dollars (called the principal) invested at an annual interest rate r (expressed in decimal form) compounded n times a year for t years is
>
> $$P = P_0 \left(1 + \frac{r}{n}\right)^{nt}$$
>
> This formula also works for compounding interest on loans.
>
> Note that r is the nominal rate; for example, $r = 0.05$ when the annual rate is 5%.

Continuous Compounding Using e

Imagine increasing the number of periods, n, without limits, so that interest is computed every week, every day, every hour, every second, and so on.

The general formula for compounded saving (or debt)

$$P = P_0 (1 + r/n)^{nt}$$

works well for a known number of compounding periods, but is difficult to evaluate for continuous compounding when n becomes extremely large. To deal with this problem we convert the formula to one that is easier to use. This involves using the known fact that as m approaches infinity then $(1 + 1/m)^m$ approaches a value of 2.71828. (See Algebra Aerobic 5.7, Problem 5.)

This constant number ≈ 2.71828 is called e, named after the Swiss mathematician Euler. It is an irrational number like π, so cannot be written as a ratio or repeating decimal. Scientific calculators have convenient e keys.

The irrational number e is a fundamental mathematical constant.

$$e \approx 2.71828$$

EXPLORE & EXTEND

5.7 Exploring with e

(A graphing program and a calculator that can evaluate powers are required.)

Following the famous mathematicians, J. Bernoulli and L. Euler, you can explore what happens to the expression $(1 + 1/n)^n$ as $n \rightarrow +\infty$, and then apply your insights to continuous compounding.

Part I Getting to e

a. Predict what will happen to each function in the table below as n becomes larger and larger. Then fill in the table to confirm or adjust your predictions.

n	$y = 1/n$	$y = 1 + 1/n$	$y = (1 + 1/n)^n$
1			
10			
100			
1000			

b. What is happening to the functions $y = 1 + 1/n$ and $y = (1 + 1/n)^n$ as n gets larger and larger? What would you expect would happen to an expression raised to the n power? The function $y = (1 + 1/n)^n$ does not behave in the same way as an exponential growth function, that is, it does not increase indefinitely. Why not? How is the base $(1 + 1/n)$ different from the base of an exponential growth function?

c. Graph the three functions in the table on the same grid, adjusting the domain as needed. Describe the long-term behavior of each function. Use e to express the limiting value for $(1 + 1/n)^n$.

Part II What does e have to do with continuous compounding?

a. When interest is compounded continuously, then the number of time periods n increases without limit. What happens to $(1 + r/n)^n$ in our compounding formula as $n \rightarrow +\infty$? Examine what happens to specific interest rates, r, in the table below and compare the limiting value of each expression to $e^{0.06}$ and $e^{0.25}$, respectively.

n	$(1 + 0.06/n)^n$	$(1 + 0.25/n)^n$
1		
10		
100		
1000		

b. Summarize your results in part (a) for any r, where $0 < r < 1$ by these steps:
   i. Substitute n_1/r for n in the expression $(1 + 1/n)^n$. Given your results from Part I, what is the limiting value of this new expression?
   ii. Raise the new expression to the r power.

Continuous Compounding Formula

What happens when interest is compounded continuously, that is what happens to the expression $(1 + r/n)^n$, when n becomes larger and larger? In Explore & Extend 5.7, you can explore what happens as n becomes larger and larger, and will find that is as $n \rightarrow +\infty$, then $(1 + r/n)^n \rightarrow e^r$. For compounding interest continuously, the following formula can be used:

Compounding Continuously for *t* Years

The value P of P_0 dollars invested at an annual interest rate r (expressed in decimal form) *compounded continuously* for t years is

$$P = P_0 \cdot e^{rt}$$

Again r is the nominal rate; for example, $r = 0.08$ when the annual rate is 8%.

EXAMPLE   3   **Different compounding intervals**
If you have $250 to invest and you are quoted a nominal interest rate of 4%, construct the equations that tell you how much money you will have if the interest is compounded once a year, quarterly, once a month, or continuously. In each case, calculate the value after 10 years.

SOLUTION   Table 5.16 shows the values of investing $250 at a nominal interest rate of 4% for different compounding intervals.

Number of Compoundings per Year	$ Value after t Years	Approximate $ Value When $t = 10$ Years
1	$250 \cdot (1 + 0.04)^t = 250 \cdot (1.04)^t$	370.06
4	$250 \cdot (1 + 0.04/4)^{4t} = 250 \cdot (1.01)^{4t}$	372.22
12	$250 \cdot (1 + 0.04/12)^{12t} \approx 250 \cdot (1.0033)^{12t}$	372.71
Continuous	$250 \cdot e^{0.04t} \approx 250 \cdot (1.0408)^t$	372.96

Table 5.16

EXAMPLE   4   **Continuously compounding debt**
Suppose you have a debt on which the nominal annual interest rate (APR) is 7% compounded continuously. What is the effective interest rate (APY)?

SOLUTION   The nominal interest rate (APR) of 7% is compounded continuously, so the equation $P = P_0 e^{0.07t}$ describes the amount P that an initial debt P_0 becomes after t years. Using a calculator, we find that $e^{0.07} \approx 1.073$. The equation could be rewritten as $P = P_0 (1.073)^t$. So the effective interest rate (APY) on your debt is about 7.3% per year.

EXAMPLE   5   **What is the better interest rate?**
You have a choice between two bank accounts. One is a passbook account in which you receive interest of 5% per year, compounded once per year. The other is a 1-year certificate of deposit (CD), which pays interest at the rate of 4.9% per year, compounded continuously. Which account is the better deal?

SOLUTION   Since the interest on the passbook account is compounded once a year, the nominal and effective interest rates are both 5%. The equation $P = P_0 (1.05)^t$ can be used to describe the amount P that the initial investment P_0 is worth after t years.

The 1-year certificate of deposit has a nominal interest rate of 4.9%. Since this rate is compounded continuously, the equation $P = P_0\, e^{0.49t}$ describes the amount P that the initial investment P_0 is worth after t years. Assuming $e^{0.049} \approx 1.0502$, the equation can also be written as $P = P_0\, (1.0502)^t$, and the effective interest rate is 5.02%. So the CD is a better deal.

Exponential Functions Base e

Continuous compounding is useful in areas other than financial contexts. Any quantity that can be viewed as continuously growing (or decaying), such as *E. coli* growth, can be converted from the form $f(t) = Ca^t$ into a continuous growth (or decay) function using a power of e as the base. Assuming $a > 0$, we can always find a value for k such that

$$a = e^k$$

So we can rewrite the function f as

$$f(t) = C(e^k)^t$$
$$= Ce^{kt}$$

In general applications, we call k the *instantaneous* or *continuous growth* (or *decay*) *rate*. The value of k may be given as either a decimal or a percent.

If $k > 0$, the function represents exponential growth

Why is this true? If an exponential function represents growth, then the growth factor $a > 1$. If we rewrite a as e^k and 1 as e^0, then $e^k > e^0$, so $k > 0$.

For example, the equation $P(t) = 100\, e^{0.06t}$ could describe the growth of 100 cells with a continuous growth rate of 0.06, or 6%, per time period t.

If $k < 0$, the function represents exponential decay

For exponential decay, the decay factor a is such that $0 < a < 1$. Rewriting a as e^k and 1 as e^0, we have $0 < e^k < e^0$. We know the value of e^k is always >0 since e is a positive number. But if $e^k < e^0$, then $k < 0$.

For example, the function $Q(t) = 50\, e^{-0.03t}$ could describe the decay of 50 cells with a continuous decay rate of 0.03 or 3% per time period t.

The program "E4: $y = Ce^{nx}$ sliders" can be used to explore the difference between continuous growth and decay.

Continuous Growth and Decay

If $Q(t) = Ce^{kt}$, then k is called the *instantaneous* or *continuous growth* (or *decay*) *rate*.

For exponential growth, k is positive.

For exponential decay, k is negative.

EXAMPLE 6   Continuous growth or decay rates
Identify the continuous growth (or decay) rate for each of the following functions and graph each function.

$$f(t) = 100 \cdot e^{0.055t}$$
$$g(t) = 100 \cdot e^{0.02t}$$
$$h(t) = 100 \cdot e^{-0.055t}$$
$$j(t) = 100 \cdot e^{-0.02t}$$

SOLUTION   The function f has a continuous growth rate of 0.055, or 5.5%, and g has a continuous growth rate of 0.02, or 2%.

The function h has a continuous decay rate of 0.055, or 5.5%, and j has a continuous decay rate of 0.02, or 2%.

The graphs of these four functions are shown in Figure 5.21.

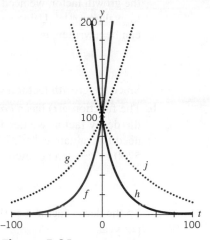

Figure 5.21   Graphs of four exponential functions.

E X A M P L E   7

Using continuous growth models

A website design company guarantees that the custom sites it creates will increase your company's sales at a continuous growth rate of 10% per month for the first year. If you currently have monthly sales of $2000, what can you expect your monthly sales, S, to be at the end of the first year?

S O L U T I O N

Since sales, S, are to grow continuously, we use the formula $S = Ce^{kt}$, where C is our initial monthly sales, k is our monthly continuous growth rate, and t is the number of months. After 12 months, sales are:

$$S = 2000e^{0.10(12)} \approx \$6640.23$$

Converting e^k into a

Using the rules of exponents, we can rewrite e^{kt} as $(e^k)^t$. When we know the value of k, we can calculate the value of e^k. For example,

$$P = 100 \cdot e^{0.06t}$$

Rule 3 of exponents $= 100 \cdot (e^{0.06})^t$

use a calculator to evaluate $e^{0.06}$ $\approx 100 \cdot 1.0618^t$

> See the program "E5: Comparing $y = Ce^x$ and $y = Ce^{rx}$."

The two functions

$$P = 100 \cdot e^{0.06t} \quad \text{and} \quad P = 100 \cdot 1.0618^t \text{ (rounded off)}$$

can be treated as equivalent. The first function (base e) suggests growth that occurs *continuously* throughout a time period, so we call 0.06, or 6%, the *continuous* growth rate per time period t. The other function suggests growth that happens all at once at the end of each time period, so 0.0618, or 6.18%, is just called the growth rate per time period t.

To do the reverse—that is, convert any base a into e^k—we need to know about logarithms base e, explained in the next chapter.

E X A M P L E   8

Growth rates and growth factors

For each of the following functions, identify the *continuous* growth (or decay) rate per year and the growth (or decay) rate based on the growth (or decay) factor. Assume t is measured in years.

a. $f(t) = 240 \, e^{0.127t}$

b. $g(t) = 5700 \, e^{-0.425t}$

SOLUTION

a. The function $f(t)$ has a *continuous* growth rate of 0.127, or 12.7%, per year. To find the growth factor, we need to convert $f(t)$ into the form $f(t) = Ca^t$. To do this we need to evaluate $e^{0.127}$. Using a calculator, we find that $e^{0.127} \approx 1.135$. So $f(t) = 240\, e^{0.127t}$ can be rewritten as

$$f(t) \approx 240 \cdot 1.135^t$$

Since the growth factor $a = 1.135$, the growth rate is 0.135, or 13.5%, per year.

b. The function $g(t)$ has a *continuous* decay rate of 0.425, or 42.5%, per year. To find the decay factor, we need to convert $g(t)$ into the form $g(t) = Ca^t$. To do this, we need to evaluate $e^{-0.425}$. Using a calculator, we find that $e^{-0.425} \approx 0.654$. So $g(t) = 5700\, e^{-0.425t}$ can be rewritten as

$$g(t) \approx 5700 \cdot 0.654^t$$

So the decay factor $a = 0.654$ and the decay rate is $1 - 0.654 = 0.346$, or 34.6%, per year.

EXAMPLE 9

Sales of bottled water over 3 decades

a. In 1976 approximately 0.28 billion gallons of bottled water were sold in the United States, according to Beverage Marketing Corp., a New York research and consulting firm. Between 1976 and 2007, the bottled water industry in the United States had a continuous growth rate of about 11.1% a year. Construct a model that represents the continuous growth of bottled water sales between 1976 and 2007.

b. Beverage Marketing later reported that in 2008 sales of bottled water peaked at approximately 8.7 billion gallons. If we extrapolate, what would the model predict for sales in 2008? How does this compare with the actual sales? Why might sales be different than predicted?

SOLUTION

a. To represent continuous growth, we construct an exponential function using base e and a continuous growth rate of 11.1% a year. If we have an initial value of 0.28 billion gallons and $t =$ number of years since 1976, then

$$f(t) = 0.28 \cdot e^{0.111t} \quad (0 \leq t \leq 31)$$

models the continuous growth of bottled water sales in the United States between 1976 and 2007.

b. If we extrapolate our model to predict sales for the year 2008, then $t = 2008 - 1976 = 32$ years and

$$f(32) = 0.28 \cdot e^{(0.111 \cdot 32)}$$
$$\approx 0.28 \cdot 34.883$$
$$\approx 9.77 \text{ billion gallons}$$

Our model's prediction of nearly 9.77 billion gallons in 2008 is somewhat over actual sales of 8.7 billion gallons. This discrepancy is a result of the first decline in bottled water sales this decade. An August 2009 *Wall Street Journal* article attributes this drop to the growth in environmental awareness and budget consciousness. We may find that in time these factors cause bottled water sales to no longer fit an exponential model at all.

You can check the current sales of bottled water at *www.beveragemarketing.com.*

EXAMPLE 10

Elimination of Tylenol

A standard dose (two tablets) of Tylenol contains 650 mg of acetaminophen. According to data at *tylenolprofessional.com,* after the medication has been fully metabolized, it is eliminated continuously from the body at a rate of approximately 23.1% per hour.

a. Create an exponential equation displaying the amount of acetaminophen (after fully metabolized) in the system after t hours.

b. How much acetaminophen is still in the system after 4 hours? After 12 hours? After 24 hours?

SOLUTION   **a.** To represent continuous decay, we construct an exponential function using base e and a continuous decay rate of 23.1% an hour. If we have an initial dose of 650 mg, A = amount of acetaminophen in the body, and t – number of hours after it has been fully metabolized, then

$$A = 650e^{-0.231t}$$

models the continuous decay of a dose of acetaminophen.

b. Substituting in values for t, we find:

$$t = 4 \text{ hours} \qquad A = 650e^{-0.231(4)} \approx 258 \text{ mg}$$
$$t = 12 \text{ hours} \qquad A = 650e^{-0.231(12)} \approx 40.65 \text{ mg}$$
$$t = 24 \text{ hours} \qquad A = 650e^{-0.231(24)} \approx 2.54 \text{ mg}$$

EXPLORE & EXTEND

5.7b   Repeated Doses of Tylenol

As we saw in Example 10, the amount of acetaminophen in the bloodstream can be viewed as continuously decaying by 23.1% per hour. A 2003 report to the American Association of Poison Control Centers stated that acetaminophen accounted for more overdoses and overdose deaths each year in the United States than any other pharmaceutical agent. In the following problem, you examine how dosages accumulate in the body. A typical dosage for arthritis sufferers recommends repeated dosages as follows:

• Take one dose (2 tablets) every 4 to 6 hours while symptoms last.

• Do not take more than 12 tablets in 24 hours.

Assume that a patient takes the maximum recommended dosage: one dose (of two tablets) every 4 hours throughout a 24-hour period, for a total of twelve 325-mg tablets.

a. Fill in the following table, displaying how acetaminophen accumulates in the body:

Time (hours)	Amount Remaining from Previous Doses (in mg)	New Doses Added (in mg)	Total Amount in Bloodstream (in mg)
0	n.a.	650	
4		650	
8		650	
12		650	
16		650	
20		650	
24		650	

Note: The model created for acetaminophen is accurate only for a limited amount of time, but it gives a fairly good approximation.

b. How many doses of acetaminophen are in the bloodstream at the end of the 24-hour period?

c. Patients are to take no more than 4000 mg of acetaminophen per day. How much acetaminophen did the above patient take?

d. What will happen to the total amount in the bloodstream if this patient continues this routine?

Algebra Aerobics 5.7

Most of these problems require a calculator that can evaluate powers of e.

1. Find the amount accumulated after 1 year on an investment of $1000 at 8.5% compounded:

   a. Annually   **b.** Quarterly   **c.** Continuously

2. Find the effective interest rate for each given nominal interest rate that is compounded continuously.

   a. 4%   **b.** 12.5%   **c.** 18%

3. Assume that each of the following describes the value of an investment, A, over t years. Identify the principal, nominal rate, effective rate, and number of interest periods per year.

   a. $A = 6000 \cdot 1.05^t$ **d.** $A = 50,000 \cdot 1.025^{2t}$

   b. $A = 10,000 \cdot 1.02^{4t}$ **e.** $A = 125e^{0.076t}$

   c. $A = 500 \cdot 1.01^{12t}$

4. Fill in the missing values, translating from e^k to a (the growth or decay factor).

   a. $5e^{0.03t} = 5(\underline{\quad})^t$

   b. $3500e^{0.25t} = 3500(\underline{\quad})^t$

   c. $660e^{1.75t} = 660(\underline{\quad})^t$

   d. $55,000e^{-0.07t} = 55,000(\underline{\quad})^t$

   e. $125,000e^{-0.28t} = 125,000(\underline{\quad})^t$

5. The value for e is often defined as the number that $(1 + 1/n)^n$ approaches as n gets arbitrarily large. Use your calculator to complete the table at the bottom of the page. Use your exponent key (x^y or y^x) to evaluate the last column. Is your value consistent with the approximate value for e of 2.71828 given in the text?

6. If a principal of $10,000 is invested at the rate of 12% compounded quarterly, the amount accumulated at the end of t years is given by the formula

$$A = 10,000 \cdot \left(1 + \frac{0.12}{4}\right)^{4t} = 10,000(1.03)^{4t}$$

The graph of this function is given at the top of the next column. Use the graph to estimate for parts (a)–(c) the amount, A, accumulated after:

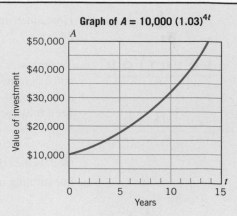

Graph of $A = 10,000\,(1.03)^{4t}$

a. 1 year

b. 5 years

c. 10 years

d. Use the graph to estimate the number of years it will take to double the original investment.

e. Use the equation to calculate the amount A after the years specified in parts (a)–(c). Use the rule of 70 to find the doubling time for A.

7. Construct a function that represents the resulting value if you invested $1000 for n years at an annually compounded interest rate of:

   a. 4%   **b.** 11%   **c.** 110%

8. At birth, Maria's parents set aside $8000 in an account designated to help pay for her college education. How much will Maria's account be worth by her 18th birthday if the interest rate was:

   a. 8% compounded quarterly?

   b. 8% compounded continuously?

   c. 8.4% compounded annually?

9. For each of the following functions, identify the continuous growth rate and then determine the effective growth rate. Assume t is measured in years.

   a. $A(t) = Pe^{0.6t}$ **b.** $N(t) = N_0 e^{2.3t}$

10. For each of the following functions, identify the continuous decay rate and then determine the effective decay rate. Assume t is measured in years.

   a. $Q(t) = Q_0 e^{-0.055t}$ **b.** $P(t) = P_0 e^{-0.15t}$

n	$1/n$	$1 + (1/n)$	$[1 + (1/n)]^n$
1	1	$1 + 1 = 2$	$2^1 = 2$
100	0.01	$1 + 0.01 = 1.01$	$(1.01)^{100} \approx 2.7048138$
1,000			
1,000,000			
1,000,000,000			

Exercises for Section 5.7

Many of these exercises require a calculator that can evaluate powers (including powers of base e) and logs. A graphing program is required for Exercise 18 and recommended for Exercise 3.

1. Construct a function that would represent the resulting value:

a. If you invested $5000 for n years at an annually compounded interest rate of:

  i 3.5%   **ii** 6.75%   **iii** 12.5%

b. If you make three different $5000 investments today at the three different interest rates listed in part (a), how much will each investment be worth in 40 years?

2. A bank compounds interest annually at 4%

a. Write an equation for the value V of $100 in t years.

b. Write an equation for the value V of $1000 in t years.

c. After 20 years will the total interest earned on $1000 be ten times the total interest earned on $100? Why or why not?

3. (Graphing program recommended.) You have a chance to invest money in a risky investment at 6% interest compounded annually. Or you can invest your money in a safe investment at 3% interest compounded annually.

a. Write an equation that describes the value of your investment after n years if you invest $100 at 6% compounded annually. Plot the function. Estimate how long it would take to double your money.

b. Write an equation that describes the value of your investment after n years if you invest $200 at 3% compounded annually. Plot the function on the same grid as in part (a). Estimate the time needed to double your investment.

c. Looking at your graph, indicate whether the amount in the first investment in part (a) will ever exceed the amount in the second account in part (b). If so, approximately when?

4. Examine the formulas below for a $3000 investment. Determine the number of interest periods, the nominal rate and the effective rate.

a. $A = 3000 \left(1 + \dfrac{0.0375}{4}\right)^{4t}$ **c.** $A = 3000(1 + 0.11)^t$

b. $A = 3000 \left(1 + \dfrac{0.082}{12}\right)^{12t}$ **d.** $A = 3000e^{0.0612t}$

5. You are looking for a safe place to put $20,000.00 for one year. You go online and find that three banks are offering the following rates for CDs. Find the effective rate for each to determine which would earn you the most interest at the end of one year.

  Bank A   2.46% continuously
  Bank B   2.48% quarterly
  Bank C   2.47% monthly

6. In 2010 the Federal Reserve was keeping interest rates low to stimulate the economy. As a result banks were posting very low interest rates on certificate of deposits (CD).

a. By law banks must state both the nominal rate and the effective (APY) rates. On August 3, 2010 the following banks listed their CD rates as follows:

Bank	Nominal Rate	Effective Rate
PNC Bank	0.60% compounded monthly	APY 0.60%
Sovereign Bank	0.60% compounded daily	APY 0.60%

Is the APY correct? Calculate the APY rates for each bank and explain why they were giving the same nominal and APY rates, since APY rates are generally higher.

b. What would be the APY for a bank whose rate was 0.60% compounded continuously?

c. Go online or check at your bank to find the current CD rates for banks in your area. Check the APY values to make sure the bank is posting it correctly.

7. Use a calculator to find the value of each expression rounded to four decimal places.

a. $\left(1 + \dfrac{0.035}{2}\right)^2$ **d.** $e^{0.035}$

b. $\left(1 + \dfrac{0.035}{4}\right)^4$ **e.** How are these values related?

c. $\left(1 + \dfrac{0.035}{12}\right)^{12}$

8. Assume $y = Pe^{rx}$ represents P dollars invested at an annual interest rate r (in decimal form) compounded continuously for x years. Then for each of the following, calculate the nominal and effective rates in percentage form.

Nominal Rate (APR)	Effective Rate (APY) (*Hint:* Evaluate at $e^k = a$)
a. $y = Pe^{0.025x}$	
b. $y = Pe^{0.039x}$	
c. $y = Pe^{0.062x}$	

9. Assume $10,000 is invested at a nominal interest rate of 8.5%. Write the equations that give the value of the money after n years and determine the effective interest rate if the interest is compounded:

a. Annually   **c.** Quarterly

b. Semiannually   **d.** Continuously

10. Assume you invest $2000 at 3.5% compounded continuously.

a. Construct an equation that describes the value of your investment at year t.

b. How much will $2000 be worth after 1 year? 5 years? 10 years?

11. Construct functions for parts (a) and (b) and compare them in parts (c) and (d).

a. $25,000 is invested at 5.75% compounded quarterly.

b. $25,000 is invested at 5.75% compounded continuously.

c. What is the amount in each account at the end of 5 years?

d. Explain, using the concept of effective rates, why one amount is larger than the other.

12. Fill in the following chart assuming that the principal is $10,000 in each case.

Nominal Interest Rate (APR)	Compounding Period	Expression for the Value of Your Account after t Years	Effective Interest Rate (APY)
5.25%	Monthly		
		$10,000(1 + 0.045/4)^{4t}$	
	Semiannually		8.16%
3.25%	Daily		
		$10,000(1.02)^t$	

13. Insert the symbol $>$, $<$, or $\approx$ to make the statement true.

a. $e^{0.045}$ ___ 1.046

b. 1.068 ___ $e^{0.068}$

c. 1.269 ___ $e^{0.238}$

d. $e^{-0.10}$ ___ 0.90

e. 0.8607 ___ $e^{-0.15}$

14. Assume that $5000 was put in each of two accounts. Account A gives 4% interest compounded semiannually. Account B gives 4% compounded continuously.

a. What are the total amounts in each of the accounts after 10 years?

b. Show that account B gives 0.04% more interest annually than account A.

15. According to Rubin and Farber's *Pathology*, "death from cancer of the lung, more than 85% of which is attributed to cigarette smoking, is today the single most common cancer death in both men and women in the United States." The accompanying graph shows the annual death rate (per thousand) from lung cancer for smokers and nonsmokers.

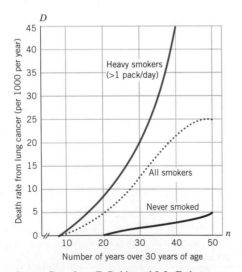

Source: Data from E. Rubin and J. L. Farber, *Pathology,* 3rd ed. (Philadelphia: Lippincott-Raven, 1998), p. 312. Copyright © 1998 by Lippincott-Raven. Reprinted by permission.

a. The death rate for nonsmokers is roughly a linear function of age. Construct a linear model for the graph of nonsmokers. Interpret your results.

b. By contrast, those who smoke more than one pack per day show an exponential rise in the death rate from lung cancer. An exponential function to model heavy smokers is $D = 1.843 \cdot e^{0.081n}$, where D is the death rate from lung cancer

(per 1,000 per year) and n = number of years over 30 years of age. Interpret the numbers in the regression equation.

c. Using the given graph, estimate a doubling time for the death rate from lung cancer and then convert to a^n where n = number of years over 30 years. How does this growth factor compare to $e^{0.081}$ in the regression equation? What is the relationship between these two numbers?

16. You want to invest money for your newborn child so that she will have $50,000 for college on her 18th birthday. Determine how much you should invest if the best annual rate that you can get on a secure investment is:

a. 6.5% compounded annually

b. 9% compounded quarterly

c. 7.9% compounded continuously

17. A city of population 1.5 million is expected to experience a 15% decrease in population every 10 years.

a. What is the 10-year decay factor? What is the *yearly* decay factor? The *yearly* decay rate?

b. Use part (a) to create an exponential population model $g(t)$ that gives the population (in millions) after t years.

c. Create an exponential population model $h(t)$ that gives the population (in millions) after t years, assuming a 1.625% continuous yearly decrease.

d. Compare the populations predicted by the two functions after 20 years. What can you conclude?

18. (Requires a graphing program.) Using technology, graph the functions $f(x) = 15,000e^{0.085x}$ and $g(x) = 100,000$ on the same grid.

a. Estimate the point of intersection. (*Hint:* Let x go from 0 to 60.)

b. If $f(x)$ represents the amount of money accumulated by investing at a continuously compounded rate (where x is in years), explain what the point of intersection represents.

19. Rewrite each continuous growth function in its equivalent form $f(t) = Ca^t$. In each case identify the continuous growth rate, and the effective growth rate. (Assume that t is in years.)

a. $P(t) = 500e^{0.02t}$

b. $N(t) = 3000e^{1.5t}$

c. $Q(t) = 45e^{0.06t}$

d. $G(t) = 750e^{0.035t}$

20. Rewrite each continuous decay function in its equivalent form $f(t) = Ca^t$. In each case identify the continuous decay rate and the effective decay rate. (Assume that t is in years.)

a. $P(t) = 600e^{-0.02t}$

b. $N(t) = 30,000e^{-0.5t}$

c. $Q(t) = 7145e^{-0.06t}$

d. $G(t) = 750e^{-0.035t}$

21. Match the function with its graph.

a. $f(x) = 10e^{0.025x}$ b. $g(x) = 10e^{0.045x}$ c. $h(x) = 10e^{0.075x}$

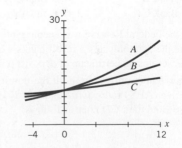

5.8 *Semi-Log Plots of Exponential Functions*

With exponential growth functions, we often face the same problem that we did in Chapter 4 when we tried to compare the size of an atom with the size of the solar system. The numbers go from very small to very large. In our *E. coli* model, for example, the number of bacteria started at 100 and grew to over 1 billion in twenty-four time periods (see Table 5.1). It is virtually impossible to display the entire data set on a standard graph. Whenever we need to graph numbers of widely varying sizes, we turn to a logarithmic scale.

Previously, using standard linear scales on both axes, we could graph only a subset of the *E. coli* data in order to create a useful graph (see Figure 5.1). However, if we convert the vertical axis to a logarithmic scale, we can plot the entire data set (Figure 5.22). When one axis uses a logarithmic scale and the other a linear scale, the graph is called a *semi-log* (or *log-linear*) plot.

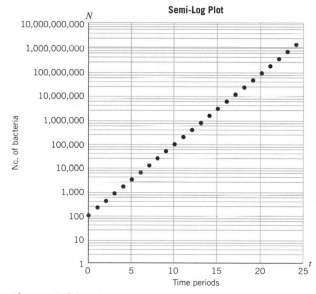

Figure 5.22 Semi-log plot of *E. coli* models data over 24 time periods.

Why Does the Graph Appear as a Straight Line?   On a semi-log plot, moving a fixed distance on the logarithmic scale is equivalent to *multiplying* by a constant factor, rather than to adding a constant factor as on a linear scale. To stay on a line with slope m, we need to move vertically m units for each unit we move horizontally. So the line tells us that each time we increase the time period by 1, the number of *E. coli* is multiplied by a constant (namely 2). That's precisely the definition of an exponential function.

In general, the graph of any exponential function on a semi-log plot will be a straight line. We'll take a closer look at why this is true in Chapter 6. Since most graphing software easily converts standard linear plots to *log-linear* or *semi-log* plots, this is one of the simplest and most reliable ways to recognize exponential growth in a data set.

> When an exponential function is plotted using a standard linear scale on the horizontal axis and a logarithmic scale on the vertical axis, its graph is a straight line. This type of graph is called a *log-linear* or *semi-log* plot.

EXAMPLE 1

To learn more about Gordon Moore's predictions, read his original paper and a more current interview in *Wired*.

Moore's Law

In 1965 Gordon Moore, cofounder of Intel, made his famous prediction that the number of transistors per integrated circuit would increase exponentially over time (doubling about every 2 years.) This became known as Moore's Law. Figure 5.23 on the next page shows the actual increase in the number of transistors over the last 40 years. Do the data justify his claim of exponential growth?

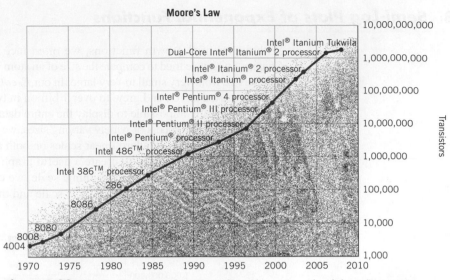

Figure 5.23 The semi-log plot of number of transistors (per circuit board) over time.
Source: Intel's website at *www.intel.com.*

SOLUTION Figure 5.23 shows a semi-log plot of the number of transistors (per integrated circuit) over time. The plot looks basically linear, indicating that the data are exponential in nature. Intel has kept up with a pace of exponential growth. But in 2005, Gordon Moore said, "It can't continue forever."

EXPLORE & EXTEND

Log-Log Graph

The following graph shows a progression of events from the beginnings of life on this planet to the present time. This type of graph extends what you learned about semi-log plots and previews what you will learn about log-log plots in Chapter 7.

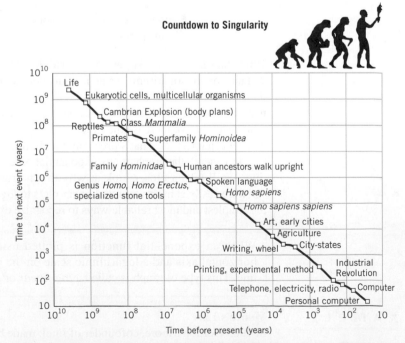

Source: *http://singularity.com.charts.* Kurzweil, R. *The Singularity is Near: When Humans Transcend Biology,* NY: Viking Press, 2005.
(Note: The time scale represents time *before* the present time in years.)

    a. What type of scale is used on the vertical and horizontal axes? Why do you think these scales were used?

    b. Pick a point on the graph, identify its coordinates, and interpret the meaning of the coordinates.

    c. What is the order of magnitude difference between the start of life on Earth and the Cambrian Explosion, measured in time before the present? Measured in time to next event?

    d. Pick two other events and describe them relative to each other.

Algebra Aerobics 5.8

1. a. Using the graph in Figure 5.22 on page 331, estimate the time it takes for the *E. coli* to increase by a factor of 10.

    b. From the original expression for the population, $N = 100 \cdot 2^t$, when does the population increase by a factor of 8?

    c. By a factor of 16?

    d. Are these three answers consistent with each other?

2. What would you expect the graph of $y = 25 \cdot 10^x$ to look like on a semi-log plot? Construct a table of values for *x* equal to 0, 1, 2, 3, 4 and 5. Plot the graph of this function on the accompanying semi-log grid. Does your graph match your prediction?

3. Use your graph for Problem 2 to estimate the values of *y* when *x* = 3.5 and when *x* = 7.

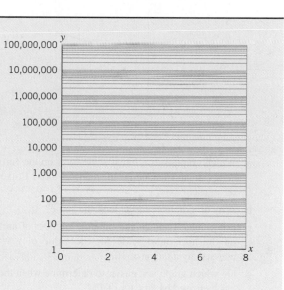

Exercises for Section 5.8

Technology for finding a best-fit function is required for Exercises 5 and 6. Graphing program is recommended for Exercise 1.

1. (Graphing program recommended.) Below is a table of values for $y = 500(3)^x$ and for log *y*.

x	*y*	log *y*
0	500	2.699
5	121,500	5.085
10	29,524,500	7.470
15	$7.17 \cdot 10^9$	9.856
20	$1.74 \cdot 10^{12}$	12.241
25	$4.24 \cdot 10^{14}$	14.627
30	$1.03 \cdot 10^{17}$	17.013

    a. Plot *y* vs. *x* on a linear scale. Remember to identify the largest number you will need to plot before setting up axis scales.

    b. Plot log *y* vs. *x* on a semi-log plot with a log scale on the vertical axis and a linear scale on the horizontal axis.

    c. Rewrite the *y*-values as powers of 10. How do these values relate to log *y*?

2. Match each function with its semi-log plot.

    a. $y = 200(1.5)^x$

    b. $y = 200(2.5)^x$

    c. $y = 200(0.9)^x$

    d. $y = 200(0.5)^x$

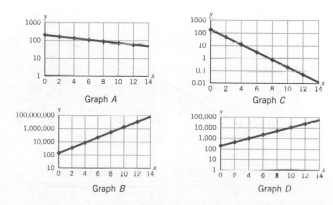

3. The three accompanying graphs are all of the same function, $y = 1000(1.5)^x$.

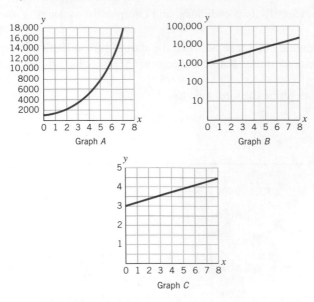

Graph A

Graph B

Graph C

a. Which graph uses a linear scale for y on the vertical axis? A power-of-10 scale on the vertical axis? Logarithms on the vertical axis?

b. Why do graphs B and C look the same?

c. For graphs A and B, estimate the number of units needed on the horizontal scale for the value of $y = 1000$ to increase by a factor of 10.

d. On which graph is it easier to determine when the function has increased by a factor of 10?

e. On which graph is it easier to determine when the function has doubled?

4. According to Rubin and Farber's *Pathology*, "Smoking tobacco is the single largest preventable cause of death in the United States, with direct health costs to the economy of tens of billions of dollars a year. Over 400,000 deaths a year—about one sixth of the total mortality in the United States—occur prematurely because of smoking." The accompanying graph compares the risk of dying for smokers, ex-smokers, and nonsmokers. It shows that individuals who have smoked for 2 years are twice

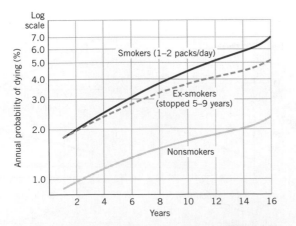

Source: E. Rubin and J. L. Farber, *Pathology*, 3rd ed. (Philadelphia: Lippincott-Raven, 1998), p. 310. Copyright © 1998 by Lippincott-Raven. Reprinted by permission from Wolters Kluwer Health.

as likely to die as a nonsmoker. Someone who has smoked for 14 years is three times more likely to die than a nonsmoker.

a. The graphs for the smokers (one to two packs per day), ex-smokers, and nonsmokers all appear roughly as straight lines on this semi-log plot. What, then, would be appropriate functions to use to model the increased probability of dying over time for all three groups?

b. The plots for smokers and ex-smokers appear roughly as two straight lines that start at the same point, but the graph for smokers has a steeper slope. How would their two function models be the same and how would they be different?

c. The plots of ex-smokers and nonsmokers appear as two straight lines that are roughly equidistant. How would their two function models be the same and how would they be different?

5. (Requires technology to find a best-fit function.)

a. Load the file CELCOUNT, which contains all of the white blood cell counts for the bone marrow transplant patient. Now graph the data on a semi-log plot. Which section(s) of the curve represent exponential growth or decay? Explain how you can tell.

b. Load the file ECOLI, which contains the *E. coli* counts for twenty-four time periods. Graph the *E. coli* data on a semi-log plot. Which section of this curve represents exponential growth or decay? Explain your answer.

6. (Requires technology to find a best-fit function.) The accompanying table shows the U.S. international trade in goods and services.

USTRADE

U.S. International Trade (Billions of Dollars)

Year	Total Exports	Total Imports
1960	25.9	22.4
1965	35.3	30.6
1970	56.6	54.4
1975	132.6	120.2
1980	271.8	291.2
1985	288.8	410.9
1990	537.2	618.4
1995	793.5	891.0
2000	1070.6	1450.4
2005	1281.5	1996.7
2008	1826.6	2522.5

Source: U.S. Department of Commerce, Bureau of Economic Analysis, U.S. Bureau of the Census, *Statistical Abstract of the United States: 2010.*

a. U.S. imports and exports both expanded rapidly between 1960 and 2005. Use technology to plot the total U.S. exports and total U.S. imports over time on the same graph.

b. Now change the vertical axis to a logarithmic scale and generate a semi-log plot of the same data as in part (a). What is the shape of the data now, and what does this suggest would be an appropriate function type to model U.S. exports and imports?

c. Construct appropriate function models for total U.S. imports and for total exports.

6. (continued)

d. The difference between the values of exports and imports is called the *trade balance*. If the balance is negative, it is called a *trade deficit*. The balance of trade has been an object of much concern lately. Calculate the trade balance for each year and plot it over time. Describe the overall pattern.

e. We have a trade deficit that has been increasing rapidly in recent years. But for quantities that are growing exponentially, the "relative difference" is much more meaningful than the simple difference. In this case the relative difference is

$$\frac{\text{exports} - \text{imports}}{\text{exports}}$$

This gives the trade balance as a fraction (or if you multiply by 100, as a percentage) of exports.

Calculate the relative difference for each year in the above table and graph it as a function of time. Does this present a more or less worrisome picture? That is, in particular over the last decade, has the relative difference remained stable or is it also rapidly increasing in magnitude?

7. The following plot shows the gross domestic product (GDP) per capita (or per person) over time.

a. What type of plot is this? Explain your answer.

b. For what time period(s) does the growth appear roughly linear? What does this tell you?

c. For what time period(s) is there a roughly linear decline? What does this tell you? What are possible explanations for the decline(s)?

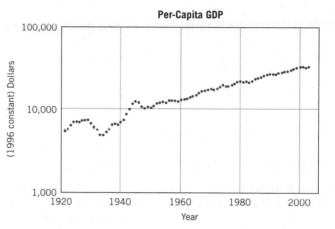

Source: http://singularity.com.charts. Kurzweil, R. *The Singularity is Near: When Humans Transcend Biology,* NY: Viking Press, 2005.

CHAPTER SUMMARY

Exponential Functions

The general form of the equation for an exponential function is

$$y = Ca^x \qquad (a > 0 \text{ and } a \neq 1), \quad \text{where}$$

C is the initial value or *y*-intercept
a is the base and is called the growth (or decay) factor

If $C > 0$ and

$a > 1$, the function represents growth
$0 < a < 1$, the function represents decay

Linear vs. Exponential Growth

Exponential growth is multiplicative, whereas linear growth is additive. Exponential growth involves multiplication by a constant factor for each unit increase in input. Linear growth involves adding a fixed amount for each unit increase in input.

In the long run, any exponential growth function will eventually overtake any linear growth function.

Graphs of Exponential Functions

For functions in the form $y = Ca^x$:
The value of *C* tells us where the graph crosses the *y*-axis.
The value of *a* affects the steepness of the graph.

Exponential growth ($a > 1$, $C > 0$). The larger the value of *a*, the more rapid the growth and the more rapidly the graph rises.

Exponential decay ($0 < a < 1$, $C > 0$). The smaller the value of *a*, the more rapid the decay and the more rapidly the graph falls.

The graphs of both exponential growth and decay functions are asymptotic to the *x*-axis.

Exponential growth: As $x \to -\infty$, $y \to 0$.
Exponential decay: As $x \to +\infty$, $y \to 0$.

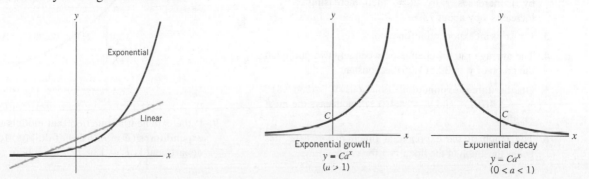

Factors, Rates, and Percents

An exponential function can be represented as a constant percent change. Growth and decay factors can be translated into constant percent increases or decreases, called *growth* or *decay rates*. For an exponential function in the form $y = Ca^x$,

$$\text{growth factor} = 1 + \text{growth rate}$$
$$a = 1 + r$$

where r is the *growth rate* in decimal form.

$$\text{decay factor} = 1 - \text{decay rate}$$
$$a = 1 - r$$

where r is the *decay rate* in decimal form.

Properties of Exponential Functions

The *doubling time* of an exponentially growing quantity is the time required for the quantity to double in size. The *half-life* of an exponentially decaying quantity is the time required for a quantity to be reduced by a half.

The *rule of 70* offers a simple way to estimate the doubling time or the half-life. If a quantity is growing at $R\%$ per year, then its doubling time is approximately $70/R$ years. If a quantity is decaying at $R\%$ per month, then $70/R$ gives its half-life in months.

Converting from n Time Units to One Time Unit

If a is the growth (or decay) factor for n time units, then $a^{1/n}$ is the growth (or decay) factor for one time unit.

Equivalent Forms of an Exponential Function

Exponential growth

$$f(t) = Ca^t = C \cdot 2^{t/n} = Ce^{kt}$$

Exponential decay

$$g(t) = Ca^t = C \cdot \left(\tfrac{1}{2}\right)^{t/n} = Ce^{kt}$$

where C = initial quantity
a = the growth (or decay) factor
t = number of time units
n = doubling (or half-life) time
k = the instantaneous growth (or decay) rate

Compounding

The value of P_0 dollars invested at an annual interest rate r (in decimal form) compounded n times a year for t years is

$$P = P_0 \left(1 + \frac{r}{n}\right)^{nt}$$

The value for r is called the nominal rate or *annual percentage rate* (APR). The actual interest rate per year is called the *effective interest rate* or the annual percentage yield (APY).

Compounding Continuously

The number $e \approx 2.71828$ is used to compute *continuous compounding*. For example, the value of P_0 dollars invested at a nominal interest rate r (expressed in decimal form) compounded continuously for t years is

$$P = P_0 e^{rt}$$

Semi-Log Plots of Exponential Functions

When an exponential function is plotted using a standard scale on the horizontal axis and a logarithmic scale on the vertical axis, its graph is a straight line. This is called a *log-linear* or *semi-log plot*.

CHECK YOUR UNDERSTANDING

I. Is each of the statements in Problems 1–22 true or false? Give an explanation for your answer.

1. If $y = f(x)$ is an exponential function and if increasing x by 1 increases y by a factor of 3, then increasing x by 2 increases y by a factor of 6.

2. If $y = f(x)$ is an exponential function and if increasing x by 1 increases y by 20%, then increasing x by 3 increases y by about 73%.

3. $y = x^3$ is an exponential function.

4. The average rate of change between any two points on the graph of $y = 32.5(1.06)^x$ is constant.

5. Of the three exponential functions $y_1 = 5.4(0.8)^x$, $y_2 = 5.4(0.7)^x$, and $y_3 = 5.4(0.3)^x$, y_3 decays the most rapidly.

6. The graph of the exponential function $y = 1.02^x$ lies below the graph of the line $y = x$ for $x > 5$.

7. If $y = 100(0.976)^x$, then as x increases by 1, y decreases by 97.6%.

8. The function in the accompanying figure represents exponential decay with an initial population of 150 and decay factor of 0.8.

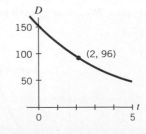

9. If the exponential function that models federal budget expenditures (E in billions of dollars) for a particular department is $E = 134(1.041)^t$, where t = number of

9. (continued)

years since 1990, then in 1990 expenditures were about $134 billion and were increasing by 4.1% per year.

10. Increasing $1000 by $100 per year for 10 years gives you more than increasing $1000 by 10% per year for 10 years.

11. The value of the dollar with inflation at 2% per month is the same as the value of the dollar with inflation at 24% per year.

12. If M dollars is invested at 6.25% compounded annually, the amount of money, A, in 14 years is $A = M(0.0625)^{14}$.

Problems 13 and 14 refer to the following graph.

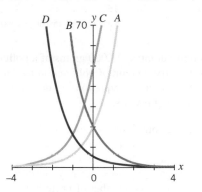

13. Of the functions plotted in the figure, graph C best describes the exponential function $f(x) = 20(3)^x$.

14. Of the functions graphed in the figure, graph B best describes the exponential function $g(x) = 20(3)^{-x}$.

15. If $B = 100(0.4)^x$, then as x increases by 1, B decreases by 60%.

16. After 30 years, the amount of interest earned on $5000 invested at 5% interest compounded annually will be half as much as the amount of interest earned on $10,000 invested at the same rate.

17. If a population behaves exponentially over time and if $\frac{\text{population in year 3}}{\text{population in year 0}} = 0.98$, then after 3 years the population will have decreased by 2%.

18. If the half-life of a substance is 10 years, then three half-lives of the substance would be 30 years.

19. If the doubling time of $100 invested at an interest rate r compounded annually is 9 years, then in 27 years the amount of the investment will be $600.

20. The doubling time for the function in the accompanying graph is approximately 10 years.

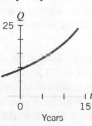

21. The amount of $100 invested at 8% compounded continuously has a doubling time of about 8.7 years.

22. The function representing the result of 500 grams of a substance doubling every 15 days for t days is $A = 500 \cdot 2^{15t}$.

II. In Problems 23–29, give an example of a function with the specified properties. Express your answer using formulas, and specify the independent and dependent variables.

23. An exponential function that has an initial population of 2.2 million people and increases 0.5% per quarter.

24. An exponential function that has an initial value of $1.4 billion and decreases 2.3% per decade.

25. Two exponential decay functions with the same initial population but with the first decaying at twice the rate of the second.

26. Two functions, one linear and one exponential, that pass through the data points $(0, 5)$ and $(3, 0.625)$.

27. An exponential function that has a doubling time of 14 years.

28. A function that describes the amount remaining of an initial amount of 200 mg of a substance with a half-life of 20 years.

29. A function (that uses e) whose graph is identical to the graph of the function $y = 100(0.974)^t$. (*Hint:* $0.974 \approx e^{-0.0263}$.)

III. Is each of the statements in Problems 30–38 true or false? If a statement is false, give a counter-example.

30. Exponential functions are multiplicative and linear functions are additive.

31. All exponential functions are increasing.

32. If $P = Ca^t$ describes a population P as a function of t years since 2000, then C is the initial population in the year 2000.

33. Quantities that increase by a constant amount represent linear growth, whereas quantities that increase by a constant percent represent exponential growth.

34. The graph of an exponential growth function plotted on a semi-log plot is concave upward.

35. For exponential functions, the growth rate is the same as the growth factor.

36. Compounding at 3% quarterly is the same as compounding at 12% annually.

37. Assuming your investment is growing, the effective interest rate is always greater than or equal to the nominal interest rate.

38. The interest rate r for continuous compounding in $y = Ce^{rx}$ is called the nominal interest rate or the instantaneous interest rate, depending on the application.

CHAPTER 5 REVIEW: PUTTING IT ALL TOGETHER

A graphing program is required for Exercise 13. Internet access is required for Exercise 9 (c). A calculator that can evaluate powers is recommended for many of the exercises.

1. In each case, generate equations that represent the population, P, as a function of time, t (in years), such that when $t = 0$, $P = 150$ and:

   a. P doubles each year.

   b. P decreases by twelve units each year.

   c. P increases by 5% each year.

   d. The annual average rate of change of P with respect to t is constant at 12.

2. Describe in words the following functions by identifying the type of function and what happens with each unit increase in x.

   a. $f(x) = 100(4)^x$ **b.** $g(x) = 100 - 4x$

3. Match each of the following functions with the appropriate graph.

   a. $y = 2^x$ **b.** $y = 3^x$ **c.** $y = 4^x$

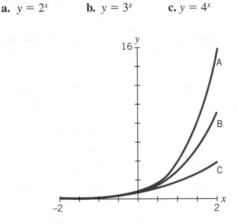

4. Sketch graphs of the following functions on the same grid:

   a. $y = 0.1^x$ **b.** $y = 0.2^x$ **c.** $y = 0.3^x$

5. Studies have shown that lung cancer is directly correlated with smoking. The accompanying graph shows for each year after a smoker has stopped smoking his or her relative risk of lung cancer. The relative risk is the number of times more likely a former smoker is to get lung cancer than someone who has never smoked. For example, if the relative risk is 4.5, then that patient is 4.5 times more likely to get lung cancer than a life-long non-smoker. Time is measured in the number of years since a smoker stopped smoking.

   a. According to the graph, how many times more likely (the relative risk) is a male who just stopped smoking to get lung cancer than a lifelong nonsmoker? How many times more likely is a female who stopped smoking 12 years ago to get lung cancer than someone who never smoked?

   b. Why is it reasonable that the relative risk is declining?

   c. What does it mean if the relative risk is 1? Would you expect the relative risk to go below 1? Explain.

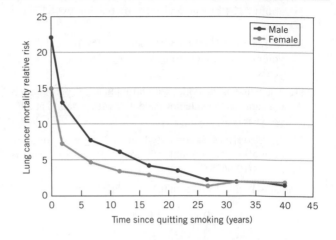

6. A polluter dumped 25,000 grams of a pollutant into a lake. Each year the amount of pollutant in the lake is reduced by 4%. Construct an equation to describe the amount of pollutant after n years.

7. You have a sore throat, so your doctor takes a culture of the bacteria in your throat with a swab and lets the bacteria grow in a Petri dish. If there were originally 500 bacteria in the dish and they grow by 50% per day, describe the relationship between $G(t)$, the number of bacteria in the dish and t, the time since the culture was taken (in days).

8. Which of the following functions are exponential and which are linear? Justify your answers.

x	$f(x)$	$g(x)$	$h(x)$	$j(x)$
0	−500	1	10.25	1
1	−200	2.5	8.25	0.5
2	100	6.25	6.25	0.25
3	400	15.625	4.25	0.125
4	700	39.0625	2.25	0.0625
5	1000	97.65625	0.25	0.03125

9. [Internet access required for part (c)]. The United Nations Department of Economic and Social Affairs reported in 1999 that "the world's population stands at 6 billion and is growing at 1.3% per year, for an annual net addition of 78 million people."

   a. This statement actually contains two contradictory descriptions for predicting world population growth. What are they?

   b. In order to decide which of the statements is more accurate, use the information provided to construct a linear model and an exponential model for world population growth.

   c. The U.S. Census Bureau has a website that gives estimates for the current world population at *http://www.census.gov/ipc/www/popclockworld.html*. Look up the current population. Which of your models is a more accurate predictor of the current world population?

10. Professional photographers consider exposure value (a combination of shutter speed and aperture) in relation to luminance (the amount of light that falls on a certain region) in setting camera controls to produce a desired effect. The data shown in the table give the relationship of the exposure value, E_v, to the luminance, L, for a particular type of film. Find a formula giving L as a function of E_v.

E_v, Exposure Value	L, Luminance (candelas/sq. meter)
0	0.125
1	0.25
2	0.5
3	1
4	2
5	4
6	8
7	16
8	32
9	64
10	128
11	256
12	512

11. According to Optimum Population Trust, the number of motor vehicles in the United Kingdom has nearly doubled every 25 years since 1925. Use the "rule of 70" to estimate the annual percent increase.

12. Mirex, a pesticide used to control fire ants, is rather long-lived in the environment. It takes 12 years for half of the original amount to break down into harmless products. Use the "rule of 70" to estimate the annual percent decrease.

13. (Graphing program required). The accompanying table gives the populations and the annual growth rates of several countries in July 2009. Assume the populations are growing exponentially at the given rate.

Country	Annual Growth Rate (%)	Population (2009)
United Arab Emirates	3.69	4,798,491
Guatemala	2.07	13,276,517
Kenya	2.69	39,002,772
United States	0.98	307,212,123

Source: www.cia.gov.

a. Write the equation of each exponential growth function in the form $y = Ca^t$, where t is the number of years from 2009. Round the initial population to the nearest thousand.

b. Create a table of values that shows the populations of each country from 2009 to 2020.

c. Using a graphing program, graph each of your functions.

d. Use your table, graphs, or calculators to estimate the number of years for each population to double.

c. Do you think the assumption of exponential growth for each of these countries is appropriate? Explain why or why not.

14. The *CIA Factbook* reported that Russia's population was 140 million in 2009 and the average annual percent change was −0.47%.

a. Is the population increasing or decreasing?

b. Construct a function to predict Russia's population after 2009. What assumptions did you make in constructing your model?

c. What does your model predict for the population in Russia in 2019?

15. The Energy Information Administration has made the following projections for energy consumption for the world:

World Energy Consumption

Year	2015	2020	2025	2030
Quadrillion Btu	563	613	665	722

Source: www.eia.doe.gov/iea/.

Would a linear or an exponential function be an appropriate model for these data? Construct a function that models the data.

16. a. Generate a series of numbers N using $N = 2^{0.5n}$, for integer values of n from 0 to 10. Use the value of $2^{0.5} \approx 1.414$ and the rules of exponents to calculate values for N.

b. Rewrite the N formula using a square root sign.

17. The accompanying table gives worldwide mobile device sales for 2008 and 2009 and forecasts for 2010.

Worldwide Mobile Device Sales to End Users (in thousands of devices)

Region	2008	2009	2010
Asia/Pacific	453,100.1	479,862.6	546,770.8
Eastern Europe	96,068.0	81,145.1	84,995.0
Japan	40,588.1	34,871.7	34,897.9
Latin America	142,323.1	119,737.5	126,772.7
Middle East and Africa	133,471.9	128,879.6	140,305.1
North America	182,245.8	182,571.6	190,130.8
Western Europe	174,455.3	186,950.5	198,498.9
Worldwide	**1,222,252.30**	**1,214,018.60**	**1,322,371.20**

Source: Gartner (December 2009).

For which region(s) does mobile terminal sales seem to be growing exponentially? Justify your answer/s.

18. In February 2007 the U.S. Intergovernmental Panel on Climate Change reported that human activities have increased greenhouse gases in the atmosphere, resulting in global warming and other climate changes. Methane (a greenhouse gas) in the atmosphere has more than doubled since pre-industrial times from about 750 parts per billion in 1850 to 1,750 in 2005. The graph on the next page shows the increase in methane gas since 1850.

18. (continued)

Methane in the Atmosphere
(parts per billion)

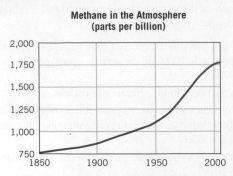

Source: *The New York Times International*, February 3, 2007.

a. Construct an exponential equation, $M(t)$ that could be used to model the increase in methane gas in the atmosphere, where t = number of years since 1850.

b. From the graph, estimate the approximate doubling time. Now calculate the approximate doubling time using the "rule of 70." How close are your answers?

c. Describe the exponential growth in methane gas since 1850 in two different ways.

19. Radioactive iodine (I-131), used to test for thyroid problems, has a half-life of 8 days. If 20 grams of radioactive iodine are used for the test, describe the relationship between $A(t)$, the amount of I-131 (in grams), and t, the time (in days) after thyroid test.

20. Xenon gas (Xe-133) is used in medical imaging to study blood flow in the heart and brain. When inhaled, it is quickly absorbed into the bloodstream, and then gradually eliminated from the body (through exhalation). About 5 minutes after inhalation, there is half as much xenon left in the body (i.e., Xe-133 has a *biological* half-life of about 5 minutes). If you originally breathe in and absorb 3 ml. of xenon gas, describe the relationship between X, the amount of xenon gas in your body (in ml), and t, the time (in minutes) since you inhaled it.

21. Match the function with its graph.

a. $f(x) = 10e^{-0.075x}$

b. $g(x) = 10e^{-0.045x}$

c. $h(x) = 10e^{-0.025x}$

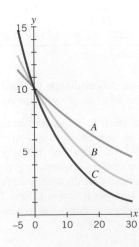

22. a. In November 2006 the Bank of America offered a 12-month certificate of deposit (CD) at a nominal rate (or annual percentage rate, APR) of 3.6%. What is the effective interest rate (or annual percentage yield, APY) if the bank compounds:

   i. Annually?

   ii. Monthly?

   iii. Continuously?

b. The Bank lists the APY as 3.66%. Which of the above compounding schedules comes closet to this?

23. At the birth of their granddaughter, the grandparents create a college fund in her name, investing $10,000 at 7% per year.

a. Construct an equation to model the growth in the account.

b. How much will be in the account in 18 years?

c. If the annual inflation stays at 3%, how much will something that cost $10,000 today cost in 18 years? Calculate the difference between your answer in part (b) and your answer in part (c).

d. Is this equivalent to the return you would have if you invested $10,000 at 4% for 18 years? If not, what does investing $10,000 at 4% represent?

24. In 2005 the Government Accountability Office issued a study of textbook prices. The report noted that in the preceding two decades textbook prices had been rising at the rate of 6% per year, roughly double the annual inflation rate of about 3% a year.

a. Assuming textbook prices continue to rise at 6% per year, if a textbook costs $80 in 2005, what will it cost in 2015?

b. A textbook that was first published in 1990 costs $120 in 2010. Since 1990 its price has increased at the rate of 6% per year. What was its price in 1990?

25. You may have noticed that when you take a branch from certain trees, the branch looks like a miniature version of the tree. When you break off a piece of the branch, that looks like the tree too. Mathematicians call this property self-similarity. The village of Bourton-on-the-Water in southwest England has a wonderful example of self-similarity: it has a 1/10 scale model of itself. Because the 1/10 scale model is a complete model of the town, it must contain a model of itself (that is, a 1/100 scale model of Bourton), and because the 1/100 scale model is also a complete model of Bourton, it also contains a scale model (that is, a 1/1000 scale model of Bourton.)

a. If A_o is the area of the actual village, how does the area of the first 1/10 scale model (where each linear dimension is one-tenth of the actual size) compare with A_o?

b. If the scale models are made of the same materials as the actual village, how does the weight of the building materials in the 1/100 scale model church compare with the weight W_o of the original church?

c. If the actual village is called a level 0 model, the 1/10 model is level 1, the 1/100 model is level 2, and so on, find a formula for the area A_n at any level n of the model as a function of A_o and for the weight W_n at any level n of the model as a function of W_o.

26. In the early 1980s Brazil's inflation was running rampant at about 10% per *month*.[6] Assuming this inflation rate continued unchecked, construct a function to describe the purchasing power of 100 cruzeiros after *n* months. (A cruzeiro is a Brazilian monetary unit.) What would 100 cruzeiros be worth after 3 months? After 6 months? After a year?

27. Lung cancer is one of the leading causes of death, especially for smokers. There are various forms of cancer that attack the lungs, including cancers that start in some other organ and metastasize to the lungs. Doubling times for lung cancers vary considerably but are likely to fall within the range of 2 to 8 months. Although individual cancers are unpredictable and may speed up or slow down, the following examples give an idea of the range of time possibilities.

  a. Two people are found to have lung cancer tumors whose volumes are each estimated at 0.5 cubic centimeters. A former asbestos worker has a tumor with an expected doubling time of 8 months. A heavy smoker has a tumor with an expected doubling time of 2 months. Write two tumor growth exponential functions, $A(t)$ and $S(t)$, for the asbestos worker and smoker, respectively, where $t =$ number of 2-month time periods.

  b. Graph $S(t)$ and $A(t)$ over 12 time periods on the same grid and compare the graphs.

  c. Assuming both tumors grow in a spherical shape, what would be the diameter of each tumor after one year? (*Note:* The volume of a sphere $V = 4/3\pi r^3$.)

28. Which of the following graphs to the right suggests a linear model for the original data? Which suggests an exponential model for the original data? Justify your answers.

[6]In such cases sooner or later the government usually intervenes. In 1985 the Brazilian government imposed an anti-inflationary wage and price freeze. When the controls were dropped, inflation soared again, reaching a high in March 1990 of 80% per month! At this level, it begins to matter whether you buy groceries in the morning or wait until that night. On August 2, 1993, the government devalued the currency by defining a new monetary unit, the cruzeiro real, equal to 1000 of the old cruzeiros. Inflation still continued, and on July 1, 1994, yet another unit, the real, was defined equal to 2740 cruzeiros reales. By 1997, inflation had slowed considerably to about 0.1% per month.

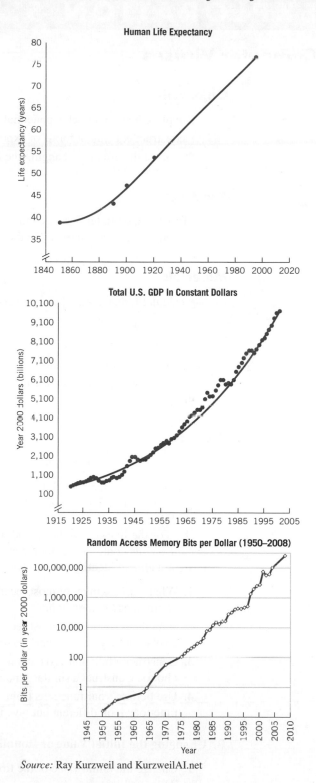

Source: Ray Kurzweil and KurzweilAI.net

Computer Viruses

Objectives

- explore how to model exponential growth of computer viruses
- construct equivalent forms of exponential functions and examine their properties
- analyze the effect of changing the initial value and the exponent of exponential functions

Readings

"Top Ten Computer Viruses and Worms," *ABC News/Technology*, September 2009. This reading and other links can be found on the course website.

Materials

- calculator that can evaluate powers or computer with spreadsheet.

Introduction

Computer viruses are a continual problem. One reason they are so devastating is because of the rate at which they spread. It is said that a virus can potentially spread worldwide in a matter of hours. Is this true?

Procedure

A. Modeling the Growth A Computer Virus

A new virus is distributed to 50 computers via a corporate email. Let $t = 0$ represent the time in hours when the first 50 computers get the infection. Fifteen minutes later another 100 computers are infected.

For the first day of the spread of the virus, assume that the number of new computers infected by the virus continues to double every 15 minutes and that the anti-virus companies and programs are not able to identify the virus or slow its growth.

1. What is the number of most recently infected computers after 1 hour? After 1 hour and 45 minutes? After 6 hours?
2. What is the number of most recently infected computers after 24 hours?
   Explain why your answer does or does not make sense.
3. Construct a function $V_h(t)$ that models the growth at each level of infection of the virus after t hours. Construct a similar function $V_m(T)$ that models the growth after T minutes.
4. Use both of your functions from above to verify your answers in number 2. Explain why the functions look different but give the same answer.

B. Changing the Initial Value or Doubling Time

1. Change the initial value of the first infected computers. How does this affect the outcome after 24 hours?
2. What if the time period of the growth changes to, say, 10 minutes or 30 minutes? Explain how the function would change and how the outcome would change.

C. Spreading of a New Virus

Assume the number of computers newly infected by a virus is given by $V(t) = 300(2)^{t/2}$, where t is the number of hours after the infection.

1. How many computers were infected initially, at $t = 0$?
2. What is the doubling time of the virus?
3. How long will it take for 800 new computers to be infected?
4. How long will it take for approximately 10,000 new computers to be infected according to this model? (*Note*: Answer this question using tables of values or a graph.)

D. Finding the Total Number of Computers Infected

1. The models constructed in parts A and B predict how many new computers are infected after a given time period. Construct a model that predicts the *total* number of computers infected after a given time period under the same conditions as in part A. [*Hint*: To find the total number of computers infected after a given time period, t, you need to add all of the previously infected computers up to and including time t.]
2. One amazing property of exponential functions is that the growth in any doubling time is greater than the total of all the preceding growth. Find out if your models confirm this property for growth after 6 hours. After 6 hours, how many new computers are infected in part A? What is the total number infected after 5 hours and 45 minutes?

Exploration-Linked Homework

Do an Internet search to find an estimate of the number of people who have computers with access to the Internet. How does this number compare to the number of computers that could potentially be infected in 24 hours with this type of computer virus? Do you think the assumptions made for this exploration are valid assumptions? Why or why not?

CHAPTER 6
LOGARITHMIC LINKS: LOGARITHMIC AND EXPONENTIAL FUNCTIONS

OVERVIEW

If we know a specific output for an exponential function, how can we find the associated input? To answer this question, we can use logarithmic functions, the close relatives of exponential functions. We return to the *E. coli* model to use logarithms to calculate the doubling time.

After reading this chapter, you should be able to
- use logarithms to solve exponential equations
- learn the rules for common and natural (base *e*) logarithms
- understand and apply the properties of logarithmic functions
- describe the relationship between logarithmic and exponential functions
- find the equation of an exponential function on a semi-log plot

6.1 *Using Logarithms to Solve Exponential Equations*

Estimating Solutions to Exponential Equations

So far, we have found a function's output from particular values of the input. For example, in Chapter 5, we modeled the growth of *E. coli* bacteria with the function

$$N = 100 \cdot 2^T$$

where N = number of *E. coli* bacteria and T = number of 20-minute intervals. If the input T is 5, then the corresponding output N is $100 \cdot 2^5 = 100 \cdot 32 = 3200$ bacteria.

Now we do the reverse, that is, find a function's input when we know the output. Starting with a value for N, the output, we find a corresponding value for T, the input. For example, at what time T, will the value for N, the number of *E. coli*, be 1000? This turns out to be a harder question to answer than it might first appear. First we will estimate a solution from a table and a graph, and then we show how to find an exact solution using logarithms.

From a data table and graph

In Table 6.1, when $T = 3$, $N = 800$. When $T = 4$, $N = 1600$. Since N is steadily increasing, if we know $N = 1000$, then the value of T is somewhere between 3 and 4.

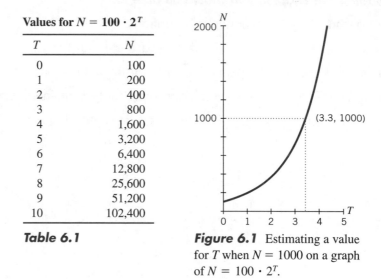

Values for $N = 100 \cdot 2^T$

T	N
0	100
1	200
2	400
3	800
4	1,600
5	3,200
6	6,400
7	12,800
8	25,600
9	51,200
10	102,400

Table 6.1

Figure 6.1 Estimating a value for T when $N = 1000$ on a graph of $N = 100 \cdot 2^T$.

We can also estimate the value for T when $N = 1000$ by looking at a graph of the function $N = 100 \cdot 2^T$ (Figure 6.1). By locating the position on the vertical axis where $N = 1000$, we can move over horizontally to find the corresponding point on the function graph. By moving from this point vertically down to the T-axis, we can estimate the T value for this point, approximately 3.3. So after about 3.3 time periods (or 66 minutes), the number of bacteria is 1000.

From an equation

An alternative strategy is to set $N = 1000$ and solve the equation for the corresponding value for T.

Start with the equation	$N = 100 \cdot 2^T$
set $N = 1000$	$1000 = 100 \cdot 2^T$
divide both sides by 100	$10 = 2^T$

Then we are left with the problem of finding a solution to the equation

$$10 = 2^T$$

We can estimate the value for T that satisfies the equation by bracketing 2^T between consecutive powers of 2. Since $2^3 = 8$ and $2^4 = 16$, then $2^3 < 10 < 2^4$. So $2^3 < 2^T < 2^4$, and therefore T is between 3 and 4, which agrees with our previous estimates. Using this strategy, we can find an approximate solution to the equation $1000 = 100 \cdot 2^T$. Strategies for finding exact solutions to such equations require the use of logarithms.

Algebra Aerobics 6.1a

1. Given the equation $M = 250 \cdot 3^T$, find values for M when:

   a. $T = 0$ **b.** $T = 1$ **c.** $T = 2$ **d.** $T = 3$

2. Use Table 6.1 to determine between which two consecutive integers the value of T lies when N is:

   a. 2000   **b.** 50,000

3. Use the graph of $y = 3^x$ below to estimate the value of x in each of the following equations.

   a. $3^x = 7$ **b.** $3^x = 0.5$

   Graph of $y = 3^x$

   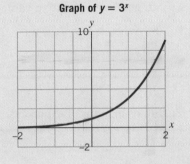

4. Use the table below to determine between which two consecutive integers the value of x in each of the following equations lies.

   a. $5^x = 73$ **b.** $5^x = 0.36$

x	5^x
-2	0.04
-1	0.2
0	1
1	5
2	25
3	125
4	625

5. For each of the following, find two consecutive integers for the exponents a and b that would make the statement true.

   a. $2^a < 13 < 2^b$ **c.** $5^a < 0.24 < 5^b$

   b. $3^a < 99 < 3^b$ **d.** $10^a < 1500 < 10^b$

Rules for Logarithms

In Chapter 4 we introduced logarithms in order to create order of magnitude scales. Recall that:

> The *logarithm base 10 of x* is the exponent of 10 needed to produce x.
>
> $$\log_{10} x = c \qquad \text{means that} \qquad 10^c = x \qquad \text{where } x > 0$$
>
> Logarithms base 10 are called *common logarithms* and $\log_{10} x$ is written as $\log x$.

So,

$$10^5 = 100,000 \qquad \text{is equivalent to saying that} \qquad \log 100,000 = 5$$

$$10^{-3} = 0.001 \qquad \text{is equivalent to saying that} \qquad \log 0.001 = -3$$

Using a calculator we can solve equations such as

$$10^x = 80$$

Rewrite it in equivalent form $\log 80 = x$

use a calculator to compute the log $1.903 \approx x$

But to solve equations such as $2^T = 10$ that involve exponential expressions where the base is not 10, we need to know more about logarithms. As the following expressions suggest, the rules of logarithms follow directly from the definition of logarithms and from the rules of exponents.

Rules for Exponents

If a is any positive real number and n and m are any real numbers, then

1. $a^n \cdot a^m = a^{(n+m)}$ **3.** $(a^m)^n = a^{(m \cdot n)}$

2. $a^n / a^m = a^{(n-m)}$ **4.** $a^0 = 1$

Corresponding Rules for Common Logarithms

If A and B are positive real numbers and p is any real number, then

1. $\log (A \cdot B) = \log A + \log B$ **3.** $\log (A^p) = p \log A$

2. $\log (A/B) = \log A - \log B$ **4.** $\log 1 = 0$ (since $10^0 = 1$)

Finding the common logarithm of a number means finding its exponent when the number is written as a power of 10. So when you see "logarithm," think "exponent," and the rules of logarithms make sense.

As we learned in Chapter 4, the log of 0 or a negative number is not defined. But when we take the log of a number, we can get 0 or a negative value. For example, $\log 1 = 0$ and $\log 0.1 = -1$.

EXPLORE & EXTEND

6.1a

Using the Rule of Exponents
Try expressing in words the rules for logarithms in terms of exponents.

We will list a rationale for each of the rules of common logarithms and prove Rule 1. We leave the other proofs as exercises.

Rule 1 $\log (A \cdot B) = \log A + \log B$

Rationale

Rule 1 of exponents says that when we multiply two terms with the same base, we keep the base and *add* the exponents; that is, $a^p \cdot a^q = a^{p+q}$.

Rule 1 of logs says that if we rewrite A and B each as a power of 10, then the exponent of $A \cdot B$ is the sum of the exponents of A and B.

Proof

If we let $\log A = x$, then	$10^x = A$
and if $\log B = y$, then	$10^y = B$
We have two equal products	$A \cdot B = 10^x \cdot 10^y$
by Rule 1 of exponents	$A \cdot B = 10^{x+y}$
Taking the log of each side	$\log(A \cdot B) = \log (10^{x+y})$
by definition of log	$\log(A \cdot B) = x + y$
substituting $\log A$ for x and $\log B$ for y	
we arrive at our desired result.	$\log(A \cdot B) = \log A + \log B$

Rule 2 $\qquad\qquad\qquad\qquad\qquad$ $\log(A/B) = \log A - \log B$

Rationale

Rule 2 of exponents says that when we divide terms with the same base, we keep the base and *subtract* the exponents, that is, $a^p/a^q = a^{p-q}$. Rule 2 of logs says that if we write A and B each as a power of 10, then the exponent of A/B equals the exponent of A *minus* the exponent of B.

E X A M P L E   1

Using the rules for logs

Indicate whether or not the statement is true using the rules of exponents and the definition of logarithms.

a. $\log(10^2 \cdot 10^3) \overset{?}{=} \log 10^2 + \log 10^3$

b. $\log(10^5 \cdot 10^{-7}) \overset{?}{=} \log 10^5 + \log 10^{-7}$

c. $\log(10^2/10^3) \overset{?}{=} \log 10^2 - \log 10^3$

d. $\log(10^3/10^{-3}) \overset{?}{=} \log 10^3/\log 10^{-3}$

S O L U T I O N

To see if an expression is a true statement, we must verify that the two sides of the expression are equal.

a. $\qquad\qquad\qquad\qquad\qquad$ $\log(10^2 \cdot 10^3) \overset{?}{=} \log 10^2 + \log 10^3$

$\qquad$ Rule 1 of exponents $\qquad\qquad\qquad$ $\log 10^5 \overset{?}{=} \log 10^2 + \log 10^3$

$\qquad$ definition of log $\qquad\qquad\qquad\qquad$ $5 \overset{?}{=} 2 + 3$

$\qquad\qquad\qquad\qquad\qquad\qquad\qquad\qquad$ $5 = 5$

Since the values on the two sides of the equation are equal, our original equation is a true statement.

b. $\qquad\qquad\qquad\qquad\qquad$ $\log(10^5 \cdot 10^{-7}) \overset{?}{=} \log 10^5 + \log 10^{-7}$

$\qquad$ Rule 1 of exponents $\qquad\qquad\qquad$ $\log 10^{-2} \overset{?}{=} \log 10^5 + \log 10^{-7}$

$\qquad$ definition of log $\qquad\qquad\qquad\qquad$ $-2 \overset{?}{=} 5 + (-7)$

$\qquad\qquad\qquad\qquad\qquad\qquad\qquad\qquad$ $-2 = -2$

Since the values on the two sides of the equation are equal, our original equation is a true statement.

c. $\qquad\qquad\qquad\qquad\qquad$ $\log(10^2/10^3) \overset{?}{=} \log 10^2 - \log 10^3$

$\qquad$ Rule 2 of exponents $\qquad\qquad\qquad$ $\log(10^{-1}) \overset{?}{=} \log 10^2 - \log 10^3$

$\qquad$ definition of log $\qquad\qquad\qquad\qquad$ $-1 \overset{?}{=} 2 - 3$

$\qquad\qquad\qquad\qquad\qquad\qquad\qquad\qquad$ $-1 = -1$

Since the values on the two sides of the equation are equal, our original equation is a true statement.

d. $\qquad\qquad\qquad\qquad\qquad$ $\log(10^3/10^{-3}) \overset{?}{=} \log 10^3/\log 10^{-3}$

$\qquad$ Rule 2 of exponents $\qquad\qquad\qquad$ $\log 10^{3-(-3)} \overset{?}{=} \log 10^3/\log 10^{-3}$

$\qquad$ combine terms in exponent $\qquad\qquad$ $\log 10^6 \overset{?}{=} \log 10^3/\log 10^{-3}$

$\qquad$ definition of log $\qquad\qquad\qquad\qquad$ $6 \overset{?}{=} 3/(-3)$

$\qquad\qquad\qquad\qquad\qquad\qquad\qquad\qquad$ $6 \neq -1$

Since the values on the two sides of the equation are not equal, our original equation is not a true statement.

Rule 3 $$\log A^p = p \log A$$

Rationale

Since

$$\log A^2 = \log(A \cdot A) = \log A + \log A = 2 \log A$$

$$\log A^3 = \log(A \cdot A \cdot A) = \log A + \log A + \log A = 3 \log A$$

$$\log A^4 = \log(A \cdot A \cdot A \cdot A) = 4 \log A$$

it seems reasonable to expect that in general

$$\log A^p = p \log A$$

Rule 4 $$\log 1 = 0$$

Rationale

Since $10^0 = 1$ by definition, the equivalent statement using logarithms is $\log 1 = 0$.

EXAMPLE 2

Simplifying expressions with logs

Simplify each expression, and if possible evaluate with a calculator.

$$\log 10^3 = 3 \log 10 = 3 \cdot 1 = 3$$

$$\log 2^3 = 3 \log 2 \approx 3 \cdot 0.301 \approx 0.903$$

$$\log \sqrt{3} = \log 3^{1/2} = \tfrac{1}{2} \log 3 \approx \tfrac{1}{2} \cdot 0.477 \approx 0.239$$

$$\log x^{-1} = (-1) \cdot \log x = -\log x$$

$$\log 0.01^a = a \log 0.01 = a \cdot (-2) = -2a$$

$$(\log 5) \cdot (\log 1) = (\log 5) \cdot 0 = 0$$

EXAMPLE 3

Expanding expressions with logs

Use the rules of logarithms to write the following expression as the sum or difference of several logs.

$$\log \left(\frac{x(y-1)^2}{\sqrt{z}} \right)$$

SOLUTION By Rule 2

$$\log \left(\frac{x(y-1)^2}{\sqrt{z}} \right) = \log x(y-1)^2 - \log \sqrt{z}$$

by Rule 1 and
exponent notation

$$= \log x + \log(y-1)^2 - \log z^{1/2}$$

by Rule 3

$$= \log x + 2 \log(y-1) - \tfrac{1}{2} \log z$$

We call this process *expanding the expression*.

EXAMPLE 4

Contracting expressions with logs

Use the rules of logarithms to write the following expression as a single logarithm.

$$2 \log x - \log(x-1)$$

SOLUTION By Rule 3

$$2 \log x - \log(x-1) = \log x^2 - \log(x-1)$$

by Rule 2

$$= \log \left(\frac{x^2}{x-1} \right)$$

We call this process *contracting the expression*.

Common Error

Probably the most common error in using logarithms stems from confusion over the division property.

$$\log A - \log B = \log\left(\frac{A}{B}\right) \quad \text{but} \quad \log A - \log B \neq \frac{\log A}{\log B}$$

For example,

$$\log 100 - \log 10 = \log\left(\frac{100}{10}\right) = \log 10 = 1$$

$$\text{but} \quad \log 100 - \log 10 \neq \frac{\log 100}{\log 10} = 2$$

E X A M P L E 5 **Solving equations that contain logs**
Solve for x in the equation $\log x + \log x^2 = 15$

S O L U T I O N Given $\log x + \log x^2 = 15$

Rule 1 for logs $\log x^3 = 15$

Rule 3 for logs $3 \log x = 15$

divide by 3 $\log x = 5$

rewrite using definition of logs $x = 10^5$ or $100{,}000$

EXPLORE & EXTEND

6.1b **Radical Thoughts**
Why does $\log\left(\sqrt{AB}\right)$ lie halfway between $\log A$ and $\log B$?

Algebra Aerobics 6.1b

1. Using only the rules of exponents and the definition of logarithm, verify that:
   a. $\log(10^5/10^7) = \log 10^5 - \log 10^7$
   b. $\log[10^5 \cdot (10^7)^3] = \log 10^5 + 3 \log 10^7$

2. Determine the rule(s) of logarithms that were used in each statement.
   a. $\log 3 = \log 15 - \log 5$
   b. $\log 1024 = 10 \log 2$
   c. $\log\sqrt{31} = \frac{1}{2}\log 31$
   d. $\log 30 = \log 2 + \log 3 + \log 5$
   e. $\log 81 - \log 27 - \log 3$

3. Determine if each of the following is true or false. For the true statements tell which rule of logarithms was used.
   a. $\log(x + y) \stackrel{?}{=} \log x + \log y$
   b. $\log(x - y) \stackrel{?}{=} \log x - \log y$
   c. $7 \log x \stackrel{?}{=} \log x^7$
   d. $\log 10^{1.6} \stackrel{?}{=} 1.6$
   e. $\dfrac{\log 7}{\log 3} \stackrel{?}{=} \log 7 - \log 3$
   f. $\dfrac{\log 7}{\log 3} \stackrel{?}{=} \log(7 - 3)$

4. Expand, using the properties of logarithms:
   a. $\log\sqrt{\dfrac{2x - 1}{x + 1}}$
   c. $\log\dfrac{x\sqrt{x + 1}}{(x - 1)^2}$
   b. $\log\dfrac{xy}{z}$
   d. $\log\dfrac{x^2(y - 1)}{y^3 z}$

5. Contract, expressing your answer as a single logarithm: $\frac{1}{3}[\log x - \log(x + 1)]$.

6. Use rules of logarithms to combine into a single logarithm (if necessary), then solve for x.
   a. $\log x = 3$
   b. $\log x + \log 5 = 2$
   c. $\log x + \log 5 = \log 2$
   d. $\log x - \log 2 = 1$
   e. $\log x - \log(x - 1) = \log 2$
   f. $\log(2x + 1) - \log(x + 5) = 0$

7. Show that $\log 10^3 - \log 10^2 \neq \dfrac{\log 10^3}{\log 10^2}$

Solving Exponential Equations Using Logarithms

Answering our original question: Solving $1000 = 100 \cdot 2^T$

Remember the question that started this chapter? We wanted to find out how many time periods it would take 100 *E. coli* bacteria to become 1000. To find an exact solution, we need to solve the equation $1000 = 100 \cdot 2^T$ or, dividing by 100, the equivalent equation, $10 = 2^T$. We now have the necessary tools.

Given	$10 = 2^T$
take the logarithm of each side	$\log 10 = \log 2^T$
use Rule 3 of logs	$\log 10 = T \log 2$
divide both sides by log 2	$\dfrac{\log 10}{\log 2} = T$
use a calculator	$\dfrac{1}{0.3010} \approx T$
divide	$3.32 \approx T$

which is consistent with our previous estimates of a value for T between 3 and 4, approximately equal to 3.3.

In our original model the time period represents 20 minutes, so 3.32 time periods represents $3.32 \cdot (20 \text{ minutes}) = 66.4$ minutes. So the bacteria would increase from the initial number of 100 to 1000 in a little over 66 minutes.

EXAMPLE 6

Using logarithms
Solve the following equation by using logarithms.

$$8^x = 2^{x+1}$$

SOLUTION

	$8^x = 2^{x+1}$
Take logs of both sides	$\log 8^x = \log 2^{x+1}$
Rule 3 of logs	$x \log 8 = (x + 1) \log 2$
distributive law	$x \log 8 = x \log 2 + \log 2$
subtract $x \log 2$ from both sides	$x \log 8 - x \log 2 = \log 2$
factor out x	$x (\log 8 - \log 2) = \log 2$
Rule 2 of logs	$x \log 4 = \log 2$
divide by log 4	$x = \dfrac{\log 2}{\log 4} = \dfrac{\log 2}{\log 2^2} = \dfrac{\log 2}{2 \log 2} = \dfrac{1}{2}$

EXAMPLE 7

Time to decay to a specified amount
In Chapter 5 we used the function $f(t) = 100(0.976)^t$ to measure the remaining amount of radioactive material as 100 milligrams (mg) of strontium-90 decayed over time t (in years). How many years would it take for there to be only 10 mg of strontium-90 left?

SOLUTION

Set $f(t) = 10$ and solve the equation.

	$10 = 100(0.976)^t$
Divide both sides by 100	$0.1 = (0.976)^t$
take the log of both sides	$\log 0.1 = \log (0.976)^t$
use Rule 3 of logs	$\log 0.1 = t \log 0.976$
divide by log 0.976	$\dfrac{\log 0.1}{\log 0.976} = t$
evaluate logs	$\dfrac{-1}{-0.0106} \approx t$
divide and switch sides	$t \approx 94 \text{ years}$

So it takes almost a century for 100 mg of strontium-90 to decay to 10 mg.

Solving for Doubling Times and Half-Lives

In the previous example on strontium-90, we specified the output (10 grams) and used logs to solve for the corresponding input (94 years). We can use the same strategy to find the doubling time or half-life of an exponential function $f(t) = C\,a^t$, where C is the initial value. Since $2C$ is double and $\frac{1}{2}C$ is half of the initial value, C. Then:

If $a > 0$, we have exponential growth, so set $2C = C\,a^t$ to solve for the doubling time.

If $0 < a < 1$, we have exponential decay, so set $(1/2)C = C\,a^t$ to solve for the half-life.

E X A M P L E 8

Doubling your money

As we saw in Chapter 5, the equation $P = 250(1.05)^n$ gives the value of $250 invested at 5% interest (compounded annually) for n years. How many years does it take for the initial $250 investment to double to $500?

S O L U T I O N

a. *Estimating the answer using the rule of 70 (Section 5.6): if $R = 5\%$ per year, then the doubling time is approximately*

$$70/R = 70/5 = 14 \text{ years}$$

b. *Calculating a more precise answer:* We can set $P = 500$ (twice the original amount of $250) and solve the equation for n.

$$500 = 250(1.05)^n$$

Divide both sides by 250	$2 = (1.05)^n$
take the log of both sides	$\log 2 = \log (1.05)^n$
use Rule 3 of logs	$\log 2 = n \log 1.05$
divide by log 1.05	$\dfrac{\log 2}{\log 1.05} = n$
evaluate with a calculator	$\dfrac{0.3010}{0.0212} \approx n$
divide and switch sides	$n \approx 14.2 \text{ years}$

So the estimate of 14 years using the rule of 70 was pretty close.

E X A M P L E 9

Half-life of epinephrine

Individuals who have severe allergies (for example, to bee stings or peanut butter) often carry an EpiPen. It is a self-injecting device that delivers a single dose of 0.3 mg of epinephrine. Epinephrine (more familiarly known as adrenaline) is a hormone that responds to stress by increasing heart rate and blood pressure and boosting the supply of oxygen and energy-giving glucose to the blood and brain. The body responds immediately to the injection. But the epinephrine decays quickly, according to the formula $E(t) = 0.3(0.794)^t$, where t is in minutes and $E(t)$ is the amount of epinephrine (in mg) remaining in the body. What is the half-life?

S O L U T I O N

Since the starting value for epinephrine is 0.3 mg, after one half-life there would be 0.15 mg. So we need to solve for time t in the equation:

$$0.15 = 0.3(0.794)^t$$

We divide by 0.3	$0.5 = (0.794)^t$
take log of both sides	$\log(0.5) = \log(0.794^t)$
apply Rule 3 of logs	$\log(0.5) = t \log(0.794)$
solve for t	$\dfrac{\log(0.5)}{\log(0.794)} = t$
evaluate with a calculator	$\dfrac{-0.3010}{-0.1002} \approx t$
switch sides and divide	$t \approx 3 \text{ minutes}$

So the half-life is about 3 minutes; that means that after 3 minutes only 50% of the epinephrine will still be active in the body.

Algebra Aerobics 6.1c

These problems require a calculator that can evaluate logs.

1. Solve the following equations for T.
   a. $60 = 10 \cdot 2^T$ **c.** $80(0.95)^T = 10$
   b. $500(1.06)^T = 2000$

2. Using the model $N = 100 \cdot 2^T$ for bacteria growth, where T is measured in 20-minute time periods, how long will it take for the bacteria count:
   a. To reach 7000?   **b.** To reach 12,000?

3. First use the rule of 70 to estimate how long it would take $1000 invested at 6% compounded annually to double to $2000. Then use logs to find a more precise answer.

4. Use the function in Example 7, p. 352, to determine how long it will take for 100 milligrams of strontium-90 to decay to 1 milligram.

5. Solve each equation for t (in years). Which of these equations asks you to find the time necessary for the initial amount to double? For the initial amount to drop to half?
   a. $30 = 60(0.95)^t$
   b. $16 = 8(1.85)^t$
   c. $500 = 200(1.045)^t$

6. Find the half-life of a substance that decays according to the following models.
   a. $A = 120(0.983)^t$ (t in days)
   b. $A = 0.5(0.92)^t$ (t in hours)
   c. $A = A_0(0.89)^t$ (t in years)

Exercises for Section 6.1

Many of these problems (and those in later sections) require a calculator that can evaluate powers and logs. Exercises 4, 21, 22, and 28 require a graphing program.

1. Use the accompanying table to estimate the number of years it would take $100 to become $300 at the following interest rates compounded annually.
   a. 3%   **b.** 7%

Compound Interest over 40 Years

Number of Years	Value of $100 at 3% ($)	Value of $100 at 7% ($)
0	100	100
10	134	197
20	181	387
30	243	761
40	326	1497

2. **a.** Generate a table of values to estimate the half-life of a substance that decays according to the function $y = 100(0.8)^x$, where x is the number of time periods, each time period is 12 hours, and y is in grams.
   b. How long will it be before there is less than 1 gram of the substance remaining?

3. The accompanying graph shows the concentration of a drug in the human body as the initial amount of 100 mg dissipates over time. Estimate when the concentration becomes:
   a. 60 mg   **b.** 40 mg   **c.** 20 mg

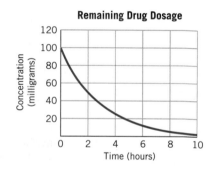

Remaining Drug Dosage

4. (Requires a graphing program.) Assume throughout that x represents time in seconds.
   a. Plot the graph of $y = 6(1.3)^x$ for $0 \le x \le 4$. Estimate the doubling time from the graph.
   b. Now plot $y = 100(1.3)^x$ and estimate the doubling time from the graph.
   c. Compare your answers to parts (a) and (b). What does this tell you?

5. Without a calculator, determine x if we know that $\log x$ equals:
   a. -3 **c.** 0   **e.** -1
   b. 6   **d.** 1

6. Given that $\log 5 \approx 0.699$, without using a calculator determine the value of:
   a. $\log 25$ **b.** $\log \dfrac{1}{25}$ **c.** $\log 10^{25}$ **d.** $\log 0.0025$
   Check your answers with a scientific calculator.

7. Expand each logarithm using only the numbers 2, 3, log 2, and log 3.

   a. log 9
   b. log 18
   c. log 54

8. Identify the rules of logarithms that were used to expand each expression.

   a. $\log 14 = \log 2 + \log 7$
   b. $\log 14 = \log 28 - \log 2$
   c. $\log 36 = 2 \log 6$
   d. $\log 9z^3 = 2 \log 3 + 3 \log z$
   e. $\log 3x^4 = \log 3 + 4 \log x$
   f. $\log \left(\dfrac{16}{3x} \right) = 4 \log 2 - (\log 3 + \log x)$

9. Use the rules of logarithms to find the value of x. Verify your answer with a calculator.

   a. $\log x = \log 2 + \log 6$
   b. $\log x = \log 24 - \log 2$
   c. $\log x^2 = 2 \log 12$
   d. $\log x = 4 \log 2 - 3 \log 2$

10. Use the rules of logarithms to show that the following are equivalent.

   a. $\log 144 = 2 \log 3 + 4 \log 2$
   b. $7 \log 3 + 5 \log 3 = 12 \log 3$
   c. $2(\log 4 - \log 3) = \log \left(\dfrac{16}{9} \right)$
   d. $-4 \log 3 + \log 3 = \log \left(\dfrac{1}{27} \right)$

11. Prove Rule 2 of logarithms:

$$\log(A/B) = \log A - \log B \quad (A, B > 0)$$

12. Expand, using the rules of logarithms.

   a. $\log \left(x^2 y^3 \sqrt{z - 1} \right)$
   c. $\log \left(t^2 \cdot \sqrt[4]{tp^3} \right)$
   b. $\log \dfrac{A}{\sqrt[3]{BC}}$

13. Contract, using the rules of logarithms, and express your answer as a single logarithm.

   a. $3 \log K - 2 \log(K + 3)$
   b. $-\log m + 5 \log(3 + n)$
   c. $4 \log T + \frac{1}{2} \log T$
   d. $\frac{1}{3}(\log x + 2 \log y) - 3(\log x + 2 \log y)$

14. For each of the following equations either prove that it is correct (by using the rules of logarithms and exponents) or else show that it is not correct (by finding numerical values for the variables that make the values of the two sides of the equation different).

   a. $\log \left(\dfrac{x}{y} \right) = \dfrac{\log x}{\log y}$
   b. $\log x - \log y = \log \left(\dfrac{x}{y} \right)$
   c. $\log(2x) = 2 \log x$
   d. $2 \log x = \log(x^2)$

   e. $\log \left(\dfrac{x + 1}{x + 3} \right) = \log(x + 1) - \log(x + 3)$
   f. $\log \left(x \sqrt{x^2 + 1} \right) = \log x + \frac{1}{2}(x^2 + 1)$
   g. $\log(x^2 + 1) = 2 \log x + \log 1$

15. Prove Rule 3 of logarithms: $\log A^p = p \log A$ (where $A > 0$).

16. Solve for t.

   a. $10^t = 4$
   d. $10^{-t} = 5$
   b. $3(2^t) = 21$
   e. $5^t = 7^{t+1}$
   c. $1 + 5^t = 3$
   f. $6 \cdot 2^t = 3^{t-1}$

17. Solve for x.

   a. $\log x = 3$
   d. $\log(x + 1) - \log x = 1$
   b. $\log(x + 1) = 3$
   e. $\log x - \log(x + 1) = 1$
   c. $3 \log x = 5$

18. Solve for x.

   a. $2^x = 7$
   d. $\log(x + 3) + \log 5 = 2$
   b. $\left(\sqrt{3} \right)^{x+1} = 9^{2x-1}$
   e. $\log(x - 1) = 2$
   c. $12(1.5)^{x+1} = 13$

19. In Exercise 1, we estimated the number of years it would take $100 to become $300 at each of the interest rates listed below, compounded annually. Now calculate the number of years by constructing and solving the appropriate equations.

   a. 3%
   b. 7%

20. Solve for the indicated variables.

   a. $825 = 275 \cdot 3^T$
   d. $100 = 25(1.18)^t$
   b. $45,000 = 15,000 \cdot 1.04^t$
   e. $32,000 = 8000(2.718)^t$
   c. $12 \cdot 10^{10} = (6 \cdot 10^8) \cdot 5^x$
   f. $8 \cdot 10^5 = 4 \cdot 10^5(2.5)^x$

21. (Requires a graphing program.) Let $f(x) = 500(1.03)^x$ and $g(x) = 4500$.

   a. Using technology, graph the two functions on the same screen.
   b. Estimate the point of intersection.
   c. Solve the equation $4500 = 500(1.03)^x$ using logarithms.
   d. Compare your answers.

22. (Requires a graphing program.) Let $f(x) = 100(0.8)^x$ and $g(x) = 10$.

   a. Using technology, graph the two functions on the same screen.
   b. Estimate the point of intersection.
   c. Solve the equation $10 = 100(0.8)^x$ using logarithms.
   d. Compare your answers.

23. Find the doubling time or half-life for each of the following functions (where x is in years).

   a. $f(x) = 100 \cdot 4^x$
   b. $g(x) = 100 \cdot \left(\frac{1}{4} \right)^x$
   c. $h(x) = A(4)^x$ (*Hint:* Set $h(x) = 2A$)
   d. $j(x) = A \cdot \left(\frac{1}{4} \right)^x$ (*Hint:* Set $j(x) = \frac{1}{2} A$)

24. The yearly per capita consumption of whole milk in the United States reached a peak of 40 gallons in 1945, at the end of World War II. By 1970 consumption was only 27.4 gallons per person. It has been steadily decreasing since 1970 at a rate of about 3.9% per year.

   a. Construct an exponential model $M(t)$ for per capita whole milk consumption (in gallons) where t = years since 1970.

   b. Use your model to estimate the year in which per capita whole milk consumption dropped to 6 gallons per person. How does this compare with the actual consumption of 6.1 gallons per person in 2007 (the most recent data available on printing)?

   c. What might have caused this decline?

25. Wikipedia is a popular, free online encyclopedia (at *en.wikipedia.org*) created in 2001. Anyone can edit the articles, so they should be taken with "a grain of salt." One Wikipedia article claims that the number of articles posted on Wikipedia has grown exponentially from 20,000 articles in 2001 to 2,000,000 in 2007.

   a. Create an exponential function $N(t)$ that represents the total number of Wikipedia articles where t = years since 2001.

   b. What is the doubling time? Interpret your answer.

   c. What does your function predict for 2010? Is your prediction above or below the actual number of 16,000,000 posted articles Wikipedia claims in 2010? The number of articles is clearly increasing, but is the rate of change also increasing or decreasing?

26. If the amount of drug remaining in the body after t hours is given by $f(t) = 100 \left(\frac{1}{2}\right)^{t/2}$ (graphed in Exercise 3), then calculate:

   a. The number of hours it would take for the initial 100 mg to become:

   i. 60 mg   **ii.** 40 mg   **iii.** 20 mg

   b. The half-life of the drug.

27. In Chapter 5 we saw that the function $N = N_0 \cdot 1.5^T$ described the actual number N of *E. coli* bacteria in an experiment after T time periods (of 20 minutes each) starting with an initial bacteria count of N_0.

   a. What is the doubling time?

   b. How long would it take for there to be ten times the original number of bacteria?

28. (Requires a graphing program.) A woman starts a training program for a marathon. She starts in the first week by doing 10-mile runs. Each week she increases her run length by 20% of the distance for the previous week.

   a. Write a formula for her run distance, D, as a function of week, W.

   b. Use technology to graph your function, and then use the graph to estimate the week in which she will reach a marathon length of approximately 26 miles.

   c. Now use your formula to calculate the week in which she will start running 26 miles.

29. The half-life of bismuth-214 is about 20 minutes.

   a. Construct a function to model the decay of bismuth-214 over time. Be sure to specify your variables and their units.

   b. For any given sample of bismuth-214, how much is left after 1 hour?

   c. How long will it take to reduce the sample to 25% of its original size?

   d. How long will it take to reduce the sample to 10% of its original size?

30. A department store has a discount basement where the policy is to reduce the selling price, S, of an item by 10% of its current price each week. If the item has not sold after the tenth reduction, the store gives the item to charity.

   a. For a $300 suit, construct a function for the selling price, S, as a function of week, W.

   b. After how many weeks might the suit first be sold for less than $150? What is the selling price at which the suit might be given to charity?

31. If you drop a rubber ball on a hard, level surface, it will usually bounce repeatedly. (See the accompanying graph below.) Each time it bounces, it rebounds to a height that is a percentage of the previous height. This percentage is called the *rebound height*.

   a. Assume you drop the ball from a height of 5 feet and that the rebound height is 60%. Construct a table of values that shows the rebound height for the first four bounces.

   b. Construct a function to model the ball's rebound height, H, on the nth bounce.

   c. How many bounces would it take for the ball's rebound height to be 1 foot or less?

   d. Construct a general function that would model a ball's rebound height H on the nth bounce, where H_0 is the initial height of the ball and r is the ball's rebound height (in decimal form).

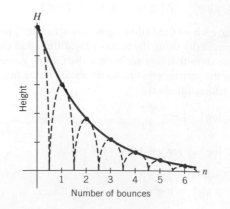

Number of bounces

32. The accompanying graph shows fish and shellfish production (in millions of pounds) by U.S. companies.

The growth appears to be exponential between 1970 and 1995. The exponential function $F(t) = 1355(1.036)^t$ models that growth, where $F(t)$ represents millions of pounds of fish and shellfish produced by U.S. companies since 1970 (where $0 \le t \le 25$).

a. Use the model to predict when fish production would reach 4000 million pounds. What year would that be? Use the graph to estimate the actual production in that year.

b. What might have caused the decline in U.S. fish production after 1995? Do you think that America's appetite for fish has waned?

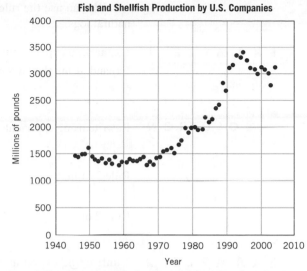

Fish and Shellfish Production by U.S. Companies

Source: U.S. Department of Agriculture, Economic Research Service, Data set on food availability, *www.ers.usda.gov.*

6.2 Using Natural Logarithms to Solve Exponential Equations Base e

If we think of the *E. coli* growth example $N = 100 \cdot 2^T$ (see Section 5.1 and Section 6.1) as growing *continuously*, then we need to represent the equivalent function using base e as

$$N = 100 \cdot (e^k)^T \qquad \text{where } e^k = 2$$

So we need to find the value of k such that $e^k = 2$. To solve this equation, we need *natural logarithms* that use e as a base.

The Natural Logarithm

The *common logarithm* uses 10 as a base. The *natural logarithm* uses e as a base and is written $\ln x$ rather than $\log_e x$. Scientific calculators have a key that computes $\ln x$.

> **The Natural Logarithm**
>
> The *logarithm base e* of x is the exponent of e needed to produce x. Logarithms base e are called *natural logarithms* and are written as $\ln x$.
>
> $$\ln x = c \qquad \text{means that} \qquad e^c = x \quad (x > 0)$$

The properties for natural logarithms (base e) are similar to the properties for common logarithms (base 10). Like the common logarithm, $\ln A$ is not defined when $A \le 0$.

> Assume A and B are positive real numbers and p is any real number.
>
> **Rules of Common Logarithms**
>
> 1. $\log(A \cdot B) = \log A + \log B$
> 2. $\log(A/B) = \log A - \log B$
> 3. $\log A^p = p \log A$
> 4. $\log 1 = 0$ (since $10^0 = 1$)
>
> **Rules of Natural Logarithms**
>
> 1. $\ln(A \cdot B) = \ln A + \ln B$
> 2. $\ln(A/B) = \ln A - \ln B$
> 3. $\ln A^p = p \ln A$
> 4. $\ln 1 = 0$ (since $e^0 = 1$)

We can use the rules of natural logarithms to manipulate expressions involving natural logs.

E X A M P L E   1

Expanding expressions with natural logs

Expand, using the laws of logarithms, the expression: $\ln\sqrt{\dfrac{x+3}{x-2}}$.

S O L U T I O N

Rewrite using exponents $\qquad \ln\sqrt{\dfrac{x+3}{x-2}} = \ln\left(\dfrac{x+3}{x-2}\right)^{1/2}$

Rule 3 of ln $\qquad\qquad\qquad\qquad = \dfrac{1}{2}\ln\left(\dfrac{x+3}{x-2}\right)$

Rule 2 of ln $\qquad\qquad\qquad\qquad = \dfrac{1}{2}\left[\ln(x+3) - \ln(x-2)\right]$

E X A M P L E   2

Contracting expressions with natural logs

Contract, expressing the answer as a single logarithm: $\frac{1}{3}\ln(x-1) + \frac{1}{3}\ln(x+1)$

S O L U T I O N

Distributive property $\quad \frac{1}{3}\ln(x-1) + \frac{1}{3}\ln(x+1) = \frac{1}{3}\left[\ln(x-1) + \ln(x+1)\right]$

Rule 1 of ln $\qquad\qquad\qquad\qquad\qquad\qquad = \frac{1}{3}\ln[(x-1)(x+1)]$

Rule 3 of ln $\qquad\qquad\qquad\qquad\qquad\qquad = \ln[(x-1)(x+1)]^{1/3}$

multiply binomials $\qquad\qquad\qquad\qquad\quad = \ln(x^2-1)^{1/3}$ or $\ln\sqrt[3]{x^2-1}$

Solving our original equation

Now we can solve our original equation:

$$e^k = 2$$

by applying ln to both sides	$\ln(e^k) = \ln 2$
using Rule 3	$k(\ln e) = \ln 2$
since $\ln e = 1$	$k = \ln 2$
evaluating $\ln 2$	$k \approx 0.6931$

Using our value for k, we can now describe the *E. coli* growing continuously as:

$$N = 100 \cdot (e^{\ln 2})^T \approx 100 \cdot (e^{0.6931})^T$$

E X A M P L E   3

Solving equations using natural logarithms

Solve the equation $10 = e^t$ for t.

S O L U T I O N

Given	$10 = e^t$
take ln of both sides	$\ln 10 = \ln e^t$
definition of ln	$\ln 10 = t$
evaluate and switch sides	$t \approx 2.303$

E X A M P L E   4

Effective vs. nominal rates

If the effective annual interest rate on an account is 5.21%, estimate the nominal annual interest rate that is compounded continuously.

S O L U T I O N

A 5.21% effective interest rate is 0.0521 in decimal form. So the equation $P = P_0(1.0521)^x$ represents the value P of P_0 dollars after x years. To find the nominal interest rate that is continuously compounded, we must convert $(1.0521)^x$ into the form e^{rx} or $(e^r)^x$. So e^r must equal 1.0521. We need to solve for r in the following equation.

$$1.0521 = e^r$$

Take ln of both sides	$\ln 1.0521 = \ln e^r$
use a calculator	$0.0508 \approx \ln e^r$
definition of ln	$0.0508 \approx r$

So $1.0521 \approx e^{0.0508}$. A nominal interest rate of 0.0508 or 5.08% compounded continuously is equivalent to an effective interest rate of 0.0521 or 5.21%.

Returning to Doubling Times and Half-Lives

In Section 6.1 we used common logs to solve for doubling times and half-lives. Now that we are dealing with exponential functions base e, we will use the natural log, abbreviated as *ln*. So to find the doubling time or half-life of an exponential function $F(t) = Ce^{kt}$:

> If $k > 0$, we have exponential growth, so we set $2C = Ce^{kt}$ to solve for the doubling time;
>
> If $k < 0$, we have exponential decay, so we set $(1/2)C = Ce^{kt}$ to solve for the half-life.

EXAMPLE 5

Half-life of atmospheric pressure

As the height above sea level increases, the atmospheric pressure continuously decreases at a rate of about 12% per 1000 meters. The atmospheric pressure is measured in atms, where 1 atm $\approx$ 15 lb/in.², the pressure at sea level.

a. Create an exponential decay function that describes the atmospheric pressure $P(h)$, where h = height above sea level (in thousands of meters). Sketch a graph of the function.

b. At what height would the atmospheric pressure be half that at sea level?

SOLUTION

a. At sea level the pressure is 1 atm. The pressure $P(h)$ (in atms) at sea level and above can be described as $P(h) = e^{-0.12\,h}$, where h is height (in thousands of meters).

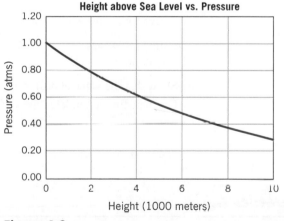

Height above Sea Level vs. Pressure

Figure 6.2

b. The pressure will be halved when $P(h) = (1/2)$ atm. To get the half-life, we can solve

$$(1/2) = e^{-0.12\,h} \qquad \text{for } h$$

Take ln of both sides	$\ln(1/2) = \ln(e^{-0.12\,h})$
evaluate, apply Rule 3 of natural logs,	$-0.6931 \approx -0.12\,h$
and solve for h	$h \approx 5.8$ thousand meters

So when the height, h, is about 5800 meters above sea level, the atmospheric pressure is half that at sea level.

EXAMPLE 6

Doubling the duckweed

The smallest flower in the world comes from the Wolffia plant. The plant itself is the size of a green sprinkle on a St. Patrick's Day donut. It's a type of duckweed, which you may have seen floating in clumps on a water's calm surface. The plant has a high growth rate. The equation $N(t) = N_0 e^{0.556t}$ gives $N(t)$, the total number of plants on day t, where N_0 is the initial number of plants. How long does it take for the duckweed to double?

SOLUTION

The population will have doubled when the number of plants reaches $2N_0$, twice the initial number, N_0. To get the doubling time, we can solve

$$2N_0 = N_0 e^{0.556\,t} \quad \text{for } t$$

Divide both sides by N_0	$2 = e^{0.556\,t}$
take ln of both sides	$\ln 2 = \ln(e^{0.556\,t})$
apply Rule 3 for natural logs	$\ln 2 = t \ln (e^{0.556})$
evaluate	$0.693 \approx 0.556\,t$
solve for t	$t \approx 1.25 \text{ days}$

So the doubling time is about 1.25 days, or 30 hours.

EXPLORE & EXTEND

6.2

Generalizing a Strategy

Finding Doubling Times

We can generalize the strategy used in Example 6 to find an expression for the doubling time. The function $g(t) = Ce^{kt}$, where $k > 0$ represents exponential growth.

a. Show that the value at $g(0) = C$.

b. One doubling period later, the value would be $2C$. To find the doubling time, set

$$2C = Ce^{kt}$$

and solve for t. This expression for t is the doubling time.

Finding Half-Lives

The function $h(t) = Ce^{kt}$, where $k < 0$ represents exponential decay.

a. Show that the value at $h(0) = C$.

b. One half-life period later, the value would be $(1/2)C$. To find the half-life, set

$$(1/2)\, C = Ce^{kt}$$

and solve for t. This expression for t is the half-life.

Algebra Aerobics 6.2a

Problems 5, 6, 7, and 9 require a calculator that can evaluate natural logs.

1. Evaluate without a calculator:

a. $\ln e^2$ **b.** $\ln 1$ **c.** $\ln \dfrac{1}{e}$ **d.** $\ln \dfrac{1}{e^2}$ **e.** $\ln\sqrt{e}$

2. Use the rules for logarithms to evaluate the following expressions.

a. $\log 10^3$ **c.** $3 \log 10^{0.09}$

b. $\log 10^{-5}$ **d.** $10^{\log 3.4}$

e. $\ln e^5$

f. $\ln e^{0.07}$

g. $\ln e^{3.02} + \ln e^{-0.27}$

h. $e^{\ln 0.9}$

3. Expand the following:

a. $\ln\sqrt{xy}$ **c.** $\ln\left((x + y)^2(x - y)\right)$

b. $\ln\left(\dfrac{3x^2}{y^3}\right)$ **d.** $\ln\dfrac{\sqrt{x + 2}}{x(x - 1)}$

4. Contract, expressing your answer as a single logarithm:

   a. $\ln x + \ln(x - 1)$ d. $\frac{1}{2}\ln(x + y)$

   b. $\ln(x + 1) - \ln x$ e. $\ln x - 2\ln(2x - 1)$

   c. $2\ln x - 3\ln y$

5. Find the nominal rate on an investment compounded continuously if the effective rate is 6.4%.

6. Determine how long it takes for $10,000 to grow to $50,000 at 7.8% compounded continuously.

7. Solve the following equations for x.

   a. $e^{x+1} = 10$ b. $e^{x-2} = 0.5$

8. Determine which of the following are true statements. If the statement is false, rewrite the right-hand side so that the statement becomes true.

   a. $\ln 81 \overset{?}{=} 4\ln 3$ d. $2\ln 10 \overset{?}{=} \ln 20$

   b. $\ln 7 \overset{?}{=} \dfrac{\ln 14}{\ln 2}$ e. $\ln\sqrt{e} \overset{?}{=} \dfrac{1}{2}$

   c. $\ln 35 \overset{?}{=} \ln 5 + \ln 7$ f. $5\ln 2 \overset{?}{=} \ln 25$

9. Use the rules of natural logarithms to contract each expression into a single logarithm (if necessary), then solve for x.

   a. $\ln 2 + \ln 6 = x$

   b. $\ln 2 + \ln x = 2.48$

   c. $\ln(x + 1) = 0.9$

   d. $\ln 5 - \ln x = -0.06$

10. If the world's current population is continuously compounding at about 1.14% per year, find the doubling time.

11. You invest $50,000 in an account where the interest is continuously compounding. If in 10 years you have $83,000, find the interest (growth) rate and the doubling time.

12. If it took a radioactive substance 5000 years to continuously decay from 2 mg to 0.3 mg, what is its decay rate? Its half-life?

Converting exponential functions from base *a* to base *e*

We can write two general equivalent forms of an exponential function as:

$$f(t) = Ca^t \quad \text{or} \quad f(t) = C(e^k)^t \quad \text{for some value of } k$$

So	$Ca^t = C(e^k)^t$
which implies that	$a = e^k$
Taking ln of both sides	$\ln a = \ln(e^k)$
using ln Rule 3	$\ln a = k\ln(e)$
since $\ln(e) = 1$, switching sides	$k = \ln a$

Now can write the *continuous* growth (or decay) function as:

$$f(t) = C(e^k)^t \qquad \text{where } k = \ln a$$

Recall from Chapter 5 (p. 324) that k is called the *continuous* or *instantaneous growth* (or *decay*) *rate*.

For exponential growth: $a > 1$, so $e^k > 1$ and k is positive
For exponential decay: $0 < a < 1$, so $0 < e^k < 1$ and k is negative

---

Exponential Functions Using Base *e*

The exponential function $y = Ca^t$ (where $a > 0$ and $a \neq 1$) can be rewritten as

$$y = C(e^k)^t \quad \text{where } k = \ln a$$

We call k the *instantaneous* or *continuous growth* (or *decay*) *rate*.

   Exponential growth: $a > 1$ and k is positive

   Exponential decay: $0 < a < 1$ and k is negative

Writing an exponential function in the form $y = Ca^t$ or $y = Ce^{kt}$ (where $k = \ln a$) is a matter of emphasis, since the graphs and functional values are identical. When we use $y = Ca^t$, we may think of the growth taking place at discrete points in time, whereas the form $y = Ce^{kt}$ emphasizes the notion of continuous growth. For example, $y = 100(1.05)^t$ could be interpreted as giving the value of $100 invested for t years at 5% compounded annually, whereas its equivalent form

$$y = 100(e^{\ln 1.05})^t \approx 100e^{0.049t}$$

suggests that the money is invested at 4.9% compounded continuously.

E X A M P L E   7

Transform from base a to base e

a. Rewrite the function $f(t) = 250(1.3)^t$ using base e.

b. Identify the growth rate (based on the growth factor) and the *continuous* growth rate, both per time period t.

S O L U T I O N

a. The function $f(t) = 250(1.3)^t$ can be rewritten using base e, by

substituting $e^{\ln 1.3}$ for 1.3	$f(t) = 250(e^{\ln 1.3})^t$
evaluating $\ln 1.3$	$\approx 250(e^{0.262})^t$
using Rule 3 of exponents	$\approx 250e^{0.262t}$

b. Since $f(t) = 250(1.3)^t$, the growth rate per time period t is 0.3 or 30%. Since we also have $f(t) \approx 250e^{0.262t}$, the *continuous* growth rate per time period t is 0.262 or 26.2%.

E X A M P L E   8

Medicare: Continuous growth

A best-fit function for Medicare expenditures is $C(n) = 11.1 \cdot (1.11)^n$ where $n =$ years since 1970 and $C(n) =$ Medicare expenditures in billions of dollars. This implies a growth rate of 11% compounded annually. Convert this function into a continuous compounding model.

S O L U T I O N

To describe the same growth in terms of continuous compounding, we need to rewrite the growth factor 1.11 as e^k, where $k = \ln 1.11 \approx 0.104$. Substituting into the original function gives us

$$C(n) \approx 11.1e^{0.104n}$$

So Medicare expenditures are continuously compounding at about 10.4% per year.

E X A M P L E   9

Strontium-90 redux: Continuous decay

In Chapter 5 (p. 304) we saw that the function $f(t) = 100(0.976)^t$ measures the amount of radioactive strontium-90 remaining as 100 milligrams (mg) decay over time t (in years). Rewrite the function using base e.

S O L U T I O N

To rewrite the function $f(t) = 100(0.976)^t$ using base e, we need to convert the base 0.976 to the form e^k, where $k = \ln 0.976$. Using a calculator to evaluate $\ln 0.976$, we get -0.0243. Substituting into our original function, we get

$$f(t) = 100(0.976)^t$$
$$\approx 100(e^{-0.0243})^t$$
$$= 100e^{-0.0243t}$$

Since the original base 0.976 is less than 1, the function represents decay. So, as we would expect, when the function is rewritten using base e, the value of k (in this case -0.0243) is negative.

Sometimes the various rates can be confusing. We provide a quick summary in the box below.

Distinguishing Different Rates

For functions in the form $f(t) = C(e^k)^t$, we call the exponent k the *continuous* growth rate (if $k > 0$) or the *continuous* decay rate (if $k < 0$).

For functions in the form $f(t) = Ca^t$, we call the base a the growth (or decay) factor.

If $a > 1$,   the growth factor $= 1 +$ growth rate

If $0 < a < 1$,   the decay factor $= 1 -$ decay rate

EXAMPLE 10

Identifying rates

a. Rewrite the function $g(t) = 340(0.94)^t$ using base e. Assume t is in years.

b. Identify the decay factor, the decay rate, and the continuous decay rate.

SOLUTION

a. Since the function $g(t) = 340(0.94)^t$ represents exponential decay, the value for k will be negative. The function g can be rewritten using base e, by

substituting $e^{\ln 0.94}$ for 0.94	$g(t) = 340(e^{\ln 0.94})^t$
evaluating $\ln 0.94$	$\approx 340(e^{-0.062})^t$
using Rule 3 of exponents	$= 340e^{-0.062t}$

b. The decay factor is 0.94. So the decay rate is $1 - 0.94 = 0.06$ or 6% per year. The *continuous* decay rate is 0.062 or 6.2% per year.

We're finally able to prove the rule of 70 for estimating the half-life or doubling time of an exponential function.

EXAMPLE 11

Proving the rule of 70

We now have the tools to prove the rule of 70 introduced in Chapter 5. Recall that the rule said if a quantity is growing (or decaying) at $R\%$ per time period, then the time it takes the quantity to double (or halve) is approximately $70/R$ time periods. For example, if a quantity increases by a rate, R, of 7% each month, the doubling time is about $70/7 = 10$ months.

SOLUTION

For growth: Let f be an exponential function of the form

$$f(t) = Ce^{rt}$$

where C is the initial quantity, r is the continuous growth rate, and t represents time. The rate r is in decimal form, and R is the equivalent amount expressed as a percentage. Since $f(t)$ represents exponential growth, $r > 0$. The doubling time for an exponential function is constant, so we need only calculate the time for any given quantity to double. In particular, we can determine the time it takes for the initial amount C (at time $t = 0$) to become twice as large; that is, we can calculate the value for t such that

	$f(t) = 2C$	
Set	$Ce^{rt} = 2C$	
divide by C	$e^{rt} = 2$	
take ln of both sides	$\ln e^{rt} = \ln 2$	
evaluate and use ln Rule 3	$rt \approx 0.693$	(1)

$R = 100r$, so $r = R/100$. If we round 0.693 up to 0.70 and

substitute in Equation (1)	$(R/100) \cdot t \approx 0.70$
multiply both sides by 100	$R \cdot t \approx 70$
divide by R we get	$t \approx 70/R$

So the time t it takes for the initial amount to double is approximately $70/R$, which is what the rule of 70 claims.

For decay: We leave the similar proof about half-lives, where $r < 0$ represents decay, to the exercises.

The rule of 70 provides reasonable estimates when $R < 10\%$, whether the growth (or decay) rate is simple, compounded, or continuous.

Algebra Aerobics 6.2b

A calculator that can evaluate logs and powers is required for Problems 2–4.

1. Identify each of the following exponential functions as representing growth or decay. (*Hint:* For parts (d)–(f) use rules for logarithms.)

   a. $M = Ne^{-0.029t}$ **d.** $P = 250\, e^{(\ln 1.056)t}$

   b. $K = 100(0.87)^r$ **e.** $A = 20\, e^{(\ln 0.834)t}$

   c. $Q = 375\, e^{0.055t}$ **f.** $y = (1.2 \cdot 10^5)(e^{\ln 0.752})^t$

2. Rewrite each of the following as a continuous growth or decay model using base e.

   a. $y = 1000(1.062)^t$ **b.** $y = 50(0.985)^t$

3. Determine the nominal and effective rates for each of the following. Assume t is measured in years. (*Hint:* To find the effective rate, convert from e^k to a.)

   a. $P = 25{,}000\, e^{(\ln 1.056)t}$ **c.** $y = 2000\, e^{(\ln 1.083)t}$

   b. $P = 10e^{(\ln 1.034)t}$ **d.** $y = (1.2 \cdot 10^5)(e^{\ln 1.295})^t$

4. Find the growth factor or decay factor for each of the following. (*Hint:* Convert from e^k to a.)

   a. $P = 50{,}000e^{0.08t}$ **b.** $y = 30e^{-0.125t}$

Exercises for Section 6.2

Some of these exercises require a calculator that can evaluate powers and logs. In Exercises 26 and 34 Internet access is optional. Exercises 26 and 36 require a graphing program.

1. Determine the rule(s) of natural logarithms that were used to expand each expression.

   a. $\ln 15 = \ln 3 + \ln 5$

   b. $\ln 15 = \ln 30 - \ln 2$

   c. $\ln 49 = 2 \ln 7$

   d. $\ln 25z^3 = 2 \ln 5 + 3 \ln z$

   e. $\ln 5x^4 = \ln 5 + 4 \ln x$

   f. $\ln\left(\dfrac{125}{3x}\right) = 3 \ln 5 - (\ln 3 + \ln x)$

2. Expand each logarithm using only the numbers 2, 5, $\ln 2$, and $\ln 5$.

   a. $\ln 25$ **b.** $\ln 250$ **c.** $\ln 625$

3. Write an equivalent expression using exponents.

   a. $n = \log 35$ **c.** $\ln x = \dfrac{3}{4}$

   b. $\ln 75 = x$ **d.** $\ln\left(\dfrac{N}{N_0}\right) = -kt$

4. Write an equivalent equation in logarithmic form.

   a. $N = 10^{-t/c}$ **c.** $e^{3x} = 27$

   b. $I = I_0 \cdot e^{-k/x}$ **d.** $\dfrac{1}{2} = e^{-kt}$

5. Use the rules of natural logarithms to find the value of x. Verify your answer with a calculator.

   a. $\ln x = \ln 2 + \ln 5$ **d.** $\ln x = 3 \ln 2 + 2 \ln 6$

   b. $\ln x = \ln 24 - \ln 2$ **e.** $\ln x = 6 \ln 2 - 2 \ln 3$

   c. $\ln x^2 = 2 \ln 11$ **f.** $\ln x = 4 \ln 2 - 3 \ln 2$

6. Use the rules of natural logarithms to contract to a single logarithm. Use a calculator to verify your answer.

   a. $2 \ln 3 + 4 \ln 2$ **c.** $2(\ln 4 - \ln 3)$

   b. $3 \ln 7 - 5 \ln 3$ **d.** $-4 \ln 3 + \ln 3$

7. Use the rules of natural logarithms to expand.

   a. $\ln\left(\sqrt{4xy}\right)$ **b.** $\ln\left(\dfrac{\sqrt[3]{2x}}{4}\right)$ **c.** $\ln\left(3 \cdot \sqrt[4]{x^3}\right)$

8. Use the rules of natural logarithms to contract to a single logarithm.

   a. $\dfrac{1}{2} \ln x - 5 \ln y$ **c.** $\ln 2 + \dfrac{1}{3} \ln x - 4 \ln y$

   b. $\dfrac{2}{5} \ln x + \dfrac{4}{5} \ln y$

9. Contract, expressing your answer as a single logarithm.

   a. $\dfrac{1}{4} \ln (x + 1) + \dfrac{1}{4} \ln (x - 3)$

   b. $3 \ln R - \dfrac{1}{2} \ln P$

   c. $\ln N - 2 \ln N_0$

10. Expand:

   a. $\ln \sqrt[3]{\dfrac{x^2 - 1}{x + 2}}$ **b.** $\ln \left[\dfrac{x}{y\sqrt{2}}\right]^2$ **c.** $\ln \left(\dfrac{K^2 L}{M + 1}\right)$

11. Solve for the indicated variable, by changing to logarithmic form. Round your answer to three decimal places.

   a. $e^r = 1.0253$ **b.** $3 = e^{0.5t}$ **c.** $\frac{1}{2} = e^{3x}$

12. Solve for x by changing to exponential form. Round your answer to three decimal places.

   a. $\ln 3x = 1$ **b.** $3 \ln x = 5$ **c.** $\ln 3 + \ln x = 1.5$

13. Solve for x.

   a. $e^x = 10$ **c.** $2 + 4^x = 7$ **e.** $\ln(x + 1) = 3$

   b. $10^x = 3$ **d.** $\ln x = 5$ **f.** $\ln x - \ln(x + 1) = 4$

14. Solve for t.

   a. $5^{t+1} = 6^t$ **d.** $\ln \left(\dfrac{t}{t - 2}\right) = 1$

   b. $e^{t^2} = 4$ **e.** $\ln t - \ln(t - 2) = 1$

   c. $5 \cdot 2^{-t} = 4$

15. Prove that $\ln(A \cdot B) = \ln A + \ln B$, where A and B are positive real numbers.

16. For each of the following equations, either prove that it is correct (by using the rules of logarithms and exponents) or else show that it is not correct (by finding numerical values for x that make the values on the two sides of the equation different).

   a. $e^{x + \ln x} = x \cdot e^x$

   b. $\ln \left(\dfrac{(x + 1)^2}{x}\right) = 2 \ln(x + 1) - \ln x$

   c. $\ln \left(\dfrac{x}{x + 1}\right) = \dfrac{\ln x}{\ln(x + 1)}$

   d. $\ln(x + x^2) = \ln x + \ln x^2$

   e. $\ln(x + x^2) = \ln x + \ln(x + 1)$

17. How long would it take $15,000 to grow to $100,000 if invested at 8.5% compounded continuously?

18. The effective annual interest rate on an account compounded continuously is 4.45%. Estimate the nominal interest rate.

19. The effective annual interest rate on an account compounded continuously is 3.38%. Estimate the nominal interest rate.

20. A town of 10,000 grew to 15,000 in 5 years. Assuming exponential growth:

   a. What is the annual growth rate?

   b. What was its annual continuous growth rate?

21. Solve the following for r.

   a. $1.025 = e^r$ **b.** $\frac{1}{2} = e^r$ **c.** $1.08 = e^r$

22. Solve the following for r

   a. $0.9 = e^r$ **b.** $2 = e^r$ **c.** $0.75 = e^r$

23. Without using a calculator, identify which functions represent growth and which decay. Then, using a calculator, find the corresponding instantaneous growth or decay rate.

   a. $f(t) = 50e^{(\ln 1.45)t}$ **d.** $j(p) = 2000e^{(\ln 0.3)p}$

   b. $g(x) = 125e^{(\ln 3.15)x}$ **e.** $k(v) = 600e^{(\ln 1.75)v}$

   c. $h(m) = 1500e^{(\ln 0.83)m}$ **f.** $l(x) = e^{(\ln 0.75)x}$

24. Use a calculator to determine the value of each expression. Then rewrite each expression in the form $e^{\ln a}$.

   a. $e^{0.083}$ **b.** $e^{-0.025}$ **c.** $e^{-0.35}$ **d.** $e^{0.83}$

25. Rewrite each of the following functions using base e.

   a. $N = 10(1.045)^t$ **c.** $P = 500(2.10)^x$

   b. $Q = (5 \cdot 10^{-7}) \cdot (0.072)^A$

26. (Requires a graphing program.) Graph each of the following. Use the equations to find, as appropriate, the doubling time or half-life.

   a. $A = 50 e^{0.025t}$ **c.** $P = (3.2 \cdot 10^6)(e^{-0.15})^t$

   b. $A = 100 e^{-0.046t}$

27. For each of the following, find the doubling time, then rewrite each function in the form $P = P_0 e^{rt}$. Assume t is measured in years.

   a. $P = P_0 2^{t/5}$ **b.** $P = P_0 2^{t/25}$ **c.** $P = P_0 2^{2t}$

28. For each of the following, find the half-life, then rewrite each function in the form $P = P_0 e^{rt}$. Assume t is measured in years.

   a. $P = P_0\left(\frac{1}{2}\right)^{t/10}$ **b.** $P = P_0\left(\frac{1}{2}\right)^{t/215}$ **c.** $P = P_0\left(\frac{1}{2}\right)^{4t}$

29. The barometric pressure, p, in millimeters of mercury, at height h, in kilometers above sea level, is given by the equation $p = 760e^{-0.128h}$. At what height is the barometric pressure 200 mm?

30. After t days, the amount of thorium-234 in a sample is $A(t) = 35e^{-0.029t}$ micrograms.

   a. How much was there initially?

   b. How much is there after a week?

   c. When is there just 1 microgram left?

   d. What is the half-life of thorium-234?

31. Assume $f(t) = Ce^{rt}$ is an exponential decay function (so $r < 0$). Prove the rule of 70 for halving times; that is, if a quantity is decreasing at $R\%$ per time period t, then the number of time periods it takes for the quantity to halve is approximately $70/R$. (*Hint:* $R = 100r$.)

32. The functions $y = 50e^{0.04t}$ and $y = 50(2)^{t/n}$ are two different ways to write the same function.

   a. What does the value 0.04 represent?

   b. Set the functions equal to each other and use rules of natural logarithms to solve for n.

   c. What does the value of n represent?

33. The functions $y = 2500e^{-0.02t}$ and $y = 2500\left(\frac{1}{2}\right)^{t/n}$ are two different ways to write the same function.

   a. What does the value -0.02 represent?

   b. Set the functions equal to each other and use rules of natural logarithms to solve for n.

   c. What does the value of n represent?

34. (Requires a graphing program.) Radioactive lead-210 decays according to the exponential formula $Q = Q_0 e^{-0.0311t}$, where Q_0 is the initial quantity in milligrams and t is in years. What is the half-life of lead-210? Verify your answer by graphing using technology.

35. Radioactive thorium-230 decays according to the formula $Q = Q_0 \left(\frac{1}{2}\right)^{t/8000}$, where Q_0 is the initial quantity in milligrams and t is in years.

a. What is the half-life of thorium-230?

b. What is the annual decay rate?

c. Translate the equation into the form $Q = Q_0 e^{rt}$. What does r represent?

36. (Internet access optional.) According to the Pet Food Institute there were 81 million cats in the U.S. in 2005, whose numbers were growing by 1.9% each year.

a. Find an exponential function using continuous growth to model the number of cats in the U.S. since 2005.

b. How many cats are predicted for 2010? For 2020?

c. *Internet search:* Find an estimate for the number of cats in 2010? Was your prediction in part (b) close?

37. Biologists believe that, in the deep sea, species density decreases exponentially with the depth. (Species density is the number of different species per a fixed area.) The accompanying graph shows data collected in the North Atlantic. Sketch an exponential decay function through the data. Then identify two points on your curve, and generate two equivalent equations that model the data, one in the form $y = Ca^t$ and the other in the form $y = Ce^{kt}$.

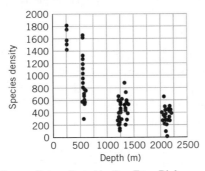

Source: Data collected by Ron Etter, Biology Department. University of Massachusetts–Boston.

38. The number of neutrons in a nuclear reactor can be predicted from the equation $n = n_0 e^{(\ln 2)t/T}$, where n = number of neutrons at time t (in seconds), n_0 = the number of neutrons at time $t = 0$, and T = the reactor period, the doubling time of the neutrons (in seconds). When $t = 2$ seconds, $n = 11$, and when $t = 22$ seconds, $n = 30$. Find the initial number of neutrons, n_0, and the reactor period, T, both rounded to the nearest whole number.

39. If an object is put in an environment at a fixed temperature, A (the "ambient temperature"), then its temperature, T, at time t is modeled by Newton's Law of Cooling:

$$T = A + Ce^{-kt}$$

where k is a positive constant. Note that T is a function of t and that as $t \to +\infty$, then $e^{-kt} \to 0$, so the temperature T gets closer and closer to the ambient temperature, A.

a. Assume that a hot cup of tea (at 160°F) is left to cool in a 75°F room. If it takes 10 minutes for it to reach 100°F, determine the constants A, C, and k in the equation for Newton's Law of Cooling. What is Newton's Law of Cooling in this situation?

b. Sketch the graph of your function.

c. What is the temperature of the tea after 20 minutes?

40. Newton's Law of Cooling (see Exercise 39) also works for objects being heated. At time $t = 0$, a potato at 70°F (room temperature) is put in an oven at 375°F. Thirty minutes later, the potato is at 220° F.

a. Determine the constants A, C, and k in Newton's Law. Write down Newton's Law for this case.

b. When is the potato at 370°F?

c. When is the potato at 374°F?

d. According to your model, when (if ever) is the potato at 375°F?

e. Sketch a graph of your function in part (a).

6.3 *Visualizing and Applying Logarithmic Functions*

The course software "E8: Logarithmic Sliders" in *Exponential & Log Functions* will help you understand the properties of logarithmic functions.

Up until now we have been dealing with logarithms of specific numbers, such as log 2 or ln 10. But since for any $x > 0$ there is a unique corresponding value of log x or ln x, we can define two logarithmic functions:

$$y = \log x \qquad \text{and} \qquad y = \ln x \qquad \text{where } x > 0$$

The Graphs of Logarithmic Functions

What will the graphs look like? We know something about the graphs from the properties of logarithms.

Properties of Logarithms

If $x > 1$, $\log x$ and $\ln x$ are both positive
If $x = 1$, $\log 1 = 0$ and $\ln 1 = 0$
If $0 < x < 1$, $\log x$ and $\ln x$ are both negative
If $x \leq 0$,   neither logarithm is defined

Table 6.2 and Figure 6.3 show some data points and the graphs of $y = \log x$ and $y = \ln x$.

Evaluating $\log x$ and $\ln x$

x	$y = \log x$	$y = \ln x$
0.001	−3.000	−6.908
0.01	−2.000	−4.605
0.1	−1.000	−2.303
1	0.000	0.000
2	0.301	0.693
3	0.477	1.099
4	0.602	1.386
5	0.699	1.609
6	0.778	1.792
7	0.845	1.946
8	0.903	2.079
9	0.954	2.197
10	1.000	2.303

Table 6.2

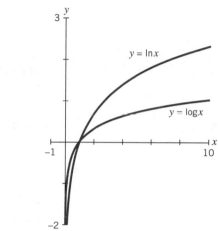

Figure 6.3 Graphs of $y = \log x$ and $y = \ln x$.

The graphs of common and natural logarithms share a distinctive shape. They are both defined only when $x > 0$ and they are both concave down. Both graphs increase throughout with no maximum or minimum value, although they grow more slowly when $x > 1$.

Logarithmic Growth

An exponential growth function grows rapidly, so a small change in the input can correspond to a large change in the output. The logarithmic function ($y = \log x$) grows slowly when $x > 1$. So a large change in the input corresponds to a small change in the output. This means that for $x > 1$ the logarithm compresses large variations, reducing large numbers to a manageable size (see Figure 6.4).

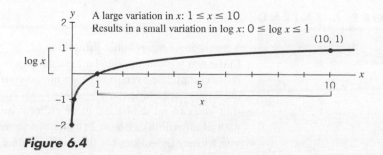

Figure 6.4

For values of x between 0 and 1, the graph of $y = \log x$ is very steep. So here, a small change in the input can correspond to a large change in the output. This means that for $0 < x < 1$, the log will expand small variations, separating them (see Figure 6.5).

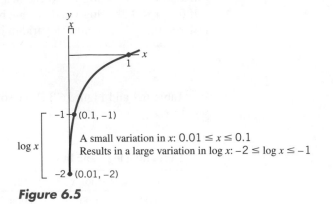

A small variation in x: $0.01 \leq x \leq 0.1$
Results in a large variation in $\log x$: $-2 \leq \log x \leq -1$

Figure 6.5

Putting the pieces together, the logarithmic function helps to separate small numbers while reducing large numbers to a manageable size.

Vertical asymptotes

The graphs of both $y = \log x$ and $y = \ln x$ are vertically asymptotic to the y-axis. For these graphs, as x approaches 0 (through positive values), the graphs come closer and closer to the vertical y-axis but never touch it. For both functions, as $x \to 0$, the values for y get increasingly negative, plunging down near the y-axis toward $-\infty$.

Domain and range

Domain: Both $y = \log x$ and $y = \ln x$ are undefined when $x \leq 0$. So the domain for both functions is restricted to all real numbers $x > 0$; that is, all x in the interval $(0, +\infty)$.

Range: The range for both functions is all real numbers; that is, all y in the interval $(-\infty, +\infty)$.

The Graphs of Logarithmic Functions

The graphs of both $y = \log x$ and $y = \ln x$

- lie to the right of the y-axis since they are defined only for $x > 0$
- share a horizontal intercept of $(1, 0)$
- are concave down
- increase throughout with no maximum or minimum
- are asymptotic to the y-axis

EXPLORE & EXTEND

6.3

Visualizing Logarithmic Functions

In the text we saw the standard graphs of $y = \log x$ and $y = \ln x$. Now we take a look at more complicated versions. Open up the course software "E8: Logarithmic Sliders" for the exponential function $y = C \log_b(ax)$. At the right-hand bottom of the screen, note that you can set the base b to be either base 10 (the common logarithm), base e (the natural logarithm), or base 2 (which we haven't worked with before). The sliders allow you to vary the values for the coefficients a and C. Since logarithms tend to be slow growing, the "max x" option allows you to expand the visible domain through powers of 10 from [0, 1] up to [0, 1000].

Part I: Working with base 10

a. Varying values for a: Set the base to 10 so that we are working with the common logarithm, and set $C = 1$ so that the equation becomes $y = \log(ax)$. Vary the values of a and the size of the domain (max x). Describe the patterns of the graphs. Is x always > 0? Do the graphs all cross the x-axis? (Be sure to extend the domain.)

b. Varying values for C: Now set $a = 1$ and vary the values of C. What are the effects on the graph? Now vary the values of a as well. What can you conclude?

c. Write a 60-second summary of your observations.

Part II: Working with base e and base 2

a. Repeat the steps in Part I for base e and base 2.

b. Summarize your general conclusions.

EXAMPLE 1

Matching graphs and functions
Match each function with the appropriate graph in Figure 6.6.

a. $y = \log(3x)$ **b.** $y = 3 + \log x$ **c.** $y = \log(x^3)$ **d.** $y = 3\log x$

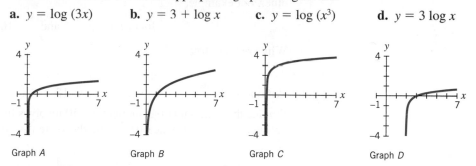

Graph A   Graph B   Graph C   Graph D

Figure 6.6 Four graphs involving logarithms.

SOLUTION

a. Graph A. When $x = 1/3$, then $y = \log(3 \cdot 1/3) = \log 1 = 0$, so the horizontal intercept is $(1/3, 0)$.

b. Graph C. When $x = 1$, then $y = 3 + \log 1 = 3 + 0 = 3$, so the graph passes through $(1, 3)$.

c. and d. Graph B. Since $\log(x^3) = 3\log x$, the graphs for functions (c) and (d) are the same. When $x = 1$, then $y = 3\log 1 = 0$, so the horizontal intercept is $(1, 0)$.

There is no match for Graph D.

EXAMPLE 2

Shifting a natural log function
Graph $y = \ln(x - 2)$. Describe its relationship to $y = \ln x$.

SOLUTION

Figure 6.7 shows the two graphs.

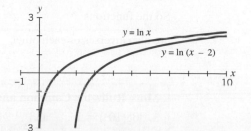

Figure 6.7 Graphs of $y = \ln x$ and $y = \ln(x - 2)$.

The function $y = \ln(x - 2)$ tells us to subtract 2 from x and then apply the function ln. So the graph of $y = \ln(x - 2)$ is the graph of $y = \ln x$ shifted two units to the right. For $y = \ln x$, the horizontal intercept is at 1 since $\ln(1) = 0$. For $y = \ln(x - 2)$, the horizontal intercept is at 3, since $\ln(3 - 2) = \ln 1 = 0$.

Inverse Functions: Logarithmic vs. Exponential

Logarithmic vs. exponential growth

Logarithmic and exponential growth are both unbounded. They both increase forever, never reaching a maximum value. But exponential growth is rapid—not only increasing, but doing so at a rate that is speeding up (accelerating). Logarithmic growth is slow—increasing, but at a rate that is slowing down (decelerating). For example, if we let $t = 1$, 3, and 5, then 10^t is respectively 10, 1000, and 100,000, but $\log(t)$ is respectively 0, 0.477, and 0.699.

Logarithmic and exponential functions are inverses of each other

What does it mean for two functions to be inverses of each other? It means that what one function does, the other undoes. Log and exponential functions are inverses of each other. For example,

$$\log(10^x) = x \qquad \text{and} \qquad 10^{\log x} = x$$

Why are these true?

Rationale for $\qquad\qquad \log(10^x) = x$

Finding the logarithm of a number base 10 involves finding the exponent of 10 needed to produce the number. Since 10^x is already written as 10 to the power x, then $\log 10^x = x$.

Rationale for $\qquad\qquad 10^{\log x} = x$

By definition, $\log x$ is the number such that when 10 is raised to that power the result is x.

Proof	Let	$y = \log x$
	rewrite using definition of logarithm	$10^y = x$
	substitute $\log x$ for y	$10^{\log x} = x$

The two functions

$$y = 10^x \qquad \text{and} \qquad y = \log x$$

are *inverses* of each other. Similarly, for natural logarithms

$$\ln(e^x) = x \qquad \text{and} \qquad e^{\ln x} = x$$

So the functions $\qquad\qquad y = e^x \qquad$ and $\qquad y = \ln x$

are also inverses of each other.

More Rules for Common and Natural Logarithms

5. $\log(10^x) = x \qquad$ and $\qquad \ln(e^x) = x$

6. $10^{\log x} = x \qquad$ and $\qquad e^{\ln x} = x$

The graphs of two inverse functions are mirror images across the diagonal line $y = x$

The graphs of two inverse functions such as $y = 10^x$ and $y = \log x$ are mirror images across the line $y = x$. If you imagine folding the graph along the dotted line $y = x$, the two curves would lie right on top of each other. See Figure 6.8.

Why are these graphs mirror images? Choose any point (a, b) on the graph of $y = \log x$. To reach that point you would need to move a units horizontally (on the x-axis) and b units vertically (on the y-axis). Now imagine folding at the dotted line $y = x$. What are the coordinates of the point's mirror image? You would need to move a units vertically (on the y-axis) and b units horizontally (on the x-axis). The mirror-image coordinates would be (b, a). So the points (a, b) and (b, a) are mirror images across the line $y = x$. Table 6.3 lists some pairs of mirror-image points that lie on the graphs of $y = 10^x$ and $y = \log x$.

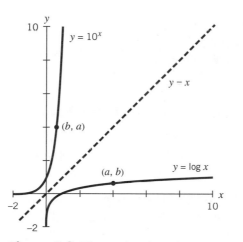

Mirror-Image Points on $y = 10^x$ and $y = \log x$

$y = 10^x$	$y = \log x$
$(0, 1)$	$(1, 0)$
$(1, 10)$	$(10, 1)$
$(2, 100)$	$(100, 2)$
$(-1, 0.1)$	$(0.1, -1)$
$(-2, 0.01)$	$(0.01, -2)$

Table 6.3

Figure 6.8 The graphs of $y = \log x$ and $y = 10^x$ are mirror images across the dotted line $y = x$.

We can use similar arguments to show that the graphs of $y = e^x$ and $y = \ln x$ are mirror images as well.

Inverse Functions and Their Graphs

The functions in each pair

$$y = 10^x \qquad \text{and} \qquad y = \log x$$
$$y = e^x \qquad \text{and} \qquad y = \ln x$$

are *inverses* of each other. What one does, the other undoes.

The graphs in each pair are mirror images across the line $y = x$.

E X A M P L E   3   **Graphs of ln x and e^x**

Graph $y = \ln x$ and $y = e^x$. Identify three pairs of points on the function graphs that are mirror images across the line $y = x$.

SOLUTION Figure 6.9 shows the graphs of $y = \ln x$ and $y = e^x$, and Table 6.4 contains some pairs of points on $y = e^x$ and $y = \ln x$.

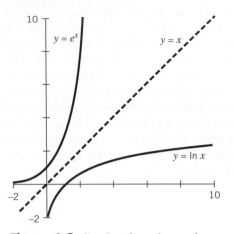

Figure 6.9 Graphs of $y = \ln x$ and $y = e^x$ are mirror images across the dotted line $y = x$.

Mirror-Image Points on $y = e^x$ and $y = \ln x$

$y = e^x$	$y = \ln x$
$(0, 1)$	$(1, 0)$
$(1, e)$	$(e, 1)$
$(-2, e^{-2})$	$(e^{-2}, -2)$

Table 6.4

Applications of Logarithmic Functions

Just as with logarithmic scales, logarithmic functions are used in dealing with quantities that vary widely in size. We'll examine two such functions used to measure acidity and noise levels.

Measuring acidity: The pH scale

Chemists use a logarithmic scale called pH to measure acidity. pH values are defined by the function

$$pH = -\log[H^+]$$

where $[H^+]$ designates the concentration of hydrogen ions. Chemists use the symbol H^+ for hydrogen ions (hydrogen atoms stripped of their one electron), and the brackets $[\ ]$ mean "concentration of." Ion concentration $[H^+]$ is measured in moles per liter, M, where one mole equals $6.022 \cdot 10^{23}$ or Avogadro's number of ions. A pH value is the negative of the logarithm of the number of moles per liter of hydrogen ions.

Table 6.5 and Figure 6.10 show a set of values and a graph for pH. The graph is the standard logarithmic graph flipped across the horizontal axis because of the negative sign in front of the log. The higher the concentration, the smaller the pH. Taking the negative log allows us to separate small increments in the input, such as translating 10^{-15} and 10^{-10} to 15 and 10.

Calculating pH Values

$[H^+]$ (moles per liter)	pH
10^{-15}	15.000
10^{-10}	10.000
1	0.000
5	−0.699
10	−1.000

Table 6.5

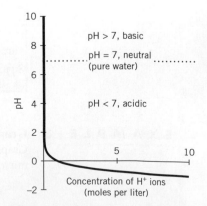

Figure 6.10 Graph of the pH function.

Typically, pH values are between 0 and 14, indicating the level of acidity. Pure water has a pH of 7.0 and is considered *neutral*. A substance with a pH < 7 is called *acidic*. A substance with a pH > 7 is called *basic* or *alkaline*.[1] The lower the pH, the more acidic the substance. The higher the pH, the less acidic and the more alkaline the substance. Table 6.6 shows approximate pH values for some common items. Most foods have a pH between 3 and 7. Substances with a pH below 3 or above 12 can be dangerous to handle with bare hands.

pH of Various Substances

Acidic (more hydrogen ions than pure water)	pH	Neutral	pH	Basic or Alkaline (fewer hydrogen ions than pure water)	pH
Gastric juice	2	Pure water	7	Egg whites, sea water	8
Coca-Cola, vinegar	3			Soap, baking soda	9
Grapes, wine	4			Detergents, toothpaste, ammonia	10
Coffee, tomatoes	5			Household cleaner	12
Bread	5.5			Caustic oven cleaner	13
Beef, chicken	6				

Table 6.6

EXAMPLE 4   **Comparing hydrogen ion concentrations**

a. Compare the hydrogen ion concentration of Coca-Cola with the hydrogen ion concentrations of coffee and ammonia.

b. Calculate the hydrogen ion concentration of Coca-Cola.

SOLUTION   **a.** Table 6.6 gives the pH of Coca-Cola as 3, coffee as 5 (so both are acidic), and ammonia as 10 (which is alkaline). Increasing the pH by 1 corresponds to decreasing the hydrogen ion concentration by a factor of 10. So Coca-Cola will have more ions than coffee, and even more hydrogen than ammonia. The hydrogen ion concentration of Coca-Cola is $10^{5-3} = 10^2 = 100$ times more than that of coffee, and $10^{10-3} = 10^7 = 10,000,000$ times more than that of ammonia!

b. To calculate the hydrogen ion concentration of Coca-Cola, we need to solve the equation

$$3 = -\log[H^+]$$

Multiply by -1 $-3 = \log[H^+]$

rewrite as powers of 10 $10^{-3} = 10^{\log[H^+]}$

evaluate and use Rule 6 of logs $0.001 = [H^+]$

So the hydrogen ion concentration of Coca-Cola is 10^{-3} or 0.001 moles per liter. That means that each liter of Coca-Cola contains $10^{-3} \cdot (6.022 \cdot 10^{23}) = 6.022 \cdot 10^{23-3} = 6.022 \cdot 10^{20}$ hydrogen ions.

EXAMPLE 5   **Calculating the pH level**

Sulfuric acid has a hydrogen ion concentration $[H^+]$ of 0.109 moles per liter. Calculate its pH.

SOLUTION   The pH of sulfuric acid equals $-\log 0.109 \approx 0.96$. This pH is so low and extremely acidic that sulfuric acid can burn flesh.

---

[1]Things that are alkaline tend to be slimy and sticky, like a bar of soap. If you wash your hands with soap and don't rinse, there will be an alkaline residue. Acids can be used to neutralize alkalinity. For example, shampoo often leaves a sticky alkaline residue. So we use acidic conditioners to neutralize the alkalinity.

The rain in many parts of the world is becoming increasingly acidic. The burning of fossil fuels (such as coal and oil) by power plants and automobile emissions release gaseous impurities into the air. The impurities combine with moisture in the air to form droplets of dilute sulfuric and nitric acids, which release hydrogen ions. High concentrations of hydrogen ions damage plants and water resources (such as the lakes of New England and Sweden) and erode structures (such as the Parthenon in Athens) by removing oxygen molecules.

EXAMPLE 6

The pH function is not linear

It is critical that health care professionals understand the nonlinearity of the pH function. An arterial blood pH of 7.35 to 7.45 for a patient is quite normal, whereas a pH of 7.1 means that the patient is severely acidotic and near death.

a. Determine the hydrogen ion concentration, $[H^+]$, first for a patient with an arterial blood pH of 7.4, then for one with a blood pH of 7.1.

b. How many more hydrogen ions are there in blood with a pH of 7.1 than in blood with a pH of 7.4?

SOLUTION

a. To find $[H^+]$ if the pH is 7.4, we must solve the equation

$$7.4 = -\log[H^+]$$

Multiply both sides by -1	$-7.4 = \log[H^+]$
then these powers of 10 are equal	$10^{-7.4} = 10^{\log[H^+]}$
use Rule 6 of logs	$10^{-7.4} = [H^+]$
evaluate and switch sides	$[H^+] \approx 0.000\,000\,040$
	$\approx 4.0 \cdot 10^{-8}\,M$

where M is in moles per liter.

Similarly, if the pH is 7.1, we can solve the equation

$$7.1 = -\log[H^+]$$

to get $\quad [H^+] \approx 7.9 \cdot 10^{-8}\,M$

b. Comparing the two concentrations gives us

$$\frac{[H^+] \text{ in blood with a pH of 7.1}}{[H^+] \text{ in blood with a pH of 7.4}} \approx \frac{7.9 \cdot 10^{-8}\,M}{4.0 \cdot 10^{-8}\,M} = \frac{7.9}{4.0} \approx 2$$

So there are almost double the number of hydrogen ions in blood with a pH of 7.1 than in blood with a pH of 7.4.

Measuring noise: The decibel scale

The decibel scale was designed to reflect the human perception of sounds.[2] When it is very quiet, it is easy to notice a small increase in sound intensity. The same increase in intensity in a noisy environment would not be noticed; it would take a much bigger change to be detected by humans. The decibel scale, like the pH scale for acidity or the Richter scale for earthquakes, is logarithmic; that is, it measures order-of-magnitude changes.

Noise levels are measured in units called *decibels,* abbreviated dB. The name is in honor of the inventor of the telephone, Alexander Graham Bell. If we designate I_0 as the intensity of a sound at the threshold of human hearing (10^{-16} watts/cm^2) and we let

---

[2]Two scientists, Weber and Fechner, studied the psychological response to intensity changes in stimuli. Their discovery, that the perceived change is proportional to the logarithm of the intensity change of the stimulus, is called the Weber-Fechner stimulus law.

I represent the intensity of an arbitrary sound (measured in watts/cm^2), then the noise level N of that sound measured in decibels (dB) is defined to be

$$N = 10 \log \left(\frac{I}{I_0} \right)$$

The unitless expression I/I_0 gives the *relative intensity* of a sound compared with the reference value of I_0. For example, if $I/I_0 = 100$, then the noise level, N, is equal to

$$N = 10 \log(100) = 10 \log(10^2) = 10(2) = 20 \text{ dB}$$

Table 6.7 shows relative intensities, the corresponding noise levels (in decibels), and how people perceive these noise levels. Note how much the relative intensity (the ratio I/I_0) of a sound source must increase for people to discern differences. Each time we *add* 10 units on the decibel scale, we *multiply* the relative intensity by 10, increasing it by one order of magnitude.

How Decibel Levels Are Perceived

Relative Intensity I/I_0	Decibels (dB)	Average Perception
1	0	Threshold of hearing
10	10	Soundproof room, very faint
100	20	Whisper, rustle of leaves
1,000	30	Quiet conversation, faint
10,000	40	Quiet home, private office
100,000	50	Average conversation, moderate
1,000,000	60	Noisy home, average office
10,000,000	70	Average radio, average factory, loud
100,000,000	80	Noisy office, average street noise
1,000,000,000	90	Loud truck, police whistle, very loud
10,000,000,000	100	Loud street noise, noisy factory
100,000,000,000	110	Elevated train, deafening
1,000,000,000,000	120	Thunder of artillery, nearby jackhammer
10,000,000,000,000	130	Threshold of pain, ears hurt

Table 6.7

E X A M P L E 7

How loud is a rock band?
What is the decibel level of a typical rock band playing with an intensity of 10^{-5} watts/cm^2? How much more intense is the sound of the band than an average conversation?

S O L U T I O N Given $I_0 = 10^{-16}$ watts/cm^2 and letting $I = 10^{-5}$ watts/cm^2 and N represent the decibel level,

by definition	$N = 10 \log \left(\dfrac{I}{I_0} \right)$
substitute for I and I_0	$= 10 \log(10^{-5}/10^{-16})$
Rule 2 of exponents	$= 10 \log(10^{11})$
Rule 5 of logs	$= 10 \cdot 11$
	$= 110$ decibels

So the noise level of a typical rock band is about 110 decibels.

According to Table 6.7, an average conversation measures about 50 decibels. So the noise level of the rock band is 60 decibels higher. Each increment of 10 decibels corresponds to a one-order-of-magnitude increase in intensity. So the sound of a rock band is about six orders of magnitude, or 10^6 (a million times), more intense than an average conversation.

EXAMPLE 8

Perceiving sound

What's wrong with the following statement? "A jet airplane landing at the local airport makes 120 decibels of noise. If we allow three jets to land at the same time, there will be 360 decibels of noise pollution."

SOLUTION

There will certainly be three times as much sound intensity, but would we perceive it that way? According to Table 6.7, 120 decibels corresponds to a relative intensity of 10^{12}. Three times that relative intensity would equal $3 \cdot 10^{12}$. So the corresponding decibel level would be

$$N = 10 \log(3 \cdot 10^{12})$$

Rule 1 of logs	$= 10(\log 3 + \log(10^{12}))$
use a calculator	$\approx 10(0.477 + 12)$
combine	$\approx 10(12.477)$
multiply and round off	≈ 125 decibels

So three jets landing will produce a decibel level of 125, not 360. We would perceive only a slight increase in the noise level.

Algebra Aerobics 6.3

A graphing program is recommended for Problem 3 and a scientific calculator for Problem 7.

1. How would the graphs of $y = \log x^2$ and $y = 2 \log x$ compare?

2. Draw a rough sketch of the graph $y = -\ln x$. Compare it with the graph of $y = \ln x$.

3. Compare the graphs of $f(x) = \log x$ and $g(x) = \ln x$. (See Figure 6.3, p. 367.)

   a. Where do they intersect?

   b. Where does each have an output value of 1? Of 2?

   c. Describe each graph for values of x such that $0 < x < 1$.

   d. Describe each graph for values of x where $x > 1$.

4. Given the accompanying graph (top of next column) of a function f, sketch the graph of its inverse.

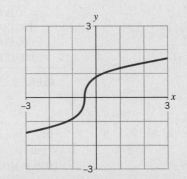

5. A typical pH value for rain or snow in the northeastern United States is about 4. Is this basic or acidic? What is the corresponding hydrogen ion concentration? How does this compare with the hydrogen ion concentration of pure water?

6. What is the decibel level of a sound whose intensity is $1.5 \cdot 10^{-12}$ watts/cm^2?

7. If the intensity of a sound increases by a factor of 100, what is the increase in the decibel level? What if the intensity is increased by a factor of 10,000,000?

Exercises for Section 6.3

Some of these exercises require a calculator that can evaluate powers and logs. Exercises 6 and 14 require a graphing program.

1. Use the rules of logarithms to explain how you can tell which graph is $y = \log x$ and which is $y = \log(5x)$.

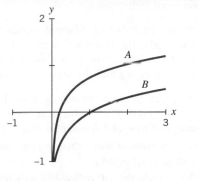

2. Use the rules of logarithms to explain how you can tell which graph is $y = \log x$ and which is $y = \log(x/5)$.

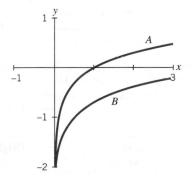

3. The functions $f(x) = \log x$, $g(x) = \log(x - 1)$, and $h(x) = \log(x - 2)$ are graphed below.
   a. Match each function with its graph.
   b. Find the value of x for each function that makes that function equal to zero.
   c. Identify the x-intercept for each function.
   d. Assuming $k > 0$, describe how the graph of $f(x)$ moves if you replace x by $(x - k)$.

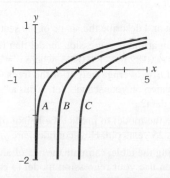

4. The functions $f(x) = \ln x$, $g(x) = \ln(x + 1)$, and $h(x) = \ln(x + 2)$ are graphed below.
   a. Match each function with its graph.
   b. Find the value of x for each function that makes that function equal to zero.
   c. Identify the coordinates of the horizontal intercept for each function.
   d. Assuming $k > 0$, describe how the graph of $f(x)$ moves if you replace x by $(x + k)$.
   e. Determine the y-intercept, if possible, for each function.

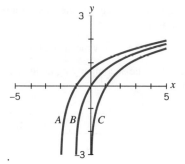

5. Examine the following graphs of four functions.
   a. Which function graphs are mirror images of each other across the y-axis?
   b. Which function graphs are mirror images of each other across the x-axis?

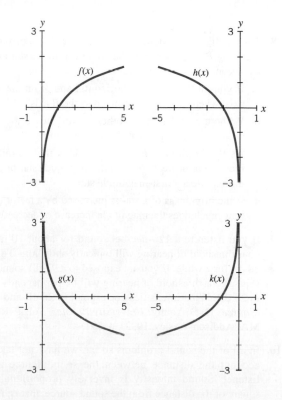

6. (Requires a graphing program.) On the same grid graph $y_1 = \ln(x)$, $y_2 = -\ln(x)$, $y_3 = \ln(-x)$ and $y_4 = -\ln(-x)$.

   a. Which pairs of function graphs are mirror images across the y-axis?

   b. Which pairs of function graphs are mirror images across the x-axis?

   c. What predictions would you make about the graphs of the functions $f(x) = a \ln x$ and $g(x) = -a \ln x$? Using technology, test your predictions for different values of a.

   d. What predictions would you make about the graphs of the functions $f(x) = a \ln x$ and $g(x) = a \ln(-x)$? Using technology, test your predictions for different values of a.

7. Logarithms can be constructed using any positive number except 1 as a base:

   $$\log_a x = y \text{ means that } a^y = x$$

   a. Complete the accompanying table and sketch the graph of $y = \log_3 x$.

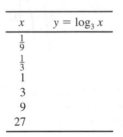

x	$y = \log_3 x$
$\frac{1}{9}$	
$\frac{1}{3}$	
1	
3	
9	
27	

   b. Now make a small table and sketch the graph of $y = \log_4 x$. (*Hint:* To simplify computations, try using powers of 4 for values of x.)

8. The stellar magnitude M of a star is approximately $-2.5 \log(B/B_0)$, where B is the brightness of the star and B_0 is a constant.

   a. If you plotted B on the horizontal and M on the vertical axis, where would the graph cross the B axis?

   b. Without calculating any other coordinates, draw a rough sketch of the graph of M. What is the domain?

   c. As the brightness B increases, does the magnitude M increase or decrease? Is a sixth-magnitude star brighter or dimmer than a first-magnitude star?

   d. If the brightness of a star is increased by a factor of 5, by how much does the magnitude increase or decrease?

9. If you listen to a 120-decibel sound for about 10 minutes, your threshold of hearing will typically shift from 0 dB up to 28 dB for a while. If you are exposed to a 92-dB sound for 10 years, your threshold of hearing will be permanently shifted to 28 dB. What intensities correspond to 28 dB and 92 dB? (*Source:* H. D. Young, *University Physics,* Vol. 1 (Reading, MA: Addison-Wesley, 1992), p. 591)

10. In all of the sound problems so far, we have not taken into account the distance between the sound source and the listener. Sound intensity is inversely proportional to the square of the distance from the sound source; that is, $I = k/r^2$,

where I is intensity, r is the distance from the sound source, and k is a constant.

Suppose that you are sitting a distance R from the TV, where its sound intensity is I_1. Now you move to a seat twice as far from the TV, a distance $2R$ away, where the sound intensity is I_2.

   a. What is the relationship between I_1 and I_2?

   b. What is the relationship between the decibel levels associated with I_1 and I_2?

11. If there are a number of different sounds being produced simultaneously, the resulting intensity is the sum of the individual intensities. How many decibels louder is the sound of quintuplets crying than the sound of one baby crying?

12. An ulcer patient has been told to avoid acidic foods. If he drinks coffee, with a pH of 5.0, it bothers him, but he can tolerate both tap water, with a pH of 5.8, and milk, with a pH of 6.9.

   a. Will a mixture of half coffee and half milk be at least as tolerable as tap water?

   b. What pH will the half coffee–half milk mixture have?

   c. In order to make 10 oz of a milk-coffee drink with a pH of 5.8, how many ounces of each are required?

13. Lemon juice has a pH of 2.1. If you make diet lemonade by mixing $\frac{1}{4}$ cup of lemon juice with 2 cups of tap water, with a pH of 5.8, will the resulting acidity be more or less than that of orange juice, with a pH of 3?

14. (Requires graphing program.) The data below show the average heights of 12 Weeping Higan cherry trees planted in Washington, D.C., as they grew over time. At the time of the planting, each of the trees was one year old and 6 feet in height.

Age of Tree (years)	Height (feet)
1	6.0
2	9.5
3	13.0
4	15.0
5	16.5
6	17.5
7	18.5
8	19.0
9	19.5
10	19.7
11	19.8

   a. Plot and describe the shape of the scatter plot data.

   b. A logarithmic regression model that represents the data is $h = 6.0182 \ln x + 6.0993$, where $x =$ age of tree (in years) and $h =$ average height of the trees (in feet). Graph this equation on your scatter plot. Is this a "good fit" to represent the data?

   c. Use the model to predict the height of a cherry tree when it is 25 years old. How confident are you in the prediction?

   d. Using the table, estimate the age of an 11-foot cherry tree. Then use your regression model to calculate the average age of an 11-foot tree.

6.4   *Using Semi-Log Plots to Construct Exponential Models for Data*

See "E ll: Semi-log Plots of $y = Ca^x$" in Logarithmic and Exponential Functions to visualize the relationship between a standard vs. a semi-log plot of an exponential function.

In Chapter 5 we learned that an exponential function appears as a straight line on a semi-log plot (where the logarithmic scale is on the vertical axis). To decide whether an exponential function is an appropriate model for a data set, we can plot the data on a semi-log plot and see if it appears to be linear. This is one of the easiest and most reliable ways to recognize exponential growth in a data set. And, as we learned in Chapters 4 and 5, a logarithmic scale has the added advantage of being able to display clearly a wide range of values.

Why Do Semi-Log Plots of Exponential Functions Produce Straight Lines?

Consider the exponential function

$$y = 3 \cdot 2^x \tag{1}$$

If we take the log of both sides	$\log y = \log (3 \cdot 2^x)$
use Rule 1 of logs	$= \log 3 + \log (2^x)$
use Rule 3 of logs	$= \log 3 + x \log 2$
evaluate logs and rearrange, we have	$\log y \approx 0.48 + 0.30x$
If we set $Y = \log y$, we have	$Y = 0.48 + 0.30x \tag{2}$

So Y (or $\log y$) is a linear function of x. Equations (1) and (2) are equivalent to each other. The graph of Equation (2) on a semi-log plot, with Y (or $\log y$) values on the vertical axis, is a straight line (see Figure 6.11). The slope is 0.30 or $\log 2$, the logarithm of the growth factor 2 of Equation (1). The vertical intercept is 0.48 or $\log 3$, the logarithm of the y-intercept 3 of Equation (1).

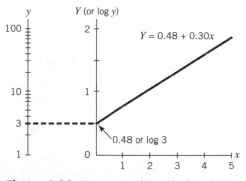

Figure 6.11 The graph of $Y = 0.48 + 0.30x$, showing the relationship between two equivalent logarithmic scales on the vertical axis.

Figure 6.11 shows two equivalent variations of logarithmic scales on the vertical axis. One plots the value of y on a logarithmic scale (using powers of 10), and the other plots the value of $\log y$ (using the exponents of the powers of 10). Spreadsheets and some graphing calculators have the ability to instantly switch axis scales between standard linear and the logarithmic scale using powers of 10. But one can in effect do the same thing by plotting $\log y$ instead of y. Notice that on the vertical log y scale, the units are now evenly spaced. This allows us to use the standard strategies for finding the slope and vertical intercept of a straight line.

In general, we can translate an exponential function in the form

$$y = Ca^x \quad \text{(where } C \text{ and } a > 0 \text{ and } a \neq 1\text{)}$$

into an equivalent linear function

$$\log y = \log C + (\log a)x \quad \text{or}$$
$$Y = \log C + (\log a)x \quad \text{where } Y = \log y$$

Finding the Equation of an Exponential Function on a Semi-Log Plot

The graph of an exponential function $y = Ca^x$ (where C and $a > 0$ and $a \neq 1$) appears as a straight line on a semi-log plot. The line's equation is

$$Y = \log C + (\log a)x \quad \text{where } Y = \log y$$

The slope of the line is $\log a$, where a is the growth (or decay) factor for $y = Ca^x$. The vertical intercept is $\log C$, where C is the vertical intercept of $y = Ca^x$.

Increase in hard drive capacity

In 1980 Seagate Technology created the first hard disk drive for microcomputers to replace the use of $5\frac{1}{4}''$ floppy disks. Hard disks became an essential part of the computer revolution, repeatedly adding more and more storage capacity. Figure 6.12 shows on a semi-log plot the astounding growth in capacity (in gigabytes or GB = one billion bytes) between 1980 and 2010.

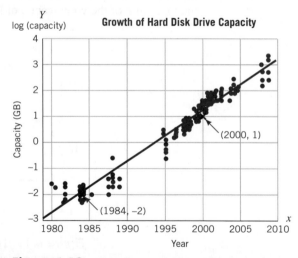

Figure 6.12 Hard disk capacity with best-fit line.

The data points lie approximately on a straight line (on the semi-log plot), so an exponential model is appropriate.

E X A M P L E   1   **Constructing an exponential function from a semi-log plot**
From Figure 6.12, estimate the annual percent growth rate in hard disk capacity between 1980 and 2010.

S O L U T I O N   We first construct a linear function for the best-fit line, then translate it to an exponential growth function to find the percent growth rate.

Figure 6.12 shows the regression line for the data. We can estimate the coordinates of two points on the regression line, *not* from the data, as (1984, −2) and (2000, 1). Then the slope or average rate of change between them is

$$\text{slope} = \frac{\text{change in } Y}{\text{change in } x} = \frac{1 - (-2)}{2000 - 1984} = \frac{3}{16} \approx 0.19$$

If we let x = number of years since 1980, the slope would stay the same. Reading off the graph, the vertical intercept is about −3. So the equation for the best-fit line is approximately

$$Y = -3 + 0.19x \qquad \text{where } Y = \log(\text{capacity}) \tag{1}$$

We can translate Equation (1) to an exponential function if we

substitute log(capacity) for Y	$\log(\text{capacity}) = -3 + 0.19\,x$
rewrite as a power of 10	$10^{\log(\text{capacity})} = 10^{(-3 + 0.19x)}$
use Rule 6 of logs and Rule 1 of exponents	$\text{capacity} = 10^{-3} \cdot 10^{0.19x}$
Rule 3 of exponents	$\text{capacity} = 10^{-3} \cdot (10^{0.19})^{x}$
Evaluate	$\text{capacity} \approx 0.001 \cdot 1.55^{x} \tag{2}$

The annual growth factor is approximately 1.55, so the annual growth rate in decimal form is 0.55. So between 1980 and 2010, hard drive capacity grew by about 55% each year!

EXPLORE & EXTEND

6.4   Death Down Under

How long do you expect to live? What's the probability you will be alive in one year? In ten years? Insurance companies determine their premiums (the amount you have to pay) by calculating the probability that you will be alive one year from now. The following are two graphs, created from "Life Tables Australia" from the Australian Bureau of Statistics, that show the same data about Australian female mortality from different perspectives. (Note: A probability of 1 means there is a 100% chance of dying that year.)

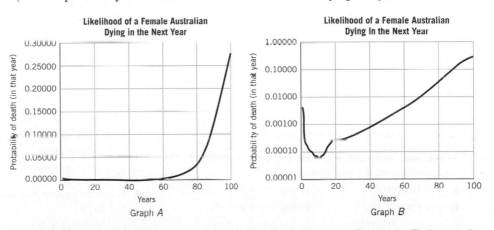

a. Identify the type of vertical scale on Graph *A* and then Graph *B*. Estimate the probability of death at 20 years from Graph *A*, then from Graph *B*. Explain why it's easier to estimate on Graph *B*.

b. Using Graph *B*, describe what happens to the probability of death between 0 and 25 years.

c. After 25 years, what does likelihood of death suggest on Graph *B*? Is this confirmed on Graph *A*?

Algebra Aerobics 6.4

Many of the problems require a calculator that can evaluate logs and exponents.

1. a. Which of the following three functions would have a straight-line graph on a standard linear plot? On a semi-log plot?

$$y = 3x + 4, \quad y = 4 \cdot 3^x, \quad \log y = (\log 3) \cdot x + \log 4$$

b. For each straight-line graph in part (a), what is the slope of the line? The vertical intercept? (*Hint:* Substitute $Y = \log y$ and $X = \log x$ as needed.)

2. Change each exponential equation to a logarithmic equation.

a. $y = 5(3)^x$ **c.** $y = 10,000(0.9)^x$

b. $y = 1000(5)^x$ **d.** $y = (5 \cdot 10^6)(1.06)^x$

3. Change each logarithmic equation to an exponential equation.

a. $\log y = \log 7 + (\log 2) \cdot x$

b. $\log y = \log 20 + (\log 0.25) \cdot x$

c. $\log y = 6 + (\log 3) \cdot x$

d. $\log y = 6 + \log 5 + (\log 3) \cdot x$

4. a. Below are linear equations of $\log y$ (or Y) in x. Identify the slope and vertical intercept for each.

   i. $\log y = \log 2 + (\log 5) \cdot x$

   ii. $\log y = (\log 0.75) \cdot x + \log 6$

   iii. $\log y = 0.4 + x \log 4$

   iv. $\log y = 3 + \log 2 + (\log 1.05) \cdot x$

b. Use the slope and intercept to create an exponential function for each equation.

5. Solve each equation using the definition of log.

a. $0.301 = \log a$ **c.** $\log a = -0.125$

b. $\log C = 2.72$ **d.** $5 = \log C$

6. Change each function to exponential form, assuming that $Y = \log y$.

a. $Y = 0.301 + 0.477x$ **c.** $Y = 1.398 - 0.046x$

b. $Y = 3 + 0.602x$

7. For each of the two accompanying graphs examine the scales on the axes. Then decide whether a linear or an exponential function would be the most appropriate model for the data.

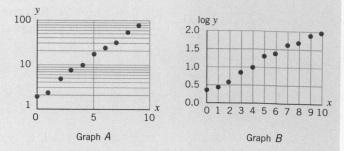

Graph *A*   Graph *B*

8. Generate an exponential function that could describe the data in the accompanying graph.

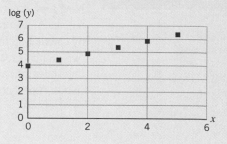

Exercises for Section 6.4

Some exercises require a calculator that can evaluate powers and logs. Exercise 10 requires a graphing program.

1. Match each exponential function in parts (a)–(d) with its logarithmic form in parts (e)–(h).

a. $y = 10,000(2)^x$ **e.** $\log y = 6.477 - 0.097x$

b. $y = 1000(1.4)^x$ **f.** $\log y = 4 + 0.301x$

c. $y = (3 \cdot 10^6)(0.8)^x$ **g.** $\log y = 3 - 0.347x$

d. $y = 1000(0.45)^x$ **h.** $\log y = 3 + 0.146x$

2. Form the exponential function from its logarithmic equivalent for each of the following.

a. $\log y = \log 1400 + (\log 1.06)x$

b. $\log y = \log (25,000) + (\log 0.87)x$

c. $\log y = 2 + (\log 2.5)x$

d. $\log y = 4.25 + (\log 0.63)x$

3. Change each exponential function to its logarithmic equivalent. Round values to the nearest thousandth.

a. $y = 30,000(2)^x$ **c.** $y = (4.5 \cdot 10^6)(0.7)^x$

b. $y = 4500(1.4)^x$ **d.** $y = 6000(0.57)^x$

4. Write the exponential equivalent for each function. Assume $Y = \log y$.

a. $Y = 2.342 + 0.123x$ **c.** $Y = 4.74 - 0.108x$

b. $Y = 3.322 + 0.544x$ **d.** $Y = 0.7 - 0.004x$

5. Match each exponential function with its semi-log plot at the top of the next page.

a. $y = 30,000(2)^x$

b. $y = 4500(1.4)^x$

c. $y = (4.5 \cdot 10^6)(0.7)^x$

d. $y = 6000(0.57)^x$

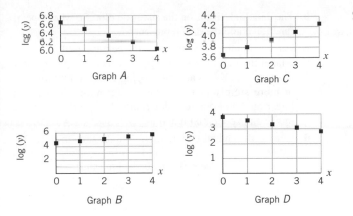

Graph A

Graph C

Graph B

Graph D

6. a. From the data in the following table, create a linear equation of Y in terms of x.

x	$\log y$ (or Y)
0	5.00000
1	5.60206
2	6.20412
3	6.80618
4	7.40824

b. Find the equivalent exponential function of y in terms of x.

7. Determine which data sets (if any) describe y as an exponential function of x, then construct the exponential function. (*Hint:* Find the average rate of change of Y with respect to x.)

a.

x	$\log y$ (or Y)
0	2.30103
10	4.30103
20	4.90309
30	5.25527
40	5.50515

b.

x	$\log y$ (or Y)
0	4.77815
10	3.52876
20	2.27938
30	1.02999
40	−0.21945

8. Construct two functions that are equivalent descriptions of the data in the accompanying graph. Describe the change in $\log y$ (or Y) with respect to x.

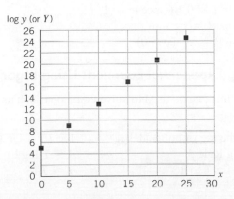

9. The accompanying graph shows the growth of a bacteria population over a 25-day period.

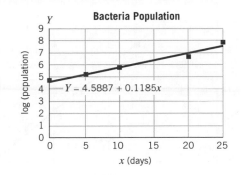

Bacteria Population

$Y = 4.5887 + 0.1185x$

a. Do the bacteria appear to be growing exponentially? Explain.

b. Translate the best-fit line shown on the graph into an exponential function and determine the daily growth rate.

10. (Requires a graphing program.) The current population of a city is 1.5 million. Over the next 40 years, the population is expected to decrease by 12% each decade.

a. Create a function that models the population decline.

b. Create a table of values at 10-year intervals for the next 40 years.

c. Graph the population of the city on a semi-log plot.

11. An experiment by a pharmaceutical company tracked the amount of a certain drug (measured in micrograms/liter) in the body over a 30-hour period. The accompanying graph shows the results.

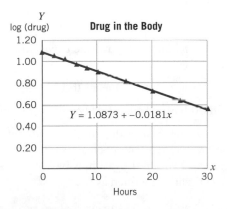

Drug in the Body

$Y = 1.0873 + -0.0181x$

a. What was the initial amount of the drug?

b. Is the amount of drug in the body decaying exponentially? If yes, state the decay rate.

c. If the amount in the body is decaying exponentially, construct an exponential model.

12. The graph at the top of the next page shows the decay of strontium-90.

a. Is the graph a semi-log plot?

b. What is the initial amount of strontium-90 for this graph?

c. According to the graph, what is the half-life of strontium-90?

d. What would be the yearly decay rate for strontium-90?

e. Find the exponential model for the decay of strontium-90.

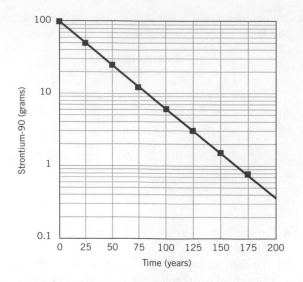

b. The data look linear on the semi-log plot. So what type of function could the data represent?

c. Let x = number of years since the beginning of 1994 and relabel the vertical axis as $Y = \log(\text{pixels}/\$)$. So the new labels for the Y-axis are log 100, log 1000, and log 10,000, or more simply just 1, 2, and 3. Then an approximate linear model of our semi-log plot is $Y = 1000\,x$. Convert this to an exponential function.

d. What is the annual growth rate? And the growth factor?

13. A pixel is the smallest point in an electronic image. Each pixel displays one color, and a collection of pixels creates an image. The more the pixels, the higher the resolution and the sharper the image. Barry Hendy of Kodak, Australia, plotted "pixels per dollar" (see graph to right) as a basic measure of value for a digital camera. (*Note:* Some have objected to the accuracy of some of the data, but not to the overall trend.)

a. About how many orders of magnitude greater in pixels/$ is the CX7300 camera (at (2004, 10000)) than the DC8460 (at (1994, 100))? Interpret your results.

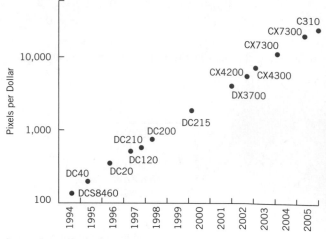

Source: Barry Hendy, Digital camera.

CHAPTER SUMMARY

Logarithms

The *logarithm base 10 of x* is the *exponent* of 10 needed to produce x.

$$\log x = c \qquad \text{means that} \qquad 10^c = x$$

Logarithms base 10 are called *common logarithms*.

The *logarithm base e of x* is the *exponent* of e needed to produce x. The number e is an irrational natural constant ≈ 2.71828. Logarithms base e are called *natural logarithms* and are written as $\ln x$:

$$\ln x = c \qquad \text{means that} \qquad e^c = x$$

If $x > 1$, $\log x$ and $\ln x$ are both positive
If $x = 1$, $\log 1 = 0$ and $\ln 1 = 0$
If $0 < x < 1$, $\log x$ and $\ln x$ are both negative
If $x \le 0$ neither logarithm is defined
and $\log 10 = 1$ and $\ln e = 1$

The rules for logarithms follow directly from the definition of logarithms and from the rules for exponents.

Rules for Logarithms

If A and B are positive real numbers and p is any real number, then:

For Common Logarithms

1. $\log(A \cdot B) = \log A + \log B$
2. $\log(A/B) = \log A - \log B$
3. $\log A^p = p \log A$
4. $\log 1 = 0$ (since $10^0 = 1$)
5. $\log 10^x = x$
6. $10^{\log x} = x$ $(x > 0)$

For Natural Logarithms

1. $\ln(A \cdot B) = \ln A + \ln B$
2. $\ln(A/B) = \ln A - \ln B$
3. $\ln A^p = p \ln A$
4. $\ln 1 = 0$ (since $e^0 = 1$)
5. $\ln e^x = x$
6. $e^{\ln x} = x$ $(x > 0)$

Solving Exponential Equations Using Logarithms

Logarithms can be used to solve equations such as $10 = 2^T$ by:

taking the log of both sides	$\log 10 = \log(2^T)$
applying Rule 3	$1 = T \log 2$
solving for T to get	$T \approx 3.32$

Logarithmic Functions

We can define two functions: $y = \log x$ and $y = \ln x$ (where $x > 0$).

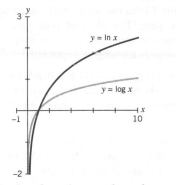

Graphs of $y = \log x$ and $y = \ln x$

When $x > 1$:   a large change in the input for $\log x$ corresponds to a small change in the output; so $\log x$ compresses large variations, reducing large numbers to a manageable size

When $0 < x < 1$:   a small change in the input for $\log x$ corresponds to a large change in the output; so $\log x$ expands small variations, separating small numbers

The graphs of these functions have similar shapes, and both

- lie to the right of the y-axis
- have a horizontal intercept of $(1, 0)$
- are concave down
- increase throughout with no maximum or minimum
- are asymptotic to the y-axis

Two well-known examples of logarithms are the pH and decibel scales.

Inverse Functions

Two functions are *inverses* of each other if what one function "does" the other "undoes." The graphs of two inverse functions are mirror images across the line $y = x$. The logarithmic and exponential functions are inverse functions.

Using Semi-Log Plots

The graph of an exponential function $y = Ca^x$ appears as a straight line on a semi-log plot. The line's equation is

$$Y = \log C + (\log a)x \qquad \text{where } Y = \log y$$

The slope of the line is $\log a$, where a is the growth (or decay) factor for $y = Ca^x$.

The vertical intercept is $\log C$, where C is the vertical intercept of $y = Ca^x$.

CHECK YOUR UNDERSTANDING

I. Is each of the statements in Problems 1–30 true or false? If false, give an explanation for your answer.

1. For the function $y = \log x$, if x is increased by a factor of 10, then y increases by 1.
2. If $\log x = c$, then x is always positive but c can be positive, negative, or zero.
3. $\ln(1.08/2) = \ln(1.08)/\ln(2)$
4. Because the function $y = \log x$ is always increasing, it has no vertical asymptotes.
5. Both $\log 1$ and $\ln 1$ equal 0.
6. $\ln a = b$ means $b^e = a$.
7. $2 < e < 3$
8. $\log(10^2) = (\log 10)^2$
9. $\log(10^3 \cdot 10^5) = 15$
10. $\log\left(\dfrac{10^7}{10^2}\right) = \dfrac{\log(10^7)}{\log(10^2)}$
11. $\log\left(\dfrac{\sqrt{AB}}{(C+2)^3}\right)$
    $= \dfrac{1}{2}(\log A + \log B) - 3\log(C+2)$
12. $5\ln A + 2\ln B = \ln(A^5 + B^2)$
13. If $(2.3)^x = 64$, then $x = \dfrac{\log 64}{\log 2.3}$.
14. If $\log 20 = t\log 1.065$, then $t = \log\left(\dfrac{20}{1.065}\right)$.
15. If $\ln 5 \approx 1.61$, then $e^{1.61} \approx 5$.

16. $\ln t$ exists for any value of t.
17. If $f(t) = 100 \cdot a^t$ and $g(t) = 100 \cdot b^t$ are exponential decay functions and $a < b$, then the half-life of f is longer than the half-life of g.
18. The graphs of the functions $y = \ln x$ and $y = \log x$ increase indefinitely.
19. The graphs of the functions $y = \ln x$ and $y = \log x$ have both vertical and horizontal asymptotes.
20. The graph of the function $y = \ln x$ lies above the graph of the function $y = \log x$ for all $x > 0$.
21. The graph of the function $y = \log x$ is always positive.
22. The graph of the function $y = \ln x$ is always decreasing.
23. The amount of $100 invested at 8% compounded continuously has a doubling time of about 8.7 years. (*Hint:* $e^{0.08} \approx 1.083$)
24. Of the functions of the form $P = P_0 e^{rt}$ graphed in the accompanying figure, graph B has the smallest growth rate, r.

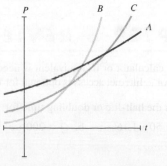

25. Of the two functions $f(x) = \ln x$ and $g(x) = \log x$ graphed in the accompanying figure, graph A is the graph of $f(x) = \ln x$.

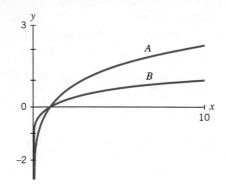

Problems 26–30 refer to the following table and semi-log plot.

U.S. Trade with China (in billions of dollars)

Year	Exports	Imports
2001	19.2	102.3
2002	22.1	125.2
2003	28.4	152.4
2004	34.7	196.7
2005	41.9	243.5
2006	50.0	263.6
2007	62.9	321.4
2008	69.7	337.8
2009	69.6	296.4

Source: U.S. Bureau of the Census, Foreign Trade Statistics, *www.census.gov/foreign/trade/balance.*

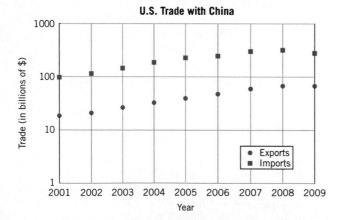

26. Between 2001 and 2009 America's trade deficit (= exports − imports) with China remained roughly constant.

27. We can think of both U.S. exports to China and imports from China (between 2001 and 2009) as functions of the year. The plots of both data sets appear to be approximately linear on a semi-log plot, so both functions are roughly exponential.

28. If $\log(\text{imports}) = 2.061 + 0.065t$ (where t = years since 2001) is the equation of the best-fit line for imports on a semi-log plot, then the annual growth rate for imports is about 6.5%.

29. If we describe the growth in U.S. imports with the equation U.S. imports $I(t) = 115(1.16)^t$ where t is in years, then the annual growth rate for imports is about 16%.

30. If equations in Problems 28 and 29 are correct, they should be equivalent.

II. In Problems 31–33, give an example of a function or functions with the specified properties. Express your answers using equations, and specify the independent and dependent variables.

31. A function whose graph is identical to the graph of the function $y = \log\sqrt{\dfrac{x}{3}}$.

32. A function (that doesn't use e) whose graph is identical to the graph of the function $y = 50.3e^{0.06t}$. (*Hint:* $e^{0.06} \approx 1.062$)

33. An exponential function $y = f(x)$ whose graph on a semi-log plot $(x, \log y)$ is the function $\log y = 2 + 0.08x$. (*Hint:* $10^{0.08} \approx 1.20$)

III. Are the statements in Problems 34–37 true or false? If a statement is true, explain how you know. If a statement is false, give a counterexample.

34. The number e used in the natural logarithm is a variable.

35. The graphs of the functions $y = Ca^x$ and $y = Ce^{rx}$, where $r = \ln a$, are identical.

36. Any substance with a pH > 7 is acidic.

37. A decibel level of 80 is 10 times greater in noise level than a decibel level of 70.

CHAPTER 6 REVIEW: PUTTING IT ALL TOGETHER

A scientific calculator or its equivalent is needed to evaluate logs and powers of e. Internet access is optional for Exercises 4 and 6.

1. What is the half-life or doubling time for:

    a. $Q = 50 \cdot 1.16^t$ **b.** $Q = 200 \cdot e^{-0.083t}$

2. a. According to the chart at the top of next page, approximately how many deaths from natural disasters in 2008 were due to floods? To severe storms? To wildfires?

    b. Why was this bar chart constructed with deaths displayed on a logarithmic scale?

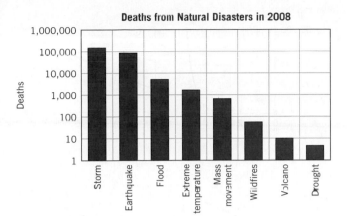

Deaths from Natural Disasters in 2008

Source: www.reliefweb.int.

3. Which of the following expressions are equivalent?

a. $\log(xy^2)$

b. $2\log(xy)$

c. $\log(xy)^2$

d. $2\log x + \log y$

e. $\log x + 2\log y$

f. $2\log x + 2\log y$

4. (Internet optional for part (d).) The Lead-Based Paint Poisoning Prevention Act (1971) was passed to reduce the toxic blood levels of lead in young children who might eat pieces of peeling lead-based paint. The following chart shows the subsequent drop in the average blood lead level of children, along with additional, related legislative acts.

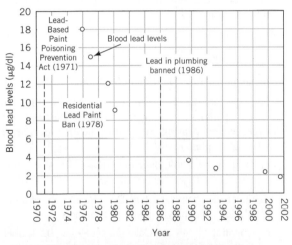

Impact of Lead Poisoning Prevention Policy on Reducing Children's Blood Lead Levels in the United States, 1971–2001

Source: Blood lead levels: National Health and Nutrition Examination Survey. National Center for Health Statistics, Centers for Disease Control and Prevention.

The best-fit exponential decay function for the data points on this graph is $L(t) = 18(0.88)^t$, where $L(t)$ is the average blood lead level (in micrograms per deciliter) of children and t — years since 1976 (the first year data are available).

a. What is the annual decay rate?

b. From the graph, estimate the half-life by estimating when the blood lead level in 1976 was reduced by 50%.

c. According to your model, when would the initial blood lead level be reduced by 50%?

d. What would your model predict for 2010? Check your answer on the Internet.

5. The equation $A(T) = 325 \cdot (0.5)^T$ and accompanying graph describe the amount of aspirin in your bloodstream at time T (measured in 20-minute periods) after you have ingested a standard aspirin of 325 milligrams (mg).

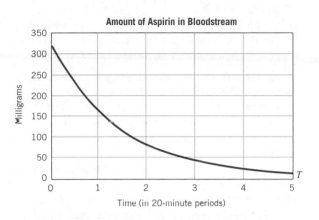

Amount of Aspirin in Bloodstream

a. Using the equation, how much aspirin will remain after one time period? After two time periods? So what is the half-life?

b. Use the graph to estimate the number of time periods it will take for the aspirin level to reach 100 mg. Now calculate the time using your equation and compare your results.

c. Many doctors suggest that their adult patients take a daily dose of "baby aspirin" (81 mg) for long-term heart protection. Construct a new function $B(T)$ to describe the amount of aspirin in your bloodstream after taking an 81-mg aspirin. (Assume the decay rate is the same as for a 325-mg aspirin.) How do the functions $A(T)$ and $B(T)$ differ, and how are they alike?

6. (Internet optional for part (d).) The Japanese government is concerned about its decreasing population. The Japanese people joke about when Japan will disappear entirely. According to a UN report, the Japanese population is believed to have peaked at 127.5 million in 2005 and, if the current rate continues, will contract to about 105 million people in 2050.

a. What is the 45-year decay factor? The annual decay factor?

b. Use the annual decay factor to construct an exponential decay function to model the Japanese population $J(x)$, where $x =$ years since 2005.

c. Using the rule of 70, estimate the half-life of the population. According to your model, what is the half-life?

d. Check on the Internet for the actual population of Japan in 2010, and compare it to the results of your model for 2010 (where $x = 5$ years since 2005).

7. India and China have the opposite problem from that of Japan (see Problem 6): a huge population that is growing. In 2005 India had an estimated population of 1.08 billion on a land area of 1.2 million square miles, and China had a population of 1.30 billion on 3.7 million square miles.

a. In 2005 did India or China have the larger population density (people per square mile)?

b. China has been trying to slow population growth with its one-child-per-family policy. As a result, in 2005 its annual population growth rate (0.6%) was considerably lower than that of India (1.6%). Assuming that each population continues to grow at the 2005 rate, construct functions $C(x)$ and $I(x)$ for the population growth in China and India, respectively, letting x = years since 2005.

c. Plot each function in part (b) on the same graph, for 30 years after 2005.

d. Looking at the graph, is there a year when India's population is projected to overtake China's? If so, use your models to predict the year and population level.

8. Polonium-210 is a toxic radioactive substance named after Poland by the Curies, who discovered it. Polonium-210 poisoning is the suspected cause of death of former Soviet spy Alexander Litvinenko, in London, November 2006.

a. Given that polonium has a half-life of approximately 138 days, construct a function to model the amount of polonium, $P(T)$, as a function of the original amount A and T, the number of half-life periods.

b. Most of the world's polonium-210 comes from Russia, which produces about 100 grams per year, which is sold commercially to the United States. How many time periods, T, would it take for 100 grams to decay until there is only 1 gram left? Translate this into days, and then into years.

9. The half-life of uranium-238 is about 5 billion years. Assume you start with 10 grams of U-238 that decays continuously.

a. Construct an equation to describe the amount of U-238 remaining after x billion years.

b. How long would it take for 10 grams of U-238 to become 5 grams?

10. Which of the following expressions are equivalent?

a. $\ln(\sqrt{x}/y)$

b. $\left(\frac{1}{2}\right)\ln(x/y)$

c. $\left(\frac{1}{2}\right)\ln x - \ln y$

d. $\left(\frac{1}{2}\right)\ln x/\ln y$

e. $\left(\frac{1}{2}\right)\ln x - \left(\frac{1}{2}\right)\ln y$

f. $\ln(\sqrt{x})/\ln y$

11. Solve each equation for t.

a. $10^t = 2.3$

b. $2\log t + \log 4 = 2$

c. $60 = 30\, e^{0.03t}$

d. $\ln(2t - 5) - \ln(t - 1) = 0$

12. Simplify these expressions without using a calculator.

a. $\ln(e^2)$

b. $e^{\ln(3)}$

c. $10^{\log(t-2)}$

d. $\ln(x^2 - x) - \ln x$

13. Convert each of the following expressions to a power of e.

a. 1.5   **b.** 0.7   **c.** 1

14. For each of the following functions identify the growth (or decay) factor, the growth (or decay) rate, and the continuous growth (or decay) rate.

a. $Q = 75(1.02)^t$ **b.** $P(x) = 50e^{-0.3t}$

15. Match each function with one (or more) of the graphs at the top of the next column.

a. $y = 2 + x$

b. $y = 2 + \log(x)$

c. $y = 100(10)^x$

d. $\log(y) = 2 + x$ (*Hint:* solve for y.)

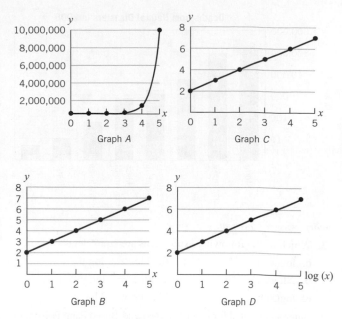

Graph A

Graph C

Graph B

Graph D

16. Of the approximately 40 million Americans suffering from hearing loss, about 10 million cases can be attributed to noise-induced hearing loss. The National Institute of Occupational Safety and Health (NIOSH), issues guidelines for safe sound exposure on a daily basis over a 40-year period. At 85 dB, a safe daily exposure is 8 hrs. For every 3 decibels over 85, the permissible safe exposure time is cut in half.

a. Create a function $T = f(d)$ where T is the number of hours of safe continuous exposure time and d is the number of decibels above 85.

b. Graph the function $f(d)$ for $0 \leq d \leq 30$.

Use your formula for f to predict how long the following rock bands can play without anyone (including the band members) experiencing hearing loss. (Assume that people are not wearing ear plugs).

c. On July 15, 2009, the band KISS reached a 136-decibel level at the Cisco Ottawa Bluesfest in Ottaway Canada.

d. Other rock bands, such as Deep Purple, Iron Maiden, and Mötley Crüe, play at about 120 decibels.

e. Counting Crows and Cake play at about 102 decibels.

17. For each function, identify the corresponding graph (here and on the next page) and describe its relationship to $y = \ln x$.

a. $y_1 = 2\ln x$

b. $y_2 = 2 + \ln x$

c. $y_3 = \ln(x + 2)$

d. $y_4 = \ln(x^2)$

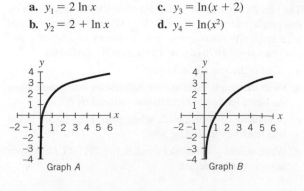

Graph A

Graph B

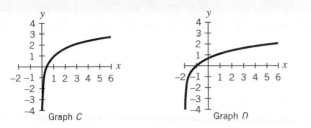

Graph *C*   Graph *D*

18. Which pair(s) of graphs show a function and its inverse?

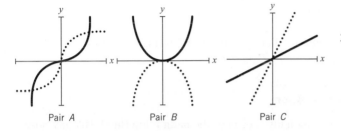

Pair *A*   Pair *B*   Pair *C*

19. If the decibel level moves from 30 (a quiet conversation) to 80 (average street noise), by how many orders of magnitude has the intensity increased? Generate your answer in two ways:

   a. Using the table in the text

   b. Using the definition of decibels

20. Beer has a pH of 4.5 and household lye a pH of 13.5.

   a. Which has the higher hydrogen ion concentration, and by how many orders of magnitude?

   b. What is the difference between the two pH values? How does this relate to your answer in part (a)?

   c. Why is it easier to use the pH number instead of $[H^+]$?

21. In Chapter 4 we encountered the Richter scale, which measures the amplitude of an earthquake. The Richter number R is defined as

$$R = \log (A/A_0)$$

where A is the amplitude of the shockwave caused by the earthquake and A_0 is the reference amplitude, the smallest earthquake amplitude that could be measured by a seismograph at the time this definition was adopted, in 1935.

Haiti suffered a magnitude 7.2 earthquake on January 12, 2010, that caused over 230,000 deaths.

Chile was hit with a magnitude 8.8 earthquake on Feb. 27, 2010, that caused about 450 deaths. On May 22, 1960, Chile had the world's largest recorded earthquake so far, measuring 9.5 on the Richter scale.

Find the ratio A/A_0 for each of the three earthquakes. How many orders of magnitude larger was each earthquake's amplitude compared with the base-level amplitude, A_0?

22. The energy magnitude, M, radiated by an earthquake measures the potential damage to man-made structures. It can be described by the formula

$$M = \left(\tfrac{2}{3}\right) \log E - 2.9$$

where the seismic energy, E, is expressed in joules. Show that for every increase in M of one unit, the associated seismic energy E is increased by about a factor of 32.

23. a. Given the following graph, generate a linear equation for $Y (= \log y)$ in terms of x.

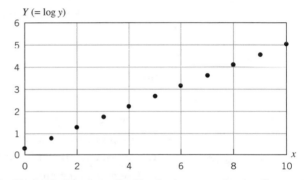

   b. Now substitute $\log y$ for Y and solve your equation for y.

   c. What type of function did you find in part (b)? What does this suggest about functions that appear linear on a semi-log graph (with the log scale on the vertical axis)?

Changing Bases

Objective

- explore the relationship between $y = \log_b x$ and $y = \log x$

Materials/Equipment

- graphing program
- graphing paper
- (optional) course software "E8: Logarithmic Sliders" and "E10: Inverse Functions: $y = a^x$ and $y = \log_a x$" in *Exponential and Logarithmic Functions*

Introduction to Changing Logarithmic Bases

Most graphing programs only use the common log ($\log x$) or the natural log ($\ln x$). But any positive number b (except 1) can be used as a base for a logarithm. The logarithm base b of x is the exponent of b needed to produce x; that is,

$$y = \log_b x \qquad \text{means that} \qquad b^y = x \quad \text{(where } x \text{ and } b > 0)$$

How would you describe the relationship between $y = \log x$ and $y = \log_b x$?

Procedure

Logarithms Base b

1. Base 2
   a. Rewrite $y = \log_2 x$ as an exponential function (base 2).
   b. Use properties of logarithms to solve the exponential function for y.
   c. Describe a formula for $\log_2 x$ in terms of $\log x$.
   d. Use a graphing program to plot $y = \log_2 x$ and $y = \log x$ on the same graph, and describe the graphical difference. Do $y = \log_2 x$ and $y = \log x$ intersect?
   When does the graph of $y = \log_2 x$ lie above the graph of $y = \log x$? Lie below it?

2. Base 3
   a.– d. Repeat the process from Part (1) using $y = \log_3 x$.

3. Base e (natural logarithms)
   a. Since $2 < e < 3$, is $\log_2 x < \ln x < \log_3 x$? Why or why not?
   b. Graph $y = \log_2 x$, $y = \ln x$, and $\log_3 x$ on the same grid to verify your answer in part (a).

4. Base b
   a.– c. Generalize the process from Parts 1 and 2, using $y = \log_b x$ where $b > 1$, including describing a general formula for $\log_b x$ in terms of $\log x$.
   d. Sketch by hand the graphs of $y = \log_b x$ and $y = \log x$ on the same graph. When would $y = \log_b x$ and $y = \log x$ intersect? When would the graph of $y = \log_b x$ lie above the graph of $y = \log x$? Lie below it?
   e. What if $b = 1$? Does b have to be an *integer* greater than 1?

Summarizing Your Results

Write a 60-second summary describing the results of changing the base in a logarithmic function.

Further Questions

1. What if $b = 1$? Rewrite $y = log_b x$ where $b = 1$ in terms of its equivalent exponential form. Then use these results and rules of logarithms to rewrite the function y. Describe the graph of $y = \log_1 x$.

2. For your general formula to work, does b have to be a positive integer >1?

3. Could you generalize $y = \log_b x$ to a value of b where $0 < b < 1$? That is, can b be a fraction such as 1/2? (*Hint:* Go through the generalization process to see if it holds when $0 < b < 1$.)

4. Why are $y = a^x$ and $y = \log_a x$ inverse functions of each other? (See "E10: Inverse Functions: $y = a^x$ and $y = \log_a x$" in *Exponential and Logarithmic Functions*.)

CHAPTER 8
QUADRATICS AND THE MATHEMATICS OF MOTION

OVERVIEW

In this chapter we add power functions to create quadratics which can be used to describe water trajectories of a fountain, model traffic flow, or predict the spread of a wildfire. We also study the mathematics of motion, deriving the equations for a freely-falling body for distance fallen and velocity.

After reading this chapter, you should be able to
- understand the behavior and construct graphs of quadratic functions
- determine the vertex and intercepts of a quadratic function
- convert quadratic functions from one form to another
- calculate the average rate of change of a quadratic function
- define and apply the equations for a freely-falling body

8.1 *An Introduction to Quadratic Functions: The Standard Form*

The Simplest Quadratic

A power function of degree 2 (in the form $y = ax^2$) is also the simplest member of a family of functions called *quadratics*. We've already encountered other quadratics such as $S = 6x^2$ (the surface area of a cube) and $A = \pi r^2$ (the area of a circle). Since $y = ax^2$ has an even integer power, its graph has the classic ∪-shaped curve called a *parabola*. (See Figure 8.1.) Certain properties are suggested by the graph:

- The parabola is concave up if $a > 0$ and concave down if $a < 0$.
- It has a minimum (or maximum) point called the *vertex*.
- It is symmetric across a vertical line called an *axis of symmetry* that runs through the vertex.

A parabola has another unique and useful property. There is an associated point off any parabola called the *focus* or *focal point*. It is located $\left| \frac{1}{4a} \right|$ units above (or below) the vertex and lies within the arms of the parabola.

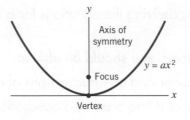

Figure 8.1 The parabolic graph of $y = ax^2$, where $a > 0$.

EXAMPLE 1

Properties of a quadratic function
Determine whether the graph of $y = 2x^2$ is concave up or down, find its vertex and focal point, and then graph the function.

SOLUTION

Since $2x^2 \geq 0$, the minimum value for $y = 2x^2$ is 0, which occurs when $x = 0$. So the function has a minimum and its vertex is at the origin (0, 0). Since $a > 0$, the graph is concave up. The coefficient, a, of x^2 is 2, so the focal point is $1/(4a) = 1/(4 \cdot 2) = 1/8$ or 0.125 units above the vertex. The coordinates of the focal point (or focus) are (0, 0.125). Figure 8.2 shows the graph of the function.

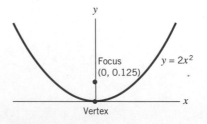

Figure 8.2 Graph of $y = 2x^2$ with vertex at (0, 0) and focal point at (0, 0.125).

The focal point merits its name. A three-dimensional parabolic bowl has vertical cross sections that are all the same parabola. If the bowl is built of a reflective material, the parabolic shape will concentrate parallel rays from the sun or a satellite TV channel at the focal point (see Figure 8.3 on the next page). This property is used to focus the sun's rays to construct primitive cooking devices or to concentrate electromagnetic waves from satellites or distant stars.

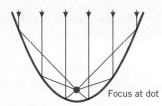

Figure 8.3 A parabola with
parallel incoming rays
concentrating at its focus.

The course software "Q10: Parabolic Reflector" gives an interactive demonstration of the reflective property of parabolas.

This process also works in reverse. For example, if you have a light bulb centered at the focus of a parabola, the light rays will bounce off a reflective parabolic surface as a set of parallel rays. This property is used to construct lamps, car headlights, or radio transmitters that send messages to robot rovers on Mars or broadcast our existence to the universe.

Designing Parabolic Devices

The easiest way to design a parabolic device is to place the vertex of the parabola at the origin (0, 0)—in other words, to construct a function of the form $y = ax^2$.

EXAMPLE 2   **Designing a cooking device**

In developing countries, such as Burkina Faso,[1] solar parabolic cooking devices provide a cheap way to cook food and pasteurize water. The cooking devices can be made any size and can even be constructed of hardboard and aluminum foil. If a large parabolic cooking device is built that is 4′ wide and 1′ deep, where should the cooking pot be centered for maximum efficiency?

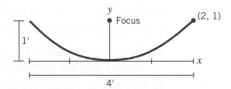

Figure 8.4 Parabolic cooking device.

SOLUTION   The cooking pot should be placed at the focal point of the parabola, where the sun's rays are concentrated. If we think of the bottom of the cooker as placed at (0, 0), we can construct a function of the form $y = ax^2$ and use it to find the focal point (see Figure 8.4). To find the value of a, we need the coordinates of a point on the parabolic cross section. Since the cooker is 4′ wide, if we move 2′ to the right of the cooker's base, and 1′ up, we will be at the point (2, 1), which lies on the rim of the cross section. Given the equation

$$y = ax^2$$

let $x = 2$ and $y = 1$ $\qquad 1 = a \cdot 2^2$

evaluate $\qquad\qquad\qquad 1 = a \cdot 4$

solve for a $\qquad\qquad\quad a = \dfrac{1}{4}$

---

[1]In 2004 the La Trame documentary film studio released *Bon Appétit, Monsieur Soleil*, a film about solar cooking in Burkina Faso. The studio has produced several versions in different languages, including English. The film covers the essential points of solar cooking, the acceptance of these devices by the population, the advantages achieved by using them (savings in money and time), and their contribution to fighting deforestation. The film is a good tool to inform people living in countries with energy problems about other ways to cook. It also serves to introduce this cooking method to decision makers in government agencies and NGOs.

So the formula for the parabolic oven cross section is $y = \left(\frac{1}{4}\right)x^2$. The focal point is at $\left|\frac{1}{4a}\right| = \left|\frac{1}{4 \cdot \left(\frac{1}{4}\right)}\right| = \frac{1}{1}$ or 1 foot above the vertex, the bottom of the cooker. Since the height of the cooker is also 1 foot, then the bottom of the pot should be suspended 1 foot above the center, at the level of the rim, for maximum efficiency.

The Standard Form of a Quadratic

A quadratic function may include a linear term added to the ax^2 term. The *general quadratic* in *standard* or *a-b-c form* is

$$y = ax^2 + bx + c \qquad (a \neq 0)$$

where a, b, and c are constants. For example, $y = 3x^2 + 7$ is a quadratic function where $a = 3$, $b = 0$, and $c = 7$.

EXPLORE & EXTEND

8.1a

Visualizing the *a-b-c* Form of a Parabola
Open the course software to "Q1: *a, b, c* Sliders" and vary the values of *a, b,* and *c* in the standard quadratic $y = ax^2 + bx + c$. Be sure to consider values for *a* that are >1, between 0 and 1, and <0. What is the effect of varying *b* while holding *a* and *c* fixed? (*Hint:* Trace the position of the vertex as you vary the value of *b*.) Describe the effect of *a, b,* and *c* on the parabola's shape in a 60-second summary.

EXAMPLE 3

A quadratic model for tuition revenue
Because of state financial problems, many publicly funded colleges are faced with large budget cuts. To raise more revenue, one state college plans to raise both tuition and student body size over the next 10 years. Its goal is to increase the current tuition of $6000 by $600 a year and increase the student population of 1200 by 40 students a year.

a. Generate two equations, one to describe the projected increases in tuition and the other to describe increases in student body size over time.

b. Generate an equation to describe the projected total tuition revenue over time.

c. Compare the current tuition revenue to the projected revenue in 10 years.

SOLUTION   **a.** If we let $S =$ student body size and $N =$ number of years after the present, then
$S = 1200 + 40N$.

If we let $T =$ cost of tuition in thousands of dollars, then $T = 6 + 0.6N$.

b. The total tuition revenue, R, is the product of the number of students, S, times the cost of tuition, T. We can find R using the FOIL technique.[2] Given:

$$R = S \cdot T$$

substitute	$= (1200 + 40N)(6 + 0.6N)$
multiply out	$= 7200 + 720N + 240N + 24N^2$
combine terms	$= 7200 + 960N + 24N^2$

---

[2] The FOIL technique: Recall that the product $(a + b)(c + d)$ is the sum of four terms: the product of the first (F) two terms, the product of the outside (O) terms, the product of the inside (I) terms, and the product of the last (L) two terms of the factors.

$$\underset{\substack{\text{F} \quad\quad \text{O} \quad\quad \text{I} \quad\quad \text{L}}}{(a + b)(c + d) = ac + ad + bc + bd}$$

So the total revenue R (in thousands of dollars) from tuition is given by the quadratic function

$$R = 7200 + 960N + 24N^2$$

c. At the present time $N = 0$, so the total tuition revenue $R = \$7200$ thousand or $\$7.2$ million. In 10 years, $N = 10$, so

$$R = 7200 + (960 \cdot 10) + (24 \cdot 10^2)$$
$$= 7200 + 9600 + 2400$$
$$= 19{,}200$$

So in 10 years the projected total tuition revenue would be $\$19{,}200$ thousand or $\$19.2$ million.

Properties of Quadratic Functions

The vertex and the axis of symmetry

The graph of the general quadratic, $y = ax^2 + bx + c$, is a parabola that is symmetric about its axis of symmetry. For any quadratic other than the simplest (of the form $y = ax^2$), the axis of symmetry is not the vertical axis. The vertex lies on the axis of symmetry and is a maximum (if the parabola is concave down) or a minimum (if the parabola is concave up) (see Figure 8.5). In Section 8.2 we'll learn how to find the coordinates of the vertex.

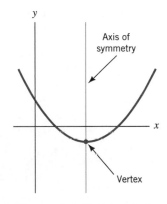

Figure 8.5 Each parabola has a vertex that lies on an axis of symmetry.

Vertical and horizontal intercepts

The general properties of vertical and horizontal intercepts are the same for all functions.

Intercepts of a Function

For any function $f(x)$:

the vertical intercept is at $f(0)$

the horizontal intercepts occur at values of x, called *zeros,* where $f(x) = 0$

For a quadratic function $f(x) = ax^2 + bx + c$, we have $f(0) = c$. So $f(x)$ has a vertical intercept at $(0, c)$. We usually shorten this to say that the vertical intercept is at c.

Because of the $\cup$ shape of its graph, a quadratic function may have no, one, or two horizontal intercepts (see Figure 8.6). If $(r, 0)$ are the coordinates of a horizontal intercept, we abbreviate this to say that the horizontal intercept is at r. These values of r are also called *zeros* of the function since $f(r) = 0$. In Section 8.4 we'll learn how to calculate the horizontal intercepts.

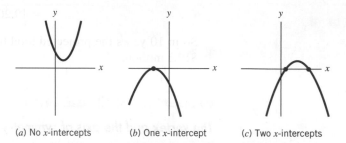

(*a*) No *x*-intercepts   (*b*) One *x*-intercept   (*c*) Two *x*-intercepts

Figure 8.6 Graphs of quadratic functions showing the three possible cases for the number of horizontal or *x*-intercepts.

Since a parabola is symmetric about its axis of symmetry, each point on one arm of the parabola has a mirror image on the other arm. In particular, if there are two horizontal intercepts at r_1 and r_2, then $(r_1, 0)$ and $(r_2, 0)$ are mirror images across the parabola's axis of symmetry. The horizontal intercepts lie at an equal distance, d, to the left and right of the axis of symmetry (see Figure 8.7). Equivalently, the axis of symmetry crosses the horizontal axis halfway between r_1 and r_2, at the point where x is $\dfrac{(r_1 + r_2)}{2}$. So the equation for the vertical line of the axis of symmetry is $x = \dfrac{(r_1 + r_2)}{2}$.

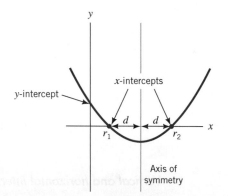

Figure 8.7 Each parabola has a vertical intercept and zero, one, or two horizontal intercepts.

The focal point

The focal point for the general parabola $y = ax^2 + bx + c$ is still $\left|\frac{1}{4a}\right|$ units above (or below) the vertex. The distance $\left|\frac{1}{4a}\right|$ from the vertex to the focal point is called the *focal length*.

EXPLORE & EXTEND

8.1b

Changing the Focus

Does increasing $|a|$ bring the focus closer to or farther away from the vertex?

The Quadratic Function and Its Graph

A *quadratic function* can be written in standard form as

$$f(x) = ax^2 + bx + c \qquad \text{(where } a \neq 0\text{)}$$

Its graph:

is called a *parabola*

is symmetric about its vertical *axis of symmetry*

has a minimum (if the parabola is concave up) or a maximum (if the parabola is concave down) point called its *vertex*

has one vertical intercept, but may have zero, one, or two horizontal intercepts

has a *focal point* $\left|\frac{1}{4a}\right|$ units above the minimum or below the maximum;

the value $\left|\frac{1}{4a}\right|$ is called the *focal length*.

Estimating the Vertex and Horizontal Intercepts

Quadratics arise naturally in area and motion problems. The earliest problems we know of that led to quadratic equations are on Babylonian tablets dating from 1700 B.C. The writings suggest a problem similar to the following example.

EXAMPLE 4   Maximizing area

What is the maximum rectangular area you can enclose within a fixed perimeter of 24 meters? What are the dimensions of the rectangle with the maximum area?

SOLUTION   If the rectangular region has length L and width W, then

$$2L + 2W = \text{perimeter}$$

Substitute	$2L + 2W = 24$
divide by 2	$L + W = 12$
subtract L from both sides	$W = 12 - L$

Since area, A, is given by	$A = L \cdot W$
substitute for W	$= L \cdot (12 - L)$
multiply through	$= 12L - L^2$

So area, A, is a quadratic function of L. Table 8.1 on the next page shows values of L from 0 to 12 meters and corresponding values for W and A.

Figure 8.8 shows a graph of area, A, versus length, L. In the equation $A = 12L - L^2$ the coefficient of L^2 is negative (-1), so the parabola is concave down. From the table and the graph, it appears that the vertex of the parabola is at $(6, 36)$; that is, at a length of 6 meters the area reaches a maximum of 36 square meters. Since $W = 12 - L$, when $L = 6$, then $W = 6$. So the rectangle has maximum area when the length equals the width, or in other words when the rectangle is a square.

Length, L (m)	Width, W (m)	Area, A (m²)
0	12	0
1	11	11
2	10	20
3	9	27
4	8	32
5	7	35
6	6	36
7	5	35
8	4	32
9	3	27
10	2	20
11	1	11
12	0	0

Table 8.1

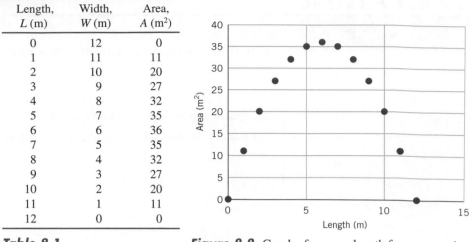

Figure 8.8 Graph of area vs. length for a rectangle.

E X A M P L E   5

The trajectory of a projectile

Figure 8.9 shows a plot of the height above the ground (in feet) of a projectile for the first 5 seconds of its trajectory. How could we:

a. Estimate the height from which the projectile was launched?

b. Estimate when the projectile will hit the ground?

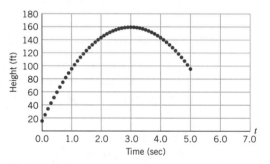

Figure 8.9 The height of a projectile.

S O L U T I O N

a. The initial height of the projectile corresponds to the vertical intercept of the parabola (where time $t = 0$). A quick look at Figure 8.9 gives us an estimate somewhere between 10 and 20 feet.

If we use technology to generate the equation for a best-fit quadratic, we get $H(t) = -16t^2 + 96t + 15$, where t = time (in seconds) and $H(t)$ = height (in feet). Since $H(0) = 15$, then 15 is the vertical intercept. In other words, when $t = 0$ seconds, the initial height is $H(0) = 15$ feet.

b. Figure 8.10 overlays on Figure 8.9 the graph of $H(t) = -16t^2 + 96t + 15$, our function model for the projectile's path. The projectile will hit the ground when the

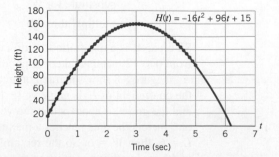

Figure 8.10 A graph of the function model for the projectile's path.

height above the ground $H(t) = 0$. This occurs at a horizontal intercept of the parabola. One horizontal intercept (not shown) would occur at a negative value of time, t, which would be meaningless here. From the graph we can estimate that the other intercept occurs at $t \approx 6.2$ seconds.

"The Mathematics of Motion" in Section 8.6 offers an extended discussion on constructing mathematical models to describe the motion of freely-falling or projected bodies.

Algebra Aerobics 8.1

1. Find the coordinates of the vertex and the focal point for each quadratic function. Then specify whether each vertex is a maximum or minimum.

   a. $y = 3x^2$ **c.** $y = \frac{1}{24}x^2$

   b. $y = -6x^2$ **d.** $y = -\frac{1}{12}x^2$

2. A designer is planning an outdoor concert place for solo performers. The stage area is to have a parabolic wall 30′ wide by 10′ deep. If the performer stands at the focal point, the sound will be reflected out in parallel sound waves directly to the audience in front. See the figure below.

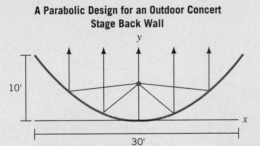

A Parabolic Design for an Outdoor Concert Stage Back Wall

   a. Where would the focal point be?

   b. What is the equation for the reflecting wall?

   c. Do you see any problem with this idea for concert acoustics? (*Hint:* What if the audience is noisy?)

3. Evaluate each of the following quadratic functions at 2, −2, 0, and z.

   a. $g(x) = x^2$ **d.** $m(s) = 5 + 2s - 3s^2$

   b. $h(w) = -w^2$ **e.** $D(r) = -(r - 3)^2 + 4$

   c. $Q(t) = -t^2 - 3t + 1$ **f.** $k(x) = 5 - x^2$

4. For each of the graphs of the quadratic functions $f(x)$, $g(x)$, $h(x)$, and $j(x)$ given at the top of the next column, determine:

   i. If the parabola is concave up or concave down

   ii. If the parabola has a maximum or a minimum

   iii. The equation of the axis of symmetry

   iv. Estimated coordinates of the vertex

   v. Estimated coordinates of the horizontal and vertical intercepts

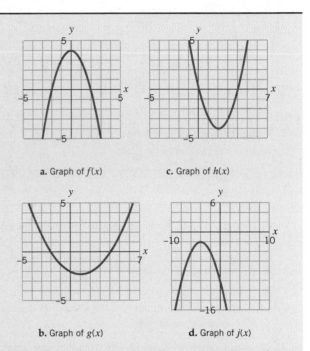

a. Graph of $f(x)$ **c.** Graph of $h(x)$

b. Graph of $g(x)$ **d.** Graph of $j(x)$

5. A manufacturer wants to make a camp stove using a parabolic reflector to concentrate sun rays at the focal point of the parabola, where the food to be cooked would be placed. The reflector must be no wider than 24″, and the depth is to be the same as the focal length, f, the distance from the focal point to the bottom. See the figure below.

A Parabolic Design for a Camp Stove

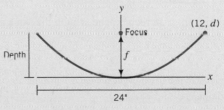

   a. What focal length would the camp stove need to have? (*Hint:* The focal length is $\left|\frac{1}{4a}\right|$)

   b. What formula is needed for manufacturing the stove reflector?

6. Put the following quadratic functions into standard a-b-c form using the FOIL technique. (See Footnote 2 on p. 470.)

$$f(x) = (x - 3)(x + 5) \qquad h(x) = 10(x - 3)(x - 5)$$
$$g(x) = (2x + 5)(x + 1) \qquad j(x) = 2(x - 3)(x + 3)$$

Exercises for Section 8.1

A graphing program is recommended for Exercises 3, 4 and 5 and required for parts of Exercises 23 and 24.

1. From the graph of each quadratic function, identify whether the parabola is concave up or down and hence whether the function has a maximum or minimum. Then estimate the vertex, the axis of symmetry, and any horizontal and vertical intercepts.

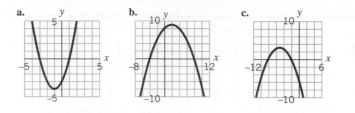

a.   **b.**   **c.**

2. Using the quadratic function $y = \frac{1}{8}x^2$:

   a. Fill in the values in the table.

x	-12	-8	-4	0	4	8	12
y							

   b. Plot the points by hand and sketch the graph of the function.

   c. Determine the coordinates of the focal point and place the focal point on the graph.

   d. What is the equation of the axis of symmetry?

   e. What are the coordinates of the vertex? Is the vertex a maximum or minimum?

3. (Graphing program recommended.) On the same graph, plot the following three functions.

   $$y_1 = x^2 \qquad y_2 = \frac{1}{4}x^2 \qquad y_3 = \frac{1}{12}x^2$$

   a. Calculate the focal point for each parabola.

   b. Compare y_2 and y_3 with y_1.

   c. Complete this sentence: As the value of $|a|$ gets smaller, the focal point gets _____ the vertex and the graph gets _____.

4. (Graphing program recommended.) On the same graph, plot the three functions.

   $$y_1 = x^2 \qquad y_2 = 4x^2 \qquad y_3 = 8x^2$$

   a. Calculate the focal point for each parabola.

   b. Compare y_3 and y_2 with y_1.

   c. Complete this sentence: As the value of $|a|$ gets larger, the focal point gets _____ to the vertex and the graph of the parabola gets _____.

5. (Graphing program recommended.) Given a point P on a parabola of the form $y = ax^2$, find the equation of the parabola. If available, use technology to verify your equations by plotting each parabola.

   a. $P = (12, 6)$ **c.** $P = (4, 12)$

   b. $P = (-12, -6)$ **d.** $P = (3, 144)$

6. Find the coordinates of the vertex for each quadratic function listed. Then specify whether each vertex is a maximum or minimum.

   a. $y = 4x^2$ **c.** $P(n) = \left(\frac{1}{12}\right)n^2$

   b. $f(x) = -8x^2$ **d.** $Q(t) = -\left(\frac{1}{24}\right)t^2$

7. Given the following focal points, write the equation of a parabola in the form $y = ax^2$ by finding a.

   a. $(0, 4)$ **c.** $\left(0, \frac{1}{16}\right)$

   b. $(0, -8)$ **d.** $\left(0, -\frac{1}{24}\right)$

8. Find the focal length (the distance from the focal point to the vertex) for each of the following.

   a. $f(x) = x^2$ **b.** $f(x) = 3x^2$ **c.** $f(x) = \frac{1}{3}x^2$

9. For each of the following functions, evaluate $f(2)$ and $f(-2)$.

   a. $f(x) = x^2 - 5x - 2$

   b. $f(x) = 3x^2 - x$

   c. $f(x) = -x^2 + 4x - 2$

10. A designer proposes a parabolic satellite dish 5 feet in diameter and 15 inches deep.

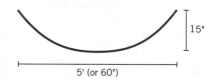

5' (or 60")

   a. What is the equation for the cross section of the parabolic dish?

   b. What is the focal length (the distance between the vertex and the focal point)?

   c. What is the diameter of the dish at the focal point?

11. An electric heater is designed as a parabolic reflector that is 5" deep. To prevent accidental burns, the centerline of the heating element, placed at the focus, must be set in 1.5" from the base of the reflector.

   a. What equation could you use to design the reflector? (*Hint:* Tip the parabolic reflector so it opens upward.)

   b. How wide would the reflector be at its rim?

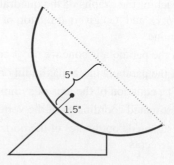

5"

1.5"

12. A parabolic reflector 3″ in diameter and 2″ deep is proposed for a spotlight.

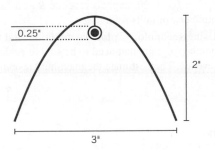

a. What formula is needed for manufacturing the reflector?

b. Where would the focus be?

c. Will a $\frac{1}{4}$″-diameter light source fit if it is centered at the focus?

13. A slimline fluorescent bulb $\frac{1}{2}$″ in diameter needs 1″ clearance top and bottom in a parabolic reflecting shade.

a. What are the coordinates of the focus for this parabola?

b. What is the equation for the parabolic curve of the reflector?

c. What is the diameter of the opening of the shade?

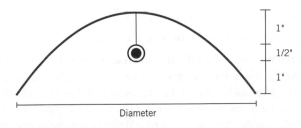

14. If we know the radius and depth of a parabolic reflector, we also know where the focus is.

a. Find a generic formula for the focal length f of a parabolic reflector expressed in terms of its radius R and depth D. The focal length $\left|\frac{1}{4a}\right|$ is the distance between the vertex and the focal point. Assume $a > 0$.

b. Under what conditions does $f = D$?

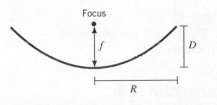

15. Construct several of your own equations of the form $y = ax^2$ and then describe in words how the focal length varies depending on how open or closed the parabolic curve is.

16. For each of the following quadratics with their respective vertices, calculate the distance from the vertex to the focal point. Then determine the coordinates of the focal point.

a. $f(x) = x^2 - 2x - 3$ with vertex at $(1, -4)$

b. $g(t) = 2t^2 - 16t + 24$ with vertex at $(4, -8)$

17. Put each of the following quadratics into standard form.

a. $f(x) = (x + 3)(x - 1)$

b. $P(t) = (t - 5)(t + 2)$

c. $H(z) = (2 + z)(1 - z)$

18. Put each of the following quadratics into standard form.

a. $g(x) = (2x - 1)(x + 3)$

b. $h(r) = (5r + 2)(2r + 5)$

c. $R(t) = (5 - 2t)(3 - 4t)$

19. Determine the dimensions for enclosing the maximum area of a rectangle if:

a. The perimeter is held constant at 200 meters.

b. The perimeter is held constant at P meters.

20. A gardener wants to grow carrots along the side of her house. To protect the carrots from wild rabbits, the plot must be enclosed by a wire fence. The gardener wants to use 16 feet of fence material left over from a previous project. Assuming that she constructs a rectangular plot, using the side of her house as one edge, estimate the area of the largest plot she can construct.

21. Which of the following are true statements for quadratic functions?

a. The vertex and focal point always lie on the axis of symmetry.

b. The graph of a parabola could have three horizontal intercepts.

c. The graph of a parabola does not necessarily have a vertical intercept.

d. If $f(2) = 0$, then f has a horizontal intercept at 2.

e. The focal point always lies above the vertex.

22. The management of a company is negotiating with a union over salary increases for the company's employees for the next 5 years. One plan under consideration gives each worker a raise of $1500 per year. The company currently employs 1025 workers and pays them an average salary of $30,000 a year. It also plans to increase its workforce by 20 workers a year.

a. Construct a function $C(t)$ that models the projected cost of this plan (in dollars) as a function of time t (in years).

b. What will the annual cost be in 5 years?

23. (Graphing program required for part (c).) A landlady currently rents each of her 50 apartments for $1250 per month. She estimates that for each $100 increase in rent, two additional apartments will remain vacant.

a. Construct a function that represents the revenue $R(n)$ as a function of the number of rent increases, n. (*Hint:* Find the rent per unit after n increases and the number of units rented after n increases.)

23. (continued)

    b. After how many rent increases will all the apartments be empty? What is a reasonable domain for this function?

    c. Using technology, plot the function. From the graph, estimate the maximum revenue. Then estimate the number of rent increases that would give you the maximum revenue.

24. (Graphing program required for part (c).)

    a. In economics, revenue R is defined as the amount of money derived from the sale of a product and is equal to the number x of units sold times the selling price p of each unit. What is the equation for revenue?

    b. If the selling price is given by the equation $p = -\frac{1}{10}x + 20$, express revenue R as a function of the number x of units sold.

    c. Using technology, plot the function and estimate the number of units that need to be sold to achieve maximum revenue. Then estimate the maximum revenue.

8.2 *Visualizing Quadratics: The Vertex Form*

In this section we show that every quadratic can be generated through transformations of the basic quadratic $y = x^2$. We start by investigating what happens to the equation of a function when we change its graph.

Stretching and Compressing Vertically

When we stretch or compress a graph we change its shape, creating a graph of a new function. To find the equation of this new function, we need to ask, "What is the relationship between the graphs of the new function and the old function?"

We can build on what we know from Chapter 7, where we learned that multiplying a power function by a constant stretches or compresses its graph. Specifically, for power functions in the form $y = kx^p$, the value of the coefficient k compresses or stretches the graph of $y = x^p$. A quadratic in the form $y = ax^2$ is a power function, so the graph of $y = ax^2$ is the graph of $y = x^2$

    vertically stretched by a factor of a, if $a > 1$

    vertically compressed by a factor of a, if $0 < a < 1$

The magnitude (or absolute value) of a tells us how much the graph of $y = x^2$ is stretched or compressed. As $|a|$ increases, a acts as a vertical stretch factor, which pulls harder and harder on the arms of the parabola anchored at the vertex, narrowing the graph. As $|a|$ decreases, a acts as a compression factor, which pulls down the graph, flattening the parabola's arms. So the value of a determines the shape of the parabola. See Figure 8.11.

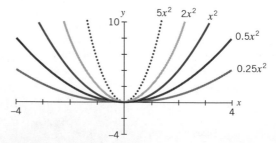

Figure 8.11 The graph of $y = x^2$ vertically compressed and stretched.

Reflecting across the Horizontal Axis

If $a < 0$, the graph of $y = ax^2$ is the graph of $y = x^2$ stretched or compressed by a factor of $|a|$ and then reflected across the x-axis. In general, the graphs of $y = ax^2$ and $y = -ax^2$ are reflections of each other across the x-axis. See Figure 8.12 on the next page.

The sign of *a* tells us whether the parabola opens up or down and if the vertex represents a maximum or a minimum value of the function. If *a* is positive, the parabola is concave up (opens upward) and the vertex is a minimum. If *a* is negative, the parabola is concave down (opens downward) and the vertex is a maximum.

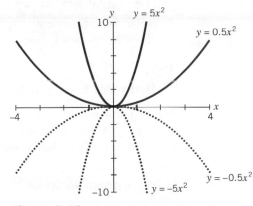

Figure 8.12 Graphs that are reflected across the *x*-axis.

E X A M P L E   1

Describing the relationship between pairs of quadratics

Describe how the following pairs of functions are related to each other and to $f(x) = x^2$. Sketch each pair and $f(x) = x^2$ on the same grid.

a. $g(x) = 3x^2$ and $h(x) = -3x^2$ **b.** $p(x) = 0.3x^2$ and $q(x) = -0.3x^2$

S O L U T I O N

a. The functions $g(x) = 3x^2$ and $h(x) = -3x^2$ are reflections of each other across the *x*-axis. For both functions, the absolute value of the coefficient *a* of x^2 is 3, so in each case $|a| > 1$. So both of the graphs of $g(x)$ and $h(x)$ are the graph of $f(x) = x^2$ stretched by a factor of 3. See Graph *A* in Figure 8.13.

b. The functions $p(x) = 0.3x^2$ and $q(x) = -0.3x^2$ are also reflections of each other across the *x*-axis. For both functions $|a| = 0.3$, so in each case $0 < |a| < 1$. So both of the graphs of $p(x)$ and $q(x)$ are the graph of $f(x) = x^2$ compressed by a factor of 0.3. See Graph *B* in Figure 8.13.

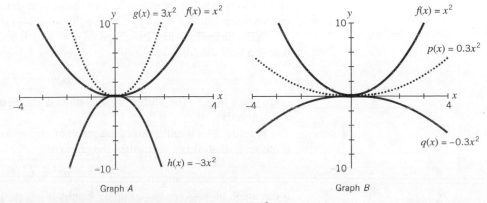

Figure 8.13 Stretching and compressing $y = x^2$.

> **Stretching, Compressing, and Reflecting the Graph of Any Function**
>
> If f is a function and a is a constant, then the graph of $af(x)$ is the graph of $f(x)$
>
> vertically stretched by a factor of a, if $a > 1$
>
> vertically compressed by a factor of a, if $0 < a < 1$
>
> vertically stretched or compressed by a factor of $|a|$
>
> reflected across the x-axis, if $a < 0$

Shifting Vertically and Horizontally

What happens when we keep the shape of the graph of $y = ax^2$ but change its position on the grid, shifting the graph vertically or horizontally? Clearly we will get a graph of a new function where the coordinates of the vertex are no longer $(0, 0)$.

Shifting a graph vertically

How does a function change when its graph is shifted up or down? Examine the graphs in Figure 8.14, Graphs A and B.

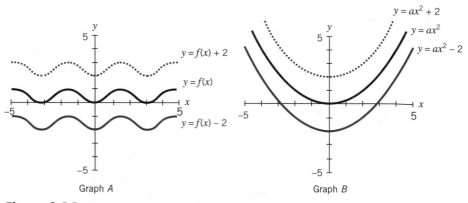

Figure 8.14 Graphs that are shifted vertically.

What happens to the output value, y, when the graph of $y = f(x)$ is shifted up two units? Every y value increases by two units. Translating this shift into equation form, our new function is $y = f(x) + 2$. (See Figure 8.14, Graph A.)

What happens when the graph of $y = f(x)$ is shifted down two units? Our new function is of the form $y = f(x) - 2$. In general, if the vertical shift is k units, then

$$y = f(x) + k$$

is the graph of $y = f(x)$ shifted up by k units if k is positive and shifted down by k units if k is negative.

Figure 8.14, Graph B shows the graph of $f(x) = ax^2$ raised or lowered by two units. If the vertical shift is k units, then the graph of

$$y = ax^2 + k$$

is the graph of $y = ax^2$ shifted up by k units if k is positive and shifted down by k units if k is negative. That also means that $(0, 0)$, the vertex of $y = ax^2$, is shifted k units (up or down) to become $(0, k)$, the vertex of $y = ax^2 + k$.

E X A M P L E 2 **Vertical shift of vertex**

If the quadratic function $f(x) = 4x^2$ is shifted down 3 units, describe the new function and its vertex.

S O L U T I O N The new function subtracts 3 from $f(x)$ to get $f(x) - 3 = 4x^2 - 3$. So the graph of $f(x)$ is lowered 3 units and the vertex is lowered from $(0, 0)$ to $(0, -3)$.

Algebra Aerobics 8.2a

In Problems 1−5, without drawing the graphs, compare the graph of part (b) to the graph of part (a).

1. **a.** $r(x) = x^2 + 2$ **b.** $s(x) = 2x^2 + 2$
2. **a.** $h(t) = t^2 + 5$ **b.** $k(t) = -t^2 + 5$
3. **a.** $f(z) = -5z^2$ **b.** $g(z) = -0.5z^2$
4. **a.** $f(x) = x^2 + 3x + 2$ **b.** $g(x) = x^2 + 3x + 8$
5. **a.** $f(t) = -3t^2 + t - 5$ **b.** $g(t) = -3t^2 + t - 2$

6. Create a quadratic equation of the form $y = ax^2 + k$ with the given values for a and the vertex. Sketch by hand the graph of each equation.

   a. $a = 3$ and the vertex is at $(0, 5)$.

   b. $a = \frac{1}{3}$ and the vertex is at $(0, -2)$.

   c. $a = -2$ and the vertex is at $(0, 4)$

7. Create new functions by performing the following transformations on $f(x) = x^2$.

   a. $g(x)$ is $f(x)$ stretched by a factor of 3.

   b. $h(x)$ is $f(x)$ stretched by a factor of 5 and reflected across the x-axis.

   c. $j(x)$ is $f(x)$ compressed by a factor of 1/2.

   d. $k(x)$ is $f(x)$ reflected across the x-axis.

8. Create a function in the form $y = ax^2 + k$ for each of the following transformations of $f(x) = x^2$:

   a. Stretched by a factor of 5 and shifted down 2 units

   b. Concave up and shifted 3 units up

   c. Multiplied by a factor of 0.5, concave down, and shifted 4.7 units down

   d. Opens up and is $f(x)$ shifted 71 units down

Shifting a graph horizontally

How does a function change when its graph is shifted to the right or left? Examine the graphs in Figure 8.15, Graphs A and B.

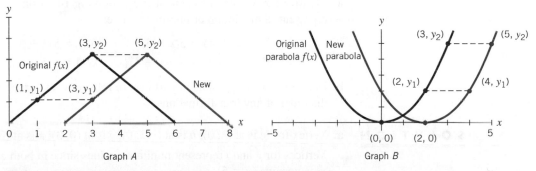

Figure 8.15 Graphs that are shifted horizontally to the right.

When we shift the graph of $y = f(x)$ two units to the right, how is the new function related to our original function? Recall that adding a constant to the output value of a function shifts its graph vertically. When we shift the graph of $f(x)$ horizontally, the input value, x, is shifted. The dotted lines in Figure 8.15 show that for any particular output value, y, the corresponding input value for the new function is $(x - 2)$. So the graph of $f(x)$ has been shifted to the *right* two units. Translating this shift into equation form, we get

$$y = f(x - 2)$$

At first this may seem counterintuitive. But remember, we are expressing the new function in terms of the original function. If you evaluate the new function at x, the same y-value of the original function $f(x)$ is now at $f(x - 2)$. For example, in Figure 8.15 if you evaluate each of the new functions in Graph A and Graph B at $x = 5$, you get an output of y_2. The same output for each of the original functions occurs at $5 - 2 = 3$.

What if we shift $y = f(x)$ to the *left* two units? We need to think about the effect of the shift and the direction of the shift. If you evaluate the new function at x, the comparable point on the graph of $f(x)$ is $f(x + 2)$. We can think of this shift as replacing x with $x - (-2)$ or $x + 2$, so our new function is

$$y = f(x + 2)$$

Figure 8.16 shows the graph of $y = 3x^2$ after being shifted to the right and then to the left 2 units. Note that the value of the x-coordinate of the vertex changed by ± 2 units.

If the horizontal shift is h units, then

$$y = a(x - h)^2$$

is the graph of $y = ax^2$ shifted right by h if h is positive, and shifted left by h if h is negative. The vertex is now at $(h, 0)$. Note that if $h = 2$, the expression $(x - h)$ becomes $x - 2$, and if $h = -2$, the expression $(x - h)$ becomes $x - (-2) = x + 2$.

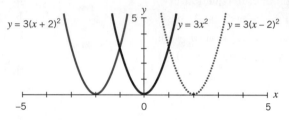

Figure 8.16 Graph of $y = 3x^2$ shifted horizontally to the left and to the right two units.

E X A M P L E   3   **Identifying the vertex**

a. Identify the vertex for each of the following functions and indicate whether it represents a maximum or minimum value:

$$g(x) = 5(x - 3)^2 \qquad j(x) = 5x^2 + 3$$
$$h(x) = -5(x + 3)^2 \qquad k(x) = -5x^2 - 3$$

b. Describe how the graphs in part (a) are related to the graph of $f(x) = 5x^2$. Specify the order of any transformations.

S O L U T I O N   **a.** Vertex for: g is at $(3, 0)$; h is at $(-3, 0)$; j is at $(0, 3)$; k is at $(0, -3)$.

Vertices for g and j represent minimum values since in both cases the coefficient for x^2 is positive ($a = 5$).

Vertices for h and k represent maximum values since in both cases the coefficient for x^2 is negative ($a = -5$).

b. The graph of $f(x) = 5x^2$

shifted horizontally to the right three units creates the graph of $g(x)$.

shifted horizontally to the left three units and then reflected across the x-axis creates the graph of $h(x)$.

shifted vertically up three units creates the graph of $j(x)$.

reflected across the x-axis and then shifted vertically down three units creates the graph of $k(x)$.

We can generalize to any function $f(x)$.

Horizontal and Vertical Shifts

The graph of $f(x) + k$ is the graph of $f(x)$ shifted vertically $|k|$ units.
   If $k > 0$, the shift is up; if $k < 0$, the shift is down.
The graph of $f(x - h)$ is the graph of $f(x)$ shifted horizontally $|h|$ units.
   If $h > 0$, the shift is to the right; if $h < 0$, the shift is to the left.

Using Transformations to Get the Vertex Form

We can use the previous transformations to generate a quadratic function in what is called the *vertex form*. We start by stretching or compressing the graph of $y = x^2$ by a factor of a. So our new function will be

$$y = ax^2$$

Next we shift the graph of the function $y = ax^2$ horizontally h units and vertically k units, to get

$$y = a(x - h)^2 + k$$

The vertex is now at (h, k). This quadratic is in the *vertex* or *a-h-k form*. Its axis of symmetry is the vertical line at $x = h$. See Figure 8.17.

"Q8: $y = ax^2$ vs. $y = a(x - h)^2 + k$," and "Q9: Finding 3 Points: *a-h-k* Form" in *Quadratic Functions* can help you visualize quadratic functions in the *a-h-k* form.

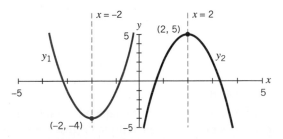

Figure 8.17 Graphs of $y_1 = 3(x + 2)^2 - 4$ and $y_2 = -3(x - 2)^2 + 5$.

The Vertex Form of a Quadratic Function

The *vertex* or *a-h-k form* of the quadratic function is

$$f(x) = a(x - h)^2 + k$$

where the vertex is at (h, k).

EXAMPLE 4 Identifying the vertex from the vertex form
For the following functions, identify the coordinates of the vertex and specify whether the vertex represents a maximum or minimum.
a. $y = -5x^2$ **b.** $y - 2(x - 3)^2 - 15$ **c.** $y = -3(x + 5)^2 + 10$

SOLUTION   **a.** The vertex is at $(0, 0)$ and represents a maximum.
   b. The vertex is at $(3, -15)$ and represents a minimum.
   c. The vertex is at $(-5, 10)$ and represents a maximum.

EXPLORE & EXTEND

8.2

Visualizing the *a-h-k* Form of a Parabola

Open the course software to "Q3: *a*, *h*, *k* Sliders" and vary the values for *a*, *h*, and *k* in the parabola $y = a(x - h)^2 + k$.

a. Confirm that the vertex of the parabola is at (*h*, *k*).

b. Are the roles for *a* in $y = ax^2 + bx + c$ and $y = a(x - h)^2 + k$ the same? What does varying *a* do to the parabola? Consider values for *a* that are > 1, between 0 and 1, and < 0.

c. Write a 60-second summary of the effects for varying *a*, *h*, and *k* on the parabola.

EXAMPLE 5

Transforming graphs

Given $f(x) = x^2$ and $g(x) = -2(x + 3)^2 + 5$, show how to transform the graph of $f(x)$ into the graph of $g(x)$.

SOLUTION

One way to transform the graph of $f(x)$ into the graph of $g(x)$ is shown in Figure 8.18.

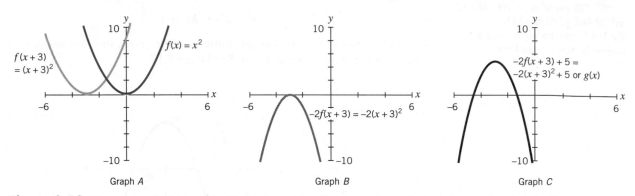

Figure 8.18 The graph of $f(x) = x^2$ in Graph A is shifted horizontally to the left three units, then in Graph B stretched vertically by a factor of 2 and reflected across the *x*-axis, and finally in Graph C shifted up vertically five units to generate $g(x) = -2(x + 3)^2 + 5$.

Domain and range

Domain: The domain for the general quadratic $y = ax^2 + bx + c$ is all real numbers; that is, all *x* in the interval $(-\infty, +\infty)$.

Range: If the quadratic is written in vertex form, $y = a(x - h)^2 + k$, it's easy to define the range, since the maximum (or minimum) of its graph is at the vertex (*h*, *k*). So if

$a > 0$, the parabola opens up, so the range is restricted to all *y* in the interval $[k, +\infty)$.

$a < 0$, the parabola opens down, so the range is restricted to all *y* in the interval $(-\infty, k]$.

EXAMPLE 6

Finding the function from its graph

Figure 8.19 on the next page shows the graph of $f(x) = 2x^2$ transformed into three new parabolas. Assume that each of the three new graphs retains the overall shape of $f(x)$.

a. Estimate the coordinates of the vertex for all four parabolas.

b. Use your estimates from part (a) to write equations for each new parabola in Figure 8.19 in vertex form.

c. Identify the domain and range of each parabola.

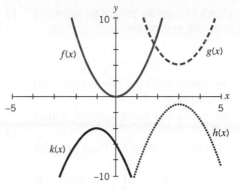

Figure 8.19 Three transformations of
$f(x) = 2x^2$

S O L U T I O N   **a.** Vertex for: $f(x)$ is at $(0, 0)$; $g(x)$ is at $(3, 4)$; $h(x)$ is at $(3, -1)$; $k(x)$ is at $(-1, -4)$

b. $g(x) = 2(x - 3)^2 + 4$; $h(x) = -2(x - 3)^2 - 1$; $k(x) = -2(x + 1)^2 - 4$

c. The domain for all four functions is $(-\infty, +\infty)$.

The range for: $f(x)$ is $[0, +\infty)$; $g(x)$ is $[4, +\infty)$; $h(x)$ is $(-\infty, -1]$; $k(x)$ is $(-\infty, -4]$.

Algebra Aerobics 8.2b

For Problems 1 to 3, without graphing, compare the positions of the vertices of parts (b) and (c) to that of part (a).

1. a. $y = x^2$

b. $y = (x + 3)^2$

c. $y = (x - 2)^2$

2. a. $f(x) = 0.5x^2$

b. $f(x) = 0.5(x - 1)^2$

c. $f(x) = 0.5(x + 4)^2$

3. a. $r = -2t^2$

b. $r = -2(t + 1.2)^2$

c. $r = -2(t - 0.9)^2$

For Problems 4 and 5, without graphing, compare the graphs of parts (b), (c), and (d) to that of part (a).

4. a. $y = x^2$

b. $y = (x - 2)^2$

c. $y = (x - 2)^2 + 4$

d. $y = (x - 2)^2 - 3$

5. a. $y = -x^2$

b. $y = -(x + 3)^2$

c. $y = -(x + 3)^2 - 1$

d. $y = -(x + 3)^2 + 4$

6. Create new functions by performing the following transformations on $f(x) = x^2$. Give the coordinates of the vertex for each new parabola.

a. $g(x)$ is $f(x)$ shifted right 2 units, stretched by a factor of 3, and then shifted down by 1 unit.

b. $h(x)$ is $f(x)$ shifted left 3 units, stretched by a factor of 2, then reflected across the x-axis, and finally shifted up by 5 units.

c. $j(x)$ is $f(x)$ shifted left 4 units, compressed by a factor of 5, and then shifted down by 3.5 units.

d. $k(x)$ is $f(x)$ shifted right 1 unit, reflected across the x-axis, and then shifted up by 4 units.

7. Give the coordinates of the vertex for each of the following functions and indicate whether the vertex represents a maximum or minimum.

a. $y = 2(x - 3)^2 - 4$ **c.** $y = -0.5(x - 4)^2$

b. $y = -3(x + 1)^2 + 5$ **d.** $y = \frac{2}{3}x^2 - 7$

Exercises for Section 8.2

A graphing program is optional for Exercise 9.

1. Which of the following quadratics have parabolic graphs that are concave up? Concave down? Explain your reasoning.

a. $y = 20 + x - x^2$ **c.** $y = -3x + 2 + x^2$

b. $y = 0.5t^2 - 2$ **d.** $y = 3(5 - x)(x + 2)$

2. Match each function to the right with its graph. Explain your reasoning for each choice.

$f(x) = x^2 + 3$ $g(x) = x^2 - 4$ $h(x) = 0.25x^2 + 3$

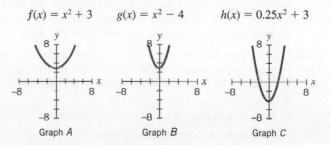

Graph A   Graph B   Graph C

3. On the same graph, sketch by hand the plots of the following functions and label each with its equation.

$y = 2x^2$ $y = -2x^2 + 3$

$y = -2x^2$ $y = -2x^2 - 3$

4. In each case sketch by hand a quadratic function $y = f(x)$ with the indicated characteristics.

   a. Does not cross the x-axis and has a negative vertical intercept.

   b. Has a vertex at (h, k) where $h < 0$ and $k > 0$ and a positive y-intercept.

   c. Crosses the x-axis at $x = 2$ and $x = -4$ and is concave up.

   d. Has the same shape as $y = x^2 - 3x$ but is raised up four units.

   e. Has the same y-intercept as $y = x^2 - 3x - 2$ but is concave down and narrower.

   f. Has the same shape as $y = \frac{x^2}{2}$, but is shifted to the left by three units.

5. Identify the stretch/compression factor and the vertex for each of the following:

   a. $y_1 = 0.3(x - 1)^2 + 8$ c. $y_3 = 0.01(x + 20)^2$

   b. $y_2 = 30x^2 - 11$ d. $y_4 = -6x^2 + 12x$

6. For each of the following functions, identify the vertex and specify whether it represents a maximum or minimum, and then sketch its graph by hand.

   a. $y = (x - 2)^2$ c. $y = -2(x + 1)^2 + 5$

   b. $y = \frac{1}{2}(x - 2)^2 + 3$ d. $y = -0.4(x - 3)^2 - 1$

7. For each of the following quadratic functions, find the vertex (h, k) and determine if it represents the maximum or minimum of the function.

   a. $f(x) = -2(x - 3)^2 + 5$ c. $f(x) = -5(x + 4)^2 - 7$

   b. $f(x) = 1.6(x + 1)^2 + 8$ d. $f(x) = 8(x - 2)^2 - 6$

8. For each quadratic function identify the vertex and specify whether it represents a maximum or minimum. Evaluate the function at 0 and then sketch a graph of the function by hand.

   a. $f(t) = 0.25(t - 2)^2 + 1$ c. $h(x) = -(x + 3)^2 + 4$

   b. $g(x) = 3 - (x - 5)^2$ d. $k(x) = \frac{(x + 5)^2}{3} - \frac{1}{3}$

9. (Graphing program optional.) Create a quadratic function in the vertex form $y = a(x - h)^2 + k$, given the specified values for a and the vertex (h, k). Then rewrite the function in the standard form $y = ax^2 + bx + c$. If available, use technology to check that the graphs of the two forms are the same.

   a. $a = 1$, $(h, k) = (2, -4)$ c. $a = -2$, $(h, k) = (-3, 1)$

   b. $a = -1$, $(h, k) = (4, 3)$ d. $a = \frac{1}{2}$, $(h, k) = (-4, 6)$

10. Transform the function $f(x) = x^2$ into a new function $g(x)$ by compressing $f(x)$ by a factor of $\frac{1}{4}$, then shifting the result horizontally left three units, and finally shifting it down by six units. Find the equation of $g(x)$ and sketch it by hand.

11. Transform the function $f(x) = 3x^2$ into a new function $h(x)$ by shifting $f(x)$ horizontally to the right four units, reflecting the result across the x-axis, and then shifting it up by five units.

   a. What is the equation for $h(x)$?

   b. What is the vertex of $h(x)$?

   c. What is the vertical intercept of $h(x)$?

12. For the following quadratic functions in vertex form, $f(x) = a(x - h)^2 + k$, determine the values for a, h, and k. Then compare each to $f(x) = x^2$, and identify which constants represent a stretch/compression factor, or a shift in a particular direction.

   a. $p(x) = 5(x - 4)^2 - 2$ c. $h(x) = -0.25\left(x - \frac{1}{2}\right)^2 + 6$

   b. $g(x) = \frac{1}{3}(x + 5)^2 + 4$ d. $k(x) = -3(x + 4)^2 - 3$

13. a. Find the equation of the parabola with a vertex of $(2, 4)$ that passes through the point $(1, 7)$.

   b. Construct two different quadratic functions both with a vertex at $(2, -3)$ such that the graph of one function is concave up and the graph of the other function is concave down.

14. For each part construct a function that satisfies the given conditions.

   a. Has a constant rate of increase of $15,000/year.

   b. Is a quadratic that opens upward and has a vertex at $(1, -4)$.

   c. Is a quadratic that opens downward and the vertex is on the x-axis.

   d. Is a quadratic with a minimum at the point $(10, 50)$ and a stretch factor of 3.

   e. Is a quadratic with a vertical intercept of $(0, 3)$ that is also the vertex.

15. If a parabola is the graph of the equation $y = a(x - 4)^2 - 5$:

   a. What are the coordinates of the vertex? Will the vertex change if a changes?

   b. What is the value of stretch factor a if the y-intercept is $(0, 3)$?

   c. What is the value of stretch factor a if the graph goes through the point $(1, -23)$?

16. Construct an equation for each of the accompanying parabolas.

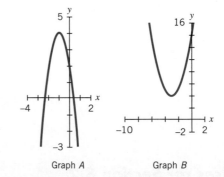

Graph A   Graph B

17. Determine the equation of the parabola whose vertex is at $(2, 3)$ and that passes through the point $(4, -1)$. Show your work, including a sketch of the parabola.

18. Students noticed that the path of water from a water fountain seemed to form a parabolic arc. They set a flat surface at the level of the water spout and measured the maximum height of the water from the flat surface as 8 inches and the distance from the spout to where the water hit the flat surface as 10 inches. Construct a function model for the stream of water.

8.3   *The Standard Form vs. the Vertex Form*

In this section we'll examine how to convert a function in standard form into vertex form by first finding the vertex. Then we'll show how to convert back, from vertex form into standard form.

Finding the Vertex from the Standard Form

See the reading "Why the Formula for the Vertex and the Quadratic Formula Work" for a derivation of this vertex formula.

What if our quadratic function is in the standard *a-b-c* form and we want to find the vertex? It's not always an easy task. If the parabola intersects the horizontal axis only once, that intersection point is the vertex. If there are two horizontal intercepts, then the horizontal coordinate of the vertex lies halfway between them. But what if there are no horizontal intercepts? The strategy for finding the vertex then is somewhat complicated. We have provided all the steps in a reading on the course website to derive the resulting formula below.

Using a formula to find the vertex

The following formula can be used to find the coordinates of the vertex of a parabola when the function is in standard form.

Formula for Finding the Vertex from the Standard Form

The vertex of a quadratic function in the form

$$f(x) = ax^2 + bx + c$$

has coordinates $\left(-\frac{b}{2a}, f\left(-\frac{b}{2a}\right)\right)$

EXAMPLE   1   **Finding the vertex**

Find the vertex and sketch the graph of $f(x) = x^2 - 10x + 100$.

SOLUTION   The function f is in standard form, $f(x) = ax^2 + bx + c$. So $a = 1$, $b = -10$, and $c = 100$. Since a is positive, the graph is concave up, so the vertex represents a minimum. Using the formula for the horizontal coordinate of the vertex, we have

$$x = -\frac{b}{2a} = -\frac{(-10)}{2(1)} = \frac{10}{2} = 5$$

To find the vertical coordinate of the vertex, we need to find the value of $f(x)$ when $x = 5$.

$$\begin{array}{lll} \text{Given} & f(x) = x^2 - 10x + 100 & \\ \text{let } x = 5 & f(5) = 5^2 - 10(5) + 100 & \\ \text{simplify} & \quad\quad = 75 & \end{array}$$

The coordinates of the vertex are (5, 75) and the *y*-intercept is at $f(0) = 100$. Figure 8.20 on the next page shows a sketch of the graph.

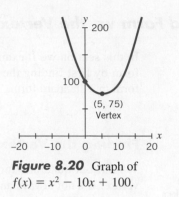

Figure 8.20 Graph of
$f(x) = x^2 - 10x + 100$.

E X A M P L E   2   **Measuring traffic flow**

Urban planners and highway designers are interested in maximizing the number of cars that pass along a section of roadway in a certain amount of time. Observations indicate that the primary variable controlling traffic flow (and hence a good choice for the independent variable) is the density of cars on the roadway: the closer each driver is to the car ahead, the more slowly he or she drives. The following quadratic relationship between traffic flow rate and density of cars was derived from observing traffic patterns in the Lincoln Tunnel, which connects New York and New Jersey:

$$t = -0.21d^2 + 34.66d$$

where t = traffic flow rate (cars/hour) and d = density of cars (cars/mile).

a. Find the coordinates of the vertex of the parabola.

b. What does the vertex represent in terms of the traffic flow rate?

c. Graph the function and estimate the value of the horizontal intercepts.

d. What do the horizontal intercepts mean in terms of the traffic flow and density of cars?

S O L U T I O N   **a.** Our equation $t = -0.21d^2 + 34.66d$ is in standard form ($t = ad^2 + bd + c$), so we can use the formula for finding the coordinates of the vertex. Since $a = -0.21$, $b = 34.66$, and $c = 0$, the horizontal coordinate of the vertex is

$$-\frac{b}{2a} = \frac{-34.66}{2(-0.21)}$$

simplify ≈ 82.52

round to the nearest integer ≈ 83

We can find the corresponding value of t by substituting $d = 83$ into our equation:

Given $t = -0.21d^2 + 34.66d$

substitute for d $t \approx -0.21(83)^2 + 34.66(83)$

simplify $= 1430.09$

round to the nearest integer ≈ 1430

So the coordinates of the vertex are approximately (83, 1430).

b. A vertex at (83, 1430) means that when the density is 83 cars/mile, the traffic flow rate reaches a maximum of 1430 cars/hour. When the density is above or below 83 cars/mile, the traffic flow rate is less than 1430 cars/hour.

c. Figure 8.21 on the next page shows the graph of $t = -0.21d^2 + 34.66d$. The horizontal intercepts are the values of d when $t = 0$. From the graph we estimate that when $t = 0$, then d is either 0 or about 165.

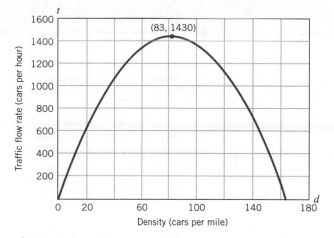

Figure 8.21 The quadratic relationship between traffic flow rate and density.
Source: Adapted from G. B. Whitman, *Linear and Nonlinear Waves* (New York: John Wiley, 1974), p. 68.

d. When the traffic flow rate is equal to zero, traffic is at a standstill. This happens when the density is either zero (when there are no cars) or approximately 165 cars per mile.

Converting between Standard and Vertex Forms

Every quadratic can be written in either standard or vertex form. In this section we examine how to convert from one form to the other.

Converting from a-h-k to a-b-c form

Every function written in the vertex or *a-h-k* form can be rewritten as a function in the standard or *a-b-c* form if we multiply out and group terms with the same power of *x*.

EXAMPLE 3

Converting from vertex to standard form
Rewrite the quadratic $f(x) = 3(x + 7)^2 - 9$ in the *a-b-c* form.

SOLUTION

The function $f(x)$ is in the *a-h-k* form where $a = 3$, $h = -7$, and $k = -9$.

Given	$f(x) = 3(x + 7)^2 - 9$
write out the factors	$= 3(x + 7)(x + 7) - 9$
multiply the factors	$= 3(x^2 + 14x + 49) - 9$
distribute the 3	$= 3x^2 + 42x + 147 - 9$
group the constant terms	$= 3x^2 + 42x + 138$

This function is now in the *a-b-c* format with $a = 3$, $b = 42$, and $c = 138$.

Converting from a-b-c to a-h-k form

 For a graphic illustration of the shift from *a-b-c* to *a-h-k* form, see "Q7: From *a-b-c* to *a-h-k* Form" in *Quadratic Functions.*

Here we examine two strategies that can be used to convert from the standard or *a-b-c* form to the vertex or *a-h-k* form.

Strategy 1: "Completing the Square."

We can convert the function $f(x) = x^2 + 14x + 9$ into *a-h-k* form using a method called *completing the square.*

When a function is in *a-h-k* form, the term $(x - h)^2$ is a perfect square; that is, $(x - h)^2$ is the product of the expression $x - h$ times itself. We can examine separately the expression $x^2 + 14x$ and ask what constant term we would need to add to it in order to make it a perfect square.

A perfect square is in the form

$$(x + m)^2 = x^2 + 2mx + m^2$$

for some number m. Notice that the coefficient of x is two times the number m in our expression for a perfect square. So to turn $x^2 + 14x$ into a perfect square $(x + m)^2$, we need to find m. Since the coefficient of x corresponds to $2m$, then $2m = 14 \Rightarrow m = 7 \Rightarrow m^2 = 49$. So if we add 49 to $x^2 + 14x$ we have a perfect square.

$$(x + 7)^2 = x^2 + 14x + 49$$

To translate $f(x) = x^2 + 14x + 9$ into the *a-k-h* form, we add 49 to make a perfect square, and then subtract 49 to preserve equality. So we have

$$f(x) = x^2 + 14x + 9$$

add and subtract 49 $= x^2 + 14x + (49 - 49) + 9$

regroup terms $= (x^2 + 14x + 49) - 49 + 9$

factor and simplify $= (x + 7)^2 - 40$

Now $f(x)$ is in *a-h-k* form. The vertex is at $(-7, -40)$.

EXAMPLE 4

Converting from standard to vertex form
Convert the function $g(t) = -2t^2 + 12t - 23$ to *a-h-k* form.

SOLUTION

This function is more difficult to convert by completing the square, since a is not 1. We first need to factor out -2 *from the t terms only*, getting

$$g(t) = -2(t^2 - 6t) - 23 \qquad (1)$$

It is the expression $t^2 - 6t$ for which we must complete the square. Since $\left(\frac{1}{2} \cdot (-6)\right)^2 = (-3)^2 = 9$, then adding 9 to $t^2 - 6t$ gives us

$$t^2 - 6t + 9 = (t - 3)^2$$

We must add the constant term 9 *inside the parentheses* in Equation (1) in order to make $t^2 - 6t$ a perfect square. Since everything inside the parentheses is multiplied by -2, we have essentially subtracted 18 from our original function, so we need to add 18 *outside the parentheses* to preserve equality. We have

Given $g(t) = -2t^2 + 12t - 23$

factor out -2 from t terms $= -2(t^2 - 6t) - 23$

add 9 inside parentheses and
    18 outside parentheses $= -2(t^2 - 6t + 9) + 18 - 23$

factor and simplify $= -2(t - 3)^2 - 5$

We now have $g(t)$ in *a-h-k* form. The vertex is at $(3, -5)$.

Strategy 2: Using the Formula for the Vertex.

We could convert $f(x) = 3x^2 - 12x + 5$ to the *a-h-k* form by using the formula for the coordinates of the vertex. Since the coefficient a is the same in both the *a-b-c* and the *a-h-k* forms, we have $a = 3$. The coordinates of the vertex of a quadratic $f(x)$ in the *a-b-c* form are given by $\left(-\frac{b}{2a}, f\left(\frac{-b}{2a}\right)\right)$.

For $f(x)$ we have $a = 3$, $b = -12$, and $c = 5$. So

$$-\frac{b}{2a} = -\frac{(-12)}{2 \cdot 3} = \frac{12}{6} = 2$$

is the horizontal coordinate of the vertex. Evaluating $f(x)$ at 2 gives

$$f(2) = (3 \cdot 2^2) - (12 \cdot 2) + 5$$
$$= 12 - 24 + 5$$
$$= -7$$

which is the vertical coordinate of the vertex. So the vertex is at $(2, -7)$. In the *a-h-k* form the vertex is at (h, k), so here $h = 2$ and $k = -7$. Substituting for a, h, and k in $f(x) = a(x - h)^2 + k$, we get

$$f(x) = 3(x - 2)^2 - 7$$

and the transformation is complete.

E X A M P L E   5   **Mystery parabola**

Find the equation of the parabola in Figure 8.22. Write it in vertex form (*a-h-k*) and in standard form (*a-b-c*).

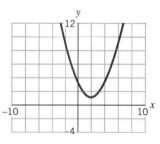

Figure 8.22 A mystery parabola.

S O L U T I O N   The vertex of the graph is at $(2, 1)$ and the graph is concave up. We can substitute the coordinates of the vertex in the *a-h-k* form of the quadratic equation to get

$$y = a(x - 2)^2 + 1 \qquad\qquad (1)$$

How can we find the value for a? If we can identify values for any other point (x, y) that lies on the parabola, we can substitute these values into Equation (1) and solve it for a. The y-intercept, estimated at $(0, 3)$, is a convenient point to pick. Setting $x = 0$ and $y = 3$, we get

$$3 = a(0 - 2)^2 + 1$$
$$3 = 4a + 1$$
$$2 = 4a$$

so $\qquad a = 0.5$

The equation in the *a-h-k* form is

$$y = 0.5(x - 2)^2 + 1$$

If we wanted it in the equivalent *a-b-c* form, we could square, multiply, and collect like terms to get

$$y = 0.5x^2 - 2x + 3$$

The Vertex of a Quadratic Function

A quadratic function in vertex form

$$f(x) = a(x - h)^2 + k \qquad \text{has a vertex at } (h, k)$$

A quadratic function in standard form

$$f(x) = ax^2 + bx + c \qquad \text{has a vertex at } \left(-\frac{b}{2a}, f\left(-\frac{b}{2a} \right) \right)$$

Whether in standard or vertex form, the value of the coefficient a determines if the parabola is relatively narrow or flat.

If $a > 0$, the parabola is concave up and has a minimum at the vertex.

If $a < 0$, the parabola is concave down and has a maximum at the vertex.

E X A M P L E 6   **Matching graphs of quadratics**

Match the following functions to the graphs in Figure 8.23. What allows you to match them easily?

a. $f(x) = x^2 - 8x + 18$

b. $g(x) = 2x^2 - 16x + 34$

c. $h(x) = -0.5x^2 + 4x - 6$

d. $j(x) = -x^2 + 8x - 14$

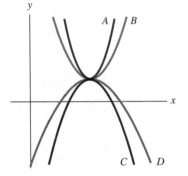

Figure 8.23 Graphs of four quadratic functions.

S O L U T I O N   All the graphs share the same vertex. Graphs A and B are concave up, so they could match with $f(x)$ or $g(x)$, which both have a positive value of a (the coefficient of x^2). Graphs C and D are concave down, so they could match with $h(x)$ or $j(x)$, which both have a negative value of a.

Graph A is narrower than B, so $g(x)$ matches with graph A and $f(x)$ matches with graph B. Graph C is narrower than D, so $j(x)$ matches with graph C and $h(x)$ matches with graph D. Reasoning: The larger the absolute value of a, the narrower the graph.

Algebra Aerobics 8.3

A graphing program is optional for Problem 3.

1. Find the vertex of the graph of each of the following quadratic functions:

   a. $f(x) = 2x^2 - 4$

   b. $g(z) = -z^2 + 6$

   c. $w = 4t^2 + 1$

2. Find the vertex of the graph of each of the following functions and then sketch the graphs on the same grid:

   a. $y = x^2 + 3$

   b. $y = -x^2 + 3$

3. In parts (a) to (d) determine the vertex and whether the graph is concave up or down. Then predict the number of x-intercepts. Graph the function to confirm your answer. Use technology if available.

   a. $f(x) = -3x^2$

   b. $f(x) = -2x^2 - 5$

   c. $f(x) = x^2 + 4x - 7$

   d. $f(x) = 4 - x - 2x^2$

4. Without drawing the graph, describe whether the graph of each of the following functions has a maximum or minimum at the vertex and is narrower or broader than $y = x^2$.

   a. $y = 2x^2 - 5$ **c.** $y = 3 + x - 4x^2$

   b. $y = 0.5x^2 + 2x - 10$ **d.** $y = -0.2x^2 + 11x + 8$

5. For each of the following functions, use the formula $h = -\frac{b}{2a}$ to find the horizontal coordinate of the vertex, and then find its vertical coordinate. Draw a rough sketch of each function.

   a. $y = x^2 + 3x + 2$ **c.** $g(t) = -t^2 - 4t - 7$

   b. $f(x) = 2x^2 - 4x + 5$

6. Find the coordinates of the vertex and the vertical intercept, and then sketch the function by hand.

   a. $y = 0.1(x + 5)^2 - 11$

   b. $y = -2(x - 1)^2 + 4$

7. Rewrite each expression in the form $(x + m)^2 - m^2$.

   a. $x^2 + 6x$ **c.** $x^2 - 30x$

   b. $x^2 - 10x$ **d.** $x^2 + x$

8. Convert the following functions to vertex form by completing the square. Identify the stretch factor and the vertex.

   a. $f(x) = x^2 + 2x - 1$

   b. $j(z) = 4z^2 - 8z - 6$

   c. $h(x) = -3x^2 - 12x$

   d. $H(t) = -16t^2 + 96t$

   e. $Q(t) = -4.9t^2 - 98t + 200$

9. Express each of the following functions in standard form $y = ax^2 + bx + c$. Identify the stretch or compression factor and the vertex.

   a. $y = 2\left(x - \frac{1}{2}\right)^2 + 5$

   b. $y = -\frac{1}{3}(x + 2)^2 + 4$

   c. $y = 10(x - 5)^2 + 12$

   d. $y = 0.1(x + 0.2)^2 + 3.8$

10. Express each of the following functions in vertex form.

    a. $y = x^2 + 6x + 7$

    b. $y = 2x^2 + 4x - 11$

Exercises for Section 8.3

A graphing program is required in Exercises 8 and 19, and optional for Exercises 5 and 6. Technology is needed in Exercises 18, 20, and 22 to generate a best-fit quadratic.

1. Convert the following quadratic functions to vertex form. Identify the coordinates of the vertex.

   a. $y = x^2 + 8x + 11$ **b.** $y = 3x^2 + 4x - 2$

2. The daily profit, f (in dollars), of a hot pretzel stand is a function of the price per pretzel, p (in dollars), given by $f(p) = -1875p^2 + 4500p - 2400$.

   a. Find the coordinates of the vertex of the parabola.

   b. Give the maximum profit and the price per pretzel that gives that profit.

3. Find the equations in vertex form of the parabolas that satisfy the following conditions. Then check your solutions by using a graphing program, if available.

   a. The vertex is at $(-1, 4)$ and the parabola passes through the point $(0, 2)$.

   b. The vertex is at $(1, -3)$ and one of the parabola's two horizontal intercepts is at $(-2, 0)$.

4. Using the strategy of "completing the square," fill in the missing numbers that would make the statement true.

   a. $x^2 + 6x + \underline{\hspace{1cm}} = (x + \underline{\hspace{1cm}})^2$

   b. $x^2 + 8x + \underline{\hspace{1cm}} = (x + \underline{\hspace{1cm}})^2$

   c. $x^2 - 4x + \underline{\hspace{1cm}} = (x - \underline{\hspace{1cm}})^2$

   d. $x^2 - 3x + \underline{\hspace{1cm}} = (x - \underline{\hspace{1cm}})^2$

   e. $2(x^2 + 2x + \underline{\hspace{1cm}}) = 2(x + \underline{\hspace{1cm}})^2$

   f. $-3(x^2 + x + \underline{\hspace{1cm}}) = -3(x + \underline{\hspace{1cm}})^2$

5. (Graphing program optional.) For each quadratic function use the method of "completing the square" to convert to the a-h-k form, and then identify the vertex. If available, use technology to confirm that the two forms are the same.

   a. $y = x^2 + 8x + 15$

   b. $f(x) = x^2 - 4x - 5$

   c. $p(t) = t^2 - 3t + 2$

   d. $r(s) = -5s^2 + 20s - 10$

   e. $z = 2m^2 + 6m - 5$

6. (Graphing program optional.) For each quadratic function convert to a-h-k form by using $h = -\frac{b}{2a}$ and then find k. If available, use technology to graph the two forms of the function to confirm that they are the same.

   a. $y = x^2 + 6x + 13$

   b. $f(x) = x^2 - 5x - 5$

   c. $g(x) = x^2 - 3x + 6$

   d. $p(r) = -3r^2 + 18r - 9$

   e. $m(z) = 2z^2 + 8z - 5$

7. Convert the following functions from the a-b-c or standard form to the a-h-k or vertex form.

   a. $f(x) = x^2 + 6x + 5$ **d.** $y = 2x^2 + 3x - 5$

   b. $g(x) = x^2 - 3x + 7$ **e.** $h(x) = 3x^2 + 6x + 5$

   c. $y = 3x^2 - 12x + 12$ **f.** $y = -x^2 + 5x - 2$

8. (Graphing program required.) Given $f(x) = -x^2 + 8x - 15$:

   a. Estimate by graphing: the x-intercepts, the y-intercept, and the vertex.

   b. Calculate the coordinates of the vertex and compare your estimate in part (a).

9. Write each of the following quadratic equations in function form (i.e., solve for y in terms of x). Find the vertex and the y-intercept using any method. Finally, using these points, draw a rough sketch of the quadratic function.

   a. $y + 12 = x(x + 1)$ **d.** $y - 8x = x^2 + 15$

   b. $2x^2 + 6x + 14.4 - 2y = 0$ **e.** $y + 1 = (x - 2)(x + 5)$

   c. $y + x^2 - 5x = -6.25$ **f.** $y + 2x(x - 6) = 20$

10. Find two different equations for a parabola that passes through the points $(-2, 5)$ and $(4, 5)$ and that opens downward. (*Hint:* Find the axis of symmetry.)

11. Without drawing the graph, list the following parabolas in order, from the narrowest to the broadest. Verify your results with technology.

    a. $y = x^2 + 20$ **d.** $y = 4x^2$

    b. $y = 0.5x^2 - 1$ **e.** $y = 0.1x^2 + 2$

    c. $y = \frac{1}{3}x^2 + x + 1$ **f.** $y = -2x^2 - 5x + 4$

12. Match each of the following graphs with one of following equations. Explain your reasoning. (*Hint:* Find the vertex.) Note that one function does not have a match.

    $f(x) = 2x^2 - 8x - 2$ $h(x) = 0.5x^2 - 2x + 3$

    $g(x) = 2x^2 - 8x + 3$ $i(x) = 0.5x^2 - 2x + 8$

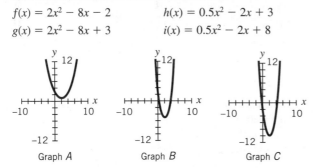

Graph *A*   Graph *B*   Graph *C*

13. Marketing research by a company has shown that the profit, $P(x)$ (in thousands of dollars), made by the company is related to the amount spent on advertising, x (in thousands of dollars), by the equation $P(x) = 230 + 20x - 0.5x^2$. What expenditure (in thousands of dollars) for advertising gives the maximum profit? What is the maximum profit?

14. Tom has a taste for adventure. He decides that he wants to bungee-jump off the Missouri River bridge. At time t (in seconds from the moment he jumps) his height $h(t)$ (in feet above the water level) is given by the function $h(t) = 20.5t^2 - 123t + 190.5$. How close to the water will Tom get?

15. A manager has determined that the revenue $R(x)$ (in millions of dollars) made on the sale of supercomputers is given by $R(x) = 48x - 3x^2$, where x represents the number of supercomputers sold. How many supercomputers must be sold to maximize revenue? According to this model, what is the maximum revenue (in millions of dollars)?

16. A Norman window has the shape of a rectangle surmounted by a semicircle of diameter equal to the width of the rectangle (see the accompanying figure at the top of next column). If the perimeter of the window is 20 feet (including the semicircle), what dimensions will admit the most light (maximize the area)? (*Hint:* Express L in terms of r. Recall that the circumference of a circle $= 2\pi r$, and the area of a circle $= \pi r^2$, where r is the radius of the circle.)

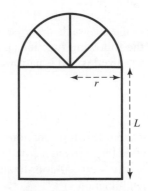

17. A pilot has crashed in the Sahara Desert. She still has her maps and knows her position, but her radio is destroyed. Her only hope for rescue is to hike out to a highway that passes near her position. She needs to determine the closest point on the highway and how far away it is.

    a. The highway is a straight line passing through a point 15 miles due north of her and another point 20 miles due east. Draw a sketch of the situation on graph paper, placing the pilot at the origin and labeling the two points on the highway.

    b. Construct an equation that represents the highway (using x for miles east and y for miles north).

    c. Now use the Pythagorean Theorem to describe the square of the distance, d, of the pilot to any point (x, y) on the highway.

    d. Substitute the expression for y from part (b) into the equation from part (c) in order to write d^2 as a quadratic in x.

    e. If we minimize d^2, we minimize the distance d. So let $D = d^2$ and write D as a quadratic function in x. Now find the minimum value for D.

    f. What are the coordinates of the closest point on the highway, and what is the distance, d, to that point?

18. (Requires technology to find a best-fit quadratic.) The accompanying figure is a plot of the data in the file BOUNCE, which shows the height of a bouncing racquetball (in feet) over time (in seconds). The path of the ball between each pair of bounces can be modeled using a quadratic function.

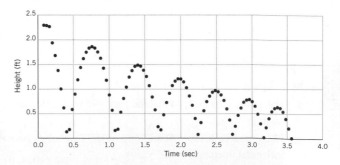

    a. Select from the file BOUNCE the subset of the data that represents the motion of the ball between the first and second bounces (that is, between the first and second times the ball hits the floor). Use technology to generate a best-fit quadratic function for this subset.

    b. From the graph, estimate the maximum height the ball reaches between the first and second bounces.

    c. Use the best-fit function to calculate the maximum height the ball reaches between the first and second bounces.

19. (Graphing program required.) At low speeds an automobile engine is not at its peak efficiency; efficiency initially rises with speed and then declines at higher speeds. When efficiency is at its maximum, the consumption rate of gas (measured in gallons per hour) is at a minimum. The gas consumption rate of a particular car can be modeled by the following equation, where G is the gas consumption rate in gallons per hour and M is speed in miles per hour:

$$G = 0.0002M^2 - 0.013M + 1.07$$

a. Construct a graph of gas consumption rate versus speed. Estimate the minimum gas consumption rate from your graph and the speed at which it occurs.

b. Using the equation for G, calculate the speed at which the gas consumption rate is at its minimum. What is the minimum gas consumption rate?

c. If you travel for 2 hours at peak efficiency, how much gas will you use and how far will you go?

d. If you travel at 60 mph, what is your gas consumption rate? How long does it take to go the same distance that you calculated in part (c)? (Recall that travel distance = speed · time traveled.) How much gas is required for the trip?

e. Compare the answers for parts (c) and (d), which tell you how much gas is used for the same-length trip at two different speeds. Is gas actually saved for the trip by traveling at the speed that gives the minimum gas consumption rate?

f. Using the function G, generate data for gas consumption rate measured in gallons per mile by completing the following table. Plot gallons per mile (on the vertical axis) vs. miles per hour (on the horizontal axis). At what speed is gallons per mile at a minimum?

Speed of Car (mph)	Measures of the Rate of Gas Consumption	
	(gal/hr)	(gal/hr)/mph = gal/mile
0		
10		
20		
30		
40		
50		
60		
70		
80		

g. Add a fourth column to the data table. This time compute miles/gal = mph/(gal/hr). Plot miles per gallon vs. miles per hour. At what speed is miles per gallon at a maximum? This is the inverse of the preceding question; we are normally used to maximizing miles per gallon instead of minimizing gallons per mile. Does your answer make sense in terms of what you found for parts (b) and (f)?

20. (Requires technology to create best-fit functions.) The following data at the top of the next column shows the average growth of the human embryo prior to birth.

Embryo Age (weeks)	Weight (g)	Length (cm)
8	3	2.5
12	36	9
20	330	25
28	1000	35
36	2400	45
40	3200	50

Source: Kimber et al., *Anatomy and Physiology* 15th Edition, © 1955, pg. 785. Reprinted by permission of Pearson Education, Inc., Upper Saddle River, NJ.

a. Plot weight (on vertical axis) vs. age (on horizontal axis). Then use technology to find the best-fit quadratic model to approximate the data.

b. According to your model, what would an average 32-week-old embryo weigh?

c. Comment on the domain for which your formula is reliable.

d. Plot length versus age; then use technology to construct an appropriate mathematical model for the length as a function of embryo age from 20 to 40 weeks.

e. Using your model, compute the age at which an embryo would be 42.5 centimeters long.

21. A shot-put athlete releases the shot at a speed of 14 meters per second, at an angle of 45 degrees to the horizontal (ground level). The height y (in meters above the ground) of the shot is given by the function

$$y = 2 + x - \frac{1}{20}x^2$$

where x is the horizontal distance the shot has traveled (in meters).

a. What was the height of the shot at the moment of release?

b. How high is the shot after it has traveled 4 meters horizontally from the release point? 16 meters?

c. Find the highest point reached by the shot in its flight.

d. Draw a sketch of the height of the shot and indicate how far the shot is from the athlete when it lands.

22. (Technology required to generate a best-fit quadratic.) The accompanying graph of the data file INCLINE shows the motion of a cart, initially at the bottom of an inclined plane, after it was given a push toward a motion detector positioned at the top of the plane. The distance (in feet) of the cart from the top of the plane is plotted vs. time (in seconds). The motion can be modeled with a quadratic function.

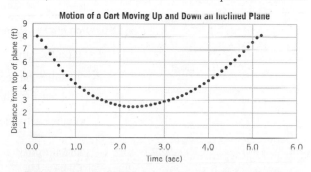

Motion of a Cart Moving Up and Down an Inclined Plane

a. Estimate the coordinates of the vertex. Describe what is happening to the cart at the vertex.

b. Use technology to generate a best-fit quadratic to the data.

c. Calculate the coordinates of the vertex.

8.4 *Finding the Horizontal Intercepts: The Factored Form*

We often want to find the exact values for horizontal intercepts when we use a function to model a real-world situation. We know that a quadratic function may have two, one, or no horizontal intercepts. The horizontal intercepts (if any) of a function $f(x)$ are the value(s) of x such that $f(x) = 0$, called the *zeros* of the function.

If $f(x) = ax^2 + bx + c$, then setting $f(x) = 0$, we have $0 = ax^2 + bx + c$. If, as in the following example, we can factor $ax^2 + bx + c$ as a product of two linear factors (each with an x term), solving for the intercepts is easy. Later on in this chapter we'll use the quadratic formula to find the horizontal intercepts when the function is not easily factored.

Using Factoring to Find the Horizontal Intercepts

EXAMPLE 1

A quadratic model for a battleship gun range

Iowa class battleships have large naval guns that have a muzzle velocity of about 2000 feet per second. If the gun is set to maximize range, then the relationship between the height of the projectile fired and the distance it travels is $h(d) = d - 8 \cdot 10^{-6}d^2$, where d = horizontal distance in feet from the battleship, and $h(d)$ = height in feet of the projectile. What is the maximum range of the battleship gunfire?

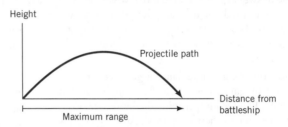

Figure 8.24 Graph of height versus distance for a projectile fired from a battleship.

SOLUTION

Assuming that the battleship and the target are at the same level, to find the maximum range we need to find where the projectile will hit the ground. At that point $h(d) = 0$, so the point represents a horizontal intercept on the graph (see Figure 8.24). To find the horizontal intercepts we set $h(d) = 0$ and solve for d.

Given the function $\qquad h(d) = d - 8 \cdot 10^{-6}d^2$

if we set $h(d) = 0$ $\qquad\qquad 0 = d - 8 \cdot 10^{-6}d^2$

and factor out d, we get $\qquad 0 = d \cdot (1 - 8 \cdot 10^{-6}d)$

So when $h(d) = 0$, then either

$d = 0$ (at the gun) $\quad or \quad 1 - 8 \cdot 10^{-6}d = 0$ (when the projectile hits the ground)

To solve the second expression for d $\qquad 1 - 8 \cdot 10^{-6}d = 0$

subtract 1 from each side $\qquad\qquad\qquad -8 \cdot 10^{-6}d = -1$

divide each side by $-8 \cdot 10^{-6}$ $\qquad\qquad\qquad d = \dfrac{-1}{(-8 \cdot 10^{-6})}$

and simplify $\qquad\qquad\qquad\qquad\qquad = 0.125 \cdot 10^6$

to get $\qquad\qquad\qquad\qquad\qquad\qquad = 125{,}000$ feet

So a projectile fired from this gun is able to hit a target that is 125,000 feet, or almost 24 miles, from the battleship.

> **Finding the Horizontal Intercepts of a Function**
> Given a function $f(x)$:
>
> > To find the horizontal or x-intercepts, set $f(x) = 0$ and solve for x.
> >
> > Every horizontal intercept is a *zero* of the function.

One of the properties of real numbers we used in the previous example was the "zero product rule"—the notion that if the product of two terms is 0, then at least one term must be 0.

> **Zero Product Rule**
> For any two numbers r and s, if the product $rs = 0$, then r or s or both must equal 0.

We'll use this rule repeatedly in our search for the horizontal intercepts of a function.

Factoring Quadratics

In the battleship example, we wrote $h(d) = d - 8 \cdot 10^{-6}d^2$, which is in standard form, as $h(d) = d(1 - 8 \cdot 10^{-6}d)$, which is in *factored form*. If a quadratic is in factored form, it's easy to find the horizontal intercepts. We set the product equal to 0 and use the zero product rule.

Factoring review

To convert $ax^2 + bx + c$ to factored form requires thinking, practice, and a few hints. It is often a trial-and-error process. We usually restrict ourselves to finding factors with integer coefficients.

First, look for common factors in all of the terms.

For example, $10x^2 + 2x$ can be factored as $2x(5x + 1)$.

Then look for factors when a = 1.

Factors are easiest to find when the coefficient of x^2 is 1. For example, to factor $x^2 + 7x + 12$, we want to rewrite it as

$$(x + m)(x + n)$$

for some m and n. Note that the coefficients of both x's in the factors equal 1, since x times x is equal to the x^2 in the original expression. If we multiply out the two factors, we get $x^2 + (m + n)x + m \cdot n$. So we need $m + n = 7$ and $m \cdot n = 12$. Since 12 is positive, m and n must have the same sign, either both positive or both negative. But since $m + n$ (=7) is positive, m and n must both be positive. So we consider pairs of positive integers whose product is 12, namely, 1 and 12, 2 and 6, or 3 and 4. We can then narrow our list of factors of 12 to those whose sum equals 7, the coefficient of the x term. Only the factors 3 and 4 fit this criterion. We can factor our polynomial as:

$$x^2 + 7x + 12 = (x + 3)(x + 4)$$

We can check that these factors work by multiplying them out, using the FOIL technique (see Footnote 2 on p. 470).

Special case: The difference of two squares.

In this case the middle terms cancel out when multiplying.

$$x^2 - 25 = (x - 5)(x + 5)$$
$$= x^2 - 5x + 5x - 25$$
$$= x^2 - 25$$
$$x^2 - n^2 = (x - n)(x + n)$$

Guidelines for Factoring $ax^2 + bx + c$ when $a = 1$

First factor out any common terms.

When $a = 1$,

find the factors of c that add to give b.

$$x^2 + bx + c = (x + m)(x + n), \quad \text{where } c = mn \text{ and } b = m + n$$

If the quadratic is the difference of two squares (so $b = 0$ and $c = n^2$ for some number n), factor it into the product of a sum and a difference.

$$x^2 - n^2 = (x - n)(x + n)$$

EXAMPLE 2

Finding factors

Put each of the following functions into factored form and then identify any horizontal intercepts.

a. $f(x) = 300x^2 + 195x$ **c.** $H(v) = -2v^2 + 18$

b. $h(t) = t^2 - 5t + 6$

SOLUTION

a. In factored form $f(x) = 15x(20x + 13)$. (You can double-check this by multiplying the factors to return to $300x^2 + 195x$.)

To find the horizontal intercepts, we need to find values for x such that $f(x) = 0$. If we set $15x(20x + 13) = 0$, then according to the zero product rule, either

$$15x = 0 \quad \text{or} \quad 20x + 13 = 0$$
$$x = 0 \qquad\qquad 20x = -13$$
$$x = \frac{-13}{20} \quad \text{or} \quad -0.65$$

So there are two horizontal intercepts, at $x = 0$ and $x = -0.65$

b. In factored form $h(t) = (t - 3)(t - 2)$. (Again we can double-check our answer by multiplying the two factors to get $t^2 - 2t - 3t + 6 = t^2 - 5t + 6$.) To find the horizontal intercepts we need to find values for t such that $h(t) = 0$. If we set $(t - 3)(t - 2) = 0$, then either

$$t - 3 = 0 \quad \text{or} \quad t - 2 = 0$$
$$t = 3 \qquad\qquad t = 2$$

So there are two horizontal intercepts, at $t = 3$ and $t = 2$.

c. In factored form $H(v) = -2(v^2 - 9)$
$$= -2(v + 3)(v - 3)$$

When $(v + 3)(v - 3) = 0$, then either

$$v + 3 = 0 \qquad \text{or} \qquad v - 3 = 0$$
$$v = -3 \qquad\qquad v = 3$$

So $H(v)$ has two horizontal intercepts, at $v = -3$ and $v = 3$.

Look for factors when a ≠ 1: The ac (or grouping) method

When $a = 1$, we factored $x^2 + bx + c = (x + m)(x + n)$ by asking: "What are the factors of c that add to give b?" Now we derive a similar method to factor a quadratic when $a \neq 1$.

We can factor a quadratic as	$ax^2 + bx + c = (sx + m)(tx + n)$
use the distributive property	$= sx(tx + n) + m(tx + n)$
multiply	$= stx^2 + snx + mtx + mn$
and regroup the middle terms	$= stx^2 + (sn + mt)x + mn$

So we get $a = st$, $b = sn + mt$ and $c = mn$, and the product $ac = stmn$.

Why would we want to do this? Notice that the coefficient of the middle term $b \, (= sn + mt)$ is the sum of certain factors of $ac \, (= stmn)$ re-arranged. So now we can ask "What are the factors of ac that add to give b?" This is called the *ac* (or *grouping*) *method* of factoring.

EXAMPLE 3

Using the *ac* method to factor

Factor $2x^2 + 5x - 12$ by using the *ac* method.

SOLUTION

Set $a = 2$, $b = 5$ and $c = -12$. Then:

Step 1: Multiply a and c to get ac $\qquad\qquad (2)(-12) = -24$

Step 2: Find the factors of ac that add to equal b (note there are many combinations, only a few are listed here)

Factors of ac	Sum of factors
$(24)(-1)$	$24 + (-1) \neq 5$
$(12)(-2)$	$12 + (-2) \neq 5$
$(8)(-3)$	$8 + (-3) = 5$

So the factors of $ac = -24$ that add to $b = 5$ are 8 and -3.

Step 3: Rewrite the middle term $(5x)$ of the original quadratic as the sum of $8x$ and $-3x$ to get: $\qquad\qquad 2x^2 + 5x - 12 = 2x^2 + 8x - 3x - 12.$

Step 4: Factor by grouping $\qquad 2x^2 + 8x - 3x - 12 = 2x(x + 4) - 3(x + 4)$
$$= (x + 4)(2x - 3)$$

So $2x^2 + 5x - 12 = (x + 4)(2x - 3)$.

EXAMPLE 4

Can a quadratic always be factored?

Factor $3x^2 - x + 5$ using the *ac* method.

SOLUTION

Set $a = 3$, $b = -1$ and $c = 5$. Then:

Step 1: Multiply to get ac $\qquad\qquad (3)(5) = 15$

Step 2: Find the factors of ac that add to equal b

Factors of ac	Sum of factors
$(15)(1)$	$15 + (1) \neq -1$
$(3)(5)$	$3 + (5) \neq -1$
$(-15)(-1)$	$-15 + (-1) \neq -1$
$(-3)(-5)$	$-3 + (-5) \neq -1$

There are no factors of 15 that add to give -1, therefore $3x^2 - x + 5$ cannot be factored using integers and is said to be *prime*. The *ac* test can always be used to determine if factoring (with integer coefficients) is possible.

Guidelines for Factoring $ax^2 + bx + c$ when $a \neq 1$: The ac Method

To factor $ax^2 + bx + c$, with no common factors, using the *ac method:*

Step 1: Multiply the leading coefficient a and the constant term c to get ac.
Step 2: Find the factors of ac that sum to get b.
Step 3: Rewrite the middle term (bx) as a sum of products of x using as coefficients the factors found in Step 2.
Step 4: Factor by grouping.

If there are no integer factors of ac that sum to get b, then $ax^2 + bx + c$ cannot be factored using integers and is said to be *prime*.

Algebra Aerobics 8.4a

1. Put the function into factored form with integer coefficients and then identify any horizontal intercepts.
   a. $y = -16t^2 + 50t$
   b. $y = t^2 - 25$
   c. $h(z) = z^2 - 3z - 4$
   d. $g(x) = 4x^2 - 9$
   e. $y = 15 - 8x + x^2$
   f. $v(x) = x^2 + 2x + 1$
   g. $p(q) = q^2 - 6q + 9$

2. When possible, put the function into factored form with integer coefficients and then identify any horizontal intercepts.
   a. $f(x) = 5 - x - 4x^2$
   b. $h(t) = 64 - 9t^2$
   c. $y = 10 - 13t - 3t^2$
   d. $z = 4w^2 - 20w + 25$
   e. $y = 2x^2 - 3x - 5$
   f. $Q(t) = 6t^2 + 11t - 10$

3. Identify which of the following quadratic functions can be factored into the product of a sum and a difference, $y = (a + b)(a - b)$, which can be factored into the square of the sum or difference, $y = (a \pm b)^2$, and which can be factored into neither.
   a. $y = x^2 - 9$
   b. $y = x^2 + 4x + 4$
   c. $y = x^2 + 5x + 25$
   d. $y = 9x^2 - 25$
   e. $y = x^2 - 8x + 16$
   f. $y = 16 - 25x^2$
   g. $y = 4 + 16x^2$

4. Find any horizontal and vertical intercepts for the following functions and explain the meaning of each within the context of the problem.
   a. An object is thrown vertically into the air at time $t = 0$ seconds. Its height, $h(t)$, in feet, is given by $h(t) = -16t^2 + 64t$.
   b. The monthly profit $P(q)$ (in dollars) is a function of q, the number of items sold. The relationship is described by $P(q) = -q^2 + 60q - 800$.

5. Match the factored form of the quadratic function with its graph.

   $$y_1 = -x(x - 2)$$
   $$y_2 = (x - 2)(x + 1)$$
   $$y_3 = (x + 4)(x + 1)$$

   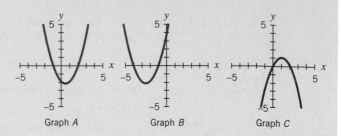

   Graph A   Graph B   Graph C

6. Given $f(x) = x^2 + x - 30$:
   a. Factor $f(x)$.
   b. Find the horizontal intercepts.
   c. Sketch the graph of $f(x)$.
   d. Describe the relationship between the factored form and the horizontal intercepts in this problem.

7. Factor each of the following using the ac method. Identify any quadratic expressions that are prime.
   a. $2x^2 + 7x - 15$
   b. $4x^2 - x - 3$
   c. $3x^2 - 13x + 14$
   d. $2t^2 - 7t + 5$

8. Factor completely, first factoring out the greatest common factor.
   a. $12x^2 - 26x + 10$
   b. $14r^3 - 21r^2 - 35r$

Using the Quadratic Formula to Find the Horizontal Intercepts

Most quadratic functions such as $f(x) = x^2 + 2x + 1$ are not easily factored. When this is the case, we still set $f(x) = 0$ and solve for x. But now we use the quadratic formula to find the solutions or *roots* to the equation $0 = ax^2 + bx + c$.

See the reading "Why the Formula for the Vertex and the Quadratic Formula Work" for a derivation of this famous formula.

The Quadratic Formula

For any quadratic equation of the form $0 = ax^2 + bx + c$ (where $a \neq 0$), the solutions, or *roots*, of the equation are given by

$$x = \frac{-b \pm \sqrt{b^2 - 4ac}}{2a}$$

The term under the radical sign, $b^2 - 4ac$, is called the *discriminant*.

The symbol $\pm$ lets us use one formula to write the two roots as

$$x = \frac{-b + \sqrt{b^2 - 4ac}}{2a} \qquad \text{and} \qquad x = \frac{-b - \sqrt{b^2 - 4ac}}{2a}$$

Note on Terminology.   The language of roots and zeros can be confusing. The numbers 3 and -3 are called the *roots* or *solutions* of the *equation* $x^2 - 9 = 0$ and the *zeros* of the *function* $f(x) = x^2 - 9$. The zeros of the function $f(x)$ are the roots of the equation $f(x) = 0$.

The discriminant

One shortcut for predicting the number of horizontal intercepts is to use the discriminant, the term $b^2 - 4ac$.

Using the Discriminant

A quadratic function $f(x) = ax^2 + bx + c$ has a discriminant of $b^2 - 4ac$.

If the discriminant > 0, there are two distinct real roots and hence two x-intercepts, at

$$x_1 = \frac{-b + \sqrt{\text{discriminant}}}{2a} \qquad \text{and} \qquad x_2 = \frac{-b - \sqrt{\text{discriminant}}}{2a}$$

If the discriminant $= 0$, there is only one distinct real root (at the vertex) and hence only one x-intercept, at

$$x = \frac{-b}{2a}$$

If the discriminant < 0, then $\sqrt{\text{discriminant}}$ is not a real number; so there are no x-intercepts and the zeros of the function are not real numbers.

E X A M P L E   5   **Identify the sign of the discriminant**

Identify whether each of the graphs in Figure 8.25 has a discriminant > 0, < 0, or equal to 0.

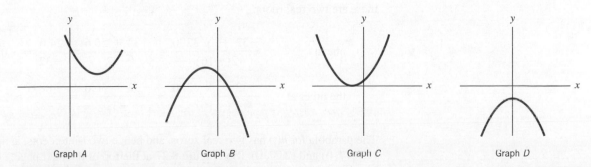

Graph *A*   Graph *B*   Graph *C*   Graph *D*

Figure 8.25 Graphs of four parabolas.

SOLUTION   Graph B has two horizontal intercepts, so the discriminant is > 0. Graph C has only one horizontal intercept, so the discriminant $= 0$. Graphs A and D have no horizontal intercepts, so the discriminant is < 0.

EXAMPLE 6

Using the discriminant

For each of the following functions, use the discriminant to predict the number of horizontal intercepts. If there are any, use the quadratic formula to find them. Then using technology, graph the function to confirm your predictions.

a. $f(z) = z^2 + 3z + 2.25$

b. $h(t) = 34 + 32t - 16t^2$

c. $g(x) = -x^2 - 6x - 10$

SOLUTION   **a.** Setting $f(z) = 0$, we have $0 = z^2 + 3z + 2.25$. Here $a = 1$, $b = 3$, and $c = 2.25$, so the discriminant is $b^2 - 4ac = 3^2 - (4 \cdot 1 \cdot 2.25) = 9 - 9 = 0$. The quadratic formula says that there is only one root, at $-b/(2a) = -3/(2 \cdot 1) = -1.5$. So $f(z)$ has one real zero, and hence one z-intercept at $(-1.5, 0)$, which must also be the vertex of the parabola. (See Figure 8.26.)

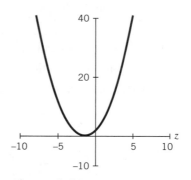

Figure 8.26 Graph of $f(z) = z^2 + 3z + 2.25$, with one horizontal intercept, at the vertex.

b. Setting $h(t) = 0$, we have $0 = 34 + 32t - 16t^2$. If we rearrange the terms as $0 = -16t^2 + 32t + 34$, it's easier to see that we should set $a = -16$, $b = 32$, and $c = 34$. The discriminant is $b^2 - 4ac = (32)^2 - (4 \cdot (-16) \cdot 34) = 1024 + 2176 = 3200$. So there are two real roots:

one at $\dfrac{-32 + \sqrt{3200}}{2(-16)} \approx \dfrac{-32 + 56.6}{-32} = \dfrac{24.6}{-32} \approx -0.77$

the other at $\dfrac{-32 - \sqrt{3200}}{2(-16)} \approx \dfrac{-32 - 56.6}{-32} = \dfrac{-88.6}{-32} \approx 2.77$

The parabola for $h(t)$ has two real zeros, and hence two t-intercepts, at approximately $(-0.77, 0)$ and $(2.77, 0)$. (See Figure 8.27 at the top of the next page.)

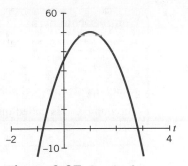

Figure 8.27 Graph of
$h(t) = 34 + 32t - 16t^2$, with
two horizontal intercepts.

c. Setting $g(x) = 0$, we have $0 = -x^2 - 6x - 10$. Here $a = -1$, $b = -6$, and $c = -10$, so the discriminant is $b^2 - 4ac = (-6)^2 - 4(-1)(-10) = 36 - 40 = -4$. The discriminant is negative, so taking its square root presents a problem. There is no real number r such that $r^2 = -4$. So, the roots at $\frac{6 \pm \sqrt{-4}}{-2}$ are not real. Since there are no real zeros for the function, there are no horizontal intercepts. (See Figure 8.28.)

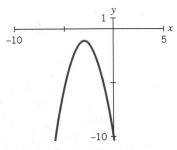

Figure 8.28 Graph of
$g(x) = -x^2 - 6x - 10$, with
no horizontal intercepts.

Imaginary and complex numbers

Mathematicians were uncomfortable with the notion that certain quadratic equations did not have solutions, so they literally invented a number system in which such equations would be solvable. In the process, they created new numbers, called *imaginary numbers*. The imaginary number i is defined as a number such that

$$i^2 = -1$$

or equivalently
$$i = \sqrt{-1}$$

A number such as $\sqrt{-4}$ is also an imaginary number. We can write $\sqrt{-4}$ as

$$\sqrt{(4)(-1)} = \sqrt{4}\sqrt{-1} = 2\sqrt{-1} = 2i$$

When a number is called imaginary, it sounds as if it does not exist. But imaginary numbers are just as legitimate as real numbers. Imaginary numbers are used to extend the real number system to a larger system called the *complex numbers*.

Complex Numbers

A *complex number* is defined as any number that can be written in the form

$$z = a + bi$$

where a and b are real numbers and $i = \sqrt{-1}$.

The number a is called the real part of z, and the number b is called the imaginary part of z.

Note that the real numbers are a subset of the complex numbers, since any real number a can be written as $a + 0 \cdot i$.

EXAMPLE 7

Representing complex numbers

Write each expression as a complex number of the form $a + bi$.

a. $-2 + 7i$ d. $\sqrt{-25}$

b. $4 + \sqrt{-9}$ e. $5 + 3i^2$

c. $13 - \sqrt{36}$

SOLUTION

a. $-2 + 7i$ is complex and already in $a + bi$ form.

b. $4 + \sqrt{-9} = 4 + \sqrt{9}\sqrt{-1} = 4 + 3i$

c. $13 - \sqrt{36} = 13 - 6 = 7 = 7 + 0 \cdot i$ (which is a real number)

d. $\sqrt{-25} = \sqrt{25}\sqrt{-1} = 5i = 0 + 5i$ (which is an imaginary number)

e. Since $i^2 = -1$, then $5 + 3i^2 = 5 + 3(-1) = 5 - 3 = 2 = 2 + 0 \cdot i$ (which is a real number)

The Factored Form

We started this section by finding the horizontal intercepts from the factored form. We now show that any quadratic function can be put into factored form, whether the zeros are real or complex. The Factor Theorem relates the zeros of a function to the factors of a function.

The Factor Theorem

Given a function $f(x)$, if $f(r) = 0$, then r is a *zero* of the function and $(x - r)$ is a factor of $f(x)$.

Using the Factor Theorem, if r_1 and r_2 are zeros of the function $f(x)$, then both $(x - r_1)$ and $(x - r_2)$ are factors of $f(x)$. So we can write a quadratic function $f(x) = ax^2 + bx + c$ in factored form as $f(x) = a(x - r_1)(x - r_2)$. Note that if you multiply out the factored form, the coefficient of x^2 is a, as it is in the standard form. The factored form is useful when we want to emphasize the zeros of a function.

The Factored Form

The quadratic function $f(x) = ax^2 + bx + c$ can be written in *factored form* as

$$f(x) = a(x - r_1)(x - r_2)$$

where r_1 and r_2 are the *zeros* of $f(x)$.

If r_1 and r_2 are real numbers, then $f(x)$ has x-intercepts at r_1 and r_2.

EXAMPLE 8   **Constructing a quadratic function with complex zeros**

a. Construct a quadratic function that is concave up and has zeros at $5 + i$ and $5 - i$. Put it into standard form.

b. What do we know about the graph of this function? Are there any other functions with the same characteristics?

SOLUTION   a. The function $f(x) = (x - (5 + i)) (x - (5 - i))$ has zeros at $5 + i$ and $5 - i$. If we multiply it out (using the FOIL technique), we get

$$(x - (5 + i))(x - (5 - i)) = x^2 - (5 - i)x - (5 + i)x + (5 + i)(5 - i)$$

use the distributive law and
   FOIL the last 2 terms $\quad = x^2 - 5x + ix - 5x - ix + (25 - 5i + 5i - i^2)$

simplify $\quad = x^2 - 10x + (25 - i^2)$

and substitute -1 for i^2 $\quad = x^2 - 10x + 26$

So we can rewrite $f(x)$ in standard form as $f(x) = x^2 - 10x + 26$. (You can double-check this by using the quadratic formula.) Since the coefficient of x^2 is 1, and hence positive, the graph is concave up.

b. Since the zeros are not real, the graph of the function has no x-intercepts.

Any function of the form $af(x) = a(x^2 - 10x + 26)$, where $a > 0$, is concave up and has zeros at $5 + i$ and $5 - i$. Since there are an infinite number of values of a, there are an infinite number of functions with these characteristics.

EXAMPLE 9   **Creating a parabola for a water jet**
A parabola has horizontal intercepts at $d = -1$ and $d = 2$, and passes through the point $(d, h) = (1.5, 1.25)$.

a. Find the equation for the parabola.

b. Graph the parabola using a dotted line. Now restrict the domain from $d = 0$ to the horizontal intercept that is positive and then color in the corresponding section of the parabola with a solid line. This section of the parabola, drawn with a solid line, describes the path of a water jet located at the center of a circular fountain, where d = horizontal distance (in feet) from the base of the fountain, and h = height (in feet) of the water.

c. At what height is the nozzle of the water jet?

d. What is the greatest height the stream of water reaches?

e. Identify the domain and range of the water jet.

SOLUTION   a. If the parabola has horizontal intercepts at $d = -1$ and $d = 2$, then the equation for the parabola is in the form $h = a(d - (-1))(d - 2)$ or $a(d + 1)(d - 2)$. The point $(1.5, 1.25)$ lies on the parabola, so it must satisfy the equation. Hence

given that $\quad\quad\quad\quad\quad\quad\quad\quad\quad\quad\quad\quad h = a(d + 1)(d - 2)$

if we substitute 1.5 for d and 1.25 for h $\quad\quad 1.25 = a (1.5 + 1)(1.5 - 2)$

simplify $\quad\quad\quad\quad\quad\quad\quad\quad\quad\quad\quad\quad 1.25 = a(2.5)(-0.5)$

multiply $\quad\quad\quad\quad\quad\quad\quad\quad\quad\quad\quad\quad 1.25 = a(-1.25)$

and divide by -1.25, we have $\quad\quad\quad\quad\quad -1 = a$

So the equation of the parabola is $\quad\quad\quad\quad\quad h = -(d + 1)(d - 2)$.

b. See Figure 8.29.

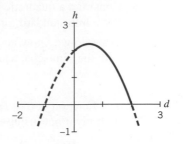

Figure 8.29 Graph of
$h = -(d + 1)(d - 2)$.

c. When $d = 0$, $h = -(0 + 1)(0 - 2) = 2$. So the water nozzle is at a height of 2 feet.

d. There are several strategies for finding the vertex. You could calculate it using the formula in Section 8.3 on p. 487. Or recall that the vertex lies on the axis of symmetry, halfway between the two horizontal intercepts (-1 and 2) at $d = \frac{(-1 + 2)}{2} = \frac{1}{2}$ or 0.5. Substituting 0.5 for d in the equation in part (a), we get

$$h = -(0.5 + 1)\,(0.5 - 2)$$
$$= -(1.5)(-1.5)$$
$$= 2.25$$

The coordinates of the vertex are (0.5, 2.25). So the maximum height of the water (at the vertex of the parabola) is 2.25 feet (or 27″), which occurs at 0.5 feet (or 6″) from the base of the fountain.

e. The domain for the water jet is all d in the interval [0, 2].

The range for the water jet is all h in the interval [0, 2.25]

The Horizontal Intercepts of a Quadratic Function

The graph of a quadratic function may have zero, one, or two horizontal intercepts, which can be found by factoring or using the quadratic formula.

The discriminant in the quadratic formula can be used to predict the number of horizontal intercepts.

Any quadratic function in standard form $f(x) = ax^2 + bx + c$ can be written in *factored form* as

$$f(x) = a(x - r_1)(x - r_2) \quad \text{where } r_1 \text{ and } r_2 \text{ are the zeros of } f(x)$$

If the zeros r_1 and r_2 are real numbers, they are the horizontal intercept(s) of $f(x)$.

Standard, Factored and Vertex Forms

A general quadratic function can be rewritten in three equivalent forms—standard, factored and vertex. Each form highlights different features of the function.

EXAMPLE 10 **Writing a quadratic function in 3 different forms**
Given the function $h(t) = t^2 - 5t + 6$ in standard form, create its equivalent in factored form, then in vertex form. Identify the advantages of each format.

SOLUTION In standard form, $h(t) = t^2 - 5t + 6$, it is easy to determine the overall shape (concave u since the coefficient of t^2 is 1) and the vertical intercept at 6 (when $h(t) = 0$). In factored form $h(t) = (t - 2)\,(t - 3)$, it's easy to identify the horizontal intercepts at $t = 2$ and 3. In vertex form $h(t) = (t - 2.5)^2 - 0.25$, the vertex is highlighted at (2.5, -0.25).

What Each Form Tells You

In each of the three quadratic forms of $f(x) = ax^2 + bx + c$, the sign of a (the coefficient of x^2) always indicates whether the parabola is concave up ($a > 0$) or concave down ($a < 0$).

The standard form: If $f(x) = ax^2 + bx + c$, the constant c is the vertical intercept.

The vertex form: If $f(x) = a(x - h)^2 + k$, the coordinates of the vertex are (h, k).

The factored form: If $f(x) = a(x - r_1)(x - r_2)$, then r_1 and r_2 are the zeros of $f(x)$. If r_1 and r_2 are real numbers, then $f(x)$ has horizontal intercepts at r_1 and r_2.

EXPLORE & EXTEND

8.4 Intersections of a Parabola and Line

We've just learned different strategies for finding where a parabola intersects with the x-axis, a horizontal line. But how would you determine the points of intersection of a parabola and a nonhorizontal line?

a. Sketch examples, then describe the possible number of intersections a parabola and a line could have.

b. Describe any intersection points of the two equations:

$$y = x^2 + 4x - 6 \quad \text{and} \quad y = 3x + 4$$

c. Now find a general expression for the intersection point(s) of a parabola of the form. $y = ax^2 + bx + c$ and a line $y = mx + b$. (*Hint:* Set the two equations equal to each other, rearrange to get a quadratic expression, then solve for x using the Quadratic Formula.) Under what conditions will there be two solutions, one solution, or none?

Algebra Aerobics 8.4b

1. Find any real numbers that satisfy the following equations.

    a. $4x + 7 = 0$

    b. $4x^2 - 7 = 0$

    c. $4x^2 - 7x = 0$

    d. $2(x + 3) = x^2$

    e. $(2x - 11)^2 = 0$

    f. $(x + 1)^2 = 81$

    g. $x = x^2 - 5$

2. Evaluate the discriminant $b^2 - 4ac$ for each of the following quadratic functions of the form $f(x) = ax^2 + bx + c$. Use the discriminant to determine the nature of the zeros of the function and the number (if any) of horizontal intercepts.

    a. $f(x) = 2x^2 + 3x - 1$ **c.** $f(x) = 4x^2 + 4x + 1$

    b. $f(x) = x^2 + 7x + 2$ **d.** $f(x) = 2x^2 + x + 5$

3. Find and interpret the horizontal and vertical intercepts for the following height equations.

    a. $h = -4.9t^2 + 50t + 80$ (h is in meters and t is in seconds)

    b. $h = 150 - 80t - 490t^2$ (h is in centimeters and t is in seconds)

    c. $h = -16t^2 + 64t + 3$ (h is in feet and t is in seconds)

    d. $h = 64t - 16t^2$ (h is in feet and t is in seconds)

4. Evaluate the discriminant and then predict the number of x-intercepts for each function. Use the quadratic formula to find all the zeros of each function and then identify the coordinates of any x-intercept(s).

    a. $y = 4 - x - 5x^2$ **d.** $y = 2x^2 - 3x - 1$

    b. $y = 4x^2 - 28x + 49$ **e.** $y = 2 - 3x^2$

    c. $y = 2x^2 + 5x + 4$

5. From the descriptions given in parts (a) and (b), determine the coordinates of the vertex, find the equation of the parabola, and then sketch the parabola.

    a. A parabola with horizontal intercepts at $x = -2$ and $x = 4$ that passes through the point $(3, 2)$.

    b. A parabola with horizontal intercepts at $x = 2$ and $x = 8$, and a vertical intercept at $y = 10$.

6. Identify the number of x-intercepts for the following functions without converting into standard form.

   a. $y = 3(x - 1)^2 + 5$

   b. $y = -2(x + 4)^2 - 1$

   c. $y = -5(x + 3)^2$

   d. $y = 3(x - 1)^2 - 2$

7. Write an equation for a quadratic function, in factored form, with the specified zeros.

   a. 2 and -3

   b. 0 and -5

   c. 8

8. For each quadratic function in the accompanying graphs, specify the number of real zeros and whether the corresponding discriminant would be positive, negative, or zero.

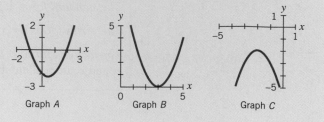

Graph A   Graph B   Graph C

Exercises for Section 8.4

Exercises 5 and 26 require the use of a graphing program. Exercises 24 and 27 recommend it.

1. Solve the following quadratic equations by factoring.

   a. $x^2 - 9 = 0$

   b. $x^2 - 4x = 0$

   c. $3x^2 = 25x$

   d. $x^2 + x = 20$

   e. $4x^2 + 9 = 12x$

   f. $3x^2 = 13x + 10$

   g. $(x + 1)(x + 3) = -1$

   h. $x(x + 2) = 3x(x - 1) - 3$

2. Find the x-intercepts for each of the following functions. Will the vertex lie above, below, or on the x-axis? Find the vertex and sketch the graph, labeling the x-intercepts.

   a. $y = (x + 2)(x + 1)$

   b. $y = 3(1 - 2x)(x + 3)$

   c. $y = -4(x + 3)^2$

   d. $y = \frac{1}{2}(x)(x - 5)$

   e. $q(x) = 2(x - 3)(x + 2)$

   f. $f(x) = -2(5 - x)(3 - 2x)$

3. Factor the quadratic expression and then sketch the graph of the function, labeling the axes and horizontal intercepts.

   a. $y = x^2 + 6x + 8$

   b. $z = 3x^2 - 6x - 9$

   c. $f(x) = x^2 - 3x - 10$

   d. $w = t^2 - 25$

   e. $r = 4s^2 - 100$

   f. $g(x) = 3x^2 - x - 4$

4. a. Construct a quadratic function with zeros at $x = 1$ and $x = 2$.

   b. Is there more than one possible quadratic function for part (a)? Why or why not?

5. (Graphing program required.) Using a graphing program, estimate the real solutions to the following equations. (*Hint:* Think of the equations as resulting from setting $f(x) = 0$.) Verify by factoring, if possible.

   a. $x^2 - 5x + 6 = 0$

   b. $3x^2 - 2x + 5 = 0$

   c. $3x^2 - 12x + 12 = 0$

   d. $-3x^2 - 12x + 15 = 0$

   e. $0.05x^2 + 1.1x = 0$

   f. $-2x^2 - x + 3 = 0$

6. Write each function in factored form, if possible, using integer coefficients.

   a. $f(x) = x^2 + 2x - 15$

   b. $g(x) = x^2 - 6x + 9$

   c. $h(a) = a^2 + 6a - 16$

   d. $k(p) = p^2 + 5p + 7$

   e. $l(s) = 5s^2 - 37s - 24$

   f. $m(t) = 5t^2 + t + 1$

7. Solve the following equations using the quadratic formula. (*Hint:* Rewrite each equation so that one side of the equation is zero.)

   a. $6t^2 - 7t = 5$

   b. $3x(3x - 4) = -4$

   c. $(z + 1)(3z - 2) = 2z + 7$

   d. $(x + 2)(x + 4) = 1$

   e. $6s^2 - 10 = -17s$

   f. $2t^2 = 3t + 9$

   g. $5 = (4x + 1)(x - 3)$

   h. $(2x - 3)^2 = 7$

8. Solve using the quadratic formula.

   a. $x^2 - 3x = 12$

   b. $3x^2 = 4x + 2$

   c. $3(x^2 + 1) = x + 2$

   d. $(3x - 1)(x + 2) = 4$

   e. $\dfrac{1}{x - 2} = \dfrac{x + 1}{x - 1}$

   f. $\dfrac{x^2}{3} + \dfrac{x}{2} - \dfrac{1}{6} = 0$

   g. $\dfrac{1}{x^2} - \dfrac{3}{x} = \dfrac{1}{6}$

9. Calculate the coordinates of the x- and y-intercepts for the following quadratics.

   a. $y = 3x^2 + 2x - 1$

   b. $y = 3(x - 2)^2 - 1$

   c. $y = (5 - 2x)(3 + 5x)$

   d. $f(x) = x^2 - 5$

10. Use the discriminant to predict the number of horizontal intercepts for each function. Then use the quadratic formula to find all the zeros. Identify the coordinates of any horizontal or vertical intercepts.

   a. $y = 2x^2 + 3x - 5$

   b. $f(x) = -16 + 8x - x^2$

   c. $f(x) = x^2 + 2x + 2$

   d. $y = 2(x - 1)^2 + 1$

   e. $g(z) = 5 - 3z - z^2$

   f. $f(t) = (t + 2)(t - 4) + 9$

11. In each part (a) to (e), graph a parabola with the given characteristics. Then write an equation of the form $y = ax^2 + bx + c$ for that parabola.

a. $a > 0$, $b^2 - 4ac > 0$, $c > 0$

b. $a > 0$, $b^2 - 4ac > 0$, $c < 0$

c. $a > 0$, $b^2 - 4ac < 0$, $b \neq 0$

d. $a < 0$, $b^2 - 4ac = 0$, $c \neq 0$

e. $b \neq 0$, $c < 0$, $b^2 - 4ac > 0$

12. For each part, draw a rough sketch of a graph of a function of the type $f(x) = ax^2 + bx + c$

a. Where $a > 0$, $c > 0$, and the function has no real zeros.

b. Where $a < 0$, $c > 0$, and the function has two real zeros.

c. Where $a > 0$ and the function has one real zero.

13. a. Construct a quadratic function $Q(t)$ with exactly one zero at $t = -1$ and a vertical intercept at -4.

b. Is there more than one possible quadratic function for part (a)? Why or why not?

c. Determine the axis of symmetry. Describe the vertex.

14. a. Construct a quadratic function $P(s)$ that goes through the point $(5, -22)$ and has two real zeros, one at $s = -6$ and the other at $s = 4$.

b. What is the axis of symmetry?

c. What are the coordinates of the vertex?

d. What is the vertical intercept?

15. Construct a quadratic function for each of the given graphs. Write the function in both factored form and standard form.

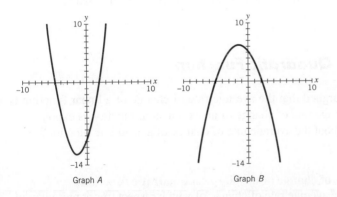

Graph A   Graph B

16. Complete the following table, and then summarize your findings.

$$i^1 = \sqrt{-1} = i \qquad i^5 = ?$$
$$i^2 = i \cdot i = -1 \qquad i^6 = ?$$
$$i^3 = i \cdot i^2 = ? \qquad i^7 = ?$$
$$i^4 = i^2 \cdot i^2 = ? \qquad i^8 = ?$$

17. Complex number expressions can be simplified by combining the real parts and then the imaginary parts. Add (or subtract) the following complex numbers and then simplify.

a. $(4 + 3i) + (-5 + 7i)$

b. $(-2 + 3i) - (-3 + 3i)$

c. $(7i - 3) + (2 - 4i)$

d. $(7i - 3) - (2 - 4i)$

18. Complex number expressions can be multiplied using the distributive property or the FOIL technique. Multiply and simplify the following. (*Note:* $i^2 = -1$.)

a. $(3 + 2i)(-2 + 3i)$

b. $(4 - 2i)(3 + i)$

c. $(2 + i)(2 - i)$

d. $(5 - 3i)(5 + 3i)$

e. $(3 - i)^2$

f. $(4 + 5i)^2$

19. A quadratic function has two complex roots, $r_1 = 1 + i$ and $r_2 = 1 - i$. Use the Factor Theorem to find the equation of this quadratic, assuming $a = 1$, and then put it into standard form.

20. The factored form of a quadratic function is $y = -2(x - (3 + i))(x - (3 - i))$. Answer the following.

a. Will the graph open up or down? Explain.

b. What are the zeros of the quadratic function?

c. Does the graph cross the x-axis? Explain.

d. Write the quadratic in standard form. (*Hint:* Multiply out; see Exercise 18.)

e. Verify your answer in part (b) by using the quadratic formula and your answer for part (d).

21. Use the quadratic formula to find the zeros of the function $f(x) = x^2 - 4x + 13$ and then write the function in factored form. Without graphing this function, how can you tell if it intersects the x-axis?

22. Let (h, k) be the coordinates of the vertex of a parabola. Since the parabola is symmetric about the vertical axis, then h is equal to the average of the two real zeros of the function (if they exist). For parts (a) and (b) use this to find h, and then construct an equation in vertex form, $y = a(x - h)^2 + k$.

a. A parabola with x-intercepts of 4 and 8, and a y-intercept of 32

b. A parabola with x-intercepts of -3 and 1, and a y-intercept of -1

c. Can you find the equation of a parabola knowing only its x-intercepts? Explain.

23. Put each of the equations into the vertex form, $y = a(x - h)^2 + k$.

a. A parabola with equation $y = x^2 + 2x - 8$

b. A parabola with equation $y = -x^2 - 3x + 4$

24. (Graphing program recommended.)

a. Write each of the following functions in both the a-b-c and the a-h-k forms. Is one form easier than the other for finding the vertex? The x- and y-intercepts?

$$y_1 = 2x^2 - 3x - 20 \qquad y_3 = 3x^2 + 6x + 3$$
$$y_2 = -2(x - 1)^2 - 3 \qquad y_4 = -(2x + 4)(x - 3)$$

b. Find the vertex and x- and y-intercepts and construct a graph by hand for each function in part (a). If you have access to a graphing program, check your work.

25. Find the equation of the graph of a parabola that has the following properties:

- The x-intercepts of the graph are at $(2, 0)$ and $(3, 0)$, *and*
- The parabola is the graph of $y = x^2$ vertically stretched by a factor of 4.

Explain your reasoning. Sketch the parabola.

26. (Graphing program required for part (b)). In Chapter 3 we determined coordinates of points where lines intersect. Once you know the quadratic formula, it's possible to determine where a line and a parabola, or two parabolas, intersect. (See Explore & Extend 8.4, p. 507.) As with two straight lines, at the point where the graphs of two functions intersect (*if* they intersect), the functions share the same x value and the same y value.

   a. Find the intersection of the parabola $y = 2x^2 - 3x + 5.1$ and the line $y = -4.3x + 10$.

   b. Plot both functions, labeling any intersection point(s).

27. (Graphing program recommended for part (b)).

   a. Find the intersection of the two parabolas $y = 7x^2 - 5x - 9$ and $y = -2x^2 + 4x + 9$.

   b. Plot both functions, labeling any intersection points.

28. Market research suggests that if a particular item is priced at x dollars, then the weekly profit $P(x)$, in thousands of dollars, is given by the function

$$P(x) = -9 + \frac{11}{2}x - \frac{1}{2}x^2$$

   a. What price range would yield a profit for this item?

   b. Describe what happens to the profit as the price increases. Why is a quadratic function an appropriate model for profit as a function of price?

   c. What price would yield a maximum profit?

29. A dairy farmer has 1500 feet of fencing. He wants to use all 1500 feet to construct a rectangle and two interior separators that together form three rectangular pens. See the accompanying figure.

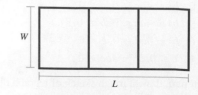

   a. If W is the width of the larger rectangle, express the length, L, of the larger rectangle in terms of W.

   b. Express the total area, $A(W)$, of the three pens as a polynomial in terms of W.

   c. What is the domain of the function $A(W)$?

   d. What are the dimensions of the larger rectangle that give a maximum area? What is the maximum area?

30. In ancient times, after a bloody defeat that made her flee her city, the queen of Carthage, Dido, found refuge on the shores of Northern Africa. Sympathetic to her plight, the local inhabitants offered to build her a new Carthage along the shores of the Mediterranean Sea. However, her city had to be rectangular in shape, and its perimeter (excluding the coastal side) could be no larger than the length of a ball of string that she could make using fine strips from only one cow hide. Queen Dido made the thinnest string possible, whose length was one mile. Dido used the string to create three non-coastal sides enclosing a rectangular piece of land (assuming the coastline was straight). She made the width exactly half the length. This way, she claimed, she would have the maximum possible area the ball of string would allow her to enclose. Was Dido right?

8.5 *The Average Rate of Change of a Quadratic Function*

In previous chapters we argued that the average rate of change of a linear function is constant, and that the average rate of change of an exponential function is exponential. What would you expect about the average rate of change of a quadratic function?

E X A M P L E 1   **Finding the average rate of change of the simplest quadratic function**
Given $y = x^2$, calculate the average rate of change of y with respect to x at unit intervals from -3 to 3. Then calculate the average rate of change of the average rate of change. What do these data points suggest?

S O L U T I O N   Column 3 in Table 8.2 (top of next page) shows the values for the average rate of change of y with respect to x. For these values the average rate of change is linear, since adding 1 to x increases the average rate of change by 2 over each interval. Column 4 computes the average rate of change of column 3 with respect to x. Since these values all are constant at 2, this confirms that column 3 (the average rate of change) is linear with respect to x. This suggests that when $y = x^2$ the average rate of change of y with respect to x is a linear function.

x	$y = x^2$	Average Rate of Change	Average Rate of Change of the Average Rate of Change
−3	9	n.a.	n.a.
−2	4	$\frac{4-9}{-2-(-3)} = -5$	n.a.
−1	1	$\frac{(1-4)}{1} = -3$	$\frac{-3-(-5)}{-1-(-2)} = 2$
0	0	$\frac{(0-1)}{1} = -1$	$\frac{(-1-(-3))}{1} = 2$
1	1	$\frac{(1-0)}{1} = 1$	$\frac{(1-(-1))}{1} = 2$
2	4	$\frac{(4-1)}{1} = 3$	$\frac{(3-1)}{1} = 2$
3	9	$\frac{(9-4)}{1} = 5$	$\frac{(5-3)}{1} = 2$

Table 8.2

E X A M P L E   2   **Finding the average rate of change of a quadratic function**
Given $y = 3x^2 - 8x - 23$, calculate the average rate of change of y with respect to x at unit intervals from −3 to 3. Then calculate the average rate of change of the average rate of change. What do the data suggest about the original function?

S O L U T I O N   Column 3 in Table 8.3 shows the values for the average rate of change. Again the values suggest that the average rate of change is linear, since adding 1 to x increases the average rate of change by 6 over each interval. This is confirmed by column 4, which shows that the average rate of change of the third column with respect to x is constant at 6. This suggests that when $y = 3x^2 - 8x - 23$, the average rate of change of y with respect to x is a linear function.

x	$y = 3x^2 - 8x - 23$	Average Rate of Change	Average Rate of Change of the Average Rate of Change
−3	28	n.a.	n.a.
−2	5	−23	n.a.
−1	−12	−17	6
0	−23	−11	6
1	−28	−5	6
2	−27	1	6
3	−20	7	6

Table 8.3

Generalizing to All Quadratic Functions

We have just seen two numerical examples that suggest that the average rate of change of a quadratic function is a linear function. We now show algebraically that this is true for *every* quadratic function.

Suppose we have $y = f(x) = ax^2 + bx + c$. In the previous examples we fixed an interval size of 1 over which to calculate the average rate of change, since it is easy to make comparisons. Now we pick a constant interval size r and for each position x compute the average rate of change of f over the interval from x to $x + r$.

First we must compute $f(x + r)$:

$$f(x + r) = a(x + r)^2 + b(x + r) + c$$

apply exponent $= a(x^2 + 2rx + r^2) + b(x + r) + c$

multiply through $= ax^2 + 2arx + ar^2 + bx + br + c$

regroup terms $= ax^2 + bx + c + (2ax + b)r + ar^2$

The program "Q12: Average Rates of Change" illustrates the relationship between distance and average velocity.

Then the average rate of change of $f(x)$ between x and $x + r$ is

$$\frac{\text{change in } f(x)}{\text{change in } x} = \frac{f(x + r) - f(x)}{(x + r) - x}$$

$$= \frac{[ax^2 + bx + c + (2ax + b)r + ar^2] - (ax^2 + bx + c)}{r}$$

$$= \frac{(2ax + b)r + ar^2}{r}$$

$$= 2ax + (b + ar)$$

The average rate of change between $(x, f(x))$ and $(x + r, f(x + r))$ is a linear function of x in its own right, where

$$\text{slope} = 2a \qquad y\text{-intercept} = b + ar$$

Note that the slope depends only on the value of a in the original equation. The y-intercept depends not only on a and b in the original equation, but also on r, the interval size over which we calculate the average rate of change. (See Figure 8.30.)

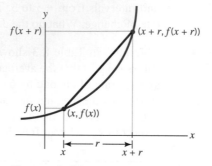

Figure 8.30 The slope of the line segment connecting two points on the parabola separated by a horizontal distance of r is $2ax + (b + ar)$.

If we took smaller and smaller values for r, then the term ar would get closer and closer to zero, and hence the y-intercept $b + ar$ would get closer and closer to b. For very small r's the average rate of change would get closer and closer to the linear expression $2ax + b$.

So given a quadratic function $f(x) = ax^2 + bx + c$, the average rate of change over small intervals around any x value can be approximated by the linear expression $2ax + b$. Thus we talk about the function $g(x) = 2ax + b$ as representing the average rate of change of $f(x)$ with respect to x. This is a central idea in calculus.

The Average Rate of Change of a Quadratic Function

Given a quadratic function $f(x) = ax^2 + bx + c$, we can think of the linear function $g(x) = 2ax + b$ as representing the average rate of change of $f(x)$ with respect to x.

E X A M P L E   3 **Construct and graph the average rate of change**
Graph each of the following functions. Construct and graph the equation for the average rate of change in each case.

a. $f(x) = x^2$

b. $f(x) = 3x^2 - 8x - 23$

SOLUTION   **a.** The average rate of change of the quadratic function $f(x) = x^2$ is the linear function $g(x) = 2x$ (see Figure 8.31).

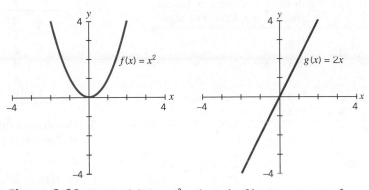

Figure 8.31 Graph of $f(x) = x^2$ and graph of its average rate of change.

b. The average rate of change of the quadratic function $f(x) = 3x^2 - 8x - 23$ is the linear function $g(x) = 6x - 8$ (see Figure 8.32).

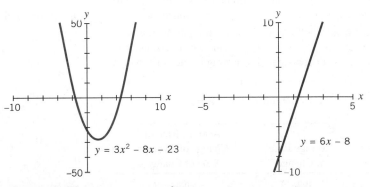

Figure 8.32 Graph of $y = 3x^2 - 8x - 23$ and graph of its average rate of change.

Algebra Aerobics 8.5

1. a. Complete the table below for the function $y = 5 - x^2$.

x	y	Average Rate of Change	Average Rate of Change of Average Rate of Change
-1	4	n.a.	n.a.
0	5	1	n.a.
1	4	-1	-2
2	1		
3			
4			

b. What do the third and fourth columns tell you?

2. Determine from each of the three following tables whether you would expect the original function $y = f(x)$ to be linear, exponential, or quadratic.

a.

x	y	Average Rate of Change	Average Rate of Change of Average Rate of Change
-1	2	n.a.	n.a.
0	1	-1	n.a.
1	6	5	6
2	17	11	6
3	34	17	6
4	57	23	6

2. (continued)

b.

x	y	Average Rate of Change	Average Rate of Change of Average Rate of Change
−1	7	n.a.	n.a.
0	4	−3	n.a.
1	1	−3	0
2	−2	−3	0
3	−5	−3	0
4	−8	−3	0

c.

x	y	Average Rate of Change	Average Rate of Change of Average Rate of Change
−1	0.5	n.a.	n.a.
0	1	0.5	n.a.
1	2	1	0.5
2	4	2	1
3	8	4	2
4	16	8	4

3. Construct a function that represents the average rate of change for the following three quadratic functions.

    a. $f(t) = t^2 + t$ **c.** $f(x) = 5x^2 + 2x + 7$

    b. $f(x) = 3x^2 + 5x$

Exercises for Section 8.5

Exercise 6 requires a graphing program.

1. Complete these sentences.

  a. For a quadratic function, the graph of its average rate of change represents a _____ function.

  b. When a quadratic function is increasing, its average rate of change is _____, and when a quadratic function is decreasing, its average rate of change is _____.

2. Complete the table for the function $y = 3 - x - x^2$.

x	y	Average Rate of Change	Average Rate of Change of Average Rate of Change
−3	−3	n.a.	n.a.
−2	1	$\frac{1 - (-3)}{-2 - (-3)} = \frac{4}{1} = 4$	n.a.
−1	3	$\frac{3 - 1}{-1 - 1(-2)} = \frac{2}{1} = 2$	$\frac{2 - 4}{-1 - (-2)} = \frac{-2}{1} = -2$
0	3	$\frac{3 - 3}{0 - (-1)} = \frac{0}{1} = 0$	$\frac{0 - 2}{0 - (-1)} =$
1	1		
2			
3			

What type of function does the average rate of change represent? What is its slope (the rate of change of the average rate of change)?

3. a. What is the average rate of change of the linear function $y = 3x - 1$? Of $y = -2x + 5$? Of $y = ax + b$? What equation could represent the average rate of change of the general linear function $y = ax + b$? Describe its graph.

  b. We have seen that the average rate of change of a quadratic function $y = ax^2 + bx + c$ can be represented by a linear function of the form $y = 2ax + b$ (where $a \neq 0$). Describe the linear graph. Could such a graph ever represent the average rate of change of a linear function?

  c. What would you predict is the equation representing the average rate of change of the cubic function $y = ax^3 + bx^2 + cx + d$?

4. a. Complete the table for the function $f(x) = 3x^2 - 2x - 5$.

x	y	Average Rate of Change	Average Rate of Change of Average Rate of Change
−3	28	n.a.	n.a.
−2	11	$\frac{11 - 28}{(-2) - (-3)} = -17$	n.a.
−1	0	$\frac{0 - 11}{(-1) - (-2)} = -11$	$\frac{(-11) - (-17)}{(-1) - (-2)} = 6$
0	−5		
1			
2			
3			

  b. What is the slope of the linear function that represents the average rate of change? How is this slope related to the average rate of change of the average rate of change?

  c. What type of function represents the average rate of change of $f(x)$?

5. Complete the table for the function $Q = 2t^2 + t + 1$.

  a. Plot $Q = 2t^2 + t + 1$. What type of function is this?

  b. What does the third column tell you about the function that represents the average rate of change for Q?

t	Q	Average Rate of Change
−3		n.a.
−2		
−1		
0		
1		
2		
3		

6. (Graphing program required.)

    a. Plot the function $h(t) = 4 + 50t - 16t^2$ for the restricted domain $0 \le t \le 3$.

    b. For what interval is this function increasing? Decreasing?

    c. Estimate the maximum point.

    d. Construct and graph a function $g(t)$ that represents the average rate of change of $h(t)$.

    e. What does $g(t)$ tell you about the function $h(t)$?

7. Match the graph of each quadratic function with the graph of its average rate of change.

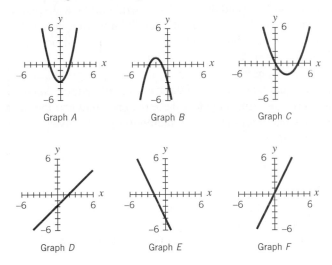

Graph A   Graph B   Graph C

Graph D   Graph E   Graph F

8. Having found the matched pairs of graphs in Exercise 7, explain the relationship between the horizontal intercept of the linear function and the vertex of the quadratic function.

9. Construct a function that represents the average rate of change for each given function.

    a. $f(t) = 3t^2 + t$

    b. $g(x) = -5x^2 + 0.4x + 3$

    c. $h(z) = 2z + 7$

10. Construct a function that represents the average rate of change for each given function.

    a. $r(t) = -3t^2 - 7$ **c.** $w(x) = -5x + 0.25$

    b. $s(v) = \frac{1}{4}v^2 - \frac{3}{2}v + 5$ **d.** $m(p) = 4p - \frac{1}{2}$

11. Determine from each of the tables whether you would expect the original function $y = f(x)$ to be linear, exponential, or quadratic.

a.

x	y	Average Rate of Change	Average Rate of Change of Average Rate of Change
−1	−2	n.a.	n.a.
0	0	2	n.a.
1	8	8	6
2	22	14	6
3	42	20	6
4	68	26	6

b.

x	y	Average Rate of Change	Average Rate of Change of Average Rate of Change
−1	9	n.a.	n.a.
0	5	−4	n.a.
1	1	−4	0
2	−3	−4	0
3	−7	−4	0
4	−11	−4	0

c.

x	y	Average Rate of Change	Average Rate of Change of Average Rate of Change
−1	0.25	n.a.	n.a.
0	1	0.75	n.a.
1	4	3	2.25
2	16	12	9
3	64	48	36
4	256	192	144

8.6 *The Mathematics of Motion*

If you are interested in learning more about how Galileo made his discoveries, read Elizabeth Cavicchi's "Watching Galileo's Learning."

Today we take for granted that scientists study physical phenomena in laboratories using sophisticated equipment. But in the early 1600s, when Galileo did his experiments on motion, the concept of laboratory experiments was unknown. In his attempts to understand nature, Galileo asked questions that could be tested directly in experiments. His use of observation and direct experimentation and his discovery that aspects of nature were subject to quantitative laws were of decisive importance, not only in science but in the broad history of human ideas.

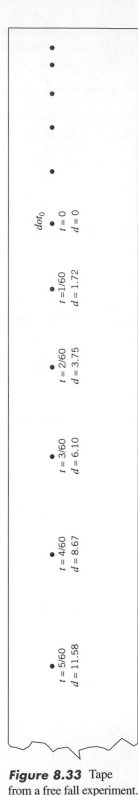

Figure 8.33 Tape from a free fall experiment.

The Scientific Method

Ancient Greeks and medieval thinkers believed that basic truths existed within the human mind and that these truths could be uncovered through reasoning, not empirical experimentation. Their method has been described as a "qualitative study of nature." Greek and medieval scientists were interested in *why* objects fall. They believed that a heavier object fell faster than a lighter one because "it has weight and it falls to the Earth because it, like every object, seeks its natural place, and the natural place of heavy bodies is the center of the Earth. The natural place of a light body, such as fire, is in the heavens, hence fire rises."[3]

Galileo changed the question from *why* things fall to *how* things fall. This question suggested other questions that could be tested directly by experiment: "By alternating questions and experiments, Galileo was able to identify details in motion no one had previously noticed or tried to observe."[4] His quantitative descriptions of objects in motion led not only to new ways of thinking about motion, but also to new ways of thinking about science. His process of careful observation and testing began the critical transformation of science from a qualitative to a quantitative study of nature.[5] Galileo's decision to search for quantitative descriptions "was the most profound and the most fruitful thought that anyone has had about scientific methodology."[6] This approach became known as the *scientific method*.

The Free Fall Experiment

Galileo sought to answer the following questions: How can we describe mathematically the distance an object falls over time? Do freely falling objects fall at a constant speed? If the speed of freely falling objects is not constant, is it increasing at a constant rate?

Deriving an Equation Relating Distance and Time

Using mathematical and technological tools not available in Galileo's time, we can describe the distance fallen over time in the free fall experiment using a "best-fit" function for our data. Galileo had to describe his finding in words. He described free fall motion first by direct measurement and then abstractly with a time-squared rule. "This discovery was revolutionary, the first evidence that motion on Earth was subject to mathematical laws."[7]

The sketch of a tape given in Figure 8.33 gives data collected by a group of students from a falling-object experiment. Each dot represents how far the object fell in each succeeding 1/60 of a second.

Since the first few dots are too close together to get accurate measurements, we start measurements at the sixth dot, which we call dot_0. At this point, the object is already in motion. This dot is considered to be the starting point, and the time, t, at dot_0 is set at 0 seconds. The next dot represents the position of the object 1/60 of a second later. Time increases by 1/60 of a second for each successive dot. In addition to assigning a time to each point, we also measure the total distance fallen, d (in cm), from the point designated dot_0. For every dot we have two values: the time, t, and the distance fallen, d. At dot_0, we set $t = 0$ and $d = 0$.

[3]M. Kline, *Mathematics for the Nonmathematician* (New York: Dover, 1967), p. 287.

[4]E. Cavicchi, "Watching Galileo's Learning," in the Anthology of Readings on the course website.

[5]Galileo's scientific work was revolutionary in terms not only of science but also of the politics of the time; his work was condemned by the ruling authorities, and he was arrested.

[6]M. Kline, op. cit., p. 288.

[7]E. Cavicchi, "Watching Galileo's Learning," in the Anthology of Readings on the course website.

Instructions for conducting the free fall experiment in a lab are on the course website. Exploration 8.1 offers a simpler version.

The time and distance measurements from the tape are recorded in Table 8.4 and plotted on the graph in Figure 8.34. Time, t, is the independent variable, and distance, d, is the dependent variable. The graph gives a representation of the data collected on distance fallen over time, not a picture of the physical motion of the object.

Time (sec)	Total Distance Fallen (cm)
0.0000	0.00
0.0167	1.72
0.0333	3.75
0.0500	6.10
0.0667	8.67
0.0833	11.58
0.1000	14.71
0.1167	18.10
0.1333	21.77
0.1500	25.71
0.1667	29.90
0.1833	34.45
0.2000	39.22
0.2167	44.22
0.2333	49.58
0.2500	55.15
0.2667	60.99
0.2833	67.11
0.3000	73.48
0.3167	80.10
0.3333	87.05
0.3500	94.23

Table 8.4

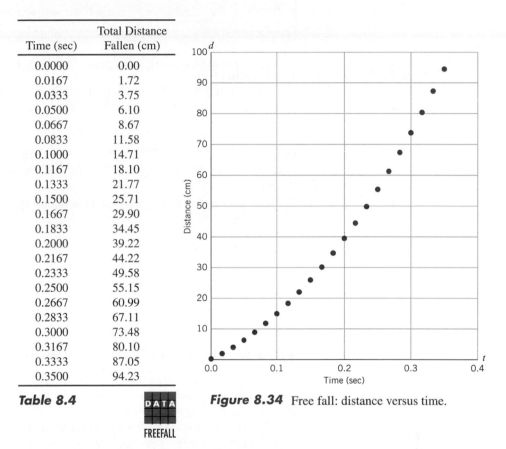

Figure 8.34 Free fall: distance versus time.

DATA

FREEFALL

   The software "Q11: Freely Falling Objects" in *Quadratic Functions* provides a simulation of the free fall experiment.

Galileo was correct in predicting a relationship between time squared and distance fallen. A best-fit function to the data in Table 8.4 (and graphed in Figure 8.34) is the quadratic,

$$d = 488t^2 + 99t \qquad (1)$$

where t = time (in sec) and d = distance fallen (in cm).

What are the units for each term of the equation? Since d is in centimeters, each term on the right-hand side of Equation (1) must also be in centimeters. Since t is in seconds, the coefficient, 488, of t^2 must be in centimeters per second squared:

$$\frac{cm}{sec^2} \cdot \frac{sec^2}{1} = cm$$

The coefficient, 99, of t must be in centimeters per second.

What do the coefficients represent? Objects fall because they are pulled down by Earth's gravity. Earth's gravitational constant, called g, is 980 cm/sec². The coefficient of the t^2 term (488) in Equation (1) is close to one-half of g. The coefficient of the t term (99) represents the initial velocity, v_0, of the object when $t = 0$. So here $v_0 = 99$ cm/sec. The value of v_0 represents how fast the object was moving at the time when we started taking measurements.

Galileo's discoveries (reconfirmed by our experiment) were the basis of the general equations relating distance fallen and time.

> The general equation for the motion of freely falling bodies that relates distance fallen, d, to time, t, is
>
> $$d = \frac{1}{2}gt^2 + v_0 t$$
>
> where v_0 is the initial velocity and g is the acceleration due to gravity on Earth.

EXAMPLE 1 **Freely falling body**

A freely falling body has an initial velocity of 125 cm/sec. Assume that $g = 980$ cm/sec^2.

a. Write an equation that relates d, distance fallen in centimeters, to t, time in seconds.

b. How far has the body fallen after 1 second? After 3 seconds?

c. If the initial velocity were 75 cm/sec, how would your equation in part (a) change?

d. Using your equation from part (c), how far has the body fallen after 1 second? After 3 seconds?

SOLUTION **a.** $d = 490t^2 + 125t$

b. After 1 second, $d = 615$ cm; after 3 seconds, $d = (490)(3^2) + (125)(3) = 4785$ cm.

c. $d = 490t^2 + 75t$

d. After 1 second, $d = 565$ cm; after 3 seconds, $d = (490)(3^2) + (75)(3) = 4635$ cm.

Velocity: Change in Distance over Time

Our data, graph, and equation confirm (as Galileo predicted) that the relationship between distance fallen and time is quadratic. But what is the relationship between the average rate of change of distance fallen and time?

Table 8.5 and Figure 8.35 show the increase in the average rate of change over time for three different pairs of points. The time interval nearest the start of the fall shows a relatively small change in the distance per time step and therefore a relatively gentle slope of 188 cm/sec. The time interval farthest from the start of the fall shows a greater change of distance per time step and a much steeper slope of 367 cm/sec. So the slope or average rate of change is increasing.

t	d	Average Rate of Change	
0.0500	6.10	$\dfrac{21.77 - 6.10}{0.1333 - 0.0500}$	≈ 188 cm/sec
0.1333	21.77		
0.0833	11.58	$\dfrac{49.58 - 11.58}{0.2333 - 0.0833}$	≈ 253 cm/sec
0.2333	49.58		
0.2167	44.22	$\dfrac{87.05 - 44.22}{0.3333 - 0.2167}$	≈ 367 cm/sec
0.3333	87.05		

Table 8.5

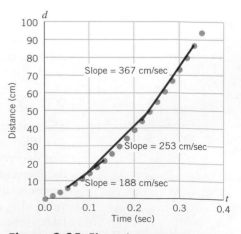

Figure 8.35 Slopes (or average velocities) between three pairs of end points.

In this experiment the average rate of change has an additional important meaning. For objects in motion, the change in distance divided by the change in time is also called the *average velocity* for that time period. For example, in the calculations in Table 8.5, the average rate of change of 188 cm/sec represents the average velocity of the falling object between 0.0500 and 0.1333 second.

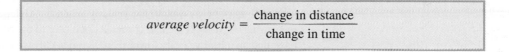

$$average\ velocity = \frac{change\ in\ distance}{change\ in\ time}$$

From Table 8.6 and Figure 8.36, we see that average velocity is not constant, but increasing. One of Galileo's questions asked whether the velocity was increasing at a constant rate?

If the rate of change of velocity is constant, then the graph of velocity vs. time should be a straight line. In Table 8.6 we calculate the average rates of change (or average velocities) for all the pairs of adjacent points in our free fall data. The results are in column 4. Since each computed velocity is the average over an interval, for increased precision we associate each velocity with the midpoint time of the interval instead of one of the end points. In Figure 8.36, we plot velocity from the fourth column against the midpoint times from the third column. The graph is strikingly linear.

See "C3: Average Velocity and Distance" in *Rates of Change*.

Time, t (sec)	Distance Fallen, d (cm)	Midpoint Time, t (sec)	Average Velocity, v (cm/sec)
0.0000	0.00		
0.0167	1.72	0.0083	103.2
0.0333	3.75	0.0250	121.8
0.0500	6.10	0.0417	141.0
0.0667	8.67	0.0583	154.2
0.0833	11.58	0.0750	174.6
0.1000	14.71	0.0917	187.8
0.1167	18.10	0.1083	203.4
0.1333	21.77	0.1250	220.2
0.1500	25.71	0.1417	236.4
0.1667	29.90	0.1583	251.4
0.1833	34.45	0.1750	273.0
0.2000	39.22	0.1917	286.2
0.2167	44.22	0.2083	300.0
0.2333	49.58	0.2250	321.6
0.2500	55.15	0.2417	334.2
0.2667	60.99	0.2583	350.4
0.2833	67.11	0.2750	367.2
0.3000	73.48	0.2917	382.2
0.3167	80.10	0.3083	397.2
0.3333	87.05	0.3250	417.0
0.3500	94.23	0.3417	430.8

Table 8.6

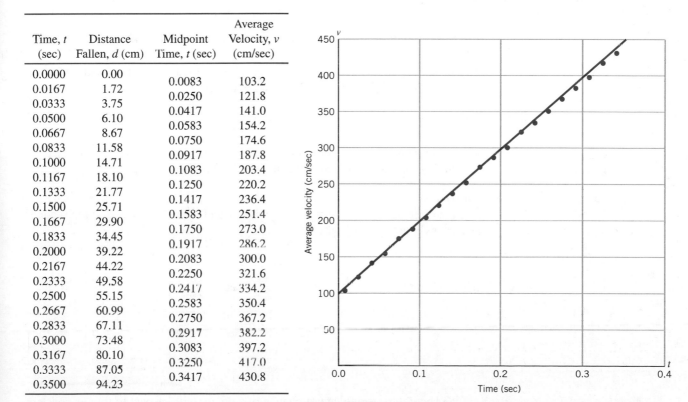

Figure 8.36 Best-fit linear function for average velocity versus time.

Generating a best-fit linear function and rounding to the nearest unit, we obtain the equation

$$average\ velocity = 977t + 98$$

where the average velocity, v, is in centimeters per second, and time t, is in seconds. The graph of this function appears in Figure 8.36. The slope of the line is constant and equals the rate of change of velocity with respect to time. This means that although the velocity is not constant, its rate of change with respect to time *is* constant.

The coefficient of t, 977, is the slope of the line and in physical terms represents g, the acceleration due to gravity. The conventional value for g is 980 cm/sec^2. So the velocity of the freely falling object increases by about 980 cm/sec during each second of free fall.

With this equation we can estimate the velocity at any given time t. When $t = 0$, then $v = 98$ cm/sec. This means that the object was already moving at about 98 cm/sec when we set $t = 0$. In our experiment, the velocity when $t = 0$ depends on where we choose to start measuring our dots. If we had chosen a dot closer to the beginning of the free fall, we would have had an initial velocity lower than 98 cm/sec. If we had chosen a dot farther away from the start, we would have had an initial velocity higher than 98 cm/sec. Note that 98 cm/sec closely matches the value of 99 cm/sec in our best-fit quadratic function (Equation 1, p. 517.)

The general equation that relates, v, the velocity of a freely falling body, to t, time, is

$$v = gt + v_0$$

where $v_0 =$ initial velocity (velocity at time $t = 0$) and g is the acceleration due to gravity.

E X A M P L E   2   **Distance and velocity**
A freely falling object has an initial velocity of 50 cm/sec.

a. Write two motion equations, one relating distance and time and the other relating velocity and time.

b. How far has the object fallen and what is its velocity after 1 second? After 2.5 seconds? Be sure to identify units in your answers.

S O L U T I O N   **a.** $d = 490t^2 + 50t$ and $v = 980t + 50$ where d is distance fallen (in cm), v is velocity (in cm/sec), and t is time (in seconds).

b. After 1 second, $d = 540$ cm and $v = 1030$ cm/sec.
After 2.5 seconds, $d = 3187.5$ cm and $v = 2500$ cm/sec.

E X A M P L E   3   **Units for velocity**
Given the velocity equation $v = gt + v_0$ where v and v_0 are in cm/sec and t in seconds, rewrite this equation using only units.

S O L U T I O N   Given $$v = gt + v_0$$

Using only units
$$\frac{\text{cm}}{\text{sec}} = \frac{\text{cm}}{\text{sec}^2} \cdot \text{sec} + \frac{\text{cm}}{\text{sec}}$$

$$\frac{\text{cm}}{\text{sec}} = \frac{\text{cm}}{\text{sec}} + \frac{\text{cm}}{\text{sec}}$$

$$\frac{\text{cm}}{\text{sec}} = \frac{\text{cm}}{\text{sec}}$$

We know the average rate of change of velocity with respect to time is constant. Now we'll find the value of that constant.

Acceleration: Change in Velocity over Time

Acceleration means a change in velocity or speed[8]. If you push the accelerator pedal in a car down just a bit, the speed of the car increases slowly. If you floor the pedal, the

[8]The difference between velocity and speed, is that velocity is associated with motion in a known direction (in the case of a falling object, straight down); speed is also distance over time, but refers to an object that could be traveling in any direction (for example, a car is driving at 60 mph, but the direction is unspecified).

speed increases rapidly. The rate of change of velocity with respect to time is called *acceleration*. Calculating the average rate of change of velocity with respect to time gives an estimate of acceleration. For example, if a car is traveling at 20 mph and 1 hour later the car has accelerated to 60 mph, then

$$\frac{\text{change in velocity}}{\text{change in time}} = \frac{(60 - 20) \text{ mph}}{1 \text{ hr}} = (40 \text{ mph})/\text{hr} = 40 \text{ mi/hr}^2$$

In 1 hour, the velocity of the car changed from 20 to 60 mph, so its average acceleration was 40 mph/hr, or 40 mi/hr^2.

$$average\ acceleration = \frac{\text{change in velocity}}{\text{change in time}}$$

Table 8.7 uses the average velocity data and midpoint time from Table 8.6 to calculate average accelerations. Figure 8.37 shows the plot of average acceleration in centimeters per second squared (the third column) versus time in seconds (the first column).

Midpoint Time, t (sec)	Average Velocity, v (cm/sec)	Average Acceleration (cm/sec^2)
0.0083	103.2	n.a.
0.0250	121.8	1116
0.0417	141.0	1152
0.0583	154.2	792
0.0750	174.6	1224
0.0917	187.8	792
0.1083	203.4	936
0.1250	220.2	1008
0.1417	236.4	972
0.1583	251.4	900
0.1750	273.0	1296
0.1917	286.2	792
0.2083	300.0	828
0.2250	321.6	1296
0.2417	334.2	756
0.2583	350.4	972
0.2750	367.2	1008
0.2917	382.2	900
0.3083	397.2	900
0.3250	417.0	1188
0.3417	430.8	828

Table 8.7

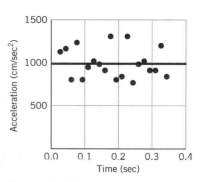

Figure 8.37 Average acceleration for free fall data.

The data lie along a roughly horizontal line. The average acceleration values vary between a low of 756 cm/sec^2 and a high of 1296 cm/sec^2 with a mean of 982.8. Rounding off, we have

$$acceleration \approx 980 \text{ cm/sec}^2$$

This is the constant we sought, which is roughly equal to the slope of the graph of velocity vs. time (Figure 8.36, p. 519). The acceleration value means that for each additional second of free fall, the velocity of the falling object increases by approximately 980 cm/sec. The longer it falls, the faster it goes. We have verified a characteristic feature of gravity near the surface of Earth: It causes objects to fall at a velocity that increases every second by about 980 cm/sec. We say that the acceleration due to gravity near Earth's surface is 980 cm/sec^2.

Converting units for g

In order to express g in feet per second squared, we need to convert 980 centimeters to feet. We start with the fact that 1 ft = 30.48 cm. So the conversion factor for centimeters to feet is (1 ft)/(30.48 cm) = 1. If we multiply 980 cm by (1 ft)/(30.48 cm) to convert centimeters to feet, we get

$$980 \text{ cm} = (980 \text{ cm})\left(\frac{1 \text{ ft}}{30.48 \text{ cm}}\right) \approx 32.15 \text{ ft} \approx 32 \text{ ft}$$

So a value of 980 cm/sec^2 for g is equivalent to approximately 32 ft/sec^2.

The numerical value used for the constant g depends on the units being used for the distance, d, and the time, t. The exact value of g also depends on where it is measured.[9]

The conventional values for g, the acceleration due to gravity near the surface of Earth, are

$$g = 32 \text{ ft/sec}^2$$

or equivalently $\qquad\qquad g = 980 \text{ cm/sec}^2 = 9.8 \text{ m/sec}^2$

EXAMPLE 4 **Using different units**

A freely falling body has an initial velocity of 2 feet per second.

a. What value would you use for g?

b. Write an equation that relates d, distance fallen, to t, time.

c. Write equation that relates v, velocity, to t, time.

SOLUTION **a.** 32 ft/sec^2

b. $d = 16t^2 + 2t$ where d = distance fallen (in feet) and t = time (in seconds).

c. $v = 32t + 2$ where v = velocity (in feet/sec) and t = time (in seconds).

Deriving an Equation for the Height of an Object in Free Fall

Assume we have the following motion equation relating distance fallen, d (in centimeters), and time, t (in seconds):

$$d = 490t^2 + 45t$$

Also assume that when $t = 0$, the height, h, of the object was 110 cm above the ground. Until now, we have considered the distance from the point the object was dropped, a value that *increases* as the object falls. How can we describe a different distance, the

---

[9]Because Earth is rotating, is not a perfect sphere, and is not uniformly dense, there are variations in g according to the latitude and elevation. The following are a few examples of local values for g.

Location	North Latitude (deg)	Elevation (m)	g (cm/sec^2)
Panama Canal	9	0	978.243
Jamaica	18	0	978.591
Denver, CO	40	1638	979.609
Pittsburgh, PA	40.5	235	980.118
Cambridge, MA	42	0	980.398
Greenland	70	0	982.534

Source: H. D. Young, *University Physics,* Vol. I, 8th ed. (Reading, MA: Addison-Wesley, 1992), p. 336.

height above ground of an object, as a function of time, a value that *decreases* as the object falls?

At time zero, the distance fallen is zero and the height above the ground is 110 centimeters. After 0.05 second, the object has fallen about 3.5 centimeters, so its height would be $110 - 3.5 = 106.5$ cm. For an arbitrary distance d, we have $h = 110 - d$. Table 8.8 gives associated values for time, t, distance fallen, d, and height above ground, h. The graphs in Figure 8.38 show distance versus time and height versus time.

Time, t (sec)	Distance Fallen, d (cm) $(d = 490t^2 + 45t)$	Height above Ground, h (cm) $(h = 110 - d)$
0.00	0.0	110.0
0.05	3.5	106.5
0.10	9.4	100.6
0.15	17.8	92.2
0.20	28.6	81.4
0.25	41.9	68.1
0.30	57.6	52.4
0.35	75.8	34.2
0.40	96.4	13.6

Table 8.8

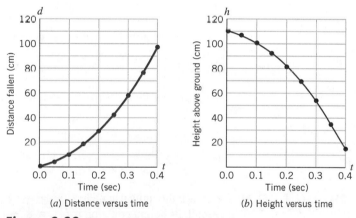

(a) Distance versus time   (b) Height versus time

Figure 8.38 Representations of free fall data.

How can we convert the equation $d = 490t^2 + 45t$, relating distance fallen and time, to an equation relating height above ground and time? We know that the relationship between height and distance is $h = 110 - d$. We can substitute the expression for d into the height equation:

$$h = 110 - d$$
$$= 110 - (490t^2 + 45t)$$
$$= 110 - 490t^2 - 45t \qquad (2)$$

Note that in Equation (2) for height, the coefficients of both t and t^2 are negative. If we consider what happens to the height of an object in free fall, this makes sense. As time increases, the height decreases. (See Table 8.8 and Figure 8.38(*b*).) When we were measuring the increasing distance an object fell, we did not take into account the direction in which it was going (up or down). We cared only about the magnitudes (the absolute values) of distance and velocity, which were positive. But when we are measuring a decreasing height or distance, we have to worry about direction. In this case we define downward motion to be negative and upward motion to be positive.

E X A M P L E   5

Worrying about up and down

Interpret the coefficients in the previous height equation $h = 110 - 45t - 490t^2$.

S O L U T I O N

The constant term is the initial height, 110 cm. The coefficient of t is the initial velocity of the object, -45 cm/sec. The term is negative because we are considering the object as moving downwards. The coefficient of t^2 is minus half of g, $-(1/2)$ 980 cm/sec^2. This term is negative as well, because now we are thinking of gravity as pulling objects downwards, decreasing the height.

Once we have introduced the notion that downward motion is negative and upward motion is positive, we can also deal with situations in which the initial velocity is upward and the acceleration is downward. The velocity equation for this situation is

$$v = -gt + v_0$$

where v_0 could be either positive or negative, depending on whether the object is thrown upward or downward, and the sign for the g term is negative because gravity accelerates downward in the negative direction.

> If we treat upward motion as positive and downward motion as negative, then the acceleration due to gravity is negative. So the general equations of motion of freely falling bodies that relate height, h, and velocity, v, to time, t, are
>
> $$h = h_0 + v_0 t - \frac{1}{2}gt^2$$
> $$v = -gt + v_0$$
>
> where h_0 = initial height, g = acceleration due to gravity, and v_0 = initial velocity (which can be positive or negative).

E X A M P L E   6

Dropping vs. being thrown down from a roof

a. A ball is dropped from a roof 8 meters from the ground. How long does it take for the ball to reach the ground?

b. A rock is thrown down from the same roof 0.6 second later. The rock and the ball hit the ground at the same time. What was the initial velocity of the rock?

S O L U T I O N

a. Since the units use meters (so $g = 9.8$ m/sec^2) and there is no initial velocity for the ball, its height equation is $h_{ball} = 8 - 4.9t^2$. To find the time it takes for the ball to hit the ground, we need to set the height $h_{ball} = 0$ and solve for t.

Given	$0 = 8 - 4.9t^2$
rearrange terms	$4.9t^2 = 8$
divide by 4.9	$t^2 \approx 1.633$
take square root	$t \approx 1.28$ sec

So it will take the ball about 1.28 seconds to hit the ground.

b. The rock will be thrown down 0.6 second after the ball is dropped. So the total time the rock falls before hitting the ground (in relation to the ball) is about $1.28 - 0.6 \approx 0.68$ second. Since the rock is thrown down, it has an initial (negative) velocity v_0, so its height equation is $h_{rock} = 8 - v_0t - 4.9t^2$ where h_{rock} is in meters. The rock hits the ground when $h_{rock} = 0$. So we can the solve the height equation for v_0 when $h_{rock} = 0$ and $t = 0.68$.

Given	$0 = 8 - v_0(0.68) - 4.9(0.68)^2$
simplify	$0 \approx 8 - 0.68v_0 - 2.27$
rearrange	$0.68v_0 \approx 5.73$
solve for v_0	$v_0 \approx 8.43$ m/sec

So the rock's initial velocity is approximately 8.43 meters/second.

Working with an Initial Upward Velocity

If we want to use the general equation to describe the height of a thrown object, we need to understand the meaning of each of the coefficients. So far we have considered downward motion. But suppose a ball is thrown upward with an initial velocity of 97 cm/sec from a height of 87 cm above the ground. Describe the relationship between the height of the ball and time with an equation.

The initial height of the ball is 87 cm when $t = 0$, so the constant term is 87 cm. The coefficient of t, or the initial-velocity term, is $+97$ cm/sec since the initial motion is upward. The coefficient of t^2, the gravity term, is -490 cm/sec^2, since gravity causes objects to fall down.

Substituting these values into the equation for height, we get

$$h = 87 + 97t - 490t^2$$

Table 8.9 gives a series of values for heights corresponding to various times. Figure 8.39 plots height above ground (cm) vs. time (sec.).

t (sec)	h (cm)
0.00	87.00
0.05	90.63
0.10	91.80
0.15	90.53
0.20	86.80
0.25	80.63
0.30	72.00
0.35	60.93
0.40	47.40
0.45	31.43
0.50	13.00

Table 8.9

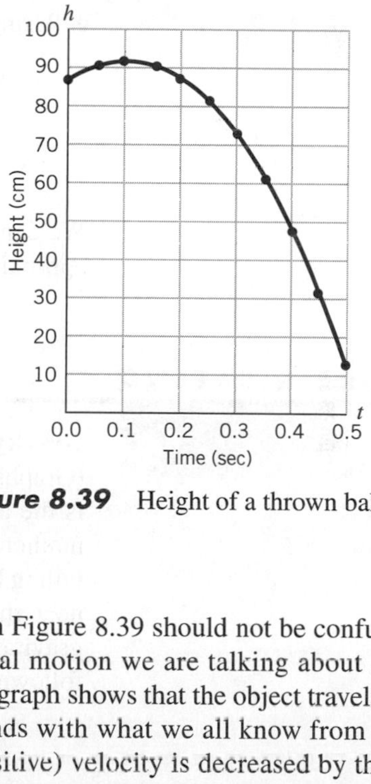

Figure 8.39 Height of a thrown ball.

The graph of the heights at each time in Figure 8.39 should not be confused with the trajectory of a thrown object. The actual motion we are talking about is purely vertical—straight up and straight down. The graph shows that the object travels up for a while before it starts to fall. This corresponds with what we all know from practical experience throwing balls. The upward (positive) velocity is decreased by the pull of gravity until the object stops moving upward and begins to fall. The downward (negative) velocity is then increased by the pull of gravity until the object strikes the ground.

E X A M P L E   7   **Shooting an arrow up**

a. Construct an equation for the height of an arrow shot vertically into the air from a height of 4 feet at an initial velocity of 176 feet/second.

b. Find the height of the arrow after 0, 2, 4, 6, and 8 seconds.

c. Sketch a graph of height vs. time.

d. Use the vertex form of the quadratic function to determine the coordinates of the maximum point.

e. Use the quadratic formula to find when the object hits the ground.

SOLUTION   **a.** $h = 4 + 176t - 16t^2$ **c.** See Figure 8.40.

b. See Table 8.10.

t	h
0	4
2	292
4	452
6	484
8	388

Table 8.10

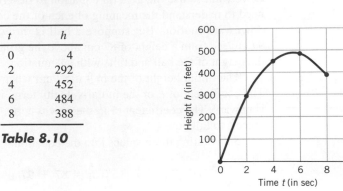

Figure 8.40   Height of an arrow.

d. The vertex form is $h = -16(t - 5.5)^2 + 488$, showing the vertex at (5.5, 488). So the arrow reaches a maximum height of 488 feet at 5.5 seconds.

e. Using the quadratic formula to solve $-16t^2 + 176t + 4 = 0$, we get

$$t = \frac{-176 \pm \sqrt{176^2 - 4(-16)4}}{2(-16)}$$

$$t \approx -0.02, 11.02$$

We can ignore the negative value for t (-0.02), so the arrow hits the ground after approximately 11 seconds.

EXPLORE & EXTEND

8.6

Gravity on Other Planets

(Graphing program required.) We have seen that the function $d = \frac{1}{2}gt^2 + v_0 t$ (where g is the acceleration due to Earth's gravity and v_0 is the object's initial velocity) is a mathematical model for the relationship between time and distance fallen for freely falling bodies near Earth's surface. This relationship also holds for freely falling bodies near the surfaces of other planets. We just replace g, the acceleration of Earth's gravitational field, with the acceleration for the planet under consideration. The following table gives the acceleration due to gravity for planets in our solar system:

Acceleration Due to Gravity

	m/sec²	ft/sec²
Mercury	3.7	12.1
Venus	8.9	29.1
Earth	9.8	32.1
Mars	3.7	12.1
Jupiter	24.8	81.3
Saturn	10.4	34.1
Uranus	8.5	27.9
Neptune	11.6	38.1

Note: Pluto is no longer considered a planet.
Source: The Astronomical Almanac, U.S. Naval Observatory, 1981.

a. Choose units of measurement (meters or feet) and three of the planets (other than Earth). For each of these planets, find an equation for the relationship between the distance an object falls and time. Fill in the table below for your chosen planets. Assume for this problem that the initial velocity of the freely falling object is 0.

Name of Planet	Function Relating Distance and Time (sec)	Units for Distance

b. Using a graphing program, plot your three functions, with time on the horizontal axis and distance on the vertical axis. What domain makes sense for your models? Why?

c. On which of your planets will an object fall the farthest in a given time? On which will it fall the least distance in a given time?

d. Examine the graphs and think about the *similarities* that they share. Describe their general shape. What happens to d as the value for t increases?

e. Think about the *differences* among the three curves. What effect does the coefficient of the t^2 term have on the shape of the graph; that is, when the coefficient gets larger (or smaller), how is the shape of the curve affected? Which graph shows d increasing the fastest compared with t?

Algebra Aerobics 8.6

Graphing program optional for Problem 5.

1. If there is no initial velocity, the distance that an object falls is directly proportional to the square of the time that it falls.

   a. Write an equation of proportionality. (Refer to Ch. 7, p. 400 for direct proportionality).

   b. In 6 seconds, an object falls 576 feet. Find the constant of proportionality. What is its unit of measure?

   c. In 6 seconds, an object falls 176.4 meters. Find the constant of proportionality. Now what is its unit of measure?

2. Construct two equations, one for the distance a freely falling object has fallen, and the other for the corresponding velocity, where $g = 980$ cm/sec² and the initial velocity is 50 cm/sec.

3. In the distance equation d for a freely falling object, $d = 16t^2 + 20t$ where $t = $ time.

   a. What are the units for d and t? How do you know this?

   b. Rewrite the equation using only units for each term and show that the units match on both sides of the equation.

4. An object is projected upwards from the ground with an initial velocity of 100 meters per second. The height h of this object after t seconds is represented by the function $h = 100t - 4.9t^2$.

   a. Fill in the table of values below.

t	0	1	6	10	11	16	20	21
h								

   b. What happens between $t = 10$ and $t = 11$? Explain why.

   c. What happens between $t = 20$ and $t = 21$? Explain why.

   d. What domain makes sense in this problem?

5. (Graphing program optional for part (d).) An object is projected upwards from a platform 50 feet high with an initial velocity of 80 feet per second. The height of this object after t seconds is represented by the function $h = 50 + 80t - 16t^2$.

   a. Fill in the table of values below.

t	0	1	2	3	4	5	6
h							

5. (continued)

b. What do you think happens between $t = 2$ and $t = 3$?

c. What happens between $t = 5$ and $t = 6$?

d. Sketch the graph of the height of the object as a function of time. Check your graph using technology.

e. What is the maximum height reached by the object? When does it reach this height?

f. Write an equation that allows you to determine the number of seconds the object travels before hitting the ground. Then solve the equation.

g. Interpret your answer to part (f) by finding the point on the graph that corresponds to that question.

6. An object is projected upwards with initial velocity v_0 and initial height h_0. The function $h = h_0 + v_0 t - (1/2)\, g t^2$ gives us the height of this object after t seconds.

a. Rewrite this equation using only units, assuming that time is in seconds, acceleration is in meters per second per second, initial velocity is in meters per second, and height is in meters. Show that the units on the left and on the right are equal.

b. Repeat part (a), now assuming time is in seconds, acceleration is in feet per second per second, velocity is in feet per second, and height is in feet.

7. The motion equations also work on other bodies in space. For instance, we can rewrite the general height function as $h = h_0 + v_0 t - (1/2)\, a t^2$ where the value of a is the acceleration due to gravity on a different planet. (See Explore & Extend 8.6.)

a. Suppose you landed on an object with negligible gravity (say a meteor), so you are virtually weightless. What would the height function become? What type of function is this?

b. If a small rock is projected upwards from the object with initial velocity, does the rock ever hit the ground? Describe the path of the rock.

Exercises for Section 8.6

A graphing program is optional for Exercises 1, 2, 11, 12, 17 and 18. Exercise 15 needs a free fall data tape.

1. (Graphing program optional.) The equation $d = 490t^2 + 50t$ describes the relationship between distance fallen, d, in centimeters, and time, t, in seconds, for a particular freely falling object.

a. Interpret each of the coefficients and specify its units of measurement.

b. Generate a table for a few values of t between 0 and 0.3 second.

c. Graph distance versus time by hand. Check your graph using a computer or graphing calculator if available.

2. (Graphing program optional.) The equation $d = 4.9t^2 + 1.7t$ describes the relationship between distance fallen, d, in meters, and time, t, in seconds, for a particular freely falling object.

a. Interpret each of the coefficients and specify its units of measurement.

b. Generate a table for a few values of t between 0 and 0.3 second.

c. Graph distance versus time by hand. Check your graph using a computer or graphing calculator if available.

d. Relate your answers to earlier results in this chapter.

3. The equation $d = \frac{1}{2}gt^2 + v_0 t$ could also be written using distances measured in meters. Rewrite the equation showing only units of measure and verify that you get meters = meters.

4. The equation $d = \frac{1}{2}gt^2 + v_0 t$ could be written using distances measured in feet. Rewrite the equation showing only units of measure and verify that you get feet = feet.

5. Complete the accompanying table. What happens to the average velocity of the object as it falls?

Time (sec)	Distance Fallen (cm)	Average Velocity (average rate of change for the previous 1/30 of a second)
0.0000	0.00	n.a.
0.0333	3.75	$\frac{3.75 - 0.00}{0.0333 - 0.0000} \approx 113$ cm/sec
0.0667	8.67	$\frac{8.67 - 3.75}{0.0667 - 0.0333} \approx 147$ cm/sec
0.1000	14.71	
0.1333	21.77	
0.1667	29.90	

6. The essay "Watching Galileo's Learning" examines the learning process that Galileo went through to come to some of the most remarkable conclusions in the history of science. Write a summary of one of Galileo's conclusions about motion. Include in your summary the process by which Galileo made this discovery and some aspect of your own learning or understanding of Galileo's discovery.

7. The data from a free fall tape generate the following equation relating distance fallen in centimeters and time in seconds:

$$d = 485.7t^2 + 7.6t$$

a. Give a physical interpretation of each of the coefficients along with its appropriate units of measurement.

b. How far has the object fallen after 0.05 second? 0.10 second? 0.30 second?

8. What would the free fall equation $d = 490t^2 + 90t$ become if d were measured in feet instead of centimeters?

9. In the equation $d = 4.9t^2 + 500t$, time is measured in seconds and distance in meters. What does the number 500 represent?

10. A freely falling object has an initial velocity of 20 ft/sec.

   a. Write one equation relating distance fallen (in feet) and time (in seconds) and a second equation relating velocity (in feet per second) and time.

   b. How many feet has the object fallen and what is its velocity after 0.5 second? After 2 seconds?

11. (Graphing program optional.) A freely falling object has an initial velocity of 12 ft/sec.

   a. Construct an equation relating distance fallen and time.

   b. Generate a table by hand for a few values of the distance fallen between 0 and 5 seconds.

   c. Graph distance vs. time by hand. Check your graph using a computer or graphing calculator if available.

12. (Graphing program optional.) Use the information in Exercise 11 to do the following:

   a. Construct an equation relating velocity and time.

   b. Generate a table by hand for a few values of velocity between 0 and 5 seconds.

   c. Graph velocity versus time by hand. If possible, check your graph using a computer or graphing calculator.

13. If the equation $d = 4.9t^2 + 11t$ represents the relationship between distance and time for a freely falling body, in what units is distance now being measured? How do you know?

14. The distance that a freely falling object with no initial velocity falls can be modeled by the quadratic function $d = 16t^2$, where t is measured in seconds and d in feet. There is a closely related function $v = 32t$ that gives the velocity, v, in feet per second at time t, for the same freely falling body.

   a. Fill in the missing values in the following table:

Time, t (sec)	Distance, d (ft)	Velocity, v (ft/sec)
1		
1.5		
2		
		80
	144	

   b. When $t = 3$, describe the associated values of d and v and what they tell you about the object at that time.

   c. Sketch both functions, distance versus time and velocity versus time, on two different graphs. Label the points from part (b) on the curves.

   d. You are standing on a bridge looking down at a river. How could you use a pebble to estimate how far you are above the water?

15. (This exercise requires a free fall data tape created using a spark timer. You could also photocopy the tape in Figure 8.33 in the text on p. 516.)

   a. Make a graph from your tape: Cut the tape with scissors crosswise at each spark dot, so you have a set of strips of paper that are the actual lengths of the distances fallen by the object during each time interval. Arrange them evenly spaced in increasing order, with the bottom of each strip on a horizontal line. The end result should look like a series of steps. You could paste or tape them down on a big piece of paper or newspaper.

   b. Use a straight edge to draw a line that passes through the center of the top of each strip. Is the line a good fit? Each separate strip represents the distance the object fell during a fixed time interval, so we can think of the series or strips as representing change in distance over time, or average velocity. Interpret the graph of the line you have constructed in terms of the free fall experiment.

16. In the height equation $h = 300 + 50t - 4.9t^2$, time is measured in seconds and height in meters.

   a. What does the number 300 represent?

   b. What does the number 50 represent? What does the fact that 50 is positive tell you?

17. (Graphing program optional.) The height of an object that was projected vertically from the ground with initial velocity of 200 m/sec is given by the equation $h = 200t - 4.9t^2$, where t is in seconds.

   a. Find the height of the object after 0.1, 2, and 10 seconds.

   b. Sketch a graph of height vs. time.

   c. Use the graph to estimate the maximum height of the projectile and the approximate number of seconds that the object traveled before hitting the ground.

18. (Graphing program optional.) The height of an object that was thrown down from a 200-meter platform with an initial velocity of 50 m/sec is given by the equation $h = -4.9t^2 - 50t + 200$, where t is in seconds and h is in meters. Sketch the graph of height versus time. Use the graph to determine the approximate number of seconds that the object traveled before hitting the ground.

19. (Graphing program optional.) Let $h = 85 - 490t^2$ be a motion equation describing height, h, in centimeters and time, t, in seconds.

   a. Interpret each of the coefficients and specify its units of measurement.

   b. What is the initial velocity?

   c. Generate a table for a few values of t between 0 and 0.3 second.

   d. Graph height versus time by hand. Check your graph using a computer or graphing calculator if possible.

20. (Graphing program optional.) Let $h = 85 + 20t - 490t^2$ be a motion equation describing height, h, in centimeters and time, t, in seconds.

   a. Interpret each of the coefficients and specify its units of measurement.

   b. Generate a table for a few values of t between 0 and 0.3 second.

   c. Graph height versus time by hand. Check your graph using a computer or graphing calculator if possible.

One screen in "Q11: Freely Falling Objects" in *Quadratic Functions* simulates the activity described in Exercise 15.

21. At $t = 0$, a ball is thrown upward at a velocity of 10 ft/sec from the top of a building 50 feet high. The ball's height is measured in feet above the ground.

    a. Is the initial velocity positive or negative? Why?

    b. Write the motion equation that describes height, h, at time, t.

22. The concepts of velocity and acceleration are useful in the study of human childhood development. The accompanying figure shows (a) a standard growth curve of weight over time, (b) the rate of change of weight over time (the *growth rate* or *velocity*), and (c) the rate of change of the growth rate over time (or acceleration). Describe in your own words what each of the graphs shows about a child's growth.

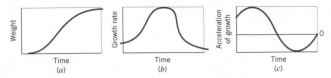

Source: Adapted from B. Bogin, "The Evolution of Human Childhood." *BioScience*, Vol. 40, p. 16.

23. The relationship between the velocity of a freely falling object and time is given by

$$v = -gt - 66$$

where g is the acceleration due to gravity and the units for velocity are centimeters per second.

    a. What value for g should be used in the equation?

    b. Generate a table of values for t and v, letting t range from 0 to 4 seconds.

    c. Graph velocity vs. time by hand and interpret your graph.

    d. What was the initial condition? Was the object dropped or thrown? Explain your reasoning.

24. A certain baseball is at height $h = 4 + 64t - 16t^2$ feet at time t in seconds. Compute the average velocity over each of the following time intervals and indicate for which intervals the baseball is rising and for which it is falling. In which interval was the average velocity the greatest?

    a. $t = 0$ to $t = 0.5$ **e.** $t = 2$ to $t = 3$

    b. $t = 0$ to $t = 0.1$ **f.** $t = 1$ to $t = 3$

    c. $t = 0$ to $t = 1$ **g.** $t = 4$ to $t = 4.01$

    d. $t = 1$ to $t = 2$

25. At $t = 0$, an object is in free fall 150 cm above the ground, falling at a rate of 25 cm/sec. Its height, h, is measured in centimeters above the ground.

    a. Is its velocity positive or negative? Why?

    b. Construct an equation that describes its height, h, at time t.

    c. What is the average velocity from $t = 0$ to $t = \frac{1}{2}$? How does it compare with the initial velocity?

26. In the Anthology Reading "Watching Galileo's Learning," Cavicchi notes that Galileo generated a sequence of odd integers from his study of falling bodies. Show that in general the odd integers can be constructed from the difference of the squares of successive integers, that is, that the terms $(n + 1)^2 - n^2$ (where $n = 0, 1, 2, 3, \ldots$) generate a sequence of all the positive odd integers.

27. Suppose an object is moving with constant acceleration, a, and its motion is initially observed at a moment when its velocity is v_0. We set time, t, equal to 0, at this point when velocity equals v_0. Then its velocity t seconds after the initial observation is $V(t) = at + v_0$. (Note that the product of acceleration and time is velocity.) Now suppose we want to find its average velocity between time 0 and time t. The average velocity can be measured in two ways. First, we can find the average of the initial and final velocities by calculating a numerical average or mean; that is, we add the two velocities and divide by 2. So, between time 0 and time t,

$$\text{average velocity} = \frac{v_0 + V(t)}{2} \qquad (1)$$

We can also find the average velocity by dividing the change in distance by the change in time. Thus, between time 0 and time t,

$$\text{average velocity} = \frac{\Delta \text{distance}}{\Delta \text{time}} = \frac{d - 0}{t - 0} = \frac{d}{t} \qquad (2)$$

If we substitute the expression for average velocity (from time 0 to time t) given by Equation (1) into Equation (2), we get

$$\frac{d}{t} = \frac{v_0 + V(t)}{2} \qquad (3)$$

We know that $V(t) = at + v_0$. Substitute this expression for $V(t)$ in Equation (3) and solve for d. Interpret your results.

28. In 1974 in Anaheim, California, Nolan Ryan threw a baseball at just over 100 mph. If he had thrown the ball straight upward at this speed, it would have risen to a height of over 335 feet and taken just over 9 seconds to fall back to Earth. Choose another planet and see what would have happened if he had been able to throw a baseball straight up at 100 mph on that planet. In your computations, use the table for the acceleration due to gravity on other planets from *Explore & Extend 8.6* on p. 526.

29. An object that is moving horizontally along the ground is observed to have (an initial) velocity of 60 cm/sec and to be accelerating at a constant rate of 10 cm/sec².

    a. Determine its velocity after 5 second, after 60 seconds, and after t seconds.

    b. Find the average velocity for the object between 0 and 5 seconds.

30. (Requires results from Exercise 29.) Find the distance traveled by the object described in the previous exercise after 5 seconds by using two different methods.

    a. Use the formula distance = rate · time. For the rate, use the average velocity found in Exercise 29(b). For time, use 5 seconds.

    b. Write an equation of motion $d = \frac{1}{2}at^2 + v_0 t$ using $a = 10$ cm/sec² and $v_0 = 60$ cm/sec and evaluate when $t = 5$. Does your answer agree with part (a)?

31. An object is observed to have an initial velocity of 200 m/sec and to be accelerating at 60 m/sec².

   a. Write an equation for its velocity after t seconds.

   b. Write an equation for the distance traveled after t seconds.

32. You may have noticed that when a basketball player or dancer jumps straight up in the air, in the middle of a blurred impression of vertical movement, the jumper appears to "hang" for an instant at the top of the jump.

   a. If a player jumps 3 feet straight up, generate equations that describe his height above ground and velocity during his jump. What initial upward velocity must the player have to achieve a 3-foot-high jump?

   b. How long does the total jump take from takeoff to landing? What is the player's downward velocity at landing?

   c. How much vertical distance is traveled in the first third of the total time that the jump takes? In the middle third? In the last third?

   d. Now explain in words why it is that the jumper appears suspended in space at the top of the jump.

33. Old Faithful, the most famous geyser at Yellowstone National Park, regularly shoots up a jet of water 120 feet high on average.

   a. At what speed must the stream of water be traveling out of the ground to go that high?

   b. How long does it take to reach its maximum height?

34. A vehicle trip is composed of the following parts:

   i. Accelerate from 0 to 30 mph in 1 minute.

   ii. Travel at 30 mph for 12 minutes.

   iii. Accelerate from 30 to 50 mph in $\frac{1}{2}$ minute.

   iv. Travel at 50 mph for 6 minutes.

   v. Decelerate from 50 to 0 mph in $\frac{1}{2}$ minute

   a. Sketch a graph of speed versus time for the trip.

   b. What are the average velocities for parts (i), (iii), and (v) of the trip?

   c. How much distance is covered in each part of the trip, and what is the total trip distance?

35. In general, for straight motion of a vehicle with constant acceleration, a, the velocity, v, at any time, t, is the original velocity, v_0, plus acceleration multiplied by time: $v = v_0 + at$. The distance traveled in time t is $d = v_0t + \frac{1}{2}at^2$.

   a. A criminal going at speed v_c passes a police car and immediately accelerates with constant acceleration a_c. If the police car has constant acceleration $a_p > a_c$, starting from 0 mph, how long will it take to pass the criminal? Give t in terms of v_c, a_p, and a_c.

   b. At what time are the police and the criminal traveling at the same speed? If they are traveling at the same speed, does it mean the police have caught up with the criminal? Explain.

CHAPTER SUMMARY

Quadratic Functions and Their Graphs

A *quadratic function* can be written in *standard form* as

$$f(x) = ax^2 + bx + c \quad \text{(where } a \neq 0\text{)}$$

Its graph

- has a distinctive ∪-shape called a *parabola*
- is symmetric across its *axis of symmetry*
- has a minimum or a maximum point called its *vertex*
- is concave up if $a > 0$ and concave down if $a < 0$
- becomes narrower as $|a|$ increases
- has a *focal point* $\left|\frac{1}{4a}\right|$ units above (or below) the vertex on the axis of symmetry

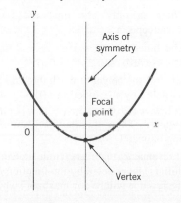

Finding the Vertex

Any quadratic function $f(x) = ax^2 + bx + c$ can also be written in *vertex* or *a-h-k* form as

$$f(x) = a(x - h)^2 + k$$

where the vertex is at $(h, k) = \left(-\frac{b}{2a}, f\left(-\frac{b}{2a}\right)\right)$. The vertex "anchors" the graph, and the coefficient a determines the shape of the parabola.

Finding the Horizontal Intercepts

Every quadratic function $f(x)$ has two, one, or no horizontal intercepts x. To find the horizontal or x-intercepts, we set $f(x) = 0$ and solve for x. The solutions are called the *zeros* of the function.

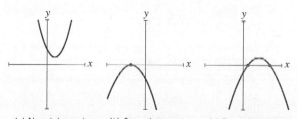

(a) No x-intercepts   (b) One x-intercept   (c) Two x-intercepts
No real zeros   One real zero   Two real zeros

The Factor Theorem says that any quadratic function $f(x) = ax^2 + bx + c$ can be written in *factored form* as

$$f(x) = a(x - r_1)(x - r_2)$$

where r_1 and r_2 are the zeros of $f(x)$.

If the zeros, r_1 and r_2, are real numbers, they are the horizontal intercept(s) of the quadratic function $f(x)$.

The Quadratic Formula

Setting the quadratic function $f(x) = ax^2 + bx + c$ equal to 0 and solving for x using the quadratic formula, gives

$$x = \frac{-b \pm \sqrt{b^2 - 4ac}}{2a}$$

The term $b^2 - 4ac$ is called the *discriminant* and can be used to predict the number of horizontal intercepts (or real zeros) of $f(x)$.

If the discriminant > 0, there are two distinct real roots and hence two x-intercepts.

If the discriminant $= 0$, there is only one distinct real root and hence only one x-intercept.

If the discriminant < 0, then the $\sqrt{b^2 - 4ac}$ is not a real number and hence there are no x-intercepts. These zeros are *complex numbers* of the form $a + bi$, where a and b are real numbers ($b \neq 0$) and $i = \sqrt{-1}$.

The Average Rate of Change of a Quadratic Function

Given a quadratic function $f(x) = ax^2 + bx + c$, the average rate of change between two points on the parabola approaches $2ax + b$ over very small intervals. We can think of the linear function $g(x) = 2ax + b$ as representing the average rate of change of $f(x)$ with respect to x.

The Mathematics of Motion

The general equations for a freely falling body are

$$d = (1/2)gt^2 + v_0 t$$
$$v = gt + v_0$$

where d = distance fallen, v = velocity, v_0 = initial velocity, t = time, and g = acceleration due to gravity.

The conventional values for g, the acceleration due to gravity near the surface of Earth, are

$$g = 32 \text{ ft/sec}^2$$

or equivalently $g = 980 \text{ cm/sec}^2 = 9.8 \text{ m/sec}^2$.

If we treat upward motion as positive and downward motion as negative, and measure the height, h, of a freely falling body, then

$$h = h_0 + v_0 t - (1/2)gt^2$$
$$v = -gt + v_0$$

where h_0 = initial height. Note that v_0, the initial velocity, can be positive (upward velocity) or negative (downward velocity).

CHECK YOUR UNDERSTANDING

I. Is each of the statements in Problems 1–20 true or false? Give an explanation for your answer.

1. If $f(t) = 2(t - 1)^2$, then $f(0) = -2$.

2. If the vertical axis is the axis of symmetry for a quadratic function $g(x)$, then $g(-2) = g(2)$.

3. The function $y = 3(x - 2)^2 + 5$ has a focal point at $\left(2\frac{1}{12}, 5\right)$.

4. The graph of the quadratic function $y = 2x^2 - 3x + 1$ is steeper than the graph of $y = 3x^2 - 3x + 1$.

5. The graph of $y = 5 - x^2$ is concave down.

6. The graph of $y = x^2 + 2x + 3$ is three units higher than the graph of $y = x^2 + 2x$.

7. The graph of $y = (x + 4)^2$ lies four units to the right of the graph of $y = x^2$.

8. A quadratic function that passes through the points $(1, 5)$ and $(7, 5)$ will have an axis of symmetry at the vertical line $x = 4$.

9. If $f(x) = x^2 - 3x - 4$, then $f(4) = 0$.

10. The function $f(x) = (x + 2)(x + 5)$ has zeros at 2 and 5.

11. The quadratic function $f(x) = 2x^2 - 3x - 1$ has a discriminant with a value of 1.

12. The function $f(x) = 3(x - 1)^2 + 2$ has an axis of symmetry at $x = 1$.

13. The function $f(x) = -2(x + 3)^2 - 1$ has a vertex at $(3, -1)$.

14. There is only one quadratic function $f(x)$ with x-intercepts at 3 and 0.

15. The function $h(t) = t^2 - 3t + 2$ has a zero at $t = 2$ because $h(0) = 2$.

16. Assume the height of a ball thrown vertically upward is modeled by the function $h(t) = -4.9t^2 + 38t + 55$, (where t is time in seconds, and $h(t)$ is the height in meters). Then the ball will hit the ground after approximately 9 seconds.

17. If revenue R (in dollars) from an item sold at price p (in dollars) is modeled by the function $R = p(100 - 5p)$, the revenue will be at a maximum when the price is \$10.

Problems 18–20 refer to the motion equation of an object, $h = 39.2 + 9.8t - 4.9\,t^2$, describing height, h, in meters and time, t, in seconds.

18. The initial velocity is 9.8 meters per second.

19. The maximum height is 39.2 meters.

20. The object hits the ground in 4 seconds.

II. For Problems 21–30 give an example of a function or functions with the specified properties. Express your answer using equations.

21. A quadratic function with vertex at the point $(0, 0)$ and with focal point $(0, -1)$.

22. A quadratic function concave down with a vertex at $(1, 3)$.

23. A quadratic function concave up with its axis of symmetry at the line $x = 3$.

24. A quadratic function concave down with vertical intercept at 2 and zeros at -2 and 2.

25. A quadratic function whose graph will be exactly the same shape as the graph of the function $r = s^2 - s$ but five units higher.

26. A quadratic function $G(x)$ whose graph will be exactly the same shape as the function $F(x) = x^2 + 2x$ but two units to the left.

27. A quadratic function whose graph will be the reflection across the t-axis of the graph of $h(t) = (t - 2)^2$.

28. Two distinct quadratic functions that intersect at the point $(1, 1)$.

29. A quadratic function with one zero at $x = -4$.

30. A motion equation of an object thrown upward from an 80-foot building with initial velocity 32 feet per second.

III. Is each of the statements in Problems 31–40 true or false? If a statement is true, explain how you know. If a statement is false, give a counterexample.

31. Quadratic functions $f(x) = ax^2 + bx + c$ always have two distinct horizontal intercepts because the equation $ax^2 + bx + c = 0$ always has two distinct solutions.

32. Quadratic functions $f(x) = ax^2 + bx + c$ always have two distinct zeros because the equation $ax^2 + bx + c = 0$ always has two roots, $x = \frac{-b \pm \sqrt{b^2 - 4ac}}{2a}$.

33. If the discriminant is 0, then the associated quadratic function has no horizontal intercept.

34. Quadratic functions that open upward have a minimum value at the vertex.

35. If $f(x) = f(-x)$, the graph of f is symmetric across the x-axis.

36. If $f(x) = -f(-x)$, the graph of f is symmetric about the origin.

37. The difference between two motion equations $F(t) = -490t^2 + 2450$ and $G(t) = -490t^2 - 1470t + 2450$ is that the first object is dropped and the second is thrown downwards with initial velocity 1470 cm/sec from the same height of 2450 cm.

Problems 38–40 refer to the motion equation of an object, $h = 80 + 64t - 16t^2$, describing the height of an object, h, in feet and time, t, in seconds.

38. The maximum height of 128 feet is reached at $t = 1$ seconds.

39. The equation of the average velocity of the object is $v = -32t + 64$.

40. The equation of the distance the object traveled is $d = 16t^2 + 64t$.

CHAPTER 8 REVIEW: PUTTING IT ALL TOGETHER

Problem 3 requires a graphing program. Problem 6 requires technology to generate a best-fit polynomial.

1. For each of the accompanying parabolas, identify the graph as concave up or down, and then estimate the minimum (or maximum) point, the axis of symmetry, any horizontal intercepts, and the domain and range.

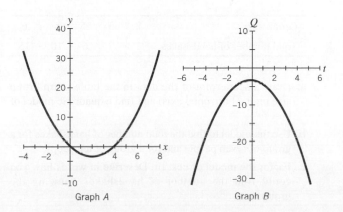

Graph *A*

Graph *B*

2. An electric heater is being designed as a parabolic reflector $6''$ deep. To prevent accidental burns, the center of the heating element is placed at the focus, which is set $1.5''$ from the vertex of the reflector.

a. What equation describes the shape of the reflector?

b. How wide will the reflector be?

c. Sketch an image of the parabolic reflector with the vertex at its origin. Put a circle at the focal point, and label the depth and width of the reflector.

d. What are reasonable values for the domain and range of your parabolic model?

3. (Requires graphing program for part (e).) A wood craftsman has created a design for a parquet floor. The pattern for an individual tile is shown in the accompanying image. The square center (x inches wide) of the tile is made from white oak hardwood and is surrounded by 1-inch strips of maple hardwood. (See image next page.)

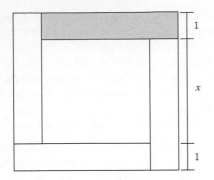

a. What is the area of the interior white oak square (in terms of x)? The area of each of the maple 1-inch strips?

b. White oak costs $2.39 per square foot; maple costs $4.49 per square foot. Calculate the cost per square inch for white oak and for maple.

c. What is the cost for the white oak in one tile of the parquet? What is the cost of the four maple strips in one tile?

d. The approximate labor cost to make each tile is $5.00. Create a cost function $C(x)$ (in dollars) for making one parquet tile. What type of function is this?

e. Use technology to graph the function $C(x)$, where x is the width of the inner white oak square. Use a domain of $0 \le x \le 15$ inches.

f. From your graph, estimate the size of a parquet tile if the total cost (including labor) is to be $7.00 or less per tile.

4. California produces nearly 95% of the processed tomatoes grown in the United States, so managing irrigation water for tomatoes is a major issue. Agricultural researchers have been able to quantify the relationship between C, the canopy coverage, and K_C, the crop coefficient.[10] Canopy cover is the percentage of the total plot covered by shade produced by the leaves of the plants. The crop coefficient is a measure of the water needed by the plants. The following graph shows the relationship between the canopy coverage and the crop coefficient.

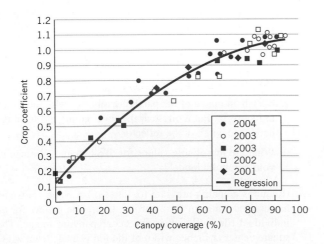

[10]B. R. Hanson and D. M. May, "New crop coefficients developed for high-yield processing tomatoes," *California Agriculture* 60(2), April–June 2006.

a. As the plants grow, will the canopy cover increase or decrease? Why?

b. As the canopy cover increases, the crop coefficient also increases—but at a decreasing rate. Why would that be true?

c. The best-fit function to the data is a quadratic:

$$K_C = 0.126 + (0.0172)C - (0.0000776)C^2$$

where $0 \le C \le 100\%$.

   i. The initial growth stage has between 0% and 10% canopy coverage. Estimate the corresponding crop coefficient for 10% canopy coverage from the graph, then calculate it using the quadratic model.

   ii. In the crop development growth stage, the canopy is between 10% and 75%. Estimate the crop coefficient when the canopy is at 75% and then calculate it. What does this number mean?

5. Construct a function for each parabola $g(x)$ and $h(x)$ in the accompanying graph.

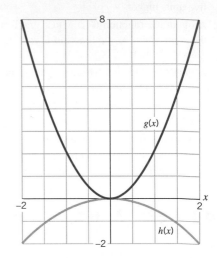

6. (Technology required to create a best-fit quadratic.) When people meet for the first time, it is customary for all people in the group to shake hands. Below is a table that shows the number of handshakes that occur depending on the size of the group.

Group size	2	3	4	5	6
Total number of handshakes	1	3	6	10	15

a. Draw a scatterplot of the data in the table (with group size on the horizontal axis) and find a quadratic model of best fit.

b. Use the model to find the total number of handshakes for a group of seven people and for a group of ten people.

c. Factor the model of best fit. Describe in words how you could find the number of handshakes knowing the group size.

7. a. Identify the coordinates of the vertex for each of the following quadratic functions.

$F(x) = x^2$, $G(x) = x^2 + 5$, $H(x) = (x + 2)^2$, and $J(x) = -(x - 1)^2 - 5$

b. Without using technology, draw a rough sketch on the same grid of all the functions for $-4 \le x \le 4$.

c. Describe how the graph of F was transformed into the graphs of G, H, and J, respectively.

8. Find any horizontal intercepts for the following functions.

a. $y = (x - 3)(2x + 1)$ **c.** $Q(t) = 2t^2 + t - 1$

b. $G(z) = 2z^2 - z + 3$

9. a. Construct a quadratic function $Q(t)$ that is concave up and has horizontal intercepts at $t = 4$ and $t = -2$. Write it in both factored and standard form. Find its vertex.

b. Construct a second function $M(t)$ that has the same horizontal intercepts as $Q(t)$ but is steeper. Write $M(t)$ in both factored and standard form. Do the two functions have the same vertex?

c. Add a term to $Q(t)$ to create a function $P(t)$ that has no horizontal intercepts.

10. Explain why you could (or couldn't) construct a parabola through any three points.

Problems 11, 12, and 13 refer to the accompanying diagram of the cross section of a swimming pool with a reflective parabolic roof.

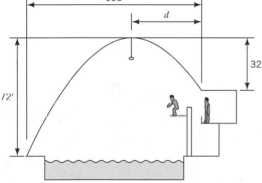

11. Find an equation for the cross section of the parabolic roof of the swimming pool in the diagram. (*Hint:* Place the origin of your coordinate system at the vertex and identify two other points on the parabola in terms of d.)

12. (Requires results of Problem 11.) The pool designer wants to mount a light source at the parabolic focus so that it sheds light evenly on the water surface below. How many feet down from the vertex must that be?

13. A diver jumps up off the high board, which is 25 feet above the surface of the water. Her height, $H(t)$, in feet above the water at t seconds, can be modeled by the function $H(t) = 25 + 12t - 16t^2$.

a. What will be the highest point above the water of her dive?

b. When will she hit the water?

14. In the United States a "heat wave" is a period of three or more consecutive days at or above 90°F. A heat wave is often accompanied by high humidity, making the air feel even hotter. The following formula combines an air temperature of 90°F with relative humidity, H, to give the apparent temperature, A, the perceived level of heat:

$$A = 86.61 - 0.132H + 0.0059H^2$$

This formula uses relative humidity as a percentage (e.g., 70% relative humidity appears in the formula as $H = 70$). Remember, this formula applies only for an air temperature of 90°F.

a. If the relative humidity is 0%, what is the apparent temperature for an air temperature of 90°F? Does the apparent temperature feel lower or higher than the air temperature of 90°F?

b. On a 90°F day, if the relative humidity is 60%, what is the apparent temperature? How much hotter do you feel?

c. An apparent temperature of 105°F or above is considered dangerous, especially for children and elders. At what relative humidity on a 90°F day is an apparent temperature of 105°F reached?

15. a. Complete the following table for the function $y = x^2 - 4x$.

x	y	Average Rate of Change	Average Rate of Change of Average Rate of Change
-1	5	n.a.	n.a.
0	0	-5	n.a.
1	-3	-3	$[-3 - (-5)] / (1 - 0) = 2$
2	-4		
3			
4			
5			

b. If you plotted the points with coordinates of the form (x, average rate of change) and connected adjacent points, what type of function would you get?

c. Does the fourth column verify your result in part (b)? Why or why not?

How Fast Are You? Using a Ruler to Make a Reaction Timer[11]

Objective

- learn about the properties of freely falling bodies and your own reaction time

Materials/Equipment

- several 12″ rulers
- narrow strips of paper and tape
- calculators

Procedure

General Description

Work in groups of two or three. Each group has a 12″ ruler and will attach a 12″ paper strip to the ruler, adding some marks (specified below). One student drops the ruler between the thumb and forefinger of a second student. The second student tries to catch the ruler as quickly as possible (see image). The reaction time of the second student can be measured by how far the ruler falls before it is caught.

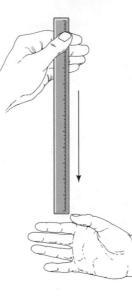

Mathematical Background

Near the surface of Earth, and neglecting air resistance, gravity causes dropped objects to fall approximately according to the formula

$$d = 16t^2$$

where d = distance fallen in feet and t = time of fall in seconds.

1. An object in free fall for 1 second will fall 16 feet, and an object in free fall for 0.5 second will fall _____ feet. (Calculate.)

2. Can you give an intuitive explanation for why the object falling for 0.5 second does not fall half as far as the object that fell for 1 second?

3. What assumptions must we make about the shape of the dropped object for it to fall according to this formula? (*Hint:* Will a sheet of paper fall 16 feet in 1 second?)

[11]This exploration was developed by Karl Schaffer, Mathematics Department, De Anza College, Cupertino, CA.

4. Do you think a heavy object will fall faster or slower or at the same rate as a light object? Explain your reasoning.

5. For each of the following times, use the formula to calculate how far a dropped object will fall.

Time, t	Distance in Feet, d	Distance in Inches
0.05 second		
0.10 second		
0.15 second		
0.20 second		
0.25 second		

6. Use tape to attach a strip of paper along the length of the ruler. Think of the ruler as measuring distance fallen. For each distance (in inches) in your previous table, put a mark on the paper that indicates the time corresponding to that distance. So you'll have unevenly spaced marks for the times 0.05 sec, 0.10 sec, up to 0.25 sec. Now the ruler is a reaction timer.

7. One member of your group holds the ruler just above the outstretched thumb and first finger of a second person. The "dropper" suddenly drops the ruler and the "catcher" tries to catch it. Use the time marks on the strip of paper to get an estimate for the reaction time of the catcher. Use the actual number of inches at the point where the catcher caught the ruler to calculate the reaction time. Record the reaction time, and average several tries. If you like, measure the reaction time for someone else in your group.

Person	Distance in Inches	Distance in Feet, d	Reaction Time, t

What was your average reaction time? How did yours compare with that of others in your group?

Further Investigations

1. A popular party trick has one person drop a dollar bill between the fingers of a second person. Usually the bill will fall through the second person's fingers without being caught. How long is a dollar bill, and what must the second person's reaction time be for the bill to be caught? Does the use of money speed up the reaction times you measured with the ruler? Does the bill fall with only negligible air resistance?

2. How do medications or drugs affect our reaction times? Test the reaction times of someone who is taking cold or flu medication or aspirin, or has just drunk a cup of coffee or glass of wine (outside of class, of course). Is his or her reaction time impaired? Is there a correlation between the amount of alcohol consumed and reaction time that might enable you to use your ruler reaction timer as a portable tester to determine whether someone who has consumed alcohol should not drive?

3. Do reaction times measured in this activity improve with practice? Why or why not? (Try it!) Do you think the catcher learns to detect subtle indications of the dropper that she or he is about to drop the ruler? How might these biases be removed from this experiment?

4. Jugglers, athletes, and dancers need to understand, either intuitively or objectively, their own reaction times. In what other occupations is reaction time important? Using the Internet, can you find the reaction times necessary in any of these areas?

5. The *Guinness Book of World Records* lists the fastest times for drawing and firing a gun. Look this up. Are these times consistent with your results?

6. What else might you investigate about reaction times and your ability to measure them?

Tables for Exploration 2.1 (p. 154) from:

The University of Massachusetts Boston

STATISTICAL PORTRAIT FALL 2010

Office of Institutional Research and Policy Studies (OIRP)

Undergraduate Admissions Summary–Fall Term

	2000	2001	2002	2003	2004	2005	2006	2007	2008	2009
FRESHMEN										
Applied	3,478	3,902	3,838	4,187	3,530	3,870	4,446	5,182	5,550	7,554
Decision Ready	2,667	2,652	2,704	2,834	2,903	3,174	3,666	4,213	4,576	6,050
Admitted	1,562	1,539	1,478	1,561	1,553	1,920	2,325	2,581	2,884	3,718
Admit Rate	58.6%	58.0%	54.7%	55.1%	53.5%	60.5%	63.4%	61.3%	63.0%	61.5%
Enrolled	706	701	576	610	565	781	974	997	1,020	987
Admitted but Deferred	32	14	23	13	30	15	8	7	3	6
Yield Rate	45.2%	45.5%	39.0%	39.1%	36.4%	40.7%	41.9%	38.6%	35.4%	26.5%
Not Admitted: Denied	1,105	1,113	1,226	1,273	1,350	1,254	1,341	1,632	1,692	2,332
Inc. Applications	811	1,250	1,134	1,353	627	696	780	969	984	1,504
TRANSFERS										
Applied	4,008	4,172	4,063	4,076	3,390	3,317	3,625	3,735	3,855	4,335
Decision Ready	3,038	2,916	2,892	2,779	2,697	2,639	2,890	2,889	3,039	3,245
Admitted	2,631	2,564	2,378	2,360	2,125	2,089	2,313	2,348	2,404	2,628
Admit Rate	86.6%	87.9%	82.2%	84.9%	78.8%	79.2%	80.0%	81.3%	79.1%	81.0%
Enrolled	1,556	1,542	1,382	1,339	1,193	1,326	1,503	1,566	1,614	1,756
Admitted but Deferred	113	45	92	20	67	51	40	38	38	34
Yield Rate	59.1%	60.1%	58.1%	56.7%	56.1%	63.5%	65.0%	66.7%	67.1%	66.8%
Not Admitted: Denied	407	352	514	419	572	550	577	541	635	617
Inc. Applications	970	1,256	1,171	1,297	693	678	735	846	816	1,090
TOTAL UNDERGRADUATES										
Applied	7,486	8,074	7,901	8,263	6,920	7,187	8,071	8,917	9,415	11,889
Decision Ready	5,705	5,568	5,596	5,613	5,600	5,813	6,556	7,102	7,615	9,295
Admitted	4,193	4,103	3,856	3,921	3,678	4,009	4,638	4,929	5,288	6,346
Admit Rate	73.5%	73.7%	68.9%	69.9%	65.7%	69.0%	70.7%	69.4%	69.4%	68.3%
Enrolled	2,262	2,243	1,958	1,949	1,758	2,107	2,477	2,563	2,634	2,743
Admitted but Deferred	145	59	115	33	97	66	48	45	41	40
Yield Rate	53.9%	54.7%	50.8%	49.7%	47.8%	52.6%	53.4%	52.0%	49.8%	43.2%
Not Admitted: Denied	1,512	1,465	1,740	1,692	1,922	1,804	1,918	2,173	2,327	2,949
Inc. Applications	1,781	2,506	2,305	2,650	1,320	1,374	1,515	1,815	1,800	2,594

Trends in New Student Race/Ethnicity in the College of Liberal Arts

	2001	2002	2003	2004	2005	2006	2007	2008	2009
Native American	0.6%	0.8%	0.4%	0.6%	0.6%	0.3%	0.8%	0.8%	0.5%
Asian/Pacific Islander	11.9%	11.9%	12.9%	12.8%	12.5%	13.3%	12.3%	11.2%	8.4%
Black	15.3%	14.0%	13.3%	15.3%	12.7%	14.6%	13.4%	14.1%	14.3%
Hispanic	7.0%	8.8%	8.0%	7.1%	11.8%	10.5%	10.0%	12.7%	12.9%
Cape Verdean	0.9%	1.0%	1.8%	1.9%	1.4%	2.0%	1.7%	2.0%	1.5%
Total minority	**35.7%**	**36.5%**	**36.4%**	**37.7%**	**39.0%**	**40.7%**	**38.2%**	**40.8%**	**37.6%**
White	55.7%	58.4%	60.1%	58.6%	59.1%	56.4%	59.2%	56.8%	59.2%
Non-resident alien	8.7%	5.1%	3.4%	3.7%	2.0%	2.8%	2.6%	2.5%	3.2%
Known race [N]	[1,509]	[1,244]	[960]	[885]	[1,034]	[1,155]	[1,162]	[1,309]	[1,172]

SAT Scores of New Freshmen by College/Program

SAT Scores of New Freshmen by College/Program, 10-Year Trend (Excluding the DSP Program, Learning Disabled and Foreign Students)

		2000	2001	2002	2003	2004	2005	2006	2007	2008	2009
College of Liberal Arts	SATVerbal	515	524	520	522	528	540	529	529	527	527
	SATMath	529	536	532	519	517	539	524	522	516	527
	Combined	1,044	1,060	1,052	1,041	1,045	1,079	1,053	1,051	1,043	1,054
	[N]	[363]	[369]	[317]	[224]	[201]	[278]	[375]	[386]	[394]	[375]
College of Management	SATVerbal	516	511	515	492	487	511	504	508	502	515
	SATMath	548	556	546	542	523	554	564	549	547	561
	Combined	1,064	1,067	1,061	1,034	1,010	1,065	1,068	1,057	1,049	1,076
	[N]	[42]	[34]	[37]	[36]	[35]	[59]	[77]	[73]	[56]	[69]

Data Dictionary for FAM1000 from March 2009 Current Population Survey

Variable Names	Definition	Unit of Measurement (Code or Allowable Range)	Variable Names	Definition	Unit of Measurement (Code or Allowable Range)
age	Age	Range: 16:85	occup	Occupation group of respondent	7 = Construction and extraction occupations
sex	Sex	1 = Male 2 = Female			8 = Installation, maintenance, and repair occupations
region	Geographic region	1 = Northeast 2 = Midwest 3 = South 4 = West			9 = Production occupations 10 = Transportation and material moving occupations 11 = Armed Forces
cencity	Metropolitan status	1 = Metropolitan 2 = Non-metropolitan 3 = Not identified	hrswork	Hours worked per week	Range: 1:99
marstat	Demographics, Marital Status	0 = Presently married 1 = Presently not married	wkswork	Weeks worked per year	Range: 1:52
			yrft	Worked full/part-time	0 = Not in universe
famsize	Number of persons in family	Range: 1:39			1 = Full year, worked full-time 2 = Full year, worked part-time
edu	Years of education from eighth grade on	0 = 8 or fewer years of education 2 = No high school diploma, with 9–12 years of education 4 = High school diploma 5 = Some college but no degree 6 = Associates degrees 8 = Bachelor's degree 10 = Master's degree (e.g., M.B.) 12 = Doctorate degree (e.g., PhD) 14 = Professional degrees (e.g., MD's)			3 = Part year, worked full-time 4 = Part year, worked part-time 5 = Nonworker
			pearnings	Total wage and salary earnings amount	Range: 0:1,000,000
			ptotinc	Personal total income	Range: −10,000:1,000,000
			faminc	Total income amount	Range: 1:1,000,000
occup	Occupation group of respondent	0 = Not in universe, or children 1 = Management, business, and financial occupations 2 = Professional and related occupations 3 = Service occupations 4 = Sales and related occupations 5 = Office and administrative support occupations 6 = Farming, fishing, and forestry occupations	race	Demographics, race of respondent	1 = White Only 2 = Black Only 3 = American Indian, Alaskan Native Only 4 = Asian Only 5 = Others
			hispanic	Demographics, detailed Hispanic recode	0 = Not in universe 1 = Mexican 2 = Puerto Rican 3 = Cuban 4 = Central/South American 5 = Other Spanish

Source: Adapted from the U.S. Bureau of the Census, Current Population Survey, March 2009, by Jie Chen, Computing Services, University of Massachusetts, Boston.

SOLUTIONS

CHAPTER 1

Section 1.1

Algebra Aerobics 1.1

1.

Fraction	Decimal	Percent
$\frac{7}{12}$	0.583	58.3%
$\frac{1}{40}$	0.025	2.5%
$\frac{1}{50}$	0.02	2%
$\frac{1}{200}$	0.005	0.5%
$\frac{7}{20}$	0.35	35%
$\frac{1}{125}$	0.008	0.8%

2. a. 500 people

   b. 38.94 or 39 students

   c. 37.5%

3. a. mean = 4; median = 5

   b. mean $= \dfrac{(-2) + 2 + 5 + 6}{4} = 2.75$; median $= \dfrac{2 + 5}{2} =$

   3.5. Because we have an even number of numbers, we need to average the two that lie in the middle when the numbers are ordered.

4. Let w represent the hourly wage of the sixth employee.

Then $\dfrac{(\$4.75 + \$5.50 + \$5.75 + \$8.00 + \$9.50 + w)}{6} =$

$\$7.50 \Rightarrow \$33.50 + w = \$45.00 \Rightarrow w = \$11.50$

5. mean = \$25,040; median = \$20,000

6. a. Frequency Count (FC) for Age 1–20 interval is 38% of total: 38% of 137 = 0.38(137) ≈ 52.

FC for Age 61–80 interval is total FC minus all the others: 137 − (52 + 35 + 28) = 137 − 115 = 22.

Relative Frequency (RF) for each interval is its FC divided by total:

RF of interval (21–40) is: $\frac{35}{137} \approx 0.255 \approx 26\%$

RF of interval (41–60) is: $\frac{28}{137} \approx 0.204 \approx 20\%$

RF of interval (61–80) is: $\frac{22}{137} \approx 0.161 \approx 16\%$

Age	Frequency Count	Relative Frequency (%)
1–20	52	38
21–40	35	26
41–60	28	20
61–80	22	16
Total	137	100

   b. 20% + 16% = 36%

7. In 2008, of the job types in the chart, home health aides had the largest number of jobs (921,700). While biomedical engineers had the smallest number of jobs (16,000), they are projected to have the greatest increase (72%). Financial examiners are projected to increase the least (41.2%). In interpreting these percentages, one needs to take into account the number of current jobs in each category.

8. Table for the histogram:

Age	Relative Frequency (%)	Frequency Count
1–20	20	(0.20)(1352) ≈ 270
21–40	35	(0.35)(1352) ≈ 473
41–60	30	(0.30)(1352) ≈ 406
61–80	15	(0.15)(1352) ≈ 203
Total	100	1352

9. a. sum = \$8750, so mean = \$8750/9 ≈ \$972.22; median = \$300

   b. sum = 4.7, so mean = 4.7 ÷ 8 ≈ 0.59; median = (0.4 + 0.5)/2 = 0.45

10. One of the values (\$6,000) is much higher than the others which forces a high value for the mean. In cases like this, the median is generally a better choice for measuring central tendency.

Exercises for Section 1.1

1. a. Solar; \$1.5 billion.

   b. Solar; it lost about 3.25 − 1.5 = 1.75 billion dollars.

   c. Transportation, energy efficiency, and smart grid gained money.

3. a. The percent of those with unmet health care needs, country for country, is higher for those with below average income.

   b. The percent disparity was largest for the U.S. It was smallest for the U.K.

   c. One would need the actual population of each country.

5. a. 45

   b. Quantitative data

   c. $\dfrac{3}{45} \approx 6.7\%$

   d.

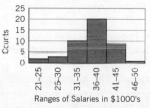

Salaries Distribution for Graduates

7. a. Mean = $386/7$ = 55.14; median = 46.

  b. Changing any entry in the list that is greater than the median to something still higher will not change the median of the list but will increase the mean. The same effect can be had if an entry less than the median is increased to a value that is still less than or equal to the median.

9. The mean annual salary is $24,700 and the median annual salary is $18,000. The mean is heavily weighted by the two high salaries. The mean salary is more attractive but is not likely to be an accurate indicator.

11. No answer is given here. (In general, when answers from students can vary quite a bit, either a typical answer or none is given.)

13. The mean age in the United States is slightly higher than the median age since there are a lot of older Americans (including the baby boomers), which pulls the mean age higher than the median. In developing countries, the mean age will be less than the median age since there are a lot of younger people and this pulls the mean lower than the median. Answers will vary by state and will depend on how the ages are distributed.

15. He is correct, provided the person leaving state A has an IQ that is below the average IQ of people in state A and above the average IQ of the people in state B.

17. The mean net worth of a group, e.g., American families, is heavily biased upward by the very high incomes of a relatively small subset of the group. The median net worth of a group such as this is not as biased, since it is in the middle of the list. The two measures would be the same if net worths were distributed symmetrically about the mean.

19. Among Internet users in the world on December 31, 2009, the number in Asia—764.4 million—dominates all other geographic regions mentioned by wide margins. They are nearly twice the number found in Europe, three times what is found in North America, and four times what is found in Latin America and the Caribbean. The counts for Africa, the Middle East, and Oceania/Australia are all below 100 million. However, proportionally, North America tops the list at 76.2% of the population, while Asia places sixth with only 20.1% of their population.

21. If we assume an estimated allowance of $5 for all students with an allowance in the $0–9 category, an estimate of $15 for all students with an allowance in the $10–19 category, and so on, then the mean $\approx$ ($5 \cdot 8 + \$15 \cdot 6 + \$25 \cdot 12 + \$35 \cdot 14 + \$45 \cdot 9)/49 = 1325/49 \approx \$27$. The median lies somewhere between $20 and $29; an estimate would be $25.

23. Answers will vary. Almost 10 hours are spent with electronic devices and only 38 minutes with print material. One question ought to be: The time adds to 10:45 per day, one wonders how the other time is spent if 8 hours are for sleep? Also the usefulness of grouping together all between the ages of 8 and 18 is questioned. The interests of 8 year olds, for example, would be quite different from the interests of those who are 18 years old. The word "Average" is not defined.

Section 1.2

Algebra Aerobics 1.2a

1. a. The median family net worth increased from 1998 to 2007. It saw sharper increases in the three-year periods 1998–2001 and then again in 2004–2007 than in the period 2001–2004.

  b. Additional information might be useful, such as inflation rate, cost of living increase or decrease, number of people in the family, number of wage earners in the family.

2. a. 1975

  b. 2030

  c. 55 years

3. a. approximately 11 billion

  b. approximately 1 billion

  c. approximately $11 - 1 = 10$ billion

  d. The total world population increased rapidly from 1950 to 2000, and is projected to continue to increase but at a decreasing rate, reaching approximately 11 billion in the year 2150.

4. a. $\frac{\text{pop } 2000}{\text{pop } 1900} \approx \frac{6}{1.5} = 4$. The population of the world in year 2000 was approximately 4 times greater than in 1900. The difference ≈ 4.5 billion.

  b. $\frac{\text{pop } 2100}{\text{pop } 2000} \approx \frac{10.5}{6} = 1.75$, so the population in 2100 is projected to be approximately 1.75 times greater than in 2000. The difference ≈ 4.5 billion.

  c. The population from 2000 to 2100 is expected to grow by 4.5 billion people, which is the same as the 4.5 billion increase from 1900 to 2000. The world population in 2000 was approximately 4 times greater than in 1900 and it is projected to be about 1.75 times greater in 2100 than in 2000. While the population continues to increase, the rate of increase is slowing down.

Algebra Aerobics 1.2b

1. a. Square the value of x, then multiply that result by 3, then subtract the value of x and add $+1$.

  b. (0, 1) is the only one of those ordered pairs that is a solution.

  c.

x	-3	-2	-1	0	1	2	3
y	31	15	5	1	3	11	25

2. a. Subtract 1 from the value of x, then square the result.

  b. (0, 1) and (1, 0) are the only ones of those ordered pairs that are solutions.

  c.

x	-3	-2	-1	0	1	2	3
y	16	9	4	1	0	1	4

3.

x	-4	-2	-1	0	1	2	4
y_1	16	10	7	4	1	-2	-8
y_2	-15	3	6	5	0	-9	-39

a. & b.

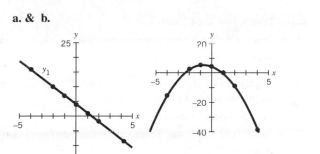

c. Yes for y_1; no for y_2.

d. No for y_1; yes for y_2.

e. No, since $4 - 3(-3) = 13$, $-2(-3)^2 - 3(-3) + 5 = -4$

4.

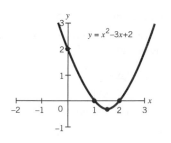

a. $y = -1/4$

b. There are many answers: $(0, 0)$ and $(-2, 5)$ are examples.

Exercises for Section 1.2

1. a. Student answers may vary. Three important facts can be culled from the given data: over time the number of diagnoses of AIDS is much larger than the number who die; the number of deaths from AIDS peaked in the early 1990s, as did the number of diagnoses; the number of diagnoses and deaths have steadily gone down over the years, and have remained stable since 1999.

b. Student answers will vary. Any answer ought to include some of the facts mentioned in part (a).

3. a.

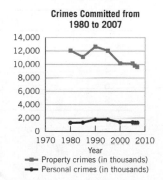

b. About nine times more in 1980 and seven times more in 2007.

c. Answers will vary. One possible sentence could be: From 1980 to 2007, property crimes went down, personal crimes stayed about the same, but during the entire time period the frequency of property crimes far exceeded that of personal crimes.

5. a. As men grow older, the risk of cancer goes up dramatically—for example, from 1 in 25,000 at age 45, to 1 in 25 at age 65, to 1 in 6 at age 80.

b.

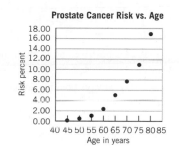

c. The ratio of risks for a 50- vs. a 45-year-old man is $0.21\%/0.004\% = 52.5$. This means that a 50-year-old man is 52.5 times more likely to have had prostate cancer than a 45-year-old man. The ratio of risks for a 55- vs. 50-year-old man is $0.83\%/0.21\% \approx 4$. So a 55-year-old man is only 4 times as likely to have had prostate cancer than a 50-year-old man. At first this may seem contradictory. But the biggest incremental risk occurs between 45 and 50. So while the absolute risk continues to rise (note that the percent risk is cumulative), the ratios of the percent risk decrease over subsequent 5-year age intervals.

d. Answers will vary. The medical profession has recommended this test for all men from age 40 up. The insurance companies may think that this is too expensive.

7. a. True; B is the newer car and it costs more than A.

b. True; A is the slower car in cruising speed and it has the larger size.

c. False; A is the larger car but it is older than B.

d. True; A carries more passengers and it is less expensive.

e. Answers will vary. You may mention that the larger range car also has the larger passenger capacity.

f. Answers will vary. Much depends on what features you value. There are many trade-offs.

9. a. Only $(1, -3)$ satisfies the formula.

b. There are many answers: e.g., $(0, 2)$, $(-1, 7)$. In general, pick a value of T and plug it into the formula to compute the corresponding value of R.

c. Here is the scatter plot for the three points given:

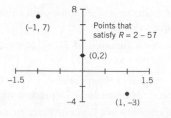

d. The plot suggests that solutions could be found on the straight line through these three points by eyeballing the line. Another source would be using the formula with other values of T to generate the corresponding values of R.

11. Answers may vary.

$(0, 1)$; $(0 - 2)^2 - 3 = 4 - 3 = 1$

$(2, -3)$; $(2 - 2)^2 - 3 = 0 - 3 = -3$

$(5, 6)$; $(5 - 2)^2 - 3 = 3^2 - 3 = 9 - 3 = 6$

13. a. $x = 0, y = 0$

  b. If $x > 0$ then x^2 is positive and is multiplied by -2, so y is negative.

  c. If $x < 0$ then x^2 is always positive and is multiplied by -2, so y is negative.

  d. For $x \neq 0$, y is always negative because x^2 is always positive and multiplying by -2 makes y always negative.

15. a. Only $(-1, 3)$ satisfies $y = 2x + 5$.

  b. $(1, 0)$ and $(2, 3)$ satisfy $y = x^2 - 1$.

  c. $(-1, 3)$ and $(2, 3)$ satisfy $y = x^2 - x + 1$

  d. Only $(1, 2)$ satisfies $y = 4/(x + 1)$.

Section 1.3

Algebra Aerobics 1.3

1. Table A represents a function. For each input value there is one and only one output value.

Table B does not represent a function. For the input of 2 there are two outputs, 7 and 8.

2. There are two output values, 5 and 7, for the input of 1. For the table to represent a function, there can be only one output value for each input. If you change 5 to 7 *or* change 7 to 5, then there will be only one output value for the input of 1.

3. Yes, it passes the vertical line test.

4. Graph *B* represents a function, while graphs *A* and *C* do not represent functions since they fail the vertical line test.

5. Neither is a function of the other. The graph fails the vertical line test, so weight is not a function of height. At height 51 inches there are two weights (115 and 120 pounds) and at height 56 inches there are two weights (135 and 140 pounds). If we reverse the axes, so that weight is on the horizontal axis and height is on the vertical axis, that graph will also fail the vertical line test, since the weight of 140 pounds has two corresponding heights of 56 inches and 58 inches.

6. a. *D* is a function of *Y*, since each value of *Y* determines a unique value of *D*.

  b. *Y* is not a function of *D*, since one value of *D*, $2.70, yields two values for *Y*, 1993 and 1997.

7. a. Tip = 15% of the meal cost, so $T = 0.15M$. Independent variable: *M* (meal cost); Dependent variable: *T* (tip).

  The equation is a function since to each value of *M* there corresponds a unique value of *T*.

  b. $T = (0.15)(\$8) = \$1.20$

  c. $T = (0.15)(\$26.42) = \$3.96$. One would probably round that up to $4.00.

Exercises for Section 1.3

1. a. Yes: each date has only one output value.

  b. No: for example, $64°$ has two output values.

3. a. function (all input values are different and thus each input has one and only one output)

  b. function (same reason as in part (a))

  c. not a function (same input values have different output values)

  d. not a function (same reason as in part (c))

5. Biotic integrity is not a function of percent of impervious area. For example, the input value of 9% has three different outputs.

7. a. The formulas are: $y = x + 5$; $y = x^2 + 1$; $y = 3$

  b. All three represent *y* as a function of *x*; each input of *x* has only one output *y*.

9. a. $S_1 = 0.90 \cdot P$; $90

  b. $S_2 = (0.90)^2 \cdot P$; $81

  c. $S_3 = (0.90)^3 \cdot P$; $72.90

  d. $S_5 = (0.90)^5 \cdot P$; $59.05; 40.95%

11. *y* is a function of *x* in parts (a), (b), and (c) but not in (d). In (d), for example, if $x = 1$, then $y = \pm 1$.

13. a. Since the dosage depends on the weight, the logical choice for the independent variable is *W* (expressed in kilograms) and for the dependent variable is *D* (expressed in milligrams).

  b. In this formula, each value of *W* determines a unique dosage *D*, so *D* is a function of *W*.

  c. The following table and graph are representations of the function. Since the points $(0, 0)$ and $(10, 500)$ are not included in the model, these points are represented with a hollow circle on the graph.

W (kg)	D (mg)
0	0
2	100
4	200
6	300
8	400

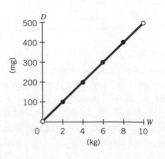

Section 1.4

Algebra Aerobics 1.4a

1. $g(0) = 0$
$g(-1) = -3$
$g(1) = 3$
$g(20) = 60$
$g(100) = 300$

2. $f(0) = (0)^2 - 5(0) + 6 = 6$, so $f(0) = 6$
$f(1) = (1)^2 - 5(1) + 6 = 2$, so $f(1) = 2$
$f(-3) = (-3)^2 - 5(-3) + 6 = 30$, so $f(-3) - 30$

3. $f(0) = \frac{2}{0-1} = -2$, so $f(0) = -2$;
$f(-1) = \frac{?}{(-1)-1} = -1$, so $f(-1) = -1$;
$f(-3) = \frac{2}{(-3)-1} = -\frac{1}{2}$, so $f(-3) = -1/2$

4. $5 - 2t = 3$, so $t = 1$; $3t - 9 = 3$, so $t = 4$; $5t - 12 = 3$, so $t = 3$.

5. $2(x - 1) - 3(y + 5) = 10 \Rightarrow 2x - 2 - 3y - 15 = 10 \Rightarrow$
$2x - 17 - 3y = 10$; $2x - 27 = 3y \Rightarrow y = \frac{2x - 27}{3}$
So y is a function of x. $f(x) = \frac{2x - 27}{3}$.

6. $x^2 + 2x - y + 4 = 0 \Rightarrow x^2 + 2x + 4 = y$.
So y is a function of x. $f(x) = x^2 + 2x + 4$

7. $7x - 2y = 5 \Rightarrow -2y = -7x + 5 \Rightarrow y = (-7x + 5)/(-2) \Rightarrow$
$y = \frac{7x - 5}{2}$
So y is a function of x. $f(x) = \frac{7x - 5}{2}$

8. $f(-4) = 2$; $f(-1) = -1$; $f(0) = -2$; $f(3) = 1$. When $x = -2$ or $x = 2$, then $f(x) = 0$.

9. $f(0) = 20$, $f(20) = 0$, $f(x) = 10$ when $x = 10$ and $x = 30$. It is a function because for every value of x, there is one and only one value of $f(x)$.

10. $f(-3) = -4$ $f(0) \approx -0.4$ $f(1) \approx -0.8$ $f(3) = 3$
$f(x) = 0 \Rightarrow x = -1$ or 2 $f(x) = 1.5 \Rightarrow x \approx 2.6$

Algebra Aerobics 1.4b

1. a. $(2, \infty)$ **b.** $[4, 20)$

2. a. $-3 \leq x < 10$

  b. $-2.5 < x \leq 6.8$

3. a. $[2.5, 3.6]$

  b. $[0.333, 1.000]$ (*Note:* The highest possible batting average is 1.000, meaning that the batter had a hit every time he has been at bat. This is commonly known as "batting a thousand.")

  c. $[35000, 50000]$

4. a. $I(n) = 1500 \cdot 0.04n$

   Domain: all integer values of n with $0 \leq n \leq 10$

   Range: $0 \leq$ dollar amounts $\leq \$600$

   Input: number of years, n

   Output: amount of interest in dollars

b. $I(6) = 360$

c. $n = 4$ if $I(n) = 240$

5. Graph A: domain: $[-2, 2]$;   range: $[0, 3]$
  Graph B: domain: $[-4, 4]$;   range: $[-2, 2]$

Exercises for Section 1.4

1. a. $T(0) = 2, T(-1) = 6, T(1) = 0, T(-5) = 42$

3. a. Tax $= 0.16 \cdot$ Income

  b. Income is the independent variable and Tax is the dependent variable.

  c. Yes, the formula represents a function: for each input there is only one output.

  d. As the Income gets closer and closer to $20,000, the Tax gets closer and closer to $3200. In fact the Tax, at some point will round off to $3200.00, even though the Income is not quite $20,000. Its domain is $[0, 20000)$ and its range is $[0, 3200]$.

5. The equation is $C = 2.00 + 0.32M$. It represents a function. The independent variable is M measured in miles. The dependent variable is C measured in dollars. Here is a table of values.

Miles	Cost ($)	Miles	Cost ($)
0	2.00	30	11.60
10	5.20	40	14.80
20	8.40	50	18.00

The graph is in the accompanying diagram. Some of the table values are marked.

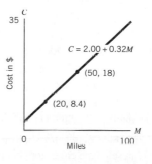

7. a. $f(2) = 4$ **c.** $f(0) = 2$
  b. $f(-1) - 4$ **d.** $f(-5) = 32$.

9. a. $p(-4) = 0.063$, $p(5) = 32$ and $p(1) = 2$.
  b. $n = 1$ only

11. a. Depth could be used as an input function but velocity cannot. Each depth has only one velocity corresponding to it but the reverse is not true since the velocity 0.59, for example, has two depths corresponding to it.

  b. The input variable is **depth**; its units are feet. The output variable is **vel**; its units are ft/sec.

c. The domain goes from 0.7 ft to 11.2 ft, and the range is from 0.22 ft/sec to 1.55 ft/sec.

d. Answers will vary. Here is one: $g(0.7) = 1.55$

13. a. $f(-2) = 5, f(-1) = 0, f(0) = -3$, and $f(1) = -4$.

b. $f(x) = -3$ if and only if $x = 0$ or 2.

c. The range of f is from -4 to $+\infty$ since we may assume that its arms extend out indefinitely.

15. $f(0) = 1, f(1) = 1$, and $f(-2) = 25$

17. a. <u>Deaths</u> is a function of <u>year</u>.

b. Input (with units) is <u>year numbers</u>; output (with units) is <u>deaths in thousands</u>.

c. Domain: <u>year numbers from 1995 to 2006</u>; range: <u>death counts go from 2,310,000 to 2,450,000</u>.

d. $D(2000) \approx \underline{2405}$ thousand deaths. This equation means that in year 2000 there were approximately 2,405,000 deaths in the United States.

e. If $D(t) = 2390$, then $t \approx$ some time in late 1998.

f. $B(2000) - D(2000) = 4050 - 2405 = 1645$. The net change in the population for the United States was 1,645,000 persons.

Section 1.5

Algebra Aerobics 1.5

1. a. "Real Median Household Income Continues to Rise"
"Are Americans Financially Better Off Today?"

b. "Newspaper Advertisement Sales Continue to Decline"
"Is the Internet Replacing the Newspaper?"

2. a. Domain: 1963 to 2009; range: 5.00% to (approximately) 14.8%

b. 6.00%

c. 1968, 1993, 1998, 1999, 2001

d. Maximum value is approximately 14.8%, which occurred in 1982; the point on the graph is (1982, 14.8%).

e. Minimum value is approximately 5%, which occurred in 2009; the point on the graph is (2009, 5%).

3. Graph B is the best match for the situation. It is the only graph that represents the child stopping at the top of the slide with a speed of zero for a few minutes.

4.

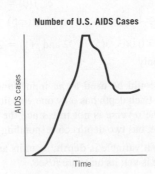

Number of U.S. AIDS Cases

AIDS cases (vertical axis)

Time (horizontal axis)

5. a. A graph that is concave up with minimum at $(-2, 1)$:

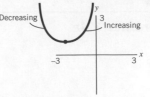

b. A graph that is concave down with maximum at $(3, -2)$:

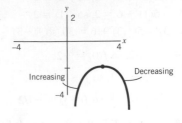

$$f(x) = 0 \Rightarrow x = -4, 0, 4$$

6. Graph A: $f(x) < 0 \Rightarrow 0 < x < 4, x < -4$
$f(x) > 0 \Rightarrow -4 < x < 0, x > 4$

$$f(x) = 0 \Rightarrow x = -3, 2$$

Graph B: $f(x) < 0 \Rightarrow -3 < x < 2$
$f(x) > 0 \Rightarrow x < -3, x > 2$

Exercises for Section 1.5

1. a. Graph B **c.** Graph A

b. Graph A **d.** Graph B

3. a. <u>Mean arctic temperature in Kelvins</u> is a function of <u>the day of the year</u>.

b. Input (with units): <u>days of the year $\{1, 2, 3 \ldots 365\}$</u>. Output (with units): <u>mean arctic temperature in Kelvins</u>.

c. Estimate—domain: <u>$\{1, 2, 3 \ldots 365\}$</u>; range: <u>$(240, 275)$</u>.

d. $T(100) \approx \underline{250 \text{ Kelvins}}$.

Estimate of the interval where $T(d) > 273.15$ is day 160 to day 230. The ice melted in the arctic between days 160 and 230 of the year or from June 10 to Aug 19.

e. The maximum value of $T(d)$ is approximately 275 Kelvins. The minimum value of $T(d)$ is approximately 242.5 Kelvins.

f. Answers will vary, but one might mention that it closely resembles the bell-shaped curve of statistics. But it is not as smooth during the first 50 days of the year.

5. a. Positive over $(-5, 0)$ and $(5, +\infty)$

b. Negative over $(0, 5)$ and $(-\infty, -5)$

c. Decreasing over $(-3, 3)$

d. Increasing over $(-\infty, -3)$ and $(3, +\infty)$

e. There is no minimum.

f. There is no maximum.

7. a. A: domain $= (-\infty, +\infty)$ and range $= (-\infty, 2)$
B: domain $= [0, +\infty)$ and range $= [0, +\infty)$
C: domain $= (-\infty, +\infty)$ and range $= (0, +\infty)$
D: domain $= (-\infty, +\infty)$ and range $= [2, +\infty)$

b. A: $(-5, -1)$, B: $(0, +\infty)$, C: $(-\infty, +\infty)$, D: $(-\infty, +\infty)$

c. A: $(-\infty, -5)$ and $(-1, +\infty)$ C: nowhere

    B: nowhere   D: nowhere

9. a. $[-6, -3)$ and $(5, 11)$

b. $(-3, 5)$ and $(11, 12]$

c. $(-6, 2)$ and $(8, 12)$

d. $(2, 8)$

e. Concave down approximately over interval $(0, 5)$; concave up approximately over interval $(5, 8)$.

f. $f(x) = 4$ when $x = 1$ and 3.

g. $f(-8)$ is not defined.

11. a. Graph B seems best. It indicates several stops, rises and falls in speed and, most importantly, it is the only one that ends with a stop.

b. Graph E is suitable. The horizontal parts on the graph indicate a time at which the bus stopped. Graphs D and F seem out of place. They would indicate that the bus went backwards.

13. a. The maximum rate is about 10%; the minimum rate is about 3%.

b. There was an overall decrease in the unemployment rate from about January 1999 to January 2000 and from June 2003 to January 2007.

c. The unemployment rate seems to be rising the fastest from January 2008 to January 2010.

d. Answers will vary, but one should look for something like the following: The unemployment rate fluctuated up and down from January 1999 to January 2010. In January 1999, it was a little over 4%, but it started to rise from January 2001 to near the middle of 2003 to about 6%. It then steadily went down from January 2004 to January 2007, going as low as about 4.5%. But it rose sharply starting in 2008 to an all-time high of 10% in January 2010.

15. a. $f(-8) = 0, f(0) = 0, f(6) = 0; f(x) > 0$ for $-8 < x < 0$ and $6 < x \le 8; f(x) < 0$ for $0 < x < 6$.

b. Maximum of $f(x)$ is at $(-4, 10)$; minimum of $f(x)$ is at $(3, -2)$.

c. $f(x)$ is increasing when $-8 < x < -4$ and $3 < x < 8$; $f(x)$ is decreasing when $-4 < x < 3$.

d. The domain is $-8 \le x \le 8$; the range is $-2 \le y \le 10$.

17. a. Range of population for Erie is from about 119 to 140 thousand; the range for Trenton is from about 90 to 125 thousand.

b. 1940 to 1960

c. 2000 to 2010

d. In 1948

19. a. From 1750 to 2100.

b. Yes, from 2100 to 2200.

c. Answers will vary; it is hoped that something is said about the dramatic increase and the slowing down noted in parts (a) and (b).

21. a. Both reached their maximum in 1992.

b. The female maximum was about 216,000; the male maximum was about 1,380,000. The female minimum was about 160,000; the male minimum was about 750,000.

c. Male enrollment started at about 1,200,000 in 1990, rose to its maximum in 1992, and then slowly decreased to below 800,000 in 2007. Female enrollment started at about 188,000 in 1990, rose to its maximum in 1992, and then went steadily down to about 173,000 in 1998. It rose a bit to about 178,000 in 2001 and then went steadily down to about 160,000 in 2007.

23.

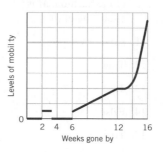

Here is one graph. Phrases like "after a while" and "for a while" were loosely interpreted. As for "levels of mobility," the idea was to give a good picture by means of a graph without being held to numerical values. Time intervals, when explicitly given, were respected. There are discontinuities in the going-to-crutches phases.

25. a. 0 and $+\infty$, respectively

b. $+\infty$ and 0, respectively

27. Clearly one cannot simulate the session here. But one can give the graphs of f and g (see the accompanying diagram) and note that the graph of g is the mirror image of the graph of f with respect to the x-axis.

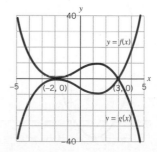

29. a. From 2000 to 2001, from 2003 to 2004, and from 2005 to 2006, the environment priority increased. From 2001 to 2003 and 2008 to 2009, the environment priority decreased. People's attitude about the importance of environmental issues varies over time.

b. Maximum value is about 62% in 2001; minimum value is about 38% in 2003. 62% of the public thought that improving the environment was a major issue in 2001. That interest dropped to 38% in 2003.

c. Answers may vary. There have been four top priorities in the public's mind from 2000 to 2009. The economy has without doubt been the most pressing issue, with jobs running second, the environment running third, and energy fourth. Toward the end of this time span, the most dominant issues were the economy, energy and jobs, and there was a decline in interest in the environment.

Ch. 1: Check Your Understanding

1. True

2. False

3. True

4. True

5. True

6. True

7. False

8. False

9. True

10. False (F is the name of the function; M is a function of q)

11. False

12. True

13. True

14. False

15. True

16. False

17. True

18. True

19. True

20. True

21. False

22. True

23. True

24. Possible answer:

z	1	2	3	4
w	10	-3	9	-3

25. Possible answer:

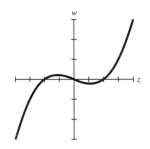

26. Possible answer

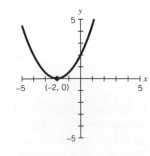

27. Possible answer:

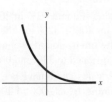

28. Possible answer:

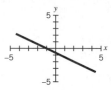

29. The number of federally prosecuted immigration cases has increased since 2000, rapidly outdistancing prosecutions of narcotics/drugs, weapons and white-collar crime cases.

30. False

31. False

32. False

33. True

34. False

35. False

36. True

37. True

38. True

39. False

40. True

Ch. 1 Review: Putting It All Together

1. a. 25%

  b. 20%

  c. 40 lb is a smaller percentage of 200 lb (his weight at age 60) than it is of 160 lb (his weight at age 20).

3. Even if teachers received no pay raise, reducing the number of lower-paid teachers would leave a smaller number of teachers with higher salaries, thereby increasing the mean salary. The money not paid to less experienced workers went toward paying the increases in the salaries of those not laid off. There is no way to tell whether the "average" was a true mean or the median. More information is needed.

5. a. Heart disease

  b.

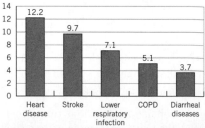

  c. Answers will vary. 62.2% of all deaths are not accounted for in this chart. Cancer is one cause; another is starvation. No disease could account for more than 3.7% because then it would be in the list of top five.

7. $E = 0.025P$

9. One title could be: "The Popularity of Emma as a Baby's Name from 1880 to 2008." That name was very popular in the late nineteenth century, when 9000 out of every million babies were given that name. But as the years went by, its popularity dwindled considerably, reaching an all-time low of about 200 per million in the 1970's. Emma's popularity has come back in the early twenty-first century, with 4500 out of every million babies being named Emma in 2008.

11. **a. i.** Function
   ii. Not a function
   iii. Function
   iv. Not a function
   b, c. Answers will vary.

13. **a.** In mid-February 2008 and again in May and June 2008.
   b. The percent of those who expected the economy to get better was more than the percent of those who expected it not to change at all until about August 2008, and the percent of those who expected it to stay the same was more than those who expected it to get worse after September 2008.
   c. Answers will vary. See part (b). One title could be: "Public opinion on what the economy would be in 6 months underwent several shifts in 2008 to 2009."

15. **a.**

D	1	2	3	4	20	100
Y	7	14	21	28	140	700

   b. $Y = 7D$. D is the independent variable and Y is the dependent variable.
   c.

Y	7	14	21	35	70	700
D	1	2	3	5	10	100

   d. $D = Y/7$ or $D = 0.14Y$; Y is the independent variable and D is the dependent variable.

17. **a.** Yes, it passes the vertical line test. Domain: [1970, 2010]; approx. range: [49, 63].
   b. Approx. 1990; approx. 63 years of age.
   c. Approx. 2003 (projected); approx. 49 years of age. AIDS is a likely candidate.
   d. Increasing from approx. 1970 to 1990 and projected to increase from 2003 to 2010; decreasing from 1990 to 2003.
   e. Concave down over interval (1985, 1995); concave up over interval (1997, 2010).
   f. In 1970 the life expectancy in Botswana was about 52 years, as opposed to about 57 years worldwide. Over the next 20 years this gap gradually closed. In 1990 the life expectancy in Botswana was about 63, and 64 was the worldwide average. However, after 1990 the life expectancy in Botswana began to rapidly decline; this decline was expected to continue until about 2003, at which time it was projected to begin to increase slowly.

19. **a.** $N(1993) \approx 3500$ juvenile arrests for murder. The coordinates are (1993, 3500). This is the maximum point.
   b. (2004, 800) is the approximate minimum point on the graph. It means that in 2004 there were about 800 juvenile arrests for murder and the count for other years is more than this amount.

c. The function is increasing from 1990 to 1993, from 2000 to 2001, and from 2004 to 2007. It is decreasing from 1993 to 2000 and from 2001 to 2004.
d. The domain is $1990 \le x \le 2007$; the range is $800 \le N(x) \le 3500$.
e. Answers may vary. Juvenile arrests went up and down over the period from 1990 to 2007. Juvenile arrests for murder peaked in 1993 and then decreased, leveling off since 2000. In 1990 there were about 2700 arrests for murder. The count rose to 3500 in 1993 and then has gradually diminished to about 1000 in 2007.

21. Answers will vary. Here are some salient features to look for: the percent share of the world's exports changed quite a bit for China and the United States from 1970 to 2010. In 1970 the United States had a 13% share, while China had less than 1%. Since then, the U.S. share has vacillated between 8% and 14%, with a definite downward turn in the 1980's and 2000's reaching a low of 8% in 2010. China's share, on the other hand, has grown steadily; it surpassed the U.S. share in 2006 and in 2010 was approximately 10%. The United States had the largest share of total world exports until 2006. However, since 1970 China's share has increased dramatically, surpassing the U.S. share in 2006.

CHAPTER 2

Section 2.1

Algebra Aerobics 2.1

1. $\dfrac{(143 - 135)\text{ lb}}{5\text{ yr}} = 1.6 \dfrac{\text{lb}}{\text{yr}}$

2. **a. i.** $\dfrac{1826.6 - 1070.6}{8} = \$94.5$ billion/year
   ii. $\dfrac{2522.5 - 1450.0}{8} \approx \$134.1$ billion/year
   iii. $\dfrac{-695.9 - (-379.4)}{8} \approx -\$39.6$ billion/year
   b. While both exports and imports were increasing, imports were increasing at a faster rate; hence, the trade balance was decreasing. The trade deficit increased by almost $40 billion/year from 2000 to 2008.

3. **a.** $\dfrac{41.9 - 52.1}{20} = -0.51$ thousand per year (or, equivalently, decreasing by 510 deaths per year)
   b. $\dfrac{43.3 - 41.9}{8} = 0.175$ thousand per year (or, equivalently, increasing by 175 deaths per year)

4. 60 mph/5 sec = 12 mph per second

5. $\dfrac{800 - 689}{9} \approx \$12.33$ per year

6. $(978 - 1056)/4 = -78/4 = -19.5$. On the average, his performance was declining by 19.5 yards per year.

7. **a.** 1972 to 1989: $\dfrac{19{,}000 - 140{,}000}{17} \approx -7{,}118$ elephants per year (losing on average over 7000 elephants annually)
   1989 to 2009: $\dfrac{30{,}000 - 19{,}000}{20} = 550$ elephants per year (gaining on average 550 elephants annually)
   b. Let n = number of years since 1989, with an initial value of 19,000; so 140,000 = 19,000 + 550n, then n = 220 years.

Exercises for Section 2.1

1. a. inches/pound   **b.** minutes/inch   **c.** pounds/inch

3. 212 miles/10.8 gal. ≈ 19.6 miles per gallon

5. $\frac{4.6 - 10.9}{2006 - 1990} = \frac{-6.3}{16} \approx -0.39$ pounds per person per year.

7. a. $\frac{499 - 498}{2008 - 2000} = \frac{1}{8}$ points per year

    b. $\frac{498 - 504}{2008 - 2000} = \frac{-6}{8}$ or $\frac{-3}{4}$ points per year

9. a. i. $\frac{13,600,000 - 630,000}{2005 - 1985} = \frac{12,970,000}{20} = 648,500$ computers per year

    ii. $\frac{4.0 - 84.1}{2005 - 1985} = \frac{-80.1}{20} \approx -4$ students per computer per year

    b. i. $\frac{14,165,000 - 13,600,000}{2006 - 2005} = \frac{565,000}{1} = 565,000$ computers per year

    ii. $\frac{3.9 - 4.0}{2006 - 2005} = \frac{-0.1}{1} = -0.1$ students per computer per year

    c. $4 - (0.1)t = 2$ or $t = 20$ years.

11. $\frac{7.5 - 6.5}{2009 - 2004} = \frac{1.0}{5} = 0.2$ hours/day/year = 12 hours/day/year

13. a. For whites: $\frac{87.1 - 26.1}{2008 - 1940} = \frac{61}{68} \approx 0.90$ percentage points per year

    For blacks: $\frac{83.0 - 7.3}{2008 - 1940} = \frac{75.7}{68} \approx 1.11$ percentage points per year

    For Asian Pacific Islanders: $\frac{88.7 - 22.6}{2008 - 1940} = \frac{66.1}{68} \approx 0.97$ percentage points per year

    For all: $\frac{86.6 - 24.5}{2008 - 1940} = \frac{62.1}{68} \approx 0.913$ percentage points per year

    b. For whites: $87.1 + 2 \cdot 0.897 \approx 88.9\%$

    For blacks: $83.0 + 2 \cdot 1.113 \approx 85.2\%$

    For Asian Pacific Islanders: $88.7 + 2 \cdot 0.972 \approx 90.6\%$

    For all: $86.6 + 2 \cdot 0.913 \approx 88.4\%$

    c. For x years since 2008:

    For whites: $100 = x \cdot 0.90 + 87.1 \Rightarrow x = (100 - 87.1)/0.90 \approx 14.3$ years

    For blacks: $100 = x \cdot 1.11 + 83.0 \Rightarrow x = (100 - 83.0)/1.11 \approx 15.3$ years

    For Asian Pacific Islanders: $100 = x \cdot 0.97 + 88.7 \Rightarrow x = (100 - 88.7)/0.97 \approx 11.6$ years

    These numbers make some sense but an absolute 100% seems a bit unrealistic.

    d. There was a major increase in the percentage of those who completed 4 years of high school or more in the 68 years since 1940 in all four categories. Over the 68-year period, the increases ranged from 61 percentage points for whites to 75.7 percentage points for blacks. Blacks had the highest rate of increase at 1.113 percentage points per year, and whites had the slowest rate at 0.897 percentage points per year. Other comments could be made.

15. a. White females had the highest life expectancy in 1900 as well as in 2010. Black males had the lowest life expectancy in 1900 as well as in 2010.

    b. For white males: $\frac{76.5 - 46.6}{2010 - 1900} = \frac{29.9}{110} \approx 0.27$ life expectancy years/year

For white females: $\frac{81.3 - 48.7}{2010 - 1900} = \frac{32.6}{110} \approx 0.30$ life expectancy years/year

For black males: $\frac{70.2 - 32.5}{2010 - 1900} = \frac{37.7}{110} \approx 0.34$ life expectancy years/year

For black females: $\frac{77.2 - 33.5}{2010 - 1900} = \frac{43.7}{110} \approx 0.40$ life expectancy years/year

Black females had the largest rate of change in life expectancy from 1900 to 2010.

    c. Life expectancies went up significantly for white and black, males and females from 1900 to 2010. The average rates of change ranged from 0.27 years of life expectancy per year for white males to 0.40 years of life expectancy per year for black females over this period.

Section 2.2

Algebra Aerobics 2.2

1. a.

World Population

Year	Total Population (in millions)	Average Rate of Change over Prior 50 yrs.
1800	980	n.a.
1850	1260	$\frac{(1260 - 980) \text{ million}}{(1850 - 1800)} = \frac{280 \text{ million}}{50 \text{ yrs}}$ = 5.6 million/yr
1900	1650	$\frac{(1650 - 1260) \text{ million}}{(1990 - 1850)} = \frac{390 \text{ million}}{50 \text{ yrs}}$ = 7.8 million/yr
1950	2520	$\frac{(2520 - 1650) \text{ million}}{(1950 - 1900)} = \frac{870 \text{ million}}{50 \text{ yrs}}$ = 17.4 million/yr
2000	6090	$\frac{(6090 - 2520) \text{ million}}{(2000 - 1950)} = \frac{3570 \text{ million}}{50 \text{ yrs}}$ = 71.4 million/yr
2050	9317	$\frac{(9317 - 6090) \text{ million}}{(2050 - 2000)} = \frac{3227 \text{ million}}{50 \text{ yrs}}$ = 64.54 million/yr

b.

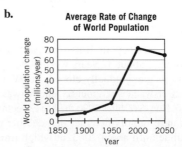

Average Rate of Change of World Population

c. During 1950–2000, average annual rate of change = 71.4 million/yr was the greatest.

d. The average rate of change increased in the time interval 1800 to 2000, yet despite a projected increase in world population from 2000 to 2050, the rate of change is projected to decrease to 64.54 million/yr.

2. From 2006 to 2007 the graph rises, then rises again from 2007 to 2008. This indicates that profits are increasing. It is flat from 2008 to 2009, meaning no increase in profits, and falls from 2009 to 2010, meaning a decrease in profits (which is not necessarily a loss.)

3. a. High School Completers

Year	Number (thousands)	Average Rate of Change (thousands/yr)
1960	1679	n.a.
1970	2757	107.8
1980	3089	33.2
1990	2355	−73.4
2000	2756	40.1
2007	2955	28.4

b. The number of individuals completing high school each year rose between 1960 and 1980, when it peaked at 3,089,000. The 1960s had the greatest rate of change, with an annual increase of 107.8 (thousand) per year. The rate slowed during the 1970s.

In the decade from 1980 to 1990, the number of high school completers each year showed a drastic decline on average of 73,400 per year. The trend reversed in the next decade (1990 to 2000), increasing on average by 40,100 per year. Between 2000 and 2007 the numbers increased on average by 28,400 per year.

c. If the average rate of change is positive, there is an increase in the number of high school completers. An example would be from 1970 to 1980, where the number of high school completers increased from 2,757,000 to 3,089,000.

d. If the average rate of change is negative, there is a decrease in the number of high school completers. An example would be from 1980 to 1990, where the number of high school completers decreased from 3,089,000 to 2,355,000.

e. The growth is slowing down.

Exercises for Section 2.2

1.

x	$f(x)$	Avg. Rate of Change
0	0	n.a.
1	1	1
2	8	7
3	27	19
4	64	37
5	125	61

a. The function is increasing.

b. The average rate of change is also increasing.

3. a.

Year	Registered Motored Vehicles (millions)	Annual Average Rate of Change (over prior decade)
1970	108	n.a.
1980	156	4.8
1990	189	3.3
2000	218	2.9
2010	270	5.2

b. The decade from 1990 to 2000 had the smallest rate of change.

c. The decade from 2000 to 2010 had the largest rate of change.

d. Student answers will vary. Their summaries might include some of the following facts. The average rate of change in the registered motor vehicle rate of change decreased each decade from 1970 to 2000, but rose sharply in the decade from 2000 to 2010. The number of registered vehicles in 2010 more than doubled from 1970 to 2010, increasing from 108 million to 270 million registered vehicles.

5.

I.

x	$f(x)$	Average Rate of Change
0	5	n.a.
10	25	2
20	45	2
30	65	2
40	85	2
50	105	2

a. $f(x)$ is increasing.

b. The average rate of change is constant.

II.

x	$g(x)$	Average Rate of Change
0	270	n.a.
10	240	−3
20	210	−3
30	180	−3
40	150	−3
50	120	−3

a. $g(x)$ is decreasing.

b. The average rate of change is constant.

7. a. Graph B

b. Graph A

c. Graph C

9. a. Table A

x	y	Average Rate of Change
0	2	n.a.
1	5	3
2	8	3
3	11	3
4	14	3
5	17	3
6	20	3

b. The average rate of change is constant.

c. Its graph is a straight line.

Table B

x	y	Average Rate of Change
0	0	n.a.
1	0.5	0.5
2	2	1.5
3	4.5	2.5
4	8	3.5
5	12.5	4.5
6	18	5.5

b. The average rate of change is increasing.

c. Its graph is concave up.

11. a. In 1920 there were $\frac{27,791,000}{106,000,000} = 0.3$ copy of newspapers printed per person (about one-third of a copy per person, or equivalently, roughly one copy for every three people).

In 2000 there were $\frac{55,800,000}{281,400,000} = 0.2$ copy of newspapers printed per person (about one-fifth of a copy per person, or equivalently, about one copy for every five people).

b.

Newspapers

Year	Average Rate of Change
1950	n.a.
1960	−0.9
1970	−1.5
1980	−0.3
1990	−13.4
2000	−13.1
2008	−9

Commercial TV Stations

Year	Average Rate of Change
1950	n.a.
1960	41.7
1970	16.2
1980	5.7
1990	35.8
2000	15.6
2008	13.1

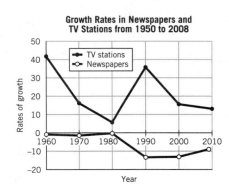

Growth Rates in Newspapers and TV Stations from 1950 to 2008

c. There would be $1353 + 13.1 \cdot (2) = 1379$ TV stations in 2010.

d. The number of U.S. newspapers decreased during this period, while the number of commercial TV stations increased. The number of TV stations increased at the largest rate for the decade from 1950 to 1960 and again from 1980 to 1990. The number of newspapers decreased by the largest rate during the decade from 1980 to 1990. While the number of TV stations increased, other news sources such as the Internet are not mentioned.

Section 2.3

Algebra Aerobics 2.3

1. a. **i.** $m = (11 - 1)/(8 - 4) = 10/4 = 2\frac{1}{2}$
  ii. $m = (6 - 6)/[2 - (-3)] = 0$
  iii. $m = [-3 - (-1)]/[0 - (-5)] = -\frac{2}{5}$
b. **i.** $m = (1 - 11)/(4 - 8) = (-10)/(-4) = 2\frac{1}{2}$
  ii. $m = (6 - 6)/(-3 - 2) = 0$
  iii. $m = [-1 - (-3)]/(-5 - 0) = 2/(-5) = -\frac{2}{5}$

2. Between 1999 and 2002 and between 2006 and 2007 the slope is approximately zero.

3. positive: 2001–2002, 2004–2008

negative: 2003–2004, 2008–2009

zero: 2000–2001, 2002–2003

4. $\frac{\Delta y}{\Delta x} = 4 \Rightarrow \frac{y - (-2)}{5 - 3} = 4 \Rightarrow \frac{y + 2}{2} = 4 \Rightarrow y + 2 = 8$
$\Rightarrow y = 6.$

5. $\frac{\Delta y}{\Delta x} = \frac{(9 + 2h) - 9}{(2 + h) - 2} = \frac{2h}{h} = 2; h \neq 0$

6. a. Points lie on the graph $\Rightarrow$ the coordinates (x, y) satisfy the equation $y = x^2$ since

$$0 = 0^2, \quad 1 = 1^2, \quad 4 = 2^2, \quad 9 = 3^2.$$

b. P_1 and P_2: $m = \frac{1 - 0}{1 - 0} = 1$; P_2 and P_3: $m = \frac{4 - 1}{2 - 1} = 3$;
P_3 and P_4: $m = \frac{9 - 4}{3 - 2} = 5.$

c. The positive slopes suggest that the function increases between $x = 0$ and $x = 3$; because the slopes increase in size as we move further to the right, the graph rises at an increasing rate.

7. $\frac{y_2 - y_1}{x_2 - x_1} = \frac{(-1)}{(-1)} \frac{(y_2 - y_1)}{(x_2 - x_1)} = \frac{-y_2 + y_1}{-x_2 + x_1} = \frac{y_1 - y_2}{x_1 - x_2}$

Exercises for Section 2.3

1. a. $m = \frac{3 - (-6)}{2 - (-5)} = \frac{9}{7}$ **b.** $m = \frac{-3 - 6}{2 - (-5)} = \frac{-9}{7}$

3. a. Graph A crosses the y-axis at $(0, -4)$ and goes through $(1, -4)$. So the slope is $\frac{(-4) - (-4)}{1 - 0} = 0.$

b. Graph B crosses the x-axis at $(2, 0)$ and the y-axis at $(0, -8)$. So the slope is $\frac{-8 - 0}{0 - 2} = 4.$

5. a. slope of segment $A = [2 - (-6)]/[-4 - (-8)] = 8/4 = 2$
slope of segment $B = [-8 - 2]/[0 - (-4)] = -10/4 = -2.5$
slope of segment $C = [-8 - (-8)]/[2 - 0] = 0/2 = 0$
slope of segment $D = [-6 - (-8)]/[6 - 2] = 2/4 = 0.5$
slope of segment $E = [6 - (-6)]/[10 - 6] = 12/4 = 3$
slope of segment $F = [-16 - 6]/[12 - 10] = -22/2 = -11$

b. The slope of segment F is steepest.

c. Segment C has a slope equal to 0.

7. a. $m = 75/10 = 7.5$

b. $70 - y = -4 \Rightarrow y = 74$

c. $32 = 4(28 - x) \Rightarrow 32 = 112 - 4x \Rightarrow x = 20$

d. $6 = 0.6(x - 10) \Rightarrow 12 = 0.6x \Rightarrow x = 20$

9. a. $-4 = (1 - t)/(-2 - 3) \Rightarrow 20 = 1 - t \Rightarrow t = -19$

b. $2/3 = (9 - 6)/(t - 5) \Rightarrow 2(t - 5) = 9 \Rightarrow t - 5 = 4.5 \Rightarrow$
$t = 9.5$

11. a. $m_1 = (7 - 3)/(4 - 2) = 4/2 = 2; m_2 = (15 - 7)/(8 - 4) = 2;$
collinear

b. $m_1 = (4 - 1)/(2 + 3) = 3/5; m_2 = (8 - 1)/(7 + 3) = 0.7;$
not collinear

13. a. $m = \frac{-2 - 0}{6 - 0} = -\frac{1}{3}$ **b.** $m = \frac{7 - 0}{-4 - 0} = -\frac{7}{4}$

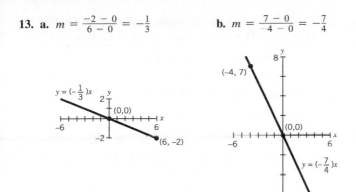

15. a. $\frac{0 - \sqrt{2}}{\sqrt{2} - 0} = -1$

b. $\frac{0 + 3/2}{-3/2 - 0} = -1$

c. $\frac{0 - b}{b - 0} = -\frac{b}{b} = -1$

d. All slopes are -1, since each pair of points is of the form $(0, a)$ and $(a, 0)$.

17. The points in parts (a) and (c).

19. a. 1/10

b. 1/12

c. old: $\frac{3}{\text{run}} = \frac{1}{10} \Rightarrow \text{run} = 30$ ft.; new: $\frac{3}{\text{run}} = \frac{1}{12} \Rightarrow \text{run} = 36$ ft.

21. Student answers will vary.

Section 2.4

Algebra Aerobics 2.4

1. a. (1960, 22.2) and (2008, 13.2)

b. (1980, 13.0) and (2008, 13.2)

c. (2000, 11.3) and (2008, 13.2)

2. a. Between 2008 and 2009 the stock price surged from $1.02 to $1.12 per share, a 9.8% increase or, equivalently, an increase of $0.10 per share. See Graph A.

b. Between 2009 and 2010 the stock price dropped drastically from $1.12 to $1.08 per share, about a 3.6% decrease or, equivalently, a decrease of $0.04 per share. See Graph B.

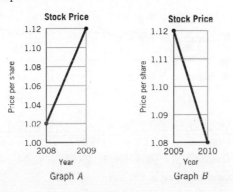

3. a. You could draw a graph that is cropped and stretched vertically, as shown in Graph A.

b. You could show a graph that is not too steep (stretched horizontally or compressed vertically) and emphasize the decline in the number of casualties in week three, as in Graph B.

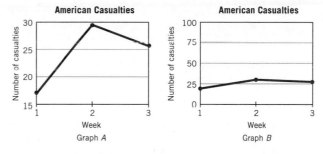

4. Some of the strategies are: use of dramatic language (the title "Gold Explodes" in big bold type, followed by "Experts Predict $1,500.00 an Ounce"; cropped vertical axis on graph (it starts at $300, not $0, making the graph look steeper); use of powerful graphics (the arrow that increases in size), suggesting dynamic growth.

Exercises for Section 2.4

1. a. From the second week of April to the beginning of May, mortgage rates have dramatically declined.

b. From the second week of March to the beginning of May, mortgage rates have "exploded."

c. From the beginning of April to the beginning of May, mortgage rates have stayed roughly the same.

In all of these cases one draws a line from the starting point on the graph to the end point on the graph; that line's slope is used to argue your point.

3. a. Graph B appears to have the steeper slope.

b. The slope of Graph A is -6 and the slope of Graph B is -4. So Graph A actually has the steeper slope.

5. a. In both grade 4 and grade 8, California math scores steadily rose over the period from 2000 to 2007. However, the California scores at both grade levels are consistently about 10 points below the national scores.

b. If one looks only at the scores from 2003 to 2007, the increases seem modest. If one looks only at the scores from 2005 to 2007, these increases seem even more modest (smaller positive slope). One could also choose different values on the vertical axis such as 0 to 500, to make the difference seem smaller.

c. One could choose smaller intervals on the vertical axis, or stretch the vertical axis. If one looks at the scores from 2000 to 2003, the scores increase more dramatically (larger positive slope).

7. **a.**

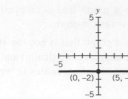

   (0, –2)   (5, –2)

b, c. The task is impossible. The slope of the line is 0, so it is not possible to make a graph through the given points appear to have a large positive slope.

9. "Precipitous," "dire," and "catastrophic" come to mind. Students will probably think of others.

11. Answers may vary.

   a. The number of persons with AIDS who are still alive has been steadily growing by approximately $\frac{575{,}000 - 200{,}000}{2008 - 1994} = \frac{375{,}000}{14} \approx 26{,}785$ persons per year. Another encouraging fact would be the steady decline in new AIDS cases from 1992 to 2008 at a rate of $\frac{40{,}000 - 75{,}000}{2008 - 1992} = \frac{-35{,}000}{16} = -2187.5$ persons per year.

   b. Despite the progress in treating AIDS and education about it, the number of cases diagnosed since 2000 has not decreased. Also, the rate of deaths from AIDS per year sharply declined from 1994 to 1996. The rate was $\frac{20{,}000 - 50{,}000}{1996 - 1994} = -15{,}000$ deaths per year; meaning a decrease of 15,000 deaths per year; however, the rate of decline has slowed down considerably since 1995 to about $\frac{15{,}000 - 20{,}000}{2008 - 1996} = \frac{-5{,}000}{12} \approx -417$ deaths per year, or a decrease of 417 deaths per year.

   c. AIDS cases increased until 1993, when new treatments, public awareness, education, and testing for the disease may have caused a decrease in the number of new cases.

13. Graph *A* gives the appearance that the percentage has declined quite a bit, while Graph *B* gives the impression that the percentage has not declined much at all. The difference is in the vertical scale: it is greatly magnified in Graph *A*.

15. **a.** From 1973 to 2008 the NAEP math scores for 17-year-old boys were higher than those scores for girls except in 1990, 1999, and 2004, when they were nearly the same. Overall, the scores have not risen or fallen much.

   b. One could change the vertical scale from, say, 290 to 310 to make the gap seem larger.

Section 2.5

Algebra Aerobics 2.5

1. The weight of a 4.5-month-old baby girl appears to be ≈ 14 lb. From the equation, the exact weight is
$W = 7.0 + 1.5(4.5) = 13.75$ lb. Our estimate is within 0.25 lb of the exact weight.

2. The age appears to be ≈ 2.5 months. Solve $11 = 7 + 1.5A \Rightarrow$ $11 - 7 = 1.5A \Rightarrow 4 = 1.5A \Rightarrow A = 2.67$ months.

3. **a.** The units for 15 are dollars/person. The units for 10 are dollars.

   b. dollars = (dollars/person)(persons) + dollars

4. **a.** The units for 1200 are dollars and for 50 are dollars/month.

   b. dollars = dollars + $\left(\frac{\text{dollars}}{\text{month}}\right)$months

5. **a.** 0.8 million dollars per year is the slope or average rate of change in the sales per year.

   b. 19 million dollars represents the sales this year.

   c. In 3 years $S = 0.8(3) + 19 = 21.4$ million dollars

   d. dollars = $\left(\frac{\text{dollars}}{\text{year}}\right)$years + dollars where all values for dollars are in millions.

6. **a.** The average cost to operate a car is $0.45 per mile, and the units are dollars per mile.

   b. $0.45(25{,}000) = \$11{,}250$

   c. dollars = $\left(\frac{\text{dollars}}{\text{mile}}\right)$miles

7. **a.** $840

   b. The beginning mortgage is $302,400.

   c. After 10 years, $B = \$302{,}400 - 840(12)(10) = \$201{,}600$;
After 20 years, $B = \$302{,}400 - 840(12)(20) = \$100{,}800$;
After 30 years, $B = \$302{,}400 - 840(12)(30) = \$0$.

8. **a.** $m = 5, b = 3$

   b. $m = 3, b = 5$

   c. $m = 5, b = 0$

   d. $m = 0, b = 3$

   e. $m = -1, b = 7.0$

   f. $m = -11, b = 10$

   g. $m = -2/3, b = 1$

   h. $m = 5, b = -3$ since $2y = 10x - 6 \Rightarrow y = 5x - 3$

9. **a.** $f(x)$ is a linear function because it is represented by an equation of the form $y = mx + b$, where here $b = 50$ and $m = -25$.

   b. $f(0) = 50 - 25(0) = 50 - 0 = 50$
$f(2) = 50 - 25(2) = 50 - 50 = 0$

   c. Since the line passes through (0, 50) and (2, 0), the slope m is: $m = \frac{50 - 0}{0 - 2} = -25$.

10. **a.** linear: $m = 3, b = 5$

   b. linear: $m = 1, b = 0$

   c. not linear

   d. linear: $m = -2/3, b = 4$

11. **a.** $y = 4 + 3x$ **b.** $y = -x$ **c.** $y = -3$

12. Graph A: $y = 3 + 4x$
Graph B: $y = -2x + 0 = -2x$
Graph C: $y = 0x + 3 = 3$
Graph D: $y = x + 0 = x$

Exercises for Section 2.5

1. a. 300, 350, 1300

  b. (0, 300), (1, 350), (20, 1300)

3. a. (300, 0) is not a solution to either equation

  b. (10, 850) is a solution to B

  c. (15, 1050) is a solution for both A and B

5. a. $D = 3.40, 3.51, 3.62, 3.73,$ and 3.84, respectively

  b. 0.11 is the slope; it represents the average rate of change of the average consumer debt per year; it is measured in thousands of dollars per year.

  c. 3.40 represents the average consumer debt when $n = 0$ years. It is measured in thousands of dollars.

7. dollars = dollars + $\frac{\text{dollars}}{\text{year}} \cdot$ years

9. a. $C(0) = \$11.00$; $C(5) = 11 + 10.50 \cdot 5 = \$63.50$; $C(10) = 11 + 10.50 \cdot 10 = \$116.00$

  b. \$11.00

  c. \$10.50 for every thousand cubic feet.

  d. $C(96) = \$1019$

11. a. 0; 4   **b.** 0; π **c.** 0; 2π **d.** -17.78; 5/9

13. a. matches **f**, since \$75 seems the most likely cost for producing one text.

  b. matches **g**, since \$0.30 seems the most likely cost to produce one CD.

  c. matches **e**, since \$800 seems the most likely cost of producing one computer.

15. a. $C(p) = 1.06 \cdot p$

  b. $C(9.50) = 1.06 \cdot 9.50 = 10.07$; $C(115.25) = 1.06 \cdot 115.25 = 122.17$ (rounded up); $C(1899) = 1.06 \cdot 1899 - 2012.94$. All function inputs and outputs are measured in dollars.

17. a. hours

  b. miles/gallon

  c. calories/grams of fat

19. a. Slope $= 0.4$, vertical intercept $= -20$

  b. Slope $= -200$, vertical intercept $= 4000$

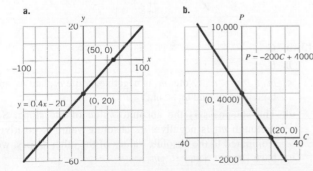

21. The equation is: $y = 3x - 2$

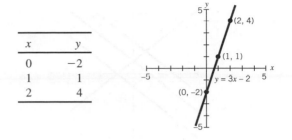

x	y
0	-2
1	1
2	4

23. The equation is $y = 0 \cdot x + 1.5 = 1.5$.

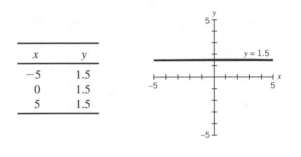

x	y
-5	1.5
0	1.5
5	1.5

Section 2.6

Algebra Aerobics 2.6

1. $0, |-1|, |-3|, 4, |-7|, 9, |-12|$

2. a. The graph of $y = 6x$ goes through the origin and is two units above the graph of $y = 6x - 2$.

  b. The graph of $y = 2 + 6x$ intersects the y-axis at (0, 2). The lines are parallel, so $y = 2 + 6x$ is four units above the graph of $y = 6x - 2$.

  c. The graph of $y = -2 + 3x$ has the same y-intercept (0, -2) but is less steep than the line $y = 6x - 2$.

  d. The graph of $y = -2 - 2x$ has the same y-intercept (0, -2) but is falling left to right, whereas $y = 6x - 2$ is rising and has a steeper slope.

3. For (a), $m = -2$, so $|m| = |-2| = 2$

  For (b), $m = -1$, so $|m| = |-1| = 1$

  For (c), $m = -3$, so $|m| = |-3| = 3$

  For (d), $m = -5$, so $|m| = |-5| = 5$

Line d is the steepest. In order from least steep to steepest, we have b, a, c, d.

4. Answers will vary depending on the original $f(x)$. For example, if $f(x) = 2 + 2x$, then we could have:

  a. $g(x) = 5 + 2x$

  b. $h(x) = -2 + 2x$

  c. $k(x) = 2 - 2x$

See graph below

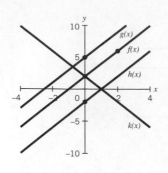

5. For $f(x) = 3x - 5$, $m = 3$.
   For $g(x) = 7 - 8x$, $m = -8$.
   $g(x)$ has a steeper slope than $f(x)$ since $|-8| > |3|$

x	$f(x)$	$g(x)$
-4	-17	39
-2	-11	23
0	-5	7
2	1	-9
4	7	-27

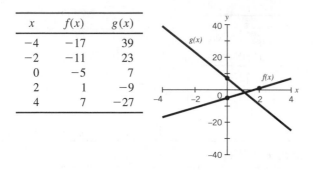

6. Answers will vary. For example,
   $f(x) = 4 - 2x$
   $g(x) = 4 - 5x$
   $h(x) = 4 - 7x$
   $h(x)$ has the steepest slope since $|-7| > |-5| > |-2|$

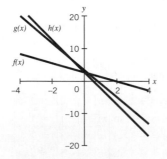

7. $f(x)$ matches Graph B, $g(x)$ matches Graph D, $h(x)$ matches Graph A, $k(x)$ matches Graph C.

Exercises for Section 2.6

1. a. Graph B **c.** Graph C
   b. Graphs A, B, and D **d.** Graphs A and D

3. Answers may vary for (b) and (c).
   a. $y = -3$ **b.** $y = x - 3$ **c.** $y = -3x + 1$

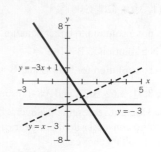

5. a. matches Graph B
   b. matches Graph D
   c. matches Graph C
   d. matches Graph A

7. Graphs may vary.
   a. Same slope   **b.** Same vertical intercept

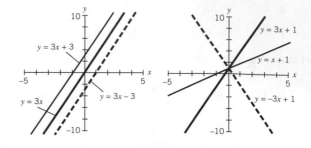

9. a. $R(t) = 5 + 13 - 5t = 18 - 5t$
   b. $S(t) = -3 + 13 - 5t = 10 - 5t$
   c. $T(t) = 13 + 2t$ (slope may vary)
   d. $U(t) = 12 - 5t$ (vertical intercept may vary)
   e. $V(t) = 15 + 5t$ (vertical intercept may vary)

11.

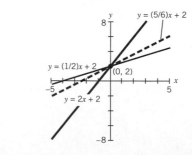

13. a. C
   b. B
   c. $m_3 < m_2 < m_1$ (Note: m_3 is negative, while m_2 and m_1 are both positive and $m_2 < m_1$.)
   d. The steepness is the absolute value of the slope. So although m_3 is negative, its absolute value is positive. Compared to the absolute values of the other slopes, we have $|m_2| < |m_1| < |m_3|$.

Section 2.7

Algebra Aerobics 2.7

1. a. $y = 1.2x - 4$

b.

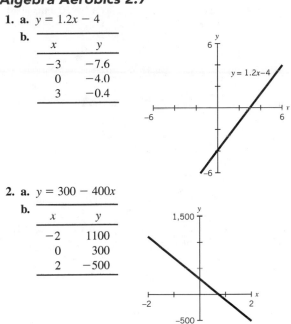

x	y
-3	-7.6
0	-4.0
3	-0.4

2. a. $y = 300 - 400x$

b.

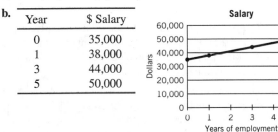

x	y
-2	1100
0	300
2	-500

3. The vertical intercept is at $(0, 1)$, so $b = 1$. The line passes through $(0, 1)$ and $(3, -5)$, so $m = \frac{1 - (-5)}{0 - 3} = -2$. The equation is: $y = -2x + 1$.

4. a. $S = \$35,000 + \$3,000x$

b.

Year	$ Salary
0	35,000
1	38,000
3	44,000
5	50,000

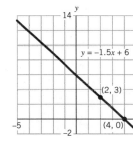

5. 3

6. a. $4; (1, 4)$ **b.** $4; (-1, 12)$

7. Graph A: $y = 6 - 2x$
Graph B: $Q = 3t + 2$

8. a. Any equation of the form $y = 6 + mx$ with three different values for m; some examples are $y = 6 + 2x$, $y = 6 - 5x$, and $y = 6 + 11x$.

b. Any equation of the form $y = -3x + b$ with three different values for b; some examples are $y = -3x + 5$, $y = -3x$, $y = -3x - 12$.

9. a. $y = 4(x + 2) + 1 \Rightarrow y = 4x + 9$

b. $y = (-2/3)(x - 3) + 5 \Rightarrow y = (-2/3)x + 7$

c. $y = -10(x + 1) + 3 \Rightarrow y = -10x - 7$

d. $y = 2(x - 1.2) + 4.5 \Rightarrow y = 2x + 2.1$

10. a. $m = \frac{11 - 5}{4 - 2} = 3$
$y = m(x - x_1) + y_1 \Rightarrow y = 3(x - 2) + 5 \Rightarrow y = 3x - 1$

b. $m = \frac{1 - 2}{6 - (-3)} = -\frac{1}{9}$
$y = m(x - x_1) + y_1 \Rightarrow y = -\frac{1}{9}(x - 6) + 1 \Rightarrow$
$y = -\frac{1}{9}x + \frac{5}{3}$

c. $m = \frac{-1 - (-7)}{4 - (-2)} = 1$
$y = m(x - x_1) + y_1 \Rightarrow y = 1(x - 4) + (-1) \Rightarrow$
$y = x - 5$

Exercises for Section 2.7

1. a. Slope $= 5$, so equation is of the form $y = 5x + b$. The line passes through $(-2, 3)$,
$\Rightarrow 3 = 5(-2) + b$
$\Rightarrow b = 13$. So $y = 5x + 13$

b. Slope $= -\frac{3}{4}$, so equation is of the form $y = -\frac{3}{4}x + b$. The line passes through $(-2, 3)$,
$\Rightarrow 3 = -\frac{3}{4}(-2) + b$
$\Rightarrow b = 1.5$. So $y = -0.75x + 1.5$

c. Slope $= 0$, so equation is of the form $y = b$. The line passes through $(-2, 3)$,
$\Rightarrow 3 = b$. So the equation is the horizontal line $y = 3$

3. a. $m = (5.1 - 7.6)/(4 - 2) = -2.5/2 = -1.25$ and
$y - 7.6 = -1.25(x - 2)$ or $y = -1.25x + 10.1$

b. $m = (16 - 12)/(7 - 5) = 4/2 = 2$ and $W - 12 = 2(A - 5)$
or $W = 2A + 2$

5. a. $y = -1.5x + 6$ **b.** $y = -0.5x + 2$

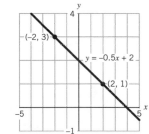

c. $y = -x + 2$ **d.** $y = 2$

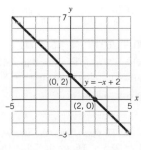

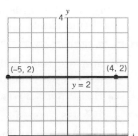

7. a. $y = -2 + \frac{2}{3}x$; the slope is 2/3 and the y-intercept is -2.

b. $y = 3 - \frac{3}{2}x$; the slope is $-3/2$ and the y-intercept is 3.

c. $y = 12 - \frac{2}{3}x$; the slope is $-2/3$ and the y-intercept is 12.

d. $y = \frac{3}{2}x$; the slope is 3/2 and the y-intercept is 0.

e. $y = \frac{3}{2}x$; the slope is 3/2 and the y-intercept is 0. So the equation is equivalent to the one in part (d).

f. $y = \frac{1}{4} + \frac{3}{4}x$; the slope is 3/4 and the y-intercept is 1/4.

9. a.

x	$f(x) = 0.10x + 10$
-100	0
0	10
100	20

Graph of $f(x)$:

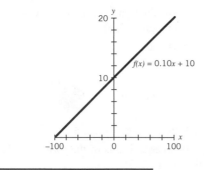

b.

x	$h(x) = 50x + 100$
-0.5	75
0	100
0.5	125

Graph of $h(x)$:

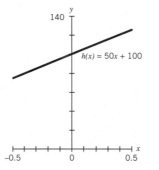

11. $C(n) = 150 + 250n$, where $C(n)$ is the cost of taking n credits measured in dollars and 250 is measured in dollars per credit.

13. $C(n) = 2.50 + 0.10n$, where $C(n)$ is the cost of cashing n checks that month.

15. a. Annual increase $= (32{,}000 - 26{,}000)/4 = \$1500$

b. $S(n) = 26{,}000 + 1500n$, where $S(n)$ is measured in dollars and n in years from the start of employment.

c. Here $0 \leq n \leq 20$ since the contract is for 20 years.

17. a. $P = 285 - 15t$

b. It will take approximately 16.33 years to make the water safe for swimming since $40 = 285 - 15t$ implies that $15t = 245$ or $t \approx 16.33$ years.

19. a. $V(t) = \frac{780{,}000}{30}(t - 1977) + 70{,}000$, where t is measured in years and $V(t)$ in dollars.

b. $V(2010) = \frac{780{,}000}{30}(2010 - 1977) + 70{,}000$; so $V(2010) = \$928{,}000$

21. The equation of the line illustrated is $y = (8/3)x - 4$. The graph in the text can be altered to appear much steeper if the graph is stretched vertically—as in the accompanying graph—where the tick marks on the x-axis are 10 units apart.

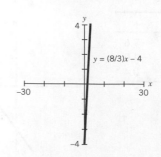

If, however, we reversed the situation and vertically compressed the graph (as in the one below, where the tick marks on the y-axis are now 10 units apart), the graph would seem less steep than the one in the text.

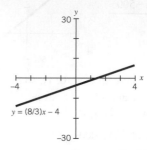

23. a. The percentage of homeowners with negative equity is $P(t) = 2.5t + 6.5$, where t is the number of quarters since the 4th quarter of 2006 and where $P(t)$ and 6.5 are measured in percentage points.

b. The domain is $0 \leq t \leq 8$, where t is the number of quarters since the 4th quarter of 2006.

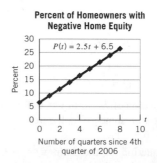

Percent of Homeowners with Negative Home Equity

c. The 4th quarter of 2008 is 8 quarters since the 4th quarter of 2006. So $P(8) = 2.5(8) + 6.5 = 26.5$ percent of homeowners had negative equity at the end of 2008.

25. The entries in the table argue that the relationship is linear. The average rate of change in salinity per degree Celsius is a constant: -0.054. Since the freezing point, P, for 0 salinity is 0 degrees Celsius, we have that $P = -0.054S$, where S is the salinity measured in ppt and P, the freezing point, is measured in degrees Celsius.

Section 2.8

Algebra Aerobics 2.8a

1. If $(0, 0)$ is on each line, then $b = 0$ in $y = mx + b$.

   a. $y = -x$.

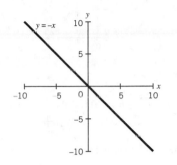

   b. $y = 0.5x$.

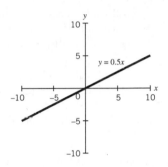

2. a. The variables x and y are directly proportional; the equation is $y = -3x$.

   b. The variables x and y are not directly proportional; the equation is $y = 3x + 5$.

3. E = euros; D = U.S. dollars

   a. $E = 0.70D$

   b. $E = 0.70 (D - 2.50)$

   c. Only (a) because it is of the form $y = mx$ for some constant m.

4. $d = 60t$. This represents direct proportionality. If the value of t doubles, the value for d also doubles. If the value for t triples, then so does the value for d.

5. Since C and N are directly proportional, $C = kN$ for some constant k. Since $50 = k \cdot 2$, then $k = \$25$, the cost per ticket. So the cost for 10 tickets is $C = 25 \cdot 10 = \$250$.

6. a. $4 = k(12) \Rightarrow k = \frac{1}{3} \Rightarrow y = \frac{1}{3}x$

   b. $300 = k(50) \Rightarrow k = 6 \Rightarrow d = 6t$

7. a. $d = kC$ **b.** $T = kI$ **c.** $t = kc$

8. a. $a = kb$; $10 = k(15) \Rightarrow \frac{2}{3} = k$; $a = \frac{2}{3}(6) = 4$

   b. $a = kb$; $4 = \frac{2}{3}b \Rightarrow b = 6$

Algebra Aerobics 2.8b

1. a. $y = -5$ **b.** $y = -3$ **c.** $y = 5$

2. a. $x = 3$ **b.** $x = 5$ **c.** $x = -3$

3. a. $y = -7$ **b.** $x = -4.3$

4. Slope is -1, y intercept is 0, so $m = -1, b = 0 \Rightarrow y = -x$.

5. $m = 360, C = 4, W = 1000$ in $W = 360C + b$. To solve for b, use the point $(4, 1000)$, so $1000 = (360)(4) + b \Rightarrow 1000 = 1440 + b \Rightarrow b = -440$. So, the equation is: $W = 360C - 440$.

6. Let m = slope of given line;

   M = slope of perpendicular line. So $M = -\frac{1}{m}$.

   a. $m = -3 \Rightarrow M = -\frac{1}{-3} = \frac{1}{3}$

   b. $m = 1 \Rightarrow M = -\frac{1}{1} = -1$

   c. $m = 3.1 \Rightarrow M = -\frac{1}{3.1} \approx -0.32$

   d. $m = -\frac{3}{5} \Rightarrow M = -\frac{1}{-3/5} = \frac{5}{3}$

7. a. Slope of $y = 2x - 4$ is 2, so line perpendicular to it has slope $-1/2$ or -0.5. Since it passes through $(3, -5)$, $x = 3$ when $y = -5$. So, $y = mx + b$ is: $-5 = (-0.5)(3) + b$. Solve it for b. $-5 = -1.5 + b \Rightarrow b = -3.5$. So, equation is: $y = -0.5x - 3.5$.

   b. Any line parallel to (but distinct from) $y = -0.5x - 3.5$ will be perpendicular to $y = 2x - 4$, but will not pass through $(3, -5)$. They have same slope m, (-0.5), but different values of b, in $y = mx + b$. Two examples are $y = -0.5x$ and $y = -0.5x - 7.5$.

   c. The three lines are parallel.

   d. Graph to verify.

8. $Ax + By = C \Rightarrow By = C - Ax \Rightarrow y = (C - Ax)/B \Rightarrow y = \frac{C}{B} - \frac{A}{B}x$, so $m = -\frac{A}{B}$

9. a. $2x + 3y = 5 \Rightarrow A = 2, B = 3 \Rightarrow m = -\frac{2}{3}$

   b. $3x - 4y = 12 \Rightarrow A = 3, B = -4 \Rightarrow m = -\frac{3}{-4} = \frac{3}{4}$

   c. $2x - y = 4 \Rightarrow A = 2, B = -1 \Rightarrow m = -\frac{2}{-1} = 2$

   d. $x = -5 \Rightarrow A = 1, B = 0 \Rightarrow m = -\frac{1}{0}$, which is undefined; this line is vertical. Note that y is not a function of x.

   e. $x - 3y = 5 \Rightarrow A = 1, B = -3 \Rightarrow m = -\frac{1}{-3} = \frac{1}{3}$

   f. $y = 4 \Rightarrow A = 0, B = 1 \Rightarrow m = -\frac{0}{1} = 0$ (this line is horizontal)

10. $2x + 3y = 5 \Rightarrow 3y = -2x + 5 \Rightarrow y = -\frac{2}{3}x + \frac{5}{3} \Rightarrow m = -\frac{2}{3}$. Parallel lines have slopes that are equal, so $m = -\frac{2}{3}$. If the line passes through the point $(0, 4)$ then $x = 0, y = 4$, the vertical intercept $b = 4 \Rightarrow y = -\frac{2}{3}x + 4$.

11. $3x + 4y = -7 \Rightarrow 4y = -7 - 3x \Rightarrow y = -\frac{7}{4} - \left(\frac{3}{4}\right)x \Rightarrow m = -\frac{3}{4}$. Because perpendicular lines have slopes that are negative reciprocals of each other, the slope of any line perpendicular to the given line is $m = -\frac{1}{-3/4} = \frac{4}{3}$. If the line passes through the point $(0, 3)$ then $x = 0, y = 3$, the vertical intercept $b = 3$. So the equation is $y = \frac{4}{3}x + 3$.

12. $4x - y = 6 \Rightarrow y = 4x - 6$. The slope of the given line is 4; therefore the slope of any line perpendicular to it is $-1/4$. If the line passes through $(2, -3) \Rightarrow x = 2, y = -3$. So $y = -\left(\frac{1}{4}\right)x + b$, and $(-3) = \frac{-1}{4}(2) + b \Rightarrow -3 = -\frac{1}{2} + b \Rightarrow b = \frac{-5}{2} \Rightarrow y = -\left(\frac{1}{4}\right)x - \frac{5}{2}$

13. a. vertical line

b. neither $\left(m = -\frac{2}{3}\right)$

c. horizontal line

14. Slope of given line $= -\frac{2}{3}$, so slope of a perpendicular line $= \frac{3}{2}$. So:

a. $y = \frac{3}{2}x + 5$

b. Substituting $(-6, 1)$, we have
$1 = \frac{3}{2}(-6) + b \Rightarrow b = 10 \Rightarrow y = \frac{3}{2}x + 10$

15. Slope $= 2$ for the given line and for lines parallel to it. So:

a. $y = 2x + 9$

b. Substituting $(4, 3)$, we have
$3 = 2(4) + b \Rightarrow b = -5 \Rightarrow y = 2x - 5$

Exercises for Section 2.8

1. a. $2 = m \cdot 10$ means $m = 0.2$

b. $0.1 = m \cdot 0.2$ means $m = 0.5$

c. $1 = m \cdot 1/4$ means $m = 4$

3. a. The slope $m = 56.92 - 42.69 = 14.23$ and since y is directly proportional to x, then $y = 14.23x$ is the equation. Thus, if $x = 5$, then $5 \cdot 14.23 = 71.15$ is the missing value.

b. The coefficient of x is the cost of a single CD.

5. a. The independent variable is the price P; the dependent variable is the sales tax T. The equation is $T = 0.065P$.

b. Independent variable is amount of sunlight S received; dependent variable is the height of the tree H. The equation is $H = kS$, where k is a constant.

c. Time t in years since 1985 is the independent variable, and salary S in dollars is the dependent variable. The equation is $S = 25,000 + 1300t$.

7. $d = 5t$; yes, d is directly proportional to t; it is more likely to be the person jogging since the rate is only 5 mph.

9. a. $m = 0$; $y = 3$

b. $m = 0$; $y = -7$

c. slope is undefined; $x = -3$

d. slope is undefined, $x = 2$

11. $B(t) = 3000$, where $t =$ years since 2007 and $0 \le t \le 3$ and $B(t)$ is measured in millions of books.

13. a. horizontal: $y = -4$; vertical: $x = 1$; line with slope 2: $-4 = 2 \cdot 1 + b \Rightarrow b = -6 \Rightarrow y = 2x - 6$.

b. horizontal: $y = 0$; vertical: $x = 2$; line with slope 2: $0 = 2 \cdot 2 + b \Rightarrow b = -4 \Rightarrow y = 2x - 4$.

15. a. The average rate of change is 10 lb per month.

b. $w(t) = 175$, where $t =$ number of months after end of spring training and 175 is measured in pounds. The graph of this function is a horizontal line.

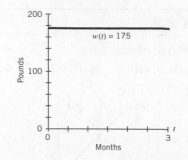

17. a. $7 = b - 3 \Rightarrow b = 10$ and therefore $y = 10 - x$.

b. $7 = b + 3 \Rightarrow b = 4$ and therefore $y = 4 + x$.

19. The lines described by: **a.** $x = 0$ **b.** $y = 0$ **c.** $y = x$.

21. a. Intercepts of one line are $(0, 4)$ and $(1, 0)$ with slope $m = -4$ and the intercepts of the other line are $(0, 4)$ and $(-1, 0)$, with slope $m = 4$. Thus the lines are not perpendicular. The equations are $f(x) = -4x + 4$ and $g(x) = 4x + 4$.

b. Intercepts of the first line are $(0, 2)$ and $(1, 0)$ with slope $m = -2$ and the intercepts of the second line are $(0, 2)$ and $(-4, 0)$, with slope $m = \frac{1}{2}$. Thus the lines are perpendicular. The equations are $f(x) = -2x + 2$ and $g(x) = \frac{1}{2}x + 2$.

23. For Graph A: both slopes are positive; same y-intercept.

For Graph B: one slope is positive, one negative; same y-intercept.

For Graph C: the lines are parallel; different y-intercepts.

For Graph D: one slope is positive, one negative; different y-intercepts.

25. a. $y = (-A/B)x + (C/B), B \ne 0$

b. The slope is $-A/B, B \ne 0$

c. The slope is $-A/B, B \ne 0$

d. The slope is $B/A, A \ne 0$

Section 2.9

Algebra Aerobics 2.9

1. a. Graph of $f(x)$

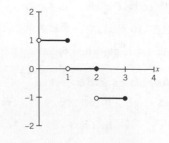

b. Graph of $g(x)$

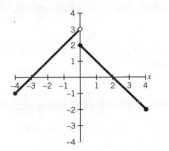

d. Federal Fund Rates (2008)

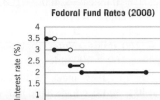
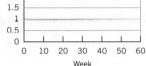

2. $Q(t) = \begin{cases} t - 2 & \text{for } -5 < t \le 0 \\ 2 - t & \text{for } \ \ 0 < t \le 5 \end{cases}$

$C(r) = \begin{cases} r + 1 & \text{for } -3 < r \le 1 \\ 5 & \text{for } \ \ 1 < r < 4 \end{cases}$

3. a. 2   **b.** 6   **c.** 2   **d.** −2   **d.** −15

4. a. $g(-3) = |-6| = 6$; $g(0) = |-3| = 3$;
$g(3) = |0| = 0$; $g(6) = |3| = 3$

b.

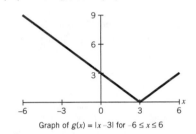

Graph of $g(x) = |x - 3|$ for $-6 \le x \le 6$

c. The graph of g is the graph of f shifted three units to the right.

d. $g(x) = \begin{cases} x - 3 & \text{if } x \ge 3 \\ 3 - x & \text{if } x < 3 \end{cases}$

5. a. $3 \le t \le 7$, which means that the values of t lie between (and include) 3 and 7.

b. $69 < Q < 81$, which means that the values of Q lie between (but exclude) 69 and 81.

6. $|T - 55°| \le 20°$, or equivalently $35° \le T \le 75°$.

7. a.

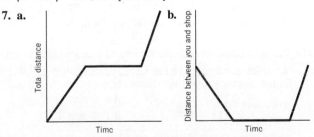

8. a. 3.5% in week 4; 2.0% in week 52; the longest period in which the rate remained the same was week 19 through week 52.

b. 2006: curbing inflation   2009: stimulating growth

c. $R(w) = \begin{cases} 3.5\% \text{ for } \ 1 \le w < \ 5 \\ 3.0\% \text{ for } \ 5 \le w < 13 \\ 2.3\% \text{ for } 13 \le w < 19 \\ 2.0\% \text{ for } 19 \le w \le 52 \end{cases}$

Exercises for Section 2.9

1. a. Graph of $f(x)$

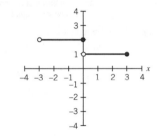

b. Graph of $g(x)$

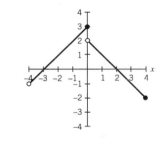

3. a. $g(x) = \begin{cases} -2 - x & \text{for } x < -2 \\ 2 + x & \text{for } x \ge -2 \end{cases}$

b. $g(x) = |2 + x|$

c. The graph of g is the same as the graph of f shifted horizontally two units to the left.

5. a. $d_A(t) = 60t$, $d_B(t) = 40t$

b. $D_{AB}(t) = 400 - d_A(t) - d_B(t) = 400 - 60t - 40t$
$= 400 - 100t$

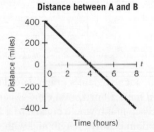

c. They will meet when the distance between them $D_{AB}(t) = 0$ miles $\Rightarrow 400 - 100t = 0 \Rightarrow 400 = 100t \Rightarrow t = 4$ hours. A will have traveled 4 hr $\cdot$ 60 miles/hr $= 240$ miles. B will have traveled 4 hr $\cdot$ 40 miles/hr $= 160$ miles. (*Note:* Together they will have traveled $240 + 160 = 400$ miles.)

d. One hour before they meet (3 hours into the trip), $D_{AB}(3) = 400 - 100 \cdot 3 = +100$ miles, which means that they are 100 miles apart and traveling *toward* each other. One hour after they meet (at 5 hours), $D_{AB}(5) = 400 - 100 \cdot 5 = -100$ miles, which means that they are 100 miles apart and traveling *away* from each other.

e. $D(t) = |400 - 100t| = \begin{cases} 400 - 100t & \text{for } 0 \leq t \leq 4 \\ 100t - 400 & \text{for } t > 4 \end{cases}$

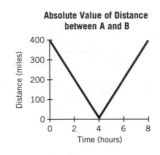

Absolute Value of Distance between A and B

7. a. All these expressions have the value of 2

b. Graph of greatest integer function $y = [x]$ for $0 \leq x < 5$

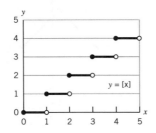

9. a. Answers will vary. This graph assumes that $1000 is in the savings account at the beginning of the month when $t = 0$ and $200 is added at the end of each week with one $300 withdrawal occurring at 2.5 weeks.

Money in a Savings Account for One Month

b. Again, each student's graph would differ but each graph would have a horizontal time axis and a vertical money axis; each graph would consist of a finite number of steps of random positive length and with random spacing as to

occurrence; each step will have its right end point excluded and its left point included; each step will be lower than the previous step by the amount of the withdrawal. The last step could occur when the day is done or the money in the ATM runs out. No graph is provided.

11. a. $T(s) = \begin{cases} 350 & \text{if } 0 \leq s < 3500 \\ 375 & \text{if } 3500 \leq s \leq 6500 \end{cases}$

where s is measured in feet above sea level and $T(s)$ is measured in degrees Fahrenheit.

b. A reasonable domain would be $0 \leq s \leq 6500$ ft above sea level, and a suitable range would be from $350 \leq T \leq 375$ degrees Fahrenheit.

13. a. $T(x) = \begin{cases} 98.6 & \text{if } 0 \leq x \leq 2 \\ 98.6 + 0.5(x - 2) & \text{if } 2 < x \leq 6 \\ 100.6 & \text{if } 6 < x \leq 12 \\ 100.6 - 0.25(x - 12) & \text{if } 12 < x \leq 20 \\ 98.6 & x > 20 \end{cases}$

where x is measured in hours from 6 A.M. and $T(x)$ is measured in temperature in degrees Fahrenheit.

Here is its graph:

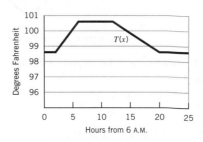

b. 8 A.M. means $t = 2$ and $T(2) = 98.6$ degrees Fahrenheit; noon means $t = 6$ and $T(6) = 100.6$ degrees Fahrenheit; 6 P.M. means $t = 12$ and $T(12) = 100.6$; 10 P.M. means $t = 16$ and $T(16) = 100.6 - 0.25(16 - 12) = 99.6$. It returns to normal when $t = 20$ since $T(20) = 98.6$.

15. The runner runs for 1 mile at 8 mph and then jogs for 3 minutes at 4 mph or for $4\frac{\text{mi}}{\text{hr}} \cdot (3 \text{ min}) \cdot \frac{1 \text{ hr}}{60 \text{ min}} = \frac{1}{5}$ mi, or 0.2 miles. So the runner's total time distribution for one repetition is

$T(d) = \begin{cases} 0.125d & \text{if } 0 \leq d \leq 1 \\ 0.125 + 0.25(d - 1) & \text{if } 1 < d \leq 1.2 \end{cases}$

where d measures distance traveled in miles and $T(d)$ is measured in hours. The graph is given below.

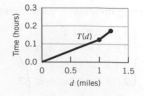

Section 2.10

Algebra Aerobics 2.10

1. a. In 1960 there were about 7.5 million college graduates, in 2009 about 29 million.

b.

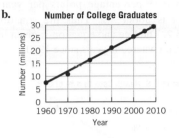

Number of College Graduates

c. Two estimated points on the line are (1960, 7.5) and (2005, 27.5).

Slope $= \frac{(27.5 - 7.5)}{(2005 - 1960)} = \frac{20}{45} \approx 0.44$ million/yr.

d. (0, 7.5) and (45, 27.5)

e. $y = 7.5 + 0.44x$

f. The number of graduates was about 7.5 million in 1960 and has been steadily increasing since—by about 0.44 million, or 440,000, persons each year.

2. a.

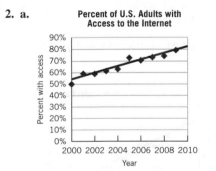

Percent of U.S. Adults with Access to the Internet

b. Points selected from the line drawn using reinitialized years: (0, 53), (9, 80). So $m = \frac{80 - 53}{9} = 3.0$. The percentage of adults with Internet access increased annually on the average by 3%.

c. (0, 53)

d. $y = 53 + 3x$ where x is the number of years after 2000.

e. The graph will become flatter because the increase will be less, since most people will already have access to the Internet.

Exercises for Section 2.10

1. Equation **a** matches with table **C**.

Equation **b** matches with table **A**.

Equation **c** matches with table **B**.

3. a. Exactly linear; $y = 1.5x - 3.5$

b. An estimated best-fit line has intercepts at (0, 6) and (2.5, 0), so the slope = $(6 - 0)/(0 - 2.5) = -2.4$. The equation is then $Q = -2.4t + 6$. (Note that the data point (0, 6.5) is not on the best-fit line.) Your answer may be somewhat different.

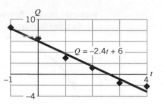

$Q = -2.4t + 6$

c. Exactly linear: $P = 3N + 35$

5. a. Exactly linear: $Y = 0.07$ billion solar units. It is a straight, horizontal line.

b. Approximately linear. Sketch a best-fit line through the data and choose any two points on the line, say (2002, 210) and (2008, 335). The slope of the line is $\frac{335 - 210}{2008 - 2002} = \frac{125}{6} \approx 21$.

See accompanying graph.

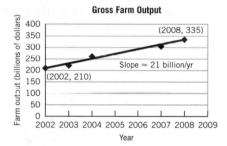

Gross Farm Output

If we reinitialize years as $x =$ years since 2002, then the vertical intercept is now about \$210 billion. Letting $y =$ gross farm output (in billions), the linear model would be $y = 210 + 21x$.

7. a. Estimates may vary. Sketch a best-fit line through the data and choose any two points on the line, say (1980, 21) and (2000, 43). The slope of the line is $\frac{43 - 21}{2000 - 1980} = \frac{22}{20} = 1.1$ percentage points per year.

See accompanying graph.

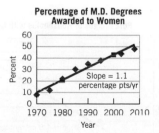

Percentage of M.D. Degrees Awarded to Women

b. If $y = 100\%$, then $100 = 10 + 1.1x$, or x is ≈ 82 years after 1970 or in 2052.

c. It is highly unlikely that this would happen. The most likely scenario is that the graph (and correspondingly the percentage of female M.D. degrees awarded) will taper off to some maximum percentage value, say at 50%. The accompanying graph fits that description. (Answers may differ somewhat.)

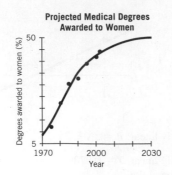

Projected Medical Degrees Awarded to Women

9. The accompanying graph gives an eyeballed best-approximation line that goes through two estimated coordinates of (0, 150) and (80, 40). The y-intercept is 150. The slope is $\frac{40 - 150}{80 - 0} = \frac{-110}{80} = -1.375$. Thus the equation of the graph can be estimated as $y = 150 - 1.375x$, where y stands for infant mortality rate (deaths per 1000 live births) and x stands for literacy rate (%). This means that on average for each 1% increase in literacy rate, the number of infant deaths drops by 1.375 per thousand.

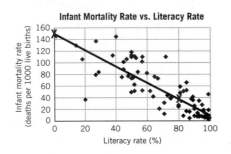

Infant Mortality Rate vs. Literacy Rate

11. **a.** Answers will vary. One estimated linear fit is given in the diagram below. Here y is measured in percents and x is measured in years. Two points are chosen: (1990, 2) and (2007, 83). Thus, the slope is $\frac{83 - 2}{2007 - 1990} = \frac{81}{17} \approx 4.8$ percentage points per year, This slope means that on average for each additional year, the percentage of households with cell phones will increase by 4.8 percentage points.

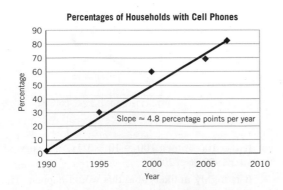

Percentages of Households with Cell Phones

Slope ≈ 4.8 percentage points per year

b. If x = years from 1990, then the vertical intercept is about 2%. So $P(x) = 2 + 4.8x$ is a linear model for $P(x)$, the percentages of households with cell phones since 1990.

c. Answers will vary. If the rate of increase continues and we extend the best-fit line, around 2010 nearly 100% of households will have cell phones.

13. **a.** Answers will vary. Choosing two points (0, 5.2) and (47, 16.2) on a best-fit line gives the slope of $\frac{16.2 - 5.2}{47 - 0} = \frac{11}{47} \approx 0.23$ percentage points per year; its vertical intercept is about 5.2. Thus the equation is $H(t) = 0.23t + 5.2$. The line and equation is given in the accompanying graph where t is years since 1960.

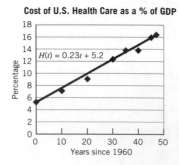

Cost of U.S. Health Care as a % of GDP

$H(t) = 0.23t + 5.2$

b. 2010 is 50 years from 1960 and $H(50) = 16.7\%$ of GDP

c. One possible reason is that the GDP has grown very large and thus the percentage for health care has not grown as much as the health care costs themselves.

15. **a.** Below are the graph of the data and a hand-drawn best-fit line.

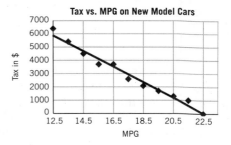

Tax vs. MPG on New Model Cars

b. The approximate coordinates of two points on this line are (17.5, 3000) and (22.5, 0) ⇒ the slope of this line is $\frac{0 - 3000}{22.5 - 17.5} = \frac{-3000}{5} = -600$ dollars per mpg. The equation of this line is $T(x) = -600x + 13,500$, where x is in mpg and $T(x)$ = tax (in dollars). The vertical intercept is very large because its value is what one would get if mpg takes the value of 0. (This value, of course, represents an impossible situation.)

c. The average rate of change for the given line is -600 dollars per mpg. As the mpg rating increases, the tax paid goes down by $600 per mpg.

17. **a.** Answers will vary. A best-fit line might include the points (2, 34) and (4, 15); its slope would be $\frac{15 - 34}{4 - 2} = \frac{-19}{2} = -9.5$ mpg/(1000 pound). This means that, on average, for every 1000-pound increase in car weight, the MPG would decrease by 9.5 gallons.

b. $M(w) = -9.5w + 53$, with $2 \le w \le 4$, where w measures the car's weight in 1000-pound units.

c. $M(3) - -9.5 \cdot (3) + 53 = 24.5$ miles per gallon.

d. For a car getting 30 mpg, $30 = -9.5 \cdot w + 53$ and then $9.5w = 53 - 30 = 23$; thus, $w = 23/9.5 = 2.4$ or about 2400 pounds. Thus, a car getting 30 mpg, according to our formula, weighs 2400 pounds.

Section 2.11

Algebra Aerobics 2.11

1. a. 0.65,   0.68,   0.07,   0.70

b. $|-0.07|$, $|0.65|$, $|-0.68|$, $|0.70|$

2. a. positive correlation: A, B, C; negative correlation: D

b. weakest correlation: A; strongest correlation: C

3. a. The slope = 6430, which means that for each additional year of education beyond grade 8, the mean personal earnings increases by $6430 per year.

b. For 4 years of education after 8th grade, median personal earnings = \$26,140 and mean personal earnings = \$31,370.

For 8 years of education after 8th grade, median personal earnings = \$51,140 and mean personal earnings = \$57,090.

Exercises for Section 2.11

1. a. Graph D shows the strongest linear relationship since its correlation coefficient (or cc) is closest in absolute value to 1. Graph C shows the weakest linear relationship since its correlation coefficient (or cc) is judged to be the farthest in absolute value from 1.

b. Graphs A and B show a positive relationship, and graphs C and D show a negative relationship.

3. a. The slope of the regression line is \$2960 per year of education; the vertical intercept is \$15,820. The correlation coefficient is 0.51.

b. The slope means that for every 1-year increase in education, the median personal earnings of a person in the South increases on average by \$2960.

c. An increase of \$2960. For ten additional years of education, those in the South can expect on average that their median personal earnings would go up by \$29,600.

5. a. The number 2310 is the slope of the regression line. It indicates that for white females, for each increase in years of education past 8 years, the median personal income on average will go up by \$2310.

b.

Years of Education Past 8	Median Personal Income
1	16350
4	23280
8	32520

The slope of the line joining (4, 23,280) and (8, 32,520) is $\frac{32520 - 23280}{8 - 4} = \frac{9240}{4} = \$2310$ per year of education beyond 8.

c. The values are the same, so the data are linear.

d. Here is the graph of the line through these three points.

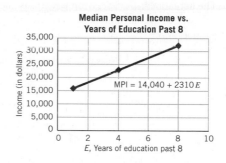

Median Personal Income vs. Years of Education Past 8

MPI = 14,040 + 2310 E

Income (in dollars) — E, Years of education past 8

e. The male income equation is $I_{male} = 3240 + 7990E$, and the female equation is $I_{female} = 14,040 + 2310E$, where E is years of education past 8. So females make more than males at 8 years of education—i.e., \$14,040 compared to \$3240. But the rate of change for males is about \$7990 versus \$2310 for females per additional year of education. So the males' median total income surpasses the females' within a few more years of education. The limitations are that these equations are for only a subset of the 1000-people sample size and the numbers of males and females are not the same. Other comments could be made.

7. a. 0.516 is the slope of the regression line, and it indicates that, on average, each increase of an inch in the height of the fathers indicates there is an increase of 0.516 inch in the mean height of the sons.

b. If $F = 64$, then $S = 33.73 + 0.516 \cdot 64 = 66.75$ in.

If $F = 73$, then $S = 33.73 + 0.516 \cdot 73 = 71.40$ in.

c. If $F = S$, then $F = 33.73 + 0.516F$ or $0.484F = 33.73$ or $S = F = 69.69$ in.

d. For each of the data points the S value represents the mean height of all the sons whose father has the given height of F. There are 17 heights listed (from 57 to 75 inches) for the 1000 fathers.

9. a. and **b.** Answers will vary depending on their eyeball estimates.

Set $x =$ years since 1985. Then sketch a best-fit for private colleges. Estimate the coordinates of two points on the line, say (2, 7000) and (20, 19000). The slope of this line is $(19000 - 7000)/(20 - 2) = 12000/18 = \$667$ per year. The vertical intercept is computed by substituting in the coordinates of the second point. Thus $19,000 = (12,000/18) \cdot 20 + b$ or $b \approx \$5667$. Thus $\text{Cost}_{private} = 667 \cdot \text{year} + 5667$ dollars. This equation is close to the actual regression line's equation. (See the following graph.)

Sketch a best-fit line for public colleges. Estimate 2 points on the line, say (5, 1500) and (20, 3500). The slope of the line through these two points is $(3500 - 1500)/(20 - 5) \approx$ $133 per year. The vertical intercept is obtained by substituting in the coordinates of the first point: $1500 = 133 \cdot 5 + b$ or $b = 835$ dollars. Thus: $\text{Cost}_{\text{public}} = 133 \cdot \text{year} + 835$ dollars. (See graph below).

The actual best-fit regression equations and their graphs are given in the diagram below. The eyeball approximations are not far off.

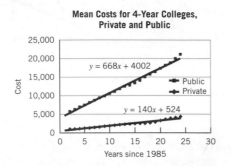

Mean Costs for 4-Year Colleges, Private and Public

$y = 668x + 4002$

$y = 140x + 524$

c. Using the equations derived by eyeballing, we see that in 2010 (25 years after 1985) the private school cost would be approximately $\text{Cost}_{\text{private}}(25) = 667 \cdot 25 + 5667 \approx$ $22,342 and the public school cost would be approximately $\text{Cost}_{\text{public}}(25) = 133 \cdot 25 + 835 \approx$ $4160.

11. a. and **b.** Student answers will vary.

The accompanying scatter plots and best-fit lines are shown below for both sexes.

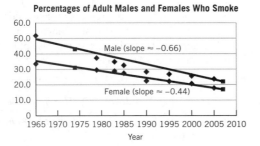

Percentages of Adult Males and Females Who Smoke

Male (slope ≈ -0.66)

Female (slope ≈ -0.44)

Estimating two points on the line for male smokers as (1975, 43) and (2007, 22), gives a slope for males as $(22 - 43)/(2007 - 1975) = -21/32 \approx -0.66$ percentage points per year. So the percentage of male smokers (out of all males) declined by 0.66 each year.

Estimating two points on the line for female smokers as (1975, 31) and (2000, 20), gives a slope for females as $(20 - 31)/(2000 - 1975) = -9/25 = -0.44$ percentage points per year. So the percentage of female smokers (out of all females) declined by 0.44 each year.

The percentage of both male and female smokers declined, but the percentage of male smokers declined more rapidly.

c. If we set $N = $ years since 1965, we can generate the graphs with regression lines and their equations and correlation coefficients (or cc's) in the diagram following. Thus, the regression line for females fits better since the

absolute value of its cc is closer to 1 than the absolute value for the male cc.

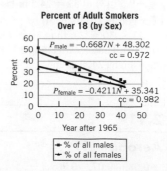

Percent of Adult Smokers Over 18 (by Sex)

$P_{\text{male}} = -0.6687N + 48.302$
$cc = 0.972$

$P_{\text{female}} = -0.4211N + 35.341$
$cc = 0.982$

Year after 1965

→ % of all males
→ % of all females

The vertical intercepts are the predictions made by the regression lines of what the percentages were for both sexes in 1965. The slopes tell much the same kind of information that was given in the answer to part (b).

d. Student answers will differ not only according to their sex but also according to their point of view. Notable factors would be that the correlations are very strong since the cc values are close to 1. Note that the percentage of smokers for both sexes is increasing. Also, the percentage of men who smoke is always larger than that of women, although in recent years these percents have come closer to each other.

13. a.

Boston Marathon Winning Times for Women

Time (minutes)

Year

b. Assume that $x = $ years since 1972. To create the first part of best-fit piecewise function by hand, we need to identify two points on that line where $0 \leq x < 10$, say (0, 190) and (5, 170). So the slope is $(170 - 190)/(5 - 0) = -20/5 = -4$. The vertical intercept is 190.

The second part of the line where $10 \leq x \leq 37$, is basically flat, so the slope is 0. The line is horizontal, so if extended it to the vertical axis, it would intersect at about 145. We can construct a piecewise function $W(x)$ where

$$W(x) = \begin{cases} 190 - 4t & \text{when } 0 \leq x < 10 \\ 145 & \text{when } 10 \leq x \leq 37 \end{cases}$$

and $W(x)$ is the winning time (in seconds) and $x = $ years since 1972.

c. Between 1972 and 1982 the women's winning time in the Boston Marathon decreased by about 4 seconds per year. After 1982 the women's record time stayed basically flat at about 145 seconds.

d. One would predict the woman's time to be about 145 seconds in 2010. Teyba Erkesso of Ethiopia won the women's marathon in a little over 146 minutes in 2010.

e. Student answers will differ. One possible description follows.

The overall trend for winning times for women at the Boston Marathon was to decrease between 1972 and 1982, but changed little after that. Between 1972 and 1982 the winning times dropped by about 4 seconds/year. After that, women's times remained flat at about 145 seconds each year.

15. a. The plot of the data and the regression line are given in the accompanying diagram. The formula relating boiling temperatures in °F to altitude is $F = 211.80 - 0.0018H$, or when suitably rounded off, $F = 212 - 0.002H$, where H is feet above sea level. The correlation coefficient is -0.9999. The answer to the second part depends on where the student lives. Other factors could be something put into the water, such as salt, or variations in the air pressure.

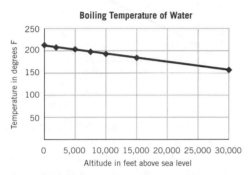

Boiling Temperature of Water

b. On Mt. McKinley water boils at $212 - 0.002 \cdot 20320 = 171.36$ °F; in Death Valley water boils at $212 - 0.002 \cdot (-285) = 212.57$ °F.

c. $32 = 212 - 0.002H$ or $-180 = -0.002H$ or $H = 90,000$ feet or approximately 17 miles. But this seems unreasonable since then the water would be outside earth's atmosphere layer (which goes to 9 miles above the earth).

17. a. Excel technology gives the regression line as $y = 3.6x + 26.3$ where x = years since 1945 and y = registrations (in millions). See the graph below.

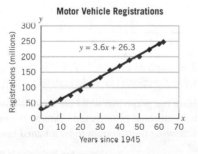

Motor Vehicle Registrations

$y = 3.6x + 26.3$

b. On average, for each additional year motor vehicle registrations have been increasing by 3.6 million per year since 1945.

c. From 1945 to 2007 is 62 years, and the equation predicted $3.6 \cdot (62) + 26.3 = 249.5$ million motor vehicle registrations. This figure is 12.1 million larger than the actual amount.

d. From 1945 to 2010 is 65 years, and $3.6 \cdot (65) + 26.3 = 260.3$ motor vehicle registrations. Compare your answer with what you find on the Internet.

19. A high correlation does not mean that there is causation involved. More studies would have to be done and in fact were done. The research leaves no doubt that cigarette smoking is indeed a cause of lung cancer. The high correlation coefficient was nevertheless an important factor.

Ch. 2: Check Your Understanding

1. False	**9.** True	**17.** True	**25.** False
2. True	**10.** True	**18.** True	**26.** False
3. False	**11.** False	**19.** False	**27.** False
4. False	**12.** True	**20.** False	**28.** False
5. True	**13.** True	**21.** True	**29.** False
6. True	**14.** False	**22.** True	**30.** True
7. True	**15.** True	**23.** False	**31.** True
8. False	**16.** True	**24.** False	

32. Possible answer: $y = -2x + 5$, y is the dependent variable, and x is the independent variable.

33. Possible answer: $D = 0$, D is the dependent variable, and p is the independent variable.

34. Possible answer: $2x - 3y = 6$, y is the dependent variable, and x is the independent variable.

35. Possible answer: $3p + 5q = -15$, q is the dependent variable, and p is the independent variable.

36. Possible answer: $T = 37d$, T is the dependent variable in minutes, and d is the independent variable in laps.

37. Possible answer: $V = 19.25 + 0.25q$, V is the dependent variable in dollars, and q is the independent variable in number of quarters from now.

38. Possible answer: $C = 2T + 3$ and $C = 2T - 1$, C is the dependent variable, and T is the independent variable.

39. Possible answer: $y = 2x + 4$, $y = 5x + 4$, $y = 4 - 2x$, $y = 4$, y is the dependent variable, and x is the independent variable.

40. Possible answer: For $m = -0.25$, $-1/m = 4$, $d = -0.25t + 3$, and $d = 4t - 1$; d is the dependent variable, and t is the independent variable.

41. Possible answer: (1990, $100), (1993, $133), (1995, $150), (2000, $175), (2009, $185). Note that this is a function given as a table rather than as an equation.

42. True	**49.** False
43. False	**50.** False
44. False	**51.** False
45. False	**52.** False
46. True	**53.** True
47. True	**54.** False
48. True	**55.** False

Ch. 2 Review: Putting It All Together

1. a. The absolute change is |$217,900 − $173,200| = $44,700.

b. The percent decrease = (173,200 − 217,900)/217,900 · 100 ≈ −20.51%.

c. This is the slope of the straight line joining the two given points: (173,200 − 217,900)/(2009 − 2007) = −44,700/2 = −22,350 dollars per year.

3. a. The function is positive between points B and E excluding the endpoints; negative between A and B (excluding point B) and between F and G (excluding point F); zero at point B and between points E and F.

b. The slope is positive between A and C; negative between D and E and between F and G; zero between C and D and between E and F.

5. a. They are both linear functions of U.S. sizes since the average rates of change are constant for both.

b. The British size equals the U.S. size plus 6.

c. If U.S. sizes are denoted by U, and French sizes by F, then $F = U + 34$.

7. a. Average rate of change = (262.7 − 109.4)/(2008 − 2000) = 153.3/8 ≈ 19.2 million cell phone subscriptions per year.

b. $C(t) = 19.2t + 109.4$, where $C(t)$ gives the number of subscriptions in millions t years from 2000.

c. $C(15) = 19.2 \cdot (15) + 109.4 = 397.4$ million subscriptions. Student responses on the reasonableness of this number will differ. There are currently more than 310 million people in the United States. It is feasible that the population would grow to 397 million in 5 years, but this would mean that almost every person has a subscription to a cell phone, which is unlikely.

9. a. Both A and B

b. Both B and C

c. C

d. A, B and D

11. $|46\% − x| \le 3\%$ and thus $43\% \le x \le 49\%$ and on the number line one would draw:

43%   49%,

where the closed circles, •, indicate that the end points of the segment are included.

13. a. (3600 people/day) · (365 days/year) = 1,314,000 people/year or, equivalently, an increase of 1.314 million people per year living in the coastal regions of the United States.

b. The coastal population $P(x)$ (in millions) can be modeled by $P(x) = 153 + 1.314x$, where x is the number of years since 2003.

c.

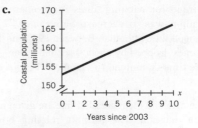

d. 2010 is 7 years from 2003 and $P(7) = 153 + 1.314 \cdot 7 \approx$ 162 million people are living in the coastal regions.

15. a.

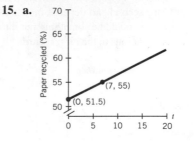

The slope = (55 − 51.5)/7 = 0.5 percentage points per year.

b. $R(t) = 51.5 + 0.5t$, where $R(t)$ is the percentage of paper recycled in t years from 2005.

c. $R(0) = 51.5\%$; $R(5) = 54\%$; $R(20) = 61.5\%$ would predict that in the year 2025, 61.5% of all paper is recycled.

17. Graph A: $y = 5$; Graph B: $x = −2$: Graph C: $y = 2x + 1$; Graph D; $y = 2x + 4$; Graph E: $y = −(1/2)x + 6$

19. a.

x	−3	−2	−1	0	1	2	3
$g(x)$	5	4	3	2	3	4	5

b.

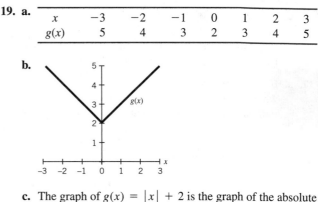

c. The graph of $g(x) = |x| + 2$ is the graph of the absolute value function $f(x) = |x|$ raised up two units.

21. a. Here is a graph that will help make the argument that voter turnout has plummeted. Answers will vary but the following points ought to be touched upon.

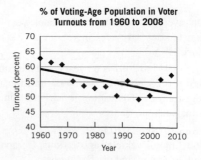

% of Voting-Age Population in Voter Turnouts from 1960 to 2008

Voter turnout has plummeted. In voter turnout as a percentage of the country's population over the years 1960–2008, there was a nearly steady decrease in this percentage. In fact, on average, this percentage decreased by 0.17 percentage point per year or by 0.56 percentage point every four years.

b. Following is a graph that will help make the argument that voter turnout has soared.

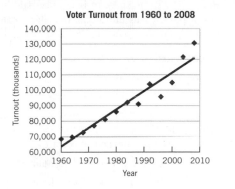

Voter turnout has soared. It is clear from the scatter plot and regression line that the voter turnout given above, on average, grew in size by about 1201 thousand voters per year from 1960 to 2008. True, there were a few times when the size of this turnout decreased from the previous election year (this happened in 1988 and 1996.) But the turnout in the next election each time was higher in both cases. One can also consider the fact that the turnout in 2008 was nearly double that in 1960. Indeed voter turnouts are soaring.

23. a. Over the last 250 years in Sweden the probability of a young child dying has steadily decreased. The child mortality rate has declined from about 40% in 1750 to less than 1% in 2000. The death rate is very similar for female and male children, though the male rates are consistently somewhat higher.

b.

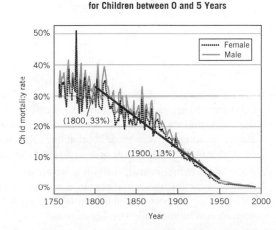

Two estimated points on the line are (1800, 33%) and (1900, 13%). So the slope of the line is $\frac{13 - 33}{1900 - 1800} = -20/100 = -0.2\%/\text{year}$. This means that on average the probability of a child between the ages of 0 to 5 dying was

decreasing by two-tenths of a percentage point each year, or equivalently, there were 2 fewer children dying per thousand.

c. If we let $t =$ years since 1800, then (0, 33%) becomes the vertical intercept. The slope remains the same. So the linear model is $P(t) = 33 - 0.2t$, where $P(t)$ gives the female child mortality rate (as a percentage) at t years after 1800.

d. After 1950 the mortality rates are very low and still declining, almost approaching 0% per year. (In fact, Sweden currently has one of the lowest child mortality rates in the world.)

25. a. The graph for income from rye shows no discernible correlation. The graph for income from corn shows a somewhat strong positive correlation. The graph for income from safflower oil also shows a negative correlation, but it is very weak.

b. Corn would most likely boost farm income, since the slope of its regression line is more positive than that for rye and the slope of the regression line for safflower oil is negative. Student answers will vary, depending on what they are able to find out on the Internet.

CHAPTER 3

Section 3.1

Algebra Aerobics 3.1

1. a. Graph A: $(1, -1)$ Graph B: $(-2, 0)$ and $(3, 2)$
Graph C: $(1.7, -0.3)$ Graph D: $(0, 1)$

b. Graph A: $y_1 > y_2$ for $x > 1$
$y_1 < y_2$ for $x < 1$

Graph B: $y_1 > y_2$ for $-2 < x < 3$
$y_1 < y_2$ for $x < -2$ or $x > 3$

Graph C: $y_1 > y_2$ for $x > 1.7$
$y_1 < y_2$ for $x < 1.7$

Graph D: $y_1 > y_2$ for $x < 0$
$y_1 < y_2$ for $x > 0$

2. Approximately (1993, 6.8 million)
This point indicates that in 1993, the United States produced and imported equal amounts of oil, namely 6.8 million barrels. From 1920 to 1993, the United States produced more oil than it imported, but after 1993, the reverse is true.

3. Gas is the cheapest system from approximately 17.5 years of operation to approximately 32.5 years of operation. Solar becomes the cheapest system after approximately 32.5 years of operation.

4. a. $(3, -1)$ is a solution since: $4(3) + 3(-1) - 9$ and $5(3) + 2(-1) = 13$. It is a solution of both equations.

b. $(1, 4)$ is not a solution since: $5(1) + 2(4) = 13$ but $4(1) + 3(4) = 16$, not 9. So it is not a solution of $4x + 3 = 9$.

5. Graph A: $(0, 2.5)$
Graph B: $(4, -4)$

Exercises for Section 3.1

1. a. (1994, 350,000)

   b. To the left of the intersection point, Pittsburgh's population is greater than that of Las Vegas and is decreasing, while Las Vegas's population is increasing. To the right of the intersection point, Pittsburgh's population is less than that of Las Vegas and continues to decrease, while Las Vegas's population continues to increase.

3. a. The two lines intersect at $(-3, 0)$ and thus $x = -3$, $y = 0$ is the estimated solution to the system of equations.

   b. The equations of the two lines are $y_1 = 2 + (2/3)x$ and $y_2 = -5 - (5/3)x$. Solving for the common x: $2 + (2/3)x = -5 - (5/3)x$ or $[(2/3) + (5/3)]x = -5 - 2$ or $(7/3)\, x = -7$ or $x = -3$. Solving for the common y gives: $y = -5 + 5$ or $y = 2 - 2$ or 0. Substituting these values solves both equations: $0 = 2 + (2/3)(-3)$ and $0 = -5 - (5/3)(-3)$. The answer to part (b), is the same as the answer to part (a).

5. a. $y = 3x + 17$.

   b. The line has the slope $m = \frac{1-0}{0-5} = -\frac{1}{5}$; since its vertical intercept is 1, it has the equation: $y = (-1/5)x + 1$.

   c. The graphs of the two lines and the point of intersection are given in the diagram below:

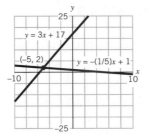

Find the common x value: $3x + 17 = -(1/5)x + 1$ or $(16/5)x = -16$ or $x = -5$. Solve for the common y value: $3 \cdot (-5) + 17 = 2$. Thus the solution is $(-5, 2)$. Check: $3 \cdot (-5) + 17 = 2$ and $-(1/5) \cdot 5 + 1 = 2$.

7. a.   **b.**

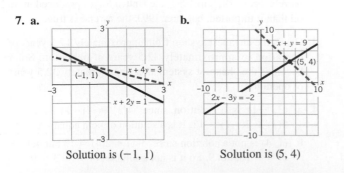

   Solution is $(-1, 1)$ Solution is $(5, 4)$

9. $y = -x - 2$ and $y = 2x - 8$ is the system, and the solution is $(2, -4)$. *Check:* $-2 - 2 = -4$ and $2 \cdot 2 - 8 = -4$, and thus the claimed solution works.

11. a. Both Africa and Latin America: During the decade of the 1980's.

   b. Between 1600 and 1700 China's population goes from being greater than that of India to being less. The population was the same around 1650 and 1725. After 1725, China's population exceeds that of India.

13. a. in 1977

   b. Before 1977, the share of insurance paid by individuals out of pocket was more than the share that was paid by private insurance. After 1977, the situation is the reverse.

15. a. After approximately 9 months.

   b. Approximately $8.60 per hour.

   c. $W_A(m) = 7.25 + 0.15m$ and $W_B(m) = 7.70 + 0.10m$

   d. The exact solution is when $W_A(m) = W_B(m)$ or $7.25 + 0.15m = 7.70 + 0.10m \Rightarrow 0.05m = 0.45 \Rightarrow m = 0.45/0.05 = 9$ months, so the estimate was correct.

   e. $W_A(9) = 7.25 + 0.15(9) = \$8.60/\text{hour}$
      $W_B(9) = 7.70 + 0.10(9) = \$8.60/\text{hour}$

   f. Before 9 months, the monthly wage rates at company B are higher; after 9 months the monthly wage rates at company A are higher. If they worked about the same number of hours at each place, one could judge the companies on accumulated wages instead of hourly rates.

17. a. The graph of the linear system is given with the intersection point marked.

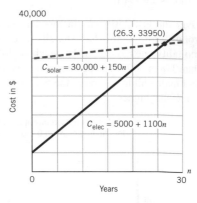

   b. 1100 and 150 are the slopes of the heating cost lines; 1100 represents the rate of change of the total cost in dollars for electric heating per year since installation; 150 is the rate of change in the total cost for solar heat in the same units.

   c. 5000 is the initial cost of installing the electric heat in dollars; 30,000 is the initial cost in dollars of installing solar heating. It cost a lot more initially to install solar heating than to install electrical heating.

   d. The point of intersection is approximately where $n = 26$ and $C = 34,000$.

   e. $n \approx 26.32$, $C \approx 33,947.37$ is a more precise answer; the values have been rounded off to two decimal places; they were obtained by setting the equations equal to each other.

f. Assuming simultaneous installation of both heating systems, the total cost of solar heat was higher than the total cost of electric heat up to year 26 (plus nearly 4 months); after that the total cost of electric heat will be greater than that of solar heat.

19. a. Setting the two equations equal gives $20{,}000 + 2500n = 25{,}000 + 2000n$ or $500n = 5000$ or $n = 10$. Plugging in that n value gives $S = 20{,}000 + 2500 \cdot 10 = 45{,}000$.

b. The graphs of the two linear equations are given in the diagram below. From inspecting the graphs it seems that the intersection occurs when $n = 10$ and $S = 45{,}000$.

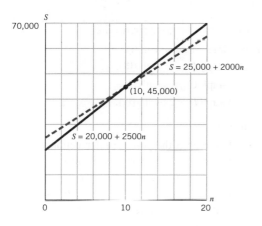

21. Let x measure years since 1977. Then the formula for the professor's assessment is: $A(x) = 70{,}000 + 26{,}000x$, where $A(x)$ is measured in dollars. The colleague's assessment formula is: $B(x) = 160{,}000 - 4{,}000x$. To find the value of x when the assessment was the same let $70{,}000 + 26{,}000x = 160{,}000 - 4000x$. So $x = 3$ or in 1980, the common assessment value was \$148,000.

Section 3.2

Algebra Aerobics 3.2a

1. a. $y = 7 - 2x$

b. $y = \dfrac{6 - 3x}{5} = \dfrac{6}{5} - \dfrac{3}{5}x$

c. $x = 2y - 1$

2. a. no solution, because the lines have the same slope but different y-intercepts, so they are parallel.

b. one solution, because the lines have different slopes.

3. a. Set $y = y \Rightarrow x + 4 = -2x + 7 \Rightarrow 3x = 3 \Rightarrow x = 1$; $y = -2(1) + 7 = 5$. *Check:* $y = (1) + 4 \Rightarrow y = 5 \Rightarrow$ solution (x, y) is $(1, 5)$.

b. Set $y = y \Rightarrow -1700 + 2100x = 4700 + 1300x \Rightarrow 800x = 6400 \Rightarrow x = 8$; $y = 4700 + 1300(8) = 15{,}100 \Rightarrow$ solution (x, y) is $(8, 15100)$.

c. Set $F = F \Rightarrow C = 32 + \frac{9}{5}C \Rightarrow 5C = 32(5) + 9C \Rightarrow -4C = 160 \Rightarrow C = -40$
$F = C \Rightarrow F = -40 \Rightarrow$ solution (C, F) is $(-40, -40)$.

4. a. Substitute $y = x + 3$ into $5y - 2x = 21 \Rightarrow 5(x + 3) - 2x = 21 \Rightarrow 5x + 15 - 2x = 21 \Rightarrow 3x = 6; x = 2;$ so, $y = (2) + 3 = 5 \Rightarrow y = 5$. Solution (x, y) is $(2, 5)$.

b. Substitute $z = 3w + 1$ into $9w + 4z = 11 \Rightarrow 9w + 4(3w + 1) = 11 \Rightarrow 9w + 12w + 4 = 11 \Rightarrow 21w = 7 \Rightarrow w = 1/3$; so $z = 3\left(\frac{1}{3}\right) + 1 = 2 \Rightarrow z = 2$. Solution (w, z) is $\left(\frac{1}{3}, 2\right)$

c. Substitute $x = 2y - 5$ into $4y - 3x = 9 \Rightarrow 4y - 3(2y - 5) = 9 \Rightarrow 4y - 6y + 15 = 9 \Rightarrow -2y = -6 \Rightarrow y = 3$; so $x = 2(3) - 5 = 1 \Rightarrow x = 1$. Solution (x, y) is $(1, 3)$.

d. Solve: $r - 2s = 5$ for r, and substitute the resulting expression for r into $3r - 10s = 13$. $r = 2s + 5 \Rightarrow 3(2s + 5) - 10s = 13 \Rightarrow 6s + 15 - 10s = 13 \Rightarrow -4s = -2 \Rightarrow s = \frac{1}{2}$; so $r = 2\left(\frac{1}{2}\right) + 5 = 6 \Rightarrow r = 6$. Solution (r, s) is $\left(6, \frac{1}{2}\right)$.

5. $2x + 3y = 9 \Rightarrow y = \dfrac{9 - 2x}{3} = 3 - \dfrac{2}{3}x$
$2x + y = 3 \Rightarrow y = 3 - 2x$

The only value of x that satisfies both equations is $x = 0$; or $(0, 3)$ is the y-intercept of both lines, so that point must satisfy both equations.

Algebra Aerobics 3.2b

1. a. By the elimination method, add equations: $2y - 5x = -1$ and $3y + 5x = 11 \Rightarrow 5y = 10 \Rightarrow y = 2$; so $3(2) + 5x = 11 \Rightarrow 5x = 5 \Rightarrow x = 1$. Solution (x, y) is $(1, 2)$.

b. Multiply the equation $(3x + 2y = 16)$ by 3 and the equation $(2x - 3y = -11)$ by $2 \Rightarrow 9x + 6y = 48$ and $4x - 6y = -22$. By the elimination method, add these equations $\Rightarrow 13x = 26 \Rightarrow x = 2$, so $3(2) + 2y = 16 \Rightarrow 2y = 10 \Rightarrow y = 5$. Solution (x, y) is $(2, 5)$.

c. By substitution of $t = 3r - 4$ into $4t + 6 = 7r \Rightarrow 4(3r - 4) + 6 = 7r \Rightarrow 12r - 16 + 6 = 7r \Rightarrow -10 + 5r = 0 \Rightarrow 5r = 10 \Rightarrow r = 2$, so $t = 3(2) - 4 = 2 \Rightarrow t = 2$. Solution (r, t) is $(2, 2)$.

d. Substitute $z = 2000 + 0.4(x - 10{,}000)$ into $z = 800 + 0.2x \Rightarrow 2000 + 0.4(x - 10{,}000) = 800 + 0.2x \Rightarrow 2000 + 0.4x - 4000 = 800 + 0.2x \Rightarrow 0.2x = 2800 \Rightarrow x = 14{,}000$; so, $z = 800 + 0.2(14{,}000) = 3600 \Rightarrow z = 3600$. Solution (x, z) is $(14000, 3600)$.

2. a. By substitution: $2x + 4 = -x + 4 \Rightarrow 3x = 0 \Rightarrow x = 0$, so $y = -(0) + 4 = 4 \Rightarrow y = 4$. So solution (x, y) is $(0, 4)$.

b. By substitution of $(y = -6x + 4)$ into $(5y + 30x = 20) \Rightarrow 5(-6x + 4) + 30x = 20 \Rightarrow -30x + 20 + 30x = 20 \Rightarrow 20 = 20$. So both equations must be equivalent. There are infinitely many solutions since both equations describe the same line.

c. $2y = 700x + 3500 \Rightarrow y = 350x + 1750$. The slopes of the lines of both the equations are 350, but the y-intercepts are different (1500 and 1750), so the lines are parallel. There is no solution.

3. In order for a system of equations to have no solutions, they must produce parallel lines with the same slope, but different y-intercepts. One example is: $y = 5x + 10$; $y = 5x + 3$.

4. a. $2x + 5y = 7 \Rightarrow y = \frac{-2x + 7}{5}$; $3x - 8y = -1 \Rightarrow$
$y = \frac{3x + 1}{8} \Rightarrow$ one solution since the lines have unequal slopes of $-\frac{2}{5}$ and $\frac{3}{8}$.

b. $3x + y = 6 \Rightarrow y = 6 - 3x$; $6x + 2y = 5 \Rightarrow y = \frac{5 - 6x}{2}$ or $y = \frac{5}{2} - 3x \Rightarrow$ no solution since the lines have the same slope of -3 and different y-intercepts of 6 and $\frac{5}{2}$.

c. $2x + 3y = 1 \Rightarrow y = \frac{1}{3} - \frac{2}{3}x$; $4x + 6y = 2 \Rightarrow y = \frac{1}{3} - \frac{2}{3}x$
$\Rightarrow$ equivalent equations and an infinite number of solutions, since the lines have same slopes and same y-intercepts.

d. $3x + y = 8 \Rightarrow y = 8 - 3x$; $3x + 2y = 8 \Rightarrow y = 4 - \frac{3}{2}x \Rightarrow$ one solution since the lines have unequal slopes of -3 and $-\frac{3}{2}$.

5. a. $6\left(\frac{x}{2} + \frac{y}{3}\right) = 6(3) \Rightarrow 3x + 2y = 18$. Substitute $y = x + 4$
$\Rightarrow 3x + 2(x + 4) = 18 \Rightarrow 5x + 8 = 18 \Rightarrow 5x = 10 \Rightarrow$
$x = 2$; so, $y = (2) + 4 = 6 \Rightarrow y = 6$. Solution (x, y) is $(2, 6)$.

b. $-60(0.5x + 0.7y) = -60(10) \Rightarrow -30x - 42y = -600$.
Add to $30x + 50y = 1000 \Rightarrow 8y = 400 \Rightarrow y = 50$, so
$30x + 50(50) = 1000 \Rightarrow 30x + 2500 = 1000 \Rightarrow$
$30x = -1500 \Rightarrow x = -50$, so the solution (x, y) is $(-50, 50)$.

6. a. $4(39) + 3q = 240 \Rightarrow 3q = 84 \Rightarrow q = 28$ gals

b. $4p + 3(20) = 240 \Rightarrow 4p = 180 \Rightarrow p = \$45$ per gal

c, d. Solve for p: $4p = 240 - 3q \Rightarrow p = 60 - 3/4q \Rightarrow p = 60 - 0.75q$.

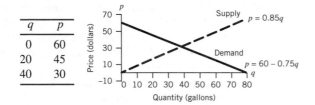

q	p
0	60
20	45
40	30

e. Solve by substitution of $p = 0.85q$ into $4p + 3q = 240 \Rightarrow$
$4(0.85q) + 3q = 240 \Rightarrow q = 37.5$; $p = 0.85(37.5) = 31.9$.
So the equilibrium point is $\sim(38, \$32)$, which means that when the price is around \$32, the demand will be around 38 gallons.

f. There is a surplus of supply because where the line $p = 39$ crosses the supply curve, it is above the demand curve, so the supply is greater than the demand.

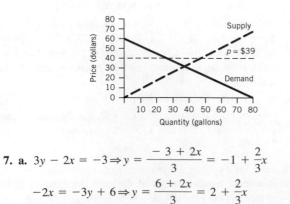

7. a. $3y - 2x = -3 \Rightarrow y = \frac{-3 + 2x}{3} = -1 + \frac{2}{3}x$

$-2x = -3y + 6 \Rightarrow y = \frac{6 + 2x}{3} = 2 + \frac{2}{3}x$

This system has no solution since no point that satisfies the first equation can satisfy the second. The system consists of two parallel lines (same slopes, different y-intercepts.)

b. $5y + x = 15 \Rightarrow y = \frac{15 - x}{5} = 3 - \frac{1}{5}x$

$10y - 30 = -2x \Rightarrow y = \frac{30 - 2x}{10} = 3 - \frac{1}{5}x$

This system has infinitely many solutions since any point that satisfies the first equation will satisfy the second. The two equations are of the same line.

Exercises for Section 3.2

1. a. Graph A goes with **(i)**.
Graph B goes with **(iii)**.
Graph C goes with **(ii)**.
Graph D goes with **(iv)**.

b. and c.

In Graph A the point of intersection visually is $(-3, 1)$.
Check: $1 = 3 - 2$ and $1 = (1/3) \cdot (-3) + 2$.

In Graph B the point of intersection visually is $(6, 4)$.
Check: $4 - 6 = -2$ and $3 \cdot 4 - 6 = 6$.

In Graph C there is no point of intersection, the lines are parallel.

In Graph D the point of intersection visually is $(3, -2)$.
Check: $3 \cdot -2 + 3 = -3$ and $-2 = -2 \cdot 3 + 4$.

3. a. The method that is easiest is often a judgment by the person solving the problem.
 i. Either is easy **iv.** Substitution
 ii. Elimination **v.** Elimination
 iii. Substitution **vi.** Substitution

b. i. Setting the y values equal to each other gives $6 = -4$. Thus there is no solution.

ii. Letting $y = 2x - 5$ in the second equation gives $5x + 2(2x - 5) = 8$ or $9x - 10 = 8$ or $9x = 18$ or $x = 2$. Thus $y = 2 \cdot 2 - 5 = -1$ and therefore the solution is $(2, -1)$. *Check:* $2(2) - (-1) = 5$; and $5 \cdot 2 + 2 \cdot (-1) = 8$.

iii. Substituting $x = 7y - 30$ into the first equation gives $3 \cdot (7y - 30) + 2y = 2$ or $21y - 90 + 2y = 2$ or $23y = 92$ or $y = 4$ and then $x = 28 - 30 = -2$. Thus the solution is $(-2, 4)$. *Check:* $3 \cdot (-2) + 2 \cdot 4 = -6 + 8 = 2$ and $7 \cdot 4 - 30 = -2$.

iv. Substituting $y = 2x - 3$ into second equation gives: $4(2x - 3) - 8x = -12$ or $8x - 12 - 8x = -12$ or $0 = 0$; thus the two equations have the same line as their graph. Thus all points on the line $y = 2x - 3$ are solutions.

v. Elimination yields $0 = -6$ and thus there is no solution.

vi. Substituting $y = 3$ into the second equation gives $x + 2 \cdot 3 = 11$ or $x = 5$. Thus the solution is $(5, 3)$. *Check:* $3 \cdot 3 = 9$ and $5 + 2 \cdot 3 = 11$.

5. a. Subtracting the first equation from the second yields $4x = -12$ or $x = -3$. Putting this value into the first equation gives $-3 + 3y = 6$ and thus $3y = 9$ and $y = 3$. Putting this value of x into the second equation gives $5(-3) + 3(3) = -6$ and thus the solution is $(-3, 3)$.

b. The graphs of the two equations and the coordinates of the intersection point are shown in the accompanying figure.

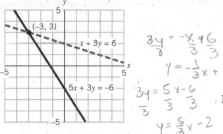

$$\frac{3y}{3} = \frac{-x + 6}{3} \frac{}{3}$$
$$y = -\frac{1}{3}x + 2$$
$$\frac{3y}{3} = \frac{5x - 6}{3} \frac{}{3}$$
$$y = \frac{5}{3}x - 2$$

7. Let $x =$ amount to be invested at 4% and let $y =$ amount to be invested at 8%. Then the system of equations to be solved is $x + y = 2000$ and $0.04x + 0.08y = 100$. Substituting $y = 2000 - x$ into the second equation gives, after simplification, $x = \$1500$ and thus $y = \$500$. [*Check*: $1500 + 500 = 2000$ and $0.04 \cdot 1500 + 0.08 \cdot 500 = 100$.]

9. a. Letting $y = x - 4/3$ from the second equation and substituting this value in the first equation, we get $\frac{x}{3} + \frac{x - 4/3}{2} = 1$. Multiplying both sides by 6 gives $2x + 3(x - 4/3) = 6$ or $5x - 4 = 6$ or $x = 2$ and thus $y = 2 - 4/3 = 2/3$. [*Check*: $2 - 2/3 = 4/3$ and $2/3 + 1/3 = 1$.]

b. Substituting $y = x/2$ from the second equation into the first equation we get $x/4 + x/2 = 9$ or $(3/4)x = 9$ or $x = 12$. Then $y = 12/2 = 6$. [*Check*: $12/4 + 6 = 9$.]

11. a. The two equations in m and b are: $-2 = 2m + b$ and $13 = -3m + b$ and the solution is $m = -3$ and $b = 4$. [*Check*: $2(-3) + 4 = -2$ and $-3(-3) + 4 = 13$.]

b. The two equations in m and b are: $38 = 10m + b$ and $-4.5 = 1.5m + b$. The solution is $m = 5$ and $b = -12$. [*Check*: $38 = 5 \cdot 10 - 12$ and $-4.5 = 5 \cdot 1.5 - 12$.]

13. a. The graphs of the supply and demand equations are in the accompanying diagram.

b. The equilibrium point is shown in the diagram. It is the spot where supply meets the demand; i.e., if the company charges $410 for a bike it will sell exactly 4000 of them and have none left over.

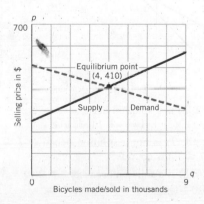

15. Two equations are equivalent if their graphs are the same, i.e., they have the same sets of solutions. An example is the system $2x + y - 1$ and $4x + 2y = 2$.

17. If we make the origin the spot on the diagram where the height (in feet) or H-axis meets the ground and let d be the distance (in feet) from the origin, then the equation of the ramp is $H = 3 - \frac{1}{12}d$. The equation to describe the rising ground is $H = \frac{1}{20}d$. Setting them equal to each other gives $3 - \frac{1}{12}d = \frac{1}{20}d$ or $\frac{36 - d}{12} = \frac{d}{20}$ or $720 - 20d = 12d$ or $720 = 32d$ or $d = 22.5$ ft and $H = 22.5/20 = 1.125$ ft. Thus the point of meeting is where $d = 22.5$ ft from the platform and $H = 1.125$ ft above ground level.

19. Possible solutions:

a. Eliminate z; $11x + 7y = 68$ (4)

b. Eliminate z; $9x + 7y = 62$ (5)

c. $x = 3$ and $y = 5$ satisfy (4) and (5)

d. Thus $z = 2 \cdot 3 + 3 \cdot 5 - 11 = 10$

e. Thus the solution is $x = 3$, $y = 5$, $z = 10$ and the check is below:

  (1) $2 \cdot 3 + 3 \cdot 5 - 10 = 11$

  (2) $5 \cdot 3 - 2 \cdot 5 + 3 \cdot 10 = 35$

  (3) $1 \cdot 3 - 5 \cdot 5 + 4 \cdot 10 = 18$

21. Answers will vary.

a. The system $y = x + 5$ and $y = x + 6$ has no solution.

b. The system $y = x + 5$ and $y = -x + 5$ has exactly one solution.

c. Algebraically: setting $x + 5 = -x + 5$ gives $2x = 0$ or $x = 0$ and thus $y = 5$; alternatively, adding the two equations together gives $2y = 10$ or $y = 5$ and thus $x = 0$. The graphs of the two lines intersecting at the point claimed is in the accompanying diagram. The answers agree.

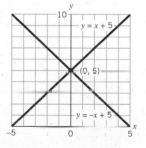

23. The system of equations has no solution if the graphs of the two equations are parallel and distinct lines. This occurs when $m_1 = m_2$ (parallel means same slope) and $b_1 \neq b_2$ (different vertical intercepts).

25. a. $y_B = 30$, $y_A = 0.625x$.

b. The common point in space that both planes will eventually occupy is where $x = 48$ and $y_B = 30$. (It is the intersection point of the graphs of $y = 30$ and $y_A = 0.625x$. These are equations for constant altitude of the flight paths of the two planes.)

c. B, in going from $(-30, 30)$ to $(48, 30)$, travels a distance of 78 miles, and this takes B 13 minutes to do (since it is traveling at 6 miles/minute). A, in traveling from $(80, 50)$ to $(48, 30)$, covers $\sqrt{(30 - 50)^2 + (48 - 80)^2} = \sqrt{400 + 1024} = \sqrt{1424} \approx 37.7$ miles, and this will take approximately 18.9 minutes (since plane A is traveling at 2 miles per minute). Thus plane A will arrive at this point nearly 6 minutes after plane B. It is a safe situation.

27. **a.** For a given price the new supply curve shows more items being made.

b. The requested graph is given in the diagram. In going from the old equilibrium point to the new one, the price goes down and the quantity made/sold goes up at the equilibrium point.

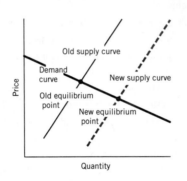

29. **a.** Higher birth rate:

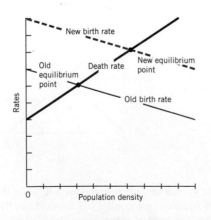

If the birth rate increases (and the death rate stays the same) the equilibrium point moves to the right and up. This means that the equilibrium point will occur at a greater population density and a higher birth rate.

b. Lower birth rate:

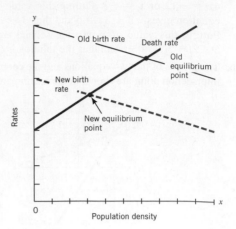

If the birth rate decreases (and the death rate stays the same) the equilibrium moves to the left and down. This means the equilibrium point will occur at a lesser population density and lower birth rate.

Section 3.3

Algebra Aerobics 3.3

1. **a.**   **d.**   **b.**   **e.**   **c.**   **f.**

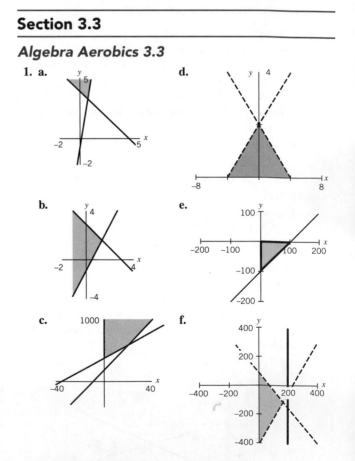

2. **a.** $(2, 3)$ is a solution because: $3 > 2(2) - 3 \Rightarrow 3 > 1$ and $3 \leq 3(2) + 8 \Rightarrow 3 \leq 14$ are true.

b. $(-4, 7)$ is not a solution because: $7 \leq 3(-4) + 8 \Rightarrow 7 \leq -4$ is not true.

c. $(0, 8)$ is a solution because: $8 > 2(0) - 3 \Rightarrow 8 > -3$ and $(8) \leq 3(0) + 8 \Rightarrow 8 \leq 8$ are true.

d. $(-4, -6)$ is a solution because: $-6 > 2(-4) - 3 \Rightarrow$ $-6 > -11$ and $-6 \leq 3(-4) + 8 \Rightarrow -6 \leq -4$ are true.

e. $(20, -8)$ is not a solution because: $-8 > 2(20) - 3 \Rightarrow$ $-8 > 37$ is not true.

f. $(1, -1)$ is not a solution because: $-1 > 2(1) - 3 \Rightarrow$ $-1 > -1$ is not true.

3. A. $y \leq 2 - x$ **B.** $y > 1 + 2x$ **C.** $y \geq -3$ **D.** $x > 4$

4. a. Approximately (100, $700). For sales of 100 books, the cost is equal to the revenue, which is $700.

b. The region between the two graphs to the left of the breakeven point.

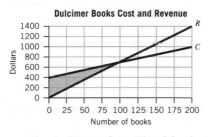

Dulcimer Books Cost and Revenue

c. $400 because that is the cost for selling 0 books (vertical intercept of the cost equation).

d. Assuming fixed costs at $400, $C_1 \geq 3x + 400$, $R_1 \leq 7x$.

5. a. $C(n) = \$500,000 + 235n$; $R(n) = 270n$.

b. $C(n) = R(n)$ at breakeven point $\Rightarrow 500,000 + 235n = 270n \Rightarrow n \approx 14,286$ tons and $C(14,286) = R(14,286) = \$3,857,143$. Selling about 14,286 tons will yield a profit of $0 since cost = revenue at the breakeven point.

c.

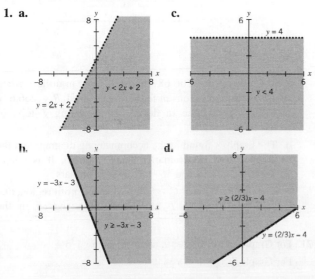

Exercises for Section 3.3

1. a.   **c.**

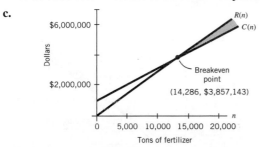

b.   **d.**

3. a. $y < \frac{2}{3}x + 2$

b. $y \geq -\frac{3}{2}x + 3$

5. a. Yes, $(0, 0)$ satisfies the inequality.

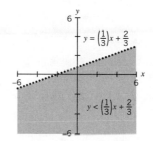

b. No, $(0, 0)$ does not satisfy the inequality.

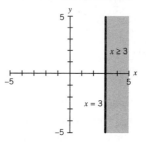

c. Yes, $(0, 0)$ satisfies the inequality.

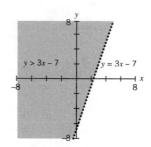

d. No, $(0, 0)$ does not satisfy the inequality.

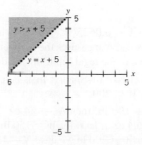

7. In this case the shaded region is above the line, since $y > \frac{3}{5}x - 3$.

Ch. 3

9. a. goes with **g.**   **c.** goes with **j.**   **e.** goes with **h.**
   b. goes with **i.**   **d.** goes with **f.**

11. a. l_1 has the equation $y = 1 + 0.25x$ and l_2 has the equation $y = 3 - 1.5x$.
   b. $3 - 1.5x \le y \le 1 + 0.25x$

13.

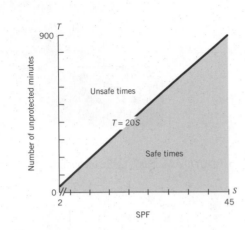

a. $T = 20S$
b. The graph is in the diagram. A suitable domain is $2 \le S \le 45$.
c. $T > 20S$ denotes unsafe times.
d. The shading and labels are found in the diagram.
e. The equation would be $T = 40S$ and its slope would be steeper.

15.

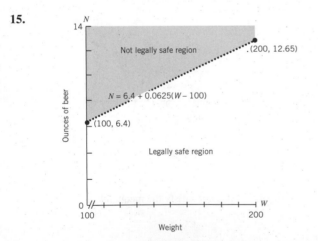

a. $N > 6.4 + 0.0625(W - 100)$, where W is measured in pounds and N measures the number of ounces of beer that gets one to the legal limit for safe driving.
b. The sketch of the shaded areas is found in the diagram for $100 \le W \le 200$.
c. If $W = 100$ lb, then $N = 6.4$ oz. Thus one may legally drink 6.4 oz or less; for $W = 150$ lb we have $N = 9.525$ oz and for $W = 200$ lb we have $N = 12.65$ oz.
d. $N = 0.0625W + 0.15$
e. The given rule of thumb translates into the formula $N = 6 + 0.05(W - 100) = 0.050W + 1$. Thus it starts

out higher and grows more slowly than the legal one. But its graph is lower from $W = 100$ to $W = 200$. It is a safe rule.

17.

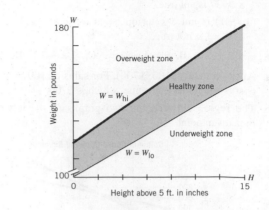

a. The two formulae and the three zones are graphed in the diagram.
b. $100 + 3.5H \le W \le 118.2 + 4.2H$ lb for $0 \le H \le 15$ in above 5 ft.
c. $W_{lo}(2) = 107$ lb and $W_{hi}(2) = 126.6$ lb. Thus the shorter woman is overweight. For the taller woman: $W_{lo}(5) = 117.5$ and $W_{hi}(5) = 139.2$. Thus the taller woman is in the healthy range.
d. $W_{hi}(4) = 135$ and $(165 - 135)/1.5 = 30/1.5 = 20$. Thus it would take 20 weeks for this woman to reach the top of the healthy range.

19.

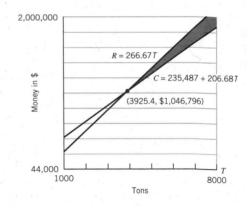

a. $C = 235,487 + 206.68T$ gives the cost in dollars when T is measured in tons of fertilizer produced. $R = 266.67T$ gives the revenue in dollars from selling T tons of fertilizer.
b. The graph is found in the accompanying diagram and the breakeven point is marked on the graph. It is where $T \approx 3925.4$ tons and $M \approx \$1,046,800$ dollars
c. The inequality $R - C > 0$ describes the profit region, and this occurs when $T > 3925.4$. It is shaded in the accompanying graph.

21. For Graph A: $x \ge 0$, $y \ge 0$, and $y < -1.5x + 3$
   For Graph B: $x + 1 \le y < 2x + 2$.

23.

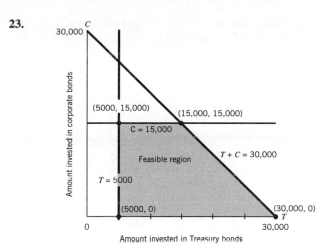

a. $0 \leq T + C \leq 30,000$, $0 \leq C \leq 15,000$ and $5000 \leq T \leq 30,000$

b. The feasible region is the shaded area of the graph.

c. Intersection points and interpretations: (5000, 0) is where $5000 is invested in T bonds; (30000, 0) is where all $30,000 is in T bonds; (5000, 15000) is where $5000 is in T bonds and $15,000 is in C bonds; and (15000, 15000) is where $15,000 is in each kind.

25. a. 12.5 is the production cost per shirt in dollars.

b. 15.5 is the selling price per shirt in dollars.

c. $15.5x = 12.5x + 360$ or $3x = 360$ or $x = 120$ units produced and sold and $C = 12.5 \cdot 120 + 360 = \$1860$. The breakeven point is (120, 1860).

d. When $x = 120$ units, then $C = 12.5 \cdot 120 + 360 = \$1860$ and $R = 15.5 \cdot 120 = \$1860$.

e.

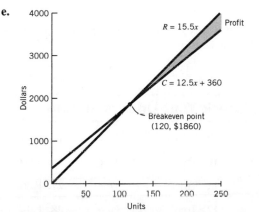

27. a. This is the area between the two graphs: above $P(t)$ and below $C(t)$. It describes the net imports during that time.

b. $P(t) \leq$ net imports $\leq C(t)$ (for $1980 \leq t \leq 2035$)

c. Student answers will vary. They should mention, for example, that energy consumption and production for the most part grew over that period from about 70 quadrillion BTU to 85 quadrillion BTU for production and 115 quadrillion BTU for consumption. Also, consumption was always more than production during that period; the gap was widest in 2007, but the gap is predicted to remain fairly constant after 2008.

29. a. It would represent those areas affected by moderate drought.

b. From the beginning of July 2008 to the middle of February 2009.

Section 3.4

Algebra Aerobics 3.4

1. a.

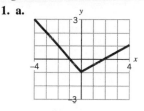

b.

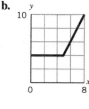

c.

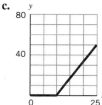

2. Graph A: $f(x) = \begin{cases} x + 3 & \text{for } x \leq 3 \\ -2x + 12 & \text{for } x > 3 \end{cases}$

     Graph B: $f(x) = \begin{cases} -2 & \text{for } x \leq 3 \\ 2x - 8 & \text{for } x > 3 \end{cases}$

3. a. $P(-5) = 3$, $P(0) = 3$, $P(2) = -3$, $P(10) = -19$

b. $W(-5) = -9$, $W(0) = -4$, $W(2) = 6$, $W(10) = 14$.

4. a. $A(i) = \begin{cases} 0.05i & \text{for } 0 \leq i \leq \$50,000 \\ 2500 + 0.15(i - 50,000) & \text{for } i > \$50,000 \end{cases}$

b. $B(i) = 0.10\,i$

c. Possible solution/s at intersection point/s:
To find if first segment of $A(i)$ intersects with $B(i)$:

$$0.10i = 0.05i$$
$$i = 0$$

To find tax, substitute $i = 0$ into $A(i)$ or $B(i)$, to get $A(0) = B(0) = 0$, so (0, 0) is a solution for this system.

To find if second segment of $A(i)$ intersects with $B(i)$:

$$0.10i = 2500 + 0.15(i - 50,000)$$
$$0.10i = 2500 + 0.15i - 7,500$$
$$-0.05i = -5,000$$
$$i = 100,000$$

To find tax, substitute $i = 100,000$ into $A(i)$ or $B(i)$, to get $A(100,000) = B(100,000) = \$10,000$, so (100000, 10000) is another solution for this system.

Ch. 3

Exercises for Section 3.4

1. Graph *B* goes with (a).

Graph *A* goes with (b).

3. a. The graph of $h(x)$ and $j(x)$:

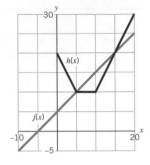

b. The estimated intersection points are (5, 10) and (15, 20).
For the first intersection let $h(x) = j(x)$ or $10 = 5 + x \Rightarrow$
$x = 5$; then $h(5) = j(5) = 10$.
For the second intersection let $h(x) = j(x)$ or
$10 + 2(x - 10) = 5 + x \Rightarrow -10 + 2x = 5 + x \Rightarrow x = 15$;
then $h(15) = j(15) = 20$.

5. The graph is given here to help one see the answers:

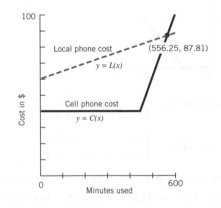

a. $C(x) = \begin{cases} 40 & \text{if } 0 \le x \le 450 \\ 40 + 0.45(x - 450) & \text{if } x > 450 \end{cases}$

$L(x) = 60 + 0.05x \quad \text{if } x \ge 0$

where x measures minutes used for long distance and $C(x)$
and $L(x)$ are measured in dollars.

The two cost functions are graphed in the diagram. They
meet at $x = 556.25$ minutes and $C(x) = L(x) \approx \$87.81$.
Thus the two plans cost the same at the point where one
uses 556.25 minutes for long distance.

c. It would be more advantageous to use the cell phone for
$0 \le x < 556.25$ minutes

d. It would be more advantageous to use the local company
plan if $x > 556.25$ minutes.

7. The graphs for parts (a) and (b) are shown in the
accompanying diagram.

a. For $0 \le T \le 20$: $D_{\text{beginner}} = (3.5/60)T$ or $0.0583T$ since
there is 1/60 of an hour in a minute; note that T is
measured in minutes and D_{beginner} is measured in miles.

b. For $0 \le T \le 10$: $D_{\text{advanced}} = (3.75/60)T$ or $0.0625T$ and for
$10 < T \le 20$ we have $D_{\text{advanced}} = 0.625 + (5.25/60)(T - 10) =$
$0.0875T - 0.25$.

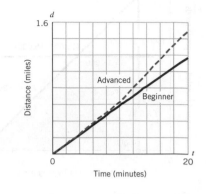

T	D_{advanced}	D_{beginner}
0	0.0000	0.0000
5	0.3125	0.2915
10	0.6250	0.5830
15	1.0625	0.8745
20	1.5000	1.1660

c. The graphs intersect only at $T = 0$, or at the beginning.

9. a. $H = 7D$ where $D \ge 0$

$H_v = \begin{cases} 10.5D & \text{for } 0 \le D \le 2 \\ 21 + 4D & \text{for } D > 2 \end{cases}$

b. Set $H = H_v$

$7D = 10.5D \Rightarrow D = 0$
$7D = 21 + 4D \Rightarrow D = 7$

The two methods give the same results when dog years = 0
and human years = 0 and when dog years = 7 years and
human years = 49 years.

Ch. 3: Check Your Understanding

1. True	**9.** False	**17.** True	**25.** True
2. False	**10.** True	**18.** True	**26.** True
3. False	**11.** True	**19.** True	**27.** False
4. True	**12.** True	**20.** False	**28.** False
5. True	**13.** False	**21.** False	**29.** True
6. False	**14.** True	**22.** True	**30.** True
7. False	**15.** False	**23.** False	
8. False	**16.** False	**24.** True	

31. Possible answer: $\begin{cases} 2x + 3y = 6 \\ 4x + 6y = 10 \end{cases}$

32. Possible answer: $\begin{cases} 2x + y = 7 \\ -6x - 3y = -21 \end{cases}$

33. Possible answer: $\begin{cases} y > 2x + 1 \\ y < 2x - 5 \end{cases}$

34. Possible answer: $\begin{cases} c = r + 1 \\ c = -r - 1 \end{cases}$

35. Possible answer: $\begin{cases} C = 25q + 2500 \\ R = 50q \end{cases}$

Ch. 3 Review: Putting It All Together

1. a. The maximum production occurred in 2008; it was about 682 MMT.
   The minimum production occurred in 2006; it was about 595 MMT.

   b. The intersection point means that consumption equaled production at the beginning of 2007.

   c. The difference was the largest in 2008. There was a surplus.

3. a. Recall that if two lines are perpendicular to each other (and neither is horizontal) then their slopes are negative reciprocals of each other.

   $y = -4x + 26$ and $y = x/4 + 9$

   b.

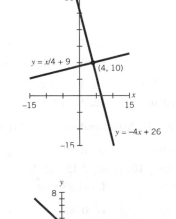

5.

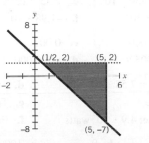

7. a. $C = 10{,}000 + 7x$; $R = 12x$, where x is the number of CDs.

   b. $C = R \rightarrow 10{,}000 + 7x - 12x \Rightarrow x = 2000$ CDs. The breakeven point is (2000, 24000).

   c. If p is the new price per CD, then $P(1600) = 10{,}000 + 7(1600) = \$21{,}200 \Rightarrow p = \$13.25$ per CD. She would need to raise the price to $13.25 for each CD.

   d. If $c = $ new fixed cost, then $c + 7(1600) = 12(1600) \Rightarrow c = \$8000$. She would need to reduce fixed costs by $2000.

9. a. Answers may vary. In 1990 oil consumption in Canada was approximately 1750 thousand barrels per day and production was approximately 1550 thousand barrels per day. The net difference between production and consumption in 1990 was a negative 200 thousand barrels per day. In 1990, Canada was consuming more oil than it was producing and would need to import oil. In 2008, Canada was producing about 2600 thousand barrels of oil per day and consuming about 2300 thousand barrels per day. In 2008, Canada was producing about 300 barrels of oil per day more than it was consuming. This may have been exported or stored.

   b. The graphs intersect in 1995, 1997, 1998, and 2001. At these points, production and consumption are the same. Hence, all the oil produced was consumed.

   c. Canada had to import oil when its consumption was more than its production. This occurred from 1986 to approximately 2001. Canada exported oil from approximately 2001 to 2008, when production was greater than consumption.

11. a. i. $x = -5$ and $y = 5/7$;   **ii.** $a = 3$ and $b = 0.5$

   b. Answers will vary. A system of two equations whose graphs are two distinct parallel lines will not have a solution.

13. a. $s + r \le 60$ minutes: $8s + 10r \ge 560$ calories; $s \ge 0$ and $r \ge 0$.

   b.

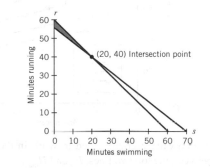

   c. There are many answers, for example: $s = 10$, $r = 50$ minutes is in the solution set and $s = 10$, $r = 40$ minutes is not in the solution set.

   d. $r + s \le 70$ minutes; $10r + 8s \ge 560$ calories; $s \ge 0$ and $r \ge 0$. The intersection point of the boundary lines changes and the shaded area representing the solution set increases.

15.

   a.

   $$A(m) = \begin{cases} 39.99 & \text{for } 0 \le m \le 450 \\ 39.99 + 0.45(m - 450) & \text{for } 450 < m \le 2500 \end{cases}$$

   b.

   $$B(m) = \begin{cases} 59.99 & \text{for } 0 \le m \le 900 \\ 59.99 + 0.40(m - 900) & \text{for } 900 < m \le 2500 \end{cases}$$

   c.

   $$C(m) = \begin{cases} 79.99 & \text{for } 0 \le m \le 1350 \\ 79.99 + 0.35(m - 1350) & \text{for } 1350 < m \le 2500 \end{cases}$$

d.

Number of Minutes Used/Month	Cost		
	Plan A	Plan B	Plan C
500	39.99 + 0.45(50) = $62.49	59.99	79.99
800	39.99 + 0.45(350) = $197.49	59.99	79.99
1000	39.99 + 0.45(550) = $287.49	59.99 + 0.40(100) = $99.99	79.99

17. Answers will vary. Answers should include some mention of the fact that real prices are always less than nominal ones; that real prices were much less than nominal ones in the 1860s, 1870s, and 1980s; that prices were relatively stable from 1879 to 1975 and again in 1989–1999, but rose after 1999 to an all-time high in 2008.

19. a. $H_b = 132 - 0.60A$

   b. $H_i = 154 - 0.70A$

   c. $H_a = 187 - 0.85A$

   d.

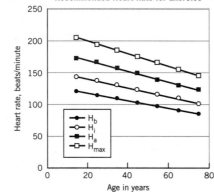

Recommended Heart Rate for Exercise

Athletes are recommended to work in the zone on and between the top two lines, H_a and H_{max}.

   e. $H_b = 120, H_i = 140, H_a = 170, H_{max} = 200$

   f. 65-year-old: $I \approx 86\%$. She is just below her $H_{max} = 220 - 65 = 155$ beats per minute.

     45-year-old: $I \approx 77\%$

     25-year-old: $I \approx 69\%$

CHAPTER 4

Section 4.1

Algebra Aerobics 4.1

1. a. 10^{10}: to express 10 billion as a power of 10, start with 1.0, then count the ten place values the decimal must be moved to the right, in order to produce 10 billion.

   b. 10^{-14}: the decimal point in 1.0 must be moved 14 place values to the left to produce 0.000 000 000 000 01.

   c. 10^5

   d. 10^{-5}

2. a. 0.000 000 01

   b. 10,000,000,000,000

   c. 0.000 1

   d. 10,000,000

3. a. 10^{-9} or 0.000 000 001 sec

   b. 10^3 or 1000 m

   c. 10^9 or 1,000,000,000 bytes (a byte is a term used to describe a unit of computer memory).

4. a. $7 \text{ cm} \cdot \dfrac{1 \text{ m}}{100 \text{ cm}} = 7 \cdot \dfrac{1}{10^2} \text{ m} = 7 \cdot 10^{-2}$ or 0.07 m

   b. $9 \text{ mm} \cdot \dfrac{1 \text{ m}}{1000 \text{ mm}} = 9 \cdot \dfrac{1}{10^3} \text{ m} = 9 \cdot 10^{-3}$ or 0.009 m

   c. $5 \text{ km} \cdot \dfrac{1000 \text{ m}}{1 \text{ km}} = 5 \cdot 10^3$ or 5000 m

5. 602,000,000,000,000,000,000,000,000

6. $3.84 \cdot 10^8$ m

7. $1 \cdot 10^{-8}$ cm

8. 0.000 000 002 m

9. a. $-705,000,000$ **c.** 5,320,000

   b. $-0.000 040 3$ **d.** 0.000 000 102 1

10. a. $-4.3 \cdot 10^7$ **c.** $5.83 \cdot 10^3$

   b. $-8.3 \cdot 10^{-6}$ **d.** $2.41 \cdot 10^{-8}$

Exercises for Section 4.1

1. a. 10^6 **c.** 10^9 **e.** 10^{13}

   b. 10^{-5} **d.** 10^{-3} **f.** 10^{-8}

3. a. $1 \cdot 10^{-1}$ m   **c.** $3 \cdot 10^{12}$ m

   b. $4 \cdot 10^3$ m   **d.** $6 \cdot 10^{-9}$ m

5. gigabyte $= 10^9$ bytes; terabyte $= 10^{12}$ bytes.

7. a. $2.9 \cdot 10^{-4}$ **d.** 10^{-11} **g.** $-4.9 \cdot 10^{-3}$

   b. $6.54456 \cdot 10^2$ **e.** $2.45 \cdot 10^{-6}$

   c. $7.2 \cdot 10^5$ **f.** $-1.98 \cdot 10^6$

9. a. 723,000   **c.** 0.001   **e.** 0.000188

   b. 0.000526   **d.** 1,500,000   **f.** 67,800,000

11. a. False; $7.56 \cdot 10^{-3}$ **d.** False; $1.596 \cdot 10^9$

   b. True   **e.** True

   c. False; $4.9 \cdot 10^7$ watts   **f.** False; $6 \cdot 10^{-12}$ second

13. a. 9   **b.** 9   **c.** 1000   **d.** -1000

Section 4.2

Algebra Aerobics 4.2a

1. a. $10^5 \cdot 10^7 = 10^{5+7} = 10^{12}$

   b. $8^6 \cdot 8^{14} = 8^{6+14} = 8^{20}$

   c. $z^5 \cdot z^4 = z^{5+4} = z^9$

   d. Cannot be simplified because bases, 5 and 6, are different.

e. $7^3 + 7^3 = 7^3(1 + 1) = 2 \cdot 7^3$

f. $5 \cdot 5^6 = 5^1 \cdot 5^6 = 5^7$

g. $3^4 + 7 \cdot 3^4 = 3^4(1 + 7) = 8 \cdot 3^4$ or $2^3 \cdot 3^4$

h. $2^3 + 2^4 = 2^3 + 2^3 \cdot 2^1 = 2^3(1 + 2^1) = 3 \cdot 2^3$

i. Cannot be simplified because bases, 2 and 5, are different.

2. a. $\dfrac{10^{15}}{10^7} = 10^{15-7} = 10^8$

b. $\dfrac{8^6}{8^4} = 8^{6-4} = 8^2$

c. $\dfrac{3^5}{3^4} = 3^{5-4} = 3^1$ or 3

d. Cannot be simplified because bases, 5 and 6, are different.

e. $\dfrac{5^1}{5^6} = 5^{-5}$ **f.** $\dfrac{3^4}{3^1} = 3^3$

g. $\dfrac{2^3 \cdot 3^4}{2^1 \cdot 3^2} = \dfrac{2^3}{2^1} \cdot \dfrac{3^4}{3^2} = 2^{3-1} \cdot 3^{4-2} = 2^2 \cdot 3^2$

h. $\dfrac{6}{2^4} = \dfrac{2 \cdot 3}{2^4 \cdot 1} = \dfrac{2^1}{2^4} \cdot \dfrac{3}{1} = 2^{1-4} \cdot 3 = 3 \cdot 2^{-3}$

3. a. $10^5 \cdot 10^6 = 10^{5+6} = 10^{11}$

b. $10^3 \cdot 10^{-6} = 10^{3+(-6)} = 10^{-3}$

c. $10^{-11} \cdot 10^{-5} = 10^{-11+(-5)} = 10^{-16}$

d. $10^9 \cdot 10^{-4} = 10^{9+(-4)} = 10^5$

e. $10^6 \cdot 10^{-(-3)} = 10^{6+3} = 10^9$

f. $10^{-5} \cdot 10^{-(-4)} = 10^{-5+4} = 10^{-1}$

g. $10^{-6} \cdot 10^{-4} = 10^{-6+(-4)} = 10^{-10}$

4. a. $10^{4(5)} = 10^{20}$

b. $7^{2(3)} = 7^6$

c. $x^{4(5)} = x^{20}$

d. $(2x)^4 = 2^4 x^4$ or $16x^4$

e. $(2a^4)^3 = 2^3(a^4)^3 = 2^3 a^{12} = 8a^{12}$

f. $(-2a)^3 = (-2)^3 a^3 = -8a^3$

g. $(-3x^2)^3 = (-3)^3(x^2)^3 = -27x^6$

h. $((x^3)^2)^4 = (x^{3\cdot2})^4 = (x^6)^4 = x^{6\cdot4} = x^{24}$

i. $(-5y^2)^3 = (-5)^3(y^2)^3 = -125y^6$

5. a. $\dfrac{(-2x)^3}{(4y)^3} = \dfrac{(-2)^3 \cdot x^3}{4^3 \cdot y^3} = \dfrac{-8x^3}{64y^3} = \dfrac{-x^3}{8y^3}$

b. $(-5)^2 = (-5)(-5) = 25$

c. $-5^2 = -(5)(5) = -25$

d. $-3(yz^2)^4 = -3(y)^4(z^2)^4 = -3y^4z^8$

e. $(-3yz^2)^4 = (-3)^4(y)^4(z^2)^4 = 81y^4z^8$

f. $(-3yz^2)^3 = (-3)^3(y)^3(z^2)^3 = -27y^3z^6$

6. $\dfrac{3.0 \cdot 10^{12}}{4.7 \cdot 10^9} \approx 0.638 \cdot 10^3$ or   638 DVDs

7. a. $(3 + 5)^3 = 8^3 = 512$

b. $3^3 + 5^3 = 27 + 125 = 152$

c. $3 \cdot 5^2 = 3 \cdot 25 = 75$

d. $-3 \cdot 5^2 = -3 \cdot 25 = -75$

Algebra Aerobics 4.2b

1. a. $(0.000\,297\,6)(43,990,000) \approx (0.000\,3)(10,000,000)$
$= 3 \cdot 10^{-4} \cdot 4 \cdot 10^7 = 12 \cdot 10^3 = 12,000$

b. $\dfrac{453,897 \cdot 2,390,702}{0.004\,38} \approx \dfrac{500,000 \cdot 2,000,000}{0.004}$

$= \dfrac{(5 \cdot 10^5)(2 \cdot 10^6)}{4 \cdot 10^{-3}} = \dfrac{10}{4} \cdot \dfrac{10^{11}}{10^{-3}}$

$= 2.5 \cdot 10^{14} \approx 3 \cdot 10^{14}$

c. $\dfrac{0.000\,000\,319}{162,000} \approx \dfrac{0.000\,000\,3}{200,000} = \dfrac{3 \cdot 10^{-7}}{2 \cdot 10^5}$

$= 1.5 \cdot 10^{-12} \approx 2 \cdot 10^{-12}$

d. $28,000,000 \cdot 7629 \approx 30,000,000 \cdot 8000$
$= 3 \cdot 10^7 \cdot 8 \cdot 10^3 = 24 \cdot 10^{10}$
$= 2.4 \cdot 10^{11} \approx 2 \cdot 10^{11}$

e. $0.000\,021 \cdot 391,000,000 \approx 0.000\,02 \cdot 400,000,000$
$= 2 \cdot 10^{-5} \cdot 4 \cdot 10^8 = 8 \cdot 10^3 = 8,000$

2. a. $(3.0 \cdot 10^3)(4.0 \cdot 10^2) = 12 \cdot 10^5$
$= 1.2 \cdot 10^6 = 1,200,000$

b. $\dfrac{(5.0 \cdot 10^2)^2}{2.5 \cdot 10^3} = \dfrac{25 \cdot 10^4}{25 \cdot 10^2} = 1.0 \cdot 10^2 = 100$

c. $\dfrac{2.0 \cdot 10^5}{5.0 \cdot 10^3} = \dfrac{20 \cdot 10^4}{5.0 \cdot 10^3} = 4 \cdot 10^1 = 40$

d. $(4.0 \cdot 10^2)^3(2.0 \cdot 10^3)^2 = (4^3 \cdot 10^6)(4 \cdot 10^6)$
$= 4^4 \cdot 10^{12} = 256 \cdot 10^{12} = 2.56 \cdot 10^{14}$
$= 256,000,000,000,000$

3. If we use 3.14 to approximate π:

a. Surface area of Jupiter $= 4\pi r^2 \approx 4\pi(7.14 \cdot 10^4\,\text{km})^2$

$= 4\pi(7.14)^2(10^4)^2\,\text{km}^2$

$\approx 4(3.14)(50.98)10^8\,\text{km}^2$

$\approx 640 \cdot 10^8\,\text{km}^2$

$\approx 6.4 \cdot 10^2 \cdot 10^8\,\text{km}^2$

$\approx 6.4 \cdot 10^{10}\,\text{km}^2$

b. Volume of Jupiter $= \frac{4}{3}\pi r^3$

$\approx (1.3) \cdot (3.14)(7.14 \cdot 10^4\,\text{km})^3$

$\approx (4.08)(7.14)^3(10^4)^3\,\text{km}^3$

$\approx (4.08)(364)10^{12}\,\text{km}^3$

$\approx 1486 \cdot 10^{12}\,\text{km}^3$

$\approx 1.486 \cdot 10^3 \cdot 10^{12}\,\text{km}^3$

$\approx 1.486 \cdot 10^{15}\,\text{km}^3$

4. If only 3/7 of the farmable land is used, the people/sq. mi. of used farmland is:

$$\dfrac{6.8 \cdot 10^9\,\text{people}}{(3/7)12 \cdot 10^6\,\text{sq. mi.}} = \left(\dfrac{7}{3}\right)\dfrac{6.8 \cdot 10^9\,\text{people}}{12 \cdot 10^6\,\text{sq. mi.}}$$

$$\approx 1.322 \cdot 10^3 \text{ or } 1322 \text{ people/sq. mi.}$$

For fractions >0, if the denominator is decreased, the value of that fraction is increased. So, one expects this ratio to be larger than the ratio of people to farmable land.

Ch. 4

Exercises for Section 4.2

1. a. 10^7

   b. $1.1 \cdot 10^4$

   c. $2 \cdot 10^3$

   d. x^{15}

   e. x^{50}

   f. $4^7 + 5^2$—this expression cannot be simplified without multiplying out the values and adding them together.

   g. z^5 **i.** 3^{-1} or $1/3$

   h. 1 or as is   **j.** 4^{11}

3. a. $16a^4$ **e.** $32x^{20}$

   b. $-2a^4$ **f.** $18x^6$

   c. $-x^{15}$ **g.** $2500a^{20}$

   d. $-8a^3b^6$ **h.** as is—nothing is simpler

5. a. $-\left(\frac{5}{8}\right)^2 = -\frac{25}{64}$

   b. $\left(\frac{3x^3}{5y^2}\right)^3 = \frac{3^3 x^9}{5^3 y^6} = \frac{27x^9}{125y^6}$

   c. $\left(\frac{-10x^5}{2b^2}\right)^4 = \frac{10^4 x^{20}}{4^2 b^8} = \frac{10{,}000x^{20}}{16b^8} = 625\frac{x^{20}}{b^8}$

   d. $\left(\frac{-x^5}{x^2}\right)^3 = -x^9$

7. a. $(2 \cdot 10^6) \cdot (4 \cdot 10^3) = 8 \cdot 10^9$

   b. $(1.4 \cdot 10^6) \div (7 \cdot 10^3) = 0.2 \cdot 10^3 = 2 \cdot 10^2$

   c. $(5 \cdot 10^{10}) \cdot (6 \cdot 10^{13}) = 30 \cdot 10^{23} = 3 \cdot 10^{24}$

   d. $(2.5 \cdot 10^{12}) \div (5 \cdot 10^5) = 0.5 \cdot 10^7 = 5 \cdot 10^6$

9. a. $x^{13}y^4$ **c.** $-8x^9y^9$ **e.** $81x^8y^{20}$

   b. $5x^4y$ **d.** $16x^{10}y^8$ **f.** $\frac{9}{25}x^4$

11. a. $10^9/10^6 = 10^3 = 1000$

   b. $1000/10 = 10^2 = 100$

   c. $1000/0.001 = 10^6 = 1{,}000{,}000$

   d. $10^{-6}/10^{-9} = 10^3 = 1000$

13. a. $\dfrac{127.4 \times 10^6 \text{ people}}{146.9 \times 10^3 \text{ square miles}} = 0.8673 \times 10^3$ people/square mile ≈ 867 people/square mile

   b. $\dfrac{309.3 \times 10^6 \text{ people}}{3720 \times 10^3 \text{ square miles}} = 0.08315 \times 10^3$ people/square mile ≈ 83.1 people/square mile

   c. $867/83.1 = 10.43$; therefore, Japan is more than 10 times as densely populated as the United States.

15. a. Worst-case scenario $-$ best-case scenario $= 4.2 \times 10^6$ gal/day $- 2.1 \times 10^5$ gal/day $= 3.99 \times 10^6$ gal/day.

   b. The worst-case scenario is $\dfrac{4.2 \times 10^6 \text{ gal/day}}{2.1 \times 10^5 \text{ gal/day}} = 20$ times larger, or one order of magnitude larger than the best-case scenario.

   c. The amount of oil spilled after 105 days for the worst case is 4.2×10^6 gal/day $\times$ 105 days $= 4.41 \times 10^8$ gallons and for the best case is 2.1×10^5 gal/day $\times$ 105 gal $= 2.205 \times 10^7$ gallons.

17. a. If a is positive, then $-a$ is negative. If n is even, then $(-a)^n$ is positive; but if n is odd, then $(-a)^n$ is negative. If a is negative, then $-a$ is positive, then $(-a)^n$ is positive whether n is even or odd.

   b. This is answered in part (a).

19. a. $\left(\frac{m^2n^3}{mn}\right)^2 = (mn^2)^2 = m^2n^4$ and $\left(\frac{m^2n^3}{mn}\right)^2 = \frac{m^4n^6}{m^2n^2} = m^2n^4$

   b. $\left(\frac{2a^2b^3}{ab^2}\right)^4 = (2ab)^4 = 16a^4b^4$ and

   $\left(\frac{2a^2b^3}{ab^2}\right)^4 = \frac{16a^8b^{12}}{a^4b^8} = 16a^4b^4$

21. $\left(\frac{2a^3}{5b^2}\right)^4 = \frac{2^4 a^{12}}{5^4 b^8} = \frac{16a^{12}}{625b^8}$

23. Two cases are distinguished:

If $n = 0$, then $(ab)^0 = 1$ and $a^0 \cdot b^0 = 1 \cdot 1 = 1$

If $n > 0$, then $a^n = a \cdot \cdots \cdot a$ (n factors) and $b^n = b \cdot \cdots \cdot b$ (n factors) and thus $a^n \cdot b^n = (a \cdot \cdots \cdot a) \cdot (b \cdot \cdots \cdot b) = (ab) \cdot \cdots \cdot (ab)$ (n factors), after rearrangement, and this product is what is meant by $(ab)^n$ when $n > 0$.

25. a. (53 terawatt-hours)/(60 $\cdot$ 10^6 persons) $= 8.83 \cdot 10^{-7}$ terawatt-hours per person in the United Kingdom and (53 terawatt-hours)/(94 $\cdot$ 10^3 sq. mi.) $= 5.64 \cdot 10^{-4}$ terawatt-hours per sq. mi. in the United Kingdom.

   b. (805 terawatt-hours)/(309 $\cdot$ 10^6 persons) $= 2.6 \cdot 10^{-6}$ terawatt-hours per person in the United States and (805 terawatt-hours)/(3.72 $\cdot$ 10^6 sq. mi.) $= 2.16 \cdot 10^{-4}$ terawatt-hours per sq. mi. in the United States.

   c. $(5.64 \cdot 10^{-4})/(2.16 \cdot 10^{-4}) = 2.61$; thus, the United Kingdom generates 2.61 times more terawatt-hours per sq. mi. than does the United States.

   d. The terawatt-hours per person generated by the United States compared to the United Kingdom is $(2.6 \cdot 10^{-6})/(8.83 \cdot 10^{-7}) = 2.94$, so the United States generates almost three times the terawatt-hours per person as the United Kingdom generates.

Section 4.3

Algebra Aerobics 4.3a

1. a. $10^{5-7} = 10^{-2} = \frac{1}{10^2}$

   b. $11^{6-(-4)} = 11^{6+4} = 11^{10}$

   c. $3^{-5-(-4)} = 3^{-5+4} = 3^{-1} = 1/3$

   d. Cannot be simplified: different bases, 5 and 6.

   e. $7^{3-3} = 7^0 = 1$

   f. $a^{-2+(-3)} = a^{-5} = \frac{1}{a^5}$

   g. $3^4 \cdot 3^3 = 3^7$

   h. $(2^2 \cdot 3) \cdot (2^6) \cdot (2^4 \cdot 3) = 2^2 \cdot 2^6 \cdot 2^4 \cdot 3 \cdot 3 = 2^{12} \cdot 3^2$

2. Time for a TV signal to travel across the
United States = (time to travel 1 kilometer) · (number of kilometers)

$$= (3.3 \cdot 10^{-6}) \text{ sec/km} \cdot (4.3 \cdot 10^3) \text{ km}$$
$$= (3.3 \cdot 4.3) \cdot (10^{-6} \cdot 10^3) \text{ sec}$$
$$= 14 \cdot 10^{-3} \text{ sec}$$
$$= 1.4 \cdot 10 \cdot 10^{-3} \text{ sec}$$
$$= 1.4 \cdot 10^{-2} \text{ sec or } 0.014 \text{ sec}$$

So it would take less than two-hundredths of a second for the signal to cross the United States.

3. a. $x^{-2}(x^5 + x^{-6}) = x^{-2}(x^5) + x^{-2}(x^{-6})$
$$= x^3 + x^{-8}$$
$$= x^3 + \frac{1}{x^8}$$

b. $-a^2(b^2 - 3ab + 5a^2)$
$$= b^2(-a^2) - 3ab(-a^2) + 5a^2(-a^2)$$
$$= -a^2b^2 + 3a^{1+2}b - 5a^{2+2}$$
$$= -a^2b^2 + 3a^3b - 5a^4$$

4. a. $10^{(4)(-5)} = 10^{-20} = \frac{1}{10^{20}}$

b. $7^{(-2)(-3)} = 7^6$

c. $\frac{1}{(2a^3)^2} = \frac{1}{4a^6}$

d. $\left(\frac{8}{x}\right)^{-2} = \left(\frac{x}{8}\right)^2 = \frac{x^2}{64}$

e. $2^{-1}x^2 = \frac{x^2}{2}$

f. $2x^2$

g. $\left(\frac{3}{2y^2}\right)^{-4} = \left(\frac{2y^2}{3}\right)^4 = \frac{2^4y^8}{3^4} = \frac{16y^8}{81}$

h. $3 \cdot (2y^2)^4 = 3 \cdot 2^4y^8 = 48y^8$

5. a. $\frac{t^{-3}(1)}{t^{-12}} = t^{-3-(-12)} = t^9$

b. $\frac{v^{-3}w^7}{v^{-6}w^{-10}} = v^{-3-(-6)}w^{7-(-10)} = v^3w^{17}$

c. $\frac{7^{-8}x^{-1}y^2}{7^{-5}x^1y^3} = 7^{(-8)-(-5)}x^{(-1)-1}y^{2-3}$
$$= 7^{-3}x^{-2}y^{-1} = \frac{1}{7^3x^2y}$$

d. $\frac{a(5b^{-1}c^3)^2}{5ab^2c^{-6}} = \frac{a^1 \cdot 5^2b^{-2}c^6}{5^1a^1b^2c^{-6}} = 5^{2-1}a^{1-1}b^{(-2)-2}c^{6-(-6)}$
$$= 5b^{-4}c^{12} = \frac{5c^{12}}{b^4}$$

Algebra Aerobics 4.3b

1. a. $\sqrt{81} = 9$

b. $\sqrt{144} = 12$

c. $\sqrt{36} = 6$

d. $-\sqrt{49} = -7$

e. not a real number

2. a. $\sqrt{9x} = 3x^{1/2}$

b. $\sqrt{\frac{x^2}{25}} = \frac{\sqrt{x^2}}{\sqrt{25}} = \frac{x}{5}$

c. $\sqrt{36x^2} = \sqrt{36}\sqrt{x^2} = 6x$

d. $\sqrt{\frac{9y^2}{25x^4}} = \frac{\sqrt{9y^2}}{\sqrt{25x^4}} = \frac{3y}{5x^2}$

e. $\sqrt{\frac{49}{x^2}} = \frac{\sqrt{49}}{\sqrt{x^2}} = \frac{7}{x}$

f. $\sqrt{\frac{4a}{169}} = \frac{\sqrt{4a}}{\sqrt{169}} = \frac{2\sqrt{a}}{13} = \frac{2a^{1/2}}{13}$

3. a. $S = \sqrt{30 \cdot 60} = \sqrt{1800} \approx 42$ mph

b. $S = \sqrt{30 \cdot 200} = \sqrt{6000} \approx 77$ mph

4. a. $\sqrt{25} < \sqrt{29} < \sqrt{36} \Rightarrow 5 < \sqrt{29} < 6$; 5 and 6

b. $\sqrt{81} < \sqrt{92} < \sqrt{100} \Rightarrow 9 < \sqrt{92} < 10$; 9 and 10

c. $\sqrt{100} < \sqrt{117} < \sqrt{121} \Rightarrow 10 < \sqrt{117} < 11$; 10 and 11.

d. $\sqrt{64} < \sqrt{79} < \sqrt{81} \Rightarrow 8 < \sqrt{79} < 9$; 8 and 9.

e. $\sqrt{36} < \sqrt{39} < \sqrt{49} \Rightarrow 6 < \sqrt{39} < 7$; 6 and 7.

5. a. $\sqrt[3]{27} = 3$

b. $\sqrt[4]{16} = 2$

c. $\frac{1}{\sqrt[3]{8}} = \frac{1}{2}$

d. $\sqrt[5]{32} = 2$

e. $\frac{1}{27^{1/3}} = \frac{1}{\sqrt[3]{27}} = \frac{1}{3}$

f. $\frac{1}{25^{1/2}} = \frac{1}{\sqrt{25}} = \frac{1}{5}$

g. $\left(\frac{8}{27}\right)^{-1/3} = \left(\frac{27}{8}\right)^{1/3}$
$$= \sqrt[3]{\frac{27}{8}} = \frac{\sqrt[3]{27}}{\sqrt[3]{8}} = \frac{3}{2}$$

h. $\sqrt{\frac{1}{16}} = \frac{1}{4}$

6. a. $(-27)^{1/3} = -3$ since $(-3)^3 = -27$

b. There is no real number solution to the fourth root of a negative number, since a negative or positive number raised to the fourth power is always positive.

c. $(-1000)^{1/3} = -10$ since $(-10)^3 = -1000$.

d. $-\sqrt[4]{16} = -2$, since $2^4 = 16 \rightarrow -\sqrt[4]{2^4} = -2$.

e. $\sqrt[3]{-8} = -2$, since $(-2)^3 = -8$.

f. $\sqrt{2500} = 50$, since $50^2 = 2500$.

7. a. $V = \frac{4}{3}\pi r^3 \Rightarrow \frac{3}{4}V = \pi r^3 \Rightarrow \frac{3V}{4\pi} = r^3 \Rightarrow$
$$r = \sqrt[3]{\frac{3 \cdot 2 \text{ feet}^3}{4\pi}} = \sqrt[3]{\frac{3 \text{ feet}^3}{2\pi}} = \sqrt[3]{0.478 \text{ feet}^3}$$
$$\approx 0.78 \text{ feet}$$

We can express 0.78 ft more meaningfully if we convert it into inches. Since 1 ft = 12 in., we can use a conversion factor of 1 = (12 in.)/(1 foot). The radius of the balloon is:

$$(0.78 \text{ ft}) \left(\frac{12 \text{ in.}}{1 \text{ ft}}\right) = 9.36 \text{ in. or } \approx 9.4 \text{ in.}$$

8. a. $\sqrt{25} = 5$

b. $-\sqrt{49} = -7$

c. -5

d. $\sqrt{45} - 3\sqrt{125} = \sqrt{9 \cdot 5} - 3\sqrt{25 \cdot 5}$
$= 3\sqrt{5} - 15\sqrt{5} = -12\sqrt{5}$

9. a. $\sqrt{36} = (6^2)^{1/2} = 6^{2/2} = 6$

b. $\sqrt[3]{27x^6} = (3^3\, x^6)^{1/3} = 3^{3/3}\, x^{6/3} = 3x^2$

c. $\sqrt[4]{81a^4\, b^{12}} = (3^4\, a^4\, b^{12})^{1/4} = 3^{4/4}\, a^{4/4}\, b^{12/4} = 3ab^3$

10. a. $r^2 = \frac{V}{\pi h} \Rightarrow r = \sqrt{\frac{V}{\pi h}}$ **d.** $a = \sqrt{c^2 - b^2}$

b. $r^2 = \frac{3V}{\pi h} \Rightarrow r = \sqrt{\frac{3V}{\pi h}}$ **e.** $x = \sqrt{\frac{S}{6}}$

c. $S = \sqrt[3]{V}$

Algebra Aerobics 4.3c

1. a. $2^{1/2} \cdot 2^{1/3} = 2^{1/2+1/3} = 2^{5/6}$

b. $5^{1/2} \cdot 5^{1/4} = 5^{1/2+1/4} = 5^{3/4}$

c. $3^{1/2} \cdot 9^{1/3} = 3^{1/2} \cdot (3^2)^{1/3} = 3^{1/2} \cdot 3^{2/3} = 3^{7/6}$

d. $x^{1/4} \cdot x^{1/3} = x^{7/12}$

e. $x^{3/4} \cdot x^{1/2} = x^{5/4}$

f. $x^{1/3} \cdot y^{2/3} \cdot x^{1/2} \cdot y^{1/2} = x^{5/6} \cdot y^{7/6}$

2. a. $\frac{2^{1/2}}{2^{1/3}} = 2^{1/2-1/3} = 2^{1/6}$

b. $\frac{2^1}{2^{1/4}} = 2^{1-1/4} = 2^{3/4}$

c. $\frac{5^{1/4}}{5^{1/3}} = 5^{1/4-1/3} = 5^{-1/12}$ or $\frac{1}{5^{1/12}}$

d. $\frac{x^{1/2}}{x^{3/4}} = x^{-\frac{1}{4}}$ or $\frac{1}{x^{1/4}}$

e. $\frac{x^{1/3} \cdot y^{2/3}}{x^{1/2} \cdot y^{1/2}} = x^{-1/6} \cdot y^{1/6}$ or $\frac{y^{1/6}}{x^{1/6}}$

3. a. $c = 17.1(0.25)^{3/8}$ cm
$c \approx 17.1(0.59)$ cm $\Rightarrow c \approx 10.2$ cm

b. $c = 17.1(25)^{3/8}$ cm
$c \approx 17.1(3.34)$ cm $\Rightarrow c \approx 57.2$ cm

4. a. $\sqrt{20x^2} = \sqrt{2^2 \cdot 5 \cdot x^2} = 2x\sqrt{5}$

b. $\sqrt{75a^3} = \sqrt{5^2 \cdot 3 \cdot a^2 \cdot a} = 5a\sqrt{3a}$

c. $\sqrt[3]{16x^3\, y^4} = \sqrt[3]{2^3 \cdot 2 \cdot x^3 \cdot y^3 \cdot y} = 2xy\sqrt[3]{2y}$

d. $\frac{\sqrt[4]{32x^4y^6}}{\sqrt[4]{81x^8y^5}} = \frac{\sqrt[4]{2^4 \cdot 2 \cdot x^4 \cdot y^4 \cdot y^2}}{\sqrt[4]{3^4 \cdot (x^2)^4 \cdot y^4 \cdot y}}$

$= \frac{2xy \cdot \sqrt[4]{2y^2}}{3x^2y \cdot \sqrt[4]{y}} = \frac{2}{3x}\sqrt[4]{\frac{2y^2}{y}} = \frac{2}{3x}\sqrt[4]{2y}$

5. a. $\sqrt{4a^2\, b^6} = (2^2\, a^2\, b^6)^{1/2} = 2^{2/2}a^{2/2}b^{6/2} = 2ab^3$

b. $\sqrt[4]{16x^4\, y^6} = (2^4\, x^4\, y^6)^{1/4} = 2^{4/4}\, x^{4/4}\, y^{6/4}$
$= 2xy^{3/2}$

c. $\sqrt[3]{8.0 \cdot 10^{-9}} = (2^3 \cdot 10^{-9})^{1/3} = 2^{3/3} \cdot 10^{-9/3}$
$= 2 \cdot 10^{-3} = 0.002$

d. $\sqrt{8a^{-4}} = (2^3 a^{-4})^{1/2} = 2^{3/2}a^{-2}$

Exercises for Section 4.3

1. a. 10^1 **c.** $10^{-6} = \frac{1}{10^6}$

b. $10^{-5} = \frac{1}{10^5}$ **d.** 10^5

3. a. xy^{12} **c.** $(x + y)^{11}$

b. $\frac{1}{x^3y^2}$ **d.** $\frac{ab^7}{c^6}$

5. a. $4.6 \cdot 10^{10}$ **d.** $5 \cdot 10^{-8}$

b. $3.7 \cdot 10^3$ **e.** $6.4 \cdot 10^{157}$

c. $2 \cdot 10^{11}$ **f.** $2.5 \cdot 10^{-21}$

7. a. $1 \cdot 10^{-1} \cdot 10^{-5} = 1 \cdot 10^{-6}$

b. $5 \cdot 10^{-5}/(5 \cdot 10^4) = 1 \cdot 10^{-9}$

c. $3/(6 \cdot 10^{-3}) = 0.5 \cdot 10^3 = 5 \cdot 10^2$

d. $8 \cdot 10^3/(8 \cdot 10^{-4}) = 1 \cdot 10^7$

e. $6.4 \cdot 10^{-3}/(8 \cdot 10^3) = 0.8 \cdot 10^{-6} = 8 \cdot 10^{-7}$

f. $5 \cdot 10^6 \cdot 4 \cdot 10^4 = 20 \cdot 10^{10} = 2 \cdot 10^{11}$

9. a. 10 **c.** $1/10 = 0.1$ **e.** -10

b. -10 **d.** $-1/10 = -0.1$ **f.** -10

11. a. $\sqrt{\frac{a^2b^4}{c^6}} = \frac{ab^2}{c^3}$ **c.** $\sqrt{\frac{49x}{y^6}} = \frac{7\sqrt{x}}{y^3}$

b. $\sqrt{36x^4y} = 6x^2\sqrt{y}$ **d.** $\sqrt{\frac{x^4y^2}{100z^6}} = \frac{x^2y}{10z^3}$

13. a. $6 \cdot 10^3$ **c.** $5 \cdot 10^5$

b. $2 \cdot 10^3$ **d.** $1.0 \cdot 10^{-2}$

15. a. 0.1 **c.** $4/3$

b. $1/5$ **d.** 0.1

17. a. $-\frac{y^3}{2^3x^9}$ **c.** $\frac{5xy^9}{3}$

b. $\frac{1}{2^2x^4}$ **d.** $\frac{36z^{24}}{x^6}$

19. a. 4 **c.** $\sqrt{2} \approx 1.414$ **e.** $2\sqrt{2} \approx 2.8284$

b. -4 **d.** not defined   **f.** 1

21. a. $\sqrt{3} + \sqrt{7} > \sqrt{3 + 7}$ **c.** $\sqrt{5^2 - 4^2} > 2$

b. $\sqrt{3^2 + 2^2} < 5$

23. 200 times longer or $1.6 \cdot 10^{-2}$ second

25. $V = (4/3)\pi r^3$ and if $V = 4$, then $r^3 = 3/\pi$. If one uses 3 as a crude estimate of π, then r is approximately 1 foot. (A more precise estimate, from using a calculator, is 0.985 ft.)

27. a. height $= 4\sqrt{3}$ cm ≈ 6.9 cm

b. area $= 16\sqrt{3}$ cm$^2 \approx 27.7$ cm^2

29. a. $2\sqrt{50} + 12\sqrt{8} = 2\sqrt{25 \cdot 2} + 12\sqrt{4 \cdot 2} = 10\sqrt{2} + 24\sqrt{2} = 34\sqrt{2}$

b. $3\sqrt{27} - 2\sqrt{75} = 3\sqrt{9 \cdot 3} - 2\sqrt{25 \cdot 3} = 9\sqrt{3} - 10\sqrt{3} = -\sqrt{3}$

c. $10\sqrt{32} - 6\sqrt{18} = 10\sqrt{16 \cdot 2} - 6\sqrt{9 \cdot 2} = 40\sqrt{2} - 18\sqrt{2} = 22\sqrt{2}$

d. $2\sqrt[3]{16} + 4\sqrt[3]{54} = 2\sqrt[3]{8 \cdot 2} + 4\sqrt[3]{27 \cdot 2} = 4\sqrt[3]{2} + 12\sqrt[3]{2} = 16\sqrt[3]{2}$

31. $5.23 \cdot 10^{-3}$, 0.00523 and $5.23/1000$

33. $\frac{1}{x^p} = \left(\frac{1}{x}\right)^p$ for any p. Here we have $p = -n$; thus

$$\frac{1}{x^{-n}} = \left(\frac{1}{x}\right)^{-n} = (x^{-1})^{-n} = x^{(-1)\cdot(-n)} = x^n$$

35. a. If $L \approx 2.67$ ft, then

$$T = 2\pi\sqrt{\frac{2.67}{32}} \approx 0.578\pi \text{ sec} \approx 1.816 \text{ sec}.$$

   b. Solving for L, we get: $L = 8T^2/(\pi^2)$. So if T is 2 seconds, then $L = 8(2)^2/(\pi^2) = 32/\pi^2 \approx 3.24$ feet.

37. a. 3673 lb   **b.** 6345 lb

Section 4.4

Algebra Aerobics 4.4

1. $2 l \cdot \frac{1 \text{ qt}}{0.946 l} \cdot \frac{32 \text{ oz}}{1 \text{ qt}} \approx 67.65$ oz

2. $120 \text{ cm} \cdot \frac{1 \text{ in}}{2.54 \text{ cm}} \approx 47.24$ in.

3. a. $12 \text{ in.} \cdot \frac{2.54 \text{ cm}}{1 \text{ in.}} = 30.48$ cm

   b. $100 \text{ yd} \cdot \frac{0.914 \text{ m}}{1 \text{ yd}} = 91.4$ m

   c. $20 \text{ kg} \cdot \frac{1 \text{ lb}}{0.4536 \text{ kg}} \approx 44.09$ lb

   d. $\frac{\$40,000}{1 \text{ year}} \cdot \frac{1 \text{ year}}{52 \text{ weeks}} \cdot \frac{1 \text{ workweek}}{40 \text{ hr}} \approx \$19.23$ per hour

   e. $\frac{24 \text{ hours}}{1 \text{ day}} \cdot \frac{60 \text{ min}}{1 \text{ hour}} \cdot \frac{60 \text{ sec}}{1 \text{ min}} = 86,400$ sec/day

   f. $1 \text{ gal} \cdot \frac{4 \text{ qt}}{1 \text{ gal}} \cdot \frac{946 \text{ ml}}{1 \text{ qt}} = 3784$ ml

   g. $\frac{1 \text{ mile}}{1 \text{ hour}} \cdot \frac{1 \text{ hour}}{3600 \text{ sec}} \cdot \frac{5280 \text{ ft}}{1 \text{ mile}} \approx 1.47$ ft/sec

4. $1 \text{ km} = 1000$ m, so the conversion factor from meters to kilometers is $\frac{1 \text{ km}}{10^3 \text{ m}}$. Hence,

$$3.84 \cdot 10^8 \text{ m} \cdot \frac{1 \text{ km}}{10^3 \text{ m}} = 3.84 \cdot 10^5 \text{ km}$$

5. $1 \text{ km} = 1000$ m, so the conversion factor from kilometers to meters is $\frac{10^3 \text{ m}}{1 \text{ km}}$. Hence

$$7.8 \cdot 10^8 \text{ km} \cdot \frac{10^3 \text{ m}}{1 \text{ km}} = 7.8 \cdot 10^8 10^3 = 7.8 \cdot 10^{11} \text{ m}$$

6. $1 \text{ km} = 0.62$ mi, so the conversion factor to convert from km to mi is $\frac{0.62 \text{ mi}}{1 \text{ km}}$. Hence

$$9.46 \cdot 10^{12} \text{ km} \cdot \frac{0.62 \text{ mi}}{1 \text{ km}} = 5.87 \cdot 10^{12} \text{ mi}$$

which is close to $5.88 \cdot 10^{12}$ mi.

7. $1 \text{ m} = 100$ cm, so the conversion factor for converting from cm to m is $\frac{1 \text{ m}}{100 \text{ cm}}$.

Hence $1 \text{ Å} = 10^{-8} \text{ cm} \cdot \frac{1 \text{ m}}{10^2 \text{ cm}} = 10^{-10}$ m.

8. $1 \text{ km} = 0.62$ mi, so the conversion factor for converting from km to miles is $\frac{0.62 \text{ mi}}{1 \text{ km}}$.

Hence, $218 \text{ km} = 218 \text{ km} \cdot \frac{0.62 \text{ mi}}{1 \text{ km}} = 135.16$ mi or ≈ 135 mi.

---

9. If a dollar bill is 6 in long, then two dollar bills/12 in = 2 dollars/ft. The number of dollars needed to reach from Earth to the sun is:

$$93,000,000 \text{ mi} \cdot 5,280 \frac{\text{ft}}{\text{mi}} \cdot \frac{2 \text{ dollars}}{\text{ft}}$$

$$\approx 9.3 \cdot 5.3 \cdot 2 \cdot 10^{7+3} \text{ dollars}$$

$$\approx 98.6 \cdot 10^{10} \approx 9.9 \cdot 10^{11}$$

or $990,000,000,000$ dollar bills, almost a trillion dollars.

10. $\frac{2,560 \text{ mi}}{4.2 \text{ hrs}} \cdot \frac{1.6 \text{ km}}{1 \text{ mi}} = \frac{4096 \text{ km}}{4.2 \text{ hrs}} = \frac{4096 \text{ km}}{4.2 \text{ hrs}} \cdot \frac{1 \text{ hr}}{60 \text{ min}}$

$= \frac{4096 \text{ km}}{252 \text{ min}}$ (or approximately 16.25 km/min)

11. $364 \text{ Smoot} \cdot \frac{5.6 \text{ ft}}{1 \text{ Smoot}} = 2038.4$ ft

Exercises for Section 4.4

1. a. $(50 \text{ miles}) \cdot \frac{1.609 \text{ km}}{1 \text{ mile}} = 80.45$ km

   b. $(3 \text{ ft}) \cdot \frac{0.305 \text{ m}}{1 \text{ ft}} \approx 0.92$ m

   c. $(5 \text{ lb}) \cdot \frac{0.4536 \text{ kg}}{1 \text{ lb}} \approx 2.27$ kg

   d. $(12 \text{ in.}) \cdot \frac{2.54 \text{ cm}}{1 \text{ in.}} = 30.48$ cm

   e. $(60 \text{ ft}) \cdot \frac{0.305 \text{ m}}{1 \text{ ft}} = 18.3$ m

   f. $(4 \text{ qt}) \cdot \frac{0.946 \text{ liters}}{1 \text{ qt}} \approx 3.78$ liters

3. a. Student estimates will vary. There are approximately 30.5 cm per foot.

   b. From conversation table, 1 ft = 0.305 m or approximately 30% of a meter.

5. Converting the units to decimeters we get:
$(100 \text{ km}) \cdot (250 \text{ m}) \cdot (25 \text{ m}) = 10^6 \cdot (2.5 \cdot 10^3) \cdot (2.5 \cdot 10^2)$ cubic decimeters $= 6.25 \cdot 10^{11}$ cubic decimeters $= (6.25 \cdot 10^{11} \text{ liters}) \cdot (10^3 \text{ droplets/liter}) = 6.25 \cdot 10^{14}$ droplets

7. $1 \text{ m} \approx 3.28$ ft and thus $9.8 \frac{\text{m}}{\text{sec}} \cdot 3.28 \frac{\text{ft}}{\text{m}} = 32.144$ ft/sec

9. a. 186,000 miles per second or one hundred and eighty-six thousand miles per second.

   b. $\frac{1.86 \cdot 10^5 \text{ mi}}{\text{sec}} \cdot \frac{1609 \text{ m}}{\text{mi}} \cdot \frac{60 \text{ sec}}{\text{min}} \cdot \frac{60 \text{ min}}{\text{hr}} \cdot \frac{24 \text{ hr}}{\text{day}} \cdot \frac{365 \text{ days}}{\text{yr}}$

     $\approx 9.438 \cdot 10^{15}$ m/yr

11. 500 seconds; 500 seconds or 8 minutes and 20 seconds from now.

13. U.S. barrel of oil = 42 gal = $42 \cdot 4$ qt = 168 qt $\approx$ $168 \cdot 0.946$ liter = 158.928 liters. Thus the British barrel of oil is larger than the U.S. barrel of oil by approximately 4.727 liters.

15. a. 1 acre = $43,560 \text{ ft}^2$ and thus is $\sqrt{43,560} \approx 208.71$ ft on each side.

   b. $150x = 1.5 \cdot 43,560$ or $x = 435.6$ ft.; perimeter = $2 \cdot (150 + 435.6) = 1171.2$ feet or 390.4 yards.

   c. 1 hectometer = 100 meters; 1 square hectometer = 10,000 square meters. 1 meter ≈ 3.28 ft. Thus 1 sq. meter ≈ 10.7584 square ft. Thus 1 acre = $43,560$ sq. ft = $43,560/10.7584 \approx 4048.929$ square meters ≈ 0.4049 square hectometer.

d. 1 hectare = 100 acres = 4,356,000 ft²/[(5280)² ft²/mi²] = 0.15625 sq. mi. Thus 1 acre = 0.0015625 sq. mile. Thus in a square mile there are 1/0.0015625 = 640 acres. This is 6.4 hectares.

17. Light travels 186,000 · 5280 ft/sec and thus it travels 186000 · 5280/10⁹ ≈ 0.98208 ft per nanosecond.

19. a. $(4.3 \text{ LY})(9.46 \cdot 10^{12} \text{ km/LY}) = 4.07 \cdot 10^{13}$ kilometers; $(4.3 \text{ LY})(5.88 \cdot 10^{12} \text{ mi/LY}) = 2.53 \cdot 10^{13}$ miles.

b. $(10^8 \text{ LY})(9.46 \cdot 10^{12} \text{ km/LY})(1000 \text{ m/km}) = 9.46 \cdot 10^{23}$ meters

c. $(1600 \text{ LY})(5.88 \cdot 10^{12} \text{ mi/LY})(5280 \text{ ft/mi}) = 4.97 \cdot 10^{19}$ feet

21. a. Volume $= \pi r^2 \cdot h = \pi \cdot 9^2 \cdot 4$ cu. ft $= 324\pi$ cu. ft ≈ 1018 cu. ft, and 1018 cu. ft · 1728 (cu. in.)/(cu. ft) = 1,759,104 cu. in., and since there are 231 gal per cubic inch, there are 1,759,104/231 ≈ 7615 gal.

b. 7615 gal/(2500 gal/hr) ≈ 3.05 hr

c. 7615 gal/(10,000 gal/lb) ≈ 0.761 lb

23. a. Obama's BMI $\approx \dfrac{(180 \text{ lb})/(2.2 \text{ lb/kg})}{(73 \text{ in.}/39.37 \text{ in./m})^2} \approx \dfrac{81.82 \text{ kg}}{1.85^2 \text{m}^2} \approx 23.9$ kg/m². He has a normal weight.

b. BMI in pounds and inches is BMI = (pounds/2.2)/(inches/39.37)² = [(39.7)²/2.2](lb/in²), or approximately 704.5 lb/in². Thus for Obama we get 704.5 · 180/73² ≈ 23.8, which is about the same.

c. Answers will vary, but note that 0.45 ≈ 1/2.2 and 0.0254 ≈ 1/39.37.

d. A kilogram, more precisely, is approximately 2.2046 lb, and 39.37²/2.2046 ≈ 703.07. Thus Kigner is correct.

25. a. $(0.5 \text{ Å}) \cdot (10^{-10} \text{ m})/(1 \text{ Å}) = 5 \cdot 10^{-1} \cdot 10^{-10} \text{ m} = 5 \cdot 10^{-11}$ m

b. $10^5 \text{ Å} \cdot (10^{-10} \text{ m})/(1 \text{ Å}) = 10^{-5}$ m

c. $10^{-5} \text{ Å} \cdot (10^{-10}) \text{m}/(1 \text{ Å}) = 10^{-15}$ m

Section 4.5

Algebra Aerobics 4.5

1. Since $\dfrac{\text{magnitude 1988 Armenia earthquake}}{\text{magnitude 1987 LA earthquake}}$

$\Rightarrow \dfrac{10^{6.9}}{10^{5.9}} = 10^1$, the Armenian earthquake had tremors about one order of magnitude larger than those in Los Angeles.

2. Since $\dfrac{\text{magnitude Chile earthquake}}{\text{magnitude 2003 Little Rock earthquake}}$

$\Rightarrow \dfrac{10^{9.5}}{10^{6.5}} = 10^3$, the maximum tremor size of the Little Rock earthquake of 2003 was 1000 times smaller (or three orders of magnitude smaller) than the maximum tremor size of the Chile earthquake.

3. Since $\dfrac{\text{your salary}}{\text{my salary of } \$100,000} = 10^1 \Rightarrow$

your salary is $1,000,000;

$\dfrac{\text{Henry's salary}}{\text{my salary } \$100,000} = 10^{-2} \Rightarrow$

$\dfrac{\text{Henry's salary}}{\text{my salary } \$100,000} = \dfrac{1}{10^2} \Rightarrow$ Henry's salary is $1000.

4. a. Since $\dfrac{\text{radius of the Milky Way}}{\text{radius of the sun}} = \dfrac{10^{21}}{10^9} = 10^{21-9} = 10^{12}$,

the radius of the Milky Way is 12 orders of magnitude larger than the radius of the sun, or equivalently, the radius of the sun is 12 orders of magnitude smaller than the radius of the Milky Way.

b. Since $\dfrac{\text{radius of a proton}}{\text{radius of the hydrogen atom}} = \dfrac{10^{-15}}{10^{-11}} = 10^{-15-(-11)} = 10^{-4}$,

the radius of the proton is four orders of magnitude smaller than the radius of the hydrogen atom, or equivalently, the radius of the hydrogen atom is four orders of magnitude larger than the radius of a proton.

5. a. $4 million since $400,000 · 10¹ = $4 million

b. $125,000 since $\dfrac{1,250,000}{10^1} = \$125,000$

6. 1 km = 1000 m = 10³ meters, so three orders of magnitude; $1 \text{ km} \cdot \dfrac{1000 \text{ m}}{1 \text{ km}} \cdot \dfrac{1000 \text{ mm}}{1 \text{ m}} = 10^6$ mm, so six orders of magnitude

7. a. $1,000,000,000 \text{ m} = 10^9 \text{ m} \Rightarrow$ plot at 10^9 m

b. $0.000\,000\,7 \text{ m} = 7 \cdot 10^{-7} \text{ m} \approx 10 \cdot 10^{-7} \text{ m} = 10^{-6} \text{ m} \Rightarrow$ plot at 10^{-6} m

c. Plot at $10^7 \cdot 10^{-2} = 10^5$ m

Exercises for Section 4.5

1. a. 12 **b.** 5 **c.** 7 **d.** 6

3. a. 52.61 is one order of magnitude larger than 5.261.

b. 5.261 is three orders of magnitude smaller than 5261.

c. 526.1 is four orders of magnitude smaller than $5.261 \cdot 10^6$.

5. a. $(50 \cdot 10^3) \cdot (3 \cdot 10^8) \cdot (365 \cdot 24 \cdot 60^2) \approx 4.7304 \cdot 10^{20} =$ number of meters in 50,000 light-years. Hence, the Milky Way is $(4.7302 \cdot 10^{20})/(0.5 \cdot 10^{-4}) = 9.4608 \cdot 10^{24}$ times larger than the first life form. Thus the Milky Way is nearly 25 orders of magnitude larger than the first living organism on Earth.

b. $(100 \cdot 10^6)/(100 \cdot 10^3) = 10^3$; thus Pleiades is three orders of magnitude older than *Homo sapiens*.

7. The raindrop is 10^{24} times heavier and this is an order of magnitude of 24.

9. a. i. Radius of the moon = 1,758,288.293 meters, or about $1.76 \cdot 10^6$ meters.

ii. Radius of Earth = 6,400,000 meters, or about $6.4 \cdot 10^6$ meters.

iii. Radius of the sun = 695,414,634.100 meters, or about $6.95 \cdot 10^8$ meters.

b. i. The surface area of a sphere is $4\pi r^2$, and thus the ratio of the surface area of Earth to that of the moon is the same as the ratio of the squares of their radii, or $[6.4 \cdot 10^6/(1.76 \cdot 10^6)]^2 \approx 13.22$. Thus the surface area of Earth is one order of magnitude bigger.

ii. The volume of a sphere is $(4/3)\pi r^3$, and thus the ratio of the volume of the sun to that of moon is the same as the ratio of the cubes of their radii, or $[6.95 \cdot 10^8/(1.76 \cdot 10^6)]^3 \approx 6.16 \cdot 10^7$. Thus the volume of the sun is seven orders of magnitude bigger than the volume of the moon.

11. Scale (a) is additive because the distances are equally spaced, and scale (b) is multiplicative or logarithmic because the distances are spaced like the logarithms of numbers.

13. a. Being 5 orders of magnitude larger than the first atoms, which is 10^{-10} m, means that it is 10^{-5} m and thus would appear at -5 on the log scale.

   b. Being 20 orders of magnitude smaller than the radius of the sun, which is 10^9 m, means that it is 10^{-11} m and thus would appear at -11 on the log scale.

15. a. 1 watt $= 10^0$ watts

   b. 10 kilowatts $= 10^4$ watts

   c. 100 billion kilowatts $= 10^2 \cdot 10^9 \cdot 10^3 = 10^{14}$ watts

   d. 1000 terawatts $= 10^{15}$ watts

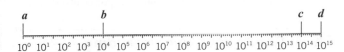

Section 4.6

Algebra Aerobics 4.6a

1. a. Since $10,000,000 = 10^7$, $\log 10,000,000 = 7$.

   b. Since $0.000\ 000\ 1 = 10^{-7}$, $\log 0.000\ 000\ 1 = -7$.

   c. Since $10,000 = 10^4$, $\log 10,000 = 4$.

   d. Since $0.0001 = 10^{-4}$, $\log 0.0001 = -4$.

   e. Since $1000 = 10^3$, $\log 1000 = 3$.

   f. Since $0.001 = 10^{-3}$, $\log 0.001 = -3$.

   g. Since $1 = 10^0$, $\log 1 = 0$.

2. a. $100,000 = 10^5$ **c.** $10 = 10^1$

   b. $0.000\ 000\ 01 = 10^{-8}$ **d.** $0.01 = 10^{-2}$

3. a. $\log N = 3 \Rightarrow N = 10^3 = 1000$

   b. $\log N = -1 \Rightarrow N = 10^{-1} = 0.1$

   c. $\log N = 6 \Rightarrow N = 10^6 = 1,000,000$

   d. $\log N = 0 \Rightarrow N = 10^0 = 1$

   e. $\log N = -2 \Rightarrow N = 10^{-2} = 0.01$

4. a. $\log 1000 = c$, $c = 3$

   b. $\log 0.001 = c$, $c = -3$

   c. $\log 100,000 = c$, $c = 5$

   d. $\log 0.000\ 01 = c$, $c = -5$

   e. $\log 1,000,000 = c$, $c = 6$

   f. $\log 0.000\ 001 = c$, $c = -6$

5. a. $x - 3 = 2$, so $x = 5$

   b. $2x - 1 = 4$, so $x = 5/2$

   c. $10^1 = x - 2$, so $x = 12$

   d. $10^{-1} = 5x$, so $x = \dfrac{0.1}{5} = 0.02$

Algebra Aerobics 4.6b

1. a. $\log 3$ is the number c such that $10^c = 3$. With a calculator you can find that: $10^{0.4} \approx 2.512$ and $10^{0.5} \approx 3.162 \Rightarrow 0.4 < \log 3 < 0.5$. A calculator gives $\log 3 \approx 0.477$.

   b. $\log 6$ is the number c such that $10^c = 6$. With a calculator you can find that: $10^{0.7} \approx 5.012$ and $10^{0.8} \approx 6.310 \Rightarrow 0.7 < \log 6 < 0.8$. A calculator gives $\log 6 \approx 0.778$.

   c. $\log 6.37$ is the number c such that $10^c = 6.37$. Since $10^{0.8} \approx 6.310$ and $10^{0.9} \approx 7.943$, then $0.8 < \log 6.37 < 0.9$. A calculator gives $\log 6.37 \approx 0.804$.

2. a. Write 3,000,000 in

  scientific notation $3,000,000 = 3 \cdot 10^6$

  and substitute $10^{0.48}$ for 3 $3,000,000 \approx 10^{0.48} \cdot 10^6$

  then combine powers $3,000,000 \approx 10^{6.48}$

  and rewrite as a logarithm $\log 3,000,000 \approx 6.48$

  A calculator gives $\log 3,000,000 \approx 6.477121255$.

   b. Write 0.006 in scientific

  notation $0.006 = 6 \cdot 10^{-3}$

  and substitute $10^{0.78}$ for 6 $0.006 \approx 10^{0.78} \cdot 10^{-3}$

  then combine powers $0.006 \approx 10^{0.78-3} \approx 10^{-2.22}$

  and rewrite as a logarithm $\log 0.006 \approx -2.22$

  A calculator gives $\log 0.006 \approx -2.22184875$.

3. a. $0.000\ 000\ 7 = 10^{\log 0.0000007} \approx 10^{-6.1549}$

   b. $780,000,000 = 10^{\log 780,000,000} \approx 10^{8.892}$

   c. $0.0042 = 10^{\log 0.0042} \approx 10^{-2.3768}$

   d. $5,400,000,000 = 10^{\log(5,400,000,000)} \approx 10^{9.732}$

4. a. You want to estimate a number x such that $10^{4.125} = x$. Since $10^4 = 10,000$, an estimate for x is 12,000. With a calculator, $x \approx 13,335$.

   b. You want to estimate a number x such that $10^{5.125} = x$. Since $10^5 = 100,000$, an estimate for x is 120,000. With a calculator, $x \approx 133,352$.

   c. You want to estimate a number x such that $10^{2.125} = x$. Since $10^2 = 100$, an estimate for x is 120. With a calculator, $x \approx 133$.

5. a. You want to estimate a value x such that $\log 250 = x$. $10^2 < 250 < 10^3 \Rightarrow 2 < x < 3$. With a calculator $x \approx 2.398$.

   b. You want to estimate a value x such that $\log 250,000 = x$. $10^5 < 250,000 < 10^6 \Rightarrow 5 < x < 6$. With a calculator, $x \approx 5.398$.

   c. You want to estimate a value x such that $\log 0.075 = x$. $10^{-2} < 0.075 < 10^{-1} \Rightarrow -2 < x < -1$. With a calculator, $x \approx -1.125$.

   d. You want to estimate a value x such that $\log 0.000\ 075 = x$. $10^{-5} < 0.000\ 075 < 10^{-4} \Rightarrow -5 < x < -4$. With a calculator, $x \approx -4.125$.

6. a. $\log 57 \approx 1.756 \Rightarrow 57 \approx 10^{1.756}$

   b. $\log 182 \approx 2.26 \Rightarrow 182 \approx 10^{2.26}$

   c. $\log 25,000 \approx 4.398 \Rightarrow 25,000 \approx 10^{4.398}$

   d. $\log 7.2 \cdot 10^9 \approx 9.857 \Rightarrow 7.2 \cdot 10^9 \approx 10^{9.857}$

Exercises for Section 4.6

1. a. $\log(10,000) = 4$
   c. $\log(1) = 0$
   b. $\log(0.01) = -2$
   d. $\log(0.00001) = -5$

3. Since $\log(375) \approx 2.574$, we have that $375 \approx 10^{2.574}$.

5. a. $\log(100) = 2$
   b. $\log(10,000,000) = 7$
   c. $\log(0.001) = -3$
   d. $10^1 = 10$
   e. $10^4 = 10,000$
   f. $10^{-4} = 0.0001$

7. a. $x = 10^{1.255} \approx 17.9887 \approx 18$
   b. $x = 10^{3.51} \approx 3235.94 \approx 3236$
   c. $x = 10^{4.23} \approx 16,982.44 \approx 16,982$
   d. $x = 10^{7.65} \approx 44,668,359.22 \approx 44,668,359$

9. a. $x = \log(12,500) \approx 4.097$
   b. $x = \log(3,526,000) \approx 6.547$
   c. $x = \log(597) \approx 2.776$
   d. $x = \log(756,821) \approx 5.879$

11. a. $x = 10^{0.82} \approx 6.607$
   b. $x = \log(0.012) \approx -1.921$
   c. $x = 10^{0.33} \approx 2.138$
   d. $x = \log(0.25) \approx -0.602$

13. a. $x - 5 = 3$ or $x = 8$
   b. $2x + 10 = 100$ or $2x = 90$ or $x = 45$
   c. $3x - 1 = -4$ or $3x = -3$ or $x = -1$
   d. $500 - 25x = 1000$ or $-25x = 500$ or $x = -20$

15. a. $1 < \log(11) < 2$ since $1 < 11 < 100$ and $\log(11) \approx 1.041$.
   b. $4 < \log(12,000) < 5$ since $10,000 < 12,000 < 100,000$ and $\log(12,000) \approx 4.079$.
   c. $-1 < \log(0.125) < 0$ since $10^{-1} = 0.1 < 0.125 < 1 = 10^0$ and $\log(0.125) \approx -0.903$.

17. a. Multiplying by 10^{-3}
   b. Multiplying by $\sqrt{10}$
   c. Multiplying by 10^2
   d. Multiplying by 10^{10}

19. a. $pH = -\log(10^{-7}) = 7$
   b. $pH = -\log(1.4 \cdot 10^{-3}) = 3 - \log(1.4) \approx 2.85$
   c. $11.5 = -\log([H^+])$; thus $[H^+] = 10^{-11.5} \approx 3.16 \cdot 10^{-12}$
   d. A higher pH means a lower hydrogen ion concentration, and one can see this because in plotting pH values one uses the numbers on the top of the given scale.
   e. Pure water is neutral, orange juice is acidic, and ammonia is basic. In plotting, one uses the top numbers to find the right spots. Thus water would be placed at the 7 mark, orange juice 85% of the way between the 2 and 3 marks, and ammonia halfway between the 11 and 12 marks.

21.

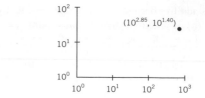

23. We measure all in seconds, the smallest time unit.
   a. 1 heartbeat $\approx 10^0$ sec ≈ 1 sec
   b. 10 minutes $= 600$ sec $\approx 10^{2.8}$ sec
   c. 7 days $= 7 \cdot 24 \cdot 3600$ sec $\approx 10^{5.8}$ sec
   d. 1 year $= 365 \cdot 24 \cdot 3600$ sec $\approx 10^{7.5}$ sec
   e. 38,000 years $= 38,000 \cdot 365 \cdot 24 \cdot 3600 \approx 10^{12.1}$ sec
   f. $2.2 \cdot 10^6 \cdot 365 \cdot 24 \cdot 3600$ sec $\approx 10^{13.8}$ sec
   (Note: the plot is omitted, but should use power-of-ten scale)

Ch. 4: Check Your Understanding

1. True	**7.** False	**13.** False	**19.** False
2. False	**8.** True	**14.** True	**20.** False
3. False	**9.** False	**15.** True	**21.** False
4. False	**10.** False	**16.** True	**22.** True
5. True	**11.** False	**17.** False	**23.** False
6. True	**12.** False	**18.** False	**24.** False

25. Possible answer: population of city A $= 583,240$ and population of city B $= 3615$.

26. Possible answer: $x = 150,000,000$.

27. Possible answer: $x = 0.45$.

28. Possible answer: b such that $b = \frac{1}{4}$.

29. $b = 1$.

30. Possible answer: $b = -3$.

31. False	**34.** True	**37.** False	**40.** True
32. False	**35.** False	**38.** False	
33. True	**36.** False	**39.** False	

Ch. 4 Review: Putting It All Together

1. a. 4200   **b.** -40 **c.** -16 **d.** $\frac{1}{10}$ **e.** 27

3. When x is an odd integer the statement is true.

5. $\frac{20 \text{ ft}}{3 \text{ sec}} \cdot \frac{60 \text{ sec}}{1 \text{ min}} \cdot \frac{60 \text{ min}}{1 \text{ hr}} \cdot \frac{1 \text{ mile}}{5280 \text{ ft}} \approx 4.55$ miles/hour

7. CEO Lawrence Culp, Jr. earns $\$141.36 \times 10^6 = \$1.4136 \times 10^8$. A minimum-wage worker earns ($\$8.25$/hour)(40 hours/week)(50 weeks/year) $= \$16,500$/year $= \$1.65 \times 10^4$/year. So, the compensation paid to the CEO is 4 orders of magnitude or 10,000 times greater than the compensation paid to a minimum-wage worker. It would take 8567 years for the minimum-wage earner to make as much as the CEO.

(Chart goes with solution 21.)

Item	Value	Value in Scientific Notation
Mass-energy of electron	0.000 000 000 000 051 J	$5.1 \cdot 10^{-14}$ J
The kinetic energy of a flying mosquito	0.000 000 160 2 J	$1.602 \cdot 10^{-7}$
An average person swinging a baseball bat	80 J	$8 \cdot 10^{1}$ J
Energy received from the sun at the Earth's orbit by one square meter in one second	1360 J	$1.360 \cdot 10^{3}$ J
Energy released by one gram of TNT	4184 J	$4.184 \cdot 10^{3}$ J
Energy released by metabolism of one gram of fat	38,000 J	$3.8 \cdot 10^{4}$ J
Approximate annual power usage of a standard clothes dryer	320,000,000 J	$3.2 \cdot 10^{8}$ J

9. Convert to same unit of measure—for example, km².

Country	Area	Scientific Notation
Russia	17,075,200 km²	$1.70752 \cdot 10^{7}$ km²
Chile	$(290{,}125 \text{ mi}^2)\left(\frac{1.609 \text{ km}}{\text{mi}}\right)^2$ $\approx 751{,}099$ km²	$7.51099 \cdot 10^{5}$ km²
Canada	$(3{,}830{,}840 \text{ mi}^2)\left(\frac{1.609 \text{ km}}{\text{mi}}\right)^2$ $\approx 9{,}917{,}589$ km²	$9.917589 \cdot 10^{6}$ km²
South Africa	1,184,825 km²	$1.184825 \cdot 10^{6}$ km²
Norway	323,895 km²	$3.23895 \cdot 10^{5}$ km²
Monaco	$0.5 \text{ mi}^2 \left(\frac{1.609 \text{ km}}{1 \text{ mi}}\right)^2$ ≈ 1.29 km²	$1.29 \cdot 10^{0}$ km²

  a. Russia is the largest in area. Monaco is the smallest in area.

  b. Russia, Canada, South Africa, Chile, Norway, Monaco

  c. Russia is seven orders of magnitude larger than Monaco.

11. False. An increase in one order of magnitude is the same as multiplying by 10. A 100% increase would only double the original amount.

13. a. Answers will vary. (0, 0) and (0.5, 8)

  b. 8/(0.5) = 16 thousand kilometers per sec/billion light-years

  c. It means that for each additional billion light-years from Earth, the recession velocity increases by 16,000 kilometers per second.

  d. $y = 16x$, where x is in billion of light years and y is in thousand km/sec.

15. Volume of a sphere $= \frac{4}{3}\pi r^3$, so the volume of Earth $=$

$$\frac{4}{3}\pi(6.3 \cdot 10^6 \text{ m})^3 = \left(\frac{4}{3}\pi(6.3)^3\right) \cdot 10^{18}\text{m}^3$$
$$\approx 1047.4 \cdot 10^{18}\text{ m}^3$$
$$\approx 1.0474 \cdot 10^{21}\text{ m}^3$$

$$\text{Density} = \frac{\text{mass}}{\text{volume}} = \frac{5.97 \cdot 10^{24}\text{ kg}}{1.0474 \cdot 10^{21}\text{ m}^3} \approx 5.7 \cdot 10^3 \text{ kg/m}^3$$

17. The patient's weight in kilograms is:

$$130 \text{ lb} \cdot \frac{0.4536 \text{ kg}}{1 \text{ lb}} \approx 59 \text{ kg}$$

$$\text{daily dosage} = \frac{5 \text{ mg}}{\text{kg}} \cdot 59 \text{ kg} \approx 295 \text{ mg}$$

Since each tablet is 100 mg, the patient should take

$$\frac{295 \text{ mg}}{100 \text{ mg/tablet}} = 2.95, \text{ or 3 tablets each day.}$$

19. a. $\log 1 = 0$

  b. $\log 1{,}000{,}000{,}000 = \log 10^9 = 9$

  c. $\log 0.000\,001 = \log 10^{-6} = -6$

21. a. See chart at top of page for solution.

  b. $\frac{3.2 \cdot 10^8 \text{ J}}{8 \cdot 10^1 \text{ J}} = 0.4 \cdot 10^7 = 4 \cdot 10^6$, or 4 million times more energy.

  c. $\frac{3.8 \cdot 10^4 \text{ J}}{1.602 \cdot 10^{-7} \text{ J}} \approx 2.372 \cdot 10^{11}$, or approximately 200 billion times more energy.

23. (Requires scientific calculator.) Converting height and weight to kilograms and centimeters, respectively:

$$W = 180 \text{ lb} \cdot \frac{0.4536 \text{ kg}}{1 \text{ lb}} \approx 81.6 \text{ kg} \quad \text{and}$$

$$H = 6 \text{ ft} \cdot \frac{30.5 \text{ cm}}{1 \text{ ft}} = 183 \text{ cm}$$

$$\text{BSA} = 71.84 \cdot 81.6^{0.425} \cdot 183^{0.725} \approx 20{,}376 \text{ cm}^2$$

Converting cm² to m²:

$$20{,}376 \text{ cm}^2 \cdot \frac{1 \text{ m}^2}{(100 \text{ cm})^2} = 2.04 \text{ m}^2$$

$$\text{daily dosage} = 15\frac{\text{mg}}{\text{m}^2} \cdot 2.04 \text{ m}^2 = 30.6 \text{ mg}$$

CHAPTER 5

Section 5.1

Algebra Aerobics 5.1

1. a. Initial value = 350
Growth factor = 5

  b. Initial value = 25,000
Growth factor = 1.5

  c. Initial value = 7000
Growth factor = 4

  d. Initial value = 5000
Growth factor = 1.025

2. a. $P = 3000 \cdot 3^t$

  b. $P = (4 \cdot 10^7)(1.3)^t$

  c. $P = 75 \cdot 4^t$

  d. $P = \$30{,}000(1.12)^t$

3. a. $P = 28 \cdot 1.065^0 = 28$ million or 28,000,000 people; the initial population

  b. $P = 28 \cdot 1.065^{10} \approx 52.6$ million or 52,600,000 people;
$P = 28 \cdot 1.065^{20} \approx 98.7$ million or 98,700,000;
$P = 28 \cdot 1.065^{30} \approx 185.2$ million or 185,200,000

  c. $t \approx 11$ years for P to double or reach 56,000,000.

4.

t	0	1	10	15	20
P	80	84.80	143.27	191.72	256.57

    a. 80

    b. 84.8

    c. The amount doubles (reaches 160) between $t = 10$ and $t = 15$

    d. $t \approx 12$ time periods.

5. $Q = 30 \cdot 2^T$

    5 years: $T = 1$, so $Q = 30 \cdot 2^1 = 60g$

    10 years: $T = 2$, so $Q = 30 \cdot 2^2 = 120g$

    15 years: $T = 3$, so $Q = 30 \cdot 2^3 = 240g$

6. a. Doubling time is approximately 1 unit of time. When $x = 0$, $y = 5$, so the initial value is 5 and $y = 5 \cdot 2^x$, where x represents the number of time units.

    b. Doubling time is approximately 2 decades ($T = 2$). When $T = 0$, $G = 300$, so the initial value is 300 and $G = 300 \cdot 2^T$, where T represents the number of 2 decades.

Exercises for Section 5.1

1. a. 275; 3   **c.** $6 \cdot 10^8$; 5   **e.** 8000; 2.718

    b. 15,000; 1.04   **d.** 25; 1.18   **f.** $4 \cdot 10^5$; 2.5

3.

Initial Value C	Growth Factor a	Exponential Function $y = Ca^x$
1600	1.05	$y = 1600 \cdot 1.05^x$
$6.2 \cdot 10^5$	2.07	$y = 6.2 \cdot 10^5 \cdot (2.07)^x$
1400	3.25	$y = 1400 \cdot (3.25)^x$

5. a. $f(x) = 6 \cdot 1.2^x$ **b.** $f(x) = 10 \cdot 2.5^x$

7. a. 1.17   **b.** 63 cells per ml

    c.

d	$C(d)$
0	63.0
1	73.7
2	86.2
3	100.9
4	118.1
5	138.1
6	161.6
7	189.1
8	221.2
9	258.8
10	302.8

    d. C doubles somewhere between $d = 4$ and $d = 5$ days.

9. a. $G(t) = 5 \cdot 10^3 \cdot 1.185^t$ cells/ml, where t is measured in hours.

    b. $G(8) = 5 \cdot 10^3 \cdot 1.185^8 = 19,440.92$ cells/ml.

11. At the top of the next column is the graph, with the point marked at which the amount owed doubles:

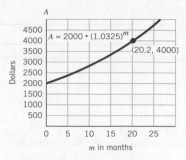

The amount owed if nothing is paid back grows very rapidly. It is not wise to not pay anything back since the amount owed doubles in about 20 months, as we see by inspecting the graph. We use the graph to solve the equation.

13. The graph looks like an exponential curve except that it does not start at a positive value for the dependent variable. Rather it starts at 0. It cannot have the form $y = Ca^t$. However, if we make time start at 2009, we see that the amount doubled in about 5 months. So an exponential model could be $y = 25(2)^T$, where T is in 5-month intervals after 2009.

15. a. *Females:* BW Teal: 360 grams; Gadwall: 750 grams; Mallard: 1050 grams

    b. *Males:* BW Teal: 370 grams; Gadwall: 800 grams; Mallard: 1150 grams.

    c. BW Teal: 80 days, Gadwall: 90 days; Mallard: 100 days; Note that after 90 days very little weight is gained by any species.

Section 5.2

Algebra Aerobics 5.2

1. a. growth   **c.** decay   **e.** decay

    b. decay   **d.** growth   **f.** growth

2. a. $y = 2300(1/3)^t$ **c.** $y = (375)(0.1)^t$

    b. $y = (3 \cdot 10^9)(0.35)^t$

3. a. Half-life is approximately 1500 years with initial value 600 g;

    $Q = 600 \cdot \left(\frac{1}{2}\right)^T$, where $T = $ number of 1500-year intervals.

    b. Half-life is $3(500) = 1500$ years with initial value of 60 g; $A = 60 \cdot \left(\frac{1}{2}\right)^T$, where $T = $ number of 1500-year intervals.

4. $A = 6 \cdot \left(\frac{1}{2}\right)^T$, where T represents an 88-year period:

    88 years: $T = 1$ $A = 6 \cdot \left(\frac{1}{2}\right)^1 = 3$ mg

    176 years: $T = 2$ $A = 6 \cdot \left(\frac{1}{2}\right)^2 = 1.5$ mg

    440 years: $T = 5$ $A = 6 \cdot \left(\frac{1}{2}\right)^5 \approx 0.19$ mg

5. $y = 12(5)^{-x}$

    $= 12(5^{-1})^x$

    $= 12(1/5)^x$ so it represents decay.

6. a. $y = 23 \cdot 2.4^{-x} \Rightarrow y = 23(2.4^{-1})^x \Rightarrow y = 23\left(\frac{1}{2.4}\right)^x \Rightarrow$
$y = 23(0.42)^x$; decay

   b. $f(x) = 8000 \cdot (0.5^{-1})^x \Rightarrow f(x) = 8000\left(\frac{1}{0.5}\right)^x \Rightarrow$
$f(x) = 8000 \cdot (2)^x$; growth

   c. $P = 52{,}000 \cdot 1.075^{-t} \Rightarrow P = 52{,}000(1.075^{-1})^t \Rightarrow$
$P = 52{,}000\left(\frac{1}{1.075}\right)^t \Rightarrow P = 52{,}000(0.93)^t$; decay

Exercises for Section 5.2

1. a. 0.43   **b.** 0.95   **c.** 1/3
For each of the functions, the decay factor represents the percentage remaining after time t.

3. a. $f(t) = 10{,}000 \cdot 0.4^t$ **c.** $f(t) = 219 \cdot 0.1^t$
   b. $f(t) = 2.7 \cdot 10^{13} \cdot 0.27^T$

5. a. This is not exponential since the base is the variable.

   b. This is exponential; the decay factor is 0.5; the vertical intercept is 100.

   c. This is exponential; the decay factor is 0.999; the vertical intercept is 1000.

7.

Initial Value, C	Decay Factor, a	Exponential Function $y = Ca^x$
500	0.95	$y = 500 \cdot 0.95^x$
$1.72 \cdot 10^6$	0.75	$y = 1.72 \cdot 10^6 \cdot (0.75)^x$
1600	0.25	$y = 1600 \cdot (0.25)^x$

9. a. matches graph B.   **b.** matches graph A.

11. In 28 years 1.25 grams would remain, so it will take 56 years to get the Strontium 90 down to 0.625 grams since that weight is ¼ of the original.

13. a. $f(t) = -4t + 20$

   b. $g(t) = 20 \cdot 0.25^t$

   c.

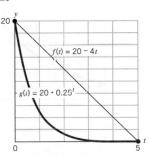

15. a. 800 grams

   b. 6 minutes

   c. $Q(t) = 800 \cdot (0.5)^T$, where $T =$ number of 6-minute periods and thus $t = 6T =$ number of minutes, or $t/6 = T$ and thus $Q(t) = 800 \cdot (0.5^{1/6})^t \approx 800 \cdot (0.891)^t$.

   d. $Q(5) \approx 449$ grams. It represents how much of substance is left after 5 minutes.

17. The graph appears to be exponential decay because the amount decreases from 100 to 50 percent in 30 years, thus it would have a half-life of thirty years. However, this would mean that in year 60 only 25 percent would remain and the graph shows that about 38% remains. Alternately using the points $(0, 100)(100, 33)$ gives the exponential form: $33 = 100(a)^{100} \Rightarrow a = (0.33)^{1/100}$, thus the equation is

$A(t) = 100(0.33)^{t/100}$. Then at year 1000, $A(1000) = 100(0.33)^{1000/100} = .00153$, which is clearly not 19 percent as stated on the graph. While the graph is clearly decaying, this graph is not exponential decay.

Section 5.3

Algebra Aerobics 5.3

1.

t	$N = 10 + 3t$	$N = 10 \cdot 3^t$
0	10	10
1	13	30
2	16	90
3	19	270
4	22	810

$N = 10 + 3t$

For every unit increase in t, N increases by 3, so the graph is linear with a slope of 3. The vertical intercept is 10.

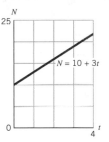

$N = 10 \cdot 3^t$

The rate of change here is not constant as in the previous problem. For example, the average rate of change between 0 and 1 is $(30 - 10)/1 = 20$ and between 1 and 2 is $(90 - 30)/1 = 60$. But the vertical intercept for the graphs of both functions is the same, 10.

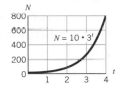

2. a.

t	$f(t) = 200 + 20t$		t	$g(t) = 200(1.20)^t$
0	200		0	200
1	220		1	240
2	240		2	288
3	260		3	345.6
4	280		4	414.72
5	300		5	497.66

   b.

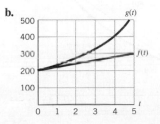

c. The function f is linear and the function g is exponential. While both functions have the same initial value, g is growing faster than f.

3. $f(x)$ linear; common difference $= 4.5$
   $g(x)$ linear; common difference $= -0.6$
   $F(x)$ exponential; common ratio ≈ 1.33
   $G(x)$ neither linear nor exponential

4. a. Linear: $m = \frac{620 - 500}{1 - 0} = 120$; $(0, 500) \Rightarrow b = 500 \Rightarrow$
   $y = 500 + 120t$
   Exponential: growth factor $= \frac{620}{500} = 1.24$; $(0, 500) \Rightarrow$
   initial amount $= 500 \Rightarrow y = 500(1.24)^t$

b. Linear: $m = \frac{3.2 - 3}{1 - 0} = 0.2$; $(0, 3) \Rightarrow b = 3 \Rightarrow y = 3 + 0.2t$
   Exponential: growth factor $= \frac{3.2}{3} \approx 1.067$; $(0, 3) \Rightarrow$ initial
   amount $= 3 \Rightarrow y = 3(1.067)^t$

5. a. growth factor $= \frac{3750}{1500} = 2.50 \Rightarrow P = 1500 \cdot 2.50^t$

b. growth factor $= \frac{82,300}{80,000} = 1.029 \Rightarrow A = 80,000 \cdot 1.029^t$

c. growth factor $= \frac{32.7}{30} = 1.09 \Rightarrow Q = 30 \cdot 1.09^t$

6. a. 2

b.

t (years)	Q
0	10
2	20
4	40
6	80
8	160

c. Let $a =$ growth factor. The 2-yr. growth factor $= 2 \Rightarrow$
   $a^2 = 2 \Rightarrow a = \sqrt{2} \approx 1.41$. The annual growth factor $\approx$
   1.41.

d. $Q = 10 \cdot 1.41^t$

Exercises for Section 5.3

1. a. linear; 5   **c.** neither; 3   **e.** exponential; 7
   b. exponential; 3   **d.** exponential; 6   **f.** linear; 0

3. a. $P(t) = 50 + 5t$

b. $Q(t) = 50 \cdot 1.05^t$

c. The graph of the two functions is in the accompanying diagram.

d. Graphing software gives that the two are equal at $(0, 50)$ and at $(26.6, 183)$. Thus the populations were both 50 people at the start and were both approximately 183 persons after 26.6 years. (Student eyeball estimates may differ.)

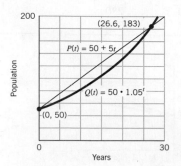

5. linear: $y = 3x + 6$; exponential: $y = 6 \cdot 1.5^x$

7. a. linear: $y = 10x + 10$; exponential: $y = 10 \cdot \left(\sqrt{3}\right)^x$

b. linear: $y = 100x + 100$; exponential $y = 100 \cdot \left(\sqrt[3]{4}\right)^x$

9. a. exponential; $f(x) = 7 \cdot 2.5^x$

b. linear: $g(x) = 0.2x + 0.5$

11.

x	$6 + 1.5x$	$6 \cdot 1.5^x$	$1.5 \cdot 6^x$
0	6.0	6	1.5
1	7.5	9	9
2	9.0	13.5	54
3	10.5	20.25	324
4	12.0	30.375	1,944
5	13.5	45.5625	11,664

The graphs are found in the accompanying diagrams.

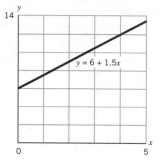

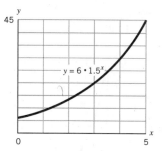

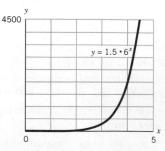

13. a. The average rate of change would be $= \frac{3000 - 14,313}{2013 - 2002} = \frac{-11,313}{11} \approx -1028$ mute swans per year since 2002.

b. If $x =$ years since 2002, we have $3000 = 14,313 \cdot a^{11}$; and thus $\frac{3000}{14,313} = a^{11}$ or $a = \left(\frac{3000}{14,313}\right)^{1/11} \approx 0.868$.

15. a.

Years after 2004	Plan A	Plan B
0	$7.00	$7.00
1	$7.35	$7.50
2	$7.72	$8.00
3	$8.10	$8.50
4	$8.51	$9.00

b. Plan A: $F_A(t) = 7 \cdot 1.05^t$ and plan B: $F_B(t) = 7 + 0.50t$

c. $F_A(21) = 7 \cdot (1.05)^{21} \approx \$19.50$ and $F_B(21) = 7 + 0.50 \cdot 21 = \$17.50$

d. Student answers will vary, but one should note that plan B is more expensive until year 15; from then on plan A would be more expensive. Going for plan A for the short term seems like a better option.

Section 5.4

Algebra Aerobics 5.4

1. a.

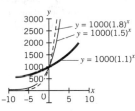

b. Yes, they intersect at (0, 1000).

c. In the first quadrant, $y = 1000(1.8)^x$ is on top, $y = 1000(1.5)^x$ is in the middle, and $y = 1000(1.1)^x$ is on the bottom.

d. In the second quadrant, $y = 1000(1.1)^x$ is on top, $y = 1000(1.5)^x$ is in the middle, and $y = 1000(1.8)^x$ is on the bottom.

2. a.

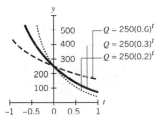

b. Yes, they intersect at (0, 250).

c. In the first quadrant: $Q = 250(0.6)^t$ is on top, $Q = 250(0.3)^t$ is in the middle, and $Q = 250(0.2)^t$ is on the bottom.

d. In the second quadrant: $Q = 250(0.2)^t$ is on top, $Q = 250(0.3)^t$ is in the middle, and $Q = 250(0.6)^t$ is on the bottom.

3. a.

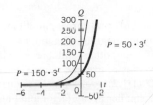

b. No, the curves do not intersect.

c. $P = 150 \cdot 3^t$ is always above $P = 50 \cdot 3^t$.

---

4. a. The horizontal axis is the horizontal asymptote.

b. The horizontal axis is the horizontal asymptote.

c. No horizontal asymptotes. The graph of this function is a line.

5. Comparing the growth factors: $1.23 > 1.092 > 1.06$, so the function h has the most rapid growth. Comparing the decay factors: $0.89 < 0.956$, so the function g has the most rapid decay.

6. a. The function g has the larger initial value since it crosses the vertical axis above the function f.

b. The function f has the larger growth factor since it is steeper, growing at a faster rate.

c. The function f approaches zero more rapidly as $x \to -\infty$.

d. The graphs intersect at approximately (3, 35); f is greater than g after the point of intersection.

Exercises for Section 5.4

1. Let $y_1 = 2^x$, $y_2 = 5^x$, and $y_3 = 10^x$.

a. $C = 1$ for each case; $a = 2$ for y_1, $a = 5$ for y_2, and $a = 10$ for y_3.

b. Each represents growth; y_1's value doubles, y_2's value is multiplied by 5, and y_3's value is multiplied by 10.

c. All three graphs intersect at (0, 1).

d. In the first quadrant the graph of y_3 is on top, the graph of y_2 is in the middle, and the graph of y_1 is on the bottom.

e. All have the graph of $y = 0$ (or the x-axis) as their horizontal asymptote.

f. Small table for each:

x	y_1	y_2	y_3
0	1	1	1
1	2	5	10
2	4	25	100

g. The graphs of the three functions are given in the accompanying diagram. They indeed confirm the answers given to the questions asked in parts (a) through (f).

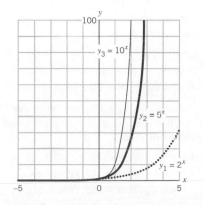

3. Let $y_1 = 3^x$, $y_2 = (1/3)^x$, and $y_3 = 3 \cdot (1/3)^x$.

a. $C = 1$ for y_1 and y_2 and $C = 3$ for y_3.

b. y_1 represents growth; y_2 and y_3 represent decay.

c. y_1 and y_3 intersect at approx. (0.5, 1.7); y_1 and y_2 intersect at (0, 1).

d. For $x > 0.5$, the graph of y_1 is on top, the graph of y_3 is in the middle, and the graph of y_2 is on the bottom. For $0 < x < 0.5$, the graph of y_3 is on top, the graph of y_1 is in the middle, and the graph of y_2 is on the bottom.

e. All the graphs have the x-axis as their horizontal asymptote.

f. A small table for each:

x	y_1	y_2	y_3
0	1	1	3
1	3	1/3	1
2	9	1/9	1/3

g. The graphs of these three functions are given in the accompanying diagram, and they confirm the answers given to the questions in parts (a) through (f).

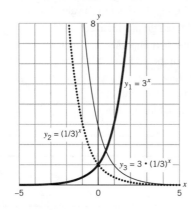

5. The function P goes with Graph C, the function Q goes with Graph A, the function R goes with Graph B, and the function S goes with Graph D.

7. a. The smaller the decay factor, the faster the descent; thus h has the smallest, then g, and finally f.

b. The point (0, 5) and no other point.

c. It will approach 0 more slowly than the other two functions.

d. h is on top and will stay on top for $x < 0$.

9. The accompanying diagram contains the graphs of $f(x) = 30 + 5x$ and $g(x) = 3 \cdot 1.6^x$ with the points of intersection marked.

a. $f(0) = 30; g(0) = 3; f(0) > g(0)$

b. $f(6) = 60; g(6) \approx 50.33; f(6) > g(6)$

c. $f(7) = 65; g(7) \approx 80.53; f(7) < g(7)$

d. $f(-5) = 5; g(-5) \approx 0.286; f(-5) > g(-5)$

e. $f(-6) = 0; g(-6) \approx 0.179; f(-6) < g(-6)$

f. f and g go to infinity; $f(x) < g(x)$ for all $x > 6.5$, approximately.

g. f goes to $-\infty$ and $g(x)$ goes to 0; thus $f(x) < g(x)$ for all $x < -6$, approximately.

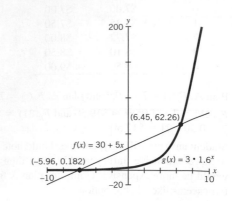

11. a. As $x \to +\infty$, $g(x)$ will dominate over $f(x)$, i.e., $g(x) > f(x)$

b. There is no one coordinate window that will display the graphs of f and g to help one see the answer to his question. Thus two displays are given: one over $2680 < x < 2700$ and another over $-0.5 < x < 0$. From these one can see that $f(x) > g(x)$ if $-0.196 < x < 2690.51$; otherwise $g(x) > f(x)$.

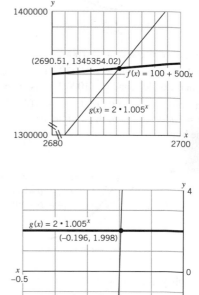

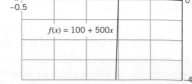

13. The graphs of the three functions are given in the accompanying diagram. The graphs of F and H are mirror images of each other and are equally steep when the absolute values of average rates of change are considered. $H(x) = 100(1/1.2)^x \approx 100(0.83)^x$ and $G(x) = 100 \cdot 0.8^x$, so $G(x)$ has a smaller decay factor and thus decays faster than $H(x)$ and is steeper than $H(x)$. Thus, the graph of G is steeper than the other two, again when considering the absolute values of the average rates of change.

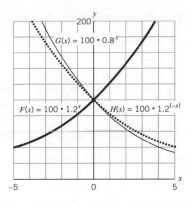

15. The graphs are given in the accompanying diagram.
    a. It is clear that the graphs of f and h are mirror images of each other with respect to the y-axis.
    b. The graphs of f and g are mirror images of each other with respect to the x-axis.
    c. The graphs of g and h are mirror images of each other with respect to the origin.
    d. The graphs of these functions are mirror images of each other with respect to the x-axis.
    e. These functions are mirror images of each other with respect to the y-axis.

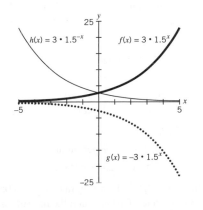

Section 5.5

Algebra Aerobics 5.5

1. **a.** 22%
   b. 106.7% or 1.067
   c. $1 - 0.972 = 0.028$ or 2.8%
   d. $1 - 0.123 = 0.877 = 87.7\%$

2.

Exponential Function	Initial Value	Growth or Decay?	Growth or Decay Factor	Growth or Decay Rate
$A = 4(1.03)^t$	4	growth	1.03	0.03 or 3%
$A = 10(0.98)^t$	10	decay	0.98	0.02 or 2%
$A = 1000(1.005)^t$	1000	growth	1.005	0.005 or 0.5%
$A = 30(0.96)^t$	30	decay	0.96	0.04 or 4%
$A = 50,000(1.0705)^t$	$50,000	growth	1.0705	0.0705 or 7.05%
$A = 200(0.51)^t$	200 g	decay	0.51	0.49 or 49%

3. **a.** growth factor = 1.06 = 106%   growth rate = 6%
   b. decay factor = 0.89 = 89%   decay rate = 11%
   c. growth factor = 1.23 = 123%   growth rate = 23%
   d. decay factor = 0.956 = 95.6%   decay rate = 4.4%
   e. growth factor = 1.092 = 109.2%   growth rate = 9.2%

4. **a.** $\frac{\$6.00}{1.65} \approx 3.64$, so 3.64 (the growth factor) = 1 + growth rate. The growth rate = 2.64 or a 264% increase in 30 years.
   b. $\frac{0.167}{0.299} \approx 0.56$, so 0.56 (the decay factor) = 1 − decay rate. The decay rate = 0.44 or a 44% decrease in one season.
   c. $\frac{23}{17} \approx 1.35$, so 1.35 = 1 + growth rate. The growth rate = 0.35 or a 35% increase in one month.

Exercises for Section 5.5

1. **a.** $Q = 1000 \cdot 3^T$ **c.** $Q = 1000 \cdot 1.03^T$
   b. $Q = 1000 \cdot 1.3^T$ **d.** $Q = 1000 \cdot 1.003^T$

3. **a.** 5%   **c.** 55%   **e.** 0.4%
   b. 18%   **d.** 34.5%   **f.** 27.5%

5.

	Initial Value	Growth or Decay?	Growth or Decay Factor	Growth or Decay Rate	Exponential Function
a.	600	growth	2.06	106%	$N(t) = 600 \cdot 2.06^t$
b.	1200	growth	3.00	200%	$N(t) = 1200 \cdot 3^t$
c.	6000	decay	0.25	75%	$N(t) = 6000 \cdot 0.25^t$
d.	$1.5 \cdot 10^6$	decay	0.75	25%	$N(t) = 1.5 \cdot 10^6 \cdot 0.75^t$
e.	$1.5 \cdot 10^6$	growth	1.25	25%	$N(t) = 1.5 \cdot 10^6 \cdot 1.25^t$
f.	7	growth	4.35	335%	$N(t) = 7 \cdot 4.35^t$
g.	60	decay	0.35	65%	$N(t) = 60 \cdot 0.35^t$

7. **a.** $A(T) = 12,000 \cdot 1.12^T$ and $B(T) = 12,000 \cdot 0.88^T$, where T = decades since 1990.
   b. If t = years since 1990, then $t = 10T$ or $T = t/10$, and thus $a(t) = 12,000 \cdot [1.12]^{t/10} = 12,000 \cdot [1.12^{0.1}]^t = 12,000 \cdot 1.0114^t$ people and $b(t) = 12,000 \cdot [0.88]^{t/10} = 12,000 \cdot [0.88^{0.1}]^t = 12000 \cdot 0.9873^t$ people where t is the number of years from 1990.
   c. $A(2) \approx 15,053$; $a(20) \approx 15,053$; $B(2) \approx 9293$ and $b(20) \approx 9293$. $A(2)$ and $B(2)$ represent the population after two decades, whereas $a(20)$ and $b(20)$ represent the population after 20 years. Note that $A(2)$ should equal $a(20)$ and $B(2)$ should equal $b(20)$. Any small differences are because of round-off error.

9. $A(x) = 500 \cdot 0.85^x$; decay rate of 15%

11. $A(x) = 225 \cdot 1.015^x$; growth rate of 1.5%

13. **a.** $P(t) = 150 \cdot 3^t$ **c.** $P(t) = 150 \cdot 0.93^t$
    b. $P(t) = 150 - 12t$ **d.** $P(t) = 150 + t$

15. In 2030 the population size will be approximately $200 \cdot 1.4^2 = 392$ million, and in 2060 it will be approximately $200 \cdot 1.4^3 = 548.8$ million. Using a graphing calculator to estimate the intersection point of the graphs of $y_1 = 200 \cdot 1.4^t$ and $y_2 = 1000$ (where y_1 and y_2 are measured in millions and t is measured in 30-year periods) gives $t \approx 4.78$. Thus, according to this model, sometime in the first part of 2113, the U.S. population will reach approximately 1 billion.

17. a. $A(n) = A_0 \cdot 0.75^n$, where n measures the number of years from the original dumping of the pollutant and A_0 represents the original amount of pollutant.

   b. We are solving $0.1 \cdot A_0 = A_0 \cdot 0.75^n$ for n and we first divide out by A_0. Then we graph $y = 0.01$ and $y = 0.75^n$ and estimate their intersection. This gives $n \approx 16$ years.

19. a. $1.007^{12} \approx 1.087$ and thus the inflation rate is about 8.7% per year.

   b. $(1 + r)^{12} = 1.05$ means that $r = (1.05)^{1/12} - 1 \approx 0.0041$. Thus the rate is about 0.4% per month.

21. a. They are equivalent, since $1.2^{0.1} = 1.0184$ to four decimal places.

   b.

x	0	5	10	15	20	25
$f(x)$	15,000.0	16431.7	18,000	19,718.0	21,600.0	23,661.6
$g(x)$	15,000.0	16431.7	18,000.1	19,718.2	21,600.3	23,662.0

   c. If x is the number of years, then $f(x) = 15,000(1.2)^{x/10}$ represents a 20% growth factor over a decade, and $g(x) = 15,000(1.0184^x)$ represents a 1.84% annual growth factor.

23. a. A graph with the data plotted and the best-fit exponential is given in the diagram.

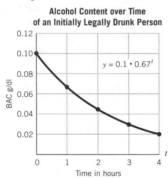

Alcohol Content over Time of an Initially Legally Drunk Person

$y = 0.1 \cdot 0.67^t$

BAC g/dl

Time in hours

   b. If one looks at the exponential graph it is easy to see that it fits the data quite well. Moreover, the ratios of successive g/dl values are nearly a constant 0.67. Another equation, easier to handle, is $y = 0.1 \cdot 0.67^t$, where t is measured in hours and y in g/dl.

   c. In this model, the g/dl decreases each hour by about 33%.

   d. A reasonable domain could be 0 to 10 hours and the corresponding range would be 0.002 to 0.100 g/dl.

   e. $0.005 = 0.1 \cdot 0.67^t$ implies that t is between 7 and 8 hours. This can be obtained by computing g/dl values for successive t values using the formula given in part (b).

25. a. Let $x =$ years since 1980. Then the rate of change over the 1980–2006 period is $\frac{33 - 94}{26 - 0} = \frac{-61}{26} \approx -2.35$ deaths per 1000 live births; a linear equation would be $d(x) = -2.35x + 94$, where $d(x) =$ number of deaths per 1000 live births x years from 1980.

   b. Solving $33 = 94a^{26}$ for a gives $a = (33/94)^{1/26}$, or $a \approx 0.9605$. So $m(x) = 94 \cdot (0.9605)^x$, where $m(x) =$ number of deaths per 1000 live births x years from 1980.

   c. For 2010, $x = 30$ years since 1980 so $m(30) = 90(0.9605)^{30} \approx 27$ deaths per 1000 live births in 2010. $d(30) = -2.35(30) + 94 \approx 24$ deaths per 1000 live births in 2010.

Section 5.6

Algebra Aerobics 5.6a

1. a. Exponential growth; doubling period is 30 days since $f(30) = 300 \cdot 2^{30/30} = 600$; Initial $= 300$;
one day later: $f(1) = 300 \cdot 2^{1/30} \approx 307.01$;
one month later: $f(30) = 300 \cdot 2^{30/30} = 600$;
one year later: $f(360) = 300 \cdot 2^{360/30} = 1,228,800$

   b. Exponential decay; half-life period is 2 days since $g(2) = 32(0.5)^{2/2} = 16$; Initial $= 32$;
one day later: $g(1) = 32 \cdot 0.5^{1/2} \approx 22.6$;
one month later: $g(30) = 32 \cdot 0.5^{30/2} = 0.00098$;
one year later: $g(360) = 32 \cdot 0.5^{360/2} =$ trace amount (almost zero)

   c. Exponential decay; half-life period is one day since $P = 32,000 \cdot 0.5^1 = 16,000$; Initial $= 32,000$;
one day later: $P = 32,000 \cdot 0.5^1 = 16,000$;
one month later: $P = 32,000 \cdot 0.5^{30} \approx 0.000\,03$;
one year later: $P = 32,000 \cdot 0.5^{360} =$ trace amount (almost zero)

   d. Exponential growth; doubling period is 360 days since $h(360) = 40,000 \cdot 2^{360/360} = 80,000$; Initial $= 40,000$;
one day later, $h(1) = 40,000 \cdot 2^{1/360} \approx 40,077$;
one month later, $h(30) = 40,000 \cdot 2^{30/360} \approx 42,379$;
one year later, $h(360) = 40,000 \cdot 2^{360/360} = 80,000$

2. a. $70/2 = 35$ years

   b. $70/0.5 = 140$ months

   c. $70/8.1 = 8.64$ yr ≈ 9 years

   d. growth factor $= 1.065 \Rightarrow$ growth rate $= 6.5\%$
$70/6.5 = 10.77 \approx 11$ years

3. a. $\frac{70}{10} = 7$; $R \approx 7\%$ per year

   b. $70/5 = 14$; $R \approx 14\%$ per minute

   c. $70/25 = 2.8$; $R \approx 2.8\%$ per second

4. a. Since $a = 0.95$, the decay rate is $r = 0.05$ or 5%, so R from the rule of 70 is 5, and the half-life is $70/5 \approx 14$ months.

   b. Since $a = 0.75$, the decay rate is $r = 0.25$ or 25%, so R from the rule of 70 is 25, and the half-life is $70/25 \approx 2.8$ seconds.

   c. The half-life is $70/35 \approx 2$ years.

Algebra Aerobics 5.6b

1. a. Initial investment is $10,000; growth factor $= 1.065$; growth rate $= 6.5\%$; doubling time: using the rule of 70, $70/6.5 \approx 10.8$ years

   b. Initial investment is $25; growth factor $= 1.08$; growth rate $= 8\%$; doubling time: using the rule of 70, $70/8 \approx 8.8$ years

   c. Initial investment is $300,000; growth factor $= 1.11$; growth rate $= 11\%$; doubling time: using the rule of 70, $70/11 \approx 6.4$ years

   d. Initial investment is $200; growth factor $= 1.092$; growth rate $= 9.2\%$; doubling time: using the rule of 70, $70/9.2 \approx 7.6$ years

2.

Function	Initial Investment	Growth Factor	Growth Rate	Amount 1 Year Later	Doubling Time (approx.)
$A = 50000 \cdot 1.072^t$	$50,000	1.072	7.2%	$53,600	9.7 years
$A = 100000 \cdot 1.067^t$	$100,000	1.067	6.7%	$106,700	10.4 years
$A = 49622 \cdot 1.058^t$	$49,622	1.058	5.8%	$52,500	12 years
$A = 3000 \cdot 1.13^t$	$3000	1.13	13%	$3390	5.4 years

3. a. $x = 5$ since $2^5 = 32$

   b. $x = 8$ since $2^8 = 256$

   c. $x = 10$ since $2^{10} = 1024$

   d. $x = 1$ since $2^1 = 2$

   e. $x = 0$ since $2^0 = 1$

   f. $x = -1$ since $2^{-1} = \frac{1}{2}$

   g. $x = -3$ since $2^{-3} = \frac{1}{2^3} = \frac{1}{8}$

   h. $x = 1/2$ since $2^{1/2} = \sqrt{2}$

4. a. 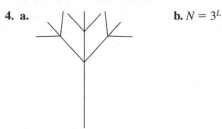   **b.** $N = 3^L$

5.

Level	# of People Called	Total Called
0	3^0	1
1	3^1	4
2	$3^2 = 9$	13
3	$3^3 = 27$	40
4	$3^4 = 81$	121
5	$3^5 = 243$	364
6	$3^6 = 729$	1093
7	$3^7 = 2187$	3280
8	$3^8 = 6561$	9841

So it will take eight levels to reach 8000 people.

Exercises for Section 5.6

1. a. Has a fixed doubling time   **d.** Has a fixed doubling time

   b. Has neither   **e.** Has neither

   c. Has a fixed half-life   **f.** Has a fixed half-life

3. b. $f(x) = 1000 \cdot 2^{t/7}$; $2^{1/7} = 1.1041$ per year; 10.41% per year

   c. $f(x) = 4 \cdot 2^{t/25}$; $2^{1/25} = 1.0281$ per minute; 2.81% per minute

   d. $f(x) = 5000 \cdot 2^{t/18}$; $2^{1/18} = 1.0393$ per month; 3.93% per month

5. a. $3 \cdot 2^{x/5} < 3(1.225)^x$, if $x > 0$; the inequality is reversed if $x < 0$.

   b. $50 \cdot (1/2)^{x/20} \approx 50 \cdot 0.9659^x$.

   c. $200 \cdot 2^{x/8} \approx 200 \cdot 1.0905^x$.

   d. $750 \cdot (1/2)^{x/165} > 750 \cdot 0.911^x$ if $x > 0$, and the inequality is reversed if $x < 0$.

7. a. and **b.** Here is the graph:

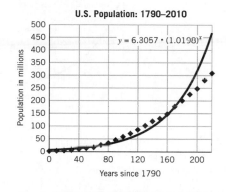

In the graph y is the U.S. population size in millions and x is the number of years since 1790. The annual growth factor is 1.0198; the annual growth rate is therefore 1.98%. The estimated initial population is 6.3 million. The two population sizes (actual and best-fit model estimate) are very close until about 1870; the actual is a bit higher than the best-fit model estimate from 1870 until 1950, and then the best-fit model estimate starts to grow much faster than the actual thereafter, exceeding the actual value by 162 million in 2010. In fact, the data look more linear after 1950.

   c. 2010 is 220 and 2025 is 235 years from 1790. The best-fit model estimates the population to be $6.3057 \cdot (1.1098)^{220} \approx$ 471 million and $6.3057(1.0198)^{235} \approx 632$ million. One can get a current estimate at any given moment by going to *http://www/popclock.html* and clicking on U.S. PopClock.

9. a. Here is the graph:

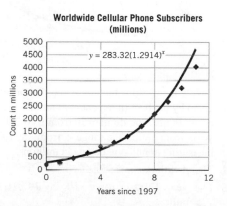

   b. The growth factor is 1.2914; the growth rate is 0.2914, or 29.14%, per year since 1997.

   c. 2010 is 13 years since 1997, and $283.32 \cdot 1.2914^{13} \approx 7870$ million cell phones. See *http://www.cbsnews.com/stories/2010/02/15business/main6209772.shtml* for a current estimate.

11. The graph in the accompanying diagram goes through $(0, 100)$ and $(12.3, 50)$, where the first coordinate is measured in years and the second is measured in grams. Using the graphing utility gives: $A(t) = 100 \cdot 0.945^t$. Using algebra we get $A(t) = 100 \cdot (0.5)^{t/12.3}$, which gives the same formula as the best-fitting exponential formula. The graph of this function is given in the accompanying diagram.

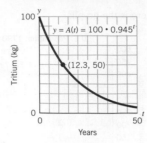

13. a. $0.5^5 = 1/32 = 0.03125$ and thus 3.125% of the original dosage is left.

  b. **i.** Approximately 2 hours.

   ii. $A(t) = 100 \cdot 0.5^{t/2}$ where t = hours after taking the drug and $A(t)$ is in mg.

   iii. $5 \cdot 2 = 10$ hours and $100 \cdot (1/2)^5 = 100/32 = 3.125$ milligrams. Also $A(10) = 100 \cdot 0.5^{10/2} = 3.125$.

   iv. Student answers will vary. They should mention the half-life and present a graph to make the drug's behavior clear to any prospective buyer.

15. a. $R = 70/5730 \approx 0.012\%$ per year

  b. $R = 70/11{,}460 \approx 0.0061\%$ per year

  c. $R = 70/5 = 14\%$ per second.

  d. $R = 70/10 = 7\%$ per second.

17. a. and **b.** The best-fit graph for the data and the best-fit exponential model are given here:

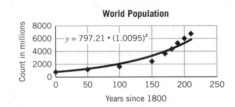

  c. 797.21 is the estimated initial population count in millions given by the best-fit exponential model. The growth factor is ≈ 1.0095. The domain is from 0 to 210: measured in years from 1800. The range is from 980 to 6850 million, measuring population counts in millions.

  d. The growth rate is 0.95%.

  e. The year 1790 is $t = -10$, so $797.21(1.0095)^{-10} \approx$ 725 million.

   The year 1920 is $t = 120$, so $797.21(1.0095)^{120} \approx$ 2479 million.

   The year 2025 is $t = 225$, so $797.21(1.0095)^{225} \approx$ 6691 million.

   The year 2050 is $t = 250$, so $797.21(1.0095)^{250} \approx$ 8475 million.

  f. From the graph estimating the intersection of population 4000 million and 8000 million, we see that the corresponding years are about 170 and 245, or the doubling time is about 75 years.

19. a. 10 half-lives gives a $0.5^{10} = 0.001$ or 0.1% of the original or a 99.9% reduction.

  b. $A(2) = 0.25A_0$, $A(3) = 0.125A_0$, $A(4) = 0.0625A_0$. After n half-lives, the amount left is $A(n) = 0.5^n \cdot A_0$.

21. a.

t, Time (months)	M, mass (g)
0	10
3	20
6	40
9	80
12	160

In general $M = 10(2^{1/3})^t = 10 \cdot 1.2599^t$ after t months.

  b. Using a calculator or a graph of the model, when $M = 2000$, then $t \approx 23$ months or nearly 2 years.

  c. $2000/10 = 200$ and thus at 2000 grams, it is 20,000% of its original size.

23. Since $1.00/1.06 = 0.943$, then the value of a dollar after t years of such inflation is given by $V(t) = 0.943^t$. Using a calculator or the "rule of 70," when $V(t) = 0.50$, gives $t \approx 12$ years.

25. a. In theory, the number recruited is $M_{\text{new}}(n) = 10^n$, where n measures the number of rounds and $M_{\text{new}}(n)$ measures the number of people participating in the nth round of recruiting. Note that this formula assumes that all who are recruited stay and that all recruits are distinct.

  b. $M_{\text{Total}}(n) = 1 + 10 + \cdots + 10^n$.

  c. $M_{\text{Total}}(5) = 111{,}111$, but only 11,110 of those stem from the originator. After 10 rounds the number recruited (not including the originator) would be 11,111,111,110, which is larger than the 2005 world population.

  d. Comments will vary, but all will probably note how fast the number of recruits needed grows and how the amounts expected are not quite what one would have thought from the advertisements. If the chain is initially successful, then you would get a large number of new recruits in a short period. However, as can be seen in part (c), it quickly becomes unrealistic for each new person on the chain to recruit 10 new people.

27. a. and **b.** Here is a graph of the data and a best-fit exponential function for that data when one chooses years from 1945 for the time variable.

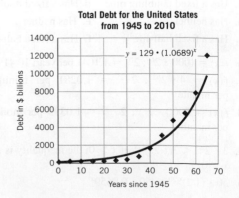

Answers will vary depending on the year chosen. In our choice, 2010 is 65 years from 1945 and the model predicts $129 \cdot (1.0689)^{65} \approx 9806$ billion dollars as the debt, which is considerably lower than the estimated debt.

c. Answers will vary depending on when they access the given Internet site.

Section 5.7

Algebra Aerobics 5.7

1. a. $\$1000(1.085) = \$1085$

  b. $\$1000\left(1 + \frac{0.085}{4}\right)^4 = \$1087.75$

  c. $\$1000e^{0.085} = \$1088.72$

2. a. $e^{0.04} = 1.0408 \Rightarrow 4.08\%$ is effective rate

  b. $e^{0.125} = 1.133 \Rightarrow 13.3\%$ is effective rate

  c. $e^{0.18} = 1.197 \Rightarrow 19.7\%$ is effective rate

3. a. Principal $= 6000$; nominal rate $= 5\%$; effective rate $= 5\%$ since $1.05^1 = 1.05$; number of interest periods $= 1$

  b. Principal $= 10,000$; nominal rate $= 8\%$; effective rate $\approx 8.24\%$ since $1.02^{4(1)} \approx 1.0824$; number of interest periods $= 4$

  c. Principal $= 500$; nominal rate $= 12\%$; effective rate $\approx 12.68\%$ since $1.01^{12(1)} \approx 1.1268$; number of interest periods $= 12$

  d. Principal $= 50,000$; nominal rate $= 5\%$; effective rate $\approx 5.06\%$ since $1.025^{2(1)} \approx 1.0506$; number of interest periods $= 2$

  e. Principal $= 125$; nominal rate $= 7.6\%$; effective rate $\approx 7.90\%$ since $e^{0.076\,(1)} \approx 1.0790$; interest is continuously compounded

4. a. $5e^{0.03t} = 5(e^{0.03})^t \approx 5(1.030)^t$

  b. $3500e^{0.25t} = 3500(e^{0.25})^t \approx 3500(1.284)^t$

  c. $660e^{1.75t} = 660(e^{1.75})^t \approx 660(5.755)^t$

  d. $55,000e^{-0.07t} = 55,000(e^{-0.07})^t \approx 55,000(0.932)^t$

  e. $125,000e^{-0.28t} = 125,000(e^{-0.28})^t \approx 125,000(0.756)^t$

5.

n	$1/n$	$1 + 1/n$	$(1 + 1/n)^n$
1	1	2	2
100	0.01	1.01	2.704 813 829
1000	0.001	1.001	2.716 923 932
1,000,000	0.000 001	1.000 001	2.718 280 469
1,000,000,000	0.000 000 001	1.000 000 001	2.718 281 827

The values for $(1 + 1/n)^n$ come closer and closer to the irrational number we define as e and are consistent with the value for e in the text of 2.71828.

6. a. about $\$12,000$

  b. about $\$18,000$

  c. about $\$33,000$

  d. about 6 years

  e. 1 year: $A = \$11,255$; 5 years: $\$18,061$; 10 years: $\$32,620$. The doubling time is approximately $70/R = 70/12 \approx 5.8$ years.

7. a. $P = \$1000(1.04)^n$ **c.** $P = \$1000(2.10)^n$

  b. $P = \$1000(1.11)^n$

8. a. $8000 \cdot \left(1 + \frac{0.08}{4}\right)^{4 \cdot (18)} = \$33,289$

  b. $8000 \cdot e^{0.08 \cdot (18)} = \$33,766$

  c. $8000 \cdot (1.084)^{18} = \$34,168$

9. a. Continuous growth rate: 0.6 or 60%; annual effective growth rate $\approx 82.2\%$ since $e^{0.6(1)} \approx 1.822$, which is the growth factor. So $1.822 - 1 = 0.822$ is the growth rate.

  b. Continuous growth rate: 2.3 or 230%; annual effective growth rate $\approx 897\%$ since $e^{2.3(1)} \approx 9.97$, which is the growth factor. So $9.97 - 1 = 8.97$ is the growth rate.

10. a. Continuous decay rate: 0.055 or 5.5%; annual effective decay rate $\approx 5.35\%$ since $e^{-0.055} \approx 0.946$, which is the decay factor. So $1 - 0.946 = 0.054$ is the decay rate.

  b. Continuous decay rate: 0.15 or 15%; annual effective decay rate $\approx 13.9\%$ since $e^{-0.15} \approx 0.861$, which is the decay factor. So $1 - 0.861 = 0.139$ is the decay rate.

Exercises for Section 5.7

1. a. i. $A(n) = 5000 \cdot (1.035)^n$

    ii. $B(n) = 5000 \cdot (1.0675)^n$

    iii. $C(n) = 5000 \cdot (1.125)^n$

  b. $A(40) = 19,796.30$ $B(40) = 68,184.45$
    $C(40) = 555,995.02$.

3. a. $V(n) = 100 \cdot 1.06^n$. Solving $V(n) = 200$ for n graphically gives $n \approx 12$ years, which is the approximate doubling time.

  b. $W(n) = 200 \cdot 1.03^n$. Solving $W(n) = 400$ for n graphically gives $n \approx 23$ years, which is the approximate doubling time.

  c. For $n > 24.14$ yrs, $W(n) < V(n)$, since $V(n)$ has a larger growth factor.

5. Bank A offers $e^{0.0246} \approx 1.02491$ and thus an effective annual rate of 2.491%.

Bank B offers $(1 + 0.0248/4)^4 \approx 1.02503$ and thus an effective annual rate of 2.503%.

Bank C offers $(1 + 0.0247/12)^{12} \approx 1.02498$ and thus an effective annual rate of 2.498%.

Thus bank B offers the most interest at the end of a year.

7. a. 1.0353   **b.** 1.0355   **c.** 1.0356   **d.** 1.0356

  e. These values represent the growth factors for compounding semi-annually, quarterly, monthly, and continuously. As the number of compoundings increase, the results come closer and closer to continuous compounding.

9. In general, if the nominal rate is 8.5%, then $A_k(n) = 10,000 \cdot (1 + 0.085/k)^{kn}$ gives the value of $\$10,000$ after n years if the interest is compounded k times per year and $A_c(n) = 10,000 \cdot e^{0.085n}$ gives that value if the interest is compounded continuously.

  a. annually: $(1 + 0.085/1)^1 = 1.0850$, and thus the effective rate is 8.50%.

  b. semi-annually: $(1 + 0.085/2)^2 \approx 1.0868$, and thus the effective rate is 8.68%.

  c. quarterly: $(1 + 0.085/4)^4 \approx 1.0877$, and thus the effective rate is 8.77%.

  d. continuously: $e^{0.085n} \approx 1.0887$, and thus the effective rate is 8.87%.

11. a. $A(t) = 25{,}000 \cdot (1 + 0.0575/4)^{4t}$, where t = number of years

b. $B(t) = 25{,}000 \cdot e^{0.0575t}$, where t = number of years

c. $A(5) = \$33{,}259.12$ and $B(5) = \$33{,}327.26$

d. The effective rate for compounding quarterly is $\left(1 + \frac{0.0575}{4}\right)^4 - 1 \approx 0.05875$, and the effective rate for compounding continuously is $e^{0.0575} - 1 \approx 0.05919$. Thus continuous compounding has a slightly greater effective rate.

13. a. $e^{0.045} \approx 1.046$

b. $1.0680 < 1.0704 \approx e^{0.068}$

c. $1.2690 > e^{0.238} \approx 1.2687$

d. $e^{-0.10} \approx 0.9048 > 0.9000$

e. $0.8607 \approx e^{-0.15}$

15. a. Answers may vary. Two points on the nonsmokers graph would be $(20, 0)$ and $(50, 5)$. The rate of change is $5/30 = 0.167$ death rate per 1000, and the straight line through these two points, where t is average age years from age 30, is $D(t) = 0.167t - 3.33$.

b. For $D = 1.843e^{0.081n}$ means that for $n = 0$, or a 30-year-old, the death rate is only 1.843 per 1000, or 18 per 10,000, due to lung cancer. The continuous growth rate is 8.1%, and an annual growth rate is $e^{0.081} - 1 \approx 1.0844 - 1 = 0.0844$, or 8.44%.

c. Choosing the points $(21, 10)$ and $(30, 20)$, the doubling time is approximately 9 years, so the exponential model is: $y = C(2)^{n/9} = C(2^{1/9})^n \approx C(1.0801)^n$, or a growth rate of 8.01%, which is close to the given rate of 8.44%.

17. a. The 10-year decay factor is 0.85. The yearly decay factor is $0.85^{1/10} \approx 0.9839$. The yearly decay rate is $1 - 0.9839 \approx 0.0161$ or 1.61%.

b. $g(t) = 1.5 \cdot 0.9839^t$, where $g(t)$ is measured in millions and t in years.

c. $h(t) = 1.5 \cdot e^{-0.01625t}$, where $h(t)$ is measured in millions and t in years.

d. $g(20) \approx 1.0842$ million and $h(20) \approx 1.0838$ million. The h function decays a bit faster than the g function, but they are quite close.

19. a. $P(t) = 500 \cdot (1.0202)^t$; the continuous growth rate is 2% and the effective annual growth rate is $= 2.02\%$.

b. $N(t) = 3000 \cdot (4.4817)^t$; the continuous growth rate is 150% and the effective growth rate is 348.17%.

c. $Q(t) = 45 \cdot (1.0618)^t$; the continuous growth rate is 6% and the effective growth rate is 6.18%.

d. $G(t) = 750 \cdot 1.0356^t$; the continuous growth rate is 3.5% and the effective growth rate is 3.56%.

21. a. The graph of $f(x)$ is the same as the graph labeled **C**.

b. The graph of $g(x)$ is the same as the graph labeled **B**.

c. The graph of $h(x)$ is the same as the graph labeled **A**.

In general, the greater the growth factor, the faster the graph grows and the larger the exponent, the greater the growth factor.

Section 5.8

Algebra Aerobics 5.8

1. a. Judging from the graph, the number of *E. coli* bacteria grows by a factor of 10 (for example, from 100 to 1000, or 100,000 to 1,000,000) in a little over three time periods.

b. From the equation $N = 100 \cdot 2^t$, we know that every three time periods, the quantity is multiplied by 2^3 or 8.

c. Every four time periods, the quantity is multiplied by 2^4 or 16.

d. The answers are consistent, since somewhere between three and four time periods the quantity should be multiplied by 10 (which is between 8 and 16).

2. The graph of $y = 25(10)^x$ will be a straight line on a semi-log plot.

x	$y = 25(10)^x$
0	25
1	250
2	2,500
3	25,000
4	250,000
5	2,500,000

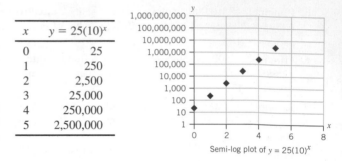

Semi-log plot of $y = 25(10)^x$

3. If $x = 3.5$, y is about 80,000. If $x = 7$, $y = 250{,}000{,}000$.

Exercises for Section 5.8

1. a. Graph of given table of values for $y = 500 \cdot 3^x$

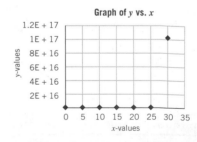

b. Graph of table for $\log(500 \cdot 3^x)$ vs. x

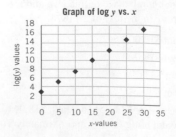

c. $y = 10^{\log(y)}$ and thus each y can be written as 10 to the power listed in the third column.

3. a. A uses the linear scale. B uses a power of 10 scale on the vertical axis and C uses a logarithm scale on the vertical axis.

b. The two graphs look the same because $\log(10^n) = n$ and the powers of 10 are spaced out like n.

c. On Graphs A and B, when $y = 1000$ is multiplied by a factor of 10 to get $y = 10,000$, x has increased by about 5.5 units.

d. On graphs B, y labels go up by factors of 10.

e. On graph A, since the scales on both axes are linear.

5. a. The accompanying graph is of the white blood cell counts on a semi-log plot. The data from October 17 to October 30 seem to be exponential, since the data in that range seem to fall along a straight line in the plot. There does not seem to be a discernible exponential decay pattern in this graph of the data.

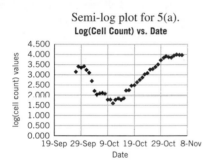

Semi-log plot for 5(a).

b. Below are two graphs of the *E. coli* counts; one has regular horizontal and vertical axes and the other is a semi-log plot. These data look very exponential from the third to the thirteenth time periods in the regular plot and fairly exponential in the semi-log plot (since that plot looks rather linear). For contrast, the regular linear plot is given as well.

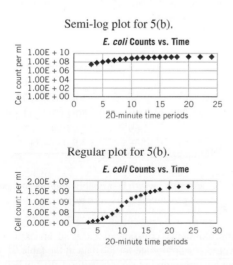

Semi-log plot for 5(b).

Regular plot for 5(b).

7. a. This is a semi-log plot. The horizontal scale is linear, and the vertical scale is a log scale.

b. The growth appears roughly linear from about 1950 to 2000, which means that function is behaving exponentially there.

c. The graph appears to decline linearly from about 1930 to 1935 and from 1945 to 1950. This means that the function describing the data is from an exponentially decaying function.

Ch. 5: Check Your Understanding

1. False	**7.** False	**13.** False	**19.** False
2. True	**8.** True	**14.** True	**20.** False
3. False	**9.** True	**15.** True	**21.** True
4. False	**10.** False	**16.** False	**22.** True
5. True	**11.** False	**17.** True	
6. False	**12.** False	**18.** True	

23. Possible answer: $P = 2.2(1.005)^{4t}$ million people, t = years.

24. Possible answer: $M = 1.4(0.977)^{\frac{t}{10}}$ billion dollars, t = years.

25. Possible answer: $y_1 = 300(0.88)^t$ and $y_2 = 300(0.94)^t$.

26. Possible answer: $y = -\frac{4.375}{3}x + 5$ and $y = 5(0.5)^x$.

27. Possible answer: $y = 5 \cdot (1.051)^t$, t = years.

28. Possible answer: $R = 200(0.966)^t$ mg, t = years.

29. Possible answer: $y = 100e^{-0.026t}$

30. True	**33.** True	**36.** False
31. False	**34.** False	**37.** True
32. True	**35.** False	**38.** True

Ch. 5 Review: Putting It All Together

1. a. $P = 150 \cdot 2^t$ **c.** $P = 150 (1.05)^t$

    b. $P = 150 - 12t$ **d.** $P = 150 + 12t$

3.

5. a. Males who just stopped smoking are about 22 times more likely (the relative risk) to get lung cancer than a lifelong nonsmoker. Females who stopped smoking 12 years ago are about three times more likely to get lung cancer than a lifelong nonsmoker.

b. One reason could be that the longer it has been since someone quit smoking, the more time the lungs have had to heal. Another reason could be that the death rate of smokers is higher than that of nonsmokers.

c. A relative risk of 1 means each group is equally likely to get lung cancer. It is highly unlikely for the relative risk of smokers vs. nonsmokers to go below 1 since that would mean smokers are less likely to get lung cancer than nonsmokers, which does not make sense.

7. Initial value = 500 and growth rate = 1.5 $\Rightarrow G(t) = 500(1.5)^t$, where $G(t)$ = number of bacteria and t = number of days.

(Note: the growth will not continue indefinitely.)

9. a. If the world population in 1999 was 6 billion people, and it grew at a rate of 1.3% per year, then it is only in the first year that there is a net addition of 78 million people. The next year the increase would be 1.3% of 6,078,000,000, which is 79,014,000. The population would continue to have an increase that becomes larger and larger than 78 million each year. The increase is a fixed amount only in a linear model.

b. Let P = world population (in millions) and t = number of years since 1999.

Linear model: $P = 6000 + 78t$

Exponential model: $6000(1.013)^t$

c. Linear model prediction for 2006:
$P = 6000 + 78(2006 - 1999) \Rightarrow P = 6546$ million

Exponential model prediction for 2006: $6000(1.013)^7 \Rightarrow P \approx 6568$ million

Answers may vary for best predictor depending on current population.

The U.S Census Bureau website gives the 2006 world population as 6.567 billion, so our exponential model is a more accurate predictor for 2006.

11. If R = the annual growth rate, then using the "rule of 70" we have $70/25 = 2.8 \Rightarrow R = 2.8\%$. So the number of motor vehicles increases by almost 3% per year.

13. In all the parts below, the independent variable t represents years since 2009 and the dependent variable represents population counts.

a. UAE: $P(t) = 4{,}798{,}000 \cdot 1.0369^t$
Guatemala: $P(t) = 13{,}277{,}000 \cdot 1.0207^t$
Kenya $P(t) = 39{,}003{,}000 \cdot 1.0269^t$
U.S.: $P(t) = 307{,}212{,}000 \cdot 1.0098^t$

It would make sense to round off the initial values to millions. Thus, we would have 4.798, 13.277, 39.003, 307.212, and 307.212 as the values in millions.

b.

Years since 2009	Population in Millions			
	UAE	Guatemala	Kenya	USA
0	4.798	13.277	39.003	307.212
1	4.975	13.552	40.052	310.223
2	5.159	13.832	41.130	313.263
3	5.349	14.119	42.236	316.333
4	5.546	14.411	43.372	319.433
5	5.751	14.709	44.539	322.563
6	5.963	15.014	45.737	325.724
7	6.183	15.325	46.967	328.917
8	6.411	15.642	48.231	332.140
9	6.648	15.966	49.528	335.395
10	6.893	16.296	50.860	338.682
11	7.148	16.633	52.229	342.001
12	7.411	16.978	53.633	345.352

c. Here are the graphs of all four on one coordinate system. Note that because the growth rates are so small, their graphs look almost linear.

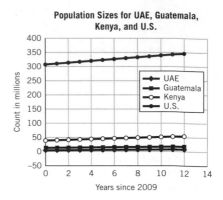

Population Sizes for UAE, Guatemala, Kenya, and U.S.

d. One can form a table in Excel or look at the graphs to do this. However, the rule of 70 might be easiest. Here is a table using the rule of 70. Note that time is measured in years since 2002.

R	Doubling Years (by Rule of 70)
3.69	70/3.69 = 19
2.07	70/2.07 = 34
2.69	70/2.69 = 26
0.98	70/.98 = 71

e. Exponential growth is probably appropriate; however, these annual growth rates may not be accurate in the long term.

15. The ratio of consecutive values over 5-year intervals is approximately constant at 1.09, so an exponential model would be appropriate for the data in the table. If we let $E(t)$ = energy consumption (in quadrillion Btu), where t = number of 5-year intervals since 2015, then an exponential function to model the data would be: $E(t) = 563 \, (1.09)^t$.

17. Here are the graphs of World Mobil Device sales for each of the regions from 2008 to 2010.

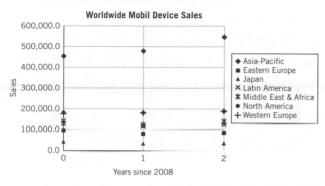

Worldwide Mobil Device Sales

Answers will vary. It would appear that perhaps the Western Europe region fits an exponential model since the data are always increasing. Using a regression equation through the points for Western Europe, one gets the equation Sales $= 174{,}724 \cdot e^{0.0646y} = 174{,}724 \cdot (1.0667)^y$, where y is years since 2008. The correlation coefficient, cc, is 0.999. Other data show much more variance, and some regions decrease and then increase.

19. Initial value $= 20$ grams gives $A(t) = 20(1/2)^{t/8}$ grams, where t is in days.

21. a. $f(x)$ matches the graph labeled **C.**

b. $g(x)$ matches the graph labeled **B.**

c. $h(x)$ matches the graph labeled **A.**

In general, the smallest exponential has the fastest decay.

23. Assuming compound interest, compounded annually:

a. $A(t) = 10{,}000 \cdot 1.07^t$, where $t =$ years from the granddaughter's birth.

b. $A(18) \approx \$33{,}800.$

c. $C(t) = 10{,}000 \cdot 1.03^t$ and thus $C(18) \approx \$17{,}024.$ The difference is $\$16{,}776$

d. No. The return of investing $\$10{,}000$ at 4% per year for 18 years is $D(t) = 10{,}000 \cdot 1.04^{18} = \$20{,}258.$ The investment formulae are not linear. So $A(t) - C(t) \neq D(t).$ It represents the return on a 4% investment and nothing more.

25. a. Each linear dimension of the model is one-tenth that of the actual village, so the area of the model (which is two-dimensional) would be $(1/10) \cdot (1/10) = 1/100$ or one-hundredth of A_o, the area of the actual village.

b. Here each linear dimension is one-hundredth that of the actual church. The weight depends on volume (which is three-dimensional) and would be $(1/100) \cdot (1/100) \cdot (1/100) = 1/1{,}000{,}000$, or one-millionth of W_o, the weight of the actual church.

c. $A_n = A_o \cdot \left(\frac{1}{10^2}\right)^n$

$W_n = W_o \cdot \left(\frac{1}{10^3}\right)^n$

27. a. $S(t) = 0.5 \cdot 2^t$, the tumor size in cubic centimeters, where $t =$ number of 2-month time periods

$A(t) = 0.5 \cdot 2^{t/4}$, the tumor size in cubic centimeters, where $t =$ number of 2-month time periods

b.

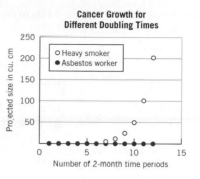

Cancer Growth for Different Doubling Times

The faster-growing cancer of the heavy smoker if untreated gets dangerously large very quickly after about 6 time periods (1 year). The slower-growing cancer of the asbestos worker after 12 time periods (2 years) is still relatively small compared to that of the smoker (the graph does not yet show an exponential curve).

c. 1 year $= 6$ time periods

The tumor size for the smoker after 1 year (or 6 time periods) is

$S(6) = 0.5 \cdot 2^6 = 32$ cubic centimeters $\Rightarrow$

$\text{Volume}_{\text{smoker}} = 32 = \left(\frac{4}{3}\right)\pi r^3 \Rightarrow \frac{32 \cdot 3}{4\pi} = r^3$

$7.64 \approx r^3$

Taking the cube root of both sides, $r \approx 1.97$ cm, or a tumor diameter of about 3.94 cm.

The tumor size for the asbestos worker after 1 year (or 6 time periods) is

$A(6) = 0.5 \cdot 2^{6/4} \approx 1.41$ cubic centimeters $\Rightarrow$

$\text{Volume}_{\text{asbestos}} = 1.41 = \left(\frac{4}{3}\right)\pi r^3 \Rightarrow \frac{1.41 \cdot 3}{4\pi} = r^3$

$0.34 \approx r^3$

Taking the cube root of both sides, $r \approx 0.70$ cm, or a tumor diameter of about 1.4 cm.

CHAPTER 6

Section 6.1

Algebra Aerobics 6.1a

1. a. When $T = 0$, $M = 250(3)^0 = 250(1) = 250.$

b. When $T = 1$, $M = 250(3)^1 = 250(3) = 750.$

c. When $T = 2$, $M = 250(3)^2 = 250(9) = 2250.$

d. When $T = 3$, $M = 250(3)^3 = 250(27) = 6750.$

2. a. $4 < T < 5$ **b.** $8 < T < 9$

3. To read each value of x from this graph, draw a horizontal line from the y-axis at a given value until it hits the curve. Then draw a vertical line to the x-axis to identify the appropriate value of x.

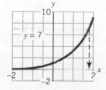

a. horizontal line $y = 7$, from y-axis to curve, then down to x-axis $\Rightarrow x \approx 1.8$

b. horizontal line $y = 0.5$, from y-axis to curve, then down to x-axis $\Rightarrow x \approx -0.5$.

4. a. $2 < x < 3$ **b.** $-1 < x < 0$

5. a. $2^3 < 13 < 2^4$ **c.** $5^{-1} < 0.24 < 5^0$

     b. $3^4 < 99 < 3^5$ **d.** $10^3 < 1500 < 10^4$

Algebra Aerobics 6.1b

1. a. $\log(10^5/10^7) = \log 10^{-2} = -2 \log 10 = -2(1) = -2$ and $\log 10^5 - \log 10^7 = 5 \log 10 - 7 \log 10 = 5 - 7 = -2$. So $\log(10^5/10^7) = \log 10^5 - \log 10^7$.

     b. $\log(10^5 \cdot (10^7)^3) = \log(10^5 \cdot 10^{21}) = \log 10^{26} = 26 \log 10 = 26$ and $\log 10^5 + 3 \log 10^7 = 5 \log 10 + 3(7) \log 10 = 5 + (3 \cdot 7) = 26$. So $\log[10^5 \cdot (10^7)^3] = \log(10^5) + 3 \log(10^7)$.

2. a. Rule 2: $\log 3 = \log \frac{15}{5} = \log 15 - \log 5$

     b. Rule 3: $\log 1024 = \log(2^{10}) = 10 \log 2$

     c. Rule 3: $\log \sqrt{31} = \log(31^{1/2}) = \frac{1}{2} \log 31$

     d. Rule 1: $\log 30 = \log(2 \cdot 3 \cdot 5) = \log 2 + \log 3 + \log 5$

     e. Rule 2: $\log 81 - \log 27 = \log\left(\frac{81}{27}\right) = \log 3$

     or Rule 3: $4 \log 3 - 3 \log 3 = \log 3$

3. a. False   **d.** True; Rule 3 and that $\log 10 = 1$

     b. False   **e.** False

     c. True; Rule 3   **f.** False

4. a. $\log \sqrt{\frac{2x-1}{x+1}} = \log\left(\frac{2x-1}{x+1}\right)^{1/2} = \frac{1}{2} \log\left(\frac{2x-1}{x+1}\right)$
$= \frac{1}{2}[\log(2x - 1) - \log(x + 1)]$

     b. $\log \frac{xy}{z} = \log(xy) - \log z = \log x + \log y - \log z$

     c. $\log \frac{x\sqrt{x+1}}{(x-1)^2} = \log \frac{x(x+1)^{1/2}}{(x-1)^2}$
$= \log x(x + 1)^{1/2} - \log(x - 1)^2$
$= \log x + \log(x + 1)^{1/2} - 2 \log(x - 1)$
$= \log x + \frac{1}{2} \log(x + 1) - 2 \log(x - 1)$

     d. $\log \frac{x^2(y-1)}{y^3z} = \log x^2(y - 1) - \log y^3 z$
$= \log x^2 + \log(y - 1) - [\log y^3 + \log z]$
$= 2 \log x + \log(y - 1) - 3 \log y - \log z$

5. $\frac{1}{3}[\log x - \log(x + 1)] = \frac{1}{3} \log \frac{x}{x+1}$
$= \log\left(\frac{x}{x+1}\right)^{1/3}$
$= \log \sqrt[3]{\frac{x}{x+1}}$

6. a. $\log x = 3 \Rightarrow 10^3 = x, x = 1000$

     b. $\log x + \log 5 = 2 \Rightarrow \log 5x = 2 \Rightarrow 10^2 = 5x \Rightarrow x = 20$

     c. $\log x + \log 5 = \log 2 \Rightarrow \log 5x = \log 2 \Rightarrow 5x = 2 \Rightarrow x = 2/5$

     d. $\log x - \log 2 = 1 \Rightarrow \log \frac{x}{2} = 1 \Rightarrow 10^1 = \frac{x}{2} \Rightarrow x = 20$

e. $\log x - \log(x - 1) = \log 2 \Rightarrow \log \frac{x}{x-1} = \log 2 \Rightarrow$
$\frac{x}{x-1} = 2 \Rightarrow x = 2(x - 1) \Rightarrow x = 2$

f. $\log(2x + 1) - \log(x + 5) = 0 \Rightarrow$
$\log \frac{2x+1}{x+5} = \log 1 \Rightarrow$
$\frac{2x+1}{x+5} = 1 \Rightarrow 2x + 1 = x + 5 \Rightarrow x = 4$

7. $\log 10^3 - \log 10^2 = 3 \log 10 - 2 \log 10 = 3 - 2 = 1$

$\frac{\log 10^3}{\log 10^2} = \frac{3 \log 10}{2 \log 10} = \frac{3}{2} = 1.5$

Since $1 \neq 1.5 \Rightarrow \log 10^3 - \log 10^2 \neq \frac{\log 10^3}{\log 10^2}$

Algebra Aerobics 6.1c

1. a. $60 = 10 \cdot 2^T \Rightarrow 6 = 2^T$ (dividing both sides by 10)
       $\log 6 = \log 2^T$ (taking the log of both sides)
       $\log 6 = T \log 2$ (using Rule 3 of logs)
       $\log 6/\log 2 = T$
       $0.7782/0.3010 \approx T$ or $T \approx 2.59$

     b. $500(1.06)^T = 2000 \Rightarrow (1.06)^T = \frac{2000}{500} \Rightarrow$
       $(1.06)^T = 4 \Rightarrow \log(1.06)^T = \log 4 \Rightarrow$
       $T(\log 1.06) = \log 4 \Rightarrow T = \frac{\log 4}{\log 1.06} = \frac{0.6021}{0.0253} \approx 23.8$

     c. $80(0.95)^T = 10 \Rightarrow (0.95)^T = 1/8 = 0.125 \Rightarrow$
       $\log(0.95)^T = \log 0.125 \Rightarrow T \log 0.95 = \log 0.125 \Rightarrow$
       $T = \frac{\log 0.125}{\log 0.95} \approx \frac{-0.9031}{-0.0223} \approx 40.5$

2. a. $7000 = 100 \cdot 2^T \Rightarrow 70 = 2^T \Rightarrow \log 70 = \log 2^T \Rightarrow$
     $\log 70 = T \log 2 \Rightarrow T = \log 70/\log 2 \approx \frac{1.845}{0.301} \approx 6.13$
     It will take 6.13 time periods or approximately $(6.13) \cdot (20)$ min $= 122.6$ min (or a little more than 2 hours) for the bacteria count to reach 7000.

     b. $12{,}000 = 100 \cdot 2^T \Rightarrow 120 = 2^T \Rightarrow \log 120 = \log 2^T \Rightarrow$
     $\log 120 = T \log 2 \Rightarrow T = \log 120/\log 2 \approx 6.907$ time periods or approximately $(6.907) \cdot (20)$ min $= 138.14$ min (or a little more than $2\frac{1}{4}$ hrs.) for the bacteria count to reach 12,000.

3. Using the rule of 70, since $R = 6\%$ per yr, then $70/R = 70/6 \approx 11.7$ yr. More precisely, we have:
$2000 = 1000(1.06)^t \Rightarrow 2 = 1.06^t \Rightarrow$
$\log 2 = \log 1.06^t \Rightarrow \log 2 = t \log 1.06 \Rightarrow$
$t = \frac{\log 2}{\log 1.06} \approx \frac{0.3010}{0.0253} \approx 11.9$ yr

4. $1 = 100(0.976)^t \Rightarrow 0.01 = (0.976)^t \Rightarrow$
$\log 0.01 = \log(0.976)^t \Rightarrow \log 0.01 = t \log 0.976 \Rightarrow$
$t = \frac{\log 0.01}{\log 0.976} = \frac{-2}{-0.0106} \approx 189$ years, or almost 2 centuries!

5. a. $30 = 60(0.95)^t \Rightarrow 0.5 = (0.95)^t \Rightarrow$
     $\log(0.5) = \log(0.95^t) \Rightarrow \log 0.5 = t \log 0.95 \Rightarrow$
     $t = \frac{\log 0.5}{\log 0.95} \approx \frac{-0.3010}{-0.0223} \approx 13.5$ years for the initial amount to drop in half.

     b. $16 = 8(1.85)^t \Rightarrow 2 = (1.85)^t \Rightarrow$
     $\log 2 = \log(1.85)^t \Rightarrow \log 2 = t \log 1.85 \Rightarrow t = \frac{\log 2}{\log 1.85} \Rightarrow$
     $t \approx \frac{0.3010}{0.2672} = 1.13$ years for the initial amount to double.

Ch. 6

c. $500 = 200(1.045)^t \Rightarrow 2.5 = (1.045)^t \Rightarrow$

$\log 2.5 = \log(1.045)^t \Rightarrow \log 2.5 = t \log 1.045 \Rightarrow$

$t = \dfrac{\log 2.5}{\log 1.045} \Rightarrow t \approx \dfrac{0.3979}{0.0191} \Rightarrow$

$t \approx 20.8$ years for the initial amount to increase from 200 to 500.

6. a. $60 = 120(0.983)^t \Rightarrow 0.5 = (0.983)^t \Rightarrow$

$\log 0.5 = t \log 0.983 \Rightarrow$

$t = \dfrac{\log 0.5}{\log 0.983} \Rightarrow t \approx \dfrac{-0.3010}{-0.0074} \Rightarrow$

$t \approx 40.6$ days

b. $0.25 = 0.5(0.92)^t \Rightarrow 0.5 = (0.92)^t \Rightarrow$

$\log 0.5 = t \log 0.92 \Rightarrow$

$t = \dfrac{\log 0.5}{\log 0.92} \Rightarrow t \approx \dfrac{-0.3010}{-0.0362} \Rightarrow$

$t \approx 8.3$ hours

c. $0.5A_0 = A_0(0.89)^t \Rightarrow 0.5 = (0.89)^t \Rightarrow$

$\log 0.5 = t \log 0.89 \Rightarrow t = \dfrac{\log 0.5}{\log 0.89} \Rightarrow t \approx \dfrac{-0.3010}{-0.0506} \Rightarrow$

$t \approx 5.9$ years

Exercises for Section 6.1

1. Student estimates will vary for each interest rate.

  a. About 38 years.   **b.** About 16 years.

3. Eyeball estimates will vary. One set of guesses is:

  a. 1.3 hrs.   **b.** 2.4 hrs.   **c.** 5 hrs.

5. a. 0.001   **b.** 10^6 **c.** 1   **d.** 10   **e.** 0.1

7. a. $2 \cdot \log 3$ **b.** $2 \cdot \log 3 + \log 2$ **c.** $3 \cdot \log 3 + \log 2$

9. a. 12   **b.** 12   **c.** 12   **d.** 2

11. If $w = \log(A)$ and $z = \log(B)$, then $10^w = A$ and $10^z = B$. Thus $A/B = 10^w/10^z = 10^{w-z}$ and therefore $\log(A/B) = \log(10^{w-z}) = w - z = \log(A) - \log(B)$, as desired.

13. a. $\log\left(\dfrac{K^3}{(K+3)^2}\right)$

  b. $\log\left(\dfrac{(3+n)^5}{m}\right)$

  c. $\log\left(T^4 \cdot \sqrt{T}\right) = \log\left(\sqrt{T^9}\right)$

  d. $\log\left(\dfrac{\sqrt[3]{xy^2}}{(xy^2)^3}\right) = \log\left([xy^2]^{-8/3}\right)$

15. Let $w = \log(A)$. Then $10^w = A$ and thus $A^p = (10^w)^p = 10^{wp}$, and then $\log(A^p) = w \cdot p = p \cdot w = p \cdot \log(A)$, as desired.

17. a. 1000

  b. 999

  c. $10^{5/3} \approx 46.42$

  d. 1/9

  e. No solution, since x can not be negative.

19. a. Solve $300 = 100 \cdot 1.03^t$ to get $t = \log(3)/\log(1.03) \approx 37.17$ years.

  b. Similarly, $\log(3)/\log(1.07) \approx 16.24$ years.

21. a. The graphs of $f(x) = 500 \cdot (1.03)^x$ and $g(x) = 4500$ are in the accompanying diagram.

b. $x = 75$ is a good eyeball estimate.

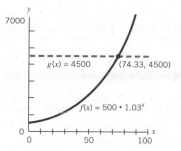

c. $\log(4500) = \log(500) + x \cdot \log(1.03)$ or $x = [\log(4500) - \log(500)]/\log(1.03) = \log(9)/\log(1.03) \approx 74.33$

d. The eyeball estimate and the logarithm-computed answer are very close.

23. a. Doubling time: $x = \log(2)/\log(4) = 0.5$ years

  b. Half-life $= \log(0.5)/\log(0.25) = 0.5$ years

  c. Doubling time $= \log(2)/\log(4) = 0.5$ years

  d. Half life $= \log(0.5)/\log(0.25) = 0.5$ years

25. a. Let $N(t) =$ number of articles posted in Wikipedia t years from 2001. The formula should be $N(t) = 20{,}000 \cdot a^t$ and $N(6) = 2{,}000{,}000 = 20{,}000 \cdot a^6$. Thus $100 = a^6$ and $(100)^{1/6} = a \approx 2.1544$. The exponential model is $N(t) = 20{,}000 \cdot (100)^{t/6} \approx 20{,}000(2.1544)^t$.

  b. To find the doubling time, we solve $40{,}000 = 20{,}000(2.1544)^t$ for t. Thus $2 = 2.1544^t$ or $\log(2) = t \log(2.1544)$ or $t = \log(2)/\log(2.1544) \approx 0.9$ years or about every 11 months.

  c. 2010 is 9 years after 2001 and $N(9) \approx 20{,}000 \cdot (100)^{9/6} = 20{,}000{,}000$ articles. It is above the actual number. Using the data for 2001, 2007, and 2010 shows that not only is the number increasing but the rate of change is also increasing since the rate of change from 2001 to 2007 was $(2{,}000{,}000 - 20{,}000)/(2007 - 2001) = 330{,}000$ articles per year, but from 2007 to 2010 the rate of change was $(16{,}000{,}000 - 2{,}000{,}000)/(2010 - 2007) \approx 4.7$ million articles per year.

27. a. $T = \log(2)/\log(1.5) \approx 1.71$ twenty-minute periods or about 34 min.

  b. $T = \log(10)/\log(1.5) \approx 5.68$ twenty-minute periods or about 114 min.

29. a. $B(t) = B_0(0.5)^{0.05t}$, where t is measured in minutes and $B(t)$ and B_0 are measured in some weight unit. None is specified in the problem.

  b. In one hour, $t = 60$ and thus $B(60) = B_0(0.5)^{0.05 \cdot 60} = 0.125 \cdot B_0$, or 12.5% is left.

  c. If its half-life is 20 minutes, then its quarter-life is 40 minutes.

  d. Solving $0.10 = 0.5^{0.05t}$ for t, we get $0.05t \cdot \log(0.5) = \log(0.10)$ or $t = \log(0.1)/[\log(0.5) \cdot 0.05] \approx 66$ minutes.

31. a. The following table shows rebound heights H (in feet) for n bounces (where $0 \leq n \leq 5$).

n	0	1	2	3	4	5
H	5.000	3.000	1.800	1.080	0.648	0.389

b. $H = 5 \cdot 0.6^n$, where H gives the height in feet above the floor at the nth bounce.

c. Looking at the table the ball is above 1 foot at the 3rd bounce and below 1 foot at the 4th bounce.

d. The general function is: $H = H_0 \cdot r^n$ where r is the rebound height in decimal form.

Section 6.2

Algebra Aerobics 6.2a

1. a. $\ln e^2 = 2 \ln e = 2(1) = 2$

b. $\ln 1 = 0$

c. $\ln \frac{1}{e} = \ln 1 - \ln e = 0 - 1 = -1$

d. $\ln \frac{1}{e^2} = \ln 1 - \ln e^2 = \ln 1 - 2 \ln e = 0 - 2(1)$
$= 0 - 2 = -2$

e. $\ln \sqrt{e} = \ln e^{1/2} = \frac{1}{2} \ln e = \frac{1}{2}(1) = 1/2$

2. a. $\log 10^3 = 3$

b. $\log 10^{-5} = -5$

c. $3 \log 10^{0.09} = 3(0.09) = 0.27$

d. $10^{\log 3.4} = 3.4$

e. $\ln e^5 = 5$

f. $\ln e^{0.07} = 0.07$

g. $\ln e^{3.02} + \ln e^{-0.27} = 3.02 - 0.27 = 2.75$

h. $e^{\ln 0.9} = 0.9$

3. a. $\ln \sqrt{xy} = \ln(xy)^{1/2} = \frac{1}{2} \ln(xy) = \frac{1}{2} [\ln x + \ln y]$

b. $\ln\left(\frac{3x^2}{y^3}\right) = \ln(3x^2) - \ln(y^3)$
$= \ln 3 + 2 \ln x - 3 \ln y$

c. $\ln\left((x + y)^2(x - y)\right) = \ln(x + y)^2 + \ln(x - y)$
$= 2 \ln(x + y) + \ln(x - y)$

d. $\ln \frac{\sqrt{x + 2}}{x(x - 1)} = \ln \frac{(x + 2)^{1/2}}{x(x - 1)} = \ln(x + 2)^{1/2} - \ln x(x - 1)$
$= \frac{1}{2} \ln(x + 2) - \ln x(x - 1)$
$= \frac{1}{2} \ln(x + 2) - [\ln x + \ln(x - 1)]$
$= \frac{1}{2} \ln(x + 2) - \ln x - \ln(x - 1)$

4. a. $\ln x(x - 1)$

b. $\ln \frac{(x + 1)}{x}$

c. $\ln x^2 - \ln y^3 = \ln \frac{x^2}{y^3}$

d. $\ln(x + y)^{1/2} = \ln \sqrt{x + y}$

e. $\ln x - 2 \ln(2x - 1) = \ln x - \ln(2x - 1)^2 = \ln \frac{x}{(2x - 1)^2}$

5. growth factor $= e^r = 1 + 0.064 = 1.064 \Rightarrow$
$\ln e^r = \ln 1.064 \Rightarrow r = \ln 1.064 = 0.062$ or 6.2%

6. $50,000 = 10,000 \, e^{0.078t} \Rightarrow 5 = e^{0.078t} \Rightarrow \ln 5 = 0.078t \Rightarrow$
$t = \ln 5/0.078 \approx 20.6$ yr

7. a. $\ln e^{x+1} = \ln 10 \Rightarrow (x + 1)\ln e = \ln 10 \Rightarrow$
$(x + 1)(1) \approx 2.30 \Rightarrow x \approx -1 + 2.30 \Rightarrow x \approx 1.30$

b. $\ln e^{x-2} = \ln 0.5 \Rightarrow (x - 2) \ln e = \ln 0.5 \Rightarrow$
$x - 2 \approx -0.69 \Rightarrow x \approx 2 - 0.69 \Rightarrow x \approx 1.31$

8. a. true, since $\ln 81 = \ln 3^4 = 4 \ln 3$

b. false; $\ln 7 = \ln \frac{14}{2} = \ln 14 - \ln 2$

c. true, since $\ln 35 = \ln(5 \cdot 7) = \ln 5 + \ln 7$

d. false; $2 \ln 10 = \ln 10^2 = \ln 100$

e. true, since $\ln e^{1/2} = \frac{1}{2} \ln e = \frac{1}{2}$

f. false: $5 \ln 2 = \ln 2^5 = \ln 32$

9. a. $\ln 2 + \ln 6 = x \Rightarrow \ln(2 \cdot 6) = x \Rightarrow \ln 12 = x \Rightarrow x \approx 2.48$

b. $\ln 2 + \ln x = 2.48 \Rightarrow \ln 2x = 2.48 \Rightarrow$
$e^{2.48} = 2x \Rightarrow x = \frac{e^{2.48}}{2} \approx 5.97$

c. $\ln(x + 1) = 0.9 \Rightarrow e^{0.9} = x + 1 \Rightarrow x = e^{0.9} - 1 \approx 1.46$

d. $\ln 5 - \ln x = -0.06 \Rightarrow \ln \frac{5}{x} = -0.06 \Rightarrow$
$e^{-0.06} = \frac{5}{x} \Rightarrow x = \frac{5}{e^{-0.06}} \approx 5.3$

10. $2 = 1e^{0.0114t} \Rightarrow 2 = e^{0.0114t} \Rightarrow \ln 2 = \ln(e^{0.0114t}) \Rightarrow$
$\ln 2 = 0.0114t \Rightarrow t = \frac{\ln 2}{0.0114} \approx 60.8$ years

11. $83,000 = 50,000e^{r(10)} \Rightarrow 1.66 = e^{10r} \Rightarrow$
$\ln(1.66) = \ln(e^{10r}) \Rightarrow \ln 1.66 = 10r \Rightarrow r = \frac{\ln 1.66}{10} \approx 0.05$
$100,000 = 50,000e^{0.05t} \Rightarrow 2 = e^{0.05t} \Rightarrow \ln 2 = \ln(e^{0.05t}) \Rightarrow$
$\ln 2 = 0.05t \Rightarrow t = \frac{\ln 2}{0.05} \approx 13.9$ years

12. $0.3 = 2e^{r(5000)} \Rightarrow 0.15 = e^{5000r} \Rightarrow \ln(0.15) = \ln(e^{5000r}) \Rightarrow$
$\ln 0.15 = 5000r \Rightarrow r = \frac{\ln 0.15}{5000} \approx -0.00038$
The half life is: $1 = 2e^{-0.00038t} \Rightarrow 0.5 = e^{-0.00038t} \Rightarrow$
$\ln 0.5 = \ln(e^{-0.00038t}) \Rightarrow \ln 0.5 = -0.00038t \Rightarrow$
$t = \frac{\ln 0.5}{-0.00038} \approx 1824$ years

Algebra Aerobics 6.2b

1. a. decay   **c.** growth   **e.** decay

b. decay   **d.** growth   **f.** decay

2. a. $e^k = 1.062 \Rightarrow k = \ln 1.062 \approx 0.060 \Rightarrow y = 1000e^{0.06t}$

b. $e^k = 0.985 \Rightarrow k = \ln 0.985 \approx -0.015 \Rightarrow y = 50e^{-0.015t}$

3. a. continuous nominal rate $\approx 5.45\%$ since $\ln 1.056 \approx 0.545$; effective rate $= 5.6\%$ since $e^{\ln 1.056} = 1.056$

b. continuous nominal rate $\approx 3.34\%$ since $\ln 1.034 \approx 0.0334$; effective rate $= 3.4\%$ since $e^{\ln 1.034} = 1.034$

c. continuous nominal rate $\approx 7.97\%$ since $\ln 1.083 \approx 0.0797$; effective rate $= 8.3\%$ since $e^{\ln 1.083} = 1.083$

d. continuous nominal rate $\approx 25.85\%$ since $\ln 1.295 \approx 0.2585$; effective rate $= 29.5\%$ since $e^{\ln 1.295} = 1.295$

4. a. growth factor $= e^{0.08} \approx 1.083$

b. decay factor $= e^{-0.125} \approx 0.883$

Exercises for Section 6.2

1. a. $\ln(A \cdot B) = \ln(A) + \ln(B)$ or Rule 1

   b. $\ln(A/B) = \ln(A) - \ln(B)$ or Rule 2

   c. $\ln(A^p) = p \cdot \ln(A)$ or Rule 3

   d. Rules 1 and 3

   e. Rules 1 and 3

   f. Rules 2 and 1

3. a. $10^n = 35$ **b.** $e^x = 75$ **c.** $e^{3/4} = x$ **d.** $N = N_0 \cdot e^{-kt}$

5. a. $x - 5 \cdot 2 = 10$ **d.** $x = 8 \cdot 36 = 288$

   b. $x = 24/2 = 12$ **e.** $x = 64/9 \approx 7.1$

   c. $x = 11$ **f.** $x = 16/8 = 2$

7. a. $\frac{1}{2}(\ln 4 + \ln x + \ln y)$ **c.** $\ln(3) + \frac{3}{4}\ln(x)$

   b. $\frac{1}{3}(\ln 2 + \ln x) - \ln(4)$

9. a. $\ln\left(\sqrt[4]{(x+1)(x-3)}\right)$ **b.** $\ln\left(\frac{R^3}{\sqrt{P}}\right)$ **c.** $\ln\left(\frac{N}{N_0^2}\right)$

11. a. $r = \ln(1.0253) \approx 0.025$ **c.** $x = \ln(0.5)/3 \approx -0.231$

   b. $t = \ln(3)/0.5 \approx 2.197$

13. a. $x = \ln(10) \approx 2.303$

   b. $x = \log(3) \approx 0.477$

   c. $x = \log(5)/\log(4) \approx 1.161$

   d. $x = e^5 \approx 148.413$

   e. $x = e^3 - 1 \approx 19.086$

   f. There is no solution since $e^4 = \frac{x}{x+1} \Rightarrow x = \frac{e^4}{1 - e^4}$ which is always negative and $\ln x$ is undefined for negative real numbers.

15. Let $w = \ln(A)$ and $z = \ln(B)$. Thus $e^w = A$ and $e^z = B$. Therefore $A \cdot B = e^{w+z}$ and therefore $\ln(A \cdot B) = w + z = \ln(A) + \ln(B)$, as desired.

17. $t = [\ln(100,000) - \ln(15,000)]/0.085 \approx 22.3$ years

19. $1.0338 = e^r$, and thus the nominal continuous interest rate $r = \ln(1.0338) \approx 0.0332 = 3.32\%$.

21. a. $r = \ln(1.025) \approx 0.0247$ **c.** $r = \ln(1.08) \approx 0.0770$

   b. $r = \ln(0.5) \approx -0.6931$

23. a. growth; 37%   **c.** decay; 19%   **e.** growth; 56%

   b. growth; 115%   **d.** decay; 120%   **f.** decay; 29%

25. a. $N = 10\, e^{\ln(1.045)t} = 10 \cdot e^{0.0440t}$

   b. $Q = 5 \cdot 10^{-7} \cdot e^{\ln(0.072)A} = 5 \cdot 10^{-7} \cdot e^{-2.631A}$

   c. $P = 500 \cdot e^{\ln(2.10)x} = 500 \cdot e^{0.742x}$

27. a. $t = 5$ years; $P = P_0 e^{0.1386t}$

   b. $t = 25$ years; $P = P_0 e^{0.0277t}$

   c. $t = 1/2$ year; $P = P_0 e^{1.3863t}$

29. If $200 = 760 \cdot e^{-0.128h}$, then $h = \ln(200/760)/(-0.128) \approx$ 10.43 km.

31. The half-life is $\ln(1/2)/(-r) = [\ln(1) - \ln(2)]/(-r) = [0 - \ln(2)]/(-r) = \ln(2)/r = 100 \cdot \ln(2)/R \approx 69.3137/R$, which is approximately $70/R$.

33. a. It is a 2% nominal continuous decay rate.

   b. $2500 \cdot e^{-0.02t} = 2500 \cdot 0.5^{t/n} \Rightarrow -0.02t = \left(\frac{t}{n}\right) \cdot \ln(0.5)$
$\Rightarrow n = \ln(0.5)/(-0.02) \approx 34.66$.

   c. It represents the half-life in whatever time units t is measured in.

35. a. Since $Q(8000) = 0.5Q_0$, the half-life is 8000 years.

   b. The annual decay rate is $1 - 0.5^{1/8000} \approx 0.000\,086\,639\,6$.

   c. Solving $e^r = 0.5^{1/8000}$ for r gives: $r \approx -0.000\,086\,643\,4$ as the nominal continuous decay rate.

37. Answers will vary. Look for an exponential curve that goes roughly through the middle of each cluster. One such curve contains the points (500, 1000) and (2000, 300). Then the best-fit exponential through these two points is $y = 1494(0.9992)^d$, where d measures depth in meters and y measures species density in an unknown unit. (Note that in base e, one gets $y = 1494e^{-0.0008d}$.)

39. a. Newton's Law here is of the form $T = 75 + Ce^{-kt}$ where $A = 75°$, the ambient temperature, and T is the temperature of the object at time t. When $t = 0$, $T = 160°$, so
$160 = 75 + Ce^0 \Rightarrow C = 85$. So the equation becomes $T = 75 + 85e^{-kt}$. When $t = 10$, $T = 100°$ so
$100 = 75 + 85e^{-10k} \Rightarrow e^{-10k} = \frac{25}{85} \approx 0.2941 \Rightarrow$
$\ln(e^{-10k}) = \ln(0.2941) \Rightarrow$
$-10k \approx -1.224 \Rightarrow k \approx 0.1224$.

So the Law of Cooling in this situation is:
$T = 75 + 85e^{-0.1224t}$

   b.

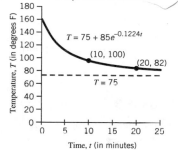

   c. When $t = 20$ minutes, then the temperature of the tea $T = 75 + 85e^{-0.1224 \cdot 20} \approx 82°$.

Section 6.3

Algebra Aerobics 6.3

1. a. Since $\log x^2 = 2 \log x$ (by Rule 3 of logarithms), the graphs of $y = \log x^2$ and $y = 2 \log x$ will be identical (assuming $x > 0$).

2. The graph of $y = -\ln x$ will be the mirror image of $y = \ln x$ across the x-axis.

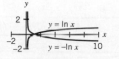

3. a. The graphs are very similar (see Figure 6.3). They intersect at $(1, 0)$, which is the x-intercept of each graph.

b. The graph of $\log x$ reaches a y value of 1 when $x = 10$ while the graph of $\ln x$ reaches a value of 1 when $x = e$ (or approximately 2.7). The graph of $\log x$ reaches 2 at $x = 100$ while the graph of $\ln x$ reaches 2 at $x = e^2 \approx 7.4$.

c. For $0 < x < 1$, both graphs lie below the x-axis and approach the y-axis asymptotically as x gets closer to 0.

d. To the right of $x = 1$, both graphs lie above the x-axis, with the $\ln x$ graph rising slightly faster than the $\log x$ graph, and thus staying above the $\log x$ graph.

4. f and its inverse f^{-1} along with the dotted line for $y = x$:

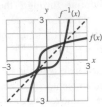

5. Acidic. If $4 = -\log[H^+]$, then $-4 = \log[H^+]$. So $[H^+] = 10^{-4}$. Since the pH is 3 less than pure water's, it will have a hydrogen ion concentration 10^3 or 1000 times higher than pure water's.

6. $N = 10 \log\left(\dfrac{1.5 \cdot 10^{-12}}{10^{-16}}\right) = 10 \log(1.5 \cdot 10^4)$

$= 10(\log 1.5 + \log 10^4) \approx 10(0.176 + 4)$

$= 10(4.176) \approx 42$ dB

7. Multiplying the intensity by $100 = 10^2$ corresponds to adding 20 to the decibel level. Multiplying the intensity by $10,000,000 = 10^7$ corresponds to adding 70 to the decibel level.

Exercises for Section 6.3

1. Since $\log(5x) = \log(5) + \log(x)$ and $\log(5) > 0$, the graph of $\log(5x)$ in Graph A is above the graph of $\log(x)$ in Graph B.

3. a. $A = \log(x)$, $B = \log(x - 1)$, and $C = \log(x - 2)$; so $f(x)$ matches A, $g(x)$ matches B, and $h(x)$ matches C.

b. $\log(1) = 0$; thus $f(1) = g(2) = h(3) = 0$.

c. f has 1 as its x-intercept; g has 2 and h has 3.

d. The graph moves from crossing the x-axis at $x = 1$ to crossing it at $x = k$. Assuming that $k > 0$, then $f(x - k)$ is the graph of f moved k units to the right.

5. a. The graphs of f and h are mirror images of each other across the y-axis, as are the graphs of g and k.

b. The graphs of f and g are mirror images of each other across the x-axis, as are the graphs of h and k.

7. The table for $\log_3(x)$ is: The table for $\log_4(x)$ is:

x	1/9	1/3	1	3	9
y	-2	-1	0	1	2

x	1/16	1/4	1	4	16
y	-2	-1	0	1	2

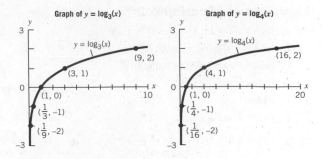

9. $dB = 10 \cdot \log(I/I_0) = 28$ implies that $I/I_0 = 10^{2.8}$ and thus $I = 10^{-13.2}$ watts/cm^2, and $dB = 92$ has $I/I_0 = 10^{9.2}$ and thus $I = 10^{-6.8}$ watts/cm^2. (Note that these answers assume, of course, that $I_0 = 10^{-16}$ watts/cm^2.)

11. If I is the intensity of one crying baby, then $5I$ is the intensity of five crying babies. Thus the perceived noise in decibels is $10 \cdot \log[5(I/I_0)] = 10 \cdot \log(5) + 10 \log(I/I_0) \approx 6.99 + 10 \cdot \log(I/I_0) \approx 7 +$ noise of one baby crying. Thus quintuplets crying are about 7 decibels louder than one baby crying.

13. Total volume $= 2\frac{1}{4} = \frac{9}{4}$ cups or equivalently 9 quarter cups. So lemon juice is $\frac{1}{9}^{th}$ of the mixture volume and water is $\frac{8}{9}^{ths}$ of the volume. Since the lemon juice has a pH of 2.1, its hydrogen ion concentration is $10^{-2.1}$ (moles per liter). Similarly, since the tap water has a pH of 5.8, its ion concentration is $10^{-5.8}$. So the hydrogen ion concentration of the mixture $= [H^+] = \left(\frac{1}{9}\right) \cdot 10^{-2.1} + \left(\frac{8}{9}\right) \cdot 10^{-5.8} \Rightarrow$ pH $= -\log[\left(\frac{1}{9}\right) \cdot 10^{-2.1} + \left(\frac{8}{9}\right) \cdot 10^{-5.8}] \approx 3.05$. Thus the mixture is slightly less acidic than orange juice (with a pH of 3).

Section 6.4

Algebra Aerobics 6.4

1. a. The graph of $y = 3x + 4$, a linear function, is a straight line on a standard linear plot. The graph of $y = 4 \cdot 3^x$, an exponential function, is a straight line graph on a semi-log plot. The equation $\log y = (\log 3) \cdot x + \log 4$ is equivalent to $y = 4 \cdot 3^x$, whose graph is a straight line on a semi-log plot.

b. The graph of $y = 3x + 4$ has slope $= 3$ and vertical intercept $= 4$. The equations $y = 4 \cdot 3^x$ and $\log y = (\log 3) \cdot x + \log 4$ are equivalent; their graphs have slope $= \log 3$ and vertical intercept $= \log 4$ on a semi-log plot.

2. a. $\log y = \log[5(3)^x] \Rightarrow \log y = \log 5 + x(\log 3)$

b. $\log y = \log[1000(5)^x] \Rightarrow \log y = \log 1000 + x \log 5$
$= 3 + x \log 5$

c. $\log y = \log[10,000(0.9)^x] \Rightarrow$
$\log y = \log 10,000 + x \log 0.9 = 4 + x \log 0.9$

d. $\log y = \log[5 \cdot 10^6(1.06)^x] \Rightarrow \log y =$
$\log 5 + \log 10^6 + x \log 1.06 = \log 5 + 6 + x \log 1.06$

3. a. $\log y = \log 7 + (\log 2)x \Rightarrow \log y = \log(7 \cdot 2^x) \Rightarrow$
$10^{\log y} = 10^{\log 7 \cdot 2^x} \Rightarrow y = 7 \cdot 2^x$

b. $\log y = \log 20 + (\log 0.25)x \Rightarrow$
$\log y = \log(20 \cdot 0.25^x) \Rightarrow$
$10^{\log y} - 10^{\log(20 \cdot 0.25^x)} \rightarrow y - 20 \cdot 0.25^x$

c. $\log y = 6 + (\log 3) \cdot x \Rightarrow 10^{\log y} = 10^{6 + \log(3^x)} \Rightarrow$
$y = 10^6 \cdot 10^{\log(3^x)} \Rightarrow y = 10^6 \cdot 3^x$

d. $\log y = 6 + \log 5 + (\log 3) \cdot x \Rightarrow$
$\log y = 6 + \log 5 + \log(3^x) \Rightarrow$
$10^{\log y} = 10^{6 + \log 5 + \log 3^x} \Rightarrow y = 10^6 \cdot 10^{\log 5} \cdot 10^{\log 3^x} \Rightarrow$
$y = 10^6(5)3^x = 5 \cdot 10^6(3)^x$

4. a. i. slope $= \log 5$; vertical intercept $= \log 2$
ii. slope $= \log 0.75$, vertical intercept $= \log 6$
iii. slope $= \log 4$; vertical intercept $= 0.4$
iv. slope $= \log 1.05$; vertical intercept $= 3 + \log 2$

b. i. $y = 2 \cdot 5^x$
ii. $y = 6 \cdot (0.75)^x$
iii. $y = 10^{0.4}(4)^x$
iv. $y = 10^3 \cdot 2 \cdot 1.05^x = 2000(1.05)^x$

5. a. $a \approx 2.00$ since $10^{0.301} \approx 2$
b. $C \approx 524.81$ since $10^{2.72} \approx 524.81$
c. $a \approx 0.75$ since $10^{-0.125} \approx 0.75$
d. $C \approx 100,000$ since $10^5 = 100,000$

6. a. $y = 10^{0.301} \cdot 10^{0.477x} \Rightarrow y = 2 \cdot 3^x$
b. $y = 10^3 \cdot 10^{0.602x} \Rightarrow y = 1000(4)^x$
c. $y = 10^{1.398} \cdot 10^{-0.046x} \Rightarrow y = 25 \cdot (0.90^x)$

7. Exponential functions for both since the graphs on semi-log plots are approximately linear.

8. Using the point $(0, 4)$, the vertical intercept, and $(2, 5)$, the slope is $\frac{5-4}{2} = 0.5$. So $\log y = 4 + 0.5x \Rightarrow y = 10^{4 + 0.5x} \Rightarrow$
$y = 10^4 \cdot 10^{0.5x} \approx 10^4(3.16)^x$.

Exercises for Section 6.4

1. a. goes with **f.**   **c.** goes with **e.**
b. goes with **h.**   **d.** goes with **g.**

3. a. $\log(y) = 4.477 + 0.301x$
b. $\log(y) = 3.653 + 0.146x$
c. $\log(y) = 6.653 - 0.155x$
d. $\log(y) = 3.778 - 0.244x$

5. a. goes with Graph **B.**   **c.** goes with Graph **A.**
b. goes with Graph **C.**   **d.** goes with Graph **D.**

7. Here are the average rates of change:

a.

x	$Y = \log(x)$	Avg. Rt. of Ch.
0	2.30103	n.a.
10	4.30103	0.20000
20	4.90309	0.06021
30	5.25527	0.03522
40	5.50515	0.02499

b.

x	$Y = \log(x)$	Avg. Rt. of Ch.
0	4.77815	n.a.
10	3.52876	-0.12494
20	2.27938	-0.12494
30	1.02999	-0.12494
40	-0.21945	-0.12494

a. This is *not* exponential; the average rate of change between consecutive points keeps decreasing.

b. The average rate of change over each decade is -0.12494. Thus the plot of Y vs. x is linear and therefore y is exponential. The exponential function is $y = 60,000(0.75)^x$.

9. a. The scatter plot of points $(x, \log(y))$, where $Y = \log(y)$, is nearly linear; thus the growth is very close to being exponential.

b. The equation of the best-fit line is given in the graph and the corresponding exponential equation is approximately $y = 38.788 \cdot 1.3137^x$. The daily growth rate is 31.37%.

11. a. Approximately 12.2 micrograms/liter.

b. Yes, because its log graph is a straight line with negative slope. The decay factor is $10^{-0.0181} \approx 0.959$, and thus the decay rate is $1 - 0.959 = 0.041$ or 4.1%.

c. The exponential model is approximately $y = 12.2 \cdot 0.959^x$.

13. a. The CX7300 is 2 orders or magnitude greater in pixels/\$ than the DC8460, i.e., it gives 100 times more pixels/\$ than the DC8460.

b. A straight line in a semi-log plot suggests an exponential function.

c. Since $Y = \log(\text{pixel}/\$) = 1000x$, then the exponential form uses powers of 10. So $P = \text{pixels}/\$ = 10^Y = 10^{1000x}$, where x measures years since 1994.

d. The growth factor is 10^{1000}. The growth rate comes from $(1 + r) = 10^{1000}$ or $r = 10^{1000} - 1$.

Ch. 6: Check Your Understanding

1. True	**8.** False	**15.** True	**22.** False
2. True	**9.** False	**16.** False	**23.** True
3. False	**10.** False	**17.** False	**24.** False
4. False	**11.** True	**18.** True	**25.** True
5. True	**12.** False	**19.** False	**26.** False
6. False	**13.** True	**20.** False	**27.** True
7. True	**14.** False	**21.** False	**28.** True

29. True

30. True

31. Possible answer: $y = \frac{1}{2}(\log x - \log 3)$

32. Possible answer: $y = 50.3(1.062)^t$

33. Possible answer: $y = 100(1.20)^x$

34. False   **35.** True   **36.** False   **37.** False

Ch. 6 Review: Putting It All Together

1. a. Doubling time: $100 = 50 \cdot 1.16^t \Rightarrow 2 = 1.16^t \Rightarrow$
$\ln(2) = t \cdot \ln(1.16) \Rightarrow t = \ln(2)/\ln(1.16) \approx 4.67$ units.

b. Half life: $100 = 200 \cdot e^{-0.083t} \Rightarrow 0.5 = e^{-0.083t} \Rightarrow$
$\ln(0.5) = -0.083t \Rightarrow t = \ln(0.5)/(-0.083) \approx 8.35$ units.

3. Expressions (a) and (e) are equivalent and (b), (c), and (f) are equivalent. Expression (d) does not match any other.

5. a. After one time period (20 minutes), $A(1) = 325 \cdot (0.5)^1 = 162.5$ mg (or half the original amount). After two time periods (40 minutes), $A(2) = 325 \cdot (0.5)2 = 81.25$ mg (or one-quarter of the original amount). Clearly the half-life is 20 minutes.

b. Using the graph, it appears that after about 1.7 time periods (34 minutes) $A(T) = 100$. Using the equation, we have $100 = 325 \cdot (0.5)^T \Rightarrow 100/325 = (0.5)^T \Rightarrow 0.3077 = (0.5)^T \Rightarrow \log(0.3077) = T \cdot \log(0.5) \Rightarrow T \approx \log(0.3077)/\log(0.5) \Rightarrow T \approx 1.70$ time periods, or 1.7 twenty-minute periods = 34 minutes. So our estimate is accurate.

c. $B(T) = 81 \cdot (0.5)^T$, so $A(T)$ and $B(T)$ have different initial amounts, but the decay rate is the same.

7. a. Density for India $= (1.08 \cdot 10^9)/(1.2 \cdot 10^6) = 0.9 \cdot 10^3 = 900$ people per square mile. Density for China $= (1.30 \cdot 10^9)/(3.7 \cdot 10^6) \approx 0.35 \cdot 10^3 = 350$ people per square mile. So India's population density is about $900/350 \approx 2.6$ times larger than China's.

b. $C(x) = 1.30(1.006)^x$ and $I(x) = 1.08(1.016)^x$, where $C(x)$ and $I(x)$ are in billions and $x =$ years since 2005.

c.

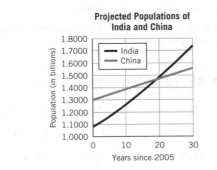

Note that India's exponential growth rate is so much higher than that for China, the graph for the Chinese population appears almost linear in comparison.

d. The projected populations are the same roughly 18 years after 2005, or in 2023. Using the models, we need the point at which $C(x) = I(x) \Rightarrow 1.30(1.006)^x = 1.08(1.016)^x \Rightarrow 1.204(1.006)^x \approx (1.016)^x \Rightarrow \log(1.204) + x \log(1.006) \approx x \log(1.016) \Rightarrow \log(1.204) \approx x [\log(1.016) - \log(1.006)] \Rightarrow 0.081 \approx x \log[(1.016)/(1.006)] \Rightarrow 0.081 \approx 0.0043x \Rightarrow x \approx 18.8$ years after 2005, that is, in late 2023 our model predicts the two populations will be the same. Evaluating $C(18.8)$, we get $1.30(1.006)^{18.8} \approx 1.455$ billion people. (To double-check you could calculate $I(18.8) \approx 1.456$ billion, with the difference due to rounding.)

9. a. Let $U(x) = 10\left(\frac{1}{2}\right)^{\frac{x}{5}}$ the exponential decay function for 10 grams of uranium-235 where x is in billions of years. For continuous decay let $\frac{1}{2} = e^k \Rightarrow k = \ln(\frac{1}{2})$ so $U(x) = 10 \, e^{(\ln\frac{1}{2})(\frac{x}{5})} = 10 \, e^{-0.1386x}$.

b. 5 billion years. (Note: one can verify this by solving $5 = 10 \, e^{-0.1386x}$)

11. a. $t = \log(2.3) \approx 0.362$

b. $t^2 \cdot 4 = 10^2 \Rightarrow t^2 = 25 \Rightarrow t = 5$ (*Note:* $t = -5$ is not a solution here, since $\log(-5)$ is not defined.)

c. $2 = e^{0.03t} \Rightarrow \ln 2 = 0.03t \Rightarrow t = (\ln 2)/0.03 \approx 0.693/0.03 \approx 23.1$

d. $\ln [(2t - 5)/(t - 1)] = 0 \Rightarrow (2t - 5)/(t - 1) = 1 \Rightarrow 2t - 5 = t - 1 \Rightarrow t = 4$

13. a. $1.5 = e^{\ln 1.5} \approx e^{0.405}$

b. $0.7 = e^{\ln 0.7} \approx e^{-0.357}$

c. $1 = e^0$

15. a. Matches Graph C

b. Matches Graph D

c, d. Both match Graphs A and B since $y = 100(10)^x \Rightarrow \log y = \log 100 + x \log 10 \Rightarrow \log y = 2 + x$.

17. a. y_1 matches Graph B since when $x = 1$ then $2 \ln(x) = 2 \ln(1) = 2 \cdot 0 = 0$. So the horizontal intercept is $(1, 0)$. This graph stretches the graph of $y = \ln(x)$ by a factor of 2.

b. y_2 matches Graph A since when $x = 1$, $2 + \ln(x) = 2 + \ln(1) = 2 + 0 = 2$. So that graph passes through $(1, 2)$. This graph is the graph of $y = \ln(x)$ shifted up by 2 units.

c. y_3 matches Graph D since when $x = 0$, $\ln(x + 2) = \ln(2) \approx 0.693 > 0$. So the vertical intercept is above the origin at approximately $(0, 0.693)$, Also, when $x = -1$, $\ln(-1 + 2) = \ln(1) = 0$, so the horizontal intercept is $(-1, 0)$. This graph is the graph of $y = \ln(x)$ shifted 2 units to the left.

d. y_4 matches Graph B since the functions $y_1 = 2 \ln(x)$ and $y_4 = \ln(x^2)$ are the same for $x > 0$. This graph stretches the graph of $y = \ln(x)$ by a factor of 2.

19. a. Using the table in the text, five orders of magnitude.

b. Using the function definition, if $30 = 10 \log(I_{30}/I_0)$, where $I_{30} =$ intensity level corresponding to 30 decibels, and $80 = 10 \log(I_{80}/I_0)$, where $I_{80} =$ intensity level corresponding to 80 decibels. Subtracting the two equations gives:

$$80 - 30 = 10 \log(I_{80}/I_0) - 10 \log(I_{30}/I_0) \Rightarrow$$
$$50 = 10 \log[(I_{80}/I_0)/(I_{30}/I_0)] \Rightarrow$$
$$5 = \log[(I_{80}/I_0) \cdot (I_0/I_{30})] \Rightarrow 5 = \log[(I_{80}/I_{30})] \Rightarrow$$
$$10^5 = I_{80}/I_{30},$$

so I_{80} is five orders of magnitude larger than I_{30}.

21. The ratios of A/A_0 are respectively $10^{7.2}$, $10^{8.8}$, and $10^{9.5}$. So A is respectively $10^{7.2} \cdot A_0$, $10^{8.8} \cdot A_0$ and $10^{9.5} \cdot A_0$. So the values of A are respectively 7.2, 8.8, and 9.5 orders of magnitude larger in amplitude than the base value of A_0.

23. a. The data appear to be exactly linear, and two estimated points on the best-fit line are $(0, 0.3)$ and $(10, 5)$. The slope $= (5 - 0.3)/(10 - 0) = 0.47$. The equation is then $Y = 0.3 + 0.47x$.

b. Substituting $\log y$ for Y gives $\log y = 0.3 + 0.47x \Rightarrow y = 10^{0.3+0.47x} \Rightarrow y = 10^{0.3}10^{0.47x} \approx 2 \cdot 3^x$.

c. The function is exponential, suggesting that the graph of exponential functions is a straight line on a semi-log plot (where the log scale is on the vertical axis).

CHAPTER 7

Section 7.1

Algebra Aerobics 7.1

1. a. $l = \frac{V}{wh}$ **c.** $w = \frac{P - 2l}{2}$

    b. $b = \frac{2A}{h}$ **d.** $h = \frac{S - 2x^2}{4x}$

2. a. $S(r) = 4\pi r^2$; $S(2r) = 4\pi(2r)^2 = 4 \cdot (4\pi r^2) = 4S(r)$; $S(3r) = 4\pi(3r)^2 = 9 \cdot (4\pi r^2) = 9S(r)$

    b. When the radius is doubled, the surface area is multiplied by 4. When the radius is tripled, the surface area is multiplied by 9.

3. a. $V(r) = (4/3)\pi r^3$; $V(2r) = (4/3)\pi(2r)^3 = 8 \cdot ((4/3)\pi r^3) = 8V(r)$; $V(3r) = (4/3)\pi(3r)^3 = 27 \cdot ((4/3)\pi r^3) = 27V(r)$

    b. When the radius is doubled, the volume is multiplied by 8. When the radius is tripled, the volume is multiplied by 27.

4. The volume grows faster than the surface area, so as the radius of a sphere increases, the ratio of (surface area)/volume decreases.

5. a. The volume of the sphere is equal to the volume of the cube when $\frac{4}{3}\pi r^3 = r^3$, which is true only if $r = 0$.

    b. The volume of the cube is greater than the volume of the sphere if $r^3 > \frac{4}{3}\pi r^3 \Rightarrow 1 > \frac{4}{3}\pi$, which is never true. The volume of the cube is less than the volume of the sphere for all $r > 0$ since $r^3 < \frac{4}{3}\pi r^3 \Rightarrow 1 < \frac{4}{3}\pi$.

6. No; since if the radius is doubled from 5 to 10, the volume is increased by a factor of 4. That is, given $V = \pi r^2 h$, if $r = 5$ then $V = \pi 5^2 \cdot 25 = 625\pi$ cubic feet and if $r = 10$ then $V = \pi 10^2 \cdot 25 = 2500\pi$ cubic feet, which is a factor of 4 larger.

7. Yes; since $V = \pi r^2 h$, if the height is doubled, the volume is doubled. That is, for this example, if $r = 5$, then $V = \pi r^2 h \Rightarrow V = \pi 12^2 \cdot 5 = 720\pi$ cubic feet and if $r = 10$, then $V = \pi 12^2 \cdot 10 = 1440\pi$ cubic feet, which is twice the volume.

8. a. The area B of the triangular prism base $= \frac{1}{2}ab \Rightarrow V = \left(\frac{1}{2}ab\right)h = \frac{1}{2}abh$.

    b. If the height is doubled, the volume is doubled, since $\left(\frac{1}{2}ab\right)2h = 2\left(\frac{1}{2}ab\right)h = 2V$.

    c. If all dimensions are doubled, the volume is increased by a factor of 8, since $\left(\frac{1}{2}(2a)(2b)\right)(2h) = 2^3 \cdot \left(\frac{1}{2}abh\right) = 8V$.

Exercises for Section 7.1

1. a. $C(r) = 2\pi r$, so $C(2r) = 2 \cdot C(r)$, $C(4r) = 4 \cdot C(r)$, and $C(4r) = 2C(2r)$. Thus the circumference is respectively doubled or quadrupled.

    b. $A(r) = \pi r^2$, so $A(2r) = 4A(r)$ and $A(4r) = 16A(r)$. Thus the area is respectively multiplied by 4 or by 16.

    c. The area grows faster than the circumference. Since $C(r)/A(r) = (2\pi r)/(\pi r^2) = 2/r$, we see that the ratio decreases as the radius increases.

3. a. $S \approx 1.26 \cdot 10^{-19}$ m^2 **c.** $S/V \approx 3.0 \cdot 10^{10}$ m^{-1}

    b. $V \approx 4.19 \cdot 10^{-30}$ m^3 **d.** ratio $= 3/r$; decreases

---

5. a. S is multiplied by 16; V is multiplied by 64.

    b. S is multiplied by n^2; V is multiplied by n^3.

    c. S is divided by 9; V is divided by 27.

    d. S is divided by n^2; V is divided by n^3.

7. a. $V = 2632.5$ cm^3

    b. quadrupled

    c. Quadruple either the length or the width or double both.

9. a. The volume doubles if the height is doubled.

    b. The volume is multiplied by 4 if the radius is doubled.

11. a. $V = 3\pi r^3$; $S = 8\pi r^2$

    b. Volume eventually grows faster since $V/S = 3r/8$; as $r \to +\infty$, V/S increases without bound.

13. a. For $x =$ radius and 3 ft $=$ height, the volume is $V = $ (area of the base)(height) $= 3\pi x^2$ ft^3.
Surface area $S = $ (circumference)(height) $= 6\pi x$ ft^2.

    b. $V/S = (3\pi x^2)/(6\pi x) = x/2$, and this increases as x increases. Thus the volume increases faster than the surface area.

Section 7.2

Algebra Aerobics 7.2

1.

Power Function	Independent Variable	Dependent Variable	Constant of Proportionality	Power
a. yes	r	A	π	2
b. yes	z	y	1	5
c. no	—	—	—	—
d. no	—	—	—	—
e. yes	x	y	3	5

2. a. y is directly proportional to x^2.

    b. y is not directly proportional to x^2.

    c. y is not directly proportional to x.

3. $g(x) = 5x^3$

    a. $g(2) = 5(2)^3$ $g(4) = 5(4)^3$
          $= 5(8)$ $= 5(64)$
          $= 40$ $= 320$ So $g(4)$ is eight times larger than $g(2)$

    b. $g(5) = 5(5)^3$ $g(10) = 5(10)^3$
          $= 5(125)$ $= 5(1000)$
          $= 625$ $= 5000$ So $g(10)$ is eight times larger than $g(5)$

    c. $g(2x) = 5(2x)^3$
          $= 5(8x^3)$
          $= 8(5x^3)$
          $= 8g(x)$ So $g(2x)$ is eight times larger than $g(x)$

    d. $g\left(\frac{1}{2}x\right) = 5\left(\frac{1}{2}x\right)^3$
          $= 5\left(\frac{1}{8}x^3\right)$
          $= \frac{1}{8}(5x^3)$
          $= \frac{1}{8}g(x)$ So $g\left(\frac{1}{2}x\right)$ is one-eighth the size of $g(x)$

4. a. $h(2) = 0.5(2)^2 = 2$ and $h(6) = 0.5(6)^2 = 18$, an increase by a factor of 9.

  b. $h(5) = 0.5(5)^2 = 12.5$ and $h(15) = 0.5(15)^2 = 112.5$, an increase by a factor of 9.

  c. $h(x)$ will increase by a factor of 9 since $h(3x) = 0.5(3x)^2 = 9 \cdot [0.5x^2] = 9 \cdot h(x)$.

  d. $h(x)$ will decrease by a factor of 9 since $h\left(\frac{1}{3}x\right) = 0.5\left(\frac{1}{3}x\right)^2 = \frac{1}{9} \cdot [0.5x^2] = \frac{1}{9} \cdot h(x)$.

5. a. $V = k \cdot r^3$ **c.** $f = k \cdot c$

  b. $V = k \cdot l \cdot w \cdot h$

6. a. y is equal to 3 times the fifth power of x.

  b. y is equal to 2.5 times the cube of x.

  c. y is equal to one-fourth of the fifth power of x.

7. a. y is directly proportional to x^5 with a proportionality constant of 3.

  b. y is directly proportional to x^3 with a proportionality constant of 2.5.

  c. y is directly proportional to x^5 with a proportionality constant of 1/4.

8. a. $P = aR^2 \Rightarrow \frac{P}{a} = \frac{aR^2}{a} \Rightarrow \frac{P}{a} = R^2 \Rightarrow \sqrt{\frac{P}{a}} = \sqrt{R^2} \Rightarrow$
$R = \pm\sqrt{\frac{P}{a}}$, unless R is nonnegative.

  b. $V = \left(\frac{1}{3}\right)\pi r^2 h \Rightarrow 3(V) = 3\left(\frac{1}{3}\right)\pi r^2 h \Rightarrow$
$3V = \pi r^2 h \Rightarrow \frac{3V}{\pi r^2} = \frac{\pi r^2 h}{\pi r^2} \Rightarrow h = \frac{3V}{\pi r^2}$

9. a. $f(2x) = 0.1(2x)^3 = 0.8x^3;$ $2f(x) = 2(0.1x^3) = 0.2x^3$
$f(2x) = 4(2f(x))$

  b. $f(3x) = 0.1(3x)^3 = 2.7x^3;$ $3f(x) = 3(0.1x^3) = 0.3x^3$
$f(3x) = 9(3f(x))$

  c. $f(2 \cdot 5) = 0.1(10)^3 = 100;$ $2f(5) = 2(0.1 \cdot 5^3) = 25$

Exercises for Section 7.2

1. a. $f(2) = 3 \cdot 2^2 = 12;$ $f(-2) = 3 \cdot (-2)^2 = 12$

  b. $g(2) = 3 \cdot 2^3 = 24;$ $g(-2) = 3 \cdot (-2)^3 = -24$

  c. $h(2) = -3 \cdot 2^2 = -12;$ $h(-2) = -3 \cdot (-2)^2 = -12$

  d. $k(2) = -3 \cdot 2^3 = -24;$ $k(-2) = -3 \cdot (-2)^3 = 24$

3. a. $k = 1/2; P = 288$ **c.** $k = 4\pi; T = 1728\pi$

  b. $k = 14; M = 84$ **d.** $k = \sqrt[3]{18}; N = 6$

5. a. $y = \frac{1}{5}x;$ 4

  b. $p = 6\sqrt{s};$ 24

  c. $A = \frac{4}{3}\pi r^3;$ 64π

  d. $P = \frac{1}{2}m^2;$ quartered, i.e., P is divided by 4.

7. a. $Y = kX^3$

  b. $k = 1.25$

  c. Increased by a factor of 125

  d. Divided by 8

  e. $X = \sqrt[3]{\frac{Y}{k}}$. So X is not directly proportional to Y. It is directly proportional to $\sqrt[3]{Y}$.

9. a. L is multiplied by 32.   **b.** M is multiplied by 2^p.

11. In all the formulas below, k is the constant of proportionality.

  a. $x = k \cdot y \cdot z^2$ **c.** $w = k \cdot x^2 \cdot y^{1/3}$

  b. $V = k \cdot l \cdot w \cdot h$ **d.** $V = k \cdot h \cdot r^2$

13. a. $C = kAt$

  b. $k = 0.06$ dollars per sq. ft, per in.

  c. The area of the four walls (without the door) $=$ $2(15 \cdot 8) + 2(20 \cdot 8) = 240 + 320 = 560$ sq. ft. The area of the door is $3 \cdot 7 = 21$ sq. ft. Thus the total wall area $=$ $560 - 21 = 539$ sq. ft.

  d. $\$129.36; \$194.04$

15. a. and **b.**

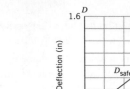

The graphs of $D_{\text{deflection}}$ and D_{safe} (with the deflections measured in inches and the plank length L measured in feet) are given in the accompanying diagram.

  c. The safety deflections are well above the actual deflections for all values of L between 0 and 20. It would cease to be safe if the plank were longer than about 22 ft. (This is the L value where D_{safe} and $D_{\text{deflection}}$ meet.)

17. In all the formulas below, k is the constant of proportionality.

  a. $d = k \cdot t^2$ **d.** $R = k \cdot [O_2] \cdot [NO]^2$

  b. $E = k \cdot m \cdot c^2$ **e.** $v = k \cdot r^2$

  c. $A = k \cdot b \cdot h$

19. a. V is quadrupled; is multiplied by 9

  b. V is doubled; is tripled

  c. V is multiplied by n^2

  d. V is multiplied by n

21. a. $h(2) = 2; h(6) = 18;$ the latter is nine times the former.

  b. $h(5) = 12.5; h(15) = 112.5;$ the latter is nine times the former.

  c. h's value is multiplied by 9.

  d. h's value is divided by 9.

Section 7.3

Algebra Aerobics 7.3

1. a. $f(2) = 4(2)^3 = 4(8) = 32$

  b. $f(-2) = 4(-2)^3 = 4(-8) = -32$

  c. $f(s) = 4s^3$

  d. $f(3s) = 4(3s)^3 = 4(27)s^3 = 108s^3$

2. a. $g(2) = -4(2)^3 = -4(8) = -32$

  b. $g(-2) = -4(-2)^3 = -4(-8) = 32$

c. $g\left(\frac{1}{2}t\right) = -4\left(\frac{1}{2}t\right)^3 = -4\left(\frac{1}{8}\right)t^3 = -\frac{1}{2}t^3$

d. $g(5t) = -4(5t)^3 = -4(125)t^3 = -500t^3$

3. a. $f(4) = 3(4)^2 = 48$ **e.** $f(2s) = 3(2s)^2 = 12s^2$

     b. $f(-4) = 3(-4)^2 = 48$ **f.** $f(3s) = 3(3s)^2 = 27s^2$

     c. $f(s) = 3s^2$ **g.** $f\left(\frac{s}{2}\right) = 3\left(\frac{s}{2}\right)^2 = \frac{3}{4}s^2$

     d. $2f(s) = 2(3s^2) = 6s^2$ **h.** $f\left(\frac{s}{4}\right) = 3\left(\frac{s}{4}\right)^2 = \frac{3}{16}s^2$

4. a.

x	$f(x) = 4x^2$	$g(x) = 4x^3$
-4	64	256
-2	16	-32
0	0	0
2	16	32
4	64	256

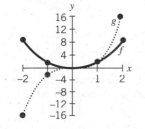

b. As $x \to +\infty$, both $f(x)$ and $g(x) \to +\infty$.

c. As $x \to -\infty$, $f(x) \to +\infty$ but $g(x) \to -\infty$.

d. The domain of both functions is all the real numbers, and the range of g is also all the real numbers; however, the range of f is only all the nonnegative real numbers.

e. They intersect at the origin and at the point $(1, 4)$.

f. All values greater than $x = 1$.

5.

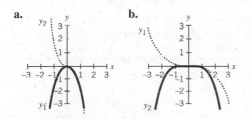

a. $f(x) = g(x)$ for $x = 0$ and $x = 1$.

b. $f(x) > g(x)$ for $0 < x < 1$ and for $x < 0$

c. $f(x) < g(x)$ for $x > 1$

6.   **a.**   **b.**

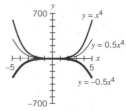

7. a. The graphs are similar; both have similar end behavior; they intersect at the origin; for positive values of x, $y_1 < y_2$; for negative values of x, $y_1 > y_2$.

     b. The graphs are similar; both have similar end behavior; they intersect at the origin; for all nonzero values of x, $y_1 > y_2$.

8. a. p is even and a is positive; $a = 3$ since $(1, 3)$ is on the graph of the function.

     b. p is odd and a is negative; $a = -2$ since $(1, -2)$ is on the graph of the function.

9.   **a.**   **b.**

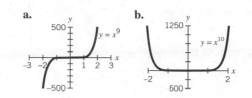

10. a. $y = x$ and $y = 4x$ and $y = -4x$:

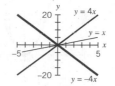

b. $y = x^4$ and $y = 0.5x^4$ and $y = -0.5x^4$

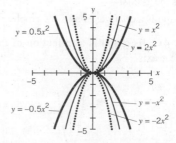

Exercises for Section 7.3

1. The graphs of the six functions are labeled in the accompanying diagram. They are all parabolas since they are graphs of functions of the form $y = ax^2$, for various values of a. The differences are due to the value of a. If $a > 0$, then the graph is concave up. If $a < 0$, then the graph is concave down. The larger the absolute value of a, the narrower the opening of the graph.

3. f goes with Graph C; g goes with Graph A; h goes with Graph D, j goes with Graph B. The graphs in B and D are mirror images of each other across the x-axis; the graph in D is steeper than that in C, the graph in A is flatter than those in C

and D for $-1 < x < +1$, but the Graph in A is steeper as $x \to \pm\infty$. The range of f is $[0, \infty)$; the range of g is $[0, \infty)$; the range of h is $[0, \infty)$; the range of j is $(-\infty, 0]$.

5. a. $g(x) = 6 \cdot x^4$

  b. $h(x) = 0.5 \cdot x^4$

  c. $j(x) = -2x^4$

7. a. $f(0) = g(0) = 0$

  b. If $0 < x < 1$, then $f(x) > g(x)$

  c. $f(1) = g(1) = 1$

  d. If $x > 1$, then $f(x) < g(x)$

  e. If $x < 0$, then $f(x) > g(x)$

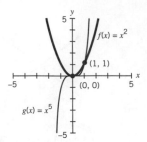

9. a. n is even   **e.** $-\infty$

  b. $k < 0$ **f.** $-\infty$

  c. Yes, $f(-2) = f(2)$ **g.** The range of f is $(-\infty, 0]$.

  d. Yes, $f(-x) = f(x)$

11.

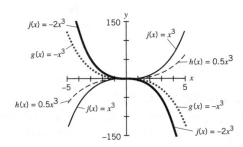

  a. The constant of proportionality for f is 1; for g it is -1; for h it is $1/2$; for j it is -2.

  b. g's graph is the reflection of f's graph across the x-axis.

  c. j's graph is both a stretch and a reflection of f's graph across the x-axis.

  d. h's graph is a compression of f's graph.

13. $g(x) = -4x^2$. Its graph is a reflection of the graph of $f(x)$ across the x-axis.

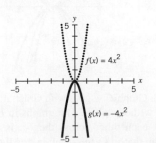

Section 7.4

Algebra Aerobics 7.4

1. a.

x	$y = 4^x$	$y = x^3$
0	1	0
1	4	1
2	16	8
3	64	27
4	256	64
5	1024	125

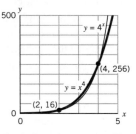

  b. $y = 4^x$ dominates $y = x^3$.

    In this case, $4^x > x^3$ for all values of x, but as $x \to +\infty$, the values grow farther and farther apart.

2. Graphs A and D are likely to be power functions; Graphs B and C are likely to be exponential functions.

3. a. $y = 2^x$ eventually dominates.

  b. $y = (1.000\ 005)^x$ eventually dominates.

Exercises for Section 7.4

1. a. The two graphs intersect at $(2, 16)$ and $(4, 256)$. Thus, the functions are equal for $x = 2$ and $x = 4$. (See the accompanying figure.)

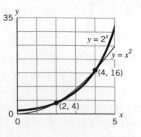

  b. As x increases, both functions grow. For $0 \le x < 2$ we have that $x^4 < 4^x$; at $x = 2$ both have a y value of 16. From $2 < x < 4$, we have $4^x < x^4$. For $x = 4$ they are again equal. For $x > 4$ we have that $x^4 < 4^x$. As x keeps on increasing 4^x will continue to grow faster than x^4.

  c. The graph of $y = 4^x$ dominates.

3. a. $f(x) > g(x)$ as $x \to -\infty$

  b. $f(x) < g(x)$ as $x \to +\infty$

  c. $f(x) > g(x)$ for x in $(-\infty, -1)$

  d. $f(x) < g(x)$ for x in $(-0.5, 1)$

  e. $f(x) > g(x)$ for x in $(1, 6)$

  f. $f(x) < g(x)$ for x in $(7, +\infty)$

5.   **Graphs of $y = 2^x$ and $y = x^2$**

For $x \geq 0$, the graphs of $y = 2^x$ and $y = x^2$ on bottom of previous page, intersect at $x = 2$ and $x = 4$. The graph of $y = 2^x$ lies above the graph of $y = x^2$ for $0 \leq x < 2$ and for $x > 4$.

Graphs of $y = 3^x$ and $y = x^3$

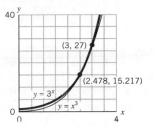

For $x \geq 0$, the graphs of $y = 3^x$ and $y = x^3$, above, intersect at $x = 3$ and at $x \approx 2.478$ The latter intersection point can only be seen if we use a graphing device and zoom in. The graph of $y = 3^x$ is above the graph of $y = x^3$ over the intervals $[0, 2.478)$ and $(3, +\infty)$. It is below that graph over the interval $(2.478, 3)$.

7. $3 \cdot 2^x > 3 \cdot x^2$ if $0 < x < 2$ and if $x > 4$. Also, $3 \cdot 2^x < 3 \cdot x^2$ if $2 < x < 4$.

9. a. $h(x)$ goes with C   **c.** $j(x)$ goes with B
    b. $i(x)$ goes with A

11. The functions $f(ax)$ and $g(ax)$ are the functions $f(x)$ and $g(x)$, respectively, where the input x has been multiplied by a. Since $f(ax) = (ax)^n = a^n \cdot f(x)$, then $f(ax)$ is $f(x)$ multiplied by a^n. Since $g(ax) = b^{ax} = (b^x)^a = [g(x)]^a$, then $g(ax)$ is $g(x)$ raised to the ath power.

13. Linear function: $m = \frac{32 - 0.5}{4 - 1} = 10.5 \Rightarrow y = 10.5x - 10$.

Exponential function: $y = \frac{1}{8} \cdot 4^x$

Power function: $y = 0.5x^3$

15. a. True   **b.** True   **c.** False   **d.** False

Section 7.5

Algebra Aerobics 7.5

1. a. x is directly proportional to y and z.

    b. y is inversely proportional to the square of x.

    c. D is directly proportional to the square root of y and inversely proportional to the cube of z.

2. a. The domain is all real numbers except 0; that is, the domain is all x in the interval $(-\infty, 0) \cup (0, +\infty)$. The range is all numbers except 0; that is, the range is all y in the interval $(-\infty, 0) \cup (0, +\infty)$.

    b. The domain is all $x > 0$; that is, the domain is all x in the interval $(0, +\infty)$. Since $\sqrt{x}$ must be positive, then $y = \frac{5}{\sqrt{x}}$ must be positive. So the range is all $y > 0$; that is, the domain is all y in the interval $(0, +\infty)$.

3. (a) and (d) represent direct proportionality; (b) and (c) represent inverse proportionality; (e) does not represent direct or inverse proportionality. It is an exponential function where x, the input, becomes an exponent in the output.

4. The volume will increase by a factor of 4, from 1 ft³ to 4 ft³. (See Table 7.6, p. 425, in Section 7.5.)

5. a. i. $\frac{15}{x^3}$ **ii.** $\frac{-10}{x^4}$ **iii.** $\frac{3.6}{x}$ **iv.** $\frac{2}{x^2 y^3}$

    b. i. $1.5x^{-2}$ **ii.** $-6x^{-3}$ **iii.** $-\frac{2}{3}x^{-2}$ **iv.** $6x^{-3}y^{-4}z^{-1}$

6. $t = kd^{-2}$ or $t = \frac{k}{d^2}$ for some constant k

7. Intensity at 4 ft is $\frac{k}{4^2} = \frac{1}{16}k$, but at 2 ft intensity is $\frac{k}{(2)^2} = \frac{1}{4}k$. The light is four times as bright at a distance of 2 ft away than it is at 4 ft away.

8. $g(x) = \frac{3}{x^4}$

    a. $g(2x) = \frac{3}{(2x)^4} = \frac{3}{16x^4} = \frac{1}{16}\left(\frac{3}{x^4}\right)$, so $g(2x) = \frac{1}{16}g(x)$.

    b. $g\left(\frac{x}{2}\right) = \frac{3}{(x/2)^4} = 3 \div \left(\frac{x}{2}\right)^4 = 3 \div \frac{x^4}{16} = 3 \cdot \frac{16}{x^4}$
       $= 16\left(\frac{3}{x^4}\right)$, so $g\left(\frac{x}{2}\right) = 16(g(x))$.

9. $h(x) = -\frac{2}{x^3}$

    a. $h(3x) = -\frac{2}{(3x)^3} = -\frac{2}{27x^3} = \frac{1}{27} \cdot \left(-\frac{2}{x^3}\right) = \frac{1}{27} \cdot h(x)$, or one-twenty-seventh of the original amount.

    b. $h\left(\frac{x}{3}\right) = -\frac{2}{(x/3)^3} = -\frac{2}{x^3/27} = -2 \cdot \frac{27}{x^3} = 27 \cdot \left(-\frac{2}{x^3}\right) = 27h(x)$, or 27 times the original amount.

10. a. $x = kyz^2$ **b.** $a = \frac{kbc^3}{d}$ **c.** $a = \frac{kb^2\sqrt{c}}{de}$

11. a. $x = \frac{k}{\sqrt{y}}$; If $x = 3$ and $y = 4$, then $3 = \frac{k}{\sqrt{4}} \Rightarrow 3 = \frac{k}{2} \Rightarrow$
    $k = 6$, so $x = \frac{6}{\sqrt{y}}$.
    So if $y = 400$, then $x = \frac{6}{\sqrt{400}} = \frac{6}{20} = 0.3$.

    b. $x = k \cdot y \cdot z$; If $x = 3$, $y = 4$ and $z = 0.5$, then $3 = k \cdot 4 \cdot (0.5) \Rightarrow k = 1.5$, so $x = 1.5 \cdot y \cdot z$. So if $x = 6$ and $z = 2$, then $6 = 1.5y(2)$ or $y = 2$.

    c. $x = \frac{ky^2}{\sqrt[3]{z}}$; If $x = 6$, $y = 6$ and $z = 8$, then
    $6 = \frac{k(6)^2}{\sqrt[3]{8}} \Rightarrow 6 = \frac{36k}{2} \Rightarrow k = \frac{1}{3}$ so $x = \frac{1}{3}\frac{y^2}{\sqrt[3]{z}}$.
    So if $y = 9$ and $z = 0.027$, then
    $x = \frac{1}{3}\frac{9^2}{\sqrt[3]{0.027}} = \frac{1}{3} \cdot \frac{81}{0.3} = 90$.

Exercises for Section 7.5

1. a. y changes from $\frac{4}{1^3} = 4$ to $\frac{4}{4^3} = \frac{1}{16}$

    b. decreases

    c. y changes from $\frac{4}{2^3} = \frac{1}{2}$ to $\frac{4}{(1/2)^3} = 32$

    d. increases

    e. y is inversely proportional to x^3 and 4 is the constant of proportionality.

    f. The domain is all real numbers except 0.

    g. The range is all real numbers except 0.

3. a. $h(2x) = \frac{k}{8x^3}$, $h(-3x) = \frac{k}{-27x^3}$ and $h(x/3) = \frac{27k}{x^3}$

    b. $j(2x) = \frac{-k}{16x^4}$, $j(-3x) = \frac{-k}{81x^4}$ and $j(x/3) = \frac{-81k}{x^4}$

5. a. $P = \frac{0.016}{f}$; 0.0016   **c.** $S = \frac{16}{wp}$; 2

  b. $Q = \frac{54}{r^2}$; 2/3   **d.** $W = \frac{2}{3\sqrt{u}}$; 1/6

7. a. B's value is divided by 16.

  b. Z's value is divided by 2^P.

9. Given: $I(x) = k/x^2$ and $4 = I(6) = k/36$; thus $k = 144 \Rightarrow$ $I(8) = 144/64 = 2.25$ watts per square meter and $I(100) = 144/10000 = 0.0144$ watt per square meter.

11. $I(d) = k/d^2$ and thus $I(4)/I(7) = 49/16 = 3.06$. The light intensity will be more than 3 times as great.

13. a. The volume becomes 1/3 of what it was.

  b. The volume becomes $1/n$ of what it was.

  c. The volume is doubled.

  d. The volume becomes n times what it was.

15. a. $x = k \cdot \frac{y}{z}$

  b. Solving $4 = k \cdot \frac{16}{32}$ for k gives $k = 8$.

  c. $x = 8 \cdot 25/5 = 40$

17. a. Let L = wavelength of a wave and t = time between waves. Using the fact that the speed of the waves is directly proportional to the square root of the wave's length, we then have $t = \frac{L}{k \cdot \sqrt{L}} = \frac{\sqrt{L}}{k}$ and thus time is directly proportional to the square root of the wavelength of the wave.

  b. If the frequency of the waves on the second day is twice that of the first, then the waves will be four times as far apart as they were on the previous day.

19. a. $v = \frac{k}{L}$ **d.** decrease

  b. 24 in.   **e.** It was doubled in length.

  c. decrease

21. a. $H = k/P = k \cdot P^{-1}$, where k is a proportionality constant.

  b. The software gave $H = 87.19 \cdot P^{-0.99}$, which is very close. Note that the fit is quite good, as can be seen from the graph of the data and the best-fit graph as given in the accompanying diagram.

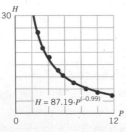

  c. The constant of proportionality is 87.19.

23. a. For a cylinder $V = \pi \cdot r^2 h$ where V = volume in cubic inches, h = height in inches, and r = radius in inches. Now $V = 13$ cubic inches. Thus $h = 13/(\pi \cdot r^2)$.

  b. Answers will vary. A reasonable domain in inches for this function is [1, 1.5], and the corresponding range in inches is approximately [1.84, 4.14]. Any smaller radius would make the can too tall and any larger radius would make the can too flat. Here is a graph:

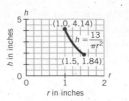

  c. Answers will vary. Selected are two points: (1.00, 4.14) and (1.5, 1.84). The first seems to give reasonable dimensions to the can. The second is less satisfactory. In actuality, a better answer is somewhere in between. If possible find a mini-coke can and measure it. The figures are below.

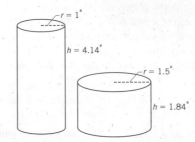

Section 7.6

Algebra Aerobics 7.6

1. a. $\frac{1}{x^2}$; $\frac{1}{(-2)^2} = \frac{1}{4}$ or 0.25

  b. $\frac{1}{x^3}$; $\frac{1}{(-2)^3} = -\frac{1}{8}$ or -0.125

  c. $\frac{4}{x^3}$; $\frac{4}{(-2)^3} = -\frac{1}{2}$ or -0.5

  d. $\frac{-4}{x^3}$; $\frac{-4}{(-2)^3} = \frac{-4}{-8} = \frac{1}{2}$ or 0.5

  e. $\frac{1}{2x^4}$; $\frac{1}{2(-2)^4} = \frac{1}{32} \approx 0.031$

  f. $\frac{-2}{x^3}$; $\frac{-2}{(-2)^3} = \frac{1}{4} = 0.25$

  g. $\frac{-2}{x^4}$; $\frac{-2}{(-2)^4} = -\frac{1}{8} = -0.125$

  h. $\frac{2}{x^4}$; $\frac{2}{(-2)^4} = \frac{1}{8} = 0.125$

2. a. Graph A: $y = \frac{a}{x^2}$ Graph B: $y = ax^3$

  b. Since the point (1, 4) lies on Graph A, $4 = \frac{a}{1^2} \Rightarrow a = 4 \Rightarrow y = \frac{4}{x^2}$.

  Since the point (2, 4) lies on Graph B, $4 = a(2)^3 \Rightarrow a = 0.5 \Rightarrow y = 0.5x^3$

3. a. $y = x^{-10}$ **b.** $y = x^{-11}$

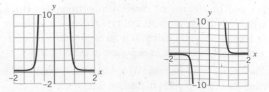

4. a. $y = x^{-2}$ and $y = x^{-3}$

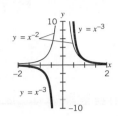

The graphs intersect at $(1, 1)$. As $x \to +\infty$ or $x \to -\infty$, both graphs approach the x-axis, but $y = x^{-3}$ is closer to the x-axis at each point after $x = 1$ and before $x = -1$.

When $x > 0$, as $x \to 0$, both graphs approach $+\infty$, but $y = x^{-3}$ is steeper.

When $x < 0$, as $x \to 0$, the graph of $y = x^{-2}$ approaches $+\infty$, and the graph of $y = x^{-3}$ approaches $-\infty$.

b. $y = 4x^{-2}$ and $y = 4x^{-3}$

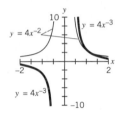

The graphs intersect at $(1, 4)$. As $x \to +\infty$ or $x \to -\infty$, both graphs approach the x-axis, but $y = 4x^{-3}$ is closer to the x-axis after $x = 1$ and before $x = -1$ than $y = 4x^{-2}$.

When $x > 0$, as $x \to 0$, both graphs approach $+\infty$.

When $x < 0$, as $x \to 0$, the graph of $y = 4x^{-2}$ approaches $+\infty$, and the graph of $y = 4x^{-3}$ approaches $-\infty$.

5. Graph A matches $f(x)$; Graph B matches $g(x)$; Graph C matches $h(x)$. The domain of all three graphs is $(-\infty, 0) \cup (0, +\infty)$. The range for Graphs A and B is $(0, +\infty)$ and for Graph C is $(-\infty, 0)$.

Exercises for Section 7.6

1. A table for r is given below.

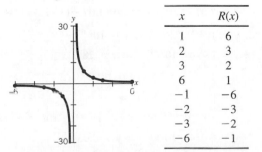

x	$R(x)$
1	6
2	3
3	2
6	1
-1	-6
-2	-3
-3	-2
-6	-1

The domain of the abstract function is all real numbers $x \neq 0$. For $x > 0$, as $x \to 0$ we have $R(x) \to +\infty$ and for $x < 0$, as $x \to 0$ we have $R(x) \to -\infty$.

3. a. Graphs of h and f:

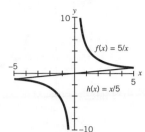

b. Answers will vary. The graphs are similar in that their domains extend from to $-\infty$ and $+\infty$, but omitting $x = 0$ for $f(x)$. They differ in many ways. The graph of h is rising, while the graph of f is always descending. The graph of f does not contain any point on either axis, but the graph of h goes through $(0, 0)$. The graph of f has both axes as its asymptotes; the graph of h does not have any asymptotes. The graph of h is a straight line; the graph of f has no straight parts at all. The graph of h is continuous (no breaking points); the graph of f is discontinuous at $x = 0$.

5. a. The graph of $y = x^2$ decreases for $x < 0$ and increases for $x > 0$ (as can be seen in the accompanying diagram).

The graph of $y = x^{-2}$ increases for $x < 0$ and decreases for $x > 0$ (as can also be seen from that diagram).

b. The two graphs intersect at $(-1, 1)$ and $(1, 1)$.

c. As x approaches $\pm\infty$, the graph of $y = x^2$ approaches $+\infty$. As x approaches $\pm\infty$, the graph of $y = x^{-2}$ approaches 0.

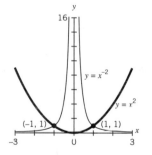

7. a. f goes with Graph A. Reason: $f(x)$ values must be positive; rapid decrease in the values of $f(x)$ when $x > 0$. The domain of f is $(-\infty, 0) \cup (0, +\infty)$. The range is $(0, +\infty)$

b. g goes with Graph C. Reason: There must be positive and negative values of $g(x)$. The domain of g is $(-\infty, 0) \cup (0, +\infty)$. The range is $(-\infty, 0) \cup (0, +\infty)$.

c. h goes with Graph D. Reason: $h(x)$ values must be positive; when $x > 0$ and $x \to +\infty$ a less rapid decrease in the values in graph D than in Graph A. The domain of h is $(-\infty, 0) \cup (0, +\infty)$. The range is $(0, +\infty)$.

d. j goes with Graph B. Reason: $j(x)$ values must all be negative. The domain of j is $(-\infty, 0) \cup (0, +\infty)$. The range is $(-\infty, 0)$.

9. a. $g(x) = 4x^{-3}$ **b.** $h(x) = \frac{1}{2}x^{-3}$ **c.** $j(x) = -3x^{-3}$

11. a. If $x > 1$, then $f(x) < g(x)$

b. If $0 < x < 1$, then $f(x) > g(x)$

c. If $x < 0$, then $f(x) < g(x)$

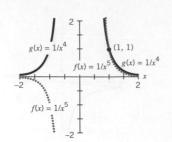

13. a. n is even

b. $k < 0$

c. $f(-1) < 0$

d. yes

e. 0

f. 0

g. The domain of f is all real numbers except 0; the range is $(-\infty, 0)$.

15. The graph of $f(x) = x^{-3}$

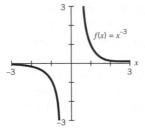

a. g has -1; h has $1/2$ and k has -2.

b. g's graph is a reflection of f's across the x-axis.

c. j's graph is a stretch and a reflection of f's across the x-axis.

d. h's graph is a compression of f's.

17. $g(x) = -4x^{-2}$

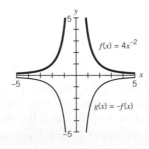

19. a. $f(-x) = \frac{1}{(-x)^4} = \frac{1}{x^4}$ and $-f(x) = -\frac{1}{x^4}$

b. $2f(x) = \frac{2}{x^4}$ and $f(2x) = \frac{1}{(2x)^4} = \frac{1}{16x^4}$

c. $g(-x) = \frac{1}{(-x)^5} = -\frac{1}{x^5}$ and $-g(x) = -\frac{1}{x^5}$

d. $2g(x) = \frac{2}{x^5}$ and $g(2x) = \frac{1}{(2x)^5} = \frac{1}{32x^5}$

e. $h(x) = -f(x) = -\frac{1}{x^4}$

f. $k(x) = -g(x) = -\frac{1}{x^5}$

21. a. $(0, 0)$ and $(1/2, 1)$ **d.** $(1, 1)$

b. $(0, 0)$ and $(1, 4)$ **e.** $(1, 4)$

c. $\left(\sqrt{\tfrac{1}{2}}, 2\right)$ and $\left(-\sqrt{\tfrac{1}{2}}, 2\right)$

Section 7.7

Algebra Aerobics 7.7

1. Let $Y = \log y$ and let $X = \log x$.

a. $\log y = \log(3 \cdot 2^x) \Rightarrow \log y = \log 3 + \log 2^x \Rightarrow$
$\log y = \log 3 + x \log 2 \Rightarrow$
$\log y = \log 3 + (\log 2)x \Rightarrow Y = 0.477 + 0.301x$

b. $\log y = \log(4x^3) \Rightarrow \log y = \log 4 + \log x^3 \Rightarrow$
$\log y = \log 4 + 3 \log x \Rightarrow Y = 0.602 + 3X$

c. $\log y = \log(12 \cdot 10^x) \Rightarrow \log y = \log 12 + \log 10^x \Rightarrow$
$\log y = \log 12 + x \log 10 \Rightarrow$
$\log y = \log 12 + (\log 10)x \Rightarrow$
$Y = 1.079 + x$ (since $\log 10 = 1$)

d. $\log y = \log(0.15x^{-2}) \Rightarrow \log y = \log 0.15 + \log x^{-2} \Rightarrow$
$\log y = \log 0.15 - 2 \log x \Rightarrow Y = -0.824 - 2X$

2. a. $y = 10^{0.067 + 1.63 \log x} \Rightarrow y = 10^{0.067} 10^{1.63 \log x} \Rightarrow$
$y = 10^{0.067} 10^{\log x^{1.63}} \Rightarrow y \approx 1.167 x^{1.63}$

b. $y = 10^{2.135 + 1.954x} \Rightarrow y = 10^{2.135} 10^{1.954x} \Rightarrow$
$y = 10^{2.135}(10^{1.954})^x \Rightarrow y \approx 136.458 \cdot 89.95^x$

c. $y = 10^{-1.963 + 0.865x} \Rightarrow y = 10^{-1.963} 10^{0.865x} \Rightarrow$
$y = 10^{-1.963}(10^{0.865})^x \Rightarrow y \approx 0.011 \cdot 7.328^x$

d. $y = 10^{0.247 - 0.871 \log x} \Rightarrow y = 10^{0.247} 10^{-0.871 \log x} \Rightarrow$
$y = 10^{0.247} 10^{\log x^{-0.871}} \Rightarrow y \approx 1.766 \cdot x^{-0.871}$

3. (a), (d), and (f) represent power functions. The graphs of their equations on a log-log plot are linear.

(b), (c), and (e) represent exponential functions. The graphs of their equations on a semi-log plot are linear.

a. $y = 10^{\log 2 + 3 \log x} \Rightarrow y = 10^{\log 2} \cdot 10^{3 \log x} \Rightarrow$
$y = 2 \cdot 10^{\log x^3} \Rightarrow y = 2x^3$

b. $y = 10^{2 + x \log 3} \Rightarrow y = 10^2 \cdot 10^{x \log 3} \Rightarrow$
$y = 100 \cdot 10^{\log 3^x} \Rightarrow y = 100 \cdot 3^x$

c. $y = 10^{0.031 + 1.25x} \Rightarrow y = 10^{0.031} \cdot 10^{1.25x} \Rightarrow$
$y = 1.07 \cdot (10^{1.25})^x \Rightarrow y = 1.07 \cdot (17.78)^x$

d. $y = 10^{2.457 - 0.732 \log x} \Rightarrow y = 10^{2.457} \cdot 10^{-0.732 \log x} \Rightarrow$
$y = 286.4 \cdot 10^{\log x^{-0.732}} \Rightarrow y = 286.4x^{-0.732}$

e. $y = 10^{-0.289 - 0.983x} \Rightarrow y = 10^{-0.289} \cdot 10^{-0.983x} \Rightarrow$
$y = 0.51 \cdot (10^{-0.983})^x \Rightarrow y = 0.51 \cdot (0.104)^x$

f. $y = 10^{-1.47 + 0.654 \log x} \Rightarrow y = 10^{-1.47} \cdot 10^{0.654 \log x} \Rightarrow$
$y = 0.034 \cdot 10^{\log x^{0.654}} \Rightarrow y = 0.034x^{0.654}$

4. a. i. $y = 4x^3$ is a power function;

ii. $y = 3x + 4$ is a linear function;

iii. $y = 4 \cdot 3^x$ is an exponential function.

Function	b. Type of Plot on Which Graph of Function Would Appear as a Straight Line	c. Slope of Straight Line
$y = 3x + 4$	Standard linear plot	$m = 3$
$y = 4 \cdot (3^x)$	Semi-log	$m = \log 3$
$y = 4 \cdot x^3$	Log-log	$m = 3$

5. This is a log-log plot, so the slope of a straight line corresponds to the exponent of a power function.

Younger ages: slope $= 1.2 \Rightarrow$ original function is of the form $y = a \cdot x^{1.2}$, where $x =$ body height in cm and $y =$ arm length in cm

Older ages: slope $= 1.0 \Rightarrow$ original function is of the form $y = a \cdot x^1$, where $x =$ body height in cm and $y =$ arm length in cm.

6. Graph *A*: Exponential function because the graph is approximately linear on a semi-log plot

Graph *B*: Exponential function because the graph is approximately linear on a semi-log plot

Graph *C*: Power function because the graph is approximately linear on a log-log plot

7. a. Estimated surface area is $10^4 = 10,000$ square cm.

b. Since $S \propto M^{2/3}$, then $S = kM^{2/3}$, where $\log k$ is the vertical intercept of the best-fit line on a log-log plot. In this case the vertical intercept $= \log 10$, so $k = 10$. So the equation is $S = 10M^{2/3}$. When $M = 70,000$g, then

$S = 10 \cdot (70,000)^{2/3} = 10 \cdot 1700 = 17,000 \text{ cm}^2$.

c. Our estimated surface area was 7000 square cm lower than the calculated area.

d. Since 1 cm $= 0.394$ in, then 1 cm$^2 = (0.394 \text{ in})^2 = 0.155 \text{ in}^2$. So $17,000 \text{ cm}^2 = (17,000)(0.155) = 2635 \text{ in}^2$. Since 1 kg $= 2.2$ lb, a weight of 70 kg translated to pounds is 70 kg $= (70 \text{ kg})\left(2.2 \frac{\text{lb}}{\text{kg}}\right) = 154$ lb.

8. a. power

b. $y = ax^{-0.23}$

c. When body mass increases by a factor of 10, heart rate is multiplied by $10^{-0.23}$ or about 0.59.

9. Estimated rate of heat production is somewhere between 10^3 and 10^4 kilocalories per day—roughly between 2000–3000 kcal/day.

Exercises for Section 7.7

1. When $x = 4$, then $100(5)^4 = 62,500$ and $100(4)^5 = 102,400$. So the exponential function $y = 100(5)^x$ in part (a) matches Graph *A*. Since an exponential function is a straight line on a semi-log plot, then the function also matches Graph *D*.

The power function $y = 100x^5$ in part (b) matches Graphs *B* and *C*, since a power function (Graph *C*) is a straight line on a log-log plot (Graph *B*).

3. For 1920–2000, the graph is a nearly straight line with negative slope on a semi-log scale. This suggests that an exponential function with a negative exponent would be the best model. For 2000–2006, the graph is flat suggesting a linear function with 0 slope.

5. a. $y = 4x^2$ **c.** $y = 1.25x^4$

b. $y = 2x^4$ **d.** $y = 0.5x^3$

7. a.

x	$y = x^5$
1	1
2	32
3	243
4	1024
5	3125
6	7776

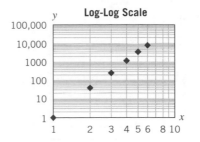

b.

x	$\log(x)$	$y = x^5$	$\log(y)$
1	0	1	0
2	0.3010	32	1.5051
3	0.4771	243	2.3856
4	0.6021	1024	3.0103
5	0.6990	3125	3.4949
6	0.7782	7776	3.8908

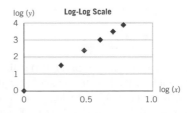

c. The graphs are exactly the same. The points lie on straight line and are plotted on equivalent scales.

9. a.

Species	Adult Mass in Grams	Egg Mass in Grams	Egg/Adult Ratio
Ostrich	113,380.0	1700.0	0.015
Goose	4,536.0	165.4	0.036
Duck	3,629.0	94.5	0.026
Pheasant	1,020.0	34.0	0.033
Pigeon	283.0	14.0	0.049
Hummingbird	3.6	0.6	0.167

Answers will vary; notable is the fact that the egg/adult mass ratio for the hummingbird is very high and it is very low for the ostrich. The other ratios are not far apart from each other.

b. The plots (where x = adult mass, y = egg mass (both in grams)) are, respectively,

   i. y vs. x

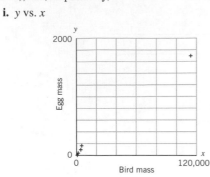

   ii. $\log y$ vs. x

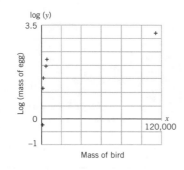

   iii. $\log y$ vs. $\log x$

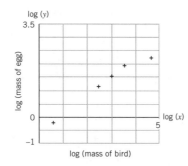

c. The log vs. log scatter diagram in part b (iii) is the most linear. Using technology the best-fit straight-line equation is: $\log(y) = \log(0.1981) + 0.7709 \cdot \log(x)$ or, in linear equation form: $Y = -0.703 + 0.7709X$ where $Y = \log(y)$ and $X = \log(x)$.

d. Regrouping, using the laws of logarithms, we get: $\log(y) = \log(0.1981 \cdot x^{0.7709})$. Thus, $y = 0.1981 \cdot x^{0.7709}$, where x = mass of the adult and y is the corresponding mass of the egg, both measured in grams.

e. If $x = 12.7$ kg (or 12,700 grams) for the weight of an adult turkey, then $y \approx 289$ grams is the predicted weight of its egg.

f. If the egg weighs 2 grams, then the adult bird is predicted to have an adult weight ≈ 20.1 grams.

11. a. Since l_1 and l_2 are both straight lines on a semi-log plot, they both represent exponential functions. Since the lines are parallel lines they have the same slope but different vertical intercepts. Thus their equations can be written as $\log(y) = mx + \log(b_1)$ and $\log(y) = mx + \log(b_2)$, with $\log(b_1) \neq \log(b_2)$. Solving each for y in terms of x, we have $y = b_1 \cdot 10^{mx}$ and $y = b_2 \cdot 10^{mx}$ with $b_1 \neq b_2$. Thus they have the same power of 10 but differ in their y-intercepts.

b. Since l_3 and l_4 are both straight lines on a log-log plot, they represent power functions. Since the lines are parallel they have the same slope but different vertical intercepts. Thus the equation of l_3 is $\log(y) = m \cdot \log(x) + \log(b_3)$ and the equation of l_4 is $\log(y) = m \cdot \log(x) + \log(b_4)$, with $\log(b_3) \neq \log(b_4)$. Solving for y in terms of x in each gives: $y = b_3 x^m$ and $y = b_4 x^m$ with $b_3 \neq b_4$. Thus they have the same power of x but differ in their coefficients of x.

13. a.

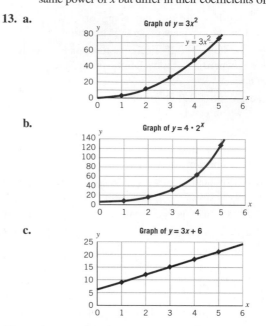

c.

15. a. A power function seems appropriate.

b. The log-log graph is linear and we get $\log(O) = 0.75 \cdot \log(m) + c$, where O stands for oxygen consumption and m stands for body mass. Thus $O = k \cdot m^{0.75}$, a power function, where $k = 10^c$.

c. A slope of 3/4 means that if the body mass, m, is multiplied by 10^4 (=10,000), then the oxygen consumption, O, is multiplied by 10^3 (=1000). If m is multiplied by 10, then O is multiplied by $10^{3/4} \approx 5.6$.

17. a.

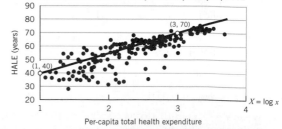

b. Two estimated points on the best-fit line are $(1, 40)$ and $(3, 70) \Rightarrow$ slope $= \frac{(70 - 40)}{(3 - 1)} = 15$. So the equation is of the form $y = b + 15X$. Substituting in $(1, 40)$ we get $40 = b + (15 \cdot 1) \Rightarrow 40 = b + 15 \Rightarrow b = 25$. So the equation is $y = 25 + 15X$. Substituting $\log x$ for X, we finally get $y = 25 + 15 \log x$. So HALE is a logarithmic function of per-capita health expenditure.

c. A logarithmic model makes sense, since there will always be a ceiling in life expectancy, so the results of adding funding will increase HALE but at a slower and slower rate.

d. It suggests that the functions are log functions.

Ch. 7: Check Your Understanding

1. True
2. False
3. False
4. False
5. True
6. False
7. False
8. True
9. True
10. False
11. False
12. True
13. True
14. True
15. False
16. False
17. False
18. True
19. True
20. False

21. True

22. Possible answer: $f(x) = 4x^8$

23. $g(x) = 3.2x^4$

24. Possible answer: $f(x) = -x^4$

25. Possible answer: $h(m) = m^{12}$

26. Possible answer: $m = \dfrac{k}{Q^3}$ for k a non-zero constant

27. Possible answer: $y = 3x^5$

28. Possible answer: $k(x) = \dfrac{-1}{x^3}$

29. Possible answer: $T(m) = 3m^4$

30. False
31. False
32. False
33. False
34. False
35. True
36. True
37. False
38. True
39. True
40. False

Ch. 7 Review: Putting It All Together

1. **a.** $V(x) = x(2x)(2x) = 4x^3$. So $V(X) = 4X^3$ and $V(2X) = 4(2X)^3 = 8(4X^3) = 8V(X) \Rightarrow$ the volume is multiplied by 8.

   b. The surface area consists of the two top and bottom sides, which are both $(2x)(2x) = 4x^2$ in area, and the other four vertical sides, each $(x)(2x) = 2x^2$ in area. So the surface area $S(x) = 2(4x^2) + 4(2x^2) = 16x^2$. Hence $S(X) = 16X^2$ and $S(2X) = 16(2X)^2 = 4(16X^2) = 4S(X) \rightarrow$ when the value of x is doubled, the surface area is multiplied by 4. So the volume grows faster than the surface area.

   c. The ratio of (surface area)/volume – $R(x) = S(x)/V(x) = (16x^2)/(4x^3) = 4/x$. If you double the value of x, from X to $2X$, since $R(X) = 4/X$, then $R(2X) = 4/(2X) = (1/2)(4/X) = (1/2)R(X)$. So if the value of x doubles, $R(x)$, the ratio of (surface area)/volume, is multiplied by 1/2—or equivalently, cut in two. This is again confirmation that the volume grows faster than the surface area.

   d. In general, as x increases, $R(x)$—the ratio of surface area to volume—will decrease. So as the rectangular solid gets larger, there will be relatively less surface area compared with the volume.

3. Think of the cake as a cylinder, with a volume of $\pi r^2 h$ where $r =$ the radius of the cake and $h -$ the height. The radii of the 10″ and 12″ cakes are 5″ and 6″, respectively. If the

corresponding heights are h_1 and h_2, then the respective cake volumes are $25\pi h_1$ and $36\pi h_2$ cubic inches. The site claims that the volume of the 12″ cake is twice that of the 10″ cake (6 quarts vs. 3 quarts). That implies that $36\pi h_2 = 2(25\pi h_1) \Rightarrow h_2/h_1 = 50/36 \approx 1.4 \Rightarrow h_2$ is about 40% larger than h_1. So for the 12″ cake to have twice the volume of the 10″ cake, the height of the 12″ cake would have to be 40% higher than the height of the 10″ cake—which seems rather unlikely.

5. **a.** $V_F(t) = 3t$; domain is $0 \le t \le 5$; represents direct proportionality.

   b. $V_D(t) = 15 - 0.5t$; domain is $0 \le t \le 30$; does not represent inverse proportionality.

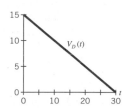

7. **a.** The friend is wrong.

   b. If the height is increased by 50% (from 4 feet to 6 feet) or, equivalently, multiplied by 1.5 (since $1.5 \cdot 4' = 6'$), the width $(2')$ and the length $(2')$ must both also be multiplied by 1.5, giving $3'$ for the new width and length. Hence the new dimensions are $6' \times 3' \times 3'$, which give a volume of 54 cubic feet. The volume of the original block of ice $4' \times 2' \times 2' = 16$ cubic feet. Since (new volume)/(old volume) $= 54/16 = 3.375$, the new volume $= 3.375 \cdot$ (old volume) or, equivalently, a 237.5% increase in volume, almost five times what your friend predicted.

   c. 10% of 54 cubic feet $= 0.10 \cdot 54 = 5.4$ cubic feet. So the volume of the melted sculpture would be $54 - 5.4 = 48.6$ cubic feet. If h is the height of the melted sculpture, then the width and length are both $0.5h$. So we have $48.6 = (0.5h)^2h = (0.5)^2h^3 \Rightarrow h^3 = (48.6)/(0.5)^2 = 194.4 \Rightarrow h \approx 5.8$ feet as the height of the melted sculpture.

9. **a.** $t = kd/r$ (the k may be needed for unit conversions)

   b. $D_1 = kD_2$

   c. $R = kV^2/W$

11. Graph C is symmetric across the y-axis. Graphs A, B, and D are rotationally symmetric about the origin. None of them has an asymptote.

13. **a.** $P(n) = 450/n$ for integer values of n between 0 and, say, 10.

   b. $P(2) = 450/2 = \$225$/person; $P(5) = 450/5 = \$90$/person. The more people, the lower the cost per person.

c. If you take the function $P(n)$ out of context and treat n as any real number ($\neq 0$), then the following graph shows the result.

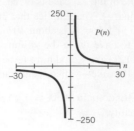

The abstract function is asymptotic to both the horizontal and vertical axes.

15. a. From the previous problem, we have $I = P/d^2$ where I is measured in foot-candles, P in candlepower, and d in feet. Setting $I = 4000$ foot-candles and $d = 3$ feet, we have $4000 = P/(3)^2 \Rightarrow P = 36{,}000$ candlepower.

b. From part (a) we know that the lamp has 36,000 candlepower. Setting $P = 36{,}000$ and $I = 2000$ foot-candles, we have $2000 = 36{,}000/d^2 \Rightarrow d^2 = 18 \Rightarrow d \approx 4.24$ feet or about 51 inches above the operating surface.

17. Graphs B and C are symmetric about the y-axis. Graph D is symmetric about the origin. Graphs A, B, and D all have the x-axis as a horizontal asymptote. Graphs B and D also have the y-axis as a vertical asymptote.

19. a. Y is a linear function of X.

b. Using the estimated points $(0, 0.3)$ and $(1, 3.3)$, we have $Y = 0.3 + 3X$.

c. Substituting $Y = \log y$ and $X = \log x$ in $Y = 0.3 + 3X$, we have $\log y = 0.3 + 3 \log x \Rightarrow \log y = 0.3 + \log(x^3)$. Rewriting using powers of 10, we have $10^{\log y} = 10^{(0.3+\log(x^3))} \Rightarrow y = 10^{0.3} \cdot 10^{\log x^3} \Rightarrow y \approx 2x^3$. Hence y is a power function and y is directly proportional to x^3.

21. To find the relationship, sketch a best-fit line to the data, where $m = $ mass (in kg) and $v = $ optimal flying speed (in meters per second). Since both axes use a logarithmic scale, we can replace the labels on the axes with $\log(m) = M$ and $\log(v) = V$, respectively. The horizontal labels will now read $-5, -4, -3, -2\ -1, 0\ (= \log 1)$, which is now the beginning of the vertical axis, followed by 1, 2, 3, 4, 5, 6; and on the vertical axis 0, 1, 2, and $2.5 \approx \log 300$.

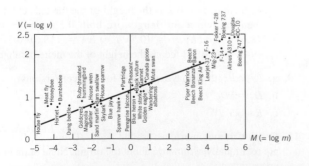

Two points on the line with coordinates of the form (M, V) are $(-1, 1)$ and $(5, 2)$. The slope is then $(2 - 1)/(5 - (-1)) = 1/6$. The vertical intercept is approximately at $(0, 1.2)$. So the

equation of the line would be $V = 1.2 + (1/6)M$. Substituting in, we get $\log(v) = 1.2 + (1/6)\log(m) = 1.2 + \log(m^{1/6})$. Rewriting both sides as powers of 10 and simplifying, we get: $v = 10^{(1.2 + \log(m^{1/6}))} = 10^{1.2} \cdot 10^{\log(m^{1/6})} \approx 16m^{1/6}$. So $v \propto m^{1/6}$ $\Rightarrow$ the cruising speed v (in meters per second) is directly proportional to the 1/6 power of body mass m (in kilograms). So Professor Bejan seems to be correct.

CHAPTER 8

Section 8.1

Algebra Aerobics 8.1

1. a. The vertex at $(0, 0)$ is the minimum point. Since $a > 0$, the focal point is $\frac{1}{4a} = \frac{1}{4(3)} = \frac{1}{12}$ units above the vertex at $\left(0, \frac{1}{12}\right)$.

b. The vertex at $(0, 0)$ is the maximum point. Since $a < 0$, the focal point is $\frac{1}{4a} = \frac{1}{4(-6)} = \frac{-1}{24}$ units below the vertex at $\left(0, -\frac{1}{24}\right)$.

c. The vertex at $(0, 0)$ is the minimum point. Since $a > 0$, the focal point is $\frac{1}{4a} = \frac{1}{4(1/24)} = 6$ units above the vertex at $(0, 6)$.

d. The vertex at $(0, 0)$ is the maximum point. Since $a < 0$, the focal point is $\frac{1}{4a} = \frac{1}{4(-1/12)} = -3$ units below the vertex at $(0, -3)$.

2. a. A point on the rim is $(15, 10)$, so if $y = ax^2$, then $10 = a(15)^2$, so $a = 2/45$. Since $a > 0$, the focal point is $\frac{1}{4a} = \frac{1}{4(2/45)} = \frac{45}{8} = 5.625$ ft from the vertex at $(0, 0)$ on the back wall.

b. $y = \frac{2}{45}x^2$

c. You could not hear well if you were more than 15 ft on either side of the stage, because the sound would be traveling in straight lines from the parabolic wall. Also, the sound would not be good if the performer moved around and did not stay at the focal point.

3. a. $g(2) = 2^2 = 4$
$g(-2) = (-2)^2 = 4$
$g(0) = 0^2 = 0$
$g(z) = z^2$

b. $h(2) = -(2)^2 = -4$
$h(-2) = -(-2)^2 = -4$
$h(0) = 0^2 = 0$
$h(z) = -z^2$

c. $Q(2) = -(2)^2 - 3(2) + 1 = -9$
$Q(-2) = -(-2)^2 - 3(-2) + 1 = 3$
$Q(0) = -(0)^2 - 3(0) + 1 = 1$
$Q(z) = -(z)^2 - 3(z) + 1 = -z^2 - 3z + 1$

d. $m(2) = 5 + 2(2) - 3(2)^2 = -3$
$m(-2) = 5 + 2(-2) - 3(-2)^2 = -11$
$m(0) = 5 + 2(0) - 3(0)^2 = 5$
$m(z) = 5 + 2z - 3z^2$

e. $D(2) = -(2 - 3)^2 + 4 = 3$
$D(-2) = -((-2) - 3)^2 + 4 = -21$
$D(0) = -((0) - 3)^2 + 4 = -5$
$D(z) = -(z - 3)^2 + 4$

f. $k(2) = 5 - 2^2 = 1$
$k(-2) = 5 - (-2)^2 = 1$
$k(0) = 5 - 0^2 = 5$
$k(z) = 5 - z^2$

4. a. $f(x)$

    i. concave down

    ii. maximum

    iii. axis of symmetry: $x = 0$

    iv. vertex: $(0, 4)$

    v. approx. horizontal intercepts. $(-2, 0)$, $(2, 0)$
    vertical intercept: $(0, 4)$

b. $g(x)$

    i. concave up

    ii. minimum

    iii. axis of symmetry: $x = 1$

    iv. approx. vertex: $(1, -2.25)$

    v. horizontal intercepts: $(-2, 0)$, $(4, 0)$
    vertical intercept: $(0, -2)$

c. $h(x)$

    i. concave up

    ii. minimum

    iii. axis of symmetry: $x = 2$

    iv. vertex: $(2, -4)$

    v. horizontal intercepts: $(0, 0)$, $(4, 0)$
    vertical intercept: $(0, 0)$

d. $j(x)$

    i. concave down

    ii. maximum

    iii. axis of symmetry: $x = -4$

    iv. vertex: $(-4, -2)$

    v. no horizontal intercepts
    vertical intercept: $(0, -10)$

5. a. Equation is of the form $y = ax^2$ where $a > 0 \Rightarrow$ focal length $= \frac{1}{4a}$. When $x = 12$, focal length $=$ depth $\Rightarrow$ $\frac{1}{4a} = a(12)^2 \Rightarrow a^2 = \frac{1}{4 \cdot 12^2} \Rightarrow a = \frac{1}{2 \cdot 12} = \frac{1}{24}$.
So focal length $= \frac{1}{4(1/24)} = \frac{1}{(1/6)} = 6$

b. $a = \frac{1}{24} \Rightarrow y = \frac{1}{24}x^2$.

6. $f(x) = x^2 + 2x - 15$ $h(x) = 10x^2 - 80x + 150$
$g(x) = 2x^2 + 7x + 5$ $j(x) = 2x^2 - 18$

Exercises for Section 8.1

1. a. concave up; minimum; vertex at $(-1, -4)$; the line $x = -1$ is axis of symmetry; $(1, 0)$ and $(-3, 0)$ are horizontal intercepts; $(0, -3)$ is vertical intercept.

b. concave down; maximum; vertex at $(2, 9)$; the line $x = 2$ is axis of symmetry; $(-4, 0)$ and $(8, 0)$ are horizontal intercepts; $(0, 8)$ is vertical intercept.

c. concave down; maximum; vertex at $(-5, 3)$; the line $x = -5$ is axis of symmetry; $(-8, 0)$ and $(-2, 0)$ are horizontal intercepts; $(0, -5)$ is vertical intercept. [Students may have a different y value for vertical intercept.]

3. The graphs of the three functions are in the accompanying diagram.

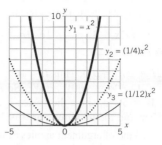

a. For y_1: $(0, 1/4)$; for y_2: $(0, 1)$; for y_3: $(0, 3)$

b. The graphs of all three functions are concave-up parabolas with vertex at $(0, 0)$, but y_1 rises faster than y_2 and y_3.

c. Farther from the vertex and the graphs get wider.

5. a. $y = (1/24)x^2$ **c.** $y = (3/4)x^2$

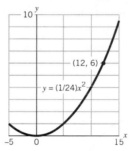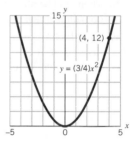

b. $y = -(1/24)x^2$ **d.** $y = 16x^2$

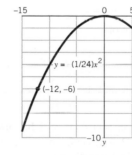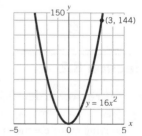

7. a. $y = (1/16)x^2$ **c.** $y = 4x^2$
   b. $y = -(1/32)x^2$ **d.** $y = 6x^2$

9. a. -8 and 12 **b.** 10 and 14 **c.** 2 and -14

11. a. $y = (1/6)x^2$

b. $5 = (1/6)x^2$ implies $x = \pm\sqrt{30}$ and thus the reflector is $2\sqrt{30} \approx 10.95$ inches wide.

13. Assuming that we rotate the parabolic model to become concave up:

a. The focus is at $(0, 1.25)$.

b. $y = 0.2x^2$

c. $2.5 = 0.2x^2$ gives $x = \pm\sqrt{12.5}$; thus the reflector should be $2\sqrt{12.5} \approx 7.07$ inches wide.

Ch. 8

15. Answers will vary. The general trend is that the smaller the focal length, the bigger the value of a and thus the narrower the opening of the parabola, since focal length $= \left|\frac{1}{4a}\right|$ when the parabola has the equation $y = ax^2$.

17. a. $f(x) = x^2 + 2x - 3$ **c.** $H(z) = -z^2 - z + 2$

 b. $P(t) = t^2 - 3t - 10$

19. a. $P = 2L + 2W = 200$; Thus $L = 100 - W$ and $A = LW = (100 - W) \cdot W = 100W - W^2$. So A has its maximum when $W = 50$ m, since the vertex is at $(50, 2500)$. Thus $L = 50$ m and the maximum area is 2500 m^2.

 b. The same kind of argument applies as in (a). $P = 2L + 2W$; thus $L = (P/2) - W$ and $A = LW = W(P/2 - W) = (P/2) \cdot W - W^2$, and this has its vertex at $(P/4, (P/4)^2)$ and the maximum area is $(P/4)^2$ m^2. Thus the dimensions are $P/4$ by $P/4$, a square.

21. a. True   **c.** False   **e.** False

 b. False   **d.** True

23. a. $R(n) = (1250 + 100n) \cdot (50 - 2n)$

 b. There will be no apartments rented when $n = 25$. Thus the domain is $0 \le n \le 25$.

 c. The graph is given in the accompanying diagram. From inspection, the practical maximum occurs when $n = 6$. At that value $R = \$70,300$.

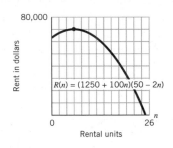

Section 8.2

Algebra Aerobics 8.2a

1. $s(x) = 2x^2 + 2$ is narrower than $r(x) = x^2 + 2$ because, looking at the coefficient of x^2, $2 > 1$. Both have a vertical intercept at $(0, 2)$ since $c = 2$ in both equations.

2. $h(t) = t^2 + 5$ is concave up since $a = 1$; $k(t) = -t^2 + 5$ is concave down since $a = -1$. They have the same shape because $|a| = 1$ in both. They both cross the vertical axis at 5 since $c = 5$ in both equations.

3. Both are concave down because both have $a < 0$. Both have a vertex at $(0, 0)$ since b and $c = 0$ in both equations. $g(z)$ is flatter than $f(z)$ because $|-0.5| < |-5|$.

4. They have the same shape and are concave up since both have $a = 1$. $g(x)$ is six units higher than $f(x)$ because the vertical intercepts are $c = 8$ and $c = 2$.

5. They have the same shape and are concave down since $a = -3$ in both equations. $g(t) = -3t^2 + t - 2$ is three units higher than $f(t) = -3t^2 + t - 5$ since $c = -2$ is three units vertically up from $c = -5$.

6. a. $y_1 = 3x^2 + 5$ **b.** $y_2 = \frac{1}{3}x^2 - 2$ **c.** $y_3 = -2x^2 + 4$

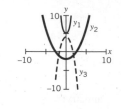

7. a. $g(x) = 3x^2$ **c.** $j(x) = 1/2x^2$

 b. $h(x) = -5x^2$ **d.** $k(x) = -x^2$

8. a. $y = 5x^2 - 2$ **c.** $y = -0.5x^2 - 4.7$

 b. $y = x^2 + 3$ **d.** $y = x^2 - 71$

Algebra Aerobics 8.2b

1. a. vertex is $(0, 0)$ **c.** vertex is $(2, 0)$

 b. vertex is $(-3, 0)$

All vertices lie on the x-axis. Vertex of (b) is three units to the left of the vertex of (a). Vertex of (c) is two units to the right of vertex of (a). All have same shape and are concave up.

2. a. vertex is $(0, 0)$ **c.** vertex is $(-4, 0)$

 b. vertex is $(1, 0)$

All vertices lie on the x-axis. The vertex of (b) is one unit to the right of the vertex of (a). The vertex of (c) is four units to the left of vertex of (a). All have the same shape and are concave up.

3. a. vertex is $(0, 0)$ **c.** vertex is $(0.9, 0)$

 b. vertex is $(-1.2, 0)$

All of the vertices lie on the horizontal axis. The vertex of (b) is 1.2 units to the left of the vertex of (a). The vertex of (c) is 0.9 units to the right of the vertex of (a). All have the same shape and are concave down.

4. $y = a(x - h)^2 + k$. The value of $h = 2$ in (b), (c), and (d), so the vertices of those parabolas are two units to the right of the vertex of (a), where $h = 0$. All are concave up with the same shape since $a = 1$ in all four equations. (c) is four units above (a) and (b); (d) is three units below (a) and (b) since $k = 0$ in (a) and (b), $k = 4$ in (c), and $k = -3$ in (d).

5. $y = a(x - h)^2 + k$. All are concave down with the same shape since $a = -1$ in all four equations. (b), (c), and (d) have vertices three units to the left of vertex of (a), since $h = -3$ in those equations, and $h = 0$ in (a). $k = 0$ in (a) and (b), but $k = -1$ in (c), so (c) is one unit below (a) and (b). $k = 4$ in (d), so (d) is four units above (a) and (b).

6. a. $g(x) = 3(x - 2)^2 - 1$, the vertex is $(2, -1)$

 b. $h(x) = -2(x + 3)^2 + 5$, the vertex is $(-3, 5)$

 c. $j(x) = \frac{1}{5}(x + 4)^2 - 3.5$, the vertex is $(-4, -3.5)$

 d. $k(x) = -(x - 1)^2 + 4$, the vertex is $(1, 4)$

7. a. The vertex at $(3, -4)$ is a minimum.

 b. The vertex at $(-1, 5)$ is a maximum.

 c. The vertex at $(4, 0)$ is a maximum.

 d. The vertex at $(0, -7)$ is a minimum.

Ch. 8

Exercises for Section 8.2

1. a. downward; the coefficient of x^2 is negative.

  b. upward; the coefficient of t^2 is positive.

  c. upward; the coefficient of x^2 is positive.

  d. downward; the coefficient of x^2 is negative.

3. The graphs of the four functions are given with their labels in the diagram below.

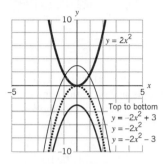

5. a. The compression factor is 0.3, the vertex is at $(1, 8)$.

  b. The expansion or stretch factor is 30 and the vertex is at $(0, -11)$.

  c. The compression factor is 0.01 and the vertex is at $(-20, 0)$.

  d. The expansion or stretch factor is -6 and the vertex is at $(1, 6)$.

7. a. $(3, 5)$; maximum

  b. $(-1, 8)$; minimum

  c. $(-4, -7)$; maximum

  d. $(2, -6)$; minimum

9. a. $y = 1(x - 2)^2 - 4 = x^2 - 4x$

  b. $y = -(x - 4)^2 + 3 = -x^2 + 8x - 13$

  c. $y = 2(x + 3)^2 + 1 = -2x^2 - 12x - 17$

  d. $y = 0.5(x + 4)^2 + 6 = 0.5x^2 + 4x + 14$

11. a. $h(x) = -3(x - 4)^2 + 5$

  b. $(4, 5)$

  c. $(0, -43)$

13. a. $y = a(x - 2)^2 + 4$ and $7 = a(1 - 2)^2 + 4$ gives $a = 3$ and thus $y = 3(x - 2)^2 + 4 = 3x^2 - 12x + 16$.

  b. If $a > 0$, the graph of $y = a(x - 2)^2 - 3$ is concave up; it is concave down if $a < 0$.

15. a. $(4, -5)$; no

  b. 0.5

  c. -2

17. Derivation: from the data we have:

$y = a(x - 2)^2 + 3$; and $-1 = a(4 - 2)^2 + 3$

This implies that $a = -1$.

Thus $y = -(x - 2)^2 + 3$ is its equation.

Check: $-(4 - 2)^2 + 3 = -4 + 3 = -1$

This is confirmed in the accompanying graph.

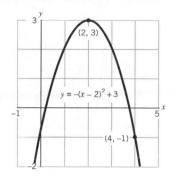

Section 8.3

Algebra Aerobics 8.3

1. a. vertex is $(0, -4)$.

  b. vertex is $(0, 6)$.

  c. vertex is $(0, 1)$.

2. a. vertex is $(0, 3)$

  b. vertex is $(0, 3)$

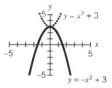

3. a. Vertex is $(0, 0)$. Parabola is concave down since $a = -3$. There is one x-intercept, $(0, 0)$, since the vertex is on the x-axis.

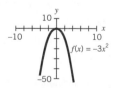

  b. Vertex is $(0, -5)$, which is below x-axis. Parabola is concave down since $a = -2$. So there are no x-intercepts.

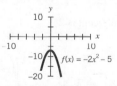

In parts (c) and (d), use the formula $h = \frac{-b}{2a}$ as the horizontal coordinate of the vertex (where $f(x) = ax^2 + bx + c$).

  c. $a = 1, b = 4 \Rightarrow h = \frac{-b}{2a} = \frac{-4}{2(1)} = -2$,

    $f(-2) = (-2)^2 + 4(-2) - 7 = 4 - 8 - 7 = -11$

The vertex is $(-2, -11)$. Parabola is concave up since $a\,(=1)$ is positive. There are two x-intercepts. They are approximately $(-5, 0)$ and $(1, 0)$.

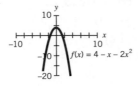

d. $f(x) = 4 - x - 2x^2 = -2x^2 - x + 4$; so $a = -2$,
$b = -1$; $h = \frac{-b}{2a} = \frac{-(-1)}{2(-2)} = \frac{1}{-4} = -\frac{1}{4}$

$f\left(-\frac{1}{4}\right) = 4 - \left(-\frac{1}{4}\right) - 2\left(-\frac{1}{4}\right)^2 = 4 + \frac{1}{4} - 2\left(\frac{1}{16}\right) =$
$4 + \frac{1}{4} - \frac{1}{8} = 4\frac{1}{8}$. So vertex is $\left(-\frac{1}{4}, 4\frac{1}{8}\right)$. Parabola is concave down since $a\,(=-2)$ is negative. There are two x-intercepts. They are approximately $(-2, 0)$ and $(1, 0)$.

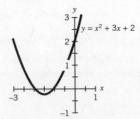

4. $y = ax^2 + bx + c$ compared with $y = x^2$ (where $a = 1$)
  a. $y = 2x^2 - 5 \Rightarrow a = 2$
    i. $a > 0 \Rightarrow$ minimum at the vertex
    ii. $|a| > 1$, so parabola is narrower than $y = x^2$.
  b. $y = 0.5x^2 + 2x - 10 \Rightarrow a = 0.5$
    i. $a > 0 \Rightarrow$ minimum at the vertex
    ii. $|a| < 1$, so parabola is broader than $y = x^2$.
  c. $y = 3 + x - 4x^2 \Rightarrow a = -4$
    i. $a < 0 \Rightarrow$ maximum at the vertex
    ii. $|a| > 1$, so parabola is narrower than $y = x^2$.
  d. $y = -0.2x^2 + 11x + 8 \Rightarrow a = -0.2$
    i. $a < 0 \Rightarrow$ maximum at the vertex
    ii. $|a| < 1$, so parabola is broader than $y = x^2$.

5. a. $a = 1$, $b = 3$; so horizontal coordinate of vertex is:
$\frac{-b}{2a} = \frac{-3}{2} = -1\frac{1}{2}$. If $x = -\frac{3}{2}, \Rightarrow$
$y = \left(-\frac{3}{2}\right)^2 + 3\left(-\frac{3}{2}\right) + 2 = \frac{9}{4} - \frac{9}{2} + 2 =$
$2\frac{1}{4} - 4\frac{2}{4} + 2 = -\frac{1}{4}$. So vertex is $\left(-1\frac{1}{2}, -\frac{1}{4}\right)$.

Vertical intercept is 2.

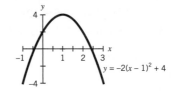

b. $a = 2$, $b = -4$, so horizontal coordinate of vertex is
$\frac{-b}{2a} = \frac{-(-4)}{2(2)} = 1$. $f(1) = 2 - 4 + 5 = 3$, so vertex is $(1, 3)$.
Since vertex is above x-axis, and $a\,(=2)$ is positive; the parabola is concave up, it does not cross x-axis, so no horizontal intercepts. Vertical intercept is 5.

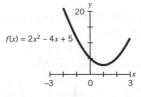

c. $a = -1$, $b = -4$; $\Rightarrow \frac{-b}{2a} = \frac{-(-4)}{2(-1)} = \frac{4}{-2} = -2$
$g(-2) = -(-2)^2 - 4(-2) - 7 = -4 + 8 - 7 = -3$, so vertex is $(-2, -3)$. Since vertex is below t-axis and $a\,(=-1)$ is negative, the parabola is concave down; it does not cross t-axis, so no horizontal intercepts. Vertical intercept $= -7$.

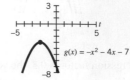

6. a. Vertex is $(-5, -11)$. y-intercept at $x = 0 \Rightarrow$
$y = 0.1(5)^2 - 11 = 0.1(25) - 11 = 2.5 - 11 = -8.5$.
So y-intercept is -8.5.

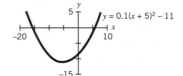

b. Vertex is $(1, 4)$. y-intercept at $x = 0 \Rightarrow y = -2(-1)^2 + 4 = -2 + 4 = 2$. So y-intercept is 2.

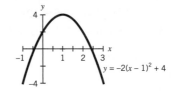

7. a. $(x + 3)^2 - 9$ **c.** $(x - 15)^2 - 225$
   b. $(x - 5)^2 - 25$ **d.** $\left(x + \frac{1}{2}\right)^2 - \frac{1}{4}$

8. a. $f(x) = x^2 + 2x - 1 = x^2 + 2x + (1 - 1) - 1$
    $= (x^2 + 2x + 1) + (-1 - 1) \Rightarrow$
    $f(x) = (x + 1)^2 - 2$
    vertex $(-1, -2)$; stretch factor 1
  b. $j(z) = 4z^2 - 8z - 6 = 4(z^2 - 2z) - 6$
    $= 4(z^2 - 2z + 1) - 4(1) - 6 \Rightarrow$
    $j(z) = 4(z - 1)^2 - 10$
    vertex $(1, -10)$; stretch factor 4
  c. $h(x) = -3x^2 - 12x = -3(x^2 + 4x)$
    $= -3(x^2 + 4x + 4) + 3(4) \Rightarrow$
    $h(x) = -3(x + 2)^2 + 12$
    vertex $(-2, 12)$; stretch factor -3

Ch. 8

d. $H(t) = -16(t^2 - 6t) \Rightarrow h(t) = -16(t^2 - 6t + 9) + 16(9)$
$$\Rightarrow h(t) = -16(t - 3)^2 + 144$$
vertex $(3, 144)$; stretch factor -16

e. $Q(t) = -4.9(t^2 + 20t) + 200$
$$\Rightarrow Q(t) = -4.9(t^2 + 20t + 10^2) + 200 + 4.9(10^2)$$
$$\Rightarrow Q(t) = -4.9(t + 10)^2 + 690$$
vertex $(-10, 690)$; stretch factor -4.9

9. a. $y = 2\left(x - \frac{1}{2}\right)^2 + 5 = 2\left(x - \frac{1}{2}\right)\left(x - \frac{1}{2}\right) + 5$
$$= 2\left(x^2 - x + \frac{1}{4}\right) + 5 = 2x^2 - 2x + \frac{1}{2} + 5 \Rightarrow$$
$$y = 2x^2 - 2x + 5\frac{1}{2}$$
vertex $(1/2, 5)$; stretch factor 2

b. $y = -\frac{1}{3}(x + 2)^2 + 4 = -\frac{1}{3}(x + 2)(x + 2) + 4$
$$= -\frac{1}{3}(x^2 + 4x + 4) + 4$$
$$= -\frac{1}{3}x^2 - \frac{4}{3}x - \frac{4}{3} + 4 \Rightarrow$$
$$y = -\frac{1}{3}x^2 - \frac{4}{3}x + 2\frac{2}{3}$$
vertex $(-2, 4)$; compression factor $-1/3$

c. $y = 10(x^2 - 10x + 25) + 12$
$$\Rightarrow y = 10x^2 - 100x + 250 + 12$$
$$\Rightarrow y = 10x^2 - 100x + 262$$
vertex $(5, 12)$; stretch factor 10

d. $y = 0.1(x^2 + 0.4x + 0.04) + 3.8$
$$\Rightarrow y = 0.1x^2 + 0.04x + 3.804$$
vertex $(-0.2, 3.8)$; compression factor 0.1

10. a. $y = x^2 + 6x + 7 = x^2 + 6x + 9 - 9 + 7 \Rightarrow$
$$y = (x + 3)^2 - 2$$

b. $y = 2x^2 + 4x - 11 = 2(x^2 + 2x) - 11$
$$= 2(x^2 + 2x + 1) - 2(1) - 11 \Rightarrow$$
$$y = 2(x + 1)^2 - 13$$

Exercises for Section 8.3

1. a. By "completing the square,"
$$y = x^2 + 8x + 11 \Rightarrow y = x^2 + 8x + 16 - 16 + 11 \Rightarrow$$
$$y = (x + 4)^2 - 5$$
So vertex is $(-4, -5)$.

b. Using the formula,
$$y = 3x^2 + 4x - 2 \Rightarrow a = 3, b = 4, c = -2 \Rightarrow$$
$$h = \frac{-b}{2a} = \frac{-4}{6} = -\frac{2}{3}$$
When $x = \frac{-2}{3}$,
$$y = 3\left(-\frac{2}{3}\right)^2 + 4\left(\frac{-2}{3}\right) - 2 \Rightarrow$$
$$y = 3\left(\frac{4}{9}\right) - \frac{8}{3} - 2 = \frac{4}{3} - \frac{8}{3} - 2 =$$
$$-\frac{4}{3} - 2 = -1\frac{1}{3} \quad 2 = -3\frac{1}{3}, \text{ so vertex is at } \left(-\frac{2}{3}, -3\frac{1}{3}\right).$$
In vertex form, $y = 3\left(x + \frac{2}{3}\right)^2 - 3\frac{1}{3}$

3. a. The vertex is $(-1, 4) \Rightarrow h = -1, k = 4 \Rightarrow$
$$y = a(x - (-1))^2 + 4 \Rightarrow y = a(x + 1)^2 + 4$$
Passing through the point $(0, 2) \Rightarrow 2 = a(0 + 1)^2 + 4 \Rightarrow$
$a = -2 \Rightarrow$ the equation is $y = -2(x + 1)^2 + 4$.

b. The vertex is $(1, -3) \Rightarrow h = 1, k = -3 \Rightarrow$
$y = a(x - 1)^2 - 3$. Passing through the point $(-2, 0) \Rightarrow$
$0 = a(-2 - 1)^2 - 3 \Rightarrow 3 = 9a \Rightarrow$ the equation is
$y = \frac{1}{3}(x - 1)^2 - 3$.

5. a. $y = (x^2 + 8x + 16 - 16) + 15 = (x^2 + 8x + 16) - 1 = (x + 4)^2 - 1; (h, k) = (-4, -1)$

b. $f(x) = (x^2 - 4x + 4 - 4) - 5 = (x^2 - 4x + 4) - 9 = (x - 2)^2 - 9; (h, k) = (2, -9)$

c. $p(t) = (t^2 - 3t + 9/4 - 9/4) + 2 = (t^2 - 3t + 2.25) - 0.25 = (t - 1.5)^2 - 0.25; (h, k) = (1.5, -0.25)$

d. $r(s) = -5(s^2 - 4s + 4 - 4) - 10 = -5(s^2 - 4s + 4) + 20 - 10 = -5(s - 2)^2 + 10; (h, k) = (2, 10)$

e. $z = 2(m^2 + 3m + 9/4 - 9/4) - 5 = 2(m^2 + 3m + 9/4) - 9/2 - 5 = 2(m + 1.5)^2 - 9.5; (h, k) = (-1.5, -9.5)$

(In each case the two graphs coincide and they are omitted here.)

7. a. $f(x) = (x + 3)^2 - 4$ **d.** $y = 2(x + 0.75)^2 - 6.125$

b. $g(x) = (x - 1.5)^2 + 4.75$ **e.** $h(x) = 3(x + 1)^2 + 2$

c. $y = 3(x - 2)^2 + 0$ **f.** $y = -(x - 2.5)^2 + 4.25$

9. a. $y = x^2 + x - 12$; its y-intercept is -12; its vertex is at $x = -0.5, y = -12.25$. Its graph is in the diagram directly below with the relevant points marked.

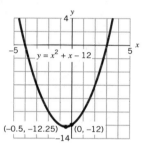

b. $y = x^2 + 3x + 7.2$; its y-intercept is 7.2; its vertex is at $x = -1.5, y = 4.95$. Its graph is directly below with the relevant points marked.

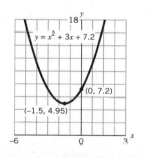

c. $y = -x^2 + 5x - 6.25$; -6.25 is its y-intercept; its vertex is at $x = 2.5, y = 0$. Its graph is directly below with the relevant points marked.

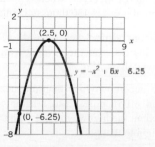

d. $y = x^2 + 8x + 15$; 15 is its y-intercept; its vertex is at $x = -4$, $y = -1$. Its graph is directly below with the relevant points marked.

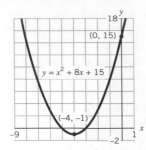

e. $y = x^2 + 3x - 11$; -11 is its y-intercept; its vertex is at $x = -1.5$, $y = -13.25$. Its graph is directly below with the relevant points marked.

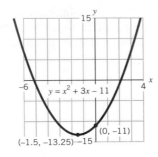

f. $y = -2x^2 + 12x + 20$; 20 is its y-intercept; its vertex is at $x = 3$, $y = 38$. Its graph is directly below with the relevant points marked.

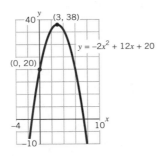

11. The larger in absolute value the coefficient of the x^2 term, the narrower the opening. Thus the order from narrow to broad is: d, f, a, b, c, and finally e. Technology confirms the principle.

13. The maximum profit occurs when $x = -20/[2 \cdot (-0.5)] = 20$; the maximum profit is 430 thousand dollars.

15. The maximum occurs when $x = -48/(2 \cdot (-3)) = \frac{-48}{-6} = 8$ super computers, and the revenue from selling 8 will be 192 million dollars.

17. a. In the accompanying diagram, next column, x and y marked off in miles, and a sample point $(7.2, 9.6)$ is marked on the highway along with the straight line from the origin to that point.

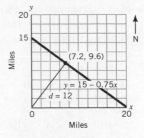

b. The highway goes through the points $(0, 15)$ and $(20, 0)$ and thus has the equation $y = 15 - 0.75x$.

c, d. The distance squared:
$$d^2 = x^2 + y^2 = x^2 + (15 - 0.75x)^2$$
$$= 1.5625x^2 - 22.5x + 225$$

e. Letting $D = d^2$, we have $D = 1.5625x^2 - 22.5x + 225$; the minimum occurs at the vertex, which is at $x = 22.5/(2 \cdot 1.5625) = 7.2$. The minimum for D is 144 and thus the minimum distance, d, is $\sqrt{144}$ or 12 miles.

f. The coordinates of the point of this shortest distance from $(0, 0)$ are $(7.2, 9.6)$. [See the graph in part (a).]

19. a, b. The graph is shown below. The minimum gas consumption rate suggested by the graph occurs when M is about 32 mph, and it is approximately 0.85 gph. (Computation on a calculator gives 32.5 for M, and the corresponding gas consumption rate is 0.86 (when rounded off).)

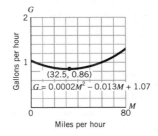

c. In 2 hours 1.72 gallons will be used and you will have traveled 65 miles.

d. If $M = 60$ mph, then $G = 1.01$ gph. It takes 1 hour and 5 minutes to travel 65 miles at 60 mph and one will have used approximately 1.094 gallons.

e. Clearly, traveling at the speed that supposedly minimizes the gas consumption rate does not conserve fuel if the trip lasts only 2 hours.

f, g.

M (mph)	G (gph)	G/M (gpm)	M/G (mpg)
0	1.07	—	0.0
10	0.96	0.09600	10.4
20	0.89	0.04450	22.5
30	0.86	0.02867	34.9
40	0.87	0.02175	46.0
50	0.92	0.01840	54.3
60	1.01	0.01683	59.4
70	1.14	0.01629	61.4
80	1.31	0.01638	61.1

For (f) we have: $G/M = (0.0002M^2 - 0.013M + 1.07)/M$. Its graph is in the accompanying diagram. Eyeballing gives the minimum gpm at $M \sim 73$ mph.

Graph for $y = G/M$

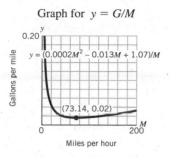

For (g) we have: $M/G = M/(0.0002M^2 - 0.013M + 1.07)$. Its graph is given in the accompanying diagram. Eyeballing gives the maximum mpg at the same M of approximately 73 mph. This is expected since maximum $- 1/$minimum.

Graph for $y = M/G$

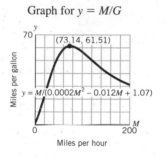

21. a. Time of release is at $x = 0$. The height then is 2 meters.

  b. At $x = 4$ m, $y = 5.2$ m; at $x = 16$ m, $y = 5.2$ m.

  c, d. The graph in the accompanying diagram shows the highest point, namely when $x = 10$ m and $y = 7$ m, and the point where the shot hits the ground, namely when x is approximately 22 m.

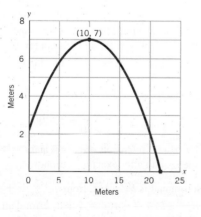

Section 8.4

Algebra Aerobics 8.4a

1. a. $y = -2t(8t - 25) = 0 \Rightarrow -2t = 0$ or $8t - 25 = 0 \Rightarrow$ $t = 0$ or $t = \frac{25}{8} \Rightarrow$ horizontal intercepts at $(0, 0)$ and $\left(\frac{25}{8}, 0\right)$

  b. $y = (t - 5)(t + 5) = 0 \Rightarrow t = 5$ or $t = -5 \Rightarrow$ horizontal intercepts at $(5, 0)$ and $(-5, 0)$

c. $h(z) = (z - 4)(z + 1) = 0 \Rightarrow z - 4 = 0$ or $z + 1 = 0 \Rightarrow$ $z = 4$ or $z = -1 \Rightarrow$ horizontal intercepts at $(4, 0)$ and $(-1, 0)$

d. $g(x) = (2x - 3)(2x + 3) = 0 \Rightarrow 2x - 3 = 0$ or $2x + 3 = 0 \Rightarrow x = 3/2$ or $x = -3/2 \Rightarrow$ horizontal intercepts at $(3/2, 0)$ and $(-3/2, 0)$

e. $y = (5 - x)(3 - x) = 0 \Rightarrow 5 - x = 0$ or $3 - x = 0 \Rightarrow$ $x = 5$ or $x = 3 \Rightarrow$ horizontal intercepts at $(5, 0)$ and $(3, 0)$

f. $v(x) = (x + 1)^2 = 0 \Rightarrow x + 1 = 0 \Rightarrow x = -1$, or one horizontal intercept at $(-1, 0)$

g. $p(q) = (q - 3)(q - 3) = 0 \Rightarrow q - 3 = 0 \rightarrow q - 3 \Rightarrow$ one horizontal intercept at $(3, 0)$

2. a. $f(x) = (5 + 4x)(1 - x) = 0 \Rightarrow 5 + 4x = 0$ or $1 - x = 0 \Rightarrow x = -5/4$ or $x = 1 \Rightarrow$ horizontal intercepts at $(-5/4, 0)$ and $(1, 0)$

b. $h(t) = (8 - 3t)(8 + 3t) = 0 \Rightarrow 8 - 3t = 0$ or $8 + 3t = 0 \Rightarrow t = \frac{8}{3}$ or $t = -\frac{8}{3} \Rightarrow$ horizontal intercepts at $\left(\frac{8}{3}, 0\right)$ and $\left(-\frac{8}{3}, 0\right)$

c. $y = (5 + t)(2 - 3t) = 0 \Rightarrow 5 + t = 0$, $2 - 3t = 0 \Rightarrow t = -5$ or $t = 2/3 \Rightarrow$ horizontal intercepts at $(-5, 0)$ and $(2/3, 0)$

d. $z = (2w - 5)(2w - 5) = 0 \Rightarrow 2w = 5 \Rightarrow w = 5/2 \Rightarrow$ one horizontal intercept $(5/2, 0)$

e. $y = (2x - 5)(x + 1) = 0 \Rightarrow 2x - 5 = 0$ or $x + 1 = 0 \Rightarrow x = \frac{5}{2}$ or $x = -1 \Rightarrow$ horizontal intercepts at $\left(\frac{5}{2}, 0\right)$ and $(-1, 0)$

f. $Q(t) = (3t - 2)(2t + 5) = 0 \Rightarrow 3t - 2 = 0$ or $2t + 5 = 0 \Rightarrow t = \frac{2}{3}$ or $t = -\frac{5}{2} \Rightarrow$ horizontal intercepts at $\left(\frac{2}{3}, 0\right)$ and $\left(-\frac{5}{2}, 0\right)$

3. Product of a sum and difference:

  a. $y = x^2 - 9 = (x + 3)(x - 3)$

  d. $y = 9x^2 - 25 = (3x + 5)(3x - 5)$

  f. $y = 16 - 25x^2 = (4 + 5x)(4 - 5x)$

  Square of sum or difference:

  b. $y = x^2 + 4x + 4 = (x + 2)^2$

  e. $y = x^2 - 8x + 16 = (x - 4)^2$

  Neither:

  c. $y = x^2 + 5x + 25$

  g. $y = 4 + 16x^2$

4. a. The vertical intercept is $(0, 0)$, which means at time $t = 0$, the object is on the ground. The horizontal intercepts: $h(t) = -16t(t - 4) = 0 \Rightarrow t = 0$ or $t = 4 \Rightarrow (0, 0)$ and $(4, 0)$, which means the object left the ground at time $t = 0$ seconds and returns to the ground at time $t = 4$ seconds.

  b. The horizontal intercepts: $P(q) = -(q^2 - 60q + 800) \Rightarrow P(q) = -(q - 20)(q - 40) = 0 \Rightarrow q = 20$ or $q = 40$. This means that if either 20 or 40 units are sold, the profit is $0 (or at breakeven). $P(q) > 0$ if $20 < q < 40$. The vertical intercept is $(0, -\$800)$, which means that if no items are sold, there is a loss of $800.

5. y_1 is Graph C because $(0, 0)$ and $(2, 0)$ are horizontal intercepts, and graph is concave down.

  y_2 is Graph A because $(2, 0)$ and $(-1, 0)$ are horizontal intercepts, and graph is concave up.

  y_3 is Graph B because $(-4, 0)$ and $(-1, 0)$ are horizontal intercepts, and graph is concave up.

6. a. $f(x) = (x - 5)(x + 6)$

b. $(5, 0)$ and $(-6, 0)$

c.

d. $f(x) = 0 \Rightarrow (x - 5)(x + 6) = 0 \Rightarrow x - 5 = 0$ or $x + 6 = 0 \Rightarrow x = 5$ or $x = -6$. So $f(x)$ has two horizontal intercepts, at $(5, 0)$ and $(-6, 0)$.

7 a. $ac = -30 \Rightarrow 2x^2 + 7x - 15 = 2x^2 + 10x - 3x - 15 = 2x(x + 5) - 3(x + 5) = (2x - 3)(x + 5)$

b. $ac = -12 \Rightarrow 4x^2 - x - 3 = 4x^2 - 4x + 3x - 3 = 4x(x - 1) + 3(x - 1) = (4x + 3)(x - 1)$

c. $ac = 42 \Rightarrow 3x^2 - 13x + 14 = 3x^2 - 7x - 6x + 14 = x(3x - 7) - 2(3x - 7) = (x - 2)(3x - 7)$

d. $ac = 10 \Rightarrow 2t^2 - 7t + 5 = 2t^2 - 5t - 2t + 5 = t(2t - 5) - 1(2t - 5) = (t - 1)(2t - 5)$

8. a. $12x^2 - 26x + 10 = 2(6x^2 - 13x + 5) = 2(6x^2 - 10x - 3x + 5) = 2(2x - 1)(3x - 5)$

b. $14r^3 - 21r^2 - 35r = 7r(2r^2 - 3r - 5) = 7r(2r - 5)(r + 1)$

Algebra Aerobics 8.4b

1. a. $4x + 7 = 0 \Rightarrow x = -7/4$

b. $4x^2 - 7 = 0 \Rightarrow 4x^2 = 7 \Rightarrow x^2 = \frac{7}{4}$

$\Rightarrow x = \pm\sqrt{\frac{7}{4}} \Rightarrow x = \pm\frac{\sqrt{7}}{2}$

c. $4x^2 - 7x = 0 \Rightarrow x(4x - 7) = 0 \Rightarrow x = 0$ or $x = \frac{7}{4}$

d. $2x + 6 = x^2 \Rightarrow$
$0 = x^2 - 2x - 6 \Rightarrow$
$x = \frac{-(-2) \pm \sqrt{(-2)^2 - 4(-6)}}{2} \Rightarrow$
$x = \frac{2 \pm \sqrt{28}}{2} \Rightarrow x = \frac{2 \pm 2\sqrt{7}}{2} \Rightarrow$
$x = \frac{2(1 \pm \sqrt{7})}{2} \Rightarrow x = 1 \pm \sqrt{7}$

e. $(2x - 11)^2 = 0 \Rightarrow 2x - 11 = 0 \Rightarrow x = \frac{11}{2}$

f. $(x + 1)^2 = 81 \Rightarrow x + 1 = \pm\sqrt{81} \Rightarrow$
$x = -1 \pm 9 \Rightarrow x = -10$ or $x = 8$

g. $0 = x^2 - x - 5 \Rightarrow$
$x = \frac{-(-1) \pm \sqrt{(-1)^2 - 4(-5)}}{2} \Rightarrow x = \frac{1 \pm \sqrt{21}}{2}$

2. a. $a = 2, b = 3, c = -1 \Rightarrow D = (3)^2 - 4(2)(-1) = 17 \Rightarrow$ two real, unequal zeros $\Rightarrow$ two horizontal intercepts

b. $a = 1, b = 7, c = 2 \Rightarrow D = (7)^2 - 4(1)(2) = 41 \Rightarrow$ two real, unequal zeros $\Rightarrow$ two horizontal intercepts

c. $a = 4, b = 4, c = 1 \Rightarrow D = (4)^2 - 4(4)(1) = 0 \Rightarrow$ one real zero (also known as a "double zero") $\Rightarrow$ one horizontal intercept

d. $a = 2, b = 1, c = 5 \Rightarrow D = (1)^2 - 4(2)(5) = -39 \Rightarrow$ no real zeros (two imaginary zeros) $\Rightarrow$ no horizontal intercepts

3. a. $h = -4.9t^2 + 50t + 80$. The vertical intercept is the initial height (at 0 seconds), which is 80 m; coordinates are $(0, 80)$. $a = -4.9, b = 50, c = 80$. So when $h = 0$ the horizontal intercepts are:

$$t = \frac{-50 \pm \sqrt{2500 - 4(-4.9)(80)}}{2(-4.9)}$$

$$= \frac{-50 \pm \sqrt{2500 + 1568}}{-9.8} = \frac{-50 \pm \sqrt{4068}}{-9.8} \Rightarrow$$

$$t = \frac{-50 + 63.8}{-9.8} \quad \text{or} \quad t = \frac{-50 - 63.8}{-9.8}$$

$$= \frac{13.8}{-9.8} \qquad\qquad = \frac{-113.8}{-9.8}$$

$$= -1.41 \qquad\qquad = 11.61 \text{ seconds}$$

Negative values of t have no meaning in a height equation, so the horizontal intercept at $(11.61, 0)$ means that the object hits the ground after 11.61 seconds.

b. $h = 150 - 80t - 490t^2$. The vertical intercept is $(0, 150)$, which means that the initial height is 150 cm. $a = -490, b = -80, c = 150 \Rightarrow$ when $h = 0$ the horizontal intercepts are at:

$$t = \frac{80 \pm \sqrt{6400 - 4(-490)(150)}}{2(-490)}$$

$$= \frac{80 \pm \sqrt{6400 + 294,000}}{-980}$$

$$= \frac{80 \pm \sqrt{300,400}}{-980} = \frac{80 \pm 548}{-980} \Rightarrow$$

$$t = \frac{80 + 548}{-980} \quad \text{or} \quad t = \frac{80 - 548}{-980}$$

$$= \frac{628}{-980} \qquad\qquad = \frac{-468}{-980}$$

$$= -0.64 \qquad\qquad = 0.48 \text{ seconds.}$$

Discard negative solution. The object hits the ground after 0.48 seconds.

c. The vertical intercept is the height in feet at $t = 0$ seconds $\Rightarrow h = 3$ feet.
If $a = -16, b = 64, c = 3 \Rightarrow$ the horizontal intercepts are:

$$t = \frac{-64 \pm \sqrt{(64)^2 - 4(-16)(3)}}{2(-16)} \Rightarrow t = \frac{-64 \pm \sqrt{4288}}{-32}$$

$\Rightarrow t = \frac{-64 \pm 65.5}{-32} \Rightarrow t \approx 4.05$, which means the object hits the ground after about 4.05 seconds or $t \approx -0.05$ (which is meaningless in this context).

d. The vertical intercept is the height in feet at $t = 0$ seconds $\Rightarrow h = 64(0) - 16(0)^2 = 0$ feet.
$a = -16, b = 64, c = 0 \Rightarrow$ horizontal intercepts are

$$t = \frac{-64 \pm \sqrt{(64)^2 - 4(-16)(0)}}{2(-16)} \Rightarrow t = \frac{-64 \pm 64}{-32} \Rightarrow$$

horizontal intercepts are $t = 0$ and $t = 4 \Rightarrow$ the object hits the ground after exactly 4 seconds.

4. Discriminant is $b^2 - 4ac$, from $ax^2 + bx + c = y$

a. $a = -5, b = -1, c = 4 \Rightarrow$ discriminant $= 1 - 4(-5)(4) = 81 > 0 \Rightarrow$ two x-intercepts. $\sqrt{81} = 9$, so roots are rational. The function has two real zeros where:

$$x = \frac{1 \pm 9}{-10} \Rightarrow$$

$$x = \frac{1 + 9}{-10} \quad \text{or} \quad x = \frac{1 - 9}{-10}$$

$$= \frac{10}{-10} \qquad\qquad = \frac{-8}{-10}$$

$$= -1 \qquad\qquad = \frac{4}{5}$$

So the x-intercepts are $(-1, 0)$ and $(4/5, 0)$.

b. $a = 4, b = -28, c = 49 \Rightarrow$ discriminant $=$
$784 - 4(4)(49) = 784 - 784 = 0 \Rightarrow$ one x-intercept.
$\sqrt{0} = 0$, so root is rational. The x-intercept is where
$x = \frac{28 \pm 0}{8} = \frac{7}{2}$;   at $\left(3\frac{1}{2}, 0\right)$.

c. $a = 2, b = 5, c = 4 \Rightarrow$ discriminant $= 25 - 4(2)(4) =$
$25 - 32 = -7$, which is negative $\Rightarrow$ no x-intercepts. The
function has two imaginary zeros at

$x = \frac{-5 \pm \sqrt{-7}}{4} = \frac{-5 \pm \sqrt{7}i}{4}$

d. $a = 2, b = -3, c = -1 \Rightarrow$ discriminant $-$
$(-3)^2 - 4(2)(-1) = 9 + 8 = 17 \Rightarrow$ two real zeros

$x = \frac{-(-3) \pm \sqrt{17}}{2(2)} \Rightarrow x = \frac{3 \pm \sqrt{17}}{4} \Rightarrow$

x-intercepts are $\left(\frac{3 + \sqrt{17}}{4}, 0\right)$ and $\left(\frac{3 - \sqrt{17}}{4}, 0\right)$

e. $a = -3, b = 0, c = 2 \Rightarrow$ discriminant $=$
$(0)^2 - 4(-3)(2) = 24 \Rightarrow$ two real zeros

$x = \frac{-0 \pm \sqrt{24}}{2(-3)} \Rightarrow x = \frac{\pm \sqrt{24}}{6} \Rightarrow x = \frac{\pm 2\sqrt{6}}{6} \Rightarrow$

$x = \pm \frac{\sqrt{6}}{3} \Rightarrow$ x-intercepts are $\left(-\frac{\sqrt{6}}{3}, 0\right), \left(+\frac{\sqrt{6}}{3}, 0\right)$

5. a. $y = a(x + 2)(x - 4)$. The point $(3, 2)$ is on the parabola $\Rightarrow$
$2 = a(3 + 2)(3 - 4) \Rightarrow a = -\frac{2}{5} \Rightarrow$

$y = -\frac{2}{5}(x + 2)(x - 4)$. The vertex is on the line of
symmetry, which lies halfway between the x-intercepts or
at $x = 1 \Rightarrow y = \frac{-2}{5}(1 + 2)(1 - 4) \Rightarrow y = \frac{18}{5} \Rightarrow$ the vertex
is at $\left(1, \frac{18}{5}\right)$.

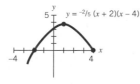

b. $y = a(x - 2)(x - 8)$. The vertical intercept is $(0, 10) \Rightarrow$
$10 = a(0 - 2)(0 - 8) \Rightarrow a = \frac{5}{8}$, so $y = \frac{5}{8}(x - 2)(x - 8)$

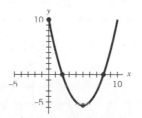

The vertex is on the line of symmetry, which lies
halfway between the x-intercepts or at $x = 5 \Rightarrow y =$
$\frac{5}{8}(5 - 2)(5 - 8) \Rightarrow y = \frac{-45}{8}$ and the vertex is $\left(5, \frac{-45}{8}\right)$.

6. a. Vertex is $(1, 5)$, above x-axis; $a = 3$ is positive, so it opens
up; so there are no x-intercepts.

b. Vertex is $(-4, -1)$, below x-axis; $a = -2$ is negative, so it
opens down; so there are no x-intercepts.

c. Vertex is $(-3, 0)$, on x-axis; $a = -5$ is negative, so it
opens down; so there is one x-intercept.

d. Vertex is $(1, -2)$, below x-axis; $a = 3$ is positive, so it
opens up; so there are two x-intercepts.

7. Answers will vary for different values of a and will be of the
form:

a. $f(x) = a(x - 2)(x + 3)$
  If $a = 1$, then
  $f(x) = (x - 2)(x + 3)$

b. $f(x) = ax(x + 5)$
  If $a = 2$, then
  $f(x) = 2x(x + 5)$

c. $f(x) = a(x - 8)^2$
  If $a = -2$, then
  $f(x) = -2(x - 8)^2$

8. Graph A: two real, unequal zeros $\Rightarrow$ discriminant is positive

Graph B: one real zero $\Rightarrow$ discriminant is equal to zero

Graph C: no real zeros $\Rightarrow$ discriminant is negative

Exercises for Section 8.4

1. a. $0 - (x - 3)(x + 3)$; thus $x = 3$ or -3

b. $0 = x(4 - x)$; thus $x = 0$ or 4

c. $0 = x(3x - 25)$; thus $x = 0$ or 25/3

d. $0 = (x + 5)(x - 4)$; thus $x = -5$ or 4

e. $0 = (2x - 3)^2$; thus $x = 3/2$ twice

f. $0 = (3x + 2)(x - 5)$; thus $x = -2/3$ or 5

g. $x^2 + 4x + 3 = -1 \Rightarrow x^2 + 4x + 4 = 0 \Rightarrow (x + 2)^2 = 0$;
thus $x = -2$ (a double zero)

h. $x^2 + 2x = 3x^2 - 3x - 3 \Rightarrow 2x^2 - 5x - 3 = 0 \Rightarrow$
$(2x + 1)(x - 3) = 0$; thus $x = -1/2$ or 3

3. a. $y = (x + 4)(x + 2)$

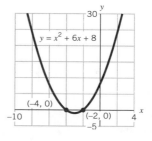

b. $z = 3(x + 1)(x - 3)$

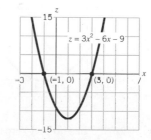

c. $f(x) = (x - 5)(x + 2)$

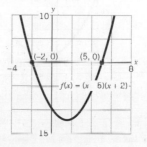

d. $w = (t - 5)(t + 5)$

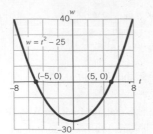

e. $r = 4(s - 5)(s + 5)$

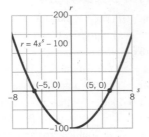

f. $g(x) = (3x - 4)(x + 1)$

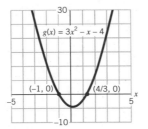

5. a. $y = x^2 - 5x + 6$ has zeros at 2 and 3, as is shown in the graph below.

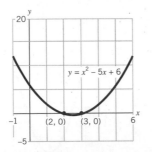

b. $y = 3x^2 - 2x + 5$ has no real zeros, as is shown in the graph below.

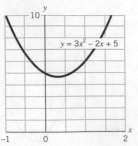

c. $y = 3x^2 - 12x + 12$ has a "double zero" at $x = 2$ since $y = 3(x - 2)^2$. See the graph below.

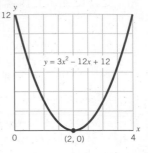

d. $y = -3x^2 - 12x + 15$ has zeros at $x = -5$ and 1; see the graph below.

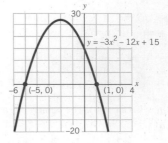

e. $y = 0.05x^2 + 1.1x$ has zeros at $x = -22$ and 0, as the graph below shows.

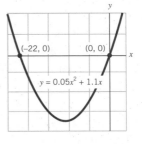

f. $y = -2x^2 - x + 3$ has roots at $x = -1.5$ and 1, as the graph below shows.

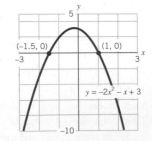

7. a. $t = \dfrac{7 \pm \sqrt{49 - 4 \cdot 6 \cdot (-5)}}{2 \cdot 6} = \dfrac{7 \pm \sqrt{169}}{12}$

$= \dfrac{7 \pm 13}{12} = \dfrac{5}{3}$ or $-\dfrac{1}{2}$

b. $x = \dfrac{12 \pm \sqrt{144 - 4 \cdot 9 \cdot 4}}{2 \cdot 9} = \dfrac{12 \pm \sqrt{0}}{18} = \dfrac{2}{3}$

c. $z = \dfrac{1 \pm \sqrt{1 - 4 \cdot 3 \cdot (-9)}}{2 \cdot 3} = \dfrac{1 \pm \sqrt{109}}{6} \approx -1.573$
or 1.907

d. $x = \dfrac{-6 \pm \sqrt{36 - 4 \cdot 1 \cdot 7}}{2 \cdot 1} = \dfrac{-6 \pm \sqrt{8}}{2} = -3 \pm \sqrt{2}$
≈ -1.586 or -4.414

e. $s = \dfrac{-17 \pm \sqrt{17^2 - 4 \cdot 6 \cdot (-10)}}{2 \cdot 6} = \dfrac{-17 \pm \sqrt{529}}{12}$
$= \dfrac{-17 \pm 23}{12} = -\dfrac{10}{3}$ or $\dfrac{1}{2}$

f. $t = \dfrac{3 \pm \sqrt{9 - 4 \cdot 2 \cdot (-9)}}{2 \cdot 2} = \dfrac{3 \pm \sqrt{81}}{4} = \dfrac{3 \pm 9}{4} = 3$
or -1.5

g. $x = \dfrac{11 \pm \sqrt{121 - 4 \cdot 4 \cdot (-8)}}{2 \cdot 4} = \dfrac{11 \pm \sqrt{249}}{8} \approx 3.347$
or -0.597

h. $x = \dfrac{12 \pm \sqrt{144 - 4 \cdot 4 \cdot 2}}{2 \cdot 4} = \dfrac{12 \pm \sqrt{112}}{8}$
$= \dfrac{12 \pm 4\sqrt{7}}{8} = \dfrac{3 \pm \sqrt{7}}{2} \approx 2.823$ or 0.177

9. a. y-intercept is 1; $y - (3x - 1)(x + 1)$ and thus x-intercepts are $1/3$ and -1.

b. y-intercept is 11; the x-intercepts are $\dfrac{6 \pm \sqrt{3}}{3} \approx 2.58$ and 1.42.

c. y-intercept is 15; x-intercepts are $5/2$ and $-3/5$.

d. The vertical intercept is $f(0) = -5$ and the x-intercepts are $\pm \sqrt{5}$.

11. Choices for values of a, b and c will vary. Here are some choices and the accompanying graphs. The equations in the form $y = ax^2 + bx + c$ are in the graph diagrams.

a. $a = 2, b = 3, c = 1$; $b^2 - 4ac = 9 - 8 = 1$

b. $a = 2, b = 3, c = -2$; $b^2 - 4ac = 9 + 16 = 25$

c. $a = 2, b = 3, c = 4$; $b^2 - 4ac = 9 - 32 = -23$

d. $a = -1, b = 2, c = -1$; $b^2 - 4ac = 4 - 4 = 0$

e. same as in (b) above.

Graph for (a)

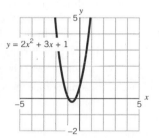

Graph for (b), (e)

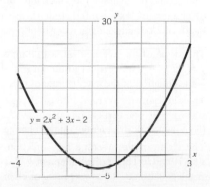

Graph for (c)

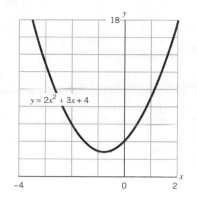

Graph for (d)

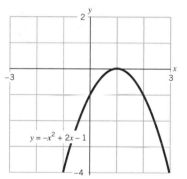

13. a. $Q(t) = -4(t + 1)^2$

b. No, since it must be of the form $Q(t) = a(t + 1)^2$ if -1 is to be a double root and a must be -4 if $Q(0) = -4$.

c. The axis of symmetry is the line whose equation is $t = -1$. The vertex is at $(-1, 0)$.

15. For Graph A: $f(x) = (x + 5)(x - 2) = x^2 + 3x - 10$
For Graph B: $g(x) = -0.5(x + 5)(x - 2) = -0.5x^2 - 1.5x + 5$

17. a. $-1 + 10i$ **b.** 1   **c.** $-1 + 3i$ **d.** $5 + 11i$

19. $f(x) = (x - (1 + i))(x - (1 - i)) =$
$x^2 - (1 - i)x - (1 + i)x + (1 + i)(1 - i) =$
$x^2 - 2x + (1 - i^2) =$
$x^2 - 2x + 2$

21. $x = \dfrac{4 \pm \sqrt{16 - 4 \cdot 1 \cdot 13}}{2 \cdot 1} = \dfrac{4 \pm \sqrt{-36}}{2}$
$= \dfrac{4 \pm 6i}{2} = 2 \pm 3i$.

Thus $f(x) = (x - 2 - 3i)(x - 2 + 3i)$. Its roots are not real and thus there are no x-intercepts.

23. a. Factoring, $y = (x + 4)(x - 2)$ and thus the roots are -4 and 2 and their average is -1. Thus $h = -1$ and $a = 1$; therefore $y = (x + 1)^2 + k = x^2 + 2x + 1 + k = x^2 + 2x - 8$ and thus $k + 1 = -8$ or $k = -9$. Thus the equation is $y = (x + 1)^2 - 9$. Or the vertex can be found by $-b/2a$ method.

b. Factoring gives $y = -(x + 4)(x - 1)$ and thus its roots are -4 and 1 and thus $h = -1.5$ and $a = -1$; therefore $y = -(x + 1.5)^2 + k = -x^2 - 3x - 2.25 + k = -x^2 - 3x + 4$ and thus $4 = k - 2.25$ or $k = 6.25$. Thus the equation is $y = -(x + 1.5)^2 + 6.25$. Or the vertex can be found by $-b/2a$ method.

25. $y = 4(x - 2)(x - 3)$. If the quadratic is to go through $(2, 0)$ and $(3, 0)$, then it must be of the form $y = a(x - 2)(x - 3)$; and if it is to stretch the graph of $y = x^2$ by a factor of 4, then $a = 4$ must hold. But this function's graph does more. It shifts the vertex of the graph of $y = x^2$ to $(2.5, -1)$. Its graph is in the accompanying diagram.

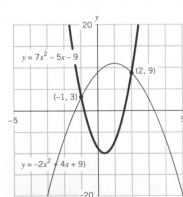

27. a. Setting the two formulas for y equal gives:

$7x^2 - 5x - 9 = -2x^2 + 4x + 9 \Rightarrow$
$0 = 9x^2 - 9x - 18 = 9(x^2 - x - 2)$
$\quad = 9(x - 2)(x + 1)$

and thus the intersection points are where $x = -1$ and $x = 2$. If $x = -1$, then $y = 3$ and if $x = 2$, then $y = 9$ (by substitution into either original equation). The graph confirming this information is given in the accompanying diagram.

b.

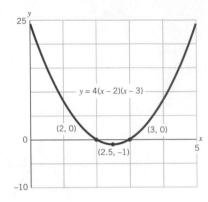

29. a. $1500 = 4W + 2L$ and thus
$L = (1500 - 4W)/2 = 750 - 2W$

b. $A(W) = W(750 - 2W) = 2W(375 - W)$

c. Domain for $A(W)$ is $0 < W < 375$

d. The area is largest when the W value is at the vertex of the parabola graph of $A(W)$, namely, when $W = 187.5$ ft. At that point $L = 375$ ft. and thus the area of the largest rectangle is 70,312.5 sq. ft. See the accompanying graph for verification.

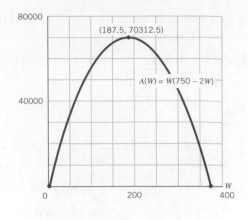

Section 8.5

Algebra Aerobics 8.5

1. a.

x	y	Average Rate of Change	Average Rate of Change of Average Rate of Change
-1	4	n.a.	n.a.
0	5	$\frac{5 - 4}{0 - (-1)} = 1$	n.a.
1	4	$\frac{4 - 5}{1 - 0} = -1$	$\frac{(-1) - 1}{1 - 0} = -2$
2	1	$\frac{1 - 4}{2 - 1} = -3$	$\frac{(-3) - (-1)}{2 - 1} = -2$
3	-4	$\frac{(-4) - 1}{3 - 2} = -5$	$\frac{(-5) - (-3)}{3 - 2} = -2$
4	-11	$\frac{(-11) - (-4)}{4 - 3} = -7$	$\frac{(-7) - (-5)}{4 - 3} = -2$

b. The third column tells us that the average rate of change of y with respect to x is decreasing at a constant rate, so the relationship is linear. The fourth column tells us that the average rate of change of the average rate of change is constant at -2.

2. a. The function is quadratic since the average rate of change is linear, that is, the average rate of change is increasing at a constant rate.

b. The function is linear since the average rate of change is constant.

c. The function is exponential since both the average rate of change and the average rate of change of the average rate of change are exponential, that is, are multiplied by a factor of 2 or increasing at a constant percentage.

3. a. The slope of the average rate of change $2a = 2$, giving $y = 2t + b$. The vertical intercept $b = 1$, so $y = 2t + 1$ is the equation of the average rate of change.

b. The slope of the average rate of change $2a = 6$, giving $y = 6x + b$. The vertical intercept $b = 5$, so $y = 6x + 5$ is the equation of the average rate of change.

c. The slope of the average rate of change is 10, giving $y = 10x + b$. The vertical intercept $b = 2$, so $y = 10x + 2$ is the equation of the average rate of change.

Exercises for Section 8.5

1. a. linear   **b.** positive, negative

3. a. $y = 3$; $y = -2$; $y = a$. It is horizontal line with slope 0 going through the point $(0, a)$.

  b. It is a straight line with slope $= 2a$ and y-intercept b. No, the slope of linear function is constant. The slope of the quadratic is a linear function.

  c. Guesses will vary. You may guess by analogy from the answers to parts (a) and (b) that its function is the quadratic $y = 3ax^2 + 2bx + c$.

5. The table and graph are given below:

t	Q	Average Rate of Change
-3	16	n.a.
-2	7	-9
-1	2	-5
0	1	-1
1	4	3
2	11	7
3	22	11

  a. The function Q is quadratic. Its graph is given in the accompanying diagram.

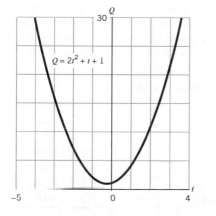

$$Q = 2t^2 + t + 1$$

  b. The third column indicates that the average rate of change of Q is linear. For each increase of 1 in t it goes up by 4.

7. Graph A goes with Graph F   Graph B goes with Graph E
Graph C goes with Graph D

9. a. $F(t) = 6t + 1$ **b.** $G(x) = -10x + 0.4$ **c.** $H(z) = 2$

11. a. quadratic   **b.** linear   **c.** exponential

Section 8.6

Algebra Aerobics 8.6

1. a. $d = kt^2$

  b. $d = kt^2 \rightarrow 576 = k(36) \rightarrow k = \frac{576}{36} = 16$ ft/sec^2

  c. $d = kt^2 \Rightarrow 176.4 = k(36) \Rightarrow k = \frac{176.4}{36} = 4.9$ m/sec^2

2. $d = 490t^2 + 50t$ $v = 980t + 50$

---

3. a. d in feet, t in seconds, since $16 = \frac{1}{2}g$, where $g = 32$ ft/sec^2

  b. feet $= \frac{\text{feet}}{\text{sec}^2}(\text{sec}^2) + \frac{\text{feet}}{\text{sec}}(\text{sec}) \Rightarrow$ feet $=$ feet

4. a.

t	0	1	6	10	11	16	20	21
h	0	95.1	423.6	510.0	507.1	345.6	40.0	-60.9

  b. Between 10 and 11 seconds, the object reaches its maximum height, then begins to fall, since h had increased from 0 to 510 meters, then decreased from 510 to 507.1 meters.

  c. Between $t = 20$ and $t = 21$, the object hits the ground, since at $t = 20$ it is still 40 meters above the ground while at $t = 21$, h has a negative value.

  d. $0 \le t \le 20.4$ since when $t = 20.4$, h is approximately 0, meaning that at $t = 20.4$ seconds the object hits the ground.

5. a.

t	0	1	2	3	4	5	6
h	50	114	146	146	114	50	-46

  b. Between $t = 2$ and $t = 3$ seconds the object reaches its maximum height and begins its descent.

  c. Between $t = 5$ and $t = 6$ the object hits the ground.

  d.

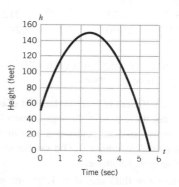

  e. It reaches its maximum height of 150 feet after 2.5 seconds.

  f. $0 = 50 + 80t - 16t^2 \Rightarrow 8t^2 - 40t - 25 = 0 \Rightarrow t = \frac{40 \pm \sqrt{1600 - 4(8)(-25)}}{2(8)} \approx \frac{40 \pm 49}{16}$

Since a negative value for t would be senseless, $t = 5.56$ seconds.

  g. The horizontal intercept is $(5.56, 0)$, which corresponds to the values $t = 5.56$ seconds and $h = 0$.

6. a. meters $=$ meters $+ \frac{\text{meters}}{\text{sec}}(\text{sec}) + \frac{\text{meters}}{\text{sec}^2}(\text{sec}^2) \Rightarrow$ meters $=$ meters

  b. feet $=$ feet $+ \frac{\text{feet}}{\text{sec}}(\text{sec}) + \frac{\text{feet}}{\text{sec}^2}(\text{sec}^2) \Rightarrow$ feet $=$ feet

7. a. If you are "weightless," then $a = 0$, so the equation is $h = h_0 + v_0 t$, which is a linear function.

  b. In this case, the object would continue to travel in a straight line.

Exercises for Section 8.6

1. For $d = 490t^2 + 50t$:

  a. 50 is measured in cm/sec; it is the initial velocity of the object falling; 490 is measured in (cm/sec)/sec and is half the acceleration due to gravity when measured in these units.

  b, c. Below is a small table of values and the graph of the equation with the table points marked on it.

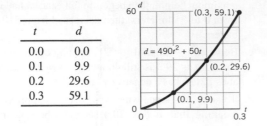

t	d
0.0	0.0
0.1	9.9
0.2	29.6
0.3	59.1

3. $m = \frac{m}{\text{sec}^2} \cdot \cancel{\text{sec}^2} + \frac{m}{\text{sec}} \cdot \cancel{\text{sec}}$

5.

Time (sec)	Distance (cm)	Avg. Vel. over Previous 1/30 sec. (cm/sec)
0.0000	0.00	n.a.
0.0333	3.75	113
0.0667	8.67	147
0.1000	14.71	181
0.1333	21.77	212
0.1667	29.90	243

The average velocity (over each 1/60 of a second) increases rapidly as time progresses.

7. **a.** The coefficient of t^2 is one-half the gravity constant. Since the coefficient of t^2 is approximately 490, distance is measured in centimeters and time in seconds, and 490 is measured in cm/sec^2. The coefficient of t is an initial velocity of 7.6 cm/sec.

  b. When $t = 0.05$ sec, $d = 1.59$ cm; when $t = 0.10$ sec, $d = 5.62$ cm, and when $t = 0.30$ sec, $d = 45.99$ cm.

9. It represents an initial velocity of the object measured in meters per second.

11. **a.** $d = 16t^2 + 12t$

  b.

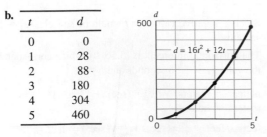

t	d
0	0
1	28
2	88
3	180
4	304
5	460

  c. The graph and table of the function are given above.

13. The distance is measured in meters if the time is measured in seconds. The use of 4.9 for half of the gravity constant is the indicator of these units.

15. **a.** Student answers will vary considerably.

  b. Since the velocity is changing at a constant rate, a straight line should be a good fit. The graph of this line is a representation of average velocity.

17. **a.** At $t = 0$ sec, $h = 0$ m; at $t = 1$ sec, $h = 195.1$ m; at $t = 2$ sec, $h = 380.4$ m; at $t = 10$ sec, $h = 1510$ m.

  b. The graph of h over t is given in the following diagram.

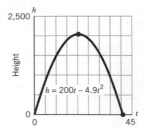

  c. The object reaches a maximum height of approximately 2000 meters after 20 seconds. It reaches the ground after approximately 40 seconds of flight.

19. For $h = 85 - 490t^2$:

  a. 85 is the height in centimeters of the falling object at the start; -490 is half the gravitational constant when measured in (cm/sec)/sec; it is negative in value since h measures height above the ground and the gravitational constant is connected with pulling objects down. This will mean subtraction from the starting height of 85 cm.

  b. The initial velocity is 0 cm/sec.

  c. Below is a table of values for this function

t	h
0.0	85.0
0.1	80.1
0.2	65.4
0.3	40.9

  d. The following diagram is the graph of the function with the table entries marked on it.

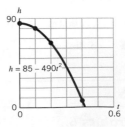

21. **a.** The initial velocity is positive since we are measuring height above ground and the object is going up at the start.

  b. The equation of motion is $h = 50 + 10t - 16t^2$, where height is measured in feet and t in seconds.

23. a. 980 cm/sec^2 since we are measuring in cm and in sec.

b.

t	v
0	-66
1	-1046
2	-2026
3	-3006
4	-3986

c. The graph is given below. The object is traveling faster and faster toward the ground. The increase in downward velocity is at a constant rate, as we can see from the constant slope of the graph. This constant acceleration, of course, is due to gravity.

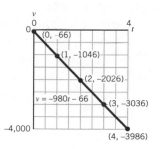

d. Ordinarily, if $t = 0$ corresponds to the actual start of the flight, then the initial condition given would indicate that the object was thrown downward at a speed of 66 meters per second. This interpretation comes from the negative sign given to the initial velocity.

25. a. Its velocity starts out negative and continues to be so since the object is falling; h is measured in cm above the ground; t is measured in seconds.

b. $h = 150 - 25t - 490t^2$; for $0 \leq t \leq 0.528$ (the second value being the approximate time in seconds it takes for the object to hit the ground).

c. The average velocity is the slope, i.e., $(15 - 150)/0.5 = -270$ cm/sec; the initial velocity is -25 cm/sec. The average velocity is 10.8 times as great in magnitude as the initial velocity.

27. Forming $\dfrac{d}{t} = \dfrac{v_0 + (v_0 + at)}{2}$ and solving for d, we get

$$d = \frac{2v_0 t + at^2}{2} = v_0 t + \frac{1}{2}at^2$$

This is very similar in form to the falling-body formula. The acceleration factor increases the velocity in a manner proportional to the square of the time traveled, and the initial velocity increases the distance in a manner proportional to the time.

29. a. After 5 seconds its velocity is 110 cm/sec; after 1 minute (or 60 seconds) its velocity is 660 cm/sec; after t seconds, its velocity is $v(t) = 60 + 10t$ cm/sec.

b. After 5 seconds its average velocity is $(60 + 110)/2 = 85$ cm/sec.

31. a. $v(t) = 200 + 60t$ meters/sec

b. $d(t) = 200t + 30t^2$ meters

33. a. Using units of feet and seconds, the equation governing the water spout is $h = -16t^2 + v_0 t$, where h is measured in feet and t, time, in seconds and where v_0 is the sought-after initial velocity. We are given that the maximum height reached is 120 ft. The maximum height is achieved at the vertex, i.e., when $t = -v_0/(-32) = \frac{v_0}{32}$. Substituting for t and h gives us

$$120 = -16\left(\frac{v_0}{32}\right)^2 + v_0\left(\frac{v_0}{32}\right) = -\frac{v_0^2}{64} + 2\,\frac{v_0^2}{64} = \frac{v_0^2}{64}$$

Thus $v_0^2 = 7680$ or $v_0 \approx 87.6$ ft per sec

b. $t = v_0/32 = 87.6/32 \approx 2.74$ sec

35. a. $d_c = v_c t + a_c t^2/2$; $d_p = a_p t^2/2$. One wants to solve for the t at which $d_c = d_p$, i.e., when $v_0 t + a_c \frac{t^2}{2} = a_p \frac{t^2}{2} \Rightarrow v_c t + a_c t^2/2 - a_p t^2/2 = 0 \Rightarrow t\,(v_c + [a_c/2 - a_p/2]t) = 0$. This occurs when $t = 0$ (when the criminal passes by the police car) and again when $t = 2v_c/(a_p - a_c)$, (when the police catch up to the criminal).

b. Now $v_c = a_c t + v_c$ and $v_p = a_p t$. One wants to solve for the t at which $v_c = v_p$, i.e., when $a_c t + v_c = a_p t$ or for $t = (a_p - a_c)/v_c$. This does not mean that the police have caught up to the criminal, but rather that the police are at that moment going as fast as the criminal is and that they are starting to go faster than the criminal.

Ch. 8: Check Your Understanding

1. False	**6.** True	**11.** False	**16.** True
2. True	**7.** False	**12.** True	**17.** True
3. False	**8.** True	**13.** False	**18.** True
4. False	**9.** True	**14.** False	**19.** False
5. True	**10.** False	**15.** False	**20.** True

21. $y = -0.25x^2$

22. Possible answer: $y = -(x - 1)^2 + 3$

23. Possible answer: $y = 2(x - 3)^2 - 5$

24. $y = \frac{-1}{2}(x - 2)(x + 2)$

25. $r = s^2 - s + 5$

26. $G(x) = (x + 2)^2 + 2(x + 2)$

27. $g(t) = -(t - 2)^2$

28. Possible answer: $y = x^2$ and $y = -(x - 1)^2 + 1$

29. Possible answer: $y = (x + 4)^2$

30. $h = -16t^2 + 32t + 80$

31. False	**34.** True	**37.** True	**40.** False
32. False	**35.** False	**38.** False	
33. False	**36.** True	**39.** True	

Ch. 8

Ch. 8 Review: Putting It All Together

1. In Graph A, the parabola is concave up with an estimated minimum at $(2, -4)$. Hence the axis of symmetry is the line $x = 2$ and there are two horizontal intercepts, at $x = 0$ and $x = 4$. Domain: $(-\infty, \infty)$; range: $[-4, +\infty)$.

   In Graph B, the parabola is concave down with an estimated maximum at $(0, -3)$. Hence the axis of symmetry is the vertical axis (the line $t = 0$) and there are no horizontal intercepts. Domain: $(-\infty, \infty)$; range: $(-\infty, -3]$.

3. **a.** Area of interior square $= x^2$ square inches; area of each of the maple strips $= (x + 1) \cdot 1 = x + 1$ square inches.

   b. Cost of white oak: ($2.39/ft^2) \cdot (1 \text{ ft}^2/144 \text{ in}^2) \approx$ $0.02/in^2$; cost of maple: ($4.49/ft^2) \cdot (1 \text{ ft}^2/144 \text{ in}^2) \approx$ $0.03/in^2$

   c. For white oak: $0.02x^2$ (in dollars); for all four maple strips: $4(0.03)(x + 1) = 0.12x + 0.12$ (in dollars)

   d. $C(x) = 0.02x^2 + 0.12x + 5.12$, a quadratic function

   e.

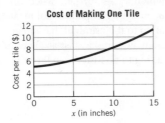

   Cost of Making One Tile

   f. Estimating from the graph, when $C(x) = \$7$, then $x \approx 7.5''$, so the width (and length) of the whole tile would be about 9.5″. To keep the cost/tile at $7 or below, the dimensions of a tile must be at most 9.5″ by 9.5″.

5. $g(x) = 2x^2$ and $h(x) = -0.5x^2$

7. **a.** Vertex for $F(x)$ is $(0, 0)$, vertex for $G(x) = (0, 5)$, vertex for $H(x)$ is $(-2, 0)$, vertex for $J(x)$ is $(1, -5)$

   b.

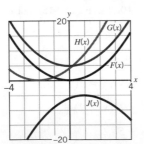

   c. The graph of $G(x)$ is the graph of $F(x)$ shifted up five units. The graph of $H(x)$ is the graph of $F(x)$ shifted left two units. The graph of $J(x)$ is the graph of $F(x)$ shifted right one unit, flipped over the x-axis, and shifted down five units.

9. **a.** One possibility is $Q(t) = (t - 4)(t + 2) = t^2 - 2t - 8$. The vertex of $Q(t)$ is at $(1, -9)$.

   b. One possibility is $M(t) = 3Q(t) = 3(t - 4)(t + 2) = 3t^2 - 6t - 24$. The vertex of $M(t)$ is at $(1, -27)$. So the vertices are not the same, though they share the same t-coordinate.

   c. One possibility is $P(t) = Q(t) + 10 = t^2 - 2t + 2$. $P(t)$ has no horizontal intercepts since the discriminant $= (-2)^2 - (4 \cdot 1 \cdot 2) = 4 - 8 = -4$, which is negative.

11. Since we have set the vertex at the origin, then the equation is of the form $y = ax^2$ (where $a < 0$). We know two points on the parabola, $(d, -32)$ and $(-(100 - d), -72) = (d - 100, -72)$. Substituting each set of points into the equation $y = ax^2$, we get the two equations

    $$-32 = ad^2 \quad \text{and} \quad -72 = a(d - 100)^2$$

    Solving both equations for a, we get

    $$a = -32/d^2 \quad \text{and} \quad a = -72/(d - 100)^2$$

    Setting both expressions for a equal and solving for d gives us

    $$-32/(d^2) = -72/(d - 100)^2$$

    cross-multiply $\quad -32(d - 100)^2 = -72d^2$

    simplify $\quad -32(d - 100)^2 + 72d^2 = 0$

    $$-32(d^2 - 200d + 10,000) + 72d^2 = 0$$

    $$40d^2 + 6400d - 320,000 = 0$$

    divide by 40 $\quad d^2 + 160d - 8000 = 0$

    Now we can solve for d either using the quadratic formula or factoring.

    Quadratic formula (letting $a = 1, b = 160, c = -8000$) gives:

    $$d = \frac{-160 \pm \sqrt{(160)^2 - 4(1)(-8000)}}{2 \cdot 1}$$

    $$= \frac{-160 \pm \sqrt{25,600 + 32,000}}{2}$$

    $$= \frac{-160 \pm \sqrt{57,600}}{2} = \frac{-160 \pm 240}{2} = -80 \pm 120$$

    So $d = 40$ feet or -200 feet. Only $d = 40$ feet makes sense.

    Factoring $d^2 + 160d - 8000 = 0$ gives $(d - 40)(d + 200) = 0$, which confirms that either $d = 40$ feet or -200 feet, where $d = 40$ feet is the only valid answer in this context.

    Substituting $d = 40$ feet into the equation $-32 = ad^2$, we get $-32 = a(40)^2 \Rightarrow a = -32/1600 = -0.02$. So the equation for the swimming pool parabolic roof is $y = -0.02x^2$, where x and y are both in feet.

13. **a.** The highest point of her dive will be at the vertex of the height function (which is concave down). Letting $a = -16, b = 12$, and $c = 25$, the t-coordinate of the vertex is at $-12/(2 \cdot (-16)) = 0.375$ seconds. Then $H(0.375) = 25 + (12 \cdot 0.375) - 16(0.375)^2 = 27.25$ feet above water will be the highest point of her dive.

   b. She will hit the water when $H(t) = 0 \Rightarrow 25 + 12t - 16t^2 = 0$. Using the quadratic formula, letting $a = -16, b = 12$, and $c = 25$, we have

    $$t = \frac{-12 \pm \sqrt{(12)^2 - 4(-16)(25)}}{2 \cdot (-16)}$$

    $$= \frac{-12 \pm \sqrt{(144 + 1600)}}{-32}$$

    $$= \frac{-12 \pm \sqrt{1744}}{-32} \approx \frac{-12 \pm 41.8}{-32}$$

    $$\approx -0.93 \text{ seconds or } 1.68 \text{ seconds.}$$

Only the positive value makes sense in this context. So about 1.68 seconds (a little under 2 seconds) after she starts her dive, she will hit the water.

15. a.

x	y	Average Rate of Change	Average Rate of Change of Average Rate of Change
-1	5	n.a.	n.a.
0	0	-5	n.a.
1	-3	-3	$\frac{-3-(-5)}{1-0}=2$
2	-4	-1	2
3	-3	1	2
4	0	3	2
5	5	5	2

b. A linear function.

c. The fourth column shows that the average rate of change (of the third column with respect to x) is constant, which means that the third column is a linear function of x.

CHAPTER 9

Section 9.1

Algebra Aerobics 9.1

1. a. $-f(x)$ matches the graph of $g(x)$ in Graph B because it is a reflection of f across the x-axis.

b. $f(-x)$ matches the graph of $h(x)$ in Graph C because it is a reflection of f across the y-axis.

c. $-f(-x)$ matches the graph of $j(x)$ in Graph A because it is a double reflection of f across both the x- and y-axes.

2. i. $f(x) = 2x - 3$

   a. $f(x + 2) = 2(x + 2) - 3 \Rightarrow f(x + 2) = 2x + 1$

   b. $\frac{1}{2}f(x) = \frac{1}{2}(2x - 3) \Rightarrow \frac{1}{2}f(x) = x - \frac{3}{2}$

   c. $-f(x) = -(2x - 3) \Rightarrow -f(x) = -2x + 3$

   d. $f(-x) = 2(-x) - 3 \Rightarrow f(-x) = -2x - 3$

   e. $-f(-x) = -(2(-x) - 3) \Rightarrow$
      $-f(-x) = -(-2x - 3) = 2x + 3$

  ii. $f(x) = 1.5^x$

   a. $f(x + 2) = 1.5^{x+2} \Rightarrow f(x + 2) = 1.5^2 \cdot 1.5^x$ or $2.25(1.5)^x$

   b. $\frac{1}{2}f(x) = \frac{1}{2}(1.5^x)$

   c. $-f(x) = -(1.5^x)$

   d. $f(-x) = 1.5^{-x} \Rightarrow f(-x) = \frac{1}{1.5^x}$

   e. $-f(-x) = -(1.5^{-x}) \Rightarrow -f(-x) = \frac{-1}{1.5^x}$

3. Graph B is symmetric across the vertical axis; Graph A is symmetric across the horizontal axis; Graph C is symmetric about the origin.

4. a. $h(t - 2) = e^{t-2}$

   b. $-h(t - 2) = -e^{t-2}$

   c. $-h(t - 2) - 1 = -e^{t-2} - 1$

5. a. $Q(t + 2) = 2 \cdot 1.06^{t+2}$

   b. $Q(t + 2) - 1 = 2 \cdot 1.06^{t+2} - 1$

   c. $-(Q(t + 2) - 1) = -(2 \cdot 1.06^{t+2} - 1) \Rightarrow$
      $-(Q(t + 2) - 1) = -2 \cdot 1.06^{t+2} + 1$

6. a. $g(x)$ is a reflection of $f(x)$ across the x-axis, since $g(x) = x^2 - 5 = -(5 - x^2) \Rightarrow g(x) = -f(x)$.

   b. $g(x)$ is a reflection of $f(x)$ across the y-axis, since $f(x) = 3 \cdot 2^x$ and $f(-x) = 3 \cdot 2^{-x} \Rightarrow f(-x) = g(x)$.

7. a. $g(x)$ is a compression of $f(x)$ by a factor of $\frac{1}{3}$ since $g(x) = \frac{1}{3x - 6} = \frac{1}{3(x - 2)} = \frac{1}{3} \cdot \frac{1}{x - 2} = \frac{1}{3}f(x)$.

   b. $g(x)$ is a vertical shift up of $f(x)$ by $\ln 3$ units since $g(x) = \ln 3x = \ln 3 + \ln x \Rightarrow g(x) = \ln 3 + f(x)$.

8. Graph A: $g(x)$
   Graph B: $f(x)$
   Graph C: $h(x)$
   Graph D: $j(x)$

Exercises for Section 9.1

1. a. g's graph is a reflection of f's graph across the x-axis followed by a stretching of the graph by a factor of 2. Thus $g(x) = -2 \cdot f(x) = -2\sqrt{x}$.

   b. g's graph is the graph of $f(x)$ shifted two units to the right. Thus $g(x) = f(x - 2) = e^{x-2}$.

   c. g's graph is the graph of $f(x)$ shifted three units to the left. Thus $g(x) = f(x + 3) = \ln(x + 3)$.

3. a.   **i.** $f(-x) = -x^3$; the original graph has been reflected across the y-axis.

     ii. $-2f(x) - 1 = -2x^3 - 1$; the original graph has been first stretched by a factor of 2, then reflected across the x-axis and then lowered one unit in the y direction.

    iii. $f(x + 2) = (x + 2)^3$; the graph has been shifted two units to the left along the x-axis.

    iv. $-f(-x) = x^3$; the graph was reflected across both axes, but the original graph is symmetric with respect to the origin and, effectively, no visual change has occurred.

The graphs are in the diagrams below, each with the graph of the original f.

Graph for (i)   Graph for (iii)

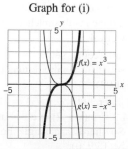

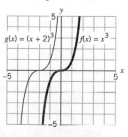

Graph for (ii)   Graph for (iv)

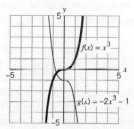

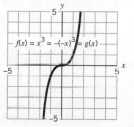

5. a. symmetric about the origin.

   b. symmetric across the x-axis

   c. symmetric across the y-axis

7. a. If $f(x) = a \cdot x^{2k}$, then $f(-x) = a(-x)^{2k} = ax^{2k} = f(x)$.

   b. If $f(x) = a \cdot x^{2k+1}$, then $f(-x) = a(-x)^{2k+1} = -ax^{2k} = -f(x)$.

   c. **i.** $f(-x) = (-x)^4 + (-x)^2 = x^4 + x^2 = f(x)$, so this is an even function.

      ii. $u(-x) = (-x)^5 + (-x)^3 = (-x^5) + (-x^3) = -(x^5 + x^3) = -u(x)$, so this is an odd function.

      iii. $h(-x) = (-x)^4 + (-x)^3 = x^4 - x^3 \neq h(x)$ and $h(-x) \neq -h(x)$, and so $h(x)$ is neither even nor odd.

      iv. $g(-x) = 10 \cdot 3^{-x} \neq g(x)$ and $\neq -g(x)$, and so $g(x)$ is neither even nor odd.

   d. The graphs of even functions are symmetric across the y-axis and the graphs of odd functions are symmetric about the origin. The graph of the function in (c)(i) has y-axis symmetry. The graph of the function in (c)(ii) has origin symmetry.

9. a. $y = 20(0.5)^{x+2} - 5$ **c.** $y = \log(x + 2)^{1/3} - 5$

   b. $y = 4(x + 2)^3 - 5$

11. a. $g(x) = f(x + 12) = -(x + 12) = -x - 12$

   b. $h(x) = f(x - 3.8) = -(x - 3.8) = -x + 3.8$

   c. $j(x) = f(x + 9) + 12 = -(x + 9) + 12 = -x + 3$

13. Graph B has the equation $y = (1/4)(x - 2)^2$. Graph C has the equation $y = (1/4)(x + 2)^2 - 1$. Graph D has the equation $y = (1/4)(x + 1)^2$.

15. a. $f(x + 2) = \ln(x + 2)$ **d.** $f(-x) = \ln(-x)$ for $x < 0$

   b. $\frac{1}{2}f(x) = \frac{1}{2}\ln(x)$ **e.** $-f(-x) = -\ln(-x)$ for $x < 0$

   c. $-f(x) = -\ln(x)$

17. a. $g(x) = -f(x) = -10 \cdot 5^x$

   b. $g(x) = f(-x) = 10 \cdot 5^{-x}$

   c. $g(x) = -f(-x) = -10 \cdot 5^{-x}$

19. a. $g(x) = 3 \ln(x + 2) - 4$, so the graph of $f(x) = \ln x$ was shifted to the left by 2, with the result stretched by a factor of 3 and then shifted down by 4.

   b. To find the vertical intercept, set $x = 0$ to get $g(0) = 3 \ln(0 + 2) - 4 \approx -1.92$. So the vertical intercept is at approximately $(0, -1.92)$.

   To find any horizontal intercepts, set $g(x) = 0$, to get $0 = 3 \ln(x + 2) - 4 \Rightarrow 4/3 = \ln(x + 2) \Rightarrow e^{4/3} = x + 2 \Rightarrow x = e^{4/3} - 2 \approx 1.79$. So there is a single horizontal intercept at approximately $(1.79, 0)$.

21. a. If we let t = time (in hours) since the corpse was discovered and T = temperature of the corpse, then since the ambient temperature is $60°$ according to Newton's Law of Cooling, $T = 60 + Ce^{-kt}$ for some constants k and C. When $t = 0$, $T = 80$, so we have $80 = 60 + C \Rightarrow C = 20$. So the equation becomes $T = 60 + 20e^{-kt}$. When $t = 2$, $T = 75$, so we have $75 = 60 + 20e^{-2k} \Rightarrow (15/20) = e^{-2k} \Rightarrow (3/4) = e^{-2k} \Rightarrow \ln(3/4) = \ln(e^{-2k}) \Rightarrow -0.288 \approx -2k \Rightarrow k \approx 0.144$. So the full equation is $T = 60 + 20e^{-0.144t}$.

b. If we assume that the normal body temperature is $98.6°$, then to find the time of death we need to solve $98.6 = 60 + 20e^{-0.144t}$ for $t \Rightarrow (38.6/20) = e^{-0.144t} \Rightarrow \ln(38.6/20) = \ln(e^{-0.144t}) \Rightarrow 0.658 \approx -0.144t \Rightarrow t \approx -4.6$ hours. So the person died about 4.6 hours before the corpse was discovered.

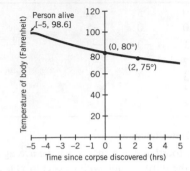

Temperature of Body Over Time

23. Graph A is $y = 2P(x)$; Graph B is $y = P(x)$; Graph C is $y = -0.5P(x)$.

Section 9.2

Algebra Aerobics 9.2

1. a. $f(2) = 2^3 = 8$

   b. $g(2) = 2(2) - 1 = 3$

   c. $h(2) = \frac{1}{2}$

   d. $(h \cdot g)(2) = h(2) \cdot g(2) = \frac{1}{2} \cdot 3 = \frac{3}{2}$

   e. $(f + g)(2) = f(2) + g(2) = 8 + 3 = 11$

   f. $\left(\frac{h}{g}\right)(2) = \frac{h(2)}{g(2)} = \frac{1/2}{3} = \frac{1}{6}$

2. a. $Q(1) + P(1) = 7 + 3 = 10$

   b. $Q(2) - P(2) = 9 - 24 = -15$

   c. $P(-1) \cdot Q(-1) = (-3 \cdot 3) = -9$

   d. $\frac{Q(3)}{P(3)} = \frac{11}{81}$

3. a. $f(t) - h(t) = (3 - 2t) - (t^2 - 1) = -t^2 - 2t + 4$

   b. $f(t) + h(t) = (3 - 2t) + (t^2 - 1) = t^2 - 2t + 2$

   c. $f(t) \cdot h(t) = (3 - 2t)(t^2 - 1) = -2t^3 + 3t^2 + 2t - 3$

   d. $\frac{h(t)}{f(t)} = \frac{t^2 - 1}{3 - 2t}$

4. a. $f(x + 1) = (x + 1)^2 + 2(x + 1) - 3 = x^2 + 4x$

   b. $f(x) + 1 = x^2 + 2x - 3 + 1 = x^2 + 2x - 2$

   c. $g(x + 1) = \frac{1}{(x + 1) - 1} = \frac{1}{x}$

   d. $g(x) + 1 = \frac{1}{x - 1} + 1 = \frac{1}{x - 1} + \frac{x - 1}{x - 1} = \frac{x}{x - 1}$

5. a. -2 **b.** 5   **c.** 15   **d.** undefined

6. a. $x = -1 \Rightarrow (f + g)(-1) = 2^{-1} + 3^{-1} = \frac{1}{2} + \frac{1}{3} = \frac{5}{6}$
   $x = 0 \Rightarrow (f + g)(0) = 2^0 + 3^0 = 1 + 1 = 2$
   $x = 2 \Rightarrow (f + g)(2) = 2^2 + 3^2 = 4 + 9 = 13$

   b. $x = -1 \Rightarrow (f - g)(-1) = 2^{-1} - 3^{-1} = \frac{1}{2} - \frac{1}{3} = \frac{1}{6}$
   $x = 0 \Rightarrow (f - g)(0) = 2^0 - 3^0 = 1 - 1 = 0$
   $x = 2 \Rightarrow (f - g)(2) = 2^2 - 3^2 = 4 - 9 = -5$

c. $x = -1 \Rightarrow (f \cdot g)(-1) = 2^{-1} \cdot 3^{-1} = \frac{1}{2} \cdot \frac{1}{3} = \frac{1}{6}$

$x = 0 \Rightarrow (f \cdot g)(0) = 2^0 \cdot 3^0 = 1 \cdot 1 = 1$

$x = 2 \Rightarrow (f \cdot g)(2) = 2^2 \cdot 3^2 = 4 \cdot 9 = 36$

d. $x = -1 \Rightarrow \left(\frac{f}{g}\right)(-1) = \frac{2^{-1}}{3^{-1}} = \frac{3}{2}$

$x = 0 \Rightarrow \left(\frac{f}{g}\right)(0) = \frac{2^0}{3^0} = 1$

$x = 2 \Rightarrow \left(\frac{f}{g}\right)(2) = \frac{2^2}{3^2} = \frac{4}{9}$

7. the percentage of people who agree with the comment or have no opinion

8. (per capita energy consumption)(population) = energy consumption of the country at time t

Exercises for Section 9.2

1. a. $3t^2 + 10t - 4$ **c.** $18t^3 + 27t^2 - 26t - 5$

   b. $-3t^2 + 2t + 6$ **d.** $\frac{3t^2 + 4t - 5}{6t + 1}$

3. a. $j(x) = 3x^5 + x^2 + x - 1$; $k(x) = 3x^5 - x^2 + x + 1$; $l(x) = 3x^7 - 3x^5 + x^3 - x$

   b. $j(2) = 101$; $k(3) = 724$; $l(-1) = 0$

5. a. $R(n) = 25n$

   b. $R(n) = 25n - 500$

   c. $R(n) = 25(n - 30) - 500 = 25n - 1250$

7. If a worker works t hours a week (where $t \geq 40$), then the worker's weekly paycheck, P (in dollars), is the sum of two terms: regular pay + overtime pay. The regular pay is $20 \cdot 40 =$ \$800 a week. The overtime pay = $30 \cdot (t - 40)$, where $t =$ total number of hours worked. So the weekly paycheck is $P = 800 + 30(t - 40)$, where $t \geq 40$.

9.

x	0	1	2	3	4	5
$h(x)$	-6	-6	-6	6	-6	-6
$j(x)$	0	-4	-16	-36	-64	-100
$k(x)$	9	5	-55	-315	-1015	-2491

11.

x	-3	-2	-1	0	1	2	3
$f(x)$	9	4	1	0	1	4	9
$g(x)$	-4	-3	-2	-1	0	1	2
$f(x) + g(x)$	5	1	-1	-1	1	5	11
$f(x) - g(x)$	13	7	3	1	1	3	7
$f(x) \cdot g(x)$	-36	-12	-2	0	0	4	18
$g(x)/f(x)$	$-4/9$	$-3/4$	-2	undefined	0	1/4	2/9

13. a. $C(n) = 500 + 40n$

   b. $P(n) = \frac{500 + 40n}{n}$

   c. $P(25) = 60$; $P(100) = 45$. As the number of people attending increases, the cost per person decreases. If only 25 attend (the minimum size), the cost would be \$60 per person. If 100 attend (the maximum size), the cost would be \$45 per person.

15. a. Graph C matches $y = (h + j)(x)$. The reasons are that $(h + j)(0) = 0 - 4 = -4$ and Graph C is the only one with that y-intercept; a quadratic plus a linear function gives us a quadratic, and Graph C is a quadratic; or $(h + j)(0)$ is defined and this eliminates Graph B.

   b. Graph A matches $y = (h \cdot j)(x)$. The reasons are that $(h \cdot j)(-2) = h(-2) \cdot j(-2) = 0 \cdot (-2) = 0$ and $(h \cdot j)(0) = h(0) \cdot j(0) = -2 \cdot 0 = 0$ and $(h \cdot j)(2) = h(2) \cdot j(2) = 0 \cdot (2) = 0$, all of which are x-intercepts of Graph A; a quadratic times a linear is a cubic, and Graph A is a cubic.

   c. Graph B matches $y = (h/j)(x)$, since $(h/j)(0) = h(0)/j(0) = 2/0$, which is not defined; $x = 2$ and -2 are both zeros for Graph B since $h(-2)/j(-2) = 0/-2 = 0$ and $h(2)/j(2) = 0/2 = 0$.

17. a. It appears that $T(q) = H(q) + O(q)$. So $T(q)$ represents the total cost of holding and ordering q items.

   b. If one orders q^* items, then ordering costs and holding costs are the same. This is where the total cost $T(q^*)$ is at a minimum.

19. a. Let $f(x)$ and $g(x)$ be two even functions. Then $f(-x) = f(x)$ and $g(-x) = g(x)$; thus if $h(x) = f(x) + g(x)$, then $h(-x) = f(-x) + g(-x) = f(x) + g(x) = h(x)$; thus h is an even function.

   b. Let $p(x)$ and $q(x)$ be two odd functions. Then $p(-x) = -p(x)$ and $q(-x) = -q(x)$; thus if $k(x) = p(x) + q(x)$, then $k(-x) = p(-x) + q(-x) = -p(x) + -q(x) = -(p(x) + q(x)) = -k(x)$; thus k is an odd function.

   c. Using the even functions in part (a): $f(-x) \cdot g(-x) = f(x) \cdot g(x)$; thus the product of two even functions is even.

   d. Using the odd functions in part (b): $p(-x) \cdot q(-x) = (-p(x)) \cdot (-q(x)) = p(x) \cdot q(x)$; thus the product of two odd functions is an even function.

   e. Using an even function from part (a) and an odd function from part (b), we get $f(-x) \cdot p(-x) = f(x) \cdot (-p(x)) = -f(x) \cdot p(x)$. This shows that the product of an even and an odd function is an odd function.

21. The function $A(t) - D(t)$ would represent the number of persons diagnosed with AIDS who did not die or those living with AIDS, assuming that those who died were all diagnosed with AIDS. One could graph $A(t) - D(t)$ or note whether the graph of $A(t)$ is above or below the graph of $D(t)$.

Section 9.3

Algebra Aerobics 9.3a

1. a. Degree is 5: $f(-1) = 11(-1)^5 + 4(-1)^3 - 11 = 11(-1) + 4(-1) - 11 = -11 - 4 - 11 = -26$

   b. Degree is 4: When $x = -1$, $y = 1 + 7(-1)^4 - 5(-1)^3$
$= 1 + 7(1) - 5(-1)$
$= 1 + 7 + 5 = 13$

   c. Degree is 4: $g(-1) = -2(-1)^4 - 20 = -2 - 20 = -22$

   d. Degree is 2: When $x = -1$, $z = 3(-1) - 4 - 2(-1)^2$
$= -3 - 4 - 2 = -9$

Ch. 9

2. a. degree $n = 5 + 3 + 2 = 10$

  b. degree $n = 2 + 12 = 14$

3. a. degree 5; a quintic polynomial function

  b. The leading term is $-2t^5$.

  c. The constant term is 0.5.

  d. $f(0) = 0.5 - 2(0)^5 + 4(0)^3 - 6(0)^2 - (0) = 0.5$
  $f(0.5) = 0.5 - 2(0.5)^5 + 4(0.5)^3 - 6(0.5)^2 - (0.5) = -1.0625$
  $f(-1) = 0.5 - 2(-1)^5 + 4(-1)^3 - 6(-1)^2 - (-1) = -6.5$

Algebra Aerobics 9.3b

1.

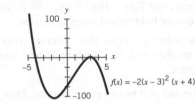

$f(x) = -2(x - 3)^2 (x + 4)$

  a. degree 3
  b. two turning points
  c. as $x \to +\infty$, $y \to -\infty$; and as $x \to -\infty$, $y \to +\infty$
  d. $y = -2x^3$
  e. The horizontal intercepts are $(3, 0)$, where the graph "touches" the x-axis and $(-4, 0)$, where the graph crosses the x-axis.
  f. $y = -72$; the vertical intercept

2. Graph A: Minimum degree = 3 because there are two turning points; positive leading coefficient

  Graph B: Minimum degree = 4 because there are three turning points; negative leading coefficient

  Graph C: Minimum degree = 5 because there are four turning points; negative leading coefficient

3. a. The y-intercept is at -3. The graph crosses the x-axis only once. It happens at about $x = 1.3$.

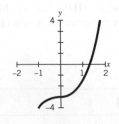

  b. The y-intercept is at 3. The graph does not intersect the x-axis.

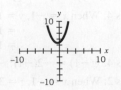

4. a. Degree is 1. x-intercept is -2.

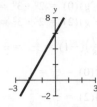

  b. Degree is 2. x-intercepts are -4 and 1.

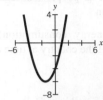

  c. Degree is 3. x-intercepts are -5, 3, and $-2\frac{1}{2}$.

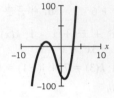

5. There are infinitely many examples of such functions. To have exactly those four x-intercepts, the functions are of the form:
  $f(x) = ax^n(x + 3)^m(x - 5)^p(x - 7)^r$, where a is a real number, and n, m, p and r are positive integers.

  (i) $f(x) = x(x + 3)(x - 5)(x - 7)$

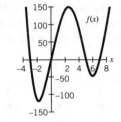

  (ii) $h(x) = 2x^2(x + 3)(x - 5)(x - 7)$

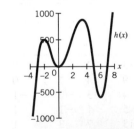

  (iii) $g(x) = -20x^2(x + 3)(x - 5)(x - 7)$

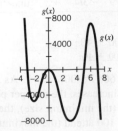

Exercises for Section 9.3

1. a. polynomial, 1   **d.** not polynomial

   b. polynomial, 3   **e.** polynomial, 5

   c. not polynomial   **f.** polynomial, 3

3. a. $\frac{1}{8}, -\frac{1}{8}$ **b.** $\frac{1}{2}, -\frac{1}{2}$ **c.** $-\frac{1}{2}, \frac{1}{2}$ **d.** $-32, 32$

5. a. goes with Graph A **c.** goes with Graph B

   b. goes with Graph C

7.   **i.** Graph A has two turning points; Graph B has three turning points; Graph C has four turning points.

   ii. Graph A has two x-intercepts; Graph B has two x-intercepts; Graph C, has five x-intercepts.

   iii. The sign of Graph A's leading term is minus; for Graph B, it is minus; for Graph C, is it plus.

   iv. The minimum degree for Graph A is three; for Graph B, it is four; for Graph C, it is five.

   v. There is no maximum value for Graph A. The maximum value for Graph B is approximately 7.5. There is no maximum value for Graph C.

   vi. None of the graphs have minimum values.

9. (a) and (e) are cubics; (b), (d), and (f) are quartics; (c) is by itself, since it is the only quintic.

11. a. Always negative; could have up to three turning points; has exactly one turning point (see accompanying graph).

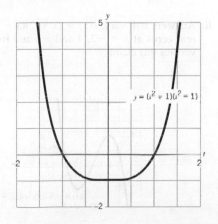

b. Always positive; in general a quartic has at most three turning points; here exactly one turning point (see accompanying graph).

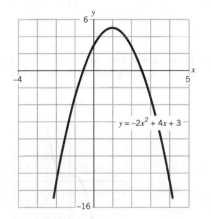

c. Negative if x is negative; positive if x is positive; at most two turning points; here none (see the accompanying graph).

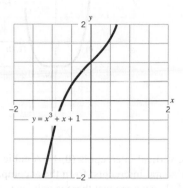

d. Positive if x is positive; negative if x is negative; in general a quintic has at most four turning points; here there are exactly four (see the accompanying graph).

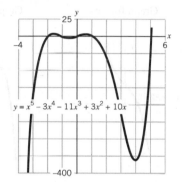

13. a. The vertical intercept is $(0, 3)$. A quadratic has at most two horizontal intercepts; from the accompanying graph we see that there are two, at $x \approx -0.6$ and 2.6.

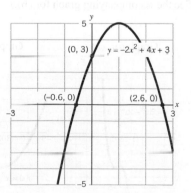

b. The vertical intercept is $(0, -1)$. A quartic at most four horizontal intercepts; from the accompanying graph, at top of next page, we see that it has two intercepts: at $t = -1$ and 1.

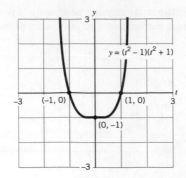

c. The vertical intercept is $(0, 1)$. A cubic has at most three horizontal intercepts; from the accompanying Graph for (c) below we see that there is only one, at $x \approx -0.682$.

d. The vertical intercept is $(0, 0)$. A quintic has at most five horizontal intercepts; from the accompanying Graph for (d) below we see that there are five, at $x = -2, -1, 0, 1,$ and 5; in fact we have that $y = (x + 2)(x + 1)x(x - 1)(x - 5)$.

Graph for (c)   Graph for (d)

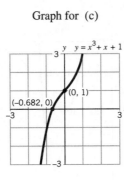   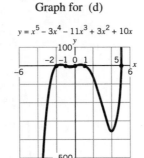

15. a. $y = 3x^3 - 2x^2 - 3$ has only one real zero at $x \approx 1.28$. (See the accompanying graph for (a).)

b. $y = x^2 + 5x + 3$ has two real zeros:

$$x = -2.5 \pm 0.5\sqrt{13} \approx -4.302 \text{ and } -0.697.$$

(See the accompanying graph for (b).)

Graph for (a)   Graph for (b)

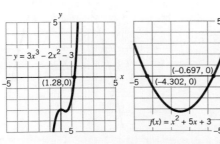

17. a. The end behavior of an odd degree polynomial as $x \to \pm\infty$ is the same as $y = a_n x^n$, so $y \to \pm\infty$. This means that the y values go from a negative to a positive or vice versa and therefore must cross the x-axis.

b. i. Answers will vary. $f(x) = x^2 + 1$ has no real zeros and its graph verifies this claim.

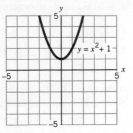

ii. Answers will vary. $g(x) = x^4 + 1$ has no real zeros and its graph verifies this claim.

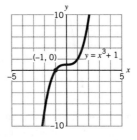

c. i. Answers will vary. $h(x) = x^3 + 1$ has one real zero at $x = -1$ and its graph verifies this claim.

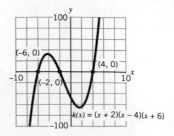

ii. Answers will vary. $k(x) = (x + 2)(x - 4)(x + 6)$ has real zeros at $x = -2, 4$ and -6 and its graph below verifies this claim.

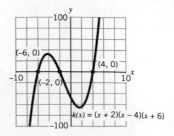

d. Answers will vary. $p(x) = (x - 4)^2 (x + 5)^2$ has two real zeros at $x = -5$ and $x = 4$. Here is its graph.

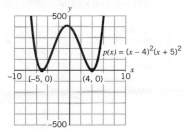

19. a. $8701 = 8n^3 + 7n^2 + 0n + 1n^0$

b. $239 = 2n^2 + 3n + 9n^0$

c. The number written in base 2 as 11001 evaluates to 25 when written in base 10 notation, since $(1 \cdot 2^4) + (1 \cdot 2^3) + (0 \cdot 2^2) + (0 \cdot 2^1) + (1 \cdot 2^0) = 25$.

d. Here is one way to find the base two equivalent: find the highest power of 2 in 35. This is $2^5 = 32$. Subtracting that from 35 leaves 3, which is easily written as $2 + 1$. Thus 35, in base 10, can be written as

$(1 \cdot 2^5) + (0 \cdot 2^4) + (0 \cdot 2^3) + (0 \cdot 2^2) + (1 \cdot 2^1) + (1 \cdot 2^0)$

and this is 100011 in base 2.

21. a. $f(x) = x^3 + 3x^2 = x^2(x + 3)$. Thus the horizontal intercepts are at $(0, 0)$ and $(-3, 0)$. A look at the accompanying graph below indicates that there are two turning points at $(0, 0)$ and $(-2, 4)$. As $x \to +\infty, f(x) \to +\infty$ and as $x \to -\infty, f(x) \to -\infty$.

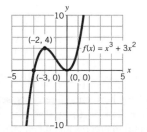

b. $g(x) = (x^4 - 81) = (x^2 + 9)(x - 3)(x + 3)$. Thus the horizontal intercepts are $(3, 0)$ and $(-3, 0)$. There is one turning point at $(0, -81)$. As $x \to +\infty, f(x) \to +\infty$ and as $x \to -\infty, f(x) \to +\infty$.

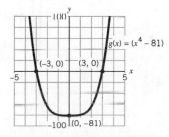

c. $h(x) = -2x^3 + x^2 + 15x = -x(2x^2 - x - 15) = x(2x + 5)(x - 3)$. Thus the x-intercepts are at $(0, 0)$, $(-5/2, 0)$, and $(3, 0)$. From the accompanying graph on next column, and the use of technology, one determines

that there are two turning points. As $x \to +\infty, f(x) \to -\infty$ and as $x \to -\infty, f(x) \to +\infty$.

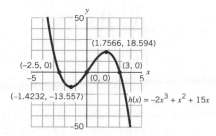

23. a. Answers will vary. The function is in the form $f(x) = a(x - 3)(x + 2)(x - 6)$. The end behavior indicates that $a > 0$; thus for $a = 1$, $f(x) = (x - 3)(x + 2)(x - 6)$. The graph confirms the asked-for behavior.

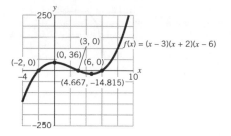

b. Answers will vary. The function is in the form $g(x) = a(x + 5)(x + 2)(x - 1)(x - 3)$. The end behavior indicates that $a < 0$. For $a = -1$, then $g(x) = -(x + 5)(x + 2)(x - 1)(x - 3)$. Its graph satisfies the requested behavior.

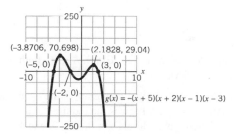

25. a. $f(x) = -2(x + 1)(x - 2)(x - 4)$

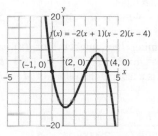

b. $g(x) = a(x - 1)(x + 1)(x - 2)(x + 2)$. At $(0, 8)$, then $8 = a(-1)(1)(-2)(2)$, so $a = 8/-4 = -2$; thus $g(x) = -2(x - 1)(x + 1)(x - 2)(x + 2)$. The accompanying graph next page, has the required properties.

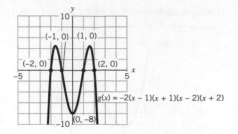

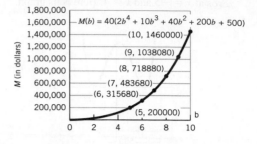

Here is the graph of $M(b)$ for $0 \le b \le 10$.

c. $h(x) = 1(x - 1)^2 (x - i)(x + i)x = (x - 1)^2(x^2 + 1)x$. Its graph confirms all but the roots at $\pm i$, which are complex roots and thus not in the real number plane.

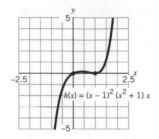

d. $k(x) = -h(x) = -(x - 1)^2(x^2 + 1)x$. The graph is given below.

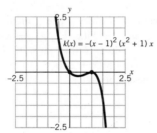

27. Graph B goes with $f(x)$; Graph A goes with $g(x)$; Graph C goes with $h(x)$.

29. a. The formula that generates the gift amounts is: $A(n) = 40 \cdot 5^n$. It gives the required amounts for the five different giving levels $n = 0, 1, 2, 3,$ and 4.

b. The amount will be $500 \cdot A(0) + 200 \cdot A(1) + 40 \cdot A(2) + 10 \cdot A(3) + 2 \cdot A(4) = 200{,}000$; thus they are planning to raise \$200,000.

c. $M(b) = 40 \cdot (2 \cdot 5^4 + 10 \cdot 5^3 + 40 \cdot 5^2 + 200 \cdot 5 + 500)$. $b = 5$ fits the school plan.

Assuming the numbers for (a) are used here then $M(b) = 40 \cdot (2 \cdot b^4 + 10 \cdot b^3 + 40 \cdot b^2 + 200 \cdot b + 500)$.

d. Here is a table of the values of M for $b = 1, 2, \ldots, 10$.

b	M	b	M
0	20000	6	315680
1	30080	7	483680
2	46880	8	718880
3	75680	9	1038080
4	123680	10	1460000
5	200000		

Section 9.4

Algebra Aerobics 9.4

1. a. horizontal intercepts at $x = 1$ and $x = -5$; vertical asymptote at $x = -3$

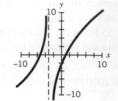

b. horizontal intercept at $x = -2/3$ and vertical asymptotes at $x = -1$ and $x = 3$

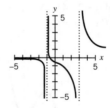

2. a. $f(r) = \dfrac{r^2 - 4r - 12}{r^2 - 4r + 3} = \dfrac{(r + 2)(r - 6)}{(r - 1)(r - 3)}$

To find any horizontal intercepts, set the numerator $= 0$ to get $(r + 2)(r - 6) = 0 \Rightarrow r = -2$ or $r = 6$. So the horizontal intercepts are at $(-2, 0), (6, 0)$.

To find any vertical asymptotes, set the denominator $= 0$ to get $(r - 1)(r - 3) = 0 \Rightarrow r = 1$ or $r = 3$. So $f(r)$ is not defined at $r = 1$ and $r = 3$. So there are two vertical asymptotes at the lines $r = 1$ and $r = 3$. The graph of $f(r)$ follows.

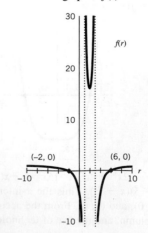

b. $g(r) = \frac{r^2 - 4r + 3}{r^2 - 4r - 12} = \frac{(r-1)(r-3)}{(r+2)(r-6)}$

To find any horizontal intercepts, set the numerator $= 0$ to get $(r-1)(r-3) = 0 \Rightarrow r = 1$ or $r = 3$. So the horizontal intercepts are at $(1, 0)$, $(3, 0)$.

To find any vertical asymptotes, set the denominator $= 0$ to get $(r+2)(r-6) = 0 \Rightarrow r = -2$ or $r = 6$. So $g(r)$ is not defined at $r = -2$ and $r = 6$. So there are two vertical asymptotes at the lines $r = -2$ and $r = 6$. The graph of $g(r)$ follows.

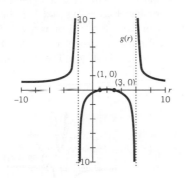

3. a. $g(x) = -\frac{1}{(x+3)^2} + 1$

b. $g(x) = -\frac{1}{(x+3)^2} + 1 \cdot \frac{(x+3)^2}{(x+3)^2}$

$= \frac{-1}{(x+3)^2} + \frac{x^2 + 6x + 9}{(x+3)^2}$

$= \frac{-1 + x^2 + 6x + 9}{x^2 + 6x + 9}$

$= \frac{x^2 + 6x + 8}{x^2 + 6x + 9} = \frac{p(x)}{q(x)}$

c. To find any horizontal intercepts, let the numerator $p(x) = 0$ to get $x^2 + 6x + 8 = (x+4)(x+2) = 0 \Rightarrow x = -4$ or $x = -2$. So the horizontal intercepts are at $(-4, 0)$ and $(-2, 0)$.

To find any vertical asymptotes, let the denominator $q(x) = 0$ to get $(x+3)^2 = 0 \Rightarrow x = -3$. So $g(x)$ is not defined when $x = -3$. So there is one vertical asymptote at the line $x = -3$.

4. a. Horizontal intercepts

b. Vertical intercepts

c. Horizontal intercept

d. Vertical asymptote

5. $f(x) = \frac{2x + 6}{x - 3}$

i. To find horizontal intercept(s): set $f(x) = 0$, which is equivalent to setting the numerator $2x + 6 = 0 \Rightarrow x = -3$. So there is one horizontal intercept at $(-3, 0)$.

To find vertical intercept: evaluate $f(0) = 6/-3 = -2$. So the vertical intercept is at $(0, -2)$.

ii. To find vertical asymptote: set $x - 3 = 0 \Rightarrow x = 3$. So the vertical asymptote is the line at $x = 3$.

iii. The end behavior as $x \to \pm\infty$ is $f(x) \to \frac{2x}{x} = 2$. So there is a horizontal asymptote at the line $y = 2$.

iv. The graph of $f(x)$:

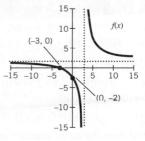

6. $g(x) = \frac{x^2 + 2x - 3}{3x - 1}$

i. To find horizontal intercept(s): set $g(x) = 0$, which is equivalent to setting $x^2 + 2x - 3 = 0 \Rightarrow (x-1)(x+3) = 0 \Rightarrow x = 1, x = -3$. So there are horizontal intercepts at $(1, 0)$ and $(-3, 0)$.

To find vertical intercept let $x = 0$, so $g(0) = 3$. The vertical intercept is at $(0, 3)$.

ii. To find vertical asymptote: set $3x - 1 = 0 \Rightarrow x = 1/3$. So the vertical asymptote is the vertical line at $x = 1/3$.

iii. The end behavior as $x \to \pm\infty$ is $g(x) \Rightarrow \frac{x^2}{3x} = \frac{1}{3}x$. So $g(x)$ is asymptotic to the line $y = \frac{1}{3}x$ (called a slant asymptote).

iv. The graph of $g(x)$:

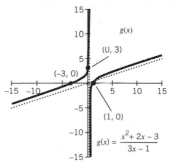

7. a. $x = 10 \Rightarrow y = \frac{1}{8} = 0.125$

$x = 100 \Rightarrow y - \frac{1}{98} \approx 0.0102$

$x = 1000 \Rightarrow y = \frac{1}{998} \approx 0.001$

The horizontal asymptote is $y = 0$.

The domain of the function is $(-\infty, 2) \cup (2, +\infty)$.

b. $x = 10 \Rightarrow y = \frac{23}{112} \approx 0.2054$

$x = 100 \Rightarrow y = \frac{203}{10,192} \approx 0.0199$

$x = 1000 \Rightarrow y = \frac{2003}{1,001,992} \approx 0.0020$

The horizontal asymptote is $y = 0$.

The domain of the function is $(-\infty, -4) \cup (-4, 2) \cup (2, +\infty)$.

8. a. $x = 10 \Rightarrow y = \frac{98}{11} \approx 8.909$

$x = 100 \Rightarrow y = \frac{9998}{101} \approx 98.990$

$x = 1000 \Rightarrow y = \frac{999,998}{1001} \approx 998.999$

As $x \to +\infty, y \to +\infty$.

b. $x = 10 \Rightarrow y = \frac{-800}{4} = -200$

$x = 100 \Rightarrow y = -\frac{980,000}{94} \approx -10,425.532$

$x = 1000 \Rightarrow y = -\frac{998,000,000}{994} \approx -1,004,024.145$

As $x \to +\infty, y \to -\infty$.

Ch. 9

Exercises for Section 9.4

1. a. $F(n) = C(n)/n = 100/n + 100$

 b. $C(10) = 1000 + 100 \cdot 10 = 2000$;
  $C(100) = 1000 + 100 \cdot 100 = 11,000$;

  $C(1000) = 1000 + 100 \cdot 1000 = 101,000$

 c. $F(10) = C(10)/10 = 2000/10 = 200$; $F(100) = C(100)/100 = 11,000/100 = 110$;

  $F(1000) = C(1000)/1000 = 101,000/1000 = 101$

 d. The cost goes up as the number of items produced goes up, but the cost per item goes down as the number of items produced goes up.

3. a. $F(10) = (30 - 2)/(30 + 5) = 28/35 = 0.8$;

  $F(100) = (300 - 2)/(300 + 5) = 298/305 \approx 0.977$;

  $F(1000) = (3000 - 2)/(3000 + 5) = 2998/3005 \approx 0.998$

  All this suggests that as $x \to \infty$, then $y \to 1$; thus $y = 1$ is the horizontal asymptote to the graph of $F(x)$.

 b. $G(10) = (200 - 8)/(100 + 10 - 6) = 192/104 \approx 1.846$;

  $G(100) = (20,000 - 8)/(10,000 + 100 - 6) = 19,998/10,094 \approx 1.981$;

  $G(1000) = (2,000,000 - 8)/(1,000,000 + 1000 - 6) = 1,999,992/1,000,994 \approx 1.998$

  All this suggests that as $x \to +\infty$, then $y \to 2$; thus the graph of $y = 2$ is the horizontal asymptote to the graph of $G(x)$.

5. a. The graph of y_1 will cross the graph of $y_2 = 3$ when $y_1 = y_2$ or $3(x^2 + 1) = 3x^2 + x + 2$; thus $3x^2 + 3 = 3x^2 + x + 2 \Rightarrow 3 = x + 2 \Rightarrow x = 1$. The domain of y_1 is all real numbers; the range of y_1 is approximately all real numbers in the interval $[1.79, 3.21]$. The accompanying graph confirms this.

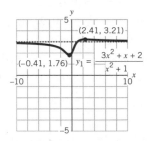

 b. If we rewrite $y_1 = \dfrac{2(x - 2)(x + 2)}{(2x - 3)(x + 2)}$, we can simplify to $y_1 = \dfrac{2(x - 2)}{(2x - 3)}$ where $x \neq -2$. The graph of this simplified y_1 will cross the graph of $y_2 = 1$ when $y_1 = y_2$ or $2x - 4 = 2x - 3$ or when $-4 = -3$, which is false, and thus the graph never crosses the horizontal asymptote. The domain of y_1 is all real numbers except $x = 1.5$ and $x = -2$; the range of y_1 is all real numbers except $y = 1$. The accompanying graph next column confirms this. Note there is a hole in the graph at $x = -2$, since the original function y_1 is not defined at $x = -2$.

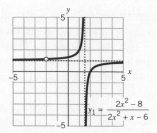

 c. The graph of y_1 will cross the graph of $y_2 = 1/4$ when $y_1 = y_2$ or $4(x^2 + 2x - 2) = 4x^2 - 1$; thus $4x^2 + 8x - 8 = 4x^2 - 1$ or $8x - 8 = -1$ or $x = 7/8$. The domain of y_1 is all real numbers except $x = \pm 0.5$. The range of y_1 is $(-\infty, 1.59] \cup [1.84, \infty)$. The accompanying graph of y_1 confirms this.

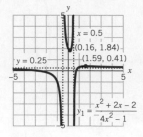

7. For Graph A, the vertical line $x = -2$ is the vertical asymptote and the horizontal line $y = -3$ is the horizontal asymptote. Its domain is all real numbers x except -2. Its range is all real numbers y except -3.

 For Graph B, the vertical line $x = 2$ is the vertical asymptote and the horizontal line $y = 4$ is the horizontal asymptote. Its domain is all real numbers x except 2. Its range is all real numbers y except 4.

 For Graph C, the vertical lines $x = -2$ and $x = 3$ are the vertical asymptotes and the horizontal line $y = -2$ is the horizontal asymptote. Its domain is all real numbers x except -2 and 3. Its range is all real numbers y except -2.

9. $g(x)$ has no horizontal intercepts. The vertical line $x = -3$ is the vertical asymptote. There is a horizontal asymptote $y = 0$. Here is the graph of $g(x)$, with the two asymptotes:

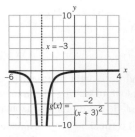

11. a. The domain of $S(x)$ is all real numbers x. The range of $S(x)$ is all y values in $[-0.5, 0.5]$.

 b. $S(x)$ has one horizontal intercept at $x = 0$. It has no vertical asymptotes.

c. As $x \to \pm\infty$, $S(x) \to 0$, so $y = 0$ is its horizontal asymptote.

d. The graph given here resembles certain snake postures.

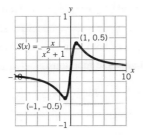

13. A possible function is $f(x) = \frac{(x + 3)(x - 4)}{(x - 1)(x + 5)}$. As $x \to \pm\infty$, then $f(x) \to 1$. The accompanying graph shows the desired properties.

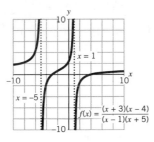

15. a. The graph of $f(x)$ is Graph C, since $f(x) = \frac{-2}{3(x + 2)} + 2 = $

$\frac{-2 + 2 \cdot 3(x + 2)}{3(x + 2)} = \frac{-2 + 6x + 12}{3(x + 2)} = \frac{6x + 10}{3(x + 2)} = \frac{2(3x + 5)}{3(x + 2)}$.

There is a horizontal intercept at $x = -5/3$, and the vertical line $x = -2$ is the vertical asymptote; no other graph has these qualities.

b. The graph of $g(x)$ is Graph B. It has vertical asymptotes at $x = -2$ and $x = 2$, and no other function does.

c. The graph of $h(x)$ is Graph A. It has y-intercepts at $x = -3$ and $x = 3$, a vertical asymptote at $x = 4$, and no other function does.

17. a. $g(x) = (1/3)f(x)$ or $f(x)$ shrinks by a factor of 1/3.

$h(x) = -2f(x - 3)$ or moves the graph 3 units to the right, then stretches $f(x)$ by a factor of 2, and reflects that function's graph across the x-axis.

$k(x) = f(-(x - 2)) + 4$ or moves the graph of $f(x)$ two units to the right, then reflects that graph across the y-axis, and then raises that graph 4 units up.

b. The domain of $g(x)$ is all real numbers x except 0; the domain of $h(x)$ is all real numbers x except 3, and the domain of $k(x)$ is all real numbers x except 2.

c. $g(x)$ has the vertical line $x = 0$ as its vertical asymptote; $h(x)$ has the vertical line $x = 3$ as its vertical asymptote; $k(x)$ has the vertical line $x = 2$ as its vertical asymptote

d. As $x \to \pm\infty$, $f(x) \to 0$, $g(x) \to 0$, and $k(x) \to 4$.

19. a. Area $= x \cdot y$, so $x \cdot y = 400$ sq^2 ft; thus $y = 400/x$.

b. $x + 2y = F$

c. $F = x + 2(400/x) = \frac{x \cdot x + 2(400)}{x} = \frac{x^2 + 800}{x}$. The graph of F shows that the approximate minimum value for F is at 56.57 ft. Note that this gives $x \approx 28.28$ ft and $y \approx 400/28.28 \approx 14.14$ ft. [*Check*: $28.28 + 2 \cdot 14.14 \approx 56.57$ ft.]

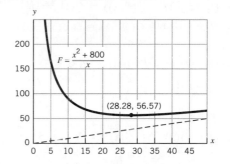

21. a. $A = $ Faculty dinners $+$ grad student dinners $=$

$(72) \cdot \frac{(g + 4)}{(g + 3)} + 12g \cdot \frac{(g + 4)}{(g + 3)} = (72 + 12g)\frac{(g + 4)}{(g + 3)}$.

b. Since $M = 24(g + 4) - A$ is the amount M paid by the department, substituting the A value from part (a) above gives:

$M = 24(g + 4) - (72 + 12g)\frac{(g + 4)}{(g + 3)}$

c. $M = 24(g + 4)\frac{(g + 3)}{(g + 3)} - (72 + 12g)\frac{(g + 4)}{(g + 3)}$

$= [24(g + 3) - (72 + 12g)]\frac{(g + 4)}{(g + 3)}$

$= [24g + 72 - 72 - 12g]\frac{(g + 4)}{(g + 3)}$

$= 12g\frac{(g + 4)}{(g + 3)} = g \cdot 12\frac{(g + 4)}{(g + 3)}$

d. The amount for grad student dinners $= 12g\frac{(g + 4)}{(g + 3)}$. If $g = 40$, then $12(40) \cdot \frac{(40 + 4)}{(40 + 3)} = \$491.16$; if $g = 10$, then $12(10) \cdot \frac{(10 + 4)}{(10 + 3)} = \$129.23$. So the cost for each of the 40 grad students is $\$491.16/40 = \$12.27$, and the cost for each of the 10 grad students is $\$129.23/10 = \$12.92$. The difference is $\$0.64$.

e. We have $g = $ the number of graduate students, and the cost of the graduate students meals is $12g \cdot \frac{(g + 4)}{(g + 3)}$. With a nonpaying guest, the denominator of the fractions is still $(g + 3)$ but the numerator is now $(g + 5)$ instead of $(g + 4)$. Thus $M = 12g\frac{(g + 5)}{(g + 3)}$.

Section 9.5

Algebra Aerobics 9.5a

1. $f(x) = 2x + 3$, $g(x) = x^2 - 4$

a. $f(g(2)) = f(0) = 2(0) + 3 = 3$

b. $g(f(2)) = g(7) = (7)^2 - 4 = 49 - 4 = 45$

c. $f(g(3)) = f(5) = 2(5) + 3 = 10 + 3 = 13$

d. $f(f(3)) = f(9) = 2(9) + 3 = 18 + 3 = 21$

e. $(f \circ g)(x) = f(g(x)) = f(x^2 - 4) = 2(x^2 - 4) + 3$
$$= 2x^2 - 8 + 3 = 2x^2 - 5$$

f. $(g \circ f)(x) = g(f(x)) = g(2x + 3) = (2x + 3)^2 - 4$
$$= 4x^2 + 12x + 9 - 4 = 4x^2 + 12x + 5$$

2. a. $(P \circ Q)(2) = P(Q(2)) = P(3(2) - 5) = P(1) = \frac{1}{1} = 1$

b. $(Q \circ P)(2) = Q(P(2)) = Q(\frac{1}{2}) = 3(\frac{1}{2}) - 5 = \frac{-7}{2}$

c. $(Q \circ Q)(3) = Q(Q(3)) = Q(3(3) - 5) = Q(4)$
$$= 3(4) - 5 = 7$$

d. $(P \circ Q)(t) = P(Q(t)) = P(3t - 5) = \frac{1}{3t - 5}$

e. $(Q \circ P)(t) = Q(P(t)) = Q(\frac{1}{t}) = 3(\frac{1}{t}) - 5 = \frac{3}{t} - 5$

3. $F(x) = \frac{2}{x - 1}$, $G(x) = 3x - 5$

a. $(F \circ G)(x) = F(G(x)) = F(3x - 5) = \frac{2}{(3x - 5) - 1}$
$$= \frac{2}{3x - 6}$$

b. $(G \circ F)(x) = G(F(x)) = G(\frac{2}{x - 1}) = 3(\frac{2}{x - 1}) - 5$
$$= \frac{6}{x - 1} - 5$$

c. $(F \circ G)(x) = \frac{2}{3x - 6} \neq (G \circ F)(x) = \frac{6}{x - 1} - 5 =$
$$\frac{6 - 5(x - 1)}{x - 1} = \frac{11 - 5x}{x - 1}$$

4. a. $f(-2) = 2$ **e.** $(g \circ f)(-2) = g(2) = 2$

b. $g(-2) = 0$ **f.** $(f \circ g)(-2) = f(0) = -2$

c. $f(0) = -2$ **g.** $(g \circ f)(0) = g(-2) = 0$

d. $g(0) = 1$ **h.** $(f \circ g)(0) = f(1) = -1$

5. a. $f(x) = x^2 - 2$

b. $g(x) = \frac{1}{2}x + 1$

c. $(g \circ f)(x) = g(x^2 - 2) = \frac{1}{2}(x^2 - 2) + 1$
$$= \frac{1}{2}x^2 - 1 + 1 = \frac{1}{2}x^2$$

d. $(f \circ g)(x) = f(\frac{1}{2}x + 1) = (\frac{1}{2}x + 1)^2 - 2$
$$= (\frac{1}{4}x^2 + x + 1) - 2 = \frac{1}{4}x^2 + x - 1$$

6. a. $(g \circ f)(-2) = \frac{1}{2}(-2)^2 = 2$ and
$$(f \circ g)(-2) = \frac{1}{4}(-2)^2 + (-2) - 1 = -2.$$

So both answers agree with the answers in Problem 4, parts (e) and (f).

7. a. $(h \circ f \circ g)(4) = h(f(g(4))) = h(f(0)) = h(3) = 4$

b. $(f \circ h \circ g)(1) = f(h(g(1))) = f(h(3)) = f(4) = 5$

8. a. $(h \circ f \circ g)(3) = h(f(g(3))) = h(f(2)) = h(-1/3) = 3$

b. $(f \circ g \circ h)(100) = f(g(h(100))) = f(g(3)) = f(2) = -1/3$

Algebra Aerobics 9.5b

1. a.

t	$g(t)$	t	$h(t)$
0	5	−1	3
1	3	1	2
2	1	3	1
3	−1	5	0

b. $(g \circ h)(3) = g(h(3)) = g(1) = 3$; $(h \circ g)(3) = h(g(3)) = h(-1) = 3$.

c. Yes, since $(g \circ h)(t) = g(h(t)) = g(\frac{5 - t}{2}) =$
$$5 - 2(\frac{5 - t}{2}) = 5 - (5 - t) = t \text{ and } (h \circ g)(t) =$$
$$h(g(t)) = h(5 - 2t) = \frac{5 - (5 - 2t)}{2} = t \text{ and the domains}$$
and ranges of both g and h are all real numbers.

d. They are inverse functions.

2. $f(x) = 2x + 1$, $g(x) = \frac{x - 1}{2}$
$$(f \circ g)(x) = f(g(x)) = f(\frac{x - 1}{2}) = 2(\frac{x - 1}{2}) + 1$$
$$= \frac{2(x - 1)}{2} + 1 = x - 1 + 1 = x$$
$$(g \circ f)(x) = g(f(x)) = g(2x + 1) = \frac{(2x + 1) - 1}{2}$$
$$= \frac{2x}{2} = x$$

3. $f(x) = \sqrt[3]{x} + 1$, $g(x) = x^3 - 1$
$$(f \circ g)(x) = f(g(x)) = f(x^3 - 1) = \sqrt[3]{(x^3 - 1) + 1}$$
$$= \sqrt[3]{x^3} = (x^3)^{1/3} = x^1 = x$$
$$(g \circ f)(x) = g(f(x)) = g(\sqrt[3]{x} + 1)$$
$$= (\sqrt[3]{x} + 1)^3 - 1 = [(x + 1)^{1/3}]^3 - 1$$
$$= (x + 1) - 1 = x$$

4. $f(f^{-1}(x)) = f(\frac{1 + x}{x}) = \frac{1}{\frac{1 + x}{x} - 1} = \frac{1}{\frac{1 + x}{x} - \frac{x}{x}}$
$$= \frac{1}{(1/x)} = x$$
$$f^{-1}(f(x)) = f^{-1}(\frac{1}{x - 1}) = \frac{1 + \frac{1}{x - 1}}{\frac{1}{x - 1}} = \frac{\frac{x - 1}{x - 1} + \frac{1}{x - 1}}{\frac{1}{x - 1}}$$
$$= \frac{\frac{x}{x - 1}}{\frac{1}{x - 1}} = \frac{x}{x - 1} \cdot \frac{x - 1}{1} = x$$

So $f(f^{-1}(x)) = x$ and $f^{-1}(f(x)) = x$.

5. Letting $f(x) = y$, we have $y = \frac{3}{x} + 5 \Rightarrow y - 5 = \frac{3}{x} \Rightarrow$ $x = \frac{3}{y - 5} \Rightarrow f^{-1}(x) = \frac{3}{x - 5}$ (using the convention of designating x as the input variable).

6. Letting $g(x) = y$, we have $y = (x - 2)^{3/2} \Rightarrow x = y^{2/3} + 2 \Rightarrow$ $g^{-1}(x) = x^{2/3} + 2$ (using the convention of designating x as the input variable).

7. Letting $h(x) = y$, we have $y = 5x^3 - 4 \Rightarrow y + 4 = 5x^3 \Rightarrow$ $x = (\frac{y + 4}{5})^{1/3} \Rightarrow h^{-1}(x) = (\frac{x + 4}{5})^{1/3}$ (using the convention of designating x as the input variable).

8. a. Saying "no"

b. Taking the bus from home, then going to class.

c. Turning off the light, leaving the room, closing the door, and then locking the door.

d. Dividing x by 5 and then adding 3

e. Subtracting 2 from z and then dividing the result by -3

9. The functions in Graphs A and D are 1-1 since they pass the horizontal line test. The functions in Graphs B and C are not 1-1, since they fail that test.

10. a. f is not one-to-one on the domain of all real numbers (since it fails the horizontal line test), so it cannot have an inverse.

b. If we restrict the domain of f to $x \geq -2$, then f has an inverse on this new domain (since it now passes the vertical test).

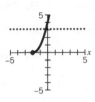

c. Letting $f(x) = y$, we have

$$y - (x + 2)^2 \Rightarrow x = +\sqrt{y} - 2 \Rightarrow f^{-1}(x) = \sqrt{x} - 2$$

(using the convention of designating x as the input variable). The graph of $f^{-1}(x)$ is shown below.

11. a. The radius $R(t)$ as a function of time t is $R(t) = 10t$.

b. $R(2) = 10 \cdot 2 = 20$, so area $A(20) = \pi(20)^2 \approx 1257$ square feet.

c. $(A \circ R)(t) = A(R(t)) \rightarrow A(10t) = \pi(10t)^2$
$$= 100\pi t^2$$

Exercises for Section 9.5

1. a. $f(g(1)) = f(0) = 2$

b. $g(f(1)) = g(1) - 0$

c. $f(g(0)) = f(1) - 1$

d. $g(f(0)) = g(2) = 3$

e. $f(f(2)) = f(3) = 0$

3. a. $g(f(2)) = g(0) = 1$

b. $f(g(-1)) = f(2) = 0$

c. $g(f(0)) = g(4) = -3$

d. $g(f(1)) = g(3) = -2$

5. a. $F(G(1)) = F(0) = 1$

b. $G(F(-2)) = G(-3) = 4$

c. $F(G(2)) = F(0.25) = 1.5$

d. $F(F(0)) = F(1) = 3$

e. $(F \circ G)(x) = 2\left(\frac{x-1}{x+2}\right) + 1$

$$= \frac{(2x - 2) + (x + 2)}{x + 2} = \frac{3x}{x + 2}$$

f. $(G \circ F)(x) = \frac{(2x + 1) - 1}{(2x + 1) + 2} = \frac{2x}{2x + 3}$

7. a. $A(r) = \pi r^2$, where r is measured in feet and $A(r)$ is measured in square feet.

b. $r = R(t) = 5t$, where t is measured in minutes and $R(t)$ is measured in feet.

c. $A(R(t)) = \pi 25t^2$, where t is measured in minutes and A is measured in square feet.

d. $A(R(10)) = \pi \cdot 25 \cdot 10^2 = 2500\pi \approx 7854$ sq. ft and $A(R(60)) = \pi \cdot 25 \cdot 60^2 \approx 282{,}743$ sq. ft.

9. $r(t) = 13t$ and thus $A(r(t)) = \pi(13t)^2 = 169\pi t^2$

11. a. $T(s) = 32 - 5s$

b. If the road is 40 feet wide, then $k = 20$ and thus $S = S(x) = \left[1 - \frac{1}{2} \cdot \left(\frac{x}{20}\right)^2\right]S_d = \left[1 - \frac{x^2}{800}\right]S_d$.

c. At the middle of the 40-ft road $x = 0$ and therefore $S(0) = \left[1 - \frac{0}{800}\right]S_d = S_d$. At the edge of the 40-ft road, $x = 20$, so $S(20) = \left[1 - \frac{20^2}{800}\right]S_d = \frac{1}{2}S_d$.

d. $T(S(x)) = T\left(\left[1 - \frac{x^2}{800}\right]S_d\right) = 32 - 5\left[1 - \frac{x^2}{800}\right]S_d$
$$= 32 - 5S_d + \frac{x^2}{160}S_d$$

e. $T(S(0)) = 32 - 5S_d$ and $T(S(20)) = 32 - 5S_d + \frac{20^2}{160}S_d = 32 - 5S_d + 2.5S_d = 32 - 2.5S_d$

13. a. $M(x) - (L \circ J \circ K)(x) = L(J(K(x))) = L(J(\log x)) = L((\log x)^3) = 1/(\log x)^3$

b. Take the log of x, cube the result, and then place it in the denominator, with 1 as the numerator.

15. If $f(x) = 4x$, $g(x) = e^x$, and $h(x) = x - 1$, then $f(g(h(x))) = f(g(x - 1)) = f(e^{x-1}) = 4e^{x-1} = j(x)$

17. $f(g(x)) = f(x^2 + 1) = \sqrt{(x^2 + 1) - 1} = \sqrt{x^2} = x$ since $x > 0$

$$g(f(x)) = g\left(\sqrt{x - 1}\right) = \left(\sqrt{x - 1}\right)^2 + 1$$
$$= (x - 1) + 1 = x$$

19. $f(g(x)) - f\left(\frac{x^3 - 5}{4}\right) = \sqrt[3]{4\left(\frac{x^3 - 5}{4}\right) + 5} = \sqrt[3]{x^3} = x$

$$g(f(x)) = g\left(\sqrt[3]{4x + 5}\right) = \frac{\left(\sqrt[3]{(4x + 5)}\right)^3 - 5}{4}$$
$$= \frac{(4x + 5) - 5}{4} = x$$

21. $F(G(t)) = F(\ln(t^{1/3})) = e^{3\ln(t^{1/3})} = e^{\ln(t)} = t$ (where $t > 0$)

$G(F(t)) = G(e^{3t}) = \ln(e^{3t})^{1/3} = \ln(e^t) = t$

23.

x	$f^{-1}(x)$
5	-2
1	-1
2	0
4	1

25. a. Yes, this is a 1-1 function since each letter is associated with a unique number. The inverse function would just consist of matching each number between 1 and 26 with its letter equivalent. The domain of the inverse function would be the integers 1 through 26.

b. "MATH RULES"

27. a. Yes, $f(x)$ has an inverse since its graph passes the horizontal line test.

b. The domain of $f(x)$ is the interval $[-4, \infty]$. The range of $f(x)$ is the interval $[0, \infty]$.

c. $f(-4) = 0$, $f(0) = 2$, and $f(5) = 3$. This means that the points $(-4, 0)$, $(0, 2)$, and $(5, 3)$ all lie on the graph of $f(x)$.

d. Given the results in part (c), the points $(0, -4)$, $(2, 0)$, and $(3, 5)$ all lie on the graph of $f^{-1}(x)$. So $f^{-1}(0) = -4$, $f^{-1}(2) = 0$, $f^{-1}(3) = 5$.

29. $Q^{-1}(x) = \frac{3x + 15}{2}$, $Q(3) = -3$, $Q^{-1}(3) = 12$

31. $Q^{-1}(x) = \frac{3}{x - 1}$, $Q(3) = 2$, $Q^{-1}(3) = 3/2$

33. a.

x (cups)	4	8	16	32
$f(x)$ (quarts)	1	2	4	8

x (quarts)	2	4	8	16
$g(x)$ (gallons)	0.5	1	2	4

b. **i.** $(g \circ f)(8) = g(2) = 0.5$ gal
   ii. $g^{-1}(2) = 8$ qt
   iii. $(f^{-1} \circ g^{-1})(1) = f^{-1}(4) = 16$ cups
   iv. $(f^{-1} \circ g^{-1})(2) = f^{-1}(8) = 32$ cups

c. $(f^{-1} \circ g^{-1})(x)$ is a function that converts gallons to cups.

35. a. $W_{men}(h) = 50 + 2.3(h - 60)$, where a reasonable domain might be $60'' \le h \le 78''$;
$W_{women}(h) = 45.5 + 2.3(h - 60)$, where a reasonable domain might be $60'' \le h \le 74''$.

b. $W_{men}(70) = 50 + 2.3(70 - 60) = 73$ kg, so 73 kg is the "ideal" weight of a $5'10''$ man.
$W_{women}(66) = 45.5 + 2.3(66 - 60) = 59.3$ kg, so 59.3 kg is the "ideal" weight of a $5'6''$ woman.

c. $W^{-1}_{men}(77.6)$ means $77.6 = 50 + 2.3(h - 60) \Rightarrow h = 72$ inches. A man with a IBW of 77.6 kg should be 72 inches or 6 ft tall.

d. $W_{newman}(h) = \frac{50 + 2.3(h - 60)}{0.4356}$;
$W_{newwomen}(h) = \frac{45.5 + 2.3(h - 60)}{0.4356}$.

e. $W^{-1}_{newwomen}(125)$ means $125 = \frac{45.5 + 2.3(h - 60)}{0.4356} \Rightarrow h \approx 63.89 \approx 64$ inches. So 125 lb is the IBW for a woman about $5'4''$ in height.

37. $F(G(x)) = F(\log_a(x)) = a^{\log_a(x)} = x$ where $x > 0$.
$G(F(x)) = G(a^x) = \log_a(a^x) = x$

Ch. 9: Check Your Understanding

1. True	**5.** True	**9.** False	**13.** True
2. False	**6.** False	**10.** False	**14.** False
3. True	**7.** True	**11.** False	**15.** True
4. False	**8.** False	**12.** True	**16.** True

17. Possible answer: $y = x^4 + x^2 + 1$

18. Possible answer: $f(x) = (x + 1)(x - 3)(x - 4)$

19. Possible answer: $h(x) = 2(x + 1)(x - 3)(x - 4)$; $g(x) = -3(x + 1)(x - 3)(x - 4)$

20. $h(t) = \frac{1}{4}(x + 2)(x + 1)(x - 2)(x - 3)$

21. Possible answer: $f(x) = x^3 + 2x^2$; $g(x) = 5x - 2$

22. $H(t) = 3t + 1$, $Q(t) = \sqrt{t}$

23. Possible answer: $f(x) = \frac{x - 2}{x(x + 3)}$

24. $h(x) = \frac{1}{2}\sqrt{x + 5}$

25. Possible answer: $f(x) = \frac{2x^2 + 1}{(x - 1)^2}$

26. True	**30.** True	**34.** False	**38.** False
27. False	**31.** True	**35.** True	**39.** False
28. False	**32.** True	**36.** False	**40.** True
29. False	**33.** True	**37.** True	

Ch. 9 Review: Putting It All Together

1. a. The graphs are:

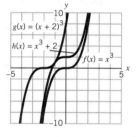

b. $g(x) = (x + 2)^3 = f(x + 2)$; $h(x) = x^3 + 2 = f(x) + 2$

3. a. True   **c.** True   **e.** True
   b. True   **d.** False   **f.** False

5. The graphs of the three functions are given below. The graph of $g(x) = 4f(x)$ or $f(x)$ is stretched by a factor of 4; the graph of $h(x) = -0.5f(x)$ or $f(x)$ is compressed by a factor of 0.5 and rotated about the x-axis.

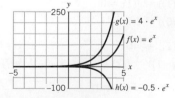

7. a. $g(x) = (2/3)f(x - 4) - 1$ or $f(x)$ is shifted to the right by 4 units, then is compressed by a factor of 2/3, and then is shifted down by 1 unit. Yes the order matters. One needs to stretch/compress before shifting up or down.

b. $g(x) = \frac{2}{3(x - 4)} - 1 = \frac{2 - 3(x - 4)}{3(x - 4)} = \frac{-3x + 14}{3x - 12}$

c. The domain of $g(x)$ is all real numbers except 4. Its graph is given on the next page.

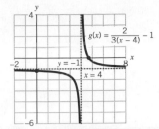

$$g(x) = \frac{2}{3(x-4)} - 1$$

$y = -1$

$x = 4$

d. The x-intercept is at $(14/3, 0)$, and the y-intercept is at $(0, -7/6)$

e. The vertical asymptote of $g(x)$ is the vertical line $x = 4$.

f. As $x \to \pm\infty$, we have $g(x) \to -1$.

9. Estimating from the graph, we have:

x	$f(x)$	$g(x)$	$f(x) + g(x)$	$f(x)/g(x)$
-2	4	-6	-2	-0.667
-1	2	-2.5	-0.5	-0.800
0	1	0	1	not defined
1	0.5	1.5	2	0.333
2	0.25	2	2.25	0.125

11. a. $S(t) = 5000 \cdot (1.04)^t$; $R(t) = 5000 \cdot (1.10)^t$; $T(t) = S(t) + R(t)$.

b. The graphs of S, R, and T are given here:

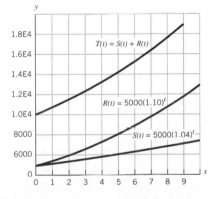

c. The money left would be just what $S(t)$ supplies, namely, $S(30) \approx \$16,217$.

d. It is not. If $A(t) = 5000 \cdot (1.14)^t$, then, e.g., at $t = 0$ we have $A(0) = 5000$ and $T(0) = S(0) + R(0) = 10000$, so $A(t) \neq T(t)$. Also, $5000(1.10)^t + 5000(1.04)^t = 5000(1.10^t + 1.04^t) \neq 5000(1.14^t)$. If we look at the graphs of A and T, we see that the two graphs are not equal most of the time.

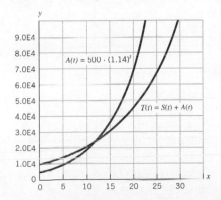

13. $g(x) = -f(x)$ and $h(x) = -f(-x)$

15. a. The graph of $j(x)$ does not match either Graph A or Graph B.

b. The graph of $g(x)$ does not match either Graph A or Graph B.

c. The graph of $f(x)$ matches Graph B, since it has vertical asymptotes $x = 1$ and $x = -1$, and a horizontal asymptote $y = 1$ and a horizontal intercept at $(0, 0)$. The domain is all real numbers except $x = 1$ and $x = -1$. The range is all real numbers except $y - 1$.

d. The graph of $h(x)$ matches Graph A, since it has a vertical asymptote $x = 1$ and a horizontal intercept at $(-5, 0)$ and a horizontal asymptote $y = 2$. The domain is all real numbers except $x = 1$. The range is all real numbers except $y = 2$.

17. a. Cost of electricity $= (0.16/kh) \cdot (500 \text{ kwh/year}) = \$80.00$ per year.

b. $C(n) = 720 + 80n$, where n is measured in years and $C(n)$ in dollars.

c. The total cost per year is $T(n) = C(n)/n = 720/n + 80$.

d. $T(5) = \$224$; $T(10) = \$152$; $T(20) = \$116$. As n increases the total cost per year, $T(n) = C(n)/n$ decreases. Thus, as the refrigerator ages, the initial cost is spread over many years and the total cost per year decreases. The graph of $T(n)$ is shown below.

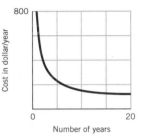

19. $f(x)$ is x^3 shifted to the right by 2 and up by 1, so it passes the horizontal line test and, yes, $f(x)$ has an inverse. We can find the inverse by solving $x = (y - 2)^3 + 1$ for y. The equation implies that $x - 1 = (y - 2)^3$; therefore, $y - 2 = \sqrt[3]{x - 1}$ or $y = 2 + \sqrt[3]{x - 1}$, which is the inverse function.

21. a. i. $762 \text{ mph} = \dfrac{762 \text{ miles}}{\text{hour}} \cdot \dfrac{5280 \text{ ft}}{\text{mile}} \cdot \dfrac{\text{hour}}{60 \text{ min}} \cdot \dfrac{\text{min}}{60 \text{ sec}}$

$= 1117.6 \text{ ft/sec}$

ii. Distance $=$ rate $\cdot$ time; thus $D(t) = 1117.6 \cdot t$, where t is measured in seconds and $D(t)$ is measured in feet.

iii. The rule of thumb seems reasonable since the sound travels approximately 1000 ft/sec

b. i. $A(r) = \pi \cdot r^2$ square feet.

ii. $A(D(t)) = \pi \cdot (1117.6t)^2$; thus $A(D(4)) = 62{,}783{,}083.49 \text{ ft}^2$. Now 1 $mi^2 = 5280^2 \text{ ft}^2$; thus $\dfrac{62{,}783{,}083.49 \text{ ft}^2}{5280^2 \text{ ft}^2/\text{mi}^2} \approx 2.25 \text{ mi}^2$

iii. If the time doubles, the distance doubles and the area quadruples.

INDEX

a-b-c form (quadratic function), 470. *See also* Standard form (quadratic functions)
Absolute value, 118
Absolute value function:
    definition of, 118, 119
    and slope, 94–96
Abstract variables, 14
Acadia National Park, night temperatures at, 91
Acceleration, 520–522
Acetaminophen, dose of, 326, 327
Acidic substances, 373
Acidity, measuring, 372–374
ac method of factoring, 499–500
Acreage, 240
Addition:
    of functions, 553
    of ordinates, 558
    of power functions, *see* Polynomial functions
Additive change, in linear functions, 278
Additive scales, 244
Aerobic exercise, 205
Aerospace industry, profits and losses in, 65
Age:
    of Earth, 218, 251
    median age of U.S. population, 10, 11, 62
Age distribution, U.S., 3–4
Aging, 161–162
a-h-k form (quadratic functions), 483. *See also* Vertex form (quadratic functions)
AIDS cases, in United States, 17, 83–84
Airplane travel, 180
Albania, pyramid schemes in, 606
Algebra, of functions, 553–557
Algeria, 300
Alkaline substances, 373
Allometric laws, 453
Allometry, 450–453
American College of Sports Medicine, 205
American Lung Association, 127
Angstrom (Å), 217, 241
Animals:
    metabolic rate vs. body mass for, 452–453
    surface area vs. body mass for, 450–452
Annual percentage rate (APR), 320
Annual percentage yield (APY), 320
Anthrax spores, 241
Antifreeze, 106
Apple Inc., 560
APR (annual percentage rate), 320
APY (annual percentage yield), 320
Arctic temperatures, mean, 43
Aspirin, 387
Asymptotes:
    of exponential functions, 289
    of logarithmic functions, 368
    of negative integer power functions, 435
    of rational functions, 577–579
Asymptotic graphs, 289
Atmospheric pressure:
    half-life of, 359
    measurement of, 431
Atomic weapons, half-life of, 276

Austin, Thomas, 301
Automobile insurance, mean cost for, 64
Average, 5. *See also* Mean; Median
Average acceleration, 521
Average cost, 576
Average rate of change, 60–64
    change in, 67–69
    definition of, 61–62
    for exponential vs. linear functions, 282
    limitations of, 62–63
    of quadratic functions, 510–513
    slope as, 72–75
Average velocity, 519
Avogadro's number, 217, 372
Axis of symmetry, 468, 471

Baby boomers, 62, 68
Babylonians, ancient, 473
BAC, *see* Blood alcohol concentration
Bank of America, 340
Bar charts, 2–3
Base:
    of an exponent, 218
    exponential functions with base e, 361–364
    on graphs of exponential functions, 286–287
    logarithmic base 10 of x, 347
Base value, *see* Initial value
Basic substance, 372
Battleship gun range, quadratic model for, 496
Beer, pH of, 388
Behavior, of polynomial functions, 565–566, 571
Bell, Alexander Graham, 374
Bernoulli, J., 322
Betelgeuse (star), 241
Big Bang theory, 216
Binary numbers, 574
Birds, masses of, 456
Birth rate, decreases in, 181
Bismuth-214, 356
Blood alcohol concentration (BAC):
    calculation of, 300
    and driving laws, 190
    and weight, 107
BMI (body mass index), 201–202, 241
Body mass, 452
    and heart rate, 454
    metabolic rate vs., 452–453
    and oxygen consumption, 457
    surface area vs., 450–452
Body mass index (BMI), 201–202, 241
Body surface area (BSA), 259
Body temperature, 123, 397
Boiling temperature, of water, 144, 246
Bombs, hydrogen, 315
Bon Appétit, Monsieur Soleil (film), 469n.1
Boston Marathon, winning times for women in, 144
Bottled water, sales of, 326
Boulding, Kenneth, 297
Bowling balls, floating of, 406

Boyle's Law of Gases, 424, 431, 432
BP (British Petroleum) oil spill, 225
Braille, print vs., 108
Brazil:
    deforestation in Brazilian Amazon, 43
    inflation in, 341
Breakeven points, 185–187
Breast cancer:
    rates of, 132
    survival rate, 615
    tumor growth, 599–601
Brewers yeast, 603–604
British Petroleum (BP) oil spill, 225
BSA (body surface area), 259

Caffeine levels, 275, 293–294
California:
    land affected by drought in, 192
    production of tomatoes in, 534
Cancer:
    breast, 132, 599–601, 615
    lung, 330, 338, 341
    prostate, 17–18
Carbon-14, 315
Carbon dioxide (CO_2) emissions, 152–153
Carlson, Tor, 603
Carrying capacity, 268, 269, 604
Cars, *see* Motor vehicles
Case, Karl, 19
Case-Shiller index, 19
Cassette sales, U.S., 33
Categorical (qualitative) data, 2
Cats, growth in population of, 366
Causation, correlation vs., 138–139
CDs (certificates of deposit), interest rates on, 329
Cell phones:
    inverse function, 591
    plans, 204
    sales of, 339
    use of, 132, 150, 295
    worldwide subscribers of, 314
Certificates of deposit (CDs), interest rates on, 329
Chan, Margaret, 605
Change(s):
    describing, using equations, 13–16
    outside and inside changes to functions, 540
    rate of change, *see* Average rate of change; Linear functions
Cherry trees, growth of, 378
Children:
    height of, 124–126, 447
    use of electronics by, 65
    weight of, 447
China:
    one-child policy in, 302, 387
    population of, 225, 387
    U.S. trade with, 386
Chlorine, 299
Cholesterol, 190
    good vs. bad, 190

Cigarettes, U.S. consumption and exports of, 66. *See also* Smoking
Civil disturbances, 74–75
Classified ads, revenue from, 17
Clean technology, venture capital investment in, 7
Closed intervals, 31
Clothing sizes, 149
CO_2 (carbon dioxide) emissions, 152–153
Coca-Cola Classic, 239
Coefficients:
    of polynomial functions, 562
    in scientific notation, 215
Coffee:
    caffeine, 275, 293–294
    cooling, 547
College completion, 65, 66
    for 1960–2009, 129
    by sex, 160
College tuition, 108, 142–143, 470
Colombia, pyramid schemes in, 606
Colorado River, 257
Combinations of functions, 553–557
Common logarithms:
    definition of, 248–249, 347
    rules for, 348, 357, 370
Comparative anatomy, 397
Complex numbers, 503–504
Composition of functions, 585–588
Compounded annually, interest, 318–320
Compounded continuously, interest, *see* Continuous compounding
Compounded *n* times a year, interest, 320–321
Compounded twice a year, interest, 320
Compounding debt, 323
Compound interest, 266–267, 318–324
    compounding at different intervals, 319–321
    continuous compounding, 321–322
    continuous compounding formula, 323–324
    simple vs., 267
Compressing:
    new functions from old, 540
    of quadratic functions, 478
Computers, 574
    viruses, 342
Computer use, data on, 65
Concavity (of a graph), 38
Conservation pricing, 196–197
Constant of proportionality, 400, 422
Constant percent change, exponential functions for, 292–297
Constant rate of change, 85
Constant terms (of polynomial functions), 562
Constructal Theory, 463
Consumer surplus, 192
Continuous compounding:
    continuous growth/decay and growth/decay rate, 325–327
    exponential functions with continuous growth/decay, 324–325
    formula for, 323–324
    of interest, 321–322
Continuous (instantaneous) decay rate, 324–327, 361–364
Continuous (instantaneous) growth rate, 324–327, 361–364
Convention on Trade in Endangered Species, 65

Conversion factors, 237–239
Cooking devices, solar parabolic, 469
Correlation, causation vs., 138–139
Correlation coefficient, 137–138
Cosmic ray bombardment, 315
Cost(s):
    average, 575–576
    of health care, 132
    holding, 560
    of inventory, 560
    marginal and fixed, 576
    mean cost for automobile insurance, 64
    ordering, 560
    of refrigerators, 614–615
    revenue vs., 186
Cotton fabrics, 191
Crime(s):
    federal prosecution of, 51
    number of personal/property, 17
    violent, 75
Crude oil prices, 205
Cryptology, 597
Cube:
    length of edge, 395
    surface area of, 394–397
    volume of, 230, 394–397
Culp, Lawrence, Jr., 257
Cylinder, volume of, 398

Dark matter, 216n3
Data:
    categorical (qualitative), 2
    choice of end points and summarizing of, 78–81
    definition of, 2
    influencing interpretations of, 78–81
    linear models of, *see* Linear models of data
    single-variable, 2–5
    two-variable, 10–16
Data Analysis of Politics and Policy (Edward Tufte), 124
Data dictionary, 134
Data points, linear vs. exponential models from two, 280–282
Data sets:
    end points for summaries of, 78–79
    functional models for, *see* Functional models
Data tables, *see* Table(s)
Death rates:
    increases in, 181
    of infants, 131, 455
    from motor vehicle accidents, 64, 66
    from natural disasters, 386–387
    of smokers, 334
    from tornadoes, 76
Debt:
    compounding, 323
    federal, *see* Federal debt
    household, 601–603
    national, 48
    net, 169
Decay, 271–276, 302
    exponential, *see* Exponential decay
    of fossil fuel emissions, 278
    of iodine-131, 271–272
    radioactive, 271–272, 304

Decay factor, 272, 286, 294
Decay rate:
    in decimal form, 294
    instantaneous, 324–327, 361–364
    in percentage form, 294
Decibel scale, 374–376
Decimal form:
    decay rate in, 294
    growth rate in, 293
Deer ticks, 317
Deflection, 403–405
Deforestation, in Brazilian Amazon, 43
Demand curve, 175
Deneb (star), 220, 240–241
Department of Energy, 206
Dependent variables:
    definition of, 21
    in function notation, 27
Depreciation:
    of a computer, 87
Devine, B. J., 598
Difference:
    of linear function output values, 279
    of two squares, 498
Digital information, 258
Direct proportionality:
    defined, 400
    and inverse proportionality, 422
    of length, surface area, and volume, 395
    of linear functions, 107–109
    with multiple variables, 404–405
    properties of, 401–404
Discriminant, 501–503
Distance:
    in equations of motion, 516–518
    and height of an object in free fall, 522–524
    and velocity, 518–520
Division:
    of functions, 553
    of logarithms, 351
    of polynomial functions, *see* Rational function[s]
DNA molecules, size of, 213, 217, 242–243
Domain(s):
    of composition functions, 586
    of a function, 31–34
    of exponential, 286
    of linear functions, 96
    of logarithmic functions, 368
    for negative integer powers, 437
    piecewise defined, 195
    of polynomial functions, 567
    for positive integer powers, 413
    of quadratic functions, 484
    of rational functions, 578
    representing with interval notation, 31
    visualizing, 32
Doubling time, 266–268, 303–309
    finding equivalent equations with, 303–305
    for lung cancer, 341
    rule of 70 for calculating, 305–306
    solving, with logarithms, 353
    solving, with natural logarithms, 359–360
Dow Jones Industrial Average, 17, 81, 252
Drinking water, access to, 153
Drought, land in California affected by, 192

Drug dosages, 259, 315
Duckweed, 360
　doubling time, 360

e (constant):
　and continuous compounding, 321–322
　in continuous compounding formula,
　　323–324
　converting, into growth/decay rate, 325–327
　definition of, 322
　exponential functions using, 324–325,
　　361–364
E. coli population growth:
　exponential function for, 292–293
　mathematical model for, 264–265
　plotting, 331
　real growth of, 268–269
Earth:
　age of, 218, 251
　distance from sun, 233, 239, 240
　gravity on, 526
　measuring, 231
　radius of, 398, 593
　radius of core of, 402–403
　size of, 221, 231, 258
Earthquakes, measurement of, 243, 244–245,
　388
Educational levels, U.S., 2–3, 77
Effective interest rate, 320
Electrical resistance (*R*), 115
Electric heating, 164, 165
Electromagnetic (EM) radiation, 246
Electronic devices, children's use of, 65
Elephants, 64–65
Elephant seals, 257
Elimination method (for linear systems),
　173–174
Elvis impersonators, 314
Embryo, growth of, 495
EM (electromagnetic) radiation, 246
End behavior, of rational functions, 579
End points:
　for average rate of change, 63
　for summaries of data sets, 78–79
Energy:
　consumption, 206, 257, 339
　consumption vs. production, 192
Energy-saving walls, 585
English system of units, 212
Epicenter, of earthquake, 243
Epinephrine, 353
EpiPen, 353
Equation(s):
　describing a function using an, 20
　describing change using, 13–16
　estimating exponential function solutions
　　from, 346–347
　graph of, 15
　of exponential functions on semi-log
　　plots, 380
　function notation with, 27–30
　of inverse functions, 593
　of linear functions, 88–90, 99–103
　of linear models of data, 126
　of power functions on log-log plots, 445
　solutions to the, 14
　solving, 27

Equilibrium point, for supply and demand, 176
Equivalent equations:
　finding, with doubling time/half-life, 303–305
　in linear systems, 175
Euler, L., 322
Even functions negative integer powers,
　434–436
Even functions positive integer powers,
　410–411
Exabytes, 258
Exchange rates (currency), 106, 108, 109,
　593, 597
Exercise, aerobic, 205
Exponents:
　common errors with, 221–222
　and converting units, 237–239
　definition of, 219–220
　fractional, 228–230, 232–234
　labeling logarithmic scales with, 252
　negative, 226–227
　and orders of magnitude, 242–245
　positive integer, 218–223
　and radicals, 230–231
　rules for, 219–221, 348
　zero and negative, 226–227
Exponential decay, 271–276, 302
　as constant percent change, 293–294
　and continuous decay rate, 361–364
　general exponential decay function,
　　272–273
　and half-life, 273–276
　of iodine-131, 271–272
Exponential equations:
　estimating solutions to, 346–347
　logarithms in solutions of, 352–353
　natural logarithms in solutions of, 357–364
Exponential functions, 263–336
　average rate of change for, 282
　for compound interest, 318–324
　for constant percent change, 292–297
　constructing, from two data points,
　　280–282
　for continuous compounding, 323–327
　in data tables, 279
　for decay, *see* Exponential decay
　for doubling time, 266–268, 303–309
　estimating solutions to, 346–347
　for fractal trees, 311–312
　graphs of, 286–289
　for growth, *see* Exponential growth
　for half-life, 273–276, 303–309
　linear vs., 278–283
　logarithmic vs., 370–372
　and Malthusian dilemma, 309–310
　power vs., 416–418
　semi-log plots of, 331–333, 379–381
　using base *e*, 361–364
Exponential growth, 264–269
　as constant percent change, 292–293
　and continuous growth rate, 361–364
　and doubling time, 266–268
　E. coli example, 264–265
　examples of, 301–302
　general exponential growth function,
　　265–266
　linear vs., 283, 295–296
　and real growth of *E. coli*, 268–269

Exponentiation, 218
Extrapolation, from linear models, 128–129

Fabrics, cotton and wool, 191
Factored form (polynomial functions), 569
Factored form (quadratic functions):
　converting to, 497–500
　and Factor Theorem, 504–506
　finding horizontal intercepts in, 496–497
　and quadratic formula, 500–504
　and standard/vertex forms, 506–507
Factoring:
　ac method of, 499–500
　of polynomial functions, 569
　of quadratic functions, 497–500
Factor Theorem, 504, 569
Falling objects:
　acceleration of, 520–522
　distance and time for, 516–518
　height of, 522–524
　velocity of, 518–520
FAM1000:
　definition, 134
　farmable land, 223, 224
Farm income, U.S., 153
Father function, 585–586
Federal debt, 63–64
　from 1945–2010, 318
　changes in average rate of change in, 68–69
　U.S., 225
Federal deficit and surplus, 12
Federal funds rate, 122
Federal minimum wage, 120–121
Federal Reserve:
　and federal funds rate, 122
　and interest rates, 329
Federal surplus, 63
Fish production, 357
Fixed cost, 576
Fixed mortgage rates, 82
Flat tax plans, graduated vs., 193–196, 207–208
Focal length, from quadratic functions, 472–473
Focal point:
　of parabola, 468
　from quadratic functions, 472–473
Focus, of parabola, 468
Food supply, population growth vs., 309–310
Fossil fuels:
　burning of, 374
　decay of emissions from, 278
Fractals, 311
Fractal trees, 311–312
Fractional exponents, 228–230, 232–234
Francium, half-life of, 278
Free fall experiment, 516. *See also* Falling
　objects
Freezing point, 106
Frequency, of radio waves, 246
Frequency, relative, 2
Frequency count, 2
Function(s), 19–23
　algebra of, 553–557
　combining, 553
　composition, 585–588
　definition of, 19
　describing, 20
　domain of a, 31–34

Function(s), (*continued*)
  evaluating, 27
  exponential, *see* Exponential functions
  horizontal intercepts of a, 497
  increasing vs. decreasing, 37
  inverse, 589–595
  linear, *see* Linear functions
  logarithmic, *see* Logarithmic Functions
  notation for, 26–30
  polynomial, *see* Polynomial functions
  power, *see* Power functions
  problem solving with, 599–610
  quadratic, *see* Quadratic functions
  range of a, 31–34
  rational, *see* Rational Functions
  relationships that are not, 21–23
  transformations of, 540–548
  visualizing, 36–40
Functional models, 443–453
  in allometry, 450–453
  introduced, 89
  height and weight data example, 447–450
  and log-log plots of power functions,
    444–447
  straight lines in, 443–444

g (constant), 522
Galileo Gaililei, 515–517, 530
Gas Guzzler Tax, 133
Gas heating, 164, 165, 170
Gasoline:
  consumption of, 117–118, 495
  prices of, 203
Gates, Bill, 5
GDP, *see* Gross domestic product
General quadratic, 470
Geometric series, 608
Global shape, of polynomial functions,
    565–566
Global warming, 613
Google, 611
Gould, Stephen Jay, 397
Government Accountability Office, 340
Graduated tax plans:
  flat vs., 193–196, 207–208
  tax rates in, 198, 199
Grain, worldwide production and consumption
    of, 202
Graph(s):
  bar charts, 2–3
  concave up vs. concave down, 38
  describing a function using an, 20
  domain and range from, 32–33
  estimating exponential function solutions
    from, 346
  estimating maximum/minimum value
    from a, 36
  estimating output of a function from a, 37
  of exponential functions, 286–289
  histograms, 3–4
  influencing data interpretations with, 79–81
  input and output values from, 28
  intersection points of nonlinear systems
    from, 160–162
  of inverse functions, 371–372
  of linear functions, 92–96, 99–103

  of linear systems, 171
  of logarithmic functions, 366–370
  with log scales, 243–245
  of negative integer power functions, 432–440
  of one-to-one functions, 594–595
  of polynomial functions, 564–567
  of positive integer power functions, 408–413
  of power functions, 408–413, 432–440,
    444–445
  of quadratic functions, *see* Vertex form
    [quadratic functions]
  of rational functions, 577–581
  scatter plots, 10–11
  time series, 11
Gravitational force, 427
Greatest integer function, 123
Greeks, ancient, 516
Greenhouse gas emissions, 152–153, 339
Gross domestic product (GDP):
  for 1915 to 2005, 341
  for 1920–2000, 335
  national debt as percentage of, 48
  net debt as percentage of, 169
Grouping method of factoring, 499–500
Growth:
  exponential, *see* Exponential growth
  linear vs. exponential, 283
  logarithmic, 367–368
  population, *see* Population growth
  of power vs. exponential functions,
    416–418
  sigmoidal curve for, 268–269
Growth factor, 265, 286, 294
Growth rate, 294
  in decimal form, 293
  instantaneous, 324–327
  in percentage form, 293
Gun ownership, 128–129
Gun range, quadratic model for, 496

H1N1 virus, progression of, 605
Haldane, J. B. S., 397
HALE (health-adjusted life expectancy), 457
Half-closed intervals, 31
Half-life, 273–276, 303–309
  of atmospheric pressure, 359
  of atomic weapons, 276
  of drugs in body, 383
  of epinephrine, 353
  of strontium-90, 304
  finding equivalent equations with,
    303–305
  of francium, 278
  of plutonium, 276, 306
  and radioactive decay, 304
  rule of 70 for calculating, 305–306
  solving, with logarithms, 353
  solving, with natural logarithms,
    359–360
Half-open intervals, 31
Harvard Bridge, 239
HDL (high-density lipoproteins), 190
Health:
  heart, 198
  and U.S. educational levels, 77
Health-adjusted life expectancy (HALE), 457

Health care:
  costs of, 132
  financing of, 169
  Medicare, 315, 362
  spending, 52
Health care gap, 7
Health insurance, in United States, 53
Hearing loss, Americans suffering from, 388
Heart health, 198, 205, 300
Heart rate:
  and body mass, 454
  target, 205
Heat, metabolic, 452
Heating:
  electric, 164, 165
  gas, 164, 165, 170
  solar, 130, 163, 165, 170, 585
Heat loss, 575, 585, 599
Heat waves, 535
Height(s):
  of an object in free fall, 522–524
  of children, 124–126, 447
  and weight, 405, 447–450
Hendy, Barry, 384
Hertz (Hz), 246
Hidden variables, 139
High-density lipoproteins (HDL), 190
High school completion, 70
Histograms, 3–4
Holdeen, Jonathan, 317
Holding costs, 560
Home ownership:
  negative equity with, 106
  rates of, 65
Horizontal asymptotes, of exponential
    functions, 289
Horizontal intercepts:
  calculating, 496–504
  estimating, 473–475
  from factored form, 496–497
  of a function, 497
  of polynomial functions, 565, 567–571
  and quadratic formula, 500–504
  of quadratic functions, 471–472
  of rational functions, 577–578
Horizontal lines, linear equations of, 110–111
Horizontal line test, 594–595
Horizontal reflections:
  across the x-axis, 541–542
  of quadratic functions, 478–480
Horizontal shifting:
  new functions from, 540
  of quadratic functions, 481–483
Household debt, 601–603
Household net worth, U.S., 40
Housing prices, U.S., 19, 149
Hubble, Edwin, 216
Hubble's Law, 257
Humans:
  growth of human embryo, 495
  history of, 217
  measurement of, 213
Hybrid cars:
  conventional vs., 209
  popularity of, 209
  sales of, 133

Hydrogen atom, radius of, 242
Hydrogen bombs, 315
Hydrogen ions, 254, 373
Hz (hertz), 246

Ice, sea, 294
Ideal body weight (IBW), 598
Imaginary numbers, 503
Income:
  and race, 84
  of typical Thai worker, 161–162
  U.S. farm, 153
Income tax, flat vs. graduated, 193–196
  rate for upper 1.15%, 187
Independent variable(s):
  definition of, 21
  in function notation, 27
  reinitializing, 127–128
India, population of, 387
Indirect proportionality, of surface area to
    volume, 396
Inequalities, linear, *see* Linear inequalities
Infants:
  median weight of female, 85, 86
  median weight of male, 85, 86
  mortality rates of, 131, 455
Infinite solutions, linear systems with, 175
Inflation, 308
  in Brazil, 341
  and household debt, 601–603
Initial upward velocity, 525–527
Initial value:
  of exponential decay function, 272
  of exponential growth function, 265
  on graphs of exponential functions,
    287–288
  of linear function, 88–89
Input values (inputs):
  from graphs and tables, 28
  of inverse functions, 589–591
Insanity, rates of, 139
Inside changes, to functions, 540
Instantaneous decay rate, 324–327, 361–364
Instantaneous growth rate, 324–327, 361–364
Insurance:
  automobile, 64
  health, 53
Intensity:
  and inverse square laws, 427–428
  of light, 426, 462
  relative, 375
Interest:
  compound, *see* Compound interest
  simple vs. compound, 267
  total, 608
Interest rates:
  on certificates of deposit, 329
  effective and nominal, 320
International Data Corporation, 258
International Telecommunications Union, 314
Internet access, 129
Internet penetration rates, by world region, 9
Interpolation, from linear models, 128–129
Intersection points:
  in linear systems, 163–166
  in nonlinear systems, 160–162

Interval notation, representing domain and
    range with, 31–32
Inventory, costs of, 560
Inverse functions, 589–595
  cell phones, 591
  equations for, 593
  horizontal line test for, 594–595
  inputs and outputs of, 589–591
  logarithmic and exponential functions as,
    370–372
  as one-to-one functions, 591–592
Inverse proportionality, 421–429
  definition of, 421
  and direct proportionality, 422
  and inverse square laws, 426–429
  properties of, 423–426
Inverse square laws, 426–429
Iodine, radioactive, 340
Iodine-131, decay of, 271–272
iPod sales, 149
Isotopes, radioactive, 271–272

Japan, population of, 225, 387
Joule (J), 258
The Joy of Cooking (Irma S. Rombauer and
    Marion Rombauer Becker), 144
Jupiter, radius of, 224
Juvenile arrests, 70–71

k (coefficient):
  and negative integer power functions,
    436–440
  and positive integer power functions,
    411–413
Kalama study, 124–126
Keillor, Garrison, 5
Kepler, Johannes, 233
Kigner, Brett, 241
kilo- (prefix), 237
Kilograms, liters vs., 239
King, Martin Luther, 75
King Kong, 405
Kleiber's Law, 452–453

Land measurement, in U.S., 240
Lake Wobebon, 5
Las Vegas, Nevada, population trends
    in, 167
LDL (low-density lipoproteins), 190
Lead-206, 313
Lead-210, 366
Lead-Based Paint Poisoning Prevention
    Act, 387
Leading terms, of polynomial functions, 562
Length:
  focal, 472–473
  of organism and population density, 457
  and surface area/volume, 395–396
License plate combinations, 418
Life expectancy, 54, 341, 380
  increases in, 151
  U.S., by race and sex, 66
Light:
  intensity of, 426, 462
  speed of, 237–238, 240
Lightning, 615

Light year, 220
Lines:
  fitting, to data, 124–126
  and functional models for data sets, 443–444
  horizontal, 110–111
  linear equalities for values above/below,
    181–182
  linear equalities for values between, 182–184
  parallel, 112, 113, 174
  perpendicular, 113
  regression, *see* Regression lines
  vertical, 111–112
Linear functions, 85–90
  and constant rate of change, 85–88
  definition of, 85
  and direct proportionality, 107–109
  exponential vs., 278–283
  finding equations from graphs of, 101–103
  finding graphs of, 99–101
  general equation for, 88–90
  and horizontal/vertical lines, 110–112
  and linear models of data, 124–129
  and parallel/perpendicular lines, 112–114
  piecewise, 117–121
  regression lines, case study using, 134–140
  from two data points, 280
  U.S. population example, 85
  visualizing, 92–96
Linear inequalities, 181–187
  and breakeven points, 185–187
  manipulating, 184–185
  for values between lines, 182–184
Linear models of data, 124–129
  fitting lines to data, 124–126
  interpolation and extrapolation from,
    128–129
  reinitializing independent variables in,
    127–128
Linear systems, 159–200
  elimination method for, 173–174
  graphs of, 171
  intersection points of, 160–166
  and linear inequalities, 181–187
  nonlinear systems vs., 160–162
  with no solution or infinite solutions,
    174–175
  piecewise linear functions in, 193–197
  solving, 171–175
  substitution method for, 172
  supply and demand example, 175–177
  with two equations, 171
Line symmetry, 439
Line tests:
  horizontal, 594–595
  vertical, 22–23
Literacy rate, and infant mortality rate, 131
Liters, kilograms vs., 239
Logarithms:
  common, *see* Common logarithms
  natural, *see* Natural logarithms
  of positive numbers, 249–252
  of powers of 10, 247–249
  properties of, 367
  rules for, 347–351
  solving exponential equations with,
    352–353

Logarithmic functions:
    and decibel scale, 374–376
    exponential vs., 370–372
    graphs of, 366–372
    and pH scale, 372–374
    and power functions, 445–446
    from power functions, 444–446
Logarithmic growth, 367–368
Logarithmic (log) scales, 242–244, 443
    plotting numbers on, 250–252
    reading, 244–245
Logistic functions, 603
Log-linear plots, 331. *See also* Semi-log plots
Log-log plots, 443–445
Log scales, *see* Logarithmic scales
Long jump, world records for women's, 145
Loss:
    in aerospace industry, 65
    regions of, 185–186
Low-density lipoproteins (LDL), 190
Lung cancer:
    doubling times for, 341
    and smoking, 330, 338
    tumor growth, 317

Madoff, Bernie, 607
Magnetic resonance imaging (MRI) system,
    cost of, 575–576
Magnitude, of slope, 94–96
Major league baseball players, average salaries
    of, 150
Mali, projected population of, 612
Malthus, Thomas Robert, 309–310
Malthusian dilemma, 309–310
Margarine, consumption of, 65
Marginal cost, 576
Margin of error, 118, 123
Marital status, of population 15 years old and
    older, 66
Mars, distance from sun, 233
Mass(es):
    of birds, 456
    body, *see* Body mass
    metabolic rate vs., 452–453
    surface area vs., 450–452
Mathematical models, 14
Maximum value, 36
MCI, 317
Mean, 4–5, 63
Median, 4–5
Medical degrees, percentage awarded to
    women, 131
Medicare, 315, 362
Men:
    in civilian work force, 168
    world records for mile of, 144
Mersenne, Marin, 429
Metabolic heat, 452
Metabolic rate, body mass vs., 452–453
Methane, 339
Metric system, 212, 214
Military reserve enlisted personnel, 46
Milk:
    per capita consumption of, 277, 356
    supply of, 176
Minimum value, 36

Minimum wages, 120–121
Mirex, 339
Missouri:
    tax plans, 198
Mobile devices:
    worldwide sales, 339
Monaco, population of, 225
Moore, Gordon, 331–332
Moore's Law, 331–332
Mortgage rates:
    fixed, 82
    in U.S., 42
Mother function, 585–586, 591
Motion, 515–527
    acceleration equations, 520–522
    distance and time in equations for, 516–518
    free fall experiment, 516–518
    height of an object in free fall, 522–524
    initial upward velocity, 525–527
    and scientific method, 516
    velocity equations, 518–520
Motor vehicles:
    fatal accidents in, 64, 66
    hybrid cars, 133, 209
    registered, 70, 145
    trade-in value of, 300
    in United Kingdom, 339
    weight of, 133–134
MRI (magnetic resonance imaging) system,
    cost of, 575–576
Multiple outputs, linear models of data for,
    128–129
Multiplication:
    of functions, 553
    and inverse proportionality, 424
Multiplicative change, in exponential
    functions, 278
Multiplicative scales, 244
Murder, juvenile arrests for, 54
Musical pitch, 308–309
Music cassette sales, U.S., 33
Mute swans, population growth of, 285

National Assessment of Educational Progress
    (NAEP) math scores, 83, 84
National Collegiate Athletic Association
    (NCAA), 23
National debt, as percentage of GDP, 48
National Institute of Occupational Safety and
    Health (NIOSH), 388
Natural disasters, deaths from, 386–387
Natural gas production, U.S., 37
Natural logarithms, 357–364
    definition of, 357
    rules for, 357–358, 370
    solving doubling time and half-life with,
    359–360
NCAA (National Collegiate Athletic
    Association), 23
Negative exponents, 226–227
Negative integer power functions:
    $f(x) = x^{-1}$ and $g(x) = x^{-2}$, 433–434
    graphs of, 432–440
    and k (coefficient), 436–440
    odd vs. even negative integer powers,
    434–436

Negative numbers, scientific notation for, 216
Negative powers, power functions with, 421–429
Neptune, distance from sun, 233–234
Net debt, 169
News communication, methods of, 72
Newspapers, 17, 72
Newton's Law of Cooling, 366, 552
New York Stock Exchange, 17
NIOSH (National Institute of Occupational
    Safety and Health), 388
Nobel Prizes, 84
Noise:
    exposure to, 316–317, 374–376, 378
    measuring levels of, 374–376
Nominal interest rate, 320
Nonlinear systems, intersection points in,
    160–162
Norman windows, shape of, 494
Notation, for functions, 26–30
nth roots, 229–230
Nuclear accidents, 304
Nuclear energy, 225
Nuclear reactors, 366
Number(s):
    binary, 574
    complex, 503–504
    imaginary, 503
    prime, 499
    square, 395, 498

Odd functions negative integer powers, 434–436
Odd functions positive integer powers, 410–411
Oil:
    British Petroleum (BP) oil spill, 225
    Canadian production and consumption, 203
    measurement of, 240
    prices of crude, 205
    U.S. production and imports, 166–167
Olympics, 132, 548
One-to-one functions, 591–592, 594–595
Ontario Association of Sport and Exercise
    Sciences, 190
Open intervals, 31
Order, of multiple transformations, 545
Ordered pairs, 11
Ordering costs, 560
Orders of magnitude, 242–245
Ordinates, addition of, 558
Organs, surface area and volume of, 397
Organic products, sales of, 25
Origin, symmetry about the, 439, 542, 543
Outliers, 5
Output values (outputs):
    estimating, from a graph, 37
    from graphs and tables, 28
    of inverse functions, 589–591
    multiple, 128–129
    ratios of, from exponential functions, 279
Outside changes, to functions, 540
Oxygen consumption, 457

Paper consumption, U.S., 151
Parabola, 468–470
Parallel lines:
    linear functions of, 112, 113
    in linear systems, 174

Pearson, Karl, 142
Pennsylvania, sales tax in, 92
Percentage form:
    decay rate in, 294
    growth rate in, 293
Percent change, constant, 292–297
Perfect squares, 228
Perpendicular lines, linear functions of, 113
Phone trees, 312
Photography, 339
pH scale, 246, 254, 372–374
Piecewise defined domains, 195
Piecewise linear functions, 117–118
    absolute value function as, 118–120
    in linear systems, 193–197
    step functions as, 120–121
    systems, 193
Pitch, musical, 308–309
Pittsburgh, Pennsylvania, population trends
    in, 167
Pixels, 384
Planets:
    patterns in position and motion of, 233
Pluto, distance from sun, 240
Plutonium, half-life of, 276, 306
Point-slope formula, 103
Poison ivy, 563
Poll figures, margin of error with, 118, 123
Polonium 210, 388
Polynomial functions:
    definition of, 562–564
    graphs of, 564–567
    horizontal intercepts of, 567–571
    quotient of, 575–581
    vertical intercept of, 567–568
Ponzi schemes, 607
Population. See also United States population
    of China, 225, 387
    of five most populous countries, 7
    food supply growth vs., 309–310
    growth of world, 13, 69, 168–169, 297, 338
    of India, 387
    of Japan, 225, 387
    of Mali, 612
    of Monaco, 225
    pyramid charts for Ghana and U.S., 9
    of Rwanda, 612
    of Saudi Arabia, 306
    in Sweden, 152
    world, 13, 46, 161, 257, 297, 316
Population density, length of organism and, 457
Population growth:
    of cats, 366
    of E. coli, 264–265, 268–269, 292–293
    food supply growth vs., 309–310
    of mute swans, 285
    of rabbits, 301
    of reindeer, 316
    of United States, 299, 604–605
    urban, 293
    world, 13, 69, 168–169, 297, 338
Positive integer exponents, 218–223
    common errors with, 221–222
    in estimations, 222–223
    exponent rules for, 219–221
    visualizing power functions with, 408

Positive integer power functions:
    $f(x) = x^2$ and $g(x) = x^3$, 408–410
    graphs of, 408–413
    and k (coefficient), 411–413
    odd vs. even, 410–411
Positive numbers, logarithms of, 249–252
Positive powers, power functions with,
    399–405
Potassium-40, 278
Poverty rate, U.S., 39, 78–79, 82
Power functions, 393–459
    definition of, 394, 400
    exponential vs., 416–418
    finding best functional model with,
        443–453
    $f(x) = x^{-1}$ and $g(x) = x^{-2}$, 433–434
    $f(x) = x^2$ and $g(x) = x^3$, 408–410
    graphs of, 408–413, 432–440
    and logarithmic functions, 445–446
    logarithmic functions from, 444–446
    log-log plots of, 444–445
    negative integer, 432–440
    with negative powers, 421–429
    positive integer, 408–413
    with positive powers, 399–405
    sum of, see Polynomial functions
    and surface area/volume, 394–397
Powers of 10:
    logarithms of, 247–249
    and the metric system, 212–214
Predator-prey models, 39
Prednisone, 563
Pressure:
    atmospheric, 359, 431
    of water, 431
Prime numbers, 499
Principal, in interest equations, 318
Print, Braille vs., 108
Producer surplus, 192
Product, of functions, 586
Profit(s):
    in aerospace industry, 65
    regions of, 185–186
Progressive tax, 193
Proportionality:
    direct, see Direct proportionality
    indirect, 396
    inverse, see Inverse proportionality
Prostate cancer, 17–18
Pyramids, 236
Pyramid schemes, 606–607

Quadratic formula, 500–504
Quadratic functions:
    average rate of change for, 510–513
    calculating vertex from, 473–475, 487–489
    factored form of, see Factored form
        [quadratic functions]
    and horizontal intercepts, 473–475,
        496–504
    horizontal reflections of, 478–480
    and parabolas, 468–470
    properties of, 471–473
    and quadratic formula, 500–504
    standard form of, see Standard form
        [quadratic functions]

    vertex form of, see Vertex form [quadratic
        functions]
    vertical and horizontal shifting of,
        480–483
    vertical stretching and compressing
        of, 478
Qualitative data, 2
Quantitative variables, 2
Queueing theory, 608
Quotient, of polynomial functions, 575–581

Rabbits, growth in population of, 301
Race, income and, 84
Radicals, 230–231
Radioactive decay, 271–272, 304
Radioactive iodine, 340
Radioactive isotopes, 271–272
Radioactive waste, 316
Radionuclides, 274
Radio waves, 246
Random access memory, 341
Range:
    of a function, 31–34
    of exponential, 286
    of linear functions, 96
    for logarithmic graphs, 368
    for negative integer powers, 437
    of polynomial functions, 567
    for positive integer powers, 413
    of quadratic functions, 484
    of rational functions, 578
    representing with interval notation, 31
    visualizing, 32
Rate of change:
    average, see Average rate of change
    constant, 85–89. See also Linear functions
Ratios:
    comparing size of objects with, 220–221
    of exponential function output values, 279
Rational function(s), 575–581
    definition of, 577
    finding average cost of MRI machine with,
        575–576
    graphs of, 577–581
Reference object, 242
Reference size, 242
Reflections:
    across both horizontal and vertical axes,
        541–542
    across horizontal axis, 478–480, 541–542
    across vertical axis, 541–542
    of quadratic functions, 478–480
Refrigerators, cost of, 614–615
Regions, terminology for, 181–182
Regression analysis, 136
Regression lines, 134–140
    interpreting, 138–139
    and nature of variables' relationship,
        139–140
    summarizing data with, 135–138
Reindeer, growth in population of, 316
Reinitializing independent variables,
    127–128
Relative frequency, 2
Relative intensity, of sound, 375
Revenue, cost vs., 186

Richter, Charles, 243
Richter scale, 243–245, 388
Rigel (star), 241
Rombauer, Irma S., 144
Rombauer Becker, Marion, 144
Roots (of the equation), 500, 501
    nth, 229
    square, 228–229
Rotational symmetry, 439
Rule of 70, 305–306, 336, 363
Rural areas, U.S. population in, 73–74
Rwanda, projected population of, 612
Ryan, Nolan, 530

Sagan, Carl, 246
Sales tax, 92, 592
Salinity, 106
San Francisco, height of tides in, 45
SAT scores, of females, 65
Saudi Arabia, population of, 306
Scaling factors, 397, 494
Scatter plots:
    linear models of data for, 128–129
    for two-variable data, 10–11
Scientific method, 516
Scientific notation, 214–217
    and metric system, 212, 214
    and powers of 10, 212–214
Scuba diving, 424
Seagate Technology, 380
Sea ice, 294
Seals, elephant, 257
Semi-log plots:
    of base 2, 605
    of exponential functions, 331–333
    exponential growth on, 444
    in exponential models of data, 379–381
Shape:
    of polynomial functions, 565–566
    and size of three-dimensional objects,
        396–397
Shellfish production, 357
Shifting:
    new functions from, 540
    of quadratic functions, 480–483
Shiller, Robert J., 19
Sigmoidal curve, for growth, 268–269, 604
Simmons, Katherine, 447
Simple interest, 266–267
    compound vs., 267
Single-variable data, 2–5
Singularity, 332
60-second summary:
    introduction to, 11
Size:
    of DNA molecules, 213, 217, 242–243
    of living cells, 242–243
    reference, 242
    and shape of three-dimensional objects,
        396–397
    of sun, 221
Skid distances, 403, 554
Sleeping bags, 183
Slope:
    as average rate of change, 72–75
    and choice of end points, 78–81

in general form of linear functions, 88
in graphs of linear functions, 93–96
in linear models of data, 125–126
Smoking:
    changes in rates of, 127–128
    and death risk, 334
    and lung cancer, 330, 338
    rate of smokers, by sex, 143
Social Security numbers, 592
Solar heating, 130, 163, 165, 170, 585
Solar parabolic cooking devices, 469
Solutions:
    to the equation, 14, 500, 501
    linear systems with no solution or infinite
        solutions, 174–175
    for linear systems with two equations, 171
    to systems of equations, 165
Sound, speed of, 258
    travel, 615
Speed. See also Velocity
    of light, 237–238, 240
    of sound, 258
Speed limit, 123
SPF (sun protection factor), 190
Spheres:
    radii of, 398
    volume of, 230
Square numbers:
    difference of, 498
    in power functions, 395
Square roots, 228–229
Standard form (quadratic functions), 470–471
    factored form vs., 506–507
    finding vertex from, 487–489
    vertex form vs., 489–492, 506–507
Step functions, 120–121
Stretching:
    new functions from, 540
    of quadratic functions, 478
Stringed instruments, 429, 461–462
Stroke, 143
Strontium-90, 277, 304, 362, 383–384
Student Statistical Portrait (UMass), 154, 617
Substitution method (for linear systems), 172
Subtraction, of functions, 553
Summaries of data sets, end points for, 78–79
Sums, of power functions, see Polynomial
    functions
Sun:
    distance from planets, 233
    mass of, 428
    radius of, 242
    size of, 221
Sunflowers, growth of, 553
Sun protection factor (SPF), 190
Suntan products, 190
Super Bowl, 7
Supply and demand, 175–177
Supply curve, 176
Surface area, 452
    body, 259
    body mass vs., 450–452
    and volume, 394–397
Surplus:
    federal, 63
    producer and consumer, 192

Swans, mute, 285
Sweden, 152
Swimming, 47, 204, 205, 299
Swine flu, progression of, 605
Symmetry:
    about the origin, 439, 542, 543
    across the y-axis, 542, 543
    axis of, 468, 471
    finding, 439–440
    of functions, 542–543
    line and rotational, 439
    of negative integer power graphs, 436
    of positive integer power graphs, 411
Systems of linear equations, 163. See also
    Linear systems

Table(s):
    describing a function using a, 20
    estimating exponential function solutions
        from, 346
    finding input and output values from, 28
    identifying exponential functions in, 279
Target heart rate, 205
Tax:
    flat tax plans, 193–196, 207–208
    graduated tax plans, 193–196, 198, 199,
        207–208
    income, 187
    sales, 92, 592
Taxi fares, in New York City, 203
Taylor series, 571
TC/HDL ratio, 190
Television signal, 227
Television stations, 72
Temperature(s):
    at Acadia National Park, 91
    body, 123, 397
    for boiling water, 144, 246
    centigrade and Fahrenheit, 91
    in heat waves, 535
    of hot coffee, 547
    January, in U.S. cities, 134
    mean Arctic, 43
    and Newton's Law of Cooling, 366, 552
    and speed of sound, 258
    wind chill, 596
Terms, of polynomial functions, 562
Texas Cancer Center, 615
Thorium-230, 366
Three-dimensional objects, size and shape of,
    396–397
Tides, 428–429
Time:
    and acceleration, 520–522
    doubling, see Doubling time
    in equations of motion, 516–518
    growth/decay factor for a time unit, 281
    and velocity, 518–520
Time series, 11
Todd, T. Wingate, 447
Tomatoes, production of, 534
Tornadoes, death rates from, 76
Toxic plumes, 587
Trade balance, 335
Trade deficit, 335, 386
Traffic flow, measuring, 488–489

Transformations:
    horizontal reflections, 478–480, 541–542
    inside and outside changes, 540
    new functions from, 540–548
    order of multiple, 545
    of quadratic functions, 478–485
    symmetry of, 542–543
    and vertex form of quadratic functions,
      483–485
    vertical and horizontal shifting, 480–483,
      540
    vertical reflections, 541–542
    vertical stretching and compressing,
      478, 540
Tritium, 315
Tufte, Edward, 124
Tuition, college, 108, 142–143, 470
Tumors:
    breast, 599–601
    lung, 317
Turning points, of polynomial functions, 565
TV signal, 227
TV stations, 72
Twitter users, 50
Two data points, linear vs. exponential models
    from, 280–282
Tylenol, dose of, 326, 327

Unemployment rate, U.S., 45
Union membership, U.S., 145
Units:
    converting, 237–239
    English system, 212
    introduced, 87
    metric system, 212, 214
United Envirotech, 299
United Nations Department of Economic and
    Social Affairs, 338
United States:
    AIDS cases in, 17
    deaths in, 36
    educational levels in, 2–3
    federal deficit of, 12
    gross domestic product of, 132
    health insurance in, 53
    household net worth in, 40
    housing prices in, 19
    international trade data for, 64, 334, 386
    live births in, 35
    music cassette sales in, 33
    natural gas production in, 37
    poverty rate in, 39
    unemployment rate in, 45
    women in military of, 83
U.S. Army, 183
U.S. Bureau of Labor Statistics, 156, 168
U.S. Bureau of the Census, 4, 8
    using data from, 4, 8
U.S. Department of Energy, 257
U.S. Geological Survey, 243
U.S. Intergovernmental Panel on Climate
    Change, 339
United States population:
    for 1790–2000, 314
    for 1830–1930, 71

    for 1970–2000, 299
    for 2000–2010, 79
    age distribution of, 3–4
    change in, over time, 60–61, 67–68, 85
    growth of, 299, 604–605
    median age of, 10, 11
    pyramid chart for, 9
    racial/ethnic composition of, 8
    in rural areas, 73–74
Universe:
    expansion of, 216
    radius of, 212
    scale and tale, 260–261
Upward velocity, initial, 525–527
Uranium-238, 313, 388
Uranus, distance from sun, 233–234
Urban areas, populations in, 293

Variables:
    abstract, 14
    dependent, 21, 27
    direct proportionality with multiple,
      404–405
    hidden, 139
    independent, 21, 27, 127–128
    nature of relationship between, 139–140
    quantitative, 2
    relationships between two, 10–16
Velocity:
    and acceleration, 520–522
    as change in distance over time,
      518–520
    initial upward, 525–527
Verhulst, P. F., 604–605
Vertex:
    of parabola, 468
    of quadratic functions, 471, 492
    from standard form of quadratic equation,
      473–475, 487–489
Vertex form (quadratic functions):
    factored form vs., 506–507
    horizontal reflections of, 478–480
    standard form vs., 489–492, 506–507
    using transformations to find, 483–485
    vertical and horizontal shifting of,
      480–483
    vertical stretching and compressing
      of, 478
Vertical asymptotes:
    of logarithmic functions, 368
    of rational functions, 577–578
Vertical compressing:
    new functions from, 540
    of quadratic functions, 478
Vertical intercepts, 89, 93
    for linear models of data, 126
    of exponential functions, 287–288
    of polynomial functions, 567–568
    of quadratic functions, 471–472
Vertical lines, linear equations and, 111–112
Vertical line test, 22–23
Vertical reflections, of functions, 541–542
Vertical shifting:
    new functions from, 540
    of quadratic functions, 480–481, 483

Vertical stretching:
    new functions from, 540
    of quadratic functions, 478
Vietnam War, 75
Violent crimes, 75
Volume:
    of cube, 230, 394–397
    of cylinder, 398
    and length/surface area, 394–397
Voter turnout, 152

Warehouse club stores, sales at, 166
Water:
    access to drinking, 153
    balancing use and resources of, 196–197
    boiling temperature of, 144, 246
    concentration of hydrogen ions in, 254
    pressure of, 431
    sales of bottled, 326
Wavelength, 246, 247
Weapons, atomic, 276
Weber-Fechner stimulus law, 374n.2
Weight:
    and blood alcohol concentration, 107
    of cars, 133–134
    of children, 447–450
    of female infants, 85, 86
    healthy zones for, 182–183, 190–191
    and height, 405, 447–450
    of male infants, 85, 86
    recommended formula for, 105
White blood cell counts, 306–307
Wikipedia, 356
Wind chill temperature, 596
Windows, Norman, 494
Wind turbines, 457, 463
Wine consumption, 131
Women:
    attitudes on role of, 18
    breast cancer rates in, 132
    clothing sizes, 149
    in civilian work force, 168
    long jump world records for, 145
    medical degrees awarded to, 131
    SAT scores of, 65
    in U.S. military, 83
Wool fabrics, 191
Words, describing a function using, 20

x-axis, reflections of functions across,
    478–480, 541–542
Xenon gas, 340

Yangtze River, 257
y-axis:
    reflections of functions across, 541–542
    symmetry across, 542, 543
y-intercept, 89, 93

Zero(es):
    as exponent, 226
    imaginary, 505, 568
    of the function, 501
    real, 504, 569
Zero product rule, 497